Nonlinear Systems

NONLINEAR SYSTEMS

The Parameter Analysis and Design

Dragoslav D. Šiljak

Associate Professor of Electrical Engineering
University of Santa Clara

John Wiley & Sons, Inc.
New York · London · Sydney · Toronto

Copyright © 1969 by John Wiley & Sons, Inc.

All rights reserved. No part of this book may
be reproduced by any means, nor transmitted,
nor translated into a machine language without the
written permission of the publisher.

Library of Congress Catalog Card Number: 68-26853
SBN 471 79168 7
Printed in the United States of America

To My Parents
Dobrilo and Ljubica

Preface

In engineering science and technology there is a need for the development of methods of nonlinear system analysis, that avoid the linear idealization of the actual system behavior. Linear mathematical models of physical systems unified by the powerful superposition principle have proved useful in numerous technical applications. From general theoretical considerations based on the superposition principle successful methods have been invented for the analysis of linear system models. Experience has indicated that a detailed study of linear differential equations that describe system models, along with appropriate computer simulation of the system, will reveal in many situations what is to be expected of the behavior of the actual nonlinear system. If a system is inherently nonlinear, however, the linear idealization may lead to both quantitatively and qualitatively erroneous results. The system analysis should therefore be performed on a nonlinear system model described by the corresponding set of nonlinear differential equations.

In contrast to linear systems there is no sufficiently powerful principle that will unify nonlinear systems. Consequently there is no general approach to nonlinear system analysis. The methods are forced at the very outset of analysis to specify a restricted class of system for their successful application. Then the analyst evaluates the method on the basis of the breadth of the corresponding class of systems, the insight the method provides in the behavior of a particular system of that class, and the usefulness of the method for guidance of computer simulation of the real system. The goal of a researcher is to extend the usefulness of known methods to a wider range of systems or to invent new methods and techniques convenient for computer implementation in relevant system analyses.

The analysis of linear systems is greatly facilitated by the fact that the general solution of linear differential equations is available. Moreover, the global behavior of linear systems when they move far away from their operating conditions can be entirely determined from their local behavior under small perturbations. This produces simplicity in the analysis but also constitutes a significant limitation in the application of linear systems. By contrast, the global behavior of nonlinear systems cannot be predicted from their local behavior and, in addition, the general solution of nonlinear differential equations is usually unavailable. Therefore, the principal task of a method for analysis of nonlinear systems is to obtain a comprehensive picture, if not quantitative then at least qualitative, of their global behavior. Due to the advent of high speed computers, it is a routine task to obtain a particular solution of nonlinear differential equations and determine the motion of a nonlinear system under stated conditions. On the other hand, there is currently a strong effort to develop methods which yield a global picture of the system motion under all admissible conditions, a picture which is broad enough to exclude confusion by irrelevant details but sharp enough to emphasize the essential system characteristics. The efficient use of computers in the simulation and analysis of nonlinear systems depends crucially on the advances made in this direction.

This book proposes a method of *parameter analysis of nonlinear systems*. The book is written for engineers by an engineer and its main purpose is to provide a tool for analysis of nonlinear systems described by high-order nonlinear differential equations that represent system models occurring in the majority of practical situations. The method and related techniques provide information about the effects on system overall behavior of changing the operating conditions and parameters. Consequently the proposed parameter method belongs to those advances in nonlinear system analysis fostered by the general availability of machine simulations and computations, yet complements rather than replaces existing methods.

Historically, the method was introduced in the last century at the Leningrad School of Theoretical and Applied Mechanics by I. A. Vishnegradsky in his classical work on the dynamics of Watt's fly-ball governor for the steam engine. When the application of automatic systems became important in modern technology and science, significant extensions of the method were made in the Russian School of Automatic Control by Yu. I. Neimark, M. V. Meerov, and others in work on stability analysis of linear systems. In a Belgrade group of Automatic Control, founded by D. Mitrović, the method was extended to the analysis and design of the system response characteristics. Mitrović made an essential advance by formulating the Vishnegradsky method not only for system stability analysis but also as a design tool in cases in which the specifications of the system performance are given in either the time or the frequency domain. In the United States G. Thaler and his associates extended the method further and used the developed techniques in significant problems of linear system design.

Rather recently the method of parameter analysis has been generalized and applied to vital problems of nonlinear systems. It has been developed for both quantitative and qualitative analysis of various important nonlinear phenomena. The present book gives a unified and self-contained exposition of the obtained results, which will hopefully encourage new applications and improvements of the method.

Although this book is essentially a *monograph*, its organization is such that the material can be conveniently used by students as a *text* for classroom presentation as well as by independent readers interested in applying the method to specific practical problems. To meet this goal its organization, as shown in the given scheme (see page x), has two significant features.

First, the level of the chapters is chosen to make the material suitable for as large an audience as possible. It will permit the reader with the usual background in linear system theory to progress rapidly toward the application of the technique. The problems associated with the chapters will provide the reader with a better understanding of the usefulness of the method through independent studies. There are computer-oriented examples, which are intended to emphasize the role of machine simulation in system analysis and design. The appendices give rigorous proofs of the statements made in the major text or complement specific procedures outlined in the corresponding chapters. They require a higher technical and mathematical background and may be omitted unless the reader wishes to acquire more insight into the validity of the related statements. Furthermore, the introductions to the chapters are intended to outline not only the problem considered in the chapter but also the references to the corresponding solution procedures proposed in the past. For a reader not interested in the history of the problem a brief review of the introduction is sufficient.

The second important characteristic of the organization of the book structure is that the various chapters have been written as independently as possible of one another. This is evident from the organizational scheme of the book; for example, if the reader is interested only in the approximate nonlinear analysis, the chapter on linear systems can be omitted completely. The necessary chapters are only the first, which describes the basis of the parameter method, and the third, which outlines the introduction to the approximate analysis. Then the subsequent chapters can be chosen at will, since the correlation between them is negligible. Other possibilities of organizing a study or classroom exposition of the given material are left to the teacher to choose on the basis of the given organizational scheme and the description of each section of the book.

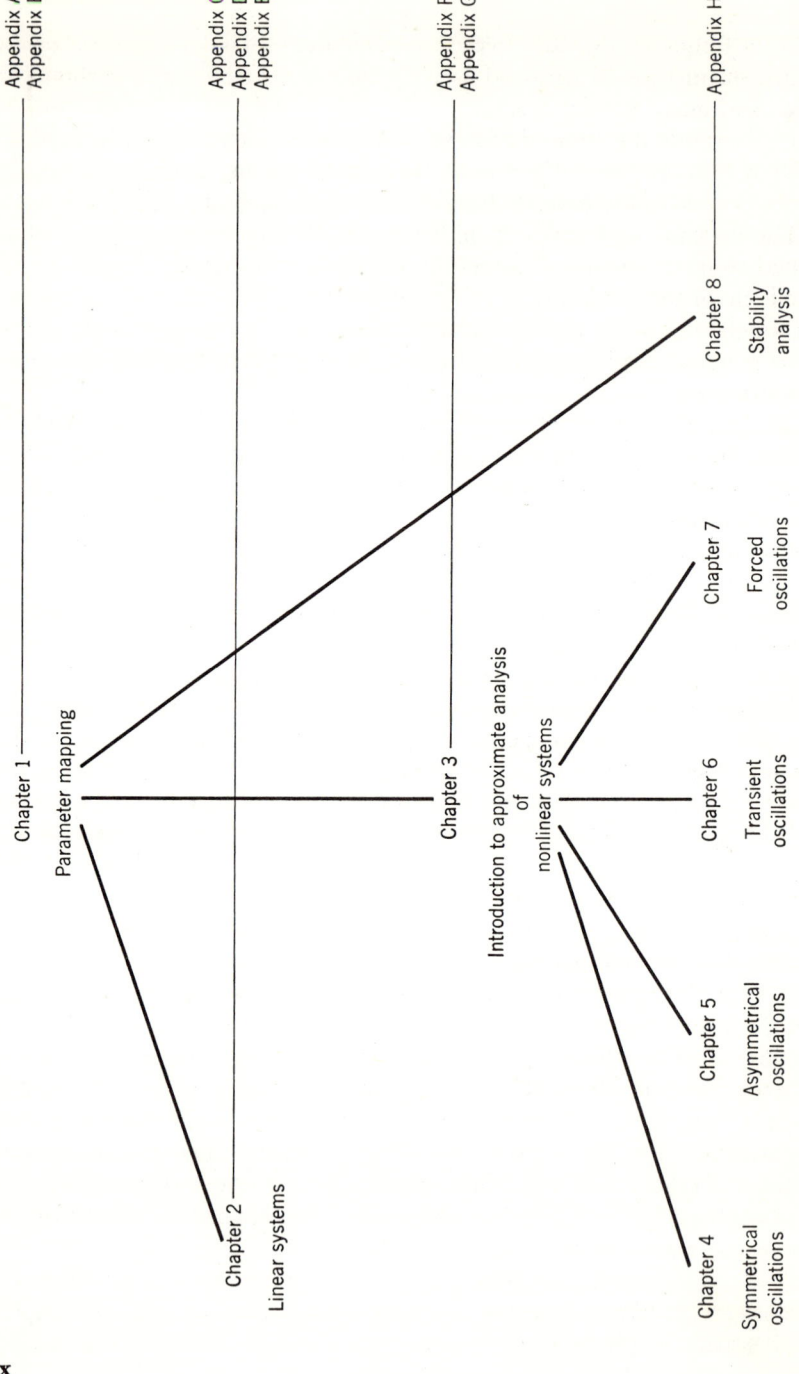

It is also important to emphasize the possibility of using the given material as an introductory text for nonlinear system analysis. One may start with Chapter 8 which introduces the state-space approach to nonlinear systems and illustrates it by a phase-plane technique elaborated in the problems associated with the chapter. Then one proceeds with the Liapunov stability theory and applies it to the absolute stability problem of Lur'e and Popov. The approximate methods of harmonic linearization can be introduced later by use of Chapter 3. The application of the linearization can be considered in Chapters 4 through 7 to the desired extent, since they are written independently of each other. The introductory material in Chapters 3 and 8 can be studied without knowledge of parameter analysis. A brief description of the contents of the chapters follows.

Chapter 1 outlines the basis of the parameter method—the parameter mapping. After some introductory examples and preliminary considerations of mapping in general, parameter mapping is defined. Basic equations and characteristic curves as a graphical interpretation of the mapping are given. They are then applied to the root distribution of relevant characteristic equations, which is closely related to the system stability, and to the root evaluation as a function of adjustable parameters, which is necessary in system transient response analysis. In addition, the root sensitivity to parameter variations is interpreted by the same concept. Appendices A and B, associated with this chapter, provide further insight into the method by related theorems, proofs, and derivations (Appendix A) or by considerations of the computational aspect of the method (Appendix B).

In Chapter 2 parameter mapping is applied to the analysis and design of linear systems. First the systems are described by a set of differential equations and the associated characteristic polynomial, the latter being the implicit mapping function of the parameter mapping. Then the stability and the system response as goals in the analysis and design are discussed. After the design is defined as a choice of system structure and parameters to obtain desired system response characteristics the design procedure is outlined. Because the procedure is carried out in terms of the pole-zero locations of the relevant system functions, a correlation between the pole-zero locations and the system response is desired. Appendix C provides necessary relations between the pole-zero locations and the time response, the latter being often the ultimate goal of the system design. The essential feature of Chapter 2, however, is the outline of a procedure for multiparameter design that is based on parameter mapping. It is illustrated by several examples in which parameter and structure adjustments are considered in multiloop feedback systems. Appendix D contains the generalized Hurwitz, Nyquist, and Mikhailov stability criteria as well as Neimark's *D*-decomposition method, all of which can be used as auxiliary tools in the system design. The *D*-decomposition method is particularly useful in the multiparameter stability analysis of linear systems. Finally an optimization procedure in the parameter plane is given in Appendix E for systems subject to statistical inputs and stability constraints.

Chapter 3 gives necessary prerequisites for nonlinear system analysis by approximate methods of harmonic linearization, usually called the describing function methods, and the parameter techniques. Attention is focused on the establishment of applicability conditions of the approximation involved in certain described classes of nonlinear systems. Standard expressions, tables, and diagrams necessary in the application of the method are given in Appendix F. The accuracy of the approximation and the applicability conditions of the linearization are extensively discussed in Appendix G using the results obtained by E. N. Rosenvasser, E. D. Garber, and others. Such an effort is necessary to obtain selective criteria for applicability of the describing function to nonlinear system analysis and to define properly a class of nonlinear systems in which the method can yield successful results.

Chapter 4 outlines the application of the harmonic linearization and parameter mapping to the investigation of symmetric periodic oscillations and their stability. The systems under investigation are free (no external forces), and the nonlinear characteristics are symmetrical to the origin so that the oscillations are symmetrical about the time axis. The stability of the oscillations is treated by the root sensitivity analysis. The overall system stability is also considered. It is pointed out that in the considerations of system stability the assumptions underlying the approximation must be satisfied. Otherwise the method may predict limit cycles that do not exist or may predict stability and fail to identify limit cycles that do exist. The systems with two nonlinearities, multivalued nonlinearities and nonlinearities with frequency-dependent and variable-pole describing functions, are also considered.

Chapter 5 presents the analysis of asymmetrical nonlinear oscillations by the parameter method and the describing function technique. The asymmetrical oscillations, which refer to the case in which the limit cycle is superimposed on a constant or slowly varying signal, are investigated in systems that have either asymmetrical nonlinear characteristics or slowly varying input signals. It is shown how the presence of a limit cycle can modify the nonlinear characteristics for the slowly varying signal. A procedure is presented by which the modified characteristic is calculated. The straightforward application of the procedure to limit-cycling control and adaptive systems is indicated. The proposed method is illustrated by examples in which the asymmetrical oscillations are investigated in multiloop feedback structures with several adjustable parameters and different nonlinear characteristics.

Chapter 6, on transient nonlinear oscillations, starts with an exposition of the extended Krylov-Bogoliubov method for the analysis of damped nonlinear oscillations in second-order systems, which is based on the work of

P. E. W. Grensted and E. P. Popov. This concept is then generalized for high-order systems by means of the parameter analysis. Because this analysis provides information about the effects on transient nonlinear oscillations of changing system parameters and choosing different initial conditions, by use of the analysis it is possible to introduce certain nonlinearities deliberately and to improve the performance of a given linear or nonlinear system.

Chapter 7 is a presentation of the analysis of forced nonlinear oscillations in the parameter plane. The basic phenomenon of jump resonance is considered. The parameter plane can display in a single diagram the whole frequency response of a nonlinear feedback system in a frequency range of interest in the analysis. Thus jump resonance can be constructed for various input amplitudes in a straightforward manner. In another approach jump resonance can be studied for various nonlinear characteristics without replotting the necessary curves each time the characteristic is changed.

Chapter 8 has the ultimate goal of presenting the parameter analysis of absolute stability in nonlinear systems. To make the exposition self-contained, first the Liapunov stability concept is outlined; then the absolute stability of nonlinear systems is defined, the "Lur'e problem" is formulated, and the Popov solution of the problem in the frequency domain is presented. The major part of the chapter is the outline of Yakubovich's results, in which the Popov solution of the "Lur'e problem" is generalized and connected with the Liapunov stability by the Yakubovich-Kalman lemma. Finally the absolute stability conditions are interpreted in the parameter space. The parameter procedure allows us to choose system parameters that will give the system a certain degree of absolute stability. The practical significance of the content of this chapter lies in the fact that a large number of nonlinear systems can be studied by the absolute stability analysis. In addition, in the framework of absolute analysis it is convenient to evaluate the linearization approach to nonlinear system analysis. This particular aspect is treated in the last section of the chapter, in which a counterexample to Aizerman's conjecture is given. This example shows that the results of the harmonic linearization should be verified before conclusions are made about the actual system behavior. Appendix H contains all the necessary proofs of stability theorems outlined in the chapter, and gives a survey of the results achieved in the absolute stability analysis of nonlinear systems, which complements the up-to-date treatment of the topic given in the book. The last section of Appendix H considers the finite regions of absolute stability.

Throughout the preparation of this manuscript I was inspired by the memory of the late Professor Dušan Mitrović. His lectures at the University of Belgrade and research activities at the Institute Michailo Pupin founded a Belgrade group of System Engineering and Automatic Control. The material presented here is in part a generalization and extension of the ideas developed by that group, and the advice and criticism of a number of its members was of invaluable assistance. In particular, I am indebted to my old friend and colleague Petar Kokotović for his help and encouragement in the preparation of this project.

During the development of the manuscript at the University of Santa Clara numerous discussions on the practical aspects of the proposed parameter analysis with Professor George Thaler were extremely useful. I am grateful to Professor Ronald Rohrer for his valuable comments on the first portion of the book, which, in particular, stimulated its educational aspect. I wish to express my gratitude to Professor Stein Weissenberger for discussions concerning the Liapunov stability theory.

I take pleasure in acknowledging that the major part of the research contained in this volume was supported and stimulated by the National Aeronautics and Space Administration under a contract monitored by Mr. Brian Doolin.

Last, but not least, I am grateful to Mrs. M. McKenna and Mrs. M. Mahaffey for typing the manuscript with patience and accuracy.

DRAGOSLAV D. ŠILJAK

Santa Clara, California
September 1968

Contents

1 **PARAMETER MAPPING** — 1
   1.1 Introduction: Historical Review — 1
   1.2 The Idea of Parameter Analysis — 4
   1.3 Parameter Mapping — 15
   1.4 Basic Equations and Characteristic Curves — 18
   1.5 Root Distribution — 32
   1.6 Root Evaluation — 45
   1.7 Root Sensitivity — 51

2 **LINEAR SYSTEMS** — 60
   2.1 Introduction — 60
   2.2 Mathematical Description of the Linear Systems — 63
   2.3 System Response and Stability — 70
   2.4 Multiparameter Adjustment. Design Constraints — 87
   2.5 Design. Compensation and Parameter Adjustment — 93

3 **INTRODUCTION TO APPROXIMATE ANALYSIS OF NONLINEAR SYSTEMS** — 107
   3.1 Introduction — 107
   3.2 Mathematical Description of Nonlinear Systems — 111
   3.3 Describing Function — 121
   3.4 Applicability Conditions — 127
   3.5 Krylov-Bogoliubov Asymptotical Method — 136
   3.6 A Comparison of Describing Function Analysis and the Krylov-Bogoliubov Asymptotical Method — 145

4 SYMMETRICAL OSCILLATIONS — 152

- 4.1 Introduction — 152
- 4.2 Solution Procedure in the Parameter Plane — 153
- 4.3 Stability and Sensitivity Analysis — 160
- 4.4 Sensitivity Analysis of Sustained Oscillations for Small Parameter Variations — 172
- 4.5 System Stability — 177
- 4.6 Two Nonlinearities — 186
- 4.7 Multivalued Nonlinearities — 194
- 4.8 Frequency-Dependent and Variable-Pole Describing Function — 198

5 ASYMMETRICAL OSCILLATIONS — 202

- 5.1 Introduction — 202
- 5.2 Basic Developments — 204
- 5.3 Asymmetrical Nonlinearities — 207
- 5.4 Constant Forcing Signals — 211
- 5.5 Slowly Varying Signals — 218
- 5.6 Conclusion — 228

6 TRANSIENT OSCILLATIONS — 232

- 6.1 Introduction — 232
- 6.2 Extension of the Krylov-Bogoliubov Method — 234
- 6.3 High-Order Systems. Parameter Plane Analysis — 245
- 6.4 Solution Procedure and Design — 251
- 6.5 Quasioptimization — 262
- 6.6 Two Nonlinearities — 265

7 FORCED OSCILLATIONS — 276

- 7.1 Introduction — 276
- 7.2 Periodic Solutions. Jump Resonance — 278
- 7.3 Stability and Sensitivity of the Periodic Solutions — 285
- 7.4 Variable Nonlinear Characteristic — 286

8 STABILITY ANALYSIS — 290

- 8.1 Introduction — 290
- 8.2 Motion of Nonlinear Systems in the State Space — 293
- 8.3 Liapunov's Stability Concept — 302
- 8.4 Stability Definitions — 307
- 8.5 Liapunov Stability Theorem. Liapunov Function — 316
- 8.6 Absolute Stability. The Lur'e Problem — 320
- 8.7 Popov's Method. Free Dynamic Systems — 326

8.8	Particular Cases	330
8.9	The Case $k = \infty$	334
8.10	Degree of Absolute Stability	336
8.11	Forced Systems	337
8.12	Nonlinear Time Varying System	341
8.13	Parameter Analysis. Sensitivity	342
8.14	Absolute Stability in the Parameter Space	352
8.15	Linearization. Aizerman's Conjecture	367

APPENDIX A	BASIC THEOREMS AND DERIVATIONS	375
APPENDIX B	COMPUTER APPLICATIONS	411
APPENDIX C	ALGEBRAIC DOMAIN VERSUS TIME DOMAIN	425
APPENDIX D	STABILITY CRITERIA FOR LINEAR SYSTEMS	448
APPENDIX E	SQUARED-ERROR OPTIMIZATION WITH STABILITY CONSTRAINTS	463
APPENDIX F	HARMONIC LINEARIZATION OF TYPICAL NONLINEAR CHARACTERISTICS	469
APPENDIX G	ACCURACY CONSIDERATIONS IN DESCRIBING FUNCTION METHOD	511
APPENDIX H	PROOFS OF THE STABILITY THEOREMS	527
PROBLEMS		573
INDEX		611

Nonlinear Systems

CHAPTER ONE

Parameter Mapping

1.1 Introduction: Historical Review

After several successful applications of feedback systems, such as Huygens' centrifugal governors for the regulation of windmills and water wheels, Airy's clock regulator, and Watt's flyball governor for the steam engine, Maxwell [1.1] postulated the problems of these systems in rigorous mathematical terms. Among the problems the foremost was system stability, which Maxwell treated by using linear differential equations as the mathematical model of a system. Maxwell reduced the stability problem to that of determining the necessary and sufficient conditions under which all roots of the system characteristic equation—an algebraic equation in complex variable—have negative real parts. He resolved it for quadratic and cubic equations and pointed out that, in case of nth-degree equations, the problem is nontrivial.

The general solution of the stability problem has been given by the stability criteria of Routh [1.2], Hurwitz [1.3], Nyquist [1.4], and Mikhailov [1.5], which provide the required conditions in different forms. On the basis of these criteria it is possible to design feedback control systems for dynamic performance specified by a certain degree of stability. However, the stability criteria do not constitute a completely satisfactory theory for the design of feedback systems. In a wide variety of control problems the designer is interested not only in stability of the system but also in the essential features of the system behavior over time. Thus there has been a strong emphasis in control theory on the development of refined techniques for the analysis and design of feedback control systems in terms of the system response to typical or test input signals.[1]

---

[1] In this introduction only the previous work related to the parameter plane concept is considered. Other methods and techniques for linear system analysis and design can be found elsewhere [1.19].

2 Parameter Mapping

The idea of investigating the system response characteristics was first introduced by Vishnegradsky [1.6] in his classical work on the dynamics of Watt's flyball governor. The algebraic approach developed by Vishnegradsky designates that the two middle coefficients of the third-degree characteristic equation be considered as variables. In the plane of the variable coefficients a diagram is plotted that enables the determination of these coefficients with respect to both the stability and the nature of the system transient response. From the work of Vishnegradsky, however, it was not possible to see how the approach could be extended to higher-degree characteristic equations.

Neimark [1.7–9], in his D-decomposition method,[2] generalized the Vishnegradsky approach for the case of nth-degree algebraic equations. Although the stability analysis assumes all the coefficients as $(n + 1)$ variables and separates the coefficient space into subspaces that correspond to a certain number of roots with negative real parts, the method is applicable only to the two-parameter stability problems. By using the D-decomposition method, we may conveniently assume two-system parameters that appear linearly in coefficients of the nth-degree characteristic equation to be variables. Then the mapping of the imaginary axis of the complex variable plane onto the plane of the variable parameters permits the designer to determine the number of the left-half-plane roots of the characteristic equation in various areas of the parameter plane. If the mapping procedure is applied to a straight line parallel to the imaginary axis of the complex plane, the method may be extended to investigations of the degree of stability; however, the plotting of the corresponding diagrams becomes a time-consuming task. Attempts to apply the D-decomposition method to the design of control systems in terms of the transient response are usually cumbersome. By applying the D-decomposition method, the designer is unable to obtain information about, or have control over, the root locations of the characteristic equation. The method has been extended to the stability analysis of systems with pure time-delay [1.8, 9], and nonlinear systems [1.12, 13]. A structural analysis of linear feedback systems and other related problems have been considered by Meerov [1.11], who used the D-decomposition method. In addition, Meerov gives an extensive list of references on the D-decomposition.

A significant extension of the Vishnegradsky approach has been proposed by Mitrović [1.14–18],[3] who emphasized the ability of the approach to relate the system adjustable parameters with the system response through the coefficients of the characteristic equation. The Mitrović method designates two last coefficients of the nth-degree characteristic equation as variables.

---

[2] A comprehensive presentation of the D-decomposition method has been given by Lanzkron and Higgins [1.10], and Meerov [1.11]. It is briefly repeated in Appendix D.

[3] The presentation of the original Mitrović method is given in the book by Thaler and Brown [1.19].

Introduction: Historical Review

By plotting the characteristic curves in the plane of the variable coefficients, the method enables the adjustments of these coefficients so that the roots of the characteristic equation may be set at any desired locations. The curves are readily plotted since explicit analytical expressions are available. After the curves are plotted, the variable coefficients can be adjusted without any calculations. All analytical and graphical operations are performed in the real domain. Since the method places in evidence all roots of the characteristic equation, it can be used in a design to satisfy both the transient and the frequency response requirements. Limitations of the method arise because only the last two coefficients can be considered as variables. The method has been extended more recently by Elliot, Thaler, and Heseltine [1.20] to several specific pairs of coefficients. The author [1.21] generalized the method so that arbitrary pairs of coefficients of the characteristic equation can be considered variable. The generalized method achieves the same degree of simplicity as does the method in its primary form. Even so, the flexibility of the method is limited because the adjustable parameters may appear in no more than two of the characteristic equation coefficients.

The Mitrović procedure has been applied to the analysis of sampled-data linear control systems [1.22–25]. The application of the method to the sensitivity analysis of linear systems, to the system design of multiloop structures with several variable parameters, and to other related problems has been performed by Thaler, Kokotović, and others [1.19–37]. A limited application of the method to the nonlinear system analysis has been presented in references 1.18 and 1.31.

A direct correlation between the system parameters and all the roots of the characteristic equation has been obtained by the author [1.38–41]. In the plane of two parameters, which can appear nonlinearly in any and all the coefficients of the characteristic equation, certain characteristic curves are obtained. These curves are constructed by mapping contours from the complex variable plane through the characteristic polynomial as a mapping function onto the parameter plane. Throughout the parameter plane map, all the roots can be determined simultaneously as numerical values without any further calculations. The mapping has been defined more precisely in reference 1.41 and is considered thoroughly in this chapter.

The parameter plane method has been applied to certain problems in sensitivity analysis of linear control systems [1.42], circuit synthesis [1.43–45], steady-state response analysis [1.46], sampled-data linear systems [1.39, 1.47–49], linear systems with pure time delay [1.50, 51], and to other related problems of linear system design [1.52–55]. Significant results have been obtained in the analysis of nonlinear systems in the parameter plane [1.40, 1.56–59]. The extension and generalization of these results is the essential goal of the sections that follow.

The major objective in this chapter is to give precise, fundamental formulation of the parameter plane analysis. It is interpreted as the mapping of the contours from the complex variable plane onto the parameter plane and is thus called the *parameter mapping*. The corresponding properties of mapping and other related features are defined. The interpretation of the resulting parameter plane diagrams is also given. This leads to a solution of the *root distribution* problem associated with the stability analysis. If a contour is specified in the complex plane and if parameter plane mapping through the characteristic equation is performed, the number of the characteristic roots relative to the contour can be read directly from the parameter plane plot as functions of the two adjustable parameters.

If the parameter plane curves are mapped back onto the complex plane—the *inverse parameter mapping*—then root loci may be obtained as functions of two variable parameters. This gives in evidence the exact root locations closely related to the *transient response* analysis.

The effect of small parameter variations on the root locations can be studied using the concept of the defined mapping. This is called the *root sensitivity* analysis.

There are two appendices associated with this chapter. Appendix A contains some theorems (the shorthand notation is Th.) and derivations (Dr.) that prove the statements outlined in the parameter plane mapping and in its applications. These theorems and derivations are not essential for the understanding of the mapping and can be omitted in the first reading of the chapter. They are important, however, in the precise definition of the mapping and its procedures.

In Appendix B, computer applications to the parameter plane analysis are discussed. Both analog and digital computer aspects are considered. It is shown that all the necessary graphical procedures in the parameter mapping can be carried out by effective digital computer programs. In the application of the mapping to the analysis and system design problems, a convergent and automatic procedure for solving algebraic equations is useful. Such a procedure is described and the necessary schemes are presented.

1.2 The Idea of Parameter Analysis

Before a precise definition of the parameter mapping is given and its properties are used in the system analysis, it is of interest to illustrate the basic idea by several simple examples.

Consider an algebraic equation of the first degree,

$$\alpha s + \beta = 0, \tag{1.1}$$

where α and β are real parameters and s is the variable. The principal task is

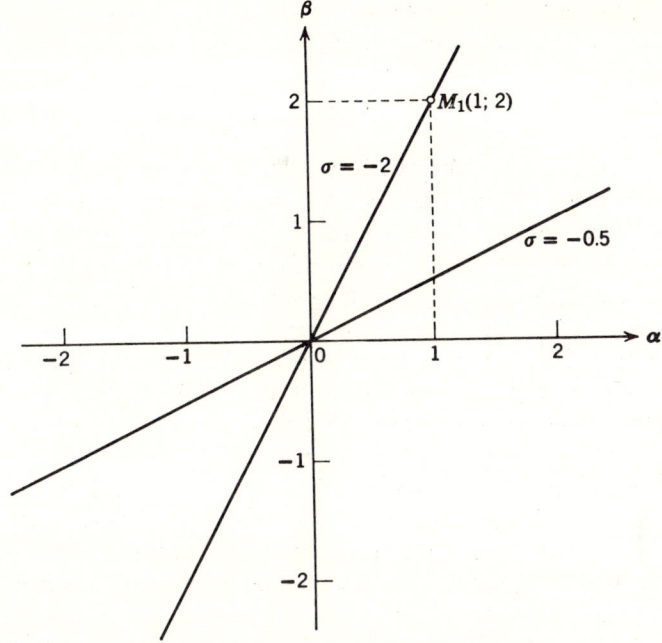

Figure 1.1 Parameter plane diagram for equation $\alpha s + \beta = 0$.

to find how the root of equation 1.1 depends on the values of the parameters α and β.

As known, for analysis purposes it is not convenient to express the roots of an algebraic equation of higher degree than two in terms of its coefficients (it is impossible for degrees higher than four). Hence the above problem must be solved by a different approach. A rectangular coordinate system with α and β as the coordinate axes can be constructed; this is called the *parameter plane*. In this plane, for any specific value of $s = \sigma$, (1.1) represents a straight line σ as shown in Figure 1.1. Now, if the straight line σ is plotted for $\sigma = -2$, then whenever a point $M(\alpha; \beta)$ is chosen on that line, the root of (1.1) will be $s = \sigma = -2$. Thus, for $M_1(1; 2)$, the root is $s = -2$, which can readily be verified from (1.1). If another σ line is plotted for $\sigma = -0.5$, the approximate parameter values for the root $-2 < s < -0.5$ can be estimated by interpolation.

Since the parameters α and β are assumed as real, (1.1) has only a real root. To consider the complex values of s, we may choose second-degree equation

$$s^2 + \alpha s + \beta = 0, \tag{1.2}$$

in which s is now either real

$$s = \sigma \tag{1.3}$$

or complex

$$s = \sigma + j\omega = -\omega_n\zeta + j\omega_n\sqrt{1-\zeta^2}, \tag{1.4}$$

where ζ $(0 \le |\zeta| \le 1)$ is referred to as the *relative damping coefficient* and ω_n is referred to as the *undamped natural frequency*.

For s *real*, the σ lines are determined again directly from the algebraic equation 1.2. By substituting (1.3) into (1.2), we obtain the equation

$$\sigma^2 + \alpha\sigma + \beta = 0, \tag{1.5}$$

which for a specified value of σ represents a straight line in the $\alpha\beta$ plane. For different values of σ, the σ lines can again be plotted in the parameter plane and applied to real root evaluation of (1.2).

To study *complex roots*, however, it is necessary to use the substitution (1.4) in (1.2). This yields

$$(-\omega_n\zeta + j\omega_n\sqrt{1-\zeta^2})^2 + (-\omega_n\zeta + j\omega_n\sqrt{1-\zeta^2})\alpha + \beta = 0. \tag{1.6}$$

Since the reals and imaginaries of (1.6) must go to zero independently, we obtain

$$\begin{aligned} -\omega_n\zeta\alpha + \beta + \omega_n^2(2\zeta^2 - 1) &= 0 \\ \alpha - 2\omega_n\zeta &= 0 \end{aligned} \tag{1.7}$$

with the assumption $\zeta \ne 1$.

Equations 1.7 may be regarded as two linear equations in two unknowns, α and β, which can be solved for α and β to obtain the well-known expressions

$$\alpha = 2\omega_n\zeta, \quad \beta = \omega_n^2 \tag{1.8}$$

since

$$\Delta = \begin{vmatrix} -\omega_n\zeta & 1 \\ 1 & 0 \end{vmatrix} = -1 \tag{1.9}$$

is different from zero.

Equations 1.8 give the necessary relationship between the parameters α and β and the complex roots $s_{1,2}(\omega_n, \zeta)$ of (1.2). This relationship may be interpreted in the $\alpha\beta$ plane. Thus, for $\zeta = 0.5$, one has from (1.8)

$$\beta = \alpha^2, \tag{1.10}$$

which represents a parabola in Figure 1.2. This parabola is a locus of points which correspond to the roots of (1.2) with $\zeta = 0.5$. Such a locus is called the ζ *curve* and, in this special case, it is the $\zeta = 0.5$ curve. Consequently, whenever a point $M(\alpha; \beta)$ is chosen to lie on this curve, equation 1.2 for the specified values of α and β will have a complex pair of roots with $\zeta = 0.5$. The value of ω_n is interpolated along the ζ curve. For example, in case of point

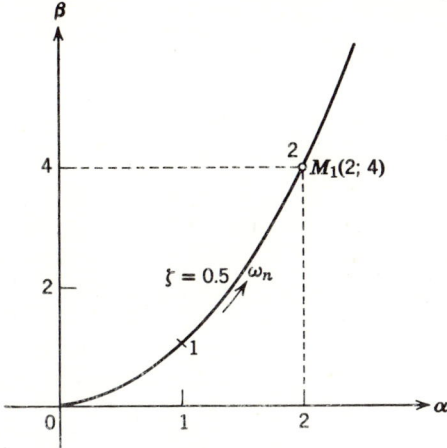

Figure 1.2 Diagram of $\zeta = 0.5$ curve.

$M_1(2; 4)$, one has $\zeta = 0.5$ and $\omega_n = 2$. Therefore, from (1.4), the root values are $s_{1,2} = -1 \pm j1.72$. Evidently, if several ζ curves are plotted in the same Figure 1.2, the complex roots can be determined for various values of α and β using the interpolation.

Another problem may now be of interest; one would like to determine a region (or regions) in the $\alpha\beta$ plane that corresponds to all the roots with a damping coefficient ζ greater than a given value. This means that the roots of (1.2) should lie inside the sector of the left half of the s plane as shown in Figure 1.3a. The roots of (1.2) can leave this sector crossing the sides of the sector (*complex root boundary*) or through the origin (*real root boundary*). To determine the region in the $\alpha\beta$ plane that corresponds to the sector of the s plane, it is necessary to map the boundaries of the sector into the parameter plane. The sides of the sector correspond clearly to the ζ curve, while the real root boundary $s = \sigma = 0$ is the $\sigma = 0$ line.

If $\zeta = 0.5$, the complex root boundary is replotted from Figure 1.2 into Figure 1.3b. The real root boundary $\sigma = 0$ is determined from equation 1.5 to get

$$\beta = 0. \tag{1.11}$$

This is actually the α axis. The desired region $R\{\zeta > 0.5\}$ is determined by inspection from Figure 1.3b.

Similarly, if the entire left half plane Re $s < 0$ is of interest, it is necessary to map the imaginary axis ($\zeta = 0$) from the s plane into the $\alpha\beta$ plane. The $\zeta = 0$ curve is the positive part of the β axis as seen from (1.8) and shown in

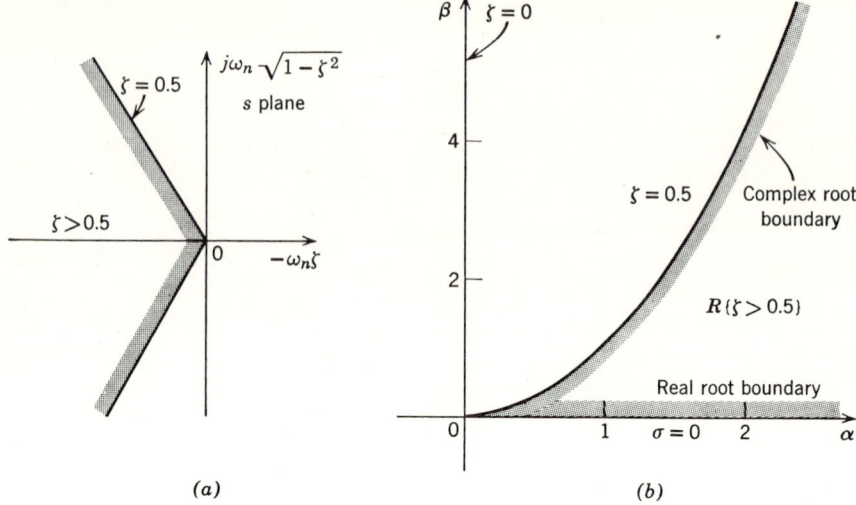

Figure 1.3 Mapping of the region $R\{\zeta > 0.5\}$.

Figure 1.3b. The region $R\{\zeta > 0\}$ is the entire first quadrant of the $\alpha\beta$ plane (the real root boundary $\sigma = 0$ remains unchanged).

Consider now an example that will illustrate an application of the parameter plane diagrams to linear system analysis. The system is shown in Figure 1.4. It is a linear feedback system with a transfer function $G(s)$ given as

$$G(s) = \frac{1}{s(s^2 + \alpha s + \beta)}. \tag{1.12}$$

The analysis task is to determine the system parameters α and β so that the system meets given performance specifications. The specifications may be given in terms of the output response $c(t)$ to a unit-step function as the input $r(t)$ by assigning the values of the maximum percent overshoot, settling time, rise time, and so on. The performance may also be specified in terms of the system transfer function

$$\frac{C}{R}(s) = \frac{1}{s^3 + \alpha s^2 + \beta s + 1} \tag{1.13}$$

when the bandwidth, peak amplitude ratio, and so on are prescribed.

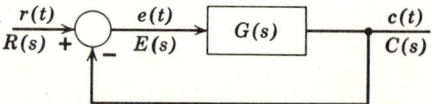

Figure 1.4 System block diagram.

In both cases, which are closely correlated, an essential role is played by the characteristic equation of the system[4]

$$F(s) \equiv s^3 + \alpha s^2 + \beta s + 1 = 0. \tag{1.14}$$

The principal task is to find how the roots of (1.14) depend on the parameters α and β. In other words, it is necessary to perform a factorization

$$(s^2 + 2\omega_n \zeta s + \omega_n^2)(s - \sigma) = 0 \tag{1.15}$$

of (1.14) and find ζ, ω_n, and σ as functions of the parameters α and β.

The parameter analysis solves the above problem graphically by constructing ζ, ω_n, and σ constant curves in the $\alpha\beta$ plane. Let us first consider the complex roots, that is, ζ and ω_n curves. They are determined in the same way as in the previous example of a second-degree equation. After substituting s in (1.14) by the second expression of (1.4), two linear equations in α and β are obtained whose solution is

$$\begin{aligned} \alpha &= \frac{1}{\omega_n^2}(2\omega_n^3 \zeta + 1) \\ \beta &= \frac{1}{\omega_n}(\omega_n^3 + 2\zeta). \end{aligned} \tag{1.16}$$

Using these equations, we can plot the parameter plane diagram of Figure 1.5. From this diagram, values of ζ and ω_n can be read as functions of system parameters α and β. Thus, two conjugate complex roots of characteristic equation 1.14 can be found by inspection for any specified pair of system parameters (α, β).

The real root σ is determined in the usual way by plotting the σ straight lines directly from the characteristic equation. For $s = \sigma$, (1.14) becomes

$$\sigma^3 + \alpha \sigma^2 + \beta \sigma + 1 = 0. \tag{1.17}$$

These σ lines are plotted on the same $\alpha\beta$ plane diagram as shown in Figure 1.6, and the real root σ of (1.14) can be evaluated for different values of system parameters.

A general computer program, which will be shown later, is available for plotting the parameter plane diagrams for nth-degree characteristic equations. This program can be utilized to map the parameter plane curves back into

---

[4] This equation was used by Vishnegradsky [1.6] when he introduced for the first time the parameter analysis in a system design. The equation was again considered recently by Towill [1.37], and his results will be used here to illustrate the general method outlined in the following section.

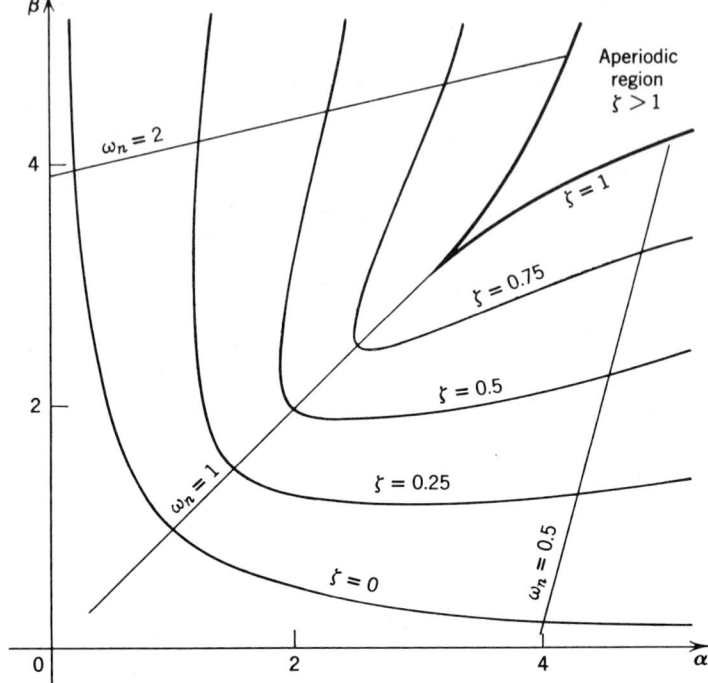

Figure 1.5 Constant ζ and ω_n curves (complex roots).

the complex plane. In the specific example under consideration, two-parameter root loci are shown in Figures 1.7a (complex roots) and 1.7b (real root). These inverse parameter plots may be more illustrative than the direct plots of Figures 1.5 and 1.6.

Once the roots of the characteristic equation are determined, the output response $c(t)$ to a unit-step function can be calculated from

$$c(t) = 1 - A_1 e^{\sigma t} + A_2 e^{-\omega_n \zeta t} \cos\left[(\omega_n \sqrt{1-\zeta^2})t + \phi\right], \quad (1.18)$$

where the constants A_1, A_2, and ϕ are dependent on α and β. From (1.18) it is clear that the nature of the response depends inherently upon the values of ζ, ω_n, and σ. This is displayed in Figure 1.8, which shows sample transient responses plotted on the parameter plane.

The philosophy of system design is to establish a simple correlation between the system parameters and the characteristic roots so that the roots may be set at desired locations by adjusting the system parameters. This is available by the parameter plane analysis. Then the roots are interpreted in terms of the time response either by calculating the actual response or by simulating the system on a computer. In the first case the response is evaluated for typical situations of the root distributions once and for all and is given by charts and diagrams.

The second case in which the computer simulation is employed is of particular interest. One may gain the impression that the simulation can be used to establish a direct correspondence between the system parameters and its response so that no intermediate steps are necessary. However, in most practical cases this is an inefficient approach. The computer simulation is limited in its ability to predict the performance for conditions not actually tested. Therefore, an intelligent method is desirable to guide the computer simulation and find the system elements and parameters that will result in desired system performance characteristics. The parameter plane method presented in the following section is just such a method; it provides an efficient computer-oriented approach to system analysis and design.

Before we conclude this section, other aspects of the parameter analysis may be illustrated. For example the steady-state response of the system in Figure 1.4 can be displayed on the parameter plane diagram.

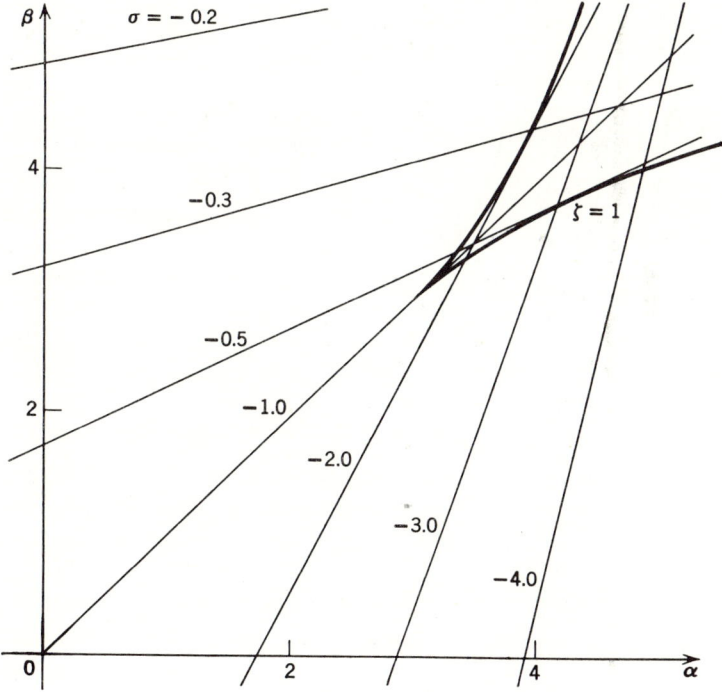

Figure 1.6 Constant σ curves (real roots).

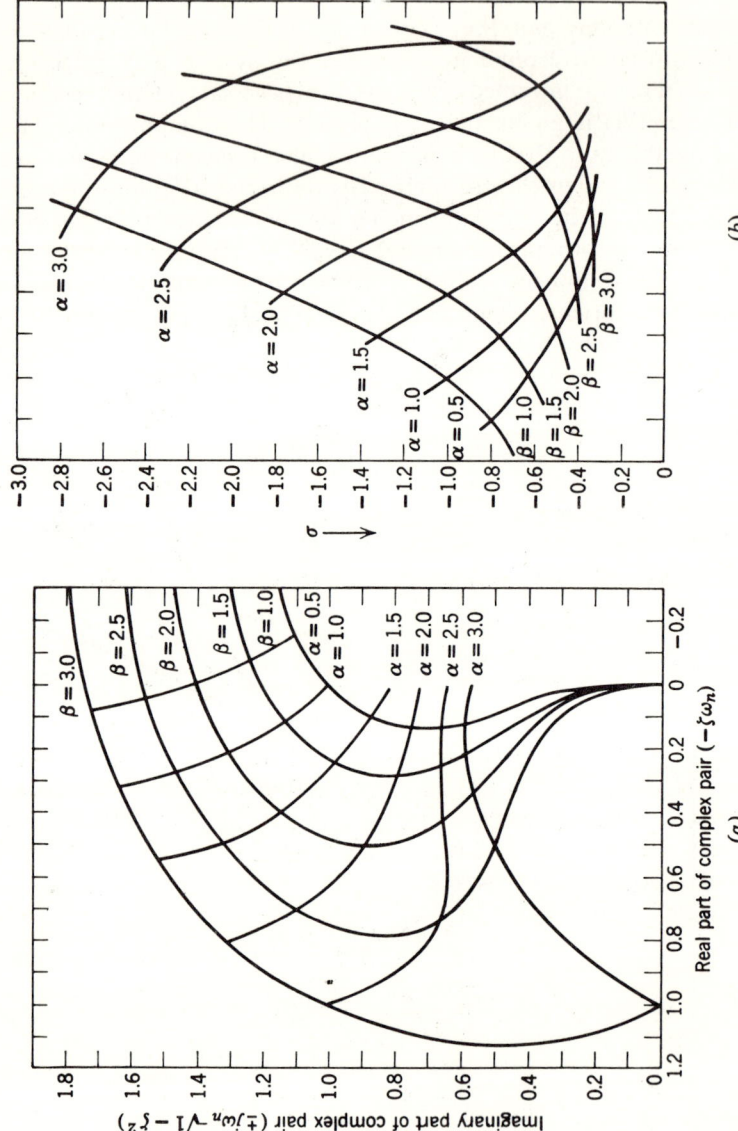

Figure 1.7 Inverse parameter diagrams in the complex plane: (a) complex roots; (b) real root.

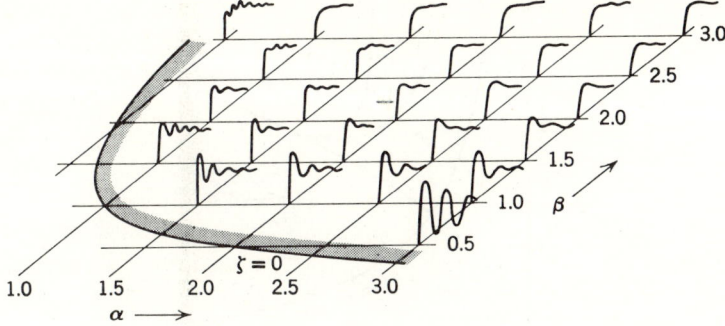

Figure 1.8 Sample transient responses.

If a unit-ramp function input $r(t) = t$ is applied to the system of Figure 1.4, the steady-state error e_{ss} is calculated by use of the error transfer function

$$\frac{E}{R}(s) = 1 - \frac{C}{R}(s) = \frac{s(s^2 + \alpha s + \beta)}{s^3 + \alpha s^2 + \beta s + 1}. \tag{1.19}$$

$$e_{ss} = \frac{1}{K_v}, \tag{1.20}$$

where the velocity error constant K_v has the value

$$K_v = \frac{1}{\beta}. \tag{1.21}$$

The K_v constant contours can be directly plotted on the parameter plane along with the ζ, ω_n, and σ curves so that both the transient and the steady-state response can be analyzed simultaneously.

Similarly, the frequency response characteristics can be given in the parameter plane. The amplitude ratio of the system specified by (1.12) is

$$\left|\frac{C}{R}(j\omega)\right| = \frac{1}{[(1 - \alpha\omega^2)^2 + (\beta\omega - \omega^3)^2]^{1/2}}. \tag{1.22}$$

Different types of the frequency response defined by (1.22) can exist, and they are shown in Figure 1.9. More detailed diagrams of the bandwidth ω_b and the peak amplitude ratio M_p are given in Figures 1.10a and 1.10b, respectively.

It may now be concluded that the parameter analysis enables the designer to adjust the system parameters so that the system has the desired performance characteristics. The performance may be specified in both the time and the

14 Parameter Mapping

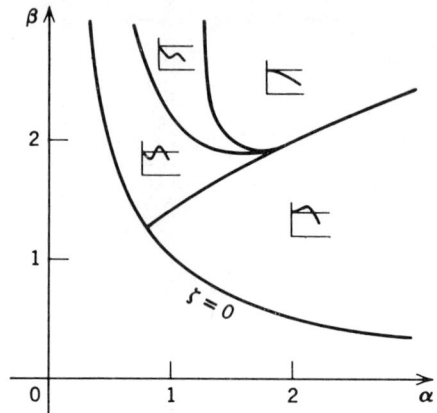

Figure 1.9 Types of the frequency response.

frequency domains. Once the parameter plane diagram is obtained, no additional calculations are necessary.

The foregoing simple examples do not by any means represent all the possibilities offered by the parameter analysis even in the area of linear systems. However, in order to utilize this approach in its full power and apply it to vital problems of both linear and nonlinear systems, it is necessary first to postulate a precise definition of the mathematical operations involved, and second to learn more about the esssential properties of these operations. The

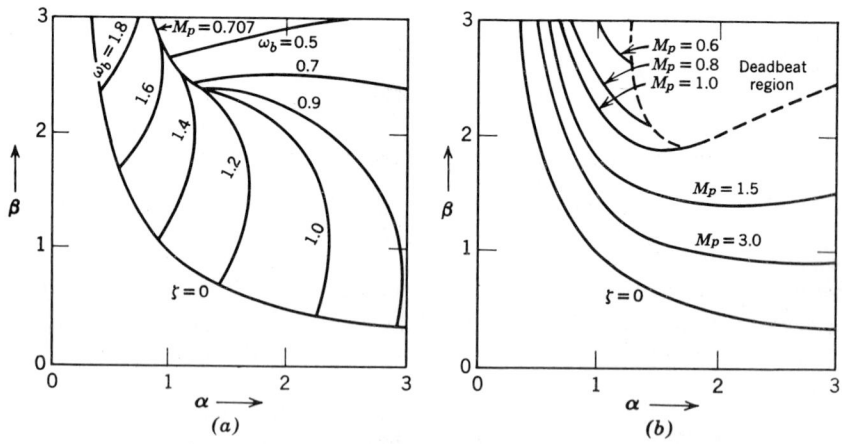

Figure 1.10 Bandwidth ω_b and M_p contours.

following sections of this chapter are devoted to a detailed study of the theoretical basis of parameter analysis.

1.3 Parameter Mapping

The parameter analysis, as introduced in the previous section, is based on a mapping procedure that transforms points from the complex $s(\sigma, \omega)$ plane onto the parameter $\alpha\beta$ plane. In the general case the implicit mapping function is nth-degree algebraic equation

$$F(s) \equiv \sum_{k=0}^{n} a_k s^k = 0, \qquad (1.23)$$

where s is the complex variable

$$s = \sigma + j\omega \qquad (1.24)$$

and a_k are the coefficients (real or complex) that are functions of two real parameters α and β,

$$a_k = a_k(\alpha, \beta), \qquad (k = 0, 1, \ldots, n). \qquad (1.25)$$

If, to a given pair of numbers (σ_1, ω_1), there corresponds a pair of numbers (α_1, β_1) so that mapping equation $F(s) = 0$ is satisfied, then for α_1 and β_1 the equation has a root $s_1 = \sigma_1 + j\omega_1$. Therefore the defined mapping can be regarded as a transformation of the points (σ, ω) from the complex s plane, which represent root values of the algebraic equation $F(s) = 0$, into certain points of the parameter $\alpha\beta$ plane. This is referred to as the *parameter mapping*.[5]

Algebraic equation 1.23 represents an implicit relationship between the parameters (α, β) and the corresponding roots expressed by the pairs of values (σ, ω). As such, it results in an implicit definition of the parameter mapping. It is then of interest to consider the possibility of an explicit formulation of the parameter mapping whereby the parameters α and β are expressed explicitly in terms of σ and ω. Then the parameter mapping will provide a useful correlation between the parameters and the roots of the algebraic equations.

If we substitute the complex variable s in (1.23) by the expression of (1.24) and apply the condition that the summations of the reals and the imaginaries must go to zero independently, then the algebraic equation can be rewritten as a system of two simultaneous equations

$$\begin{aligned} R(\sigma, \omega, \alpha, \beta) &= 0 \\ I(\sigma, \omega, \alpha, \beta) &= 0. \end{aligned} \qquad (1.26)$$

---

[5] It will be observed in the following development that the parameter mapping differs significantly from the *conformal mapping* in the calculus of complex variables.

If, in (1.26), σ and ω are considered as independent variables and α and β as two unknowns, then the equations may be solved for α and β as

$$\alpha = \alpha(\sigma, \omega)$$
$$\beta = \beta(\sigma, \omega), \qquad (1.27)$$

provided that the Jacobian $J = J[(R, I)/(\alpha, \beta)]$, which is

$$J = \begin{vmatrix} \dfrac{\partial R}{\partial \alpha} & \dfrac{\partial R}{\partial \beta} \\ \dfrac{\partial I}{\partial \alpha} & \dfrac{\partial I}{\partial \beta} \end{vmatrix} = \frac{\partial R}{\partial \alpha}\frac{\partial I}{\partial \beta} - \frac{\partial R}{\partial \beta}\frac{\partial I}{\partial \alpha}, \qquad (1.28)$$

exists and is different from zero. Since the equations of (1.26) are nonlinear in the unknowns α and β, the explicit solution in the form of (1.27) may not be possible. The sufficient conditions for the existence of solution, however, can be exactly specified by applying the Fundamental Implicit Theorem A.3 as demonstrated by Theorem A.4 of Appendix A.

In some cases the solution of (1.26) is possible by explicit algebraic processes and the functions $\alpha(\sigma, \omega)$ and $\beta(\sigma, \omega)$ can be expressed in terms of familiar elementary functions. These cases are of practical value and will be discussed in detail in Derivation A.2 of Appendix A. It is important to note that a nonzero Jacobian for (1.28) not only provides a sufficient condition for the existence of the explicit solution, but also plays a significant role in parameter mapping of points from the s plane onto the $\alpha\beta$ plane.

Because the parameter mapping is defined as a solution of nonlinear simultaneous equations, it is not generally one-to-one mapping.[6] A pair of values (σ, ω) may have several, or even infinite, corresponding images $M(\alpha; \beta)$ in the $\alpha\beta$ plane. The infinite number of images M occurs when the system of equations 1.27 degenerates to a single equation, as is discussed in the following developments.

---

[6] Consider two sets of points S_1 and S_2 in the s plane and $\alpha\beta$ plane, respectively. If, to every point $s(\sigma; \omega)$ of a set S_1, there is assigned a point $M(\alpha; \beta)$ in a set S_2, we say that we have *a mapping T of S_1 into S_2* and we can write $T: S_1 \to S_2$. If M is the point in S_2 assigned by the mapping T to the point s in S_1, we say that s is mapped into M, and we may write $M = T(s)$. The point M is called the *image* of s, and the point s is called an *antecedent* of M. The transformation T is then a *point transformation T* whereby, to certain pairs of numbers (σ, ω), there correspond certain pairs of numbers (α, β). If each element of the set S_2 appears to be an image under T of some point s of the set S_1, then the transformation T is said to be *onto*, or the transformation T maps the set S_1 *onto* the set S_2. If any two distinct points in S_1 are assigned distinct image points in S_2, then T is said to be a *one-to-one mapping* of S_1 into S_2.

The parameter mapping as defined by (1.27) is *continuous*[7] if the individual functions $\alpha(\sigma, \omega)$ and $\beta(\sigma, \omega)$ are continuous functions of their arguments σ and ω. The mapping is *differentiable* if these functions are differentiable, and *continuously differentiable* if the functions have continuous partial derivatives.

Now we may attempt the problem of defining the inverse parameter mapping.[8] Consider the transformation (1.27). It may be possible to solve the equations defining the transformation for σ and ω in terms of α and β so that we get equations of the form

$$\sigma = \sigma(\alpha, \beta)$$
$$\omega = \omega(\alpha, \beta). \qquad (1.29)$$

In most practical cases the functions $\sigma(\alpha, \beta)$ and $\omega(\alpha, \beta)$ cannot be derived in an analytical form. However, a graphical procedure is possible to evaluate pairs of values (σ, ω) that correspond to given pairs (α, β) for certain limited variations of the parameters α and β.

Evidently transformation (1.29) cannot be an inverse transformation since the parameter mapping, in general, is not a one-to-one mapping. Furthermore it is obvious that to each pair of values (α, β) there are n and only n pairs of values (σ, ω) which determine all n roots of the algebraic equation 1.23. Therefore, corresponding to the above terminology used for parameter mapping, the transformation (1.29) can be regarded as one-to-n point transformation of the points from the parameter $\alpha\beta$ plane into the s plane. For brevity we shall use the term *inverse parameter mapping* to denote the transformation (1.29), being aware that this is incorrect in the exact mathematical sense of inverse transformation. This can be motivated by the fact that the notion of the Riemann surfaces[9] may be introduced to obtain one-to-one correspondence and continuity of transformation in the sense suggested by the construction of the surfaces.

The existence of the inverse transformation is formulated by Theorem A.5 of Appendix A, which is closely correlated with the Fundamental Implicit

---

[7] In general, a mapping T of a set S_1 into a set S_2 is said to be *continuous* if, for each point s in S_1 and any *neighborhood* of its image M in S_2, there exists a neighborhood of s_0 that is mapped by T into the neighborhood of M_0. A neighborhood of a point $(\sigma_0; \omega_0)$ means the set of points inside a circle having a center $(\sigma_0; \omega_0)$ and radius δ; it is then said that the neighborhood is of radius δ. Each point $(\sigma; \omega)$ of the neighborhood satisfies the inequality $(\sigma - \sigma_0)^2 + (\omega - \omega_0)^2 < \delta^2$, where δ is a positive number.

[8] If T is one-to-one mapping, we can form a mapping which assigns to every point M in S_2 its unique antecedent s under the mapping T. This latter mapping is called the *inverse* of T and is denoted by T^{-1} in order to write $s = T^{-1}(M)$. In other words, the inverse transformation T^{-1} maps the set S_2 one-to-one onto the set S_1.

[9] Construction of the Riemann surfaces is explained, for example, in Z. Nehari, *Conformal Mapping*, McGraw-Hill, New York, 1952.

18   Parameter Mapping

Theorem. It is shown that a sufficient condition for the transformation (1.29) is that the Jacobian J of equation 1.28 does not vanish. In general, in studying the inverse transformation, we stay away from points at which the Jacobian of the transformation is equal to zero.

Formally, the coefficients of the algebraic equation can be considered as functions of several parameters. This leads to a definition of the parameter mapping into a parameter space, which is discussed in Section D.5 of Appendix D. After completing the reading of this chapter, the reader may easily apply the parameter plane mapping to the multiparameter analysis by using those formal extensions. There is yet another possible approach with more practical aspects, as shown in Chapter 2, on linear system analysis and design.

1.4 Basic Equations and Characteristic Curves

Let us consider more closely the parameter transformation defined by (1.27). Two cases will be discussed separately: the case in which the coefficients a_k of the algebraic equation are real, and the case in which they are complex numbers. It will be shown that the basic equations 1.27 represent, in both cases, a convenient relationship between the roots of the algebraic equation 1.23 and the parameters α and β. This relationship has to be developed for the complex roots and the real roots separately. It is then interpreted graphically in the parameter $\alpha\beta$ plane by studying the position of the operating point $M(\alpha; \beta)$ in relation to the characteristic curves plotted in the parameter plane. The characteristic curves are the images in the $\alpha\beta$ plane of specific Jordan curves[10] of the s plane, and they will be defined and discussed in the following pages.

Consider first the case in which the coefficients a_k of the algebraic equations are real and the relationship between the complex roots and the parameters is desired. Then, after the substitution of expression $s = \sigma + j\omega$ into algebraic equation 1.23, equations 1.26 have been obtained, and they can be written as

$$R \equiv \sum_{k=0}^{n} a_k X_k = 0 \qquad (1.30a)$$

$$I \equiv \sum_{k=0}^{n} a_k Y_k = 0, \qquad (1.30b)$$

---

[10] The set of points $(\sigma; \omega)$ where $\sigma = \sigma(\alpha)$, $\omega = \omega(\alpha)$, and $\sigma(\alpha)$ and $\omega(\alpha)$ are continuous functions of α in an interval $I(\alpha_1 \leq \alpha \leq \alpha_2)$ is called a *continuous arc L*. In the complex plane a notation $s = s(\alpha) = \sigma(\alpha) + j\omega(\alpha)$ for an arc can be used. The arc L is *differentiable* if $s'(\alpha)$ exists and is continuous. If, in addition $s'(\alpha) \neq 0$, the arc L is said to be *regular*. An arc L is *simple*, or a *Jordan arc*, if $s(\alpha_1) = s(\alpha_2)$ only for $\alpha_1 = \alpha_2$. In other words, a Jordan arc L is a continuous arc that does not cut itself. A continuous arc C with only one multiple point, which is a double point corresponding to the end points α_1 and α_2 of the interval $I(\alpha_1 \leq \alpha \leq \alpha_2)$, is called a *simple closed curve C*.

where a_k are the coefficients of (1.23) as defined in (1.25) and

$$X_k = X_k(\sigma, \omega)$$
$$Y_k = Y_k(\sigma, \omega) \qquad (1.31)$$

are functions defined by the expression $s = \sigma + j\omega$ and

$$s^k = X_k + jY_k. \qquad (1.32)$$

The functions X_k and Y_k can be obtained by applying the recurrence formulas

$$X_{k+1} - 2\sigma X_k + (\sigma^2 + \omega^2)X_{k-1} = 0$$
$$Y_{k+1} - 2\sigma Y_k + (\sigma^2 + \omega^2)Y_{k-1} = 0, \qquad (1.33)$$

where $X_0 \equiv 1$, $X_1 \equiv \sigma$, $Y_0 \equiv 0$, $Y_1 \equiv \omega$. As can be seen from (1.33), the functions X_k and Y_k are polynomials in two variables, σ and ω. These polynomials are treated in more detail in Dr. A.1 of Appendix A, where their relation to Chebyshev's functions is also given.

Equations 1.30, which are considered as two equations in the two unknowns α and β, may be solved for α and β to obtain equations

$$\alpha = \alpha(\sigma, \omega)$$
$$\beta = \beta(\sigma, \omega). \qquad (1.27)$$

The possibility of this solution depends entirely on the functions $a_k = a_k(\alpha, \beta)$ as shown in connection with Th. A.3 of Appendix A. In the following developments, these functions will be polynomials in α and β, and therefore continuous functions of their arguments.

In the rectangular coordinate system with α and β as axes, (1.27) may represent the loci of points corresponding to the complex roots with constant real part σ or constant frequency ω by assigning a constant value to the variable σ or ω, respectively. In general these loci of points are called the *characteristic curves* and, specifically, they are named the Σ and ω curves.

The Σ and ω curves as defined above represent the images of the Jordan arcs, which are straight lines parallel to the real axis and to the imaginary axis of the s plane, respectively, as shown in Figure 1.11a. Since the parameter mapping is not a one-to-one mapping, the Σ and ω curves are not simple curves. However, the curves are continuous in the sense of Th. A.8 of Appendix A as long as the corresponding Jacobian does not vanish. Th. A.8 considers the continuity of the zeros of algebraic polynomials with respect to parameter variations.

It is important to note here that a general computer program can be readily written for calculation and plotting of the characteristic curves. This is considered in Appendix B, where computer applications are discussed.

To interpret more specifically the relationship of equations 1.27, consider the case in which the coefficients a_k of (1.25) are *linear functions* [1.38] of

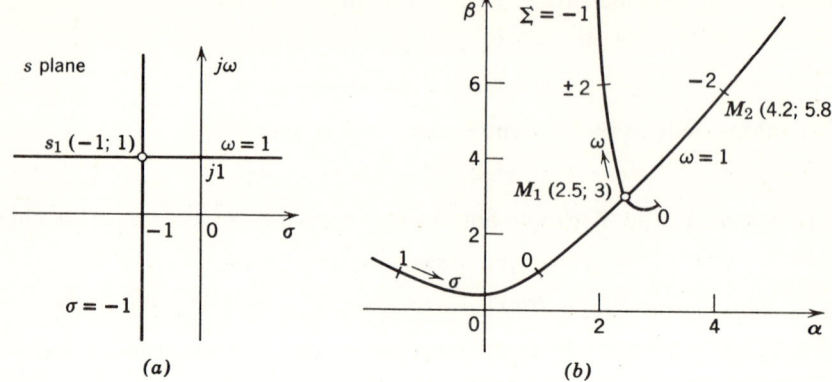

Figure 1.11 Parameter mapping: (a) s plane arcs; (b) parameter plane curves.

parameters α and β; for example,

$$a_k = b_k\alpha + c_k\beta + d_k, \quad (k = 0, 1, \ldots, n). \tag{1.34}$$

Other cases of a_k functions, which are of practical interest, are discussed in Dr. A.2 in Appendix A.

According to (1.34), (1.30) can be rewritten as

$$\begin{aligned}\alpha B_1(\sigma, \omega) + \beta C_1(\sigma, \omega) + D_1(\sigma, \omega) &= 0 \\ \alpha B_2(\sigma, \omega) + \beta C_2(\sigma, \omega) + D_2(\sigma, \omega) &= 0,\end{aligned} \tag{1.35}$$

where

$$\begin{aligned} B_1 &= \sum_{k=0}^{n} b_k X_k & B_2 &= \sum_{k=0}^{n} b_k Y_k \\ C_1 &= \sum_{k=0}^{n} c_k X_k & C_2 &= \sum_{k=0}^{n} c_k Y_k \\ D_1 &= \sum_{k=0}^{n} d_k X_k & D_2 &= \sum_{k=0}^{n} d_k Y_k. \end{aligned} \tag{1.36}$$

Equations 1.35 may be solved for unknowns α and β

$$\begin{aligned}\alpha &= \frac{C_1 D_2 - C_2 D_1}{B_1 C_2 - B_2 C_1} \\ \beta &= \frac{B_2 D_1 - B_1 D_2}{B_1 C_2 - B_2 C_1}\end{aligned} \tag{1.37}$$

provided that the determinant

$$\Delta = \begin{vmatrix} B_1 & C_1 \\ B_2 & C_2 \end{vmatrix} = B_1 C_2 - B_2 C_1 \tag{1.38}$$

Basic Equations and Characteristic Curves 21

is different from zero. This requirement concerning the determinant Δ is equivalent to that of the Jacobian J, since the linear case of a_k defined in (1.34) is a special case of the nonlinear equations 1.25 and the Jacobian J reduces to the determinant Δ of (1.38).

To illustrate the application of the basic equations, consider again the third-degree algebraic equation

$$s^3 + \alpha s^2 + \beta s + 1 = 0, \tag{1.14}$$

which we investigated in Section 1.2. Let us map the $\sigma = -1$ and $\omega = 1$ straight lines from the s plane of Figure 1.11a into the corresponding images $\Sigma = -1$ and $\omega = 1$ curves in the parameter plane.

According to (1.14), (1.36) and (1.38) are

$$\begin{aligned}
B_1 &= \sigma^2 - \omega^2 & B_2 &= 2\sigma\omega \\
C_1 &= \sigma & C_2 &= \omega \\
D_1 &= \sigma^3 - 3\sigma\omega^2 + 1 & D_2 &= 3\sigma^2\omega - \omega^3 \\
J &= \Delta = B_1 C_2 - B_2 C_1 = -\omega(\sigma^2 + \omega^2)
\end{aligned} \tag{1.39}$$

and equations 1.37 become

$$\begin{aligned}
\alpha &= \frac{1 - 2\sigma(\sigma^2 + \omega^2)}{\sigma^2 + \omega^2} \\
\beta &= \frac{(\sigma^2 + \omega^2)^2 - 2\sigma}{\sigma^2 + \omega^2}.
\end{aligned} \tag{1.40}$$

If a specific value $\sigma = -1$ is used, equations 1.40 determine the $\Sigma = -1$ curve by

$$\begin{aligned}
\alpha &= \frac{3 + 2\omega^2}{1 + \omega^2} \\
\beta &= \frac{(1 + \omega^2)^2 + 2}{1 + \omega^2}.
\end{aligned} \tag{1.41}$$

By assigning different values to ω, we plot the $\Sigma = -1$ curve in Figure 1.11b. If ω is kept equal to 1 in (1.40), the characteristic $\omega = 1$ curve is defined by

$$\begin{aligned}
\alpha &= \frac{1 - 2\sigma(\sigma^2 + 1)}{\sigma^2 + 1} \\
\beta &= \frac{(\sigma^2 + 1)^2 - 2\sigma}{\sigma^2 + 1}
\end{aligned} \tag{1.42}$$

and also plotted in Figure 1.11b. The values of σ are interpolated along the $\omega = 1$ curve.

22   Parameter Mapping

From the definition of the characteristic Σ and ω curves, it follows that whenever an operating point $M(\alpha; \beta)$ is placed on a particular Σ or ω curve, the algebraic equation has a complex root $s = \sigma + j\omega$ with the value of σ or ω for which the curve is plotted. Therefore, for the image $M_1(2.5; 3)$ of the point $s_1(-1; 1)$, which lies at the intersection of the $\Sigma = -1$ and $\omega = 1$ curve in Figure 1.11b, the algebraic equation 1.14 will have a complex root $s_1 = -1 + j1$. Similarly, for the parameter values $\alpha = 4.2$ and $\beta = 5.8$, which determine the point $M_2(4.2; 5.8)$ lying on the curve $\omega = 1$ at $\sigma = -2$, one root of (1.14) is $s_1 = -2 + j1$. Since the algebraic equation 1.14 has real coefficients only, the complex roots appear in conjugate pairs (see Th. A.9). This explains why equations 1.41 for the $\Sigma = -1$ curve do not change if ω changes in sign, and why the values $\pm \omega$ are interpolated along the $\Sigma = -1$ curve of Figure 1.11b. Thus, in general, whenever the coefficients a_k are real (not necessarily linear functions of the parameters α and β), the conditions $\alpha(\sigma, \omega) = \alpha(\sigma, -\omega)$, $\beta(\sigma, \omega) = \beta(\sigma, -\omega)$ are valid for the Σ curves, and they need to be plotted only for positive values of the argument ω.

Besides Σ and ω curves, it is possible to define other useful characteristic curves relating the complex roots and parameters of an algebraic equation. If the substitution

$$s = -\omega_n \zeta + j\omega_n \sqrt{1 - \zeta^2} \qquad (1.43)$$

is used where ω_n is the undamped natural frequency and ζ is the relative damping coefficient, the equation

$$F(s) \equiv \sum_{k=0}^{n} a_k s^k = 0 \qquad (1.23)$$

can be rewritten as

$$\sum_{k=0}^{n} a_k X_k = 0 \qquad (1.30a)$$

$$\sum_{k=0}^{n} a_k Y_k = 0, \qquad (1.30b)$$

where the functions $X_k = X_k(\omega_n, \zeta)$, $Y_k = Y_k(\omega_n, \zeta)$ are obtained by recurrence formulas

$$\begin{aligned} X_{k+1} + 2\omega_n \zeta X_k + \omega_n^2 X_{k-1} &= 0 \\ Y_{k+1} + 2\omega_n \zeta Y_k + \omega_n^2 Y_{k-1} &= 0, \end{aligned} \qquad (1.44)$$

with $X_0 \equiv 1$, $X_1 \equiv -\omega_n \zeta$, $Y_0 \equiv 0$, $Y_1 \equiv \omega_n \sqrt{1 - \zeta^2}$. Equations 1.44 are equivalent to (1.33) since

$$\begin{aligned} \sigma &= -\omega_n \zeta \\ \omega &= \omega_n \sqrt{1 - \zeta^2}. \end{aligned} \qquad (1.45)$$

Because, in (1.30), the coefficients a_k are functions of the parameters α and β, these equations may be considered as two simultaneous equations in two unknowns, α and β, which may be solved for α and β as

$$\begin{aligned}\alpha &= \alpha(\omega_n, \zeta) \\ \beta &= \beta(\omega_n, \zeta)\end{aligned} \quad (1.46)$$

provided that the Jacobian J of (1.28) does not vanish. Equations 1.46 may represent the loci of points in the $\alpha\beta$ plane that correspond to the roots with a specific value of the relative damping coefficient ζ or the undamped natural frequency ω_n. These loci of points are called the ζ and ω_n *curves*.

It is of particular interest to consider the case in which the damping coefficient $\zeta = 1$. The $\zeta = 1$ curve represents the loci of points that correspond to double real roots $s = -\omega_n = \sigma$ according to (1.43) and $s = \sigma \pm j\omega$. The $\zeta = 1$ curve is significant in the evaluation of real roots of relevant algebraic equations in the parameter plane, as will be shown later. However, the curve $\zeta = 1$ cannot be obtained from (1.30), since for $\zeta = 1$, $Y_k \equiv 0$ for all k, and (1.30b) automatically vanishes. To include the case $\zeta = 1$, an alternative formulation Y_k^* of the functions Y_k is introduced by

$$Y_k = \omega_n \sqrt{1 - \zeta^2} Y_k^* \quad (1.47)$$

and

$$Y_{k+1}^* + 2\omega_n \zeta Y_k^* + \omega_n^2 Y_{k-1}^* = 0 \quad (1.48)$$

where $Y_0^* \equiv 0$, $Y_1^* \equiv 1$. The functions $Y_k^*(\omega_n, \zeta)$ are polynomials in ω_n and ζ as is the case with the functions $X_k(\omega_n, \zeta)$. Since, in functions Y_k^*, the term $(1 - \zeta^2)^{1/2}$ does not appear, they do not vanish for $\zeta = 1$. Furthermore, there exists a relationship (see Appendix A)

$$X_k = -\omega_n \zeta Y_k^* - \omega_n^2 Y_{k-1}^* \quad (1.49)$$

that enables (1.30) to be rewritten in terms of Y_k^* polynomials only as

$$\sum_{k=0}^{n} a_k Y_{k-1}^* = 0 \quad (1.50a)$$

$$\sum_{k=0}^{n} a_k Y_k^* = 0. \quad (1.50b)$$

In the derivation of (1.50), it was assumed that $\zeta \neq 1$, $\omega_n \neq 0$. However, in Dr. A.3 of Appendix A, it is shown that equations 1.50 are valid for $\zeta = 1$. The case $\omega_n = 0$ corresponds to $s = 0$, which is treated in the considerations of the real roots of algebraic equations.

To illustrate the application of (1.50), consider again the algebraic equation 1.14 and map the $\zeta = 0, 0.5, 1$ and $\omega_n = 0.5, 1, 2$ contours of the s plane. Since, in dealing with algebraic equations that have real coefficients, it is

24 Parameter Mapping

sufficient to consider only the upper half of the s plane, Figure 1.12 displays only the second quadrant of the s plane.

The corresponding mapping equations are

$$\alpha B_1(\omega_n, \zeta) + \beta C_1(\omega_n, \zeta) + D_1(\omega_n, \zeta) = 0$$
$$\alpha B_2(\omega_n, \zeta) + \beta C_2(\omega_n, \zeta) + D_2(\omega_n, \zeta) = 0, \quad (1.51)$$

where

$$\begin{aligned}
B_1 &= Y_1^* & B_2 &= Y_2^* \\
C_1 &= Y_0^* & C_2 &= Y_1^* \\
D_1 &= Y_2^* + Y_{-1}^* & D_2 &= Y_3^* + Y_0^*.
\end{aligned} \quad (1.52)$$

By calculating the corresponding function Y_k^* from (1.48) and thereby solving (1.51) for α and β, we obtain again the equations

$$\alpha = \frac{1}{\omega_n^2}(2\omega_n^3 \zeta + 1)$$

$$\beta = \frac{1}{\omega_n}(\omega_n^3 + 2\zeta). \quad (1.16)$$

Equations 1.16 can readily be used to plot the required parameter plane curves shown in Figure 1.13.

The parameter plane curves of Figure 1.13 are interpreted in the same fashion as that of Figure 1.11b. For example, if the point $M_1(2; 2)$ is chosen as an operating point, (1.14) has two complex roots, $s_{1,2} = -0.5 \pm j0.865$,

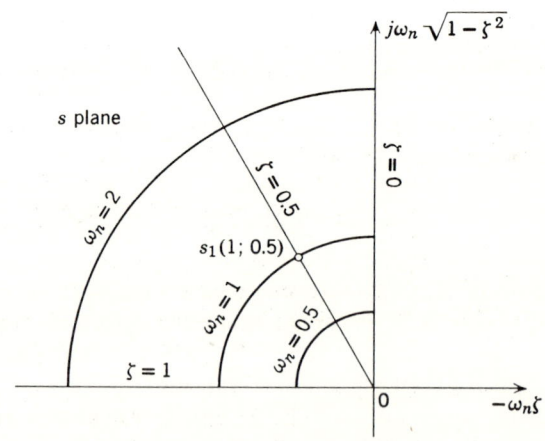

Figure 1.12   The s plane contours.

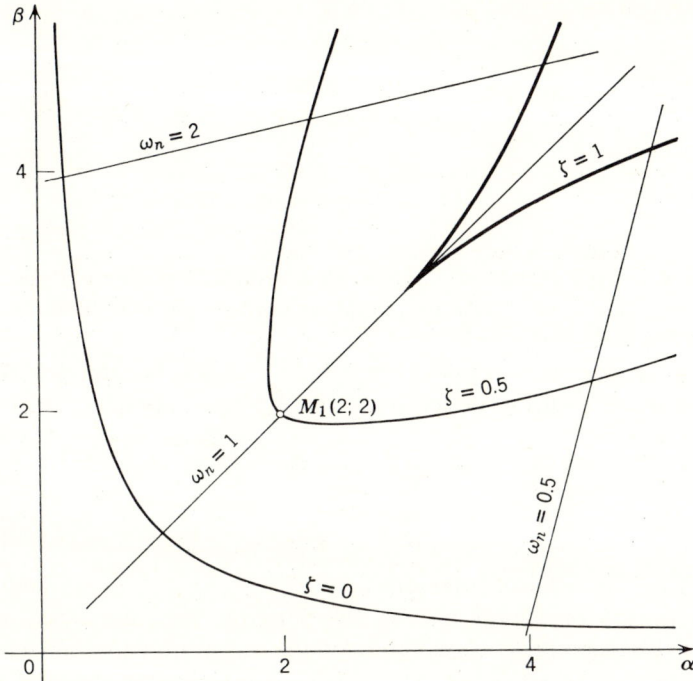

Figure 1.13 Parameter plane curves.

determined by (1.43) for $\omega_n = 1$ and $\zeta = 0.5$.[11] Furthermore, by interpolating between the parameter plane curves of Figure 1.13, it is possible to determine the complex roots of (1.14) for various values of the parameters α and β.

In the following discussion, the idea of root evaluation in the parameter plane is generalized for nth degree algebraic equations with more complex distribution of parameters than that of (1.14).

Real roots of an algebraic equation are related to the parameters directly through the equation itself. If the complex variable s is replaced in the

---

[11] It is of interest to note also that the $\zeta = 0$ curve of Figure 1.13 has been first plotted by Vishnegradsky [1.6] and then interpreted in the stability analysis of linear dynamic systems. Vishnegradsky showed that whenever the parameters α and β are chosen so that the operating point $M(\alpha; \beta)$ lies above the $\zeta = 0$ curve, algebraic equation 1.14 has all its roots with negative real parts and the corresponding system is stable. Moreover, Vishnegradsky indicated the fact that if parameters α and β are chosen so that the operating point $M(\alpha; \beta)$ lies in the region determined by the $\zeta = 1$ curve, (1.14) will have all its roots real. This corresponds to aperiodic system response.

algebraic equation 1.23 by
$$s = \sigma, \tag{1.3}$$
(1.23) becomes
$$F(\sigma, \alpha, \beta) = 0, \tag{1.53}$$
or, more specifically,
$$F \equiv \sum_{k=0}^{n} a_k \sigma^k = 0, \tag{1.54}$$
where $a_k = a_k(\alpha, \beta)$.

Equation 1.53 or equation 1.54 can be considered as one equation in two unknowns, α and β, and one independent variable σ. Therefore (1.53) may be solved for one of the unknowns in terms of the other and the variable σ under the conditions formulated by the Fundamental Implicit Theorem (Th. A.3). If the solution exists, it can be explicitly obtained as one of the following expressions:
$$\alpha = \alpha(\sigma, \beta) \tag{1.55}$$
or
$$\beta = \beta(\sigma, \alpha) \tag{1.56}$$
depending on the form of functions $a_k(\alpha, \beta)$.

Equation 1.55 or equation 1.56 represent the loci of points in the $\alpha\beta$ plane that correspond to a real root of the algebraic equation having the specified value σ. These loci of points are called the σ curves. By definition of the parameter mapping, the σ curve represents an infinite set of points that is an image of the corresponding point lying on the real axis of the s plane. Then, whenever the parameters α and β are chosen so that the operating point $M(\alpha; \beta)$ lies on a σ curve, the corresponding algebraic equation has a real root $s = \sigma$.

The σ curves may be interpreted in a slightly different way. When $\omega = 0$ in the substitution $s = \sigma + j\omega$, the functions Y_k are all identically equal to zero and $X_k = \sigma^k$ for all k as shown in Derivation A.1. This means that for $\omega = 0$, the imaginary part I of the polynomial $F(s)$, given in (1.30b), is identically equal to zero. The system of two simultaneous equations degenerates to a single equation 1.30a, which has the form of (1.53) or, more specifically, (1.54). Therefore, the $\omega = 0$ curve, which represents the real axis of the s plane, does not exist as a curve but as an infinite set of σ curves corresponding to points on the real axis.

As a consequence, the loci of points in the $\alpha\beta$ plane, corresponding to constant σ but with $\omega \neq 0$, are called the Σ curves to distinguish them from the images of the points from the real axis of the s plane, represented by σ curves, for which $\omega = 0$. In fact, a Σ curve together with the corresponding σ curve may be regarded as the image of the related σ straight line from the s plane; for example, $\sigma = -1$ line of Figure 1.11a. Then the intersection of the σ line with the real axis may be considered as a singular point in parameter mapping, which results in an image curve rather than in an image point. In further applications, it is found convenient to distinguish between the image of the σ constant line for $\omega \neq 0$, which is the Σ curve, and the image of the intersection point of the σ straight line with the real axis when $\omega = 0$, the σ curve.

It should be noted that besides the parameter mapping of the real values of the variable s, there are other singular cases due to a zero value of the related Jacobian. These singular cases occur rarely and are treated in Dr. A.4.

In the specific case of the coefficients a_k as linear functions of the parameters α and β, the σ curves are straight lines determined by

$$\alpha B(\sigma) + \beta C(\sigma) + D(\sigma) = 0, \qquad (1.57)$$

where

$$B = \sum_{k=0}^{n} b_k \sigma^k, \quad C = \sum_{k=0}^{n} c_k \sigma^k, \quad D = \sum_{k=0}^{n} d_k \sigma^k. \qquad (1.58)$$

Equations 1.57 and 1.58 are obtained from (1.23) by substituting a_k with the expression of (1.34). These equations may be used to solve for the real roots of the specific equation 1.14 as functions of the parameters α, β and thereby illustrate the application of the σ curves.

If we substitute $s = \sigma$ in (1.14), the σ curves (or the σ straight lines) are determined by

$$\alpha \sigma^2 + \beta \sigma + \sigma^3 + 1 = 0. \qquad (1.17)$$

By this equation, the points on the real axis of the s plane corresponding to $\sigma = -0.3, -0.5, -1,$ and -2, as shown in Figure 1.14, are mapped into σ straight lines in the parameter $\alpha\beta$ plane as illustrated in Figure 1.15. By definition of the σ curve, the parameter values, $\alpha = 2$ and $\beta = 2$, which determine the location of the operating point $M_1(2; 2)$ on the $\sigma = -1$ line, yield a real root $s = -1$ of (1.14).

The σ curves as shown in Figure 1.15 complete Figure 1.11b for the numerical evaluation of all the roots of the algebraic equation 1.14. By interpolating between the ζ, ω_n, and σ curves of Figure 1.15, both the complex and the real roots can be determined as functions of the parameters α and β. For example, at point $M_2(1; 3)$, the complex conjugate pair of roots $s_{1,2}$ has $\zeta \simeq 0.2$, $\omega_n \simeq 1.7$, and the real root is $s_3 \simeq -0.37$. To improve the accuracy of the obtained results, additional characteristic curves should be plotted in the neighborhood of the point M_2. Therefore Figure 1.15 as the image of Figure 1.14 under the parameter transformation illustrates the main property of the parameter plane mapping—*numerical evaluation of all roots of the algebraic equation as functions of the related parameters*. A more detailed treatment of this evaluation will be given in a following section.

28 Parameter Mapping

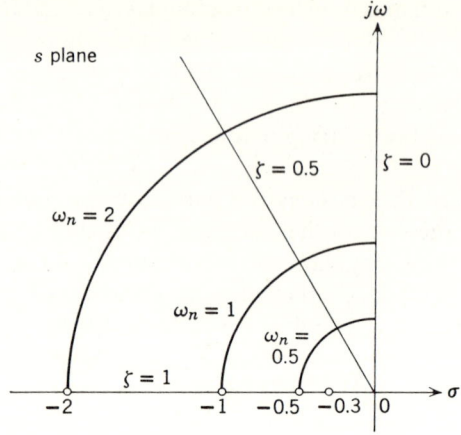

Figure 1.14 The s plane contours and points.

If the algebraic equation

$$F(s) \equiv \sum_{k=0}^{n} a_k s^k = 0 \qquad (1.23)$$

has complex coefficients a_k[12]

$$a_k = (b_k + jc_k)\alpha + (d_k + je_k)\beta + f_k + jh_k, \qquad (k = 0, 1, \ldots, n), \qquad (1.59)$$

the substitution $s = \sigma + j\omega$ into (1.23) yields the two simultaneous equations

$$\begin{aligned} R &\equiv B_1\alpha + C_1\beta + D_1 = 0 \\ I &\equiv B_2\alpha + C_2\beta + D_2 = 0 \end{aligned} \qquad (1.60)$$

in which

$$B_1 = \sum_{k=0}^{n}(b_k X_k - c_k Y_k), \qquad B_2 = \sum_{k=0}^{n}(b_k Y_k + c_k X_k)$$

$$C_1 = \sum_{k=0}^{n}(d_k X_k - e_k Y_k), \qquad C_2 = \sum_{k=0}^{n}(d_k Y_k + e_k X_k) \qquad (1.61)$$

$$D_1 = \sum_{k=0}^{n}(f_k X_k - h_k Y_k), \qquad D_2 = \sum_{k=0}^{n}(f_k Y_k + h_k X_k).$$

In equations 1.61, functions $X_k = X_k(\sigma, \omega)$ and $Y_k = Y_k(\sigma, \omega)$ are as defined by (1.33).

---

[12] In certain cases of nonlinear system analysis in which a multivalued nonlinear characteristic is present, the characteristic equation of the corresponding linearized system has complex coefficients (see Section 4.7).

Basic Equations and Characteristic Curves   29

Equations 1.60 are two simultaneous linear equations in two unknowns, α and β, which can be solved for α and β as

$$\alpha = \alpha(\sigma, \omega)$$
$$\beta = \beta(\sigma, \omega), \quad (1.27)$$

provided that the determinant

$$\Delta = \begin{vmatrix} B_1 & C_1 \\ B_2 & C_2 \end{vmatrix} = B_1 C_2 - B_2 C_1 \quad (1.38)$$

does not vanish.

Equations 1.27 represent a relationship between the complex roots of (1.23) and the parameters α and β in much the same manner as in the real coefficient case. The essential difference, however, between the real and the complex coefficient cases is that the roots corresponding to the latter do not appear necessarily in conjugate pairs. An immediate consequence is that when we map a Jordan arc from the s plane onto the $\alpha\beta$ plane, the whole arc has to

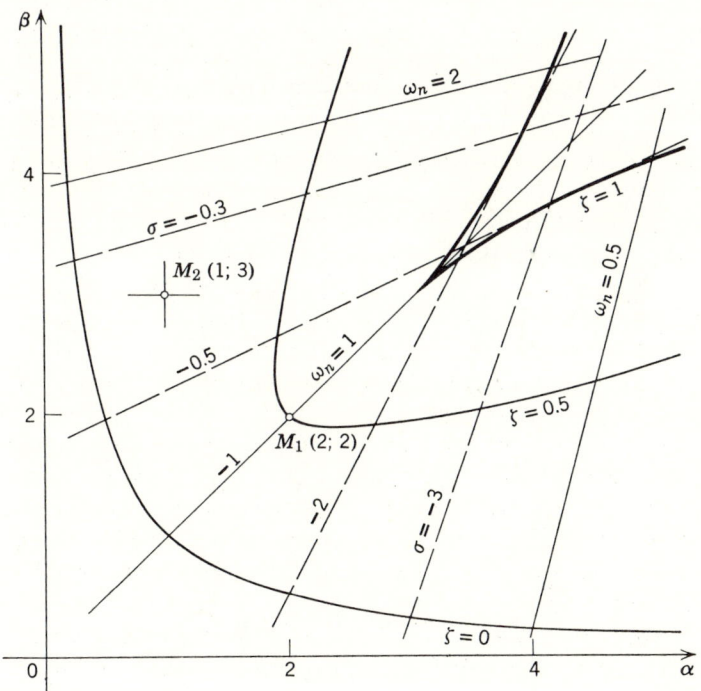

Figure 1.15   Parameter plane curves.

30   Parameter Mapping

be considered; nevertheless, the arc is symmetric with respect to the real axis of the s plane. A formal difference between the real and the complex coefficient cases is that in the latter, equations 1.61 should be used rather than (1.36). This, however, does not imply any significant difficulty in calculating the characteristic Σ and ω curves. A similar relationship between the complex roots of an algebraic equation with complex coefficients and the parameters α and β can be readily derived in terms of the characteristic ζ and ω_n curves.

To illustrate the application of the foregoing relationship, consider an equation of the form

$$s^3 + (1 - j)\alpha s^2 - j\beta s + 1 - j = 0 \tag{1.62}$$

and map the imaginary axis from the s plane onto the $\alpha\beta$ plane. In the parameter plane the imaginary axis is represented by the $\Sigma = 0$ curve. By applying the equations 1.60 and 1.61 to equation 1.62, the $\Sigma = 0$ curve, which corresponds to the substitution $s = j\omega$ and is the image of the imaginary axis under parameter mapping, is determined by

$$\alpha = \frac{\omega^3 + 1}{\omega^2}$$
$$\beta = \omega^2. \tag{1.63}$$

By assigning different values to ω, the $\Sigma = 0$ curve is plotted in Figure 1.16 by solid lines. It can be observed immediately that $\alpha(\sigma, \omega) \neq (\sigma, -\omega)$,

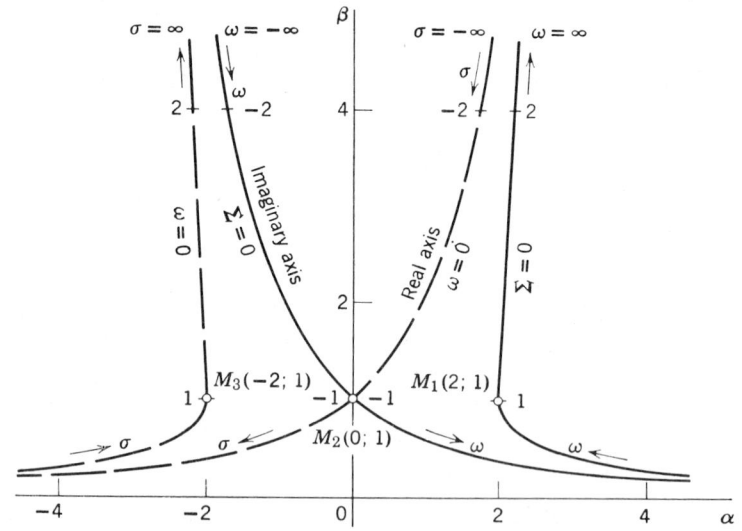

Figure 1.16   Parameter plane curves.

$\beta(\sigma, \omega) \neq \beta(\sigma, -\omega)$, and the $\Sigma = 0$ curve has to be plotted for both the positive and the negative values of the argument ω. Consequently, in an algebraic equation with complex coefficients, the complex roots are not necessarily in conjugate pairs. That is illustrated by the operating point $M_1(2; 1)$ for which (1.62) has a root $s = j$. The conjugate root $s = -j$ appears in (1.62) if the operating point is moved from $M_1(2; 1)$ to $M_2(0; 1)$. Since these two points do not coincide, (1.62) cannot have a conjugate pair of roots $s_{1,2} = \pm j$ for any combinations of the parameters α and β. Furthermore, no combination of α and β can guarantee a complex conjugate pair of roots $s_{1,2} = \pm j\omega$ for any value of ω between $-\infty$ and $+\infty$ since the curve $\Sigma = 0$ does not intersect itself for any value of ω.

Interesting results can be obtained by considering real roots of an algebraic equation with complex coefficients. For $s = \sigma$ ($\omega = 0$), the function X_k and Y_k expressed as (see Dr. A.1 of Appendix A)

$$X_k = \sigma X_{k-1} - \omega Y_{k-1}$$
$$Y_k = \sigma Y_{k-1} + \omega X_{k-1}, \quad (1.64)$$

with $X_0 \equiv 1$, $X_1 \equiv \sigma$, $Y_0 \equiv 0$, and $Y_1 \equiv \omega$, are reduced to

$$X_k = \sigma X_{k-1} = \sigma^k$$
$$Y_k \equiv 0 \quad (1.65)$$

and equations 1.61 become

$$B_1 = \sum_{k=0}^{n} b_k \sigma^k, \quad B_2 = \sum_{k=0}^{n} c_k \sigma^k$$
$$C_1 = \sum_{k=0}^{n} d_k \sigma^k, \quad C_2 = \sum_{k=0}^{n} e_k \sigma^k \quad (1.66)$$
$$D_1 = \sum_{k=0}^{n} f_k \sigma^k, \quad D_2 = \sum_{k=0}^{n} h_k \sigma^k.$$

According to (1.60) and (1.66), we obtain

$$\alpha = \alpha(\sigma)$$
$$\beta = \beta(\sigma). \quad (1.67)$$

A significant difference between the real and complex coefficient cases for real roots representation is that, for a given value of σ, there is a single pair of parameter values (α, β), resulting from the fact that by substitution of $s = \sigma(\omega = 0)$ into algebraic equation 1.23 with complex coefficients a_k, the summation of imaginaries is not necessarily identical to zero as it was in the case with real coefficients. Thus, in the complex coefficient case, the image in the parameter plane of a point on the real axis in the s plane is generally a single point rather than a σ curve.

32 Parameter Mapping

To illustrate the above statement, consider again (1.62) and plot the $\omega = 0$ curve, which is the image of the real axis of the s plane. In accordance with (1.62) and (1.66), equations 1.67 are

$$\alpha = -\frac{\sigma^3 + 1}{\sigma^2}$$
$$\beta = \sigma^2. \tag{1.68}$$

By assigning different values to the variable σ, the $\omega = 0$ curve is plotted by the dashed lines in Figure 1.16. If, for example, the operating point is located at $M_3(-2; 1)$, then (1.62) will have a real root $s = 1$ since, at the point M_2, the value of σ is 1. It is of interest to note that from the diagram of $\omega = 0$ curve, it follows that (1.62) cannot have two real roots, whatever the parameter values of α and β, since the $\omega = 0$ curve does not intersect itself. For (1.62) to have a real root, the operating point $M(\alpha; \beta)$ must be located on the $\omega = 0$ curve.

1.5 Root Distribution

By applying the parameter mapping to closed, simple Jordan curves of the s plane, the parameter $\alpha\beta$ plane can be decomposed into a finite number of *regions*[13] denoted by R_i that correspond to the algebraic equation having i *roots* inside the specified Jordan contour, and $(n - i)$ *roots* outside the contour, where n is the degree of the algebraic equation. In general, the regions are bounded by the complex root and real root boundaries. The complex root boundaries are the images in the $\alpha\beta$ plane of the whole contour except its intersections with the real axis of the s plane. These intersections are mapped in the parameter plane as real root boundaries. To determine the *region index i*, that is, the number of roots in various regions of the $\alpha\beta$ plane after the root boundaries are plotted, rules and graphical techniques have been developed [1.38, 41] which are similar to that used in the conformal mapping procedure. It will be observed, however, that the parameter mapping

---

[13] Before the notion of the region R is given, certain preliminary consideration of the point sets is necessary. As known, a set of points S means an aggregate, class, or collection of points which is formed according to a certain rule. A set S is called *open* if every point of the set S has a neighborhood lying wholly within the set. A set S is called *closed* if the points which are not in the set S form another set S_1 that is an open set. An open set S is called a *connected open set* or *domain D* if, besides being open, it has the property that any two points $(\sigma_0; \omega_0)$ and $(\sigma_1; \omega_1)$ of the set can be joined by a broken line lying wholly within the set S. Consequently, a domain D cannot be formed of two nonoverlapping open sets. The term *region R* is used to represent a set consisting of a domain plus, perhaps, some or all of its boundary points. If all boundary points are included, the region is called a *closed region*; it then necessarily forms a closed set. A domain may be considered as an *open region*.

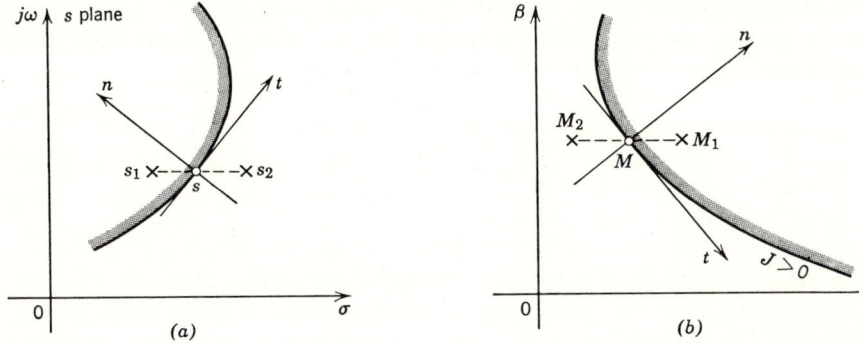

Figure 1.17 Complex root boundaries: (a) s plane boundary; (b) αβ plane boundary.

of the contours from the s plane onto the αβ plane is quite distinct from the conformal mapping.

Consider first the complex root boundaries. In Figure 1.17 are shown the corresponding complex root boundaries in the s and αβ plane. From the definition of parameter plane curves, when the operating point $M(\alpha; \beta)$ is located on the curve plotted for σ and ω, the related algebraic equation has a root value $s = \sigma + j\omega$, which is the antecedent point of M located on the contour C of the s plane. Let M_1 and M_2 be two neighboring points of M, which are located at two different sides of the boundary. Now, as the operating point crosses the complex root boundary in going from one side of the boundary (position M_1) to the other side of the boundary (position M_2), at least one complex root will go across the s plane boundary going either from s_1 to s_2 or vice versa. To determine the correspondence between the crossings in the two planes, the boundaries have to be properly oriented. If a right-handed rectangular system (t, n) of tangent and normal directions of the Jordan curve C is defined at the crossover point s, as shown in Figure 1.17a, the boundary by convention is shaded on the side of the positive n direction, or to the left facing the positive t direction. Similarly, the αβ plane boundary is shaded according to the positive n direction as illustrated in Figure 1.17b. Then, if the operating point M crosses the boundary in the αβ plane from the shaded to the unshaded side, a complex root of the related algebraic equation simultaneously goes across the s plane boundary in the same manner—from the shaded to the unshaded side, and vice versa. A word must be said here regarding the relative orientation of the two mutually perpendicular t and n directions. In the s plane, the two directions form a rectangular right-handed coordinate system similar to the (σ, ω) system; that is, a rotation of $\pi/2$ transforms the positive t axis into the positive n axis, while the inverse transformation requires a rotation of $-\pi/2$. This relative orientation of (t, n)

system is postulated by convention. The orientation of t and n directions in the $\alpha\beta$ plane and the side of the parameter plane curve to be shaded are determined by the sign of the Jacobian $J = J[(R, I)/(\alpha, \beta)]$

$$J = \frac{\partial R}{\partial \alpha}\frac{\partial I}{\partial \beta} - \frac{\partial R}{\partial \beta}\frac{\partial I}{\partial \alpha} \tag{1.28}$$

as stated in the following theorem:

If the Jacobian $J[(R, I)/(\alpha, \beta)]$ is positive (negative), then the positive t and n directions at the point $M(\alpha; \beta)$ on the $\alpha\beta$ plane curve form a right- (left-) handed rectangular coordinate system, and the curve should be shaded to the left (right) facing the positive t direction.

In order to prove this theorem, note that the polynomial function $F = R + jI$ is an entire function in the s plane and, consequently, the Cauchy-Riemann equations, $\partial R/\partial \sigma = \partial I/\partial \omega$, $\partial R/\partial \omega = -\partial I/\partial \sigma$, can be reformulated with respect to two arbitrary perpendicular directions that do not necessarily have to be parallel to the σ and ω axes of the coordinate (σ, ω) system. Thus, for a (t, n) system,

$$\frac{\partial R}{\partial t} = \frac{\partial I}{\partial n}, \quad \frac{\partial R}{\partial n} = -\frac{\partial I}{\partial t}. \tag{1.69}$$

We note that the existence of the four partial derivatives in (1.69) is implied by the existence of $F'(s)$. By using (1.69), we can write the Jacobian $J_1 = J_1[(R, I)/(t, n)]$ of R and I with respect to t and n as

$$J_1 = \frac{\partial R}{\partial t}\frac{\partial I}{\partial n} - \frac{\partial R}{\partial n}\frac{\partial I}{\partial t} = \left(\frac{\partial R}{\partial t}\right)^2 + \left(\frac{\partial I}{\partial t}\right)^2 = |F'(s)|^2, \tag{1.70}$$

which is always positive. (The case $J_1 = 0$ is excluded for obvious reasons.)

Now, define a vector $\mathbf{J}_2 = J_2\mathbf{k}$ as the cross product

$$\mathbf{J}_2 = \Delta\mathbf{t} \times \Delta\mathbf{n}, \tag{1.71}$$

where

$$\begin{aligned}\Delta\mathbf{t} &= \frac{\partial t}{\partial \alpha}\mathbf{i} + \frac{\partial t}{\partial \beta}\mathbf{j} \\ \Delta\mathbf{n} &= \frac{\partial n}{\partial \alpha}\mathbf{i} + \frac{\partial n}{\partial \beta}\mathbf{j}.\end{aligned} \tag{1.72}$$

The vectors $\mathbf{i}$ and $\mathbf{j}$ are unit vectors in the α and β directions, and the vector $\mathbf{k}$ is a unit vector so that the vectors $\mathbf{i}, \mathbf{j}, \mathbf{k}$ form a right-handed rectangular system.[14] Note that the side of the parameter plane curve to be shaded is determined by the positive Δn direction. This direction relative to the vector Δt in the t direction is determined from (1.71) by the sign of $J_2 = J_2[(t, n)/(\alpha, \beta)]$ given as

$$J_2 = \frac{\partial t}{\partial \alpha}\frac{\partial n}{\partial \beta} - \frac{\partial t}{\partial \beta}\frac{\partial n}{\partial \alpha}. \tag{1.73}$$

Now all that is left to be proved is that the sign of J_2 is the same as that of J given in (1.28). It is a matter of elementary operations to verify the relationships

$$J_2\left(\frac{t, n}{\alpha, \beta}\right) \cdot J_3\left(\frac{\alpha, \beta}{t, n}\right) = 1 \tag{1.74}$$

$$J\left(\frac{R, I}{\alpha, \beta}\right) \cdot J_3\left(\frac{\alpha, \beta}{t, n}\right) = J_1\left(\frac{R, I}{t, n}\right), \tag{1.75}$$

where $J_3 = J_3[(\alpha, \beta)/(t, n)]$ is defined as

$$J_3 = \frac{\partial \alpha}{\partial t}\frac{\partial \beta}{\partial n} - \frac{\partial \alpha}{\partial n}\frac{\partial \beta}{\partial t}. \tag{1.76}$$

From (1.74) it follows that J_2 and J_3 have always the same sign. Since J_1 is always positive as shown in (1.70), J has the same sign as J_3 and, therefore, as J_2. This proves the theorem.

In relation to Figure 1.17, it can be concluded that the Jacobian J is greater than zero and the (t, n) system is a right-handed system in the $\alpha\beta$ plane. The parameter plane curve is shaded to the left facing the positive t direction. Then, when the operating point M moves across the boundary from the shaded side (position M_1) to the unshaded side (position M_2), a complex root simultaneously crosses the boundary in the s plane from the shaded side (position s_1) to the unshaded side (position s_2).

To discuss the crossing of the complex boundaries more specifically, consider again the equation

$$s^3 + (1-j)\alpha s^2 - j\beta s + 1 - j = 0 \tag{1.62}$$

and map the s plane contour, which consists of the imaginary ω axis and the infinite semicircle enclosing the left half of the s plane as shown in Figure 1.18a. The image of this contour is the $\Sigma = 0$ curve replotted from Figure

---

[14] If, in equations 1.72, the vectors $\Delta\alpha = \Delta\alpha\mathbf{i}$ and $\Delta\beta = \Delta\beta\mathbf{j}$ are used instead of $\mathbf{i}$ and $\mathbf{j}$ to make the vector

$$\Delta t = \frac{\partial t}{\partial \alpha}\Delta\alpha + \frac{\partial t}{\partial \beta}\Delta\beta$$

to be the vector of the total differential Δt in the t direction, the final result will be the same since sign $J_2 = \text{sign } \Delta\alpha\Delta\beta J_2$.

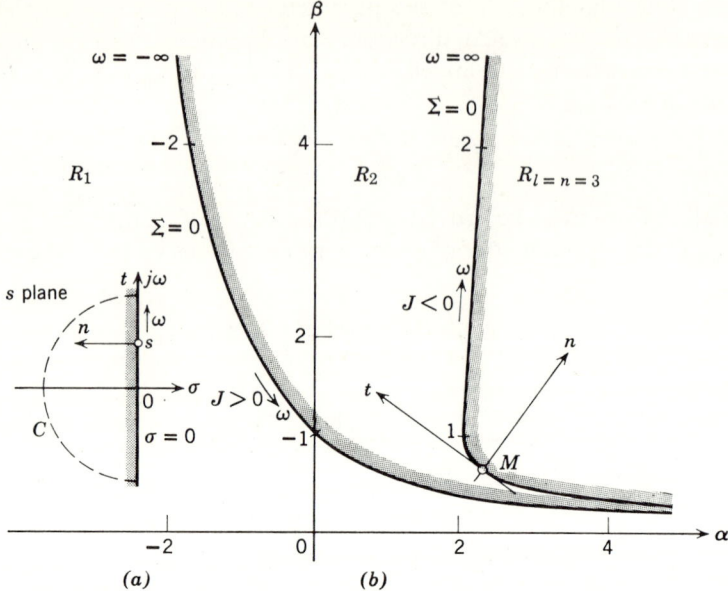

Figure 1.18 Mapping of complex root boundaries: (a) s plane contour; (b) αβ plane diagram.

1.16 onto Figure 1.18b. Since the polynomial on the left side of (1.62) behaves as s^3 on the infinite semicircle of the s plane, and this polynomial term is independent of α and β, the infinite semicircle need not be considered in parameter mapping. Similarly, since the intersection of the contour C with the real axis is at the origin of the s plane ($s = 0$), the origin has no image in the $\alpha\beta$ plane and, consequently, there are no real root boundaries in the parameter plane.

On the imaginary axis and $\Sigma = 0$ curve, $s = j\omega$, the real and imaginary parts corresponding to (1.62) are

$$R = -\alpha\omega^2 + \beta\omega + 1$$
$$I = -\omega^3 + \alpha\omega^2 - 1 \tag{1.77}$$

and the Jacobian from (1.28) is

$$J = \frac{\partial R}{\partial \alpha}\frac{\partial I}{\partial \beta} - \frac{\partial R}{\partial \beta}\frac{\partial I}{\partial \alpha} = -\omega^3. \tag{1.78}$$

In the case of the contour C, along the imaginary axis the (t, n) system is equivalent to the $(\omega, -\sigma)$ system. The $\Sigma = 0$ curve is shaded for negative values of ω to the left facing the direction in which ω decreases, since J is greater than zero. On the second part of the curve $\Sigma = 0$, the Jacobian reverses sign, and for positive values of ω, the $\Sigma = 0$ curve is shaded to the right facing the direction in which ω increases.

After the root boundaries are plotted and properly shaded, it is possible to determine the number of roots that lie inside the specified contour C and that can be determined in each of the regions of the decomposed parameter plane. The region R_l, which corresponds to the maximum number l of the roots lying in C, can be found by inspection using the rules of crossing the complex root boundaries. The region R_l may be the region R_n that corresponds to the case in which all the roots of the related algebraic equation lie inside the s plane contour. It is necessary, but not sufficient, that the region R_n is the region R_l, since R_n may be an empty set; that is, there is no combination of parameters α and β for which all the roots lie inside a specified contour.[15]

To check that the condition $R_l = R_n$ is being satisfied, it is sufficient to know the distribution of the roots at only one point in the parameter plane. In a majority of cases this distribution of the roots at certain points of the $\alpha\beta$ plane can be found by inspection. Therefore, at the origin of the $\alpha\beta$ plane of Figure 1.18b, (1.46) reduces to $s^3 + 1 - j = 0$, which has the three roots $s_1 = 2^{1/6} \exp(j\pi/4)$, $s_2 = 2^{1/6} \exp(j11\pi/12)$, $s_3 = 2^{1/6} \exp(j17\pi/4)$ with s_2 in the left-half plane (or inside the contour C). Thus the origin of the $\alpha\beta$ plane corresponds to the region R_1. The other regions, R_2 and R_3, are determined by inspection.

The real root boundaries are the σ curves which correspond to the points of intersection of the specified contour and the real axis of the s plane. If the intersection points are mapped onto the $\alpha\beta$ plane and the operating point $M(\alpha; \beta)$ crosses one of the plotted σ curves, a real root in the s plane crosses simultaneously the specified s plane contour along the real axis. To determine the direction of the real root motion along the axis, the σ curves should be properly shaded. After the complex root boundaries are shaded in accordance with the sign of the Jacobian, the real root boundaries are simply oriented with respect to the shade of the complex root boundaries. Then, when the point M crosses the real root boundary from the shaded to the unshaded side of the σ curve, the corresponding real root crosses the s plane contour in the same manner from the shaded to the unshaded side along the real axis, and vice versa.

To illustrate the above statements, consider again the algebraic equation

$$s^3 + \alpha s^2 + \beta s + 1 = 0 \qquad (1.14)$$

---

[15] Moreover, as will be shown later, the region R_n, or any other region in the parameter plane, is not necessarily either a bounded or a connected region.

Parameter Mapping

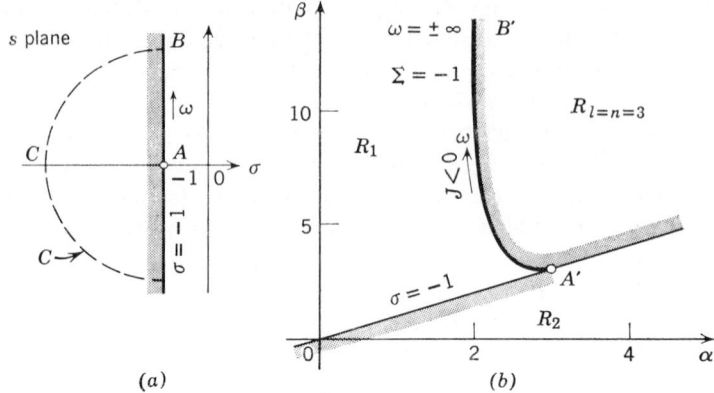

Figure 1.19 Mapping of real root boundaries: (*a*) *s* plane contour; (*b*) $\alpha\beta$ plane curves.

and map the contour C of Figure 1.19a onto the $\alpha\beta$ plane. The image of this contour in the $\alpha\beta$ plane is shown in Figure 1.19b. The complex root boundary is the $\Sigma = -1$ curve replotted from Figure 1.11b. The curve should be double-shaded because the Jacobian reverses the sign when ω changes the sign, but the direction in which ω increases is also reversed at $\omega = 0$. This double shading is true whenever the parameter mapping is applied to algebriac equations with real coefficients and the contours to be mapped are symmetric with respect to the real axis of the *s* plane. For simplicity in such cases, the complex root boundaries will be single-shaded, although we must remember that whenever the operating point crosses the complex root boundary, two complex conjugate roots simultaneously cross the *s* plane contour according to the shading rule.

The real root boundaries are the σ curves corresponding to the points of intersection of the specified *s* plane contour with the real axis of the *s* plane. These points are the point A and the point C which lies at infinity where the infinite semicircle of Figure 1.19a intersects the real axis. Since, for $s \to \infty$, the dominant polynomial term is s^3, which is independent of α and β, the real root boundary $\sigma = -\infty$ does not exist in the $\alpha\beta$ plane. The other real root boundary is $\sigma = -1$ at the point A. By using (1.17) for $\sigma = -1$, the σ curve is plotted in Figure 1.19b. In general, the side of real root boundaries to be shaded is determined at the points in the $\alpha\beta$ plane, which correspond to the intersections of the *s* plane contour with the real axis. In the case of Figure 1.19b, the competent point is A', which is related to A of Figure 1.19a. At the point A', the shade on the real root boundary must be on the same side as that on the corresponding complex root boundary. This is illustrated in Figure 1.19b.

A question may now arise whether the described way of shading is unique or there are some other possibilities of shading real root boundaries that would conflict with the described one. All the other possibilities are illustrated in Figure 1.20. As can be seen from this figure, none of the shading, other than that of Figure 1.19b, is consistent. In encircling the point A' and counting the number of roots in each of the regions starting with an arbitrary number l at one of the regions, it can be concluded that in each case of Figure 1.20 a contradiction is obtained. After encircling the point A' once around, the starting and terminal region indices are not equal, and the statement about shading of the real root boundaries is proved by eliminating all possible alternatives.

Although the parameter mapping can be applied to any simple closed Jordan contour in the s plane, the usual contours to be mapped into the $\alpha\beta$ plane are the ones shown in Figure 1.21 or a combination of them. Then the root boundaries in the $\alpha\beta$ plane are the characteristic curves defined previously. The application of the parameter mapping to the contour of Figure 1.21b will be illustrated by the following example.

Consider the algebraic equation

$$0.04\alpha s^4 + (0.36\alpha + 0.04)s^3 + (2\alpha + 0.36)s^2$$
$$+ (2\alpha + 0.4\beta + 2)s + 2\beta + 2 = 0. \quad (1.79)$$

It is required to determine values of the parameters α and β for which all the roots of (1.79) are located inside the relative damping contour of Figure 1.21b when $\zeta = 0.5$.

To solve the above root distribution problem in the $\alpha\beta$ plane, note that the variable parameters appear linearly in the coefficients of (1.79). Then, with the help of equations 1.34, 1.36, and 1.37, the parameter plane equations are

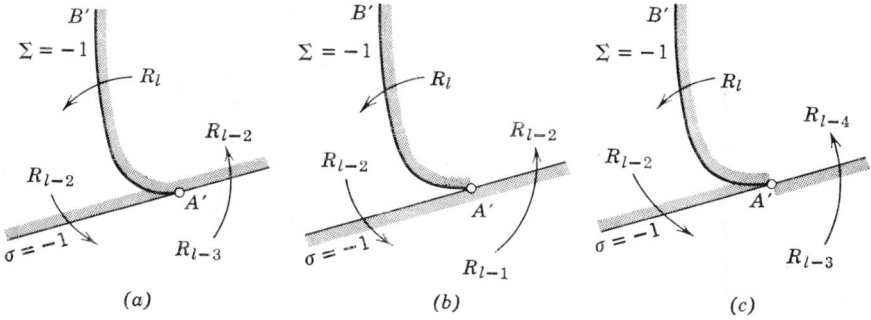

Figure 1.20 Incorrect shading of the real root boundaries.

derived as

$$\alpha = \frac{-0.016\omega_n^3 + 0.144\omega_n^2 - 0.72\omega_n + 3.2}{0.064\omega_n^3 - 0.8\omega_n^2 + 4\omega_n - 4}$$

$$\beta = \frac{0.0016\omega_n^6 - 0.0144\omega_n^5 + 0.05\omega_n^4 - 0.56\omega_n^3 + 3.28\omega_n^2 - 4\omega_n + 4}{0.064\omega_n^3 - 0.8\omega_n^2 + 4\omega_n - 4}.$$

(1.80)

These equations determine the $\zeta = 0.5$ curve in the parameter plane, which represents the complex root boundary corresponding to the specified s plane contour. Assigning different values to ω_n between 0 and $+\infty$ along the upper

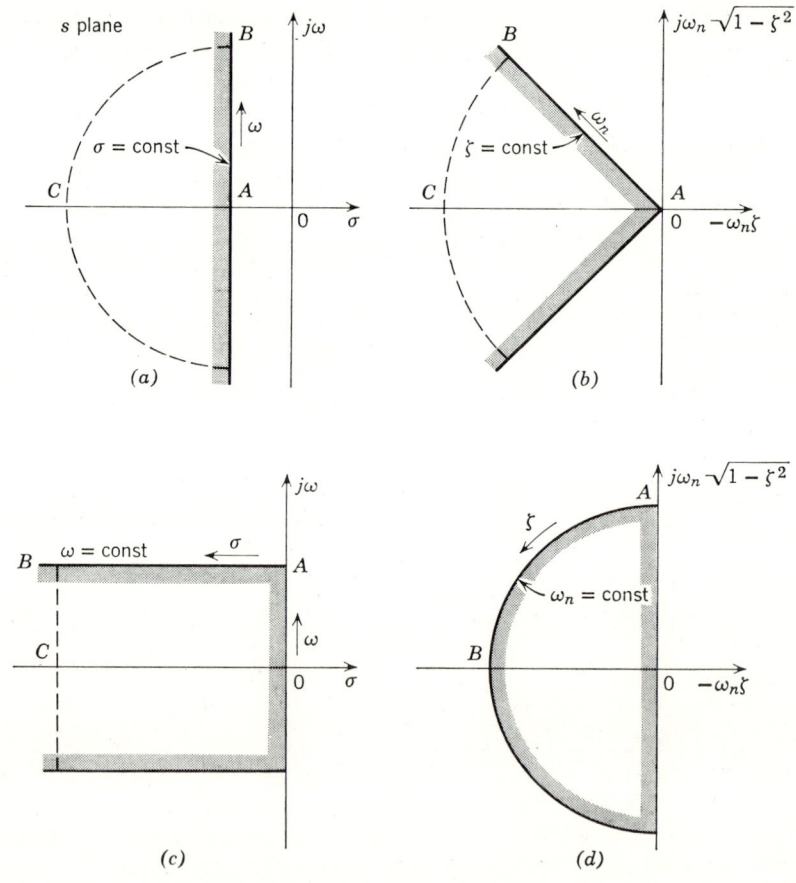

Figure 1.21 The s plane contours.

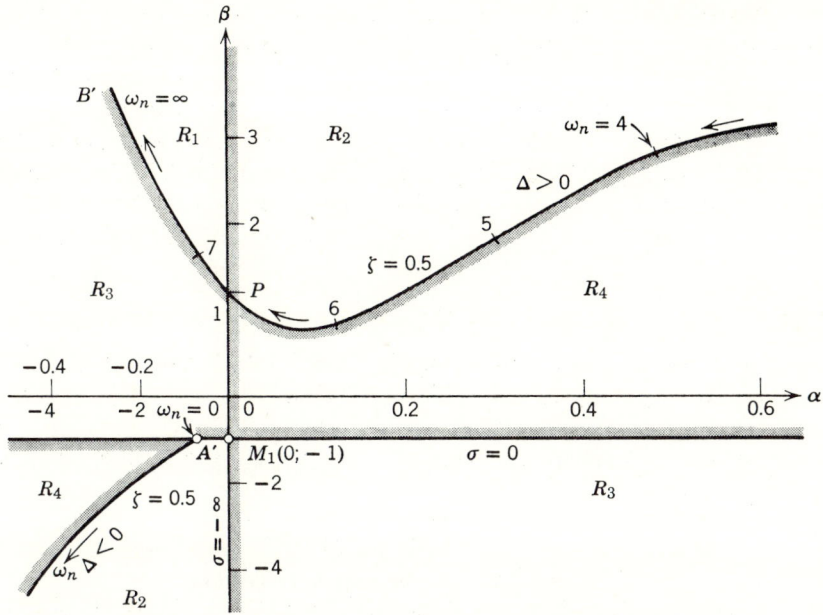

Figure 1.22 Root distribution in the parameter plane.

constant damping line AB of Figure 1.21b, and using (1.80), yields the $\zeta = 0.5$ curve. The curve consists of two branches as plotted in Figure 1.22.[16]

From Figure 1.21b it is seen that the real roots may leave the contour along the real axis at the origin and infinity of the s plane. Thus two real root boundaries, $\sigma = 0$ and $\sigma = -\infty$, can exist. Putting the value $s = \sigma = 0$ into (1.79), the corresponding boundary is

$$\beta = -1. \tag{1.81}$$

In general, the real root boundary $\sigma = 0$ that corresponds to the zero root $s = 0$ is obtained by equating to zero the absolute term a_0 of the algebraic equation.

The possibility for a root to leave an s plane contour through infinity requires a more detailed study of the boundary $\sigma = -\infty$ with the introduction of the Riemann sphere. Consider the sphere ψ of Figure 1.23 with a radius $\frac{1}{2}$ and the center at $(x; y; z) = (0; 0; \frac{1}{2})$, which is tangent to the s plane ($z = 0$) at the point $0(0; 0; 0)$ diametrically opposite the point $V(0; 0; 1)$. The line joining a point $S(x/(1-z); y/(1-z); 0)$, to the point V will

---

[16] Note again that the calculation and plotting of the characteristic curves can be performed by the use of a general computer program as shown in Appendix B.

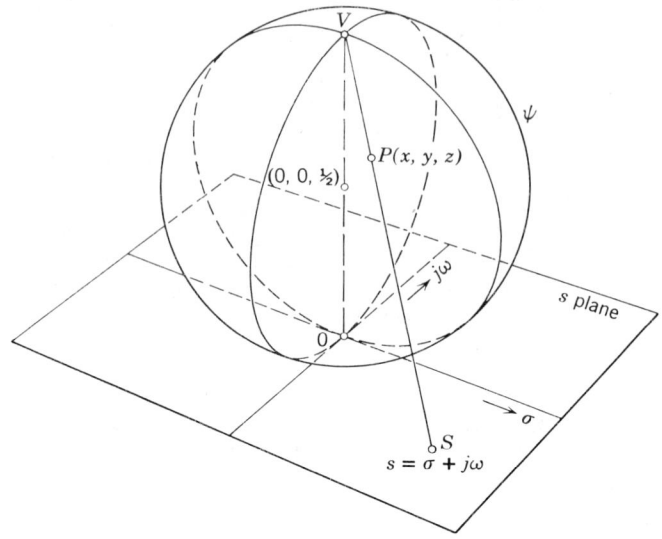

Figure 1.23 Stereographic projection.

pierce the sphere ψ at the point $P(x; y; z)$, which is a projection of the point S on the sphere ψ with the center of projection at the point V. The mapping of points from the s plane onto the sphere ψ is a continuous one-to-one mapping. If the point S tends to infinity in any direction, the projection P will approach the point V, and every complex number s, including $s = \infty$, is represented by a point on the sphere ψ. Therefore, the projection of a contour from Figure 1.21, which has a point at infinity, is a closed contour on the sphere ψ. The projected contour on ψ contains the point V, and the roots may leave the contour through that point.

To find the real root boundary $\sigma = -\infty$, note that by the transformation $s = 1/w$ the algebraic equation $F(s) = 0$ is transformed to $F_1(w) = 0$, which has a zero root $w = 0$ corresponding to infinity root $s = \pm\infty$ of equation $F(s) = 0$. The condition for zero root $w = 0$ of $F_1(w) = 0$ is that the absolute term of $F_1(w)$ is equal to zero. According to the substitution $s = 1/w$, the absolute term of $F_1(w)$ is the coefficient a_n of $F(s) = 0$. Thus, by equating to zero $a_4 = 0.04\alpha$ of (1.79), the real root boundary $\sigma = -\infty$ is found to be

$$\alpha = 0. \tag{1.82}$$

By using the rules for shading the complex root boundaries, which are outlined above, one branch of the $\zeta = 0.5$ curve in Figure 1.22 is shaded on the left and the other branch on the right, since the determinant $\Delta = 0.064\omega_n^3 - 0.8\omega_n^2 + 4\omega_n - 4$ in the denominators of (1.80) changes the sign passing from one branch to the other. After the complex root boundaries are shaded, the real root boundary $\sigma = 0$ is simply oriented at the point A' as explained previously. On the other hand, the point B' is in infinity and there is no change in the shading on the corresponding real root boundary $\sigma = -\infty$ in the finite $\alpha\beta$ plane. It is to be noted that the other intersections between the real root and complex root boundaries (for example, point P of Figure 1.22) should not be considered.

If the root boundaries are appropriately shaded, the number of roots that lie inside the specified contour can be determined in each of the regions of the parameter plane. First, regions R_l, which correspond to the maximum number l of the roots, may be found by inspection using the rules for crossing the root boundaries. These regions may relate to the case in which all n roots are located inside the s plane contour. As mentioned before, to find $R_l = R_n$, it is sufficient to determine the index i of any region R_i in the $\alpha\beta$ plane. Unfortunately, there is no simple and general way to determine the index i from the parameter plane plot by avoiding the factorization of the polynomial at a point on the $\alpha\beta$ plane. In the foregoing section we saw how the factorization can be performed by plotting several parameter plane curves.[17]

In a majority of practical cases the number of roots at a point on the parameter plane can be found by inspection. In the example under investigation, for point $M_1(0; -1)$ of Figure 1.22 where $\alpha = 0$ and $\beta = -1$, (1.79) reduces to

$$0.04s^3 + 0.36s^2 + 1.6s = 0, \qquad (1.83)$$

which has its roots readily obtained as $s_1 = 0$, $s_{2,3} = -4.5 \pm j4.45$ ($\omega_n = 6.32$; $\zeta = 0.713$), and $s_4 = -\infty$. Since, in the neighborhood of the point M_1, the complex roots $s_{2,3}$ remain in the specified contour ($\zeta \geq 0.5$), by moving M_1 slightly inside the region with l roots, the two real roots will enter the contour because of the shading rule. Therefore from Figure 1.22 the values of the parameters α and β, which yield the algebraic equation 1.79 having all the roots with the damping coefficients $\zeta = 0.5$, may be found without any calculations. The parameter values corresponding to $\zeta \geq 0.5$ should be chosen from R_4. Note that the region R_4 is neither a bounded nor a connected region and that there are two distinct areas in the parameter plane for which all the roots are located inside the specified s plane contour.

The outlined graphical techniques may be applied to a mapping of more complicated s plane contours and can be used to solve the root distribution problem in a general manner. To illustrate this statement, consider the

---

[17] Actually, the factorization can be easily avoided if the s plane contours are mapped onto the $F(s)$ plane and the obtained plot is interpreted by applying the Cauchy argument principle as shown in Th. A.6.

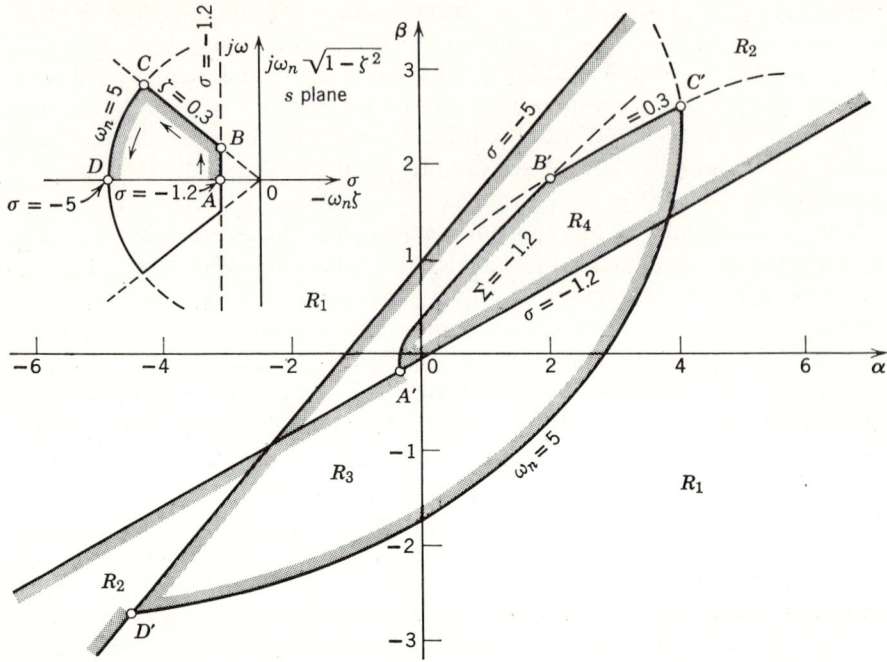

Figure 1.24 The s plane contour and parameter plane diagram.

algebraic equation

$$0.04s^4 + 0.34s^3 + (0.2\alpha + 1.12)s^2 + (0.5\alpha + \beta + 1.7)s + 2\beta + 1 = 0 \tag{1.84}$$

and determine the values of α and β so that (1.84) has all its roots within the s plane contour as shown in the upper left corner of Figure 1.24.

The specified s plane contour is a combination of the contours shown in Figure 1.21. Each part of the contour can be mapped onto the $\alpha\beta$ plane separately by equations

$$\alpha = \frac{C_1 D_2 - C_2 D_1}{B_1 C_2 - B_2 C_1}$$
$$\beta = \frac{B_2 D_1 - B_1 D_2}{B_1 C_2 - B_2 C_1}, \tag{1.37}$$

where

$$\begin{aligned}
&B_1 = 0.2Y_1^* + 0.5Y_0^*, \qquad B_2 = 0.2Y_2^* + 0.5Y_1^*, \\
&C_1 = Y_0^* + 2Y_{-1}^*, \qquad\quad C_2 = Y_1^* + 2Y_0^*, \\
&D_1 = 0.04Y_3^* + 0.34Y_2^* + 1.12Y_1^* + 1.7Y_0^* + Y_{-1}^*, \\
&D_2 = 0.04Y_4^* + 0.34Y_3^* + 1.12Y_2^* + 1.7Y_1^* + Y_0^*.
\end{aligned} \tag{1.85}$$

By using these equations, the complex root boundaries $\Sigma = -1.2$, $\zeta = 0.3$, and $\omega_n = 5$ curves are plotted in Figure 1.24.

The real root boundaries, which correspond to $\sigma = -1.2$ and $\sigma = -5$, are respectively

$$-0.312\alpha + 0.8\beta + 0.07 = 0 \tag{1.86}$$

$$2.5\alpha - 3\beta + 3 = 0. \tag{1.87}$$

Applying the outlined graphical techniques, we determine region R_4, which corresponds to the case in which all the roots are inside the specified contour of the s plane. Note that R_4 is a connected region.

1.6 Root Evaluation

It has been pointed out that equations

$$\begin{aligned}\sigma &= \sigma(\alpha, \beta) \\ \omega &= \omega(\alpha, \beta)\end{aligned} \tag{1.29}$$

may represent an inverse parameter mapping. This mapping is one-to-n point transformation since, for each point in the $\alpha\beta$ plane, there are n points in the s plane that represent n roots of the corresponding algebraic equation. It was also pointed out that the functions $\sigma(\alpha, \beta)$ and $\omega(\alpha, \beta)$ cannot be obtained analytically. However, in connection with Figure 1.15, it was shown that all the roots can be evaluated by interpolating between the characteristic curves. This point transformation from the parameter plane into the s plane can be facilitated by the use of the $\zeta = 1$ curve.

Consider again the algebraic equation

$$0.04\alpha s^4 + (0.36\alpha + 0.04)s^3 + (2\alpha + 0.36)s^2 \\ + (2\alpha + 0.4\beta + 2)s + 2\beta + 2 = 0 \tag{1.79}$$

and note that the parameters α and β appear linearly. Thus the σ curves are straight lines tangent to the $\zeta = 1$ curve. With the help of (1.80) the $\zeta = 1$ and $\zeta = 0.5$ curves are plotted in Figure 1.25. The point $M_1(0.105; 0.75)$ is located on the $\zeta = 0.5$ curve at the point where $\omega_n = 6.1$. Thus the complex pair of roots of (1.79) for $\alpha = 0.105$ and $\beta = 0.75$ is given by $s_{1,2} = -3.05 \pm j5.28$. Real roots are evaluated from σ lines. Since the point M_1 lies on the $\sigma = -10$ line, the third root is $s_3 = \sigma = -10$. The last fourth root may be determined by interpolating between the $\sigma = -3$ and $\sigma = -2$ lines (it has been evaluated as $s_4 = -2.25$). Similarly, if $M_2(0.175; 1.1)$ is considered, the complex roots are computed from the $\zeta = 0.5$ curve. The real root $s_3 = -\omega_n = \sigma = -3$ is evaluated from the point of the $\zeta = 1$ curve at which a straight line drawn from M_2 is tangent to the curve (see Dr. A.3).

46 Parameter Mapping

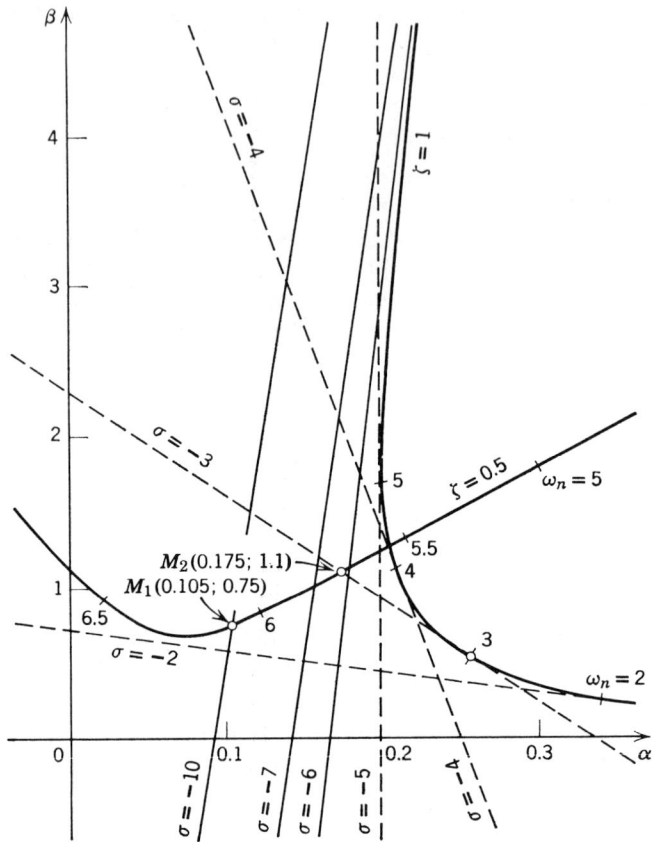

Figure 1.25 Root evaluation.

The fourth root is $s_4 = -6.3$. It is important to note that in choosing a point $M(\alpha; \beta)$ to the left of the $\zeta = 1$ curve, we can draw two tangents to the curve and there are always two real roots of (1.79). The other two are complex. On the right side of the $\zeta = 1$ curve, (1.79) has two pairs of complex conjugate roots since no tangents can be drawn to the curve. By moving along the $\zeta = 0.5$ curve and calculating the corresponding root values, a root locus is sketched as shown in Figure 1.26. On the root locus, three points, a, b, and c, are indicated that correspond to the parameter values of $(\alpha; \beta)$, and they are taken as (0.25; 1.45), (0.19; 1.08), and (0.12; 0.8), respectively. Note that the locus is plotted by varying both parameters so that two complex conjugate roots are forced to lie on the $\zeta = 0.5$ damping line in the s plane.

Another example of inverse parameter transformation is a mapping of a straight line from the $\alpha\beta$ plane onto the s plane. In other words, from the parameter plane diagram the roots are calculated along a given straight line. If the straight line is parallel to any of the parameter plane axes, then the inverse transformation results in a root locus corresponding to the parameter whose axis is parallel to the mapped line. This situation is illustrated in Figure 1.27, which represents the parameter plane diagram of Section 1.5 (Figure 1.25) with a few more ζ and ω_n curves. The parameter is chosen to be $\beta = 3$ and the parameter α is varied along the straight line AB. By calculating the roots along the line AB, a root locus of Figure 1.28 is obtained. The corresponding values of α are interpolated along the locus. A family of root loci for different values of β may be obtained from Figures 1.25 and 1.27 in a straightforward manner.

By using the technique displayed in the above example, another problem of inverse mapping can be solved. It is possible to map a closed contour from the $\alpha\beta$ plane onto the s plane. The result will be root areas rather than a root locus, which indicates root locations of the corresponding algebraic equation for all possible pairs (α, β) chosen inside the closed contour of the $\alpha\beta$ plane.

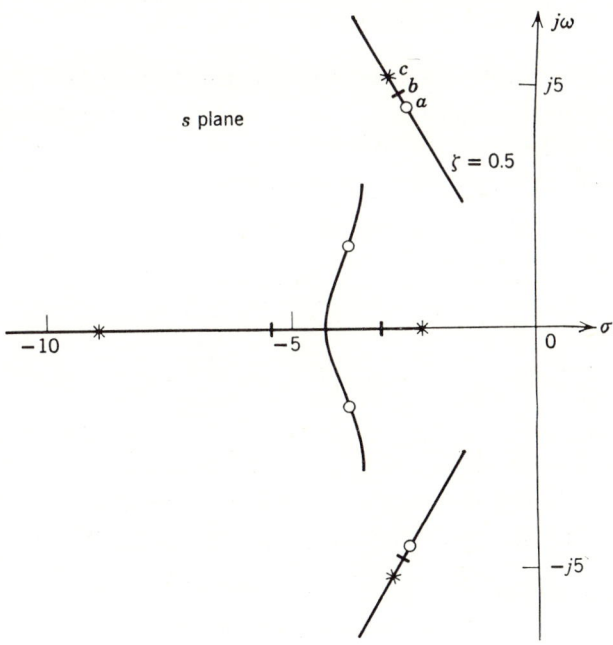

Figure 1.26 Two-parameter root locus.

Parameter Mapping

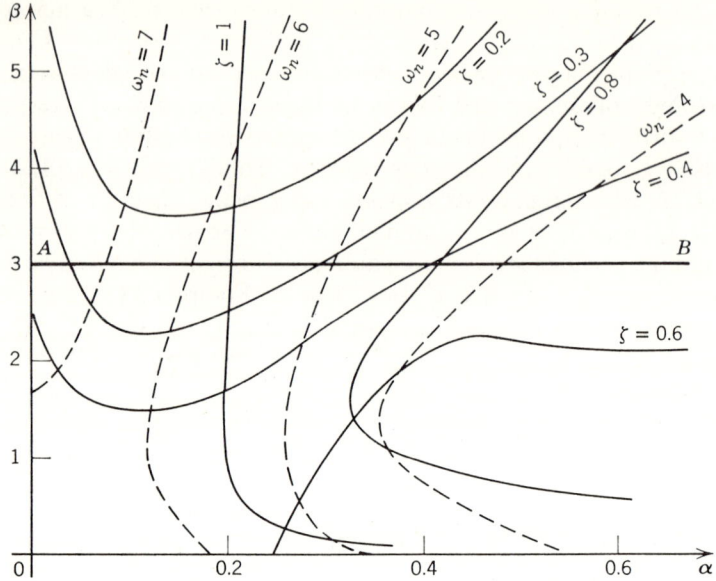

Figure 1.27 Parameter plane diagram.

The rules of the mapping and the mapping procedure itself will be best shown by the following example.

Consider again the algebraic equation

$$0.04s^4 + 0.3s^3 + (0.2\alpha + 1.12)s^2 + (0.5\alpha + \beta + 1.7)s + 2\beta + 1 = 0. \quad (1.84)$$

By using

$$\alpha = \frac{C_1 D_2 - C_2 D_1}{B_1 C_2 - B_2 C_1}$$

$$\beta = \frac{B_2 D_1 - B_1 D_2}{B_1 C_2 - B_2 C_1}, \quad (1.37)$$

where

$$\begin{aligned}
&B_1 = 0.2 Y_1^* + 0.5 Y_0^*, \qquad B_2 = 0.2 Y_2^* + 0.5 Y_1^*, \\
&C_1 = Y_0^* + 2 Y_{-1}^*, \qquad\quad C_2 = Y_1^* + 2 Y_0^*, \\
&D_1 = 0.04 Y_3^* + 0.34 Y_2^* + 1.12 Y_1^* + 1.7 Y_0^* + Y_{-1}^*, \\
&D_2 = 0.04 Y_4^* + 0.34 Y_3^* + 1.12 Y_2^* + 1.7 Y_1^* + Y_0^*,
\end{aligned} \quad (1.85)$$

the net of ζ, ω_n, and σ curves are plotted in Figure 1.29. This net enables a numerical evaluation of all roots of (1.84) for various values of system parameters. The pair of complex conjugate roots is found in the usual fashion by interpolating between the ζ and ω_n curves. The real roots are evaluated by interpolating between the σ curves, which are straight lines because the parameters α and β appear linearly in (1.84). It should be noted that the dotted σ lines determine one real root and the other real root is evaluated by interpolating between the solid σ lines.

The closed contour to be mapped from the $\alpha\beta$ plane onto the s plane is a parallelogram $ABCD$ that includes all possible combinations of the parameters α and β in the given range

$$4 \leq \alpha \leq 10, \quad 1 \leq \beta \leq 3. \tag{1.88}$$

The parallelogram $ABCD$ is plotted in Figure 1.29. By performing a factorization of (1.84) along the contour $ABCD$, it is possible to obtain a root area diagram as shown in Figure 1.30. As a result of inverse mapping of the closed contour from the $\alpha\beta$ plane onto the s plane, a disconnected set of roots is

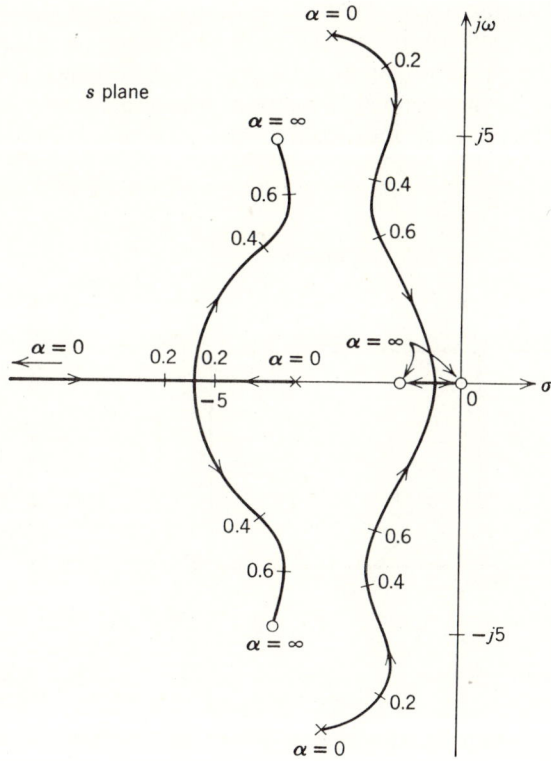

Figure 1.28 Root locus for $\beta = 3$.

obtained. To each of the four roots, however, there is associated a connected closed set of points indicating various positions of the corresponding root for the range (1.88) of values (α, β). It should be noted that the sets may be infinite and/or overlapping each other. In the latter case the concept of Riemann surfaces may be used to interpret the location of the roots. The number of roots is always n; nevertheless, two or more of them may come from overlapping sets.

As expected, it can be seen that the shading rule in the inverse mapping applies. Of course, the shading rule is valid only for the mapping of the complex roots. We can verify the rule by noting that in enclosing the parallelogram $ABCD$ as indicated by the arrows in the counterclockwise direction, the contour $A'B'C'D'$ of the s plane in Figure 1.30 is enclosed in the opposite direction. This occurs because the Jacobian of the inverse transformation has the same sign as the Jacobian of the direct transformation (see equation 1.74), the latter being negative. (It can be observed from the relative location of the ζ and ω_n curves that the Jacobian of the direct transformation is negative. For example, by following any particular ζ curve in the direction in which ω_n increases, the right side of the curve should be shaded, since it corresponds to an increase in the value of ζ and thus a negative Jacobian is indicated.)

In this section, large parameter variations of α and β have been considered, and their effects on root locations of the algebraic equation have been

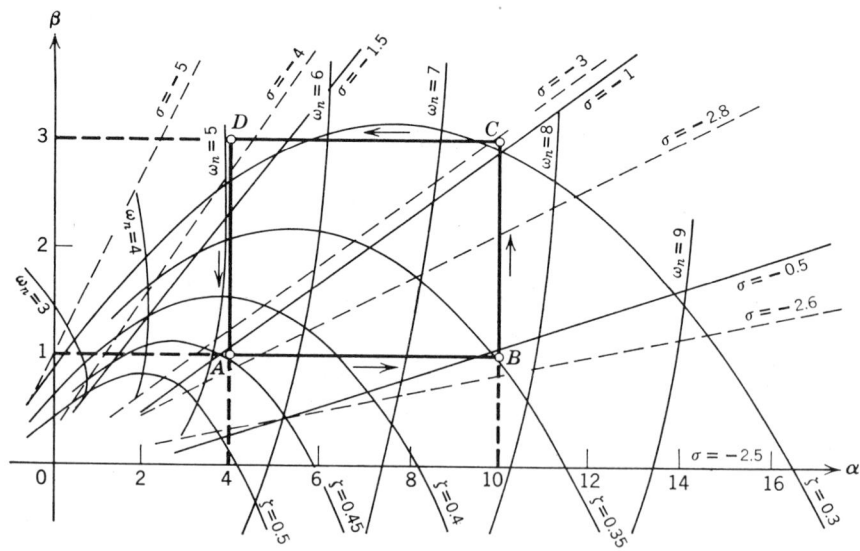

Figure 1.29 Parameter plane diagram for inverse mapping.

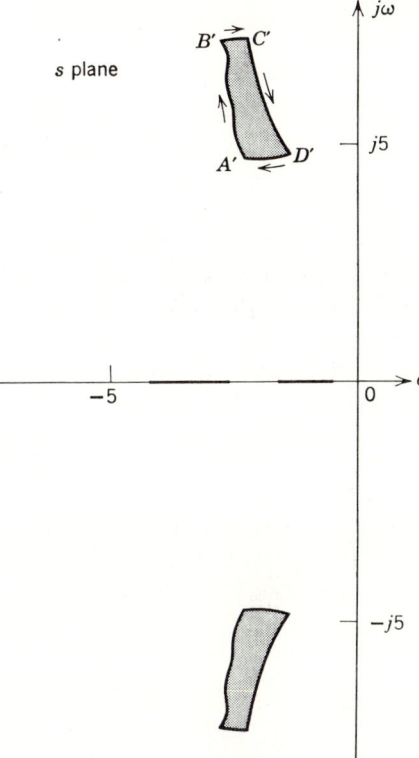

Figure 1.30 Root area diagram of inverse mapping.

evaluated. In the section that follows, the variations of the root locations due to *small parameter variations* are investigated.

1.7 Root Sensitivity

The continuous property of parameter mapping may be utilized to determine how small changes in the parameters effect the root values, this being called the *root sensitivity* to parameter variations. To give a mathematical formulation of the root sensitivity problem, consider the algebraic equation

$$F(s) \equiv \sum_{k=0}^{n} a_k s^k = 0, \qquad (1.23)$$

where the coefficients a_k ($k = 0, 1, \ldots, n$) are real and continuous functions

52   Parameter Mapping

of the parameters p_r $(r = 1, 2, \ldots, m)$

$$a_k = a_k(p_r), \qquad (k = 0, 1, \ldots, n). \tag{1.89}$$

For the case in which a set of numerical values of the parameters p_r^0 $(r = 1, 2, \ldots, m)$ is given, (1.89) yields the numerical values of the coefficients a_k^0 $(k = 0, 1, \ldots, n)$ that determine the corresponding root locations of the algebraic equation 1.23. Thus any variations of the parameters will result in a change of the root locations. In the preceding section the change of root locations has been investigated for large variations in the parameters α and β on the basis of explicit formulation of the parameter mapping by the functions $\alpha = \alpha(\sigma, \omega)$, $\beta = \beta(\sigma, \omega)$ or $\alpha = \alpha(\omega_n, \zeta)$, $\beta = \beta(\omega_n, \zeta)$. To study the small variations of parameters, it is not necessary to have the explicit expression, since the implicit definition of parameter mapping through the algebraic equation 1.23 is sufficient.

In order to evaluate numerically the change of root locations due to differential variations of parameters, the complex roots and the real roots are considered separately, and attention is focused on each individual root and each individual parameter independently. Therefore the definitions of complex and real root sensitivity to a parameter variation are to be established.

To define the complex root sensitivity, express the ith pair of complex roots of (1.23) as

$$s_{i,i+1} = \sigma_i \pm j\omega_i. \tag{1.90}$$

The sensitivity of complex roots $s_{i,i+1}$ can now be investigated in the real domain by introducing the sensitivity of the real part σ_i and the frequency ω_i.

The sensitivity of the real part σ_i with respect to a differential change in a parameter p_r, which will be denoted by $S_{i,r}^\sigma$, may be defined as the logarithmic derivative of σ_i with respect to the logarithm of the parameter p_r as

$$S_{i,r}^\sigma \equiv \frac{\partial \ln \sigma_i}{\partial \ln p_r}. \tag{1.91}$$

Similarly, the sensitivity of the frequency ω_i denoted by $S_{i,r}^\omega$ is defined as

$$S_{i,r}^\omega \equiv \frac{\partial \ln \omega_i}{\partial \ln p_r}. \tag{1.92}$$

If the jth real root of (1.23) is denoted by s_j (the notation σ_j is reserved for complex roots and therefore s_j is used to avoid confusion), the real root sensitivity $S_{j,r}^s$ is defined as

$$S_{j,r}^s \equiv \frac{\partial \ln s_j}{\partial \ln p_r}. \tag{1.93}$$

Root Sensitivity

In order to determine complex root sensitivity it is necessary to rewrite (1.23) as two simultaneous equations

$$\sum_{k=0}^{n} a_k X_k = 0$$
$$\sum_{k=0}^{n} a_k Y_k = 0$$
(1.30)

as shown previously. Since $a_k = a_k(p_r)$, $X_k = X_k(\sigma, \omega)$, $Y_k = Y_k(\sigma, \omega)$, equations 1.30 represent an implicit relationship between the variables p_r, σ, and ω. Thus, differentiating equations 1.30 with respect to a parameter p_r, we obtain

$$\frac{\partial \ln \sigma}{\partial \ln p_r} \frac{\sigma}{p_r} \sum_{k=0}^{n} k a_k X_{k-1} - \frac{\partial \ln \omega}{\partial \ln p_r} \frac{\omega}{p_r} \sum_{k=0}^{n} k a_k Y_{k-1} + \sum_{k=0}^{n} \frac{\partial a_k}{\partial p_r} X_k = 0$$
$$\frac{\partial \ln \sigma}{\partial \ln p_r} \frac{\sigma}{p_r} \sum_{k=0}^{n} k a_k Y_{k-1} + \frac{\partial \ln \omega}{\partial \ln p_r} \frac{\omega}{p_r} \sum_{k=0}^{n} k a_k X_{k-1} + \sum_{k=0}^{n} \frac{\partial a_k}{\partial p_r} Y_k = 0.$$
(1.94)

In deriving equations 1.94 we use the relationship

$$\frac{\partial X_k}{\partial \sigma} = k X_{k-1}, \qquad \frac{\partial X_k}{\partial \omega} = -k Y_{k-1}$$
$$\frac{\partial Y_k}{\partial \sigma} = k Y_{k-1}, \qquad \frac{\partial Y_k}{\partial \omega} = k X_{k-1},$$
(1.95)

which is obtained by differentiating equation

$$s^k = X_k + j Y_k$$
(1.32)

with respect to σ and ω. From (1.95) it follows that the Cauchy-Riemann conditions

$$\frac{\partial X_k}{\partial \sigma} = \frac{\partial Y_k}{\partial \omega}, \qquad \frac{\partial X_k}{\partial \omega} = -\frac{\partial Y_k}{\partial \sigma},$$
(1.96)

corresponding to the function s^k, are satisfied. The conditions of (1.95) together with (1.96) represent another important relationship between the functions X_k and Y_k.

Now, if the numerical values a_k^0 ($k = 0, 1, \ldots, n$) corresponding to a specific set of parameters p_r^0 ($r = 1, 2, \ldots, m$) are given, attention is focused on the ith pair of complex roots, and equations 1.94 become

$$P_1 S_{i,r}^\sigma + Q_1 S_{i,r}^\omega + R_1 = 0$$
$$P_2 S_{i,r}^\sigma + Q_2 S_{i,r}^\omega + R_2 = 0,$$
(1.97)

where

$$P_1 = \sigma_i \sum_{k=0}^{n} k a_k X_{k-1}, \qquad P_2 = \sigma_i \sum_{k=0}^{n} k a_k Y_{k-1}$$

$$Q_1 = -\omega_i \sum_{k=0}^{n} k a_k Y_{k-1}, \qquad Q_2 = \omega_i \sum_{k=0}^{n} k a_k X_{k-1} \qquad (1.98)$$

$$R_1 = p_r^{\,0} \sum_{k=0}^{n} \frac{\partial a_k}{\partial p_r^{\,0}} X_k, \qquad R_2 = p_r^{\,0} \sum_{k=0}^{n} \frac{\partial a_k}{\partial p_r^{\,0}} Y_k.$$

In (1.98) the functions X_{k-1} and Y_{k-1} are calculated for $\sigma = \sigma_i$, $\omega = \omega_i$, and $\partial a_k/\partial p_r^{\,0} = (\partial a_k/\partial p_r)|_{p_r = p_r^{\,0}}$.

Equations 1.97 are two equations in two unknowns, $S_{i,r}^{\sigma}$ and $S_{i,r}^{\omega}$, which may be solved to obtain the complex root sensitivity as

$$\begin{aligned} S_{i,r}^{\sigma} &= \frac{Q_1 R_2 - Q_2 R_1}{P_1 Q_2 - P_2 Q_1} \\ S_{i,r}^{\omega} &= \frac{P_2 R_1 - P_1 R_2}{P_1 Q_2 - P_2 Q_1}. \end{aligned} \qquad (1.99)$$

Note that in calculating $S_{i,r}^{\sigma}$ and $S_{i,r}^{\omega}$ for each pair of complex roots, the values of $P_1, P_2, Q_1,$ and Q_2 must be determined. The calculation is facilitated if the two sums

$$\sum_{k=0}^{n} k a_k X_{k-1} \quad \text{and} \quad \sum_{k=0}^{n} k a_k Y_{k-1}$$

are calculated before $P_1, P_2, Q_1 = -(\omega_i/\sigma_i) P_2$, and $Q_2 = (\omega_i/\sigma_i) P_1$. The values of P_1, P_2, Q_1, Q_2 are the same for all the parameters p_r. However, R_1 and R_2 must be calculated for each pair of roots and for each parameter separately.

Determination of the real root sensitivity can be made by considering (1.23) directly. By differentiating the equation with respect to the parameter p_r and using the definition (1.93), we obtain

$$S_{j,r}^{s} = p_r \frac{\sum_{k=0}^{n} (\partial a_k/\partial p_r) s_j}{\sum_{k=0}^{n} k a_k s_j}. \qquad (1.100)$$

For the case in which the parameters are given as numerical values $p_r^{\,0}$ ($r = 1, 2, \ldots, m$), equations 1.89 determine the corresponding values of $a_k^{\,0}$ ($k = 0, 1, \ldots, n$) and $\partial a_k/\partial p_r^{\,0}$ ($k = 0, 1, \ldots, n$). Then (1.100) permits a simple computation of the real root sensitivity $S_{j,r}^{s}$ related to the real root s_j and the parameter p_r.

It is of interest to note that the real root sensitivity cannot be applied to double-real roots of the algebraic equation. For the double-real root s_j, the derivative $F'(s_j)$ appearing as the denominator on the right side of (1.100) goes to zero and $S^s_{j,r}$ becomes infinite. The sensitivity of the double-real root should be found as the sensitivity of the pair of complex conjugate roots for which $\zeta_j = 1$ and $\omega_{n_j} = \sigma_j$. Then, by such variations of parameters, the double-real roots become two distinct real roots ($\zeta > 1$) or two complex conjugate roots ($\zeta < 1$), or they remain double-real roots ($\zeta = 1$). This case is discussed in Dr. A.5 in Appendix A, where the root sensitivity calculations are illustrated by an example.

So far we have considered the sensitivities of σ and ω. To calculate the sensitivities of ω_n and ζ, which are defined as

$$S^{\omega_n}_{i,r} \equiv \frac{\partial \ln \omega_{ni}}{\partial \ln p_r}$$

$$S^{\zeta}_{i,r} \equiv \frac{\partial \ln \zeta_i}{\partial \ln p_r},$$

(1.101)

we can use the relationship

$$\sigma = -\omega_n \zeta$$
$$\omega = \omega_n \sqrt{1 - \zeta^2}.$$

(1.45)

By considering σ, ω, ω_n, and ζ as functions of the parameter p_r and differentiating (1.45) with respect to p_r, we obtain, after simple manipulations,

$$S^{\omega_n}_{i,r} = \zeta^2 S^{\sigma}_{i,r} - (1 - \zeta^2) S^{\omega}_{i,r}$$
$$S^{\zeta}_{i,r} = (1 - \zeta^2)(S^{\sigma}_{i,r} - S^{\omega}_{i,r}).$$

(1.102)

It should be noted that (1.102) cannot be used to calculate the sensitivity of double-real roots ($\zeta = 1$).

The procedure of root sensitivity analysis can be directly extended to the case in which the coefficients a_k of the algebraic equations are complex. The analysis is then based upon the relations outlined in Section 1.4. Furthermore, by using the derived expressions for root sensitivities, we can find the conditions on the parameter values that yield the minimum of sensitivity. In particular, if the conditions result in zero sensitivity, then the root invariance to small parameter changes is established. This can be readily applied to linear and nonlinear system analysis and design, which are considered in the following chapters.

References

[1.1] J. C. Maxwell, On Governors, *Proc. Royal Soc. London*, **16**, 270–283 (1866). (The paper is reproduced in R. Bellman and R. Kalaba, *Selected Papers on Mathematical Trends in Control Theory*, Dover, New York, 1964, pp. 5–17.)

[1.2] E. J. Routh, Dynamics of a System of Rigid Bodies, *Adams Prize Essay*, Macmillan, London, 1877.

[1.3] A. Hurwitz, Über die Bedinungen, unter welchen eine Gleichung nur Wurzeln mit negativen reellen Teilen besitzt, *Math. Ann.*, **46**, 273–284 (1895). (English translation: *Selected Papers on Mathematical Trends in Control Theory*, by R. Bellman and R. Kalaba, Dover, New York, 1964, pp. 72–82.)

[1.4] H. Nyquist, Regeneration Theory, *Bell System Tech. J.*, **11**, 126–147 (1932).

[1.5] A. W. Mikhailov, Methods for Harmonic Analysis in the Automatic Control System Theory (in Russian), *Avtomatika i Telemekhanika*, No. 3 (1938).

[1.6] I. A. Vishnegradsky, Sur la théorie générale des régulateurs, *Compt. Rend. Acad. Sci.*, **83**, 318–321 (1876).

[1.7] Yu. I. Neimark, On the problem of the Distribution of the Roots of Polynomials (in Russian), *Dokl. Akad. Nauk SSSR*, **58**, 357–360 (1947).

[1.8] Yu. I. Neimark, *Stability of Linearized Systems* (in Russian), LKVVIA, Leningrad, 1949.

[1.9] Yu. I. Neimark, D-decomposition of the Space of Quasipolynomials (On the Stability of Linearized Distributed Systems) (in Russian), *Appl. Math. and Mech.*, **13**, No. 4, 349–380 (1949).

[1.10] R. W. Lanzkron and T. J. Higgins, D-decomposition Analysis of Automatic Control Systems, *IRE Trans. Auto. Control*, **AC-4**, No. 4, 150–171 (December 1959).

[1.11] M. V. Meerov, Structural Synthesis of High-Accuracy Automatic Control Systems (in Russian), State Press of Physics and Mathematical Literature, Moscow, 1959. (English translation: Pergamon Press, New York, 1965.)

[1.12] H. J. Harrington, Limit-cycle Stability Study of a Feedback Control System by a New Describing-function Technique, *AIEE Trans.*, Pt. II (*Applications and Industry*), **78**, 129–134 (July 1959).

[1.13] M. L. Shooman, Stability Analysis of a Nonlinear System in the Parameter Plane, *IEEE Trans. Auto. Control*, **AC-9**, No. 3, 267–269 (July 1964).

[1.14] D. Mitrović, Graphical Conditions for All Roots of an Algebraic Equation to Have Negative Real Parts (in French), *Compt. Rend. Acad. Sci.*, **240**, 1177–1178 (1955).

[1.15] D. Mitrović, Graphical Conditions for the Arguments of All Roots of an Algebraic Equation to Lie Between $\pi/2 + \mu$ and $3\pi/2 - \mu$ (in French), *Compt. Rend., Acad. Sci.*, **243**, 831–833 (1956).

[1.16] D. Mitrović, Graphical Analysis and Synthesis of Feedback Control Systems; I—Theory and Analysis, II—Synthesis, III—Sampled-data Feedback Control Systems, *AIEE Trans.*, Pt. II (*Applications and Industry*), **77**, 476–503 (1958; January 1959).

[1.17] D. Mitrović, Automatic Plotting of Characteristic Curves and Analog Solution of Algebraic Equations, *AIEE Trans.*, Pt. I (*Communication and Electronics*), **80**, 701–703 (1961; January 1962).

[1.18] D. Mitrović, Algebraic Analysis and Synthesis of Control Systems (in Serbian), *Zavod za Izd. Udž. SRS*, University of Belgrade, 1966.

[1.19] G. J. Thaler and R. G. Brown, *Analysis and Design of Feedback Control Systems*, McGraw-Hill, New York, 1960, pp. 344–388.

[1.20] D. W. Elliot, G. J. Thaler, and J. C. W. Heseltine, Feedback Compensation Using Derivative Signals, *IEEE Trans.*, Pt. II (*Applications and Industry*), **82**, 269–274 (November 1963).

[1.21] D. Šiljak, Generalization of Mitrović's Method, *IEEE Trans.*, Pt. II (*Applications and Industry*), **83**, 314–320 (September 1964).

[1.22] D. Šiljak, Sampled-data Control Systems with Transport Lag by Mitrović's Algebraic Method, *AIEE Trans.*, Pt. II (*Applications and Industry*), **80**, 247–251 (November 1961).

[1.23] D. Šiljak, and R. I. Petrović, Stability Consideration of Sampled-data Control Systems Applied to Multi-channel Measurement, *Bull. Inst. Nucl. Sci. "Boris Kidrich"* (Belgrade), **12,** 61–72 (October 1961).

[1.24] D. Šiljak, Sampled-data Control Systems with Finite Sampling Duration by Mitrović's Method, *Publ. of EE Faculty of the Univ. of Belgrade*, Nos. 31–33, 9–17 (January 1961).

[1.25] P. Kokotović, and D. D. Šiljak, The Sensitivity Problem in Continuous and Sampled-data Linear Control Systems by Generalized Mitrović's Method, *IEEE Trans.*, Pt. II (*Applications and Industry*), **83,** 321–324 (September 1964).

[1.26] P. Kokotović, and D. D. Šiljak, Automatic Analog Solution of Algebraic Equations and the Plotting of Root Loci by Generalized Mitrović's Method, *IEEE Trans.*, Pt. II (*Application and Industry*), **83,** 324–328 (September 1964).

[1.27] T. Brkić and P. Kokotović, Design of the Voltage Regulator by Mitrović's Algebraic Method (in Serbian), *Elektroprivreda*, **15,** Nos. 6–7, 277–286 (1962).

[1.28] M. Gospić and P. Kokotović, An Approach to the Synthesis of Third-order Linear Systems by Using Mitrović's Diagram (in Serbian), *The Seventh Yugoslav Conf. of ETAN*, Novi Sad, Yugoslavia, September, 1962.

[1.29] S. Božanić and P. Kokotović, A Note on the Application of the Mitrović Procedure to the Pole-Zero Variations (in Serbian), *The Seventh Yugoslav Conf. of ETAN*, Novi Sad, Yugoslavia, September, 1962.

[1.30] O. Mikić, An Application of the Mitrović Method to the Analysis and Synthesis of Regulatory Systems with Pure Delay (in Serbian), *The Seventh Yugoslav Conf. of ETAN*, Novi Sad, Yugoslavia, September, 1962.

[1.31] P. L. Wilson, Use of Mitrović's Method in Predicting Limit Cycles in Feedback Control Systems with One and Two Gain-variable Nonlinearities, M.S. Thesis, U.S. Naval Postgraduate School, Monterey, California, 1964.

[1.32] P. Kokotović, Analog Automatum for Plotting Mitrović's Characteristic Curves (in Serbian), *Tehnika*, **19,** No. 5, 47–53 (1964).

[1.33] D. Šiljak, Generalization of the Fundamental Equations in Mitrović's Method (in Serbian), *Automatika*, No. 1, 25–26 (September 1962).

[1.34] M. Gospić, A Contribution to the Synthesis Procedure for Third-order Servosystems by Mitrović's Method (in Serbian), *Automatika*, No. 1, 34–40 (1962).

[1.35] M. Rakić, A Contribution to the Algebraic Method for Synthesis of Linear Automatic Control Systems (in Serbian), *Automatika*, **1,** No. 3, 171–175 (1965).

[1.36] H. H. Choe, T. Ohta, and G. J. Thaler, Some Extensions of Mitrović's Method: I—Analysis and Design with sketches, II—Three Parameter Analysis and Adjustment, III—Analytic Design of Compensation, IV—Frequency Response Techniques, *Automatika*, **1,** No. 2, 39–53 (1965); **2,** Nos. 1–2, 33–48 (1966). [The fourth part of the sequence has been published also in *IEEE Transactions on Automatic Control*, **AC-11,** No. 3, 569–573 (July 1966).]

[1.37] D. R. Towill, Analysis and Synthesis of Feedback Compensated Third-Order Control Systems via the Coefficient Plane, *Radio and Electron. Eng.*, **32,** No. 2 (August 1966).

[1.38] D. D. Šiljak, Analysis and Synthesis of Feedback Control Systems in the Parameter Plane, Pt. I—Linear Continuous Systems, *IEEE Trans.* (*Applications and Industry*), **83,** 449–458 (November 1964).

[1.39] D. D. Šiljak, Analysis and Synthesis of Feedback Control Systems in the Parameter Plane, Pt. II—Sampled-data Systems, *IEEE Trans.*, Pt. II (*Applications and Industry*), **83,** 458–466 (November 1964).

[1.40] D. D. Šiljak, Analysis and Synthesis of Feedback Control Systems in the Parameter Plane, Pt. III—Nonlinear Systems, *IEEE Trans.*, Pt. II (*Applications and Industry*), **83,** 466–473 (November 1964).

[1.41] D. D. Šiljak, Generalization of the Parameter Plane Method, *IEEE Trans., Auto. Control*, **AC-11**, No. 1, 63–70 (January 1966).
[1.42] A. Burzio and D. D. Šiljak, Minimization of Sensitivity with Stability Constraints in Linear Control Systems, *IEEE Trans. Auto. Control*, **AC-11**, No. 3, 567–569 (July 1966).
[1.43] F. H. Holister, Network Analysis and Design by Parameter Plane Techniques, Ph.D. Thesis, U.S. Naval Postgraduate School, Monterey, California, 1965.
[1.44] F. H. Holister and G. J. Thaler, Loaded and Null Adjusted Symmetrical Parallel-tee Network, *Proceedings of the 1965 National Electronics Conference*, Chicago, 753–758 (October 1965).
[1.45] F. H. Holister and G. J. Thaler, Symmetrical Parallel-tee Network-Parameter Plane Analysis and Synthesis, *Proceedings of the 3rd Annual Allerton Conf. on Network and System Theory*, Univ. of Illinois, Monticello, Ill., 430–438 (October 1965).
[1.46] J. B. Moore, Steady-state Response in the Parameter Plane, *Automatika*, **1**, No. 2, 55–58 (1965).
[1.47] D. D. Šiljak, An Extension of the Parameter Plane Method, *Automatika*, **1**, No. 2, 59–64 (1965).
[1.48] L. D. Nace, Application of Mitrović's Method to Feedback Compensation of Sampled-data Control Systems, M.S. Thesis, U.S. Naval Postgraduate School, Monterey, California, 1965.
[1.49] A. R. Miller, Parameter Plane Analysis of Sampled-data Systems (with Extensions for Continuous System Applicability), Ph.D. Thesis, U.S. Naval Postgraduate School, Monterey, California, 1965.
[1.50] L. Eisenberg, Stability of Linear Systems with Transport Lag, *IEEE Trans. Auto. Control*, **AC-11**, No. 2, 247–255 (April 1966).
[1.51] L. Eisenberg, Design of Feedback Control Systems with Transport Lag by Parameter Plane Techniques, Ph.D. Thesis, Newark College of Engineering Newark, New Jersey, 1966.
[1.52] R. M. Nutting, Parameter Plane Technique for Feedback Control Systems, M.S. Thesis, U.S. Naval Postgraduate School, Monterey, California, 1965.
[1.53] J. A. Paine, Application of Parameter Plane Techniques to Control System Compensation (Identical Double Section Cascade Compensation), M.S. Thesis, U.S. Naval Postgraduate School, Monterey, California, 1966.
[1.54] G. J. Thaler and A. B. Lemanski, Linear Control System Synthesis Using Algebraic Methods, *Proceedings of the First Annual Princeton Conference on Information Sciences and Systems*, Princeton University, Princeton, N.J., 268–271 (March 1967).
[1.55] J. B. Moore and R. C. Dorf, The Design of an Attitude Control System for a Space Vehicle, *Proceedings of the 1966 National Electronics Conference*, Chicago, 715–718 (October 1966).
[1.56] D. D. Šiljak and M. R. Stojić, Sensitivity Analysis of Self-excited Nonlinear Oscillations, *IEEE Trans. Auto. Control*, **AC-10**, No. 4, 413–420 (October 1965).
[1.57] M. M. Yockey, Use of Mitrović's Method in the Analysis of a Control System with Two Gain-variable Feedback Nonlinearities, M.S. Thesis, U.S. Naval Postgraduate School, Monterey, California, 1965.
[1.58] C. A. Pelegrini, An Investigation of Nonlinear Systems Performance on the Parameter Plane, M.S. Thesis, U.S. Naval Postgraduate School, Monterey, California, 1966.
[1.59] D. D. Šiljak, Analysis of Asymmetrical Nonlinear Oscillations in the Parameter Plane, *IEEE Trans. Auto. Control*, **AC-11**, No. 2, 239–247 (April 1966).
[1.60] J. B. Moore, A Convergent Algorithm for Solving Polynomial Equations, *J. Assoc. Computing Machinery*, **14**, No. 2, 311–315 (April 1967).

[1.61] D. D. Šiljak, Popov Inequality via Parameter Plane, *Proceedings of the First Annual Princeton Conference on Information Sciences and Systems*, Princeton University, Princeton, N.J., 183–187 (March 1967).

[1.62] D. D. Šiljak, Absolute Stability in the Parameter Space, *Proceedings of the First Asilomar Conference on Circuits and Systems*, Asilomar, California, 624–631 (November 1967).

[1.63] D. Mitrović, Stability Analysis of Sampled-Data Control Systems (in Serbian), *Automatika*, **1**, No. 1, 1–6 (1965).

[1.64] G. J. Thaler and A. G. Thompson, Parameter Plane Methods for the Study of the Frequency Response of Linear Time-Invariant Systems, *The 1967 Applied Mechanics Conference*, Adelaide, Australia, Paper No. 2292, 68–74 (June 1967).

[1.65] R. D. Showman, Simplified Processing of Star Tracker Commands for Satellite Attitude Control, *IEEE Trans. Auto. Control*, **AC-12**, No. 4, 353–359 (August 1967).

[1.66] S. M. Seltzer, A Launch Vehicle Stability Analysis using a Generalized Mitrović Technique, *Proceedings of the First Asilomar Conference on Circuits and Systems*, Asilomar, California, 397–406 (November 1967).

[1.67] L. Grujić, Possibilities of Linear System Design on the Basis of Conditional Optimization in Parameter Plane, Pts. I and II, *Automatika*, **2**, Nos. 1–2, 49–72 (1966).

[1.68] M. Vukobratović and R. Tonić, Parametric Invariance of Dynamic Systems with Respect to an Index of Stability (in Serbian), *Automatika*, No. 1, 17–21 (1967).

[1.69] D. D. Šiljak, On Absolute Stability and Sensitivity, *Second IFAC Symposium on System Sensitivity and Adaptivity*, Dubrovnik, Yugoslavia, August 1968.

[1.70] D. D. Šiljak, Absolute Stability and Parameter Sensitivity, *Int. J. Control* (to be published in 1968).

CHAPTER TWO

Linear Systems

2.1 Introduction

Numerous methods and techniques have been developed for the analysis and design of linear systems. In control system analysis several criteria have been formulated, by Hurwitz, Nyquist, Mikhailov, and others, that solve the classical problem of system stability. In a wide variety of control problems, however, stability is necessary, but not sufficient, to ensure a satisfactory system performance, and the designer is usually interested in the essential features of the system response. Thus there has been a strong emphasis in control theory on the development of refined techniques for the analysis and design of control systems in terms of the system response to typical or test input signals. The problem consists of determining the system structure and parameters so as to achieve certain system response characteristics. In this chapter a solution to the problem is attempted by the application of the parameter plane technique [2.1–3].

After the linear system is described by a set of differential equations with constant coefficients, the corresponding characteristic equation is obtained as an algebraic equation with the coefficients given as known functions of the system parameters. Some of these parameters are adjustable and should be determined by the design procedure to meet the system specifications given in advance. These specifications can be met by adjusting the system parameters so that the roots of the characteristic equation are set at some desired locations. Thus the graphical technique outlined in the previous chapter is used to obtain a correlation between the adjustable system parameters and the root locations of the characteristic equation.

In the following section a mathematical description of linear systems is given, which is convenient for further use of the parameter plane method

This description leads to a convenient introduction of the notions and definitions employed later in the actual design. Then the system response and its characteristics are discussed and the system stability is considered on the basis of the introduced mathematical description. Examples are then given that illustrate the use of parameter plane diagrams in determining the system time and frequency responses as well as the degree of system stability as functions of adjustable system parameters. Once the pole-zero locations of the closed-loop transfer function are evaluated, the corresponding transient response can be estimated from the charts given in Appendix C. The charts are given for typical pole-zero configurations and can be used conveniently in the design process to guide the search for regions of satisfactory parameter values. At the final stage of the process, when the ranges of the parameter values are found that may yield satisfactory design specifications, the best values can be chosen by computer simulation or by direct experiments on the actual system. The frequency and steady-state responses can be evaluated directly from the parameter plane. The same is true for the degree of system stability. This is all conveniently accomplished in the parameter plane when two or at most three system parameters are adjustable and the other parameters have specified numerical values.

Multiparameter adjustment is attempted by using several techniques sequentially. Once the starting design specifications are given, we should choose the *system structure* and possibly introduce additional blocks, usually called *compensators*, so that free parameters are available for adjustment in meeting the specifications. The choice of the structures and compensators is guided by the distribution of the adjustable parameters among the coefficients of the corresponding characteristic equation. A convenient distribution of the parameters is achieved if there is a sufficient number of free parameters in all the characteristic equation coefficients. Then the approximate truncation technique is used for a rough estimation of the root locations and possible parameter values. The design specification usually given in the time domain can then be checked by reference to Appendix C, where charts of the transient responses are given in relation to the typical root configurations. These typical configurations are based on the dominancy conditions of certain characteristic roots and therefore are the goals in adjusting the parameters in the truncated characteristic equation. If the structure and compensators are not chosen properly and the specifications cannot be met for any admissible parameter values, the truncation procedure can be repeated with another structure and/or compensator. At this stage of the design previous experience, as well as the use of well-known linear methods such as the Bode and Nichols frequency response techniques, the Evans root-locus method, and others, can be useful.

Once a convenient structure is chosen and the free system parameters are placed at the desired locations, constraint equations and inequalities relating these parameters can be formulated directly from the design specifications. Constraints resulting from the characteristic root specifications, prescribed zeros of the closed-loop transfer function, dominant-root parameter sensitivity specifications, and error-constant values given in advance are incorporated analytically in the multiparameter design. The relative stability and dominancy constraints are considered, using either approximate analytical methods or graphical techniques in the parameter and complex planes. By solving available constraint equations and using a method of factoring polynomials (Appendix B), along with a plotting of the obtained results and any relevant system characteristic on both the parameter and the complex plane, the adjustable system parameters can be chosen in a systematic manner. The procedure is illustrated by examples.

It should be noted that the presented method and the procedure are oriented to facilitate a computer simulation that should be used as a final step in the system design. The limitation of the simulation in predicting the system performance for conditions not actually tested can be removed. The computer simulation can be guided by the proposed design techniques toward a system structure and a set of system parameters that result in the desired system performance characteristics.

There are three appendices associated with this chapter. Appendix C gives the charts that provide the transient responses to unit-step input functions, which correspond to typical pole-zero configurations of the relevant closed-system transfer functions. Thus the charts permit a direct conversion between the time and the algebraic domains in the design process. In this appendix a recursive system is also given, which relates the time-domain description of the linear systems with the algebraic domain. The recursive system enables the calculation of the characteristic polynomial from a specified state-space vector equation.

Appendix D outlines the generalized stability criteria of Hurwitz, Nyquist, and Mikhailov. The generalization is in the sense of the degree of stability that is important in the application of these criteria to linear system design as well as stability analysis of nonlinear systems (see Chapter 8). Particular attention is devoted in this appendix to the D-decomposition method, which attempts the stability of systems with several adjustable parameters.

In Appendix E the optimization of linear systems subject to stochastic (or deterministic) inputs is discussed. A procedure for squared-error optimization is presented with stability constraints. This procedure yields a compromise optimization solution that does not violate a necessary degree of system stability. The procedure that uses the parameter plane technique is illustrated by an example.

2.2 Mathematical Description of the Linear Systems

In general the analysis of a physical system is concerned with the problem of investigating the change in the state of the system caused by external forces applied to the system. The quantities that determine the state of the system are called the *generalized coordinates* of the system and are functions of time. Since physical systems can be composed of mechanical systems, electrical networks, chemical plants, and so forth, the generalized coordinates may represent such physical quantities as position, electric current, temperature, and so on. The number of coordinates necessary and sufficient to determine uniquely the state of the system is called the *number of system degrees of freedom*.

The values of the generalized coordinates, however, are not sufficient to describe the change of the system or its motion in a dynamic sense. From these values it is not possible to predict the system state at any future time, and the corresponding rates of change of the generalized coordinates must be simultaneously given.[1] If all the generalized coordinates and their first derivatives are known at the same instant of time, the future state of the system can be, in principle, entirely predicted. In mathematical terms this means that the second derivatives of the generalized coordinates are uniquely determined once the generalized coordinates and their first derivatives are known. The relationship among the generalized coordinates, their first and second derivatives, and the external forces acting on the system, which at any instant entirely describes the state of the system, is given by a set of differential equations usually called the *equations of motion*. The number of independent equations of motion is equal to the number of system degrees of freedom.

If a system is subject to an external force, usually called the *input*, and the change in one of the generalized coordinates designated as the *output* of the system (or *system response*) is of interest, a single differential equation relating the input and the output of the system can be obtained by eliminating the rest of the system coordinates. If the input and output functions are denoted by $f(t)$ and $x(t)$, respectively, the system establishes, in a mathematical sense, a certain relationship between the functions $f(t)$ and $x(t)$, which can be symbolically represented as

$$D[x(t)] = f(t), \qquad (2.1)$$

where D is a certain operator. The form of the operator D depends on the system under investigation and may be a function of $x(t), f(t)$, and t.

---

[1] If only the values of the generalized coordinates are given, the system can have arbitrary velocities and its position at the next instant, after an infinitesimal interval of time, is also arbitrary.

64   Linear Systems

A *linear* physical system for which the *superposition principle* is valid (that is, its response to several external forces acting together is the sum of the responses obtained for each of the forces acting alone) is described by (2.1), in which the operator D is linear. For the operator D to be linear, it should satisfy the following relationship:

$$D\left[\sum_k x_k(t)\right] = \sum_k D[x_k(t)] \tag{2.2a}$$

$$D[cx(t)] = cD[x(t)], \quad c = \text{const} \tag{2.2b}$$

for all admissible $x_k(t)$ and $x(t)$, respectively.

Equation 2.2a is significant in that it enables a general solution of the linear equations of motion to be found as a summation of all particular solutions determined separately. The relationship 2.2b is sometimes called the *principle of homogeneity*. It may be concluded that *a system is linear if and only if it satisfies both the principle of superposition and the principle of homogeneity*.

A system is *time invariant* or fixed if the operator D is independent of time; that is,

$$D[x(t - \tau)] = f(t - \tau), \quad \tau = \text{const} \tag{2.3}$$

for all admissible $x(t)$ and $f(t)$. Otherwise, the system is called *time varying*.

If the operator D is a function of the input $f(t)$ or the output $x(t)$, it is a nonlinear operator for which the superposition and homogeneity principles are not valid, and the corresponding system is *nonlinear*. Whether the system is linear or nonlinear time varying depends also upon the validity of these principles.

The classical formulation of differential equations for a physical system is now illustrated by a simple example in which the linear *RLC* electric circuit shown in Figure 2.1 is considered.

The *RLC* circuit is a physical system with one degree of freedom that is described by a second-order linear differential equation

$$L\frac{d^2q}{dt^2} + R\frac{dq}{dt} + \frac{1}{C}q = e, \tag{2.4}$$

where $q = q(t)$ is a generalized coordinate that is the electric charge on the

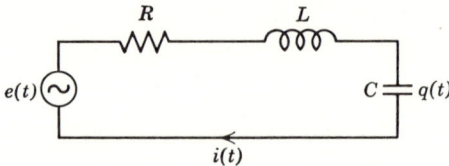

Figure 2.1   *RLC* circuit.

capacitor plates, $e = e(t)$ is the external force that is the voltage of the electric generator, and R, L, and C are resistance, inductance, and capacitance, which are called the *system parameters*. If the function $e(t)$ is given, the equation of motion is (2.4), which can be solved for the function $q(t)$. The function $e(t)$ is the input, and the function $q(t)$ is the output of the system of Figure 2.1, which can be symbolically represented by (2.1), in which the operator D is given as

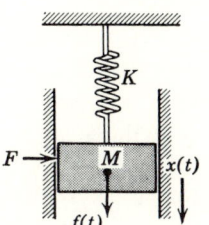

$$D \equiv u\frac{d^2}{dt^2} + v\frac{d}{dt} + w. \qquad (2.5)$$

The coefficients u, v, and w are functions of the system parameters, namely $u = L$, $v = R$, and $w = 1/C$. The operator D is a linear operator with constant coefficients independent of time.

Figure 2.2 Mass, damper, and spring system.

Similarly, the differential equation for a linear mechanical mass, damper, and spring system shown in Figure 2.2 is .

$$M\frac{d^2x}{dt^2} + F\frac{dx}{dt} + Kx = f, \qquad (2.6)$$

where the position $x = x(t)$ of the center of the mass M is the generalized coordinate, the force $f = f(t)$ is the external force, and the mass M, damping F, and the stiffness constant K are the system parameters. The operator D in this case has the same form as (2.5) except that $u = M$, $v = F$, and $w = K$.

In general, when a linear system is composed of several different electrical, mechanical, and other elements, the equations of motion have the following form.[2]

$$\begin{aligned} D_{11}x_1 + D_{12}x_2 + \cdots + D_{1l}x_l &= f_1 \\ D_{21}x_1 + D_{22}x_2 + \cdots + D_{2l}x_l &= f_2 \\ &\vdots \\ D_{l1}x_1 + D_{l2}x_2 + \cdots + D_{ll}x_l &= f_l, \end{aligned} \qquad (2.7)$$

where $x_i = x_i(t)$, $(i = 1, 2, \ldots, l)$, are generalized coordinates of the system; $f_i = f_i(t)$, $(i = 1, 2, \ldots, l)$, are external forces applied to the system; and

---

[2] Equations of motion (2.7) can be written in the state-space form, which is a convenient system mathematical description for numerous methods in system analysis. This is discussed in Section C.3 of Appendix C, where a transition from the operational form (2.7) to the state-space description is given. In addition, the section also gives concepts of controllability and observability that are significant in the analysis of dynamic systems. The state-space description of nonlinear systems is considered in Section 8.2.

66 Linear Systems

the operator D_{ij} has the form

$$D_{ij} \equiv u_{ij}\frac{d^2}{dt^2} + v_{ij}\frac{d}{dt} + w_{ij}. \tag{2.8}$$

If the system parameters entering the coefficients u_{ij}, v_{ij}, and w_{ij} of the operator D_{ij} are not functions of the generalized coordinates, their derivatives, or time, then equations 2.7 form a set of linear differential equations with constant coefficients. If all the external forces f_i are identically zero, we say that the system is *free (unforced)*. If, in addition, the operators D_{ij} are all independent of time and the system is both free and time invariant, it is said that the system is *autonomous*. Otherwise the system is said to be *nonautonomous*.

To find the solution of (2.7) and determine all functions $x(t)$ which make expression (2.7) an identity, it is necessary to know all the external forces as functions of time and the initial state of the system. It is normally assumed that the external forces are applied to the system at $t = 0$, and that the initial state of the system is determined by the value of the generalized coordinates and their first derivatives at $t = 0$. Since equations 2.7 are linear and the superposition principle is valid, the influence of each external force on any of the generalized coordinates can be considered separately, and the obtained results added to determine the entire effect. Therefore, without a loss in generality, we may consider the case in which all the external forces in (2.7) are zero except f_i, and equations 2.7 have the form

$$\begin{aligned}
D_{11}x_1 + D_{12}x_2 + \cdots + D_{1l}x_l &= 0 \\
&\vdots \\
D_{i-1,1}x_1 + D_{i-1,2}x_2 + \cdots + D_{i-1,l}x_l &= 0 \\
D_{i1}x_1 + D_{i2}x_2 + \cdots + D_{il}x_l &= f_i \\
D_{i+1,1}x_1 + D_{i+1,2}x_2 + \cdots + D_{i+1,l}x_l &= 0 \\
&\vdots \\
D_{l1}x_1 + D_{l2}x_2 + \cdots + D_{ll}x_l &= 0.
\end{aligned} \tag{2.9}$$

Attention may now be focused on any of the generalized coordinates, for example, x_j. Then the equation relating the external force f_i and the coordinate x_j is obtained from (2.9) by Cramer's rule as

$$x_j = \frac{\Delta_j}{\Delta} = \frac{\Delta_{ij}}{\Delta} f_i, \tag{2.10}$$

where the determinants Δ and Δ_j are

$$\Delta = \begin{vmatrix} D_{11} & D_{12} & \cdots & D_{1l} \\ D_{21} & D_{22} & \cdots & D_{2l} \\ \cdot & \cdot & & \cdot \\ \cdot & \cdot & & \cdot \\ \cdot & \cdot & & \cdot \\ D_{l1} & D_{l2} & \cdots & D_{ll} \end{vmatrix}$$

(2.11)

$$\Delta_j = \begin{vmatrix} D_{11} & D_{12} & \cdots & D_{1,j-1} & 0 & D_{1,j+1} & \cdots & D_{1l} \\ \cdot & \cdot & & \cdot & & \cdot & & \cdot \\ \cdot & \cdot & & \cdot & & \cdot & & \cdot \\ \cdot & \cdot & & \cdot & & \cdot & & \cdot \\ D_{i-1,1} & D_{i-1,2} & \cdots & D_{i-1,j-1} & 0 & D_{i-1,j+1} & \cdots & D_{i-1,l} \\ D_{i1} & D_{i2} & \cdots & D_{i,j-1} & f_i & D_{i,j+1} & \cdots & D_{il} \\ D_{i+1,1} & D_{i+1,2} & \cdots & D_{i+1,j-1} & 0 & D_{i+1,j+1} & \cdots & D_{i+1,l} \\ \cdot & \cdot & & \cdot & & \cdot & & \cdot \\ \cdot & \cdot & & \cdot & & \cdot & & \cdot \\ \cdot & \cdot & & \cdot & & \cdot & & \cdot \\ D_{l1} & D_{l2} & \cdots & D_{l,j-1} & 0 & D_{l,j+1} & \cdots & D_{ll} \end{vmatrix}$$

and Δ_{ij} is the cofactor corresponding to the ith row and the jth column of the determinant Δ_j. Since Δ and Δ_{ij} are linear operators that can be denoted by $\Delta \equiv D$ and $\Delta_{ij} \equiv D_i$, the input $f_i = f$ and the output $x_j = x$ are related by a single differential equation

$$Dx = D_i f, \qquad (2.12)$$

where

$$D \equiv a_n \frac{d^n}{dt^n} + a_{n-1} \frac{d^{n-1}}{dt^{n-1}} + \cdots + a_1 \frac{d}{dt} + a_0 \qquad (2.13a)$$

$$D_i \equiv b_m \frac{d^m}{dt^m} + b_{m-1} \frac{d^{m-1}}{dt^{m-1}} + \cdots + b_1 \frac{d}{dt} + b_0 \qquad (2.13b)$$

and $n \leq 2l$, l being the number of system degrees of freedom. In physical systems, also, $n \geq m$. The coefficients a_k and b_k of (2.13) are all real functions of the system parameters.

When the external force $f = f(t)$ is given, the right side $D_i f$ of (2.12) is a known function of time. Then the solution $x = x(t)$ as the system output is determined uniquely if the initial conditions

$$x(0), \left[\frac{dx}{dt}\right]_{t=0}, \left[\frac{d^2x}{dt^2}\right]_{t=0}, \ldots, \left[\frac{d^{n-1}x}{dt^{n-1}}\right]_{t=0} \quad (2.14)$$

are given in advance.

A linear system can be conveniently expressed in an operational form if a differential operator

$$s \equiv \frac{d}{dt} \quad (2.15)$$

is defined to give

$$s^k[x(t)] \equiv \frac{d^k}{dt^k}[x(t)], \quad (k = 0, 1, \ldots). \quad (2.16)$$

The operational form of (2.12) is

$$D(s)x = D_i(s)f, \quad (2.17)$$

where $D(s)$ and $D_i(s)$ are polynomials in s given as

$$D(s) = \sum_{k=0}^{n} a_k s^k, \quad D_i(s) = \sum_{k=0}^{m} b_k s^k. \quad (2.18)$$

The polynomial $D(s)$ is referred to as the *characteristic polynomial* and $D(s) = 0$ is called the *characteristic equation*. In the presentation of the parameter plane method [2.1–3] the characteristic polynomial will be denoted by $F(s)$, and the roots s_k of the equation $F(s) = 0$ will be called the *characteristic roots*.

The relationship 2.17 leads directly to the expression of the *transfer function* of the system

$$G(s) = \frac{D_i(s)}{D(s)}, \quad (2.19)$$

where $D(s)$ and $D_i(s)$ are given in (2.18).

The transfer function concept is extensively used in the analysis of networks and systems along with *block diagrams* and *signal-flow graphs*. While the transfer functions are used to describe mathematically the parts of the system, the block diagrams and signal-flow graphs display the *system structure*; that is, they show how the parts are mutually interconnected.

The transfer function representation of linear systems is based on the *Laplace transformation*.[3] In Figure 2.3, the basic structure of feedback control

---

[3] It is assumed that the reader is familar with the Laplace transformation and its application to the linear system analysis. The definition of the transformation and the basic concept of its use are repeated in Appendix C. A rigourous and extensive study of the transformation is given in references 2.4, 5.

In addition, the fundamental notions and the related terms of the linear system analysis are only listed here for completeness. For further explanations, references 2.6, 7 can be recommended.

Mathematical Description of the Linear Systems

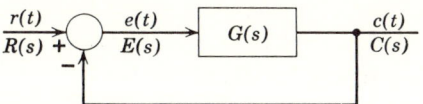

Figure 2.3 Block diagram of the basic feedback control system.

systems is shown. The function $G(s)$ is usually called the *open-loop transfer function* and is defined as[4]

$$G(s) = \frac{C(s)}{E(s)}, \tag{2.20}$$

where $E(s) = \mathscr{L}[e(t)]$, $C(s) = \mathscr{L}[c(t)]$ are the Laplace transforms of the *error signal* $e(t)$ and the *output signal* $c(t)$, respectively. The symbol $\mathscr{L}$ is used to denote the Laplace transformation. The *closed-loop transfer function* $[C/R](s) = C(s)/R(s)$ relates the Laplace transforms $R(s)$ and $C(s)$ of the input $r(t)$ and the output $c(t)$ and, for the system of Figure 2.3, it is

$$\frac{C}{R}(s) = \frac{G(s)}{1 + G(s)}, \tag{2.21}$$

where equation

$$1 + G(s) = 0 \tag{2.22}$$

is another form of the characteristic equation. The block $G(s)$ is in the so-called *direct path*. The line connecting the output and the summing point around the block $G(s)$ is called the *feedback path*. More complicated system structures can usually be reduced to that of Figure 2.3 by the algebra of block diagrams or signal flow-graph techniques.

It should also be mentioned that the output, or the system response, $c(t)$ can be calculated from the function $C(s)$ by performing the inverse Laplace transformation. Symbolically, $c(t) = \mathscr{L}^{-1}[C(s)]$, where $\mathscr{L}^{-1}$ represents the inverse operation (see Appendix C). Thus the operation designated by the symbol $\mathscr{L}$ is sometimes called the direct Laplace transformation. For functions that are common for linear system analysis, both operations, $\mathscr{L}$ and $\mathscr{L}^{-1}$, are affected by tables [2.4, 5].

---

[4] The transfer function of (2.20) is identical to the definition of (2.19) if the original differential equation 2.17 describes the block $G(s)$ of Figure 2.3 only, and the pair of functions (x, f) are replaced by (c, e). Of course, a differential equation such as (2.17) can be written for the whole system of Figure 2.3. Then the corresponding characteristic equations are equivalent, $F(s) \equiv 1 + G(s) = 0$.

Moreover, in describing linear systems by transfer functions, we normally assume that the systems have no internal energy prior to the application of the input. This provides no loss in generality, since any initial energy stored in the system can be represented by added inputs.

70   Linear Systems

In a wide class of theoretical and practical problems the description of linear systems in state space is used. This time-domain description is considered in references 2.6 and 2.7, in which the corresponding analysis techniques are outlined. It is of interest to relate the state space description with the operator description, which is done in Appendix C.

This consideration of the description of linear systems is assumed to be sufficient for the linear analysis presented in this chapter. For more detailed treatment of the subject, references 2.6, 7 are recommended.

2.3 System Response and Stability

Once the linear system is described by a differential equation or a transfer function and the input function is specified, we are interested in determining the output function. In the case of a differential equation the output function represents the solution, whereas in the transfer function description it is referred to as the system response.

To find the solution of a differential equation, two methods are used: *classical* and *operational*. Both methods are well-known techniques for solving linear differential equations with constant coefficients, and they will be discussed briefly only to introduce the necessary terms and definitions. For further study of the techniques, reference 2.8 is recommended for classical methods and 2.5, 6 for operational techniques.

Although the classical method is less practical than the operational techniques, it is used first to introduce the basic terms of linear analysis. Thus consider the differential equation

$$\sum_{k=0}^{n} a_k \frac{d^k x}{dt^k} = \sum_{k=0}^{m} b_k \frac{d^k f}{dt^k}, \qquad n \geq m \tag{2.12}$$

As known, the general solution $x = x(t)$ of (2.12) which makes this expression an identity can be written as

$$x(t) = x_t(t) + x_s(t), \tag{2.23}$$

where $x_t = x_t(t)$ is the general solution of the corresponding homogeneous equation

$$\sum_{k=0}^{n} a_k \frac{d^k x_t}{dt^k} = 0 \tag{2.24}$$

and $x_s = x_s(t)$ is the particular solution of (2.12).

The general solution x_t of (2.24) is composed of n linearly independent solutions

$$x_t = \sum_{k=1}^{n} C_k e^{s_k t}, \tag{2.25}$$

where C_k are arbitrary constants and s_k are the roots of the characteristic equation

$$F(s) \equiv \sum_{k=0}^{n} a_k s^k = 0. \qquad (2.26)$$

The form of solution (2.25) is valid on the condition that all the roots s_k of (2.26) are real and distinct. If the characteristic equation has a multiple real root $s_i = \sigma_i$ of the pth order, then there are p linearly independent solutions $t^k e^{\sigma_i t}$, $(k = 0, 1, \ldots, p - 1)$ in the summation of (2.25). When $s_{i,i+1} = \sigma_i \pm j\omega_i$ is of order p, we obtain linearly independent solutions by replacing the corresponding $2p$ functions by $t^k e^{\sigma_i t} \cos \omega_i t$, $t^k e^{\sigma_i t} \sin \omega_i t$, $(k = 0, 1, \ldots, p - 1)$.

The solution x_t is general in the sense that every solution of (2.24) can be obtained from (2.25) by proper choice of the constant C_k. The constants C_k, which are called the *integration constants*, are defined uniquely for every given solution x_t. The constants C_k are evaluated from the initial conditions (equation 2.14), as will be shown later.

The stability problem, which arises in the theory and a whole series of applications of linear differential equations, is formulated in terms of the behavior of the solution $x_t(t)$ as $t \to \infty$. A linear system is called stable if

$$x_t \to 0 \quad \text{when} \quad t \to \infty. \qquad (2.27)$$

When this condition is not satisfied and the solution x_t increases without limit as $t \to \infty$, the system is unstable. Since in stable systems the solution $x_t(t)$ disappears as $t \to \infty$ and is significant only in the transition from one state of the system to another, it is called the *transient solution*.

The stability condition (2.27) is related only to the roots of the characteristic equation and is satisfied if and only if all the roots have negative real parts. To prove this, denote the roots of (2.26) as

$$s_i = \sigma_i + j\omega_i, \quad (i = 1, 2, \ldots, n) \qquad (2.28)$$

if $\sigma_i < 0$, $(i = 1, 2, \ldots, n)$, then there exists a positive number a such that

$$\sigma_i < -a, \quad (i = 1, 2, \ldots, n). \qquad (2.29)$$

In this case it can be shown that for every solution $x_t(t)$ of (2.25), a positive number M can be found so that

$$|x_t(t)| < M e^{-at}, \quad \text{for} \quad t \geq 0. \qquad (2.30)$$

To prove this inequality, consider an arbitrary term (for example, the ith term) $t^k e^{s_i t}$ in (2.25). Then we get

$$\left| \frac{t^k e^{s_i t}}{e^{-at}} \right| = t^k e^{(\sigma_i + a)t}. \qquad (2.31)$$

Since the number $\sigma_i + a$ is negative because of the inequality 2.29, the function $t^k e^{(\sigma_i + a)t}$ tends to zero as $t \to \infty$ and therefore is bounded for $t \geq 0$. Thus

$$\left| \frac{t^k e^{s_i t}}{e^{-at}} \right| < M_i, \qquad \text{for } t \geq 0, \tag{2.32}$$

or

$$|t^k e^{s_i t}| < M_i e^{-at}, \qquad \text{for } t \geq 0. \tag{2.33}$$

For the general solution $x_t(t)$ of (2.25), we obtain

$$x_t(t) \leq (|C_1| M_1 + |C_2| M_2 + \cdots + |C_n| M_n) e^{-at} = M e^{-at} \tag{2.34}$$

and the inequality 2.30 is thereby proved. In case one of the roots s_i of $F(s) = 0$ has a positive real part $\sigma_i > 0$, there exists a solution $e^{s_i t}$ of (2.24) that increases without bound and the system is unstable. Linear systems that have the roots of the characteristic equation on the imaginary axis are of little significance since they cannot exist in practice, and they may be considered stable or unstable according to convention.

Since the stability of a linear system depends only on the location of the roots of the characteristic equation $F(s) = 0$, it is not affected by the values of the initial conditions and the forcing function $f(t)$. Furthermore, the characteristic polynomial $F(s) \equiv \Delta(s)$ of (2.11) has the same form irrespective of the choice of the generalized coordinate and the forcing function under consideration. Therefore the stability is an inherent characteristic of the system itself.

It should be noted that the inequality 2.30 not only assures system stability, but also gives an estimate of the rate of convergence to zero of the solution $x_t(t)$, which establishes the important notion of the *relative stability*. The number a that is defined in (2.29) is a measure of the degree of relative stability. In geometrical terms the relative stability is related to the case in which all the roots of the characteristic equation are located to the left of a straight line that is drawn in the left half of the complex s plane parallel to the imaginary axis at a distance a.[5] This problem was considered in Chapter 1, Section 1.5 on root distribution, where the distribution of the characteristic roots was discussed relative to several s plane contours of Figure 1.21. In particular, the inequality 2.30 is satisfied if the operating point $M(\alpha; \beta)$ is located inside the region R_n bounded by the $\Sigma = -a$ curve in the parameter $\alpha\beta$ plane. Other specifications of the relative stability may be related to the

---

[5] In particular, when all the roots of the characteristic equation are to the left of the imaginary axis ($a = 0$), the system is sometimes called *absolutely stable*. The notion of absolute stability is not in the Liapunov sense as introduced in Chapter 8, on stability of nonlinear systems. Here the term *absolute* comes as a consequence of the relative stability definition.

contours of Figure 1.21 or a combination of these contours. This was discussed rather extensively in the preceding chapter, and the applications will be illustrated by examples.

The particular solution $x_s(t)$ of (2.12), which is called the *steady-state solution*, represents the response as $t \to \infty$. It depends on the form of the forcing function $f(t)$ and is not affected by the choice of initial conditions.

In linear system analysis the class of forcing functions $f(t)$ is normally such that $f(t) \equiv 0$ for $t < 0$, and for $t \geq 0$

$$f(t) = \sum_{i=1}^{m} \mu_i e^{\lambda_i t}, \qquad (2.35)$$

where μ_i are constants or polynomials of t; λ_i are negative real numbers (in special cases, they may be zero) or pairs of complex conjugate numbers with real parts negative or zero; and m is an integer.[6]

If (2.35) is substituted into (2.12) after simple mathematical operations, we obtain

$$\sum_{k=0}^{n} a_k \frac{d^k x}{dt^k} = \sum_{i=1}^{m} L_i e^{\lambda_i t}, \qquad (2.36)$$

where L_i are known functions of μ_i, λ_i, b_k, and t. The steady-state solution $x_s(t)$ now has the following form:

$$x_s(t) = \sum_{i=1}^{m} N_i e^{\lambda_i t}. \qquad (2.37)$$

Since λ_i are known, to determine the solution $x_s(t)$ it is necessary to evaluate N_i, ($i = 1, 2, \ldots, m$), which are constants or polynomials of t. Consider the case when none of the constants λ_i are equal to any of the characteristic roots. In this case the N_i are constants determined by the identity

$$\sum_{i=1}^{m} N_i e^{\lambda_i t} \sum_{k=0}^{n} a_k \lambda_i^k \equiv \sum_{i=1}^{m} L_i e^{\lambda_i t}. \qquad (2.38)$$

The identity 2.38 is obtained from (2.36) by substituting $x(t)$ with $x_s(t)$ of (2.37). The identity 2.38 yields the constants N_i as

$$N_i = \frac{L_i}{\sum_{k=0}^{n} a_k \lambda_i^k}, \qquad (2.39)$$

---

[6] For example, the step function for which $f(t) \equiv 0$ when $t < 0$ and $f(t) = \text{const} \neq 0$ when $t \geq 0$, is obtained from (2.35) for $m = 1$, $\lambda_1 = 0$, and $\mu_1 = \text{const} \neq 0$. In particular, when $\mu_1 = 1$, the function $f(t)$ is the unit step function $u(t)$. Similarly, if $m = 2$, $\lambda_1 = j\omega$, $\lambda_2 = -j\omega$, and μ_1 and μ_2 are two complex conjugate numbers, the functions $\sin \omega t$ and $\cos \omega t$ are obtained for the function $f(t)$. The function $f(t)$ of (2.35) is sometimes called the *quasipolynomial* [2.8].

74   Linear Systems

and the particular corresponding solution $x_s(t)$ is determined as

$$x_s(t) = \sum_{i=1}^{m} \frac{L_i}{\sum_{k=0}^{n} a_k \lambda_i^k} e^{\lambda_i t}. \qquad (2.40)$$

The obtained steady-state solution $x_s(t)$ is finite because the constants λ_i are not equal to any of the characteristic roots. Even if some of λ_i are equal to some of the roots, it can be shown that $x_s(t)$ will have a finite form [2.8].

It is of particular interest to note that in order to determine the steady-state solution $x_s(t)$, we need not solve the characteristic equation. The solution $x_s(t)$ can be found without any principal difficulties—directly from the given differential equation. In feedback systems the equations are often written with respect to the error signal $e(t)$ (see Figure 2.3) and $x_s(t)$ represents the steady-state error. Thus for typical test input signals (equation 2.35), the coefficients a_k and b_k of (2.12) can readily be determined to yield a sufficiently small value of the steady-state error at any instant of time. This is normally accomplished by adjusting the corresponding error constants defined in terms of the given transfer functions, as will be shown later.

Since the steady-state solution $x_s(t)$ is determined, the whole solution $x(t)$ is obtained as the algebraic sum of the transient solution $x_t(t)$ of (2.25) and the steady-state solution $x_s(t)$ of (2.40) as stated in (2.23). Thus

$$x(t) = \sum_{k=1}^{n} C_k e^{s_k t} + \sum_{i=1}^{m} \frac{L_i}{\sum_{k=0}^{n} a_k \lambda_i^k} e^{\lambda_i t}. \qquad (2.41)$$

When the initial conditions listed in (2.14) are given, the integration constants C_k are determined from the corresponding set of n algebraic equations in n unknowns C_k, ($k = 1, 2, \ldots, n$). The algebraic equations are obtained by forcing the solution $x(t)$ of (2.41) to satisfy n conditions imposed by (2.14). The calculation of the integration constants may become tedious, and the operational solution method based on the Laplace transform is superior to the classical method outlined here. The latter method, however, is convenient to introduce the notions of stability and transient and steady-state response.

Qualitative study of the transient solution can be performed by stability analysis. Stability criteria have been established to check both the absolute and the relative stability of linear systems. It is important to note that the Hurwitz, Nyquist, and Mikhailov stability outlined in Appendix D, as well as the parameter plane method presented in the preceding chapter, can check the stability without solving for the roots of the characteristic equation. Moreover, the Nyquist criterion and the parameter plane analysis enable the adjustment of the system parameters to achieve a desired degree of relative stability. The advantages of the parameter plane method are, first, that two system parameters can be varied at a time and, second, that the root distribution is placed in evidence. This is particularly true in relative stability analysis. Both techniques can be used in stability analysis of nonlinear systems, as will be shown later.

In linear systems, however, the degree of stability is not usually a satisfactory description of the system performance, and the final goal of the analysis is to obtain a satisfactory system response. It can be easily shown that two systems can have the same degree of stability, but the corresponding responses may differ significantly. Thus the numerical evaluation of all the characteristic roots is necessary to obtain the exact solution of the linear differential equations representing the system. Note that the root values are necessary for the evaluation of the transient response, whereas the steady-state response can be found without solving the characteristic equation, as can be concluded from the solution procedure outlined above.

It is now important to emphasize that in stability and solution problems associated with linear differential equations with constant coefficients the central problem is the algebraic problem—the evaluation of all the roots of relevant characteristic equations.

A significant fact, however, is that the numerical evaluation of all the roots should be performed for various values of adjustable system parameters. This problem can be solved by the parameter analysis as shown in Section 1.6, where the root evaluation and the inverse parameter mapping were considered. The application of the parameter plane to the root solution problem is straightforward and will be illustrated in the following examples.

The problem now is the correlation between the characteristic roots and the system response. It is quite impractical to solve the related differential equations for a set of roots calculated from the parameter plane each time a system parameter is changed. We avoid this by interpreting the design of linear systems in terms of pole-zero locations of the system transfer function. The parameter plane method gives the pole-zero locations as functions of adjustable system parameters.[7] Then a straightforward interpretation of the obtained pole-zero locations in terms of the time-domain solution is desired. This can be accomplished by displaying the system responses corresponding to the fundamental pole-zero configurations as shown in Appendix C. These fundamental configurations, based upon the well-known dominancy of control poles, are not only the most frequent distributions of system function

---

[7] When the system transfer function is a ratio of two polynomials, the parameter plane method can be applied to each of them to calculate the corresponding pole-zero locations when the system parameters are varied. Sometimes one parameter diagram can display both the zeros and the poles of the system function [2.3].

76   Linear Systems

poles and zeros, but are normally also the most desired configurations to which the parameter plane design is directed.

Of course the correlation between the algebraic domain and the time domain can be realized by computer simulation of the actual system. After suitable pole-zero locations are achieved and the parameters are fixed, the time-domain solution can be found. If the tested system response does not satisfy the given requirements, a readjustment of the pole-zero locations is necessary. This readjustment should be governed by a simple correlation between the pole-zero configuration and the time-domain response, which is given in Appendix C. Then the interchanges of the algebraic domain considerations and the time-domain solutions of the system response are performed until a satisfactory solution is achieved.

In most cases the transition from the pole-zero locations to the transient response during the readjustment of parameters is not necessary, because the design goal can be interpreted directly in terms of the pole-zero positions. This, of course, depends largely on the designer's experience in working with pole-zero locations, although the charts of Appendix C as well as the computer simulation can be both helpful and useful.

Once the pole-zero locations are displayed as a function of adjustable system parameters, the frequency characteristics can be readily sketched. Thus the parameter plane gives indirectly the frequency response of the system. By plotting the constant bandwidth curves in the parameter plane, the essential characteristic of the frequency response can be considered directly.

Let us define what is meant by *constant bandwidth curve*: A constant bandwidth curve for $|G(j\omega_b)| = M$ is a curve drawn upon the parameter plane that specifies the relation between the parameters necessary if the transfer function $G(s)$, which is a function of parameters, is to have a magnitude M at the real frequency ω_b.

Once such curves are obtained for selected values of M, the frequency response may be sketched. Alternatively, the constant bandwidth curve corresponding to some M and ω_b specification can be drawn and, from this curve, values of the parameters can be found that will guarantee that the specification is met.

Consider the following transfer function $G(s)$:

$$G(s) = \frac{Q(s)}{P(s)}, \qquad (2.42)$$

where

$$\begin{aligned} P(s) &= \sum_{k=0}^{n}(a_k\alpha + b_k\beta + c_k\alpha\beta + d_k)s^k \\ Q(s) &= \sum_{k=0}^{m}(e_k\alpha + f_k\beta + g_k\alpha\beta + h_k)s^k \end{aligned} \qquad (2.43)$$

and $n \geq m$. By substituting $s = j\omega_b$ in (2.42), we obtain $M = |G(j\omega_b)|$ as

$$M = \left(\frac{Q_1^2 + Q_2^2}{P_1^2 + P_2^2}\right)^{1/2}, \qquad (2.44)$$

where

$$\begin{aligned}
P_1 &= \sum_{\substack{k=0 \\ \text{even}}}^{n} (a_k\alpha + b_k\beta + c_k\alpha\beta + d_k)\omega_b^k \\
P_2 &= \sum_{\substack{k=1 \\ \text{odd}}}^{n} (a_k\alpha + b_k\beta + c_k\alpha\beta + d_k)\omega_b^k \\
Q_1 &= \sum_{\substack{k=0 \\ \text{even}}}^{m} (e_k\alpha + f_k\beta + g_k\alpha\beta + h_k)\omega_b^k \\
Q_2 &= \sum_{\substack{k=1 \\ \text{odd}}}^{m} (e_k\alpha + f_k\beta + g_k\alpha\beta + h_k)\omega_b^k.
\end{aligned} \qquad (2.45)$$

Thus $M = M(\alpha, \beta, \omega_b)$. If all the coefficients a_k, b_k, c_k and so on, are calculated and the value of M is specified (usually M is -3 db), then the $\omega_b = $ const loci can be plotted in the $\alpha\beta$ plane. To do this, note that (2.44) can be reduced to a quadratic equation in β with the coefficients as functions of M, ω_b, and α. Since M is specified, by assigning a value for ω_b, a curve $\omega_b = $ const can be plotted on the $\alpha\beta$ plane. The solution of the resulting quadratic equation can be readily achieved on a digital computer. Thus both the transient and the frequency responses can be obtained from the parameter $\alpha\beta$ plane.

To illustrate the above procedures, consider the symmetrical parallel-T network of Figure 2.4. This network has been the subject of several papers [2.9–11], primarily because it is possible to obtain complex zeros of the transfer function anywhere in the left-half s plane and as far as 30° from the imaginary axis into the right-half s plane by suitably choosing the network

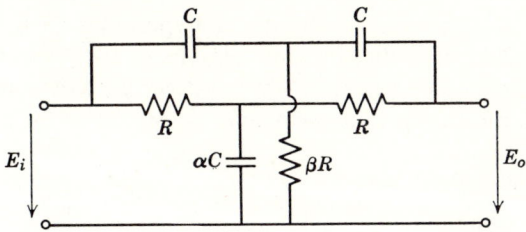

Figure 2.4 Symmetrical parallel-T network.

elements.[8] The transfer function $G(S) = E_0(S)/E_i(S)$ of the network is

$$G(S) = \frac{\alpha\beta S^3 + 2\beta S^2 + 2\beta S + 1}{\alpha\beta S^3 + (\alpha + 2\beta + 2\alpha\beta)S^2 + (\alpha + 2\beta + 2)S + 1}, \quad (2.46)$$

where $S = RCs$. By varying the parameters α and β the poles and zeros of $G(S)$ can be adjusted. The objective of the analysis is to determine in what manner the poles, zeros, and -3 db frequencies of (2.46) change as the parameters α and β are varied. This can be accomplished by preparing three parameter planes—one each for the poles, zeros, and -3 db bandwidth of (2.46).

Since the network shown in Figure 2.4 is a passive RC network, all of the poles will lie along the negative real axis in the S plane. Let σ denote the value of S at which such a pole of (2.46) occurs. We may thus assert that poles occur at those values of σ that satisfy equation

$$\alpha\beta\sigma^3 + (\alpha + 2\alpha\beta + 2\beta)\sigma^2 + (\alpha + 2\beta + 2)\sigma + 1 = 0. \quad (2.47)$$

Rewriting (2.47) in the form of a plane algebraic curve in the $\alpha\beta$ plane, we obtain

$$(\sigma^2 + \sigma)\alpha + (2\sigma^2 + 2\sigma)\beta + (\sigma^3 + 2\sigma^2)\alpha\beta + (2\sigma + 1) = 0. \quad (2.48)$$

Equation 2.48 represents the σ curves that constitute a family of hyperbolas in the $\alpha\beta$ plane with the family parameter σ. The σ curves represented by (2.48) are plotted in Figure 2.5. We enter this diagram with values of α and β and then read the values of the three poles directly. Alternatively, we may determine the values of α and β to produce some selected pole values.

The zeros of $G(S)$ are the roots of equation

$$\alpha\beta S^3 + 2\beta S^2 + 2\beta S + 1 = 0. \quad (2.49)$$

Since α and β are presumed real, it can be asserted that (2.49) has at least one real root. As before, let $S = \sigma$ in (2.49) and rewrite the resulting equation as a plane algebraic curve in the $\alpha\beta$ plane:

$$(2\sigma^2 + 2\sigma)\beta + \sigma^3\alpha\beta + 1 = 0. \quad (2.50)$$

Equation 2.50 also represents a family of hyperbolas in the $\alpha\beta$ plane, as shown in Figure 2.6. Note that in some regions of the plane it is possible to have three real zeros ($\zeta = 1$ curve). The positive real axis coincides with the $\zeta = -0.5$ curve, indicating that zeros may be placed as far as 30° from the

---

[8] Barker and Rosenstein [2.10], using the root-locus techniques, investigated this network. Their analysis, while complete and detailed, was complicated considerably by the inherent single-parameter capability of the root-locus technique.

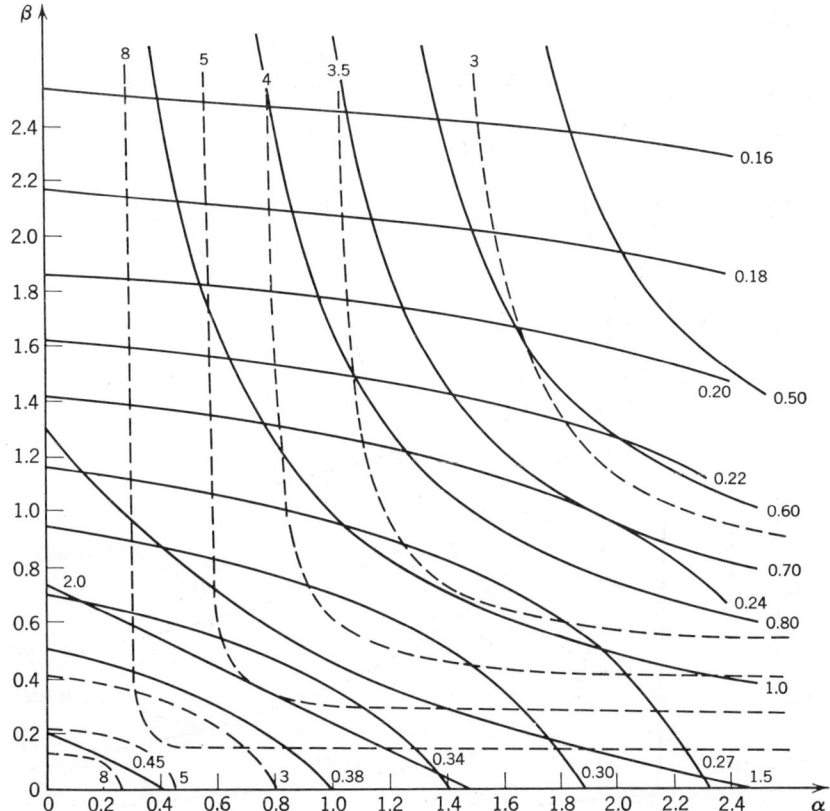

Figure 2.5 Poles of the parallel-T network (note: curve labels refer to negative root values).

imaginary axis into the right-half S plane. The $\zeta = 0$ curve is a straight line with equation $\alpha = 4\beta$. This indicates that if we are interested only in those cases in which a pair of conjugate zeros are placed on the $j\omega$ axis, then we may substitute this linear relation between the parameters into the network equation, thus reducing the problem of analysis and synthesis to one of a single parameter.

The -3 db constant bandwidth curves are plotted for the investigated network in Figure 2.7. The curves correspond to selected values of ω_b with $M = 1/\sqrt{2}$ as required for -3 db curves.

The diagrams of Figures 2.5–2.7 display both the transient and the frequency responses in the parameter plane. The pole-zero locations are obtained by Figures 2.5 and 2.6, and can be interpreted in the time domain with the

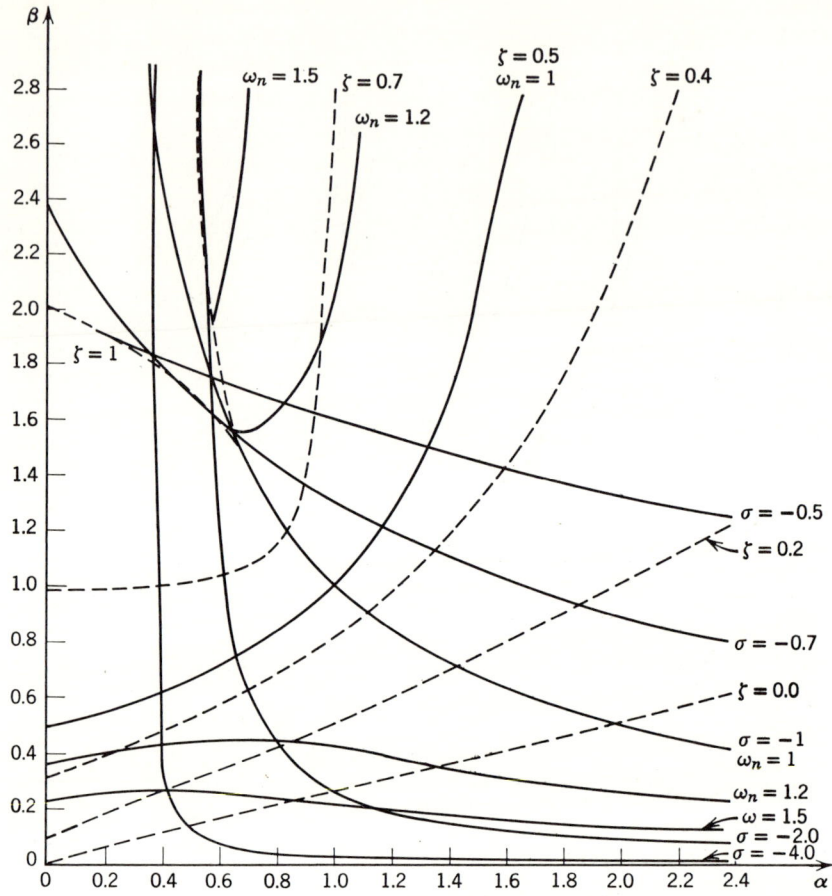

Figure 2.6   Zeros of the parallel-T network.

help of Appendix C. The constant bandwidth curves are directly interpreted from the parameter plane diagram of Figure 2.7.[9]

It should be noted here that in many cases of control system analysis and design, the zeros of the system function are in factored form, or they can be read from the same diagram displaying the corresponding pole location [2.3]. Moreover, the design is usually directed toward the dominancy of a complex pair of roots located in a certain region or on a certain contour of the s plane. Thus by focusing attention on a part of the $\alpha\beta$ plane or displaying the

---

[9] The loci of constant quality factor Q can also be plotted in the parameter plane, as indicated in reference 2.9.

System Response and Stability   81

necessary curves on the s plane using the inverse parameter plane mapping, the number of curves and the presentation of the graphical means can be greatly reduced and easily interpreted in the actual design. This, however, is based upon the familiarity of the designer with the proposed technique. In that respect the following material and examples can be helpful.

The steady-state response can be analyzed directly in the parameter plane by using the error constants. They are utilized in control system design as measures of the steady-state error because they are directly related to the error and are readily obtained from the system transfer functions. In this section error constants of control systems with simple aperiodic inputs are defined.

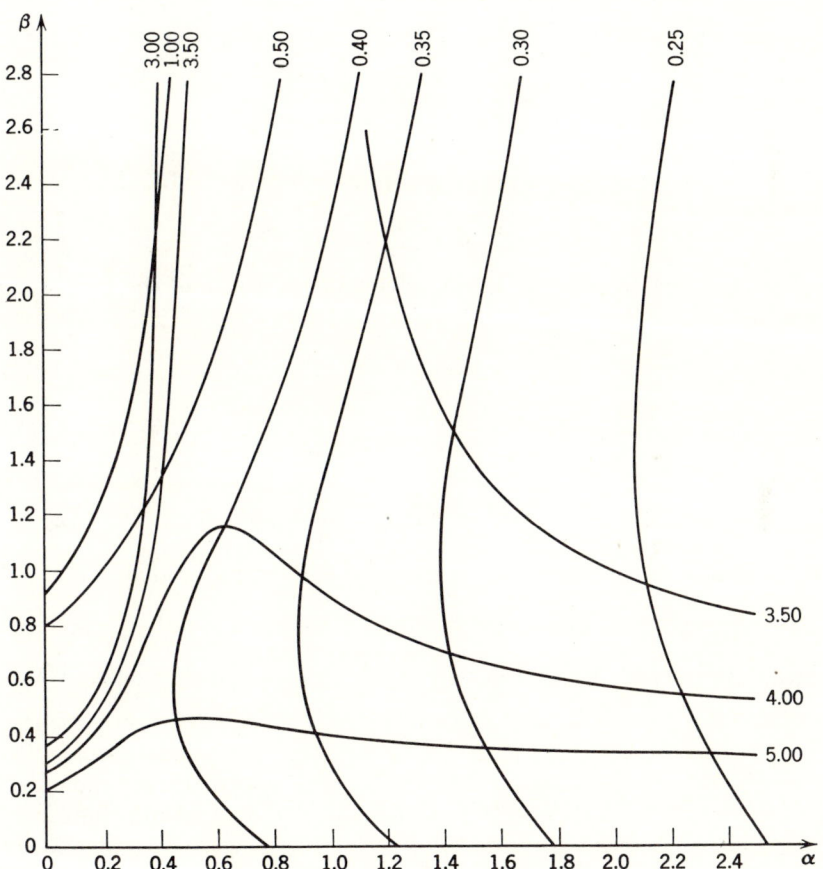

Figure 2.7 The -3 db constant bandwidth curves for the symmetrical parallel-T network (note: the curve labels refer to values of the frequency ω_b).

82 Linear Systems

The definitions of the error constants are based upon the application of the final value theorem of the Laplace transformation to the system of Figure 2.3. For the system of Figure 2.3, the error transform $E(s)$ is

$$E(s) = \frac{R(s)}{1 + G(s)}. \qquad (2.51)$$

If the system is stable and the function $sE(s)$ is analytic in the right half of the s plane and on the imaginary axis, the steady-state error is given by the final value theorem

$$e_{ss} = \lim_{s \to 0} \frac{sR(s)}{1 + G(s)}. \qquad (2.52)$$

Three particular forms of the input function $r(t)$ are considered:

1. $r(t)$ is a unit step function $[r(t) = u(t)]$; in this case $R(s) = 1/s$ and

$$e_{ss} = \lim_{s \to 0} \frac{1}{1 + G(s)} = \frac{1}{1 + \lim_{s \to 0} G(s)}. \qquad (2.53)$$

If the definition of the *positional error constant* K_p is introduced as

$$K_p = \lim_{s \to 0} G(s), \qquad (2.54)$$

a simple relationship between the steady-state error e_{ss} and the constant K_p is obtained as

$$e_{ss} = \frac{1}{1 + K_p}. \qquad (2.55)$$

2. $r(t)$ is a unit ramp function $[r(t) = u(t)t]$; in this case $R(s) = 1/s^2$ and

$$e_{ss} = \lim_{s \to 0} \frac{1/s}{1 + G(s)} = \frac{1}{\lim_{s \to 0} sG(s)}. \qquad (2.56)$$

If the definition of the *velocity error constant* K_v is introduced as

$$K_v = \lim_{s \to 0} sG(s), \qquad (2.57)$$

the steady-state error e_{ss} is expressed simply in terms of the constant K_v as

$$e_{ss} = \frac{1}{K_v}. \qquad (2.58)$$

3. $r(t)$ is a unit parabolic function $[r(t) = u(t)t^2/2]$; in this case $R(s) = 1/s^3$ and

$$e_{ss} = \lim_{s \to 0} \frac{1/s^2}{1 + G(s)} = \frac{1}{\lim_{s \to 0} s^2 G(s)}. \qquad (2.59)$$

If the definition of the *acceleration error constant* K_a is introduced as

$$K_a = \lim_{s \to 0} s^2 G(s), \tag{2.60}$$

the steady-state error e_{ss} is given by

$$e_{ss} = \frac{1}{K_a}. \tag{2.61}$$

The error constants K_p, K_v, and K_a are, therefore, measures of the steady-state error when the input is a unit step, a unit ramp, and a unit parabolic function, respectively. As a result of the given definitions of K_p, K_v, and K_a, only one of the error constants possesses a finite, nonzero value for any given system. For example, when the transfer function $G(s)$ has a simple pole at the origin of the s plane, K_p is infinite, K_v is finite and nonzero, and K_a is zero. The value of the nonzero, finite error constant depends on some of the system parameters and is a direct measure of the steady-state error. The effectiveness of the feedback in reducing sensitivity of system characteristics to variations of system parameters is determined, at least in part, by that value. From the definitions of the error constants listed above it is obvious that the higher value of the nonzero, finite error constant results in a small steady-state error. In control system design the minimum allowable value of the nonzero error constant is usually given in advance or it is necessary to choose a system adjustable parameter so as to achieve a maximum possible value of that constant.

Two graphical procedures in the parameter plane can be used to incorporate the steady-state response specifications in the system design. In the first, error-constant contours are plotted on the parameter plane enabling the steady-state error to be minimized subject to constraints imposed on the system pole-zero locations. In the second procedure the pole-zero locations are studied in the parameter plane on which the characteristic curves are plotted for a specified error constant, and are portrayed as a function of three-system parameters [2.12].

To present the first minimization procedure, denote the nonzero error constant by K_e and note that, in general, it is a function of the parameters α and β,

$$K_e = K_e(\alpha, \beta). \tag{2.62}$$

The solution of the simultaneous equations

$$\frac{\partial K_e}{\partial \alpha} = 0, \quad \frac{\partial K_e}{\partial \beta} = 0, \tag{2.63}$$

which satisfy the conditions

$$\frac{\partial^2 K_e}{\partial \alpha^2} < 0, \quad \left|\frac{\partial^2 K_e}{\partial \alpha \, \partial \beta}\right| - \frac{\partial^2 K_e}{\partial \alpha^2} \frac{\partial^2 K_e}{\partial \beta^2} < 0, \tag{2.64}$$

are the maximum values of K_e on the $\alpha\beta$ plane. The absolute maximum K_e is, therefore, readily determined. However, the absolute maximum is usually unrealizable because infinite or negative values of α and β are required. Even if the parameter values are realizable the stability and transient response associated with the maximum K_e are not acceptable. To ensure a more acceptable transient response, relative stability constraints are included in the analytical derivations as follows.

Consider the maximization of K_e subject to the constraints that all the roots of the characteristic equation have damping ratios ζ greater than, or equal to, the prescribed value ζ_0; that is, $\zeta \geq \zeta_0$. The region R_n of interest in the $\alpha\beta$ plane is therefore bounded by the $\zeta = \zeta_0$ curve, determined by $\alpha = \alpha(\zeta_0, \omega_n)$, $\beta = \beta(\zeta_0, \omega_n)$. If the maximum K_e occurs inside the region R_n, then its value may be determined from (2.63) and (2.64). If the maximum K_e does not occur inside the region R_n, then it occurs on the boundary of R_n itself and its value may be calculated from the following equation:

$$\left(\frac{\partial K_e}{\partial \omega_n}\right)_{\zeta=\zeta_0} = \left(\frac{\partial K_e}{\partial \alpha} \cdot \frac{\partial \alpha}{\partial \omega_n} + \frac{\partial K_e}{\partial \beta} \cdot \frac{\partial \beta}{\partial \omega_n}\right)_{\zeta=\zeta_0} = 0 \quad (2.65)$$

and the condition

$$\left(\frac{\partial^2 K_e}{\partial \omega_n^2}\right)_{\zeta=\zeta_0} < 0. \quad (2.66)$$

The partial derivatives $\partial \alpha/\partial \omega_n$ and $\partial \beta/\partial \omega_n$ may be obtained from the parameter plane equations $\alpha = \alpha(\zeta_0, \omega_n)$ and $\beta = \beta(\zeta_0, \omega_n)$ in a straightforward manner, but the calculation of the maxima values of K_e may become tedious as the roots of a high-degree algebraic equation in ω_n must be found and examined.

The constraints $\zeta \geq \zeta_0$ are not sufficient to ensure an acceptable transient response, but to include further constraints analytically would be difficult. Using the parameter plane, however, the problem of adjusting the system parameters to give the best compromise between steady-state and transient

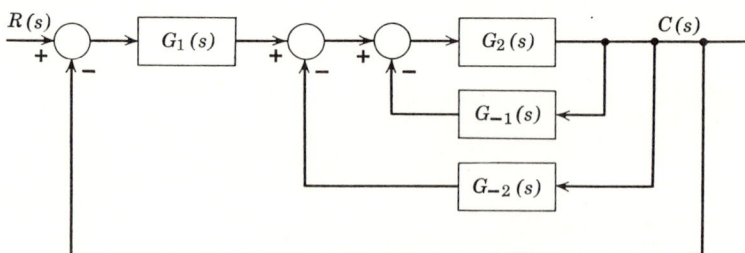

Figure 2.8 System block diagram.

System Response and Stability   85

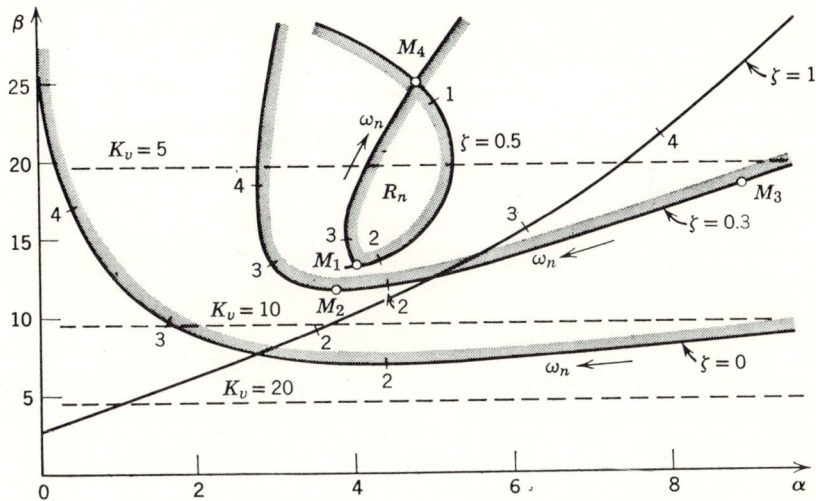

Figure 2.9 Parameter plane diagram for maximization of K_v with relative stability constraints.

response may be undertaken by plotting contours of constant K_e on the $\alpha\beta$ plane. The curves are interpreted as in the following examples.

Consider the system of Figure 2.8, where

$$G_1(s) = K_1 \frac{(s+1)}{(s+0.2)}, \qquad G_2(s) = \frac{K_2}{s(s^2+0.8s+1)}, \qquad (2.67)$$

$$G_{-1}(s) = K_{-1}s, \qquad G_{-2}(s) = K_{-2}s^2.$$

The system characteristic equation is

$$s^4 + (\alpha + 1)s^3 + (0.2\alpha + \beta + 1.16)s^2 + (0.2\beta + \gamma + 0.2)s + \gamma = 0, \qquad (2.68)$$

where $\alpha = K_2 K_{-2}$, $\beta = K_2 K_{-1}$, and $\gamma = K_1 K_2$. The error constant K_v can be expressed as

$$K_v = \frac{\gamma}{0.2(1+\beta)}. \qquad (2.69)$$

Figure 2.9 gives the parameter plane curves for the condition $\gamma = 20$. By plotting just a few ζ lines, the region R_n of interest in the $\alpha\beta$ plane is evident. The points of the relative maximums are found by inspection for the two curves $\zeta = 0.5$ and $\zeta = 0.3$. They are M_1 and M_2 with the corresponding

86   Linear Systems

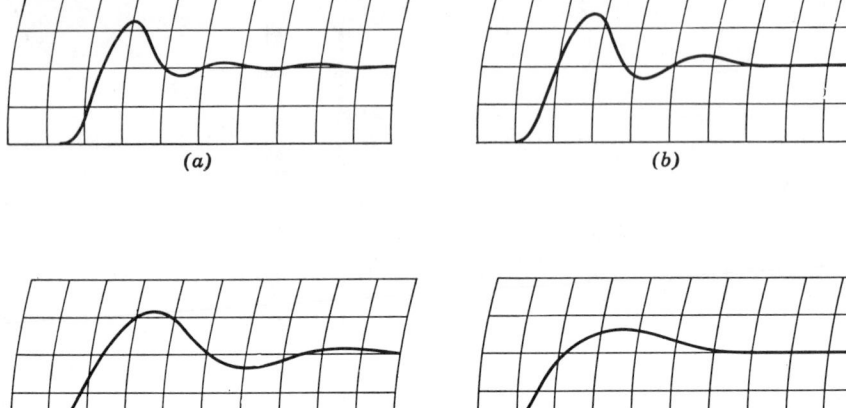

Figure 2.10   Unit-step responses (note: magnitude scale is 0.5 per line; time scale is 1 sec per line).

root and error constant values

$$M_1(4.1; 13.3) \qquad M_2(3.7; 11.3)$$
$$\zeta_1 = 0.5, \ \omega_{n1} \simeq 2.5 \qquad \zeta_1 = 0.3, \ \omega_{n1} \simeq 2.4$$
$$\zeta_2 \simeq 0.69, \ \omega_{n2} \simeq 2.15 \qquad \zeta_2 \simeq 0.86, \ \omega_{n2} \simeq 2.1$$
$$K_v \simeq 7 \qquad K_v \simeq 8.$$

Because of the poor dominancy conditions, these two points are unacceptable as shown by the related transient responses given in Figures 2.10a and 2.10b. The overshoots of the responses to a unit-step input may be considered excessive, being over 50% of the steady-state value of the responses.

To decrease the overshoot, the operating point can be moved to $M_3(9; 18.3)$ where two tangents can be drawn to the $\zeta = 1$ curve, indicating two real roots, $\sigma_3 \simeq -1.7$, $\sigma_4 \simeq -7.4$. The complex roots are determined by $\zeta_1 = 0.3$, $\omega_{n1} \simeq 1.2$. The transient response is shown in Figure 2.10c. The error constant is decreased to $K_v \simeq 5.5$. Further decreasing of the overshoot may lead to a point $M_4(4.75; 25)$ where the roots are $\zeta_1 = \zeta_2 = 0.5$, $\omega_{n1} = 0.9$, $\omega_{n2} = 6$. The dominancy of the roots with (ζ_1, ω_{n1}) is achieved and results in a transient response given in Figure 2.10d. The overshoot may be acceptable (approximately 30%), but the error constant has been decreased down to $K_v \simeq 2$, which may result in an unacceptable steady-state error.

When the error constant is specified, the pole-zero locations and the transient response can be investigated as a function of α, β, and a third parameter γ on a single-parameter plane diagram. The effect of adjusting the two parameters α and β may be studied on the $\alpha\beta$ plane, and for each point a third parameter γ is then given as a known function of α, β, and K_e. The same example is used to show that the additional degree of freedom can improve the actual design specifications.

For the system of Figure 2.8, if K_v is specified, the characteristic equation may be expressed in terms of α and β by eliminating the third parameter γ from (2.68) and (2.69). For $K_v = 10$, $\gamma = 2(1 + \beta)$ and

$$s^4 + (\alpha + 1)s^3 + (0.2\alpha + \beta + 1.16)s^2 + (2.2\beta + 2.2)s + (2\beta + 2) = 0. \tag{2.70}$$

The corresponding parameter plane diagram is shown on Figure 2.11. Whenever the operating point $M(\alpha; \beta)$ is placed on this diagram, the error constant is $K_v = 10$. For several points on the plane, which may seem to have acceptable root configuration, the unit-step response is plotted by an analog computer. The scales for magnitude and time of the responses are the same as that of Figure 2.10; that is, for magnitude 0.5 second per line and for time 1 second per line. Comparing the diagrams of Figures 2.10 and 2.11, we can conclude that the second approach can offer a better system performance for both the transient and the steady-state response characteristics.

2.4 Multiparameter Adjustment. Design Constraints

In this section the parameter plane concept and technique are extended to the selection of several adjustable parameters to achieve desired system performance characteristics. The procedure is not essentially dependent on the structure of the system and can be advantageously applied to multiloop control systems.

The effects of the structural changes of the system are important because such intentional changes introduce an additional number of adjustable parameters that are used to meet design specifications and requirements. These effects are considered in the following section, where the compensation techniques are outlined. In this section, however, it is assumed that the system structure is given and the design requirements are satisfied by parameter adjustment only.

Two-parameter limitation occurs because there are only two equations relating the two adjustable parameters; they are obtained by equating the real and imaginary part of the characteristic polynomial to zero. These two equations are then solved for the two adjustable parameters in terms of the root values. The main idea of the following extension is not to generate more equations artificially from the pure mathematical relations, since they

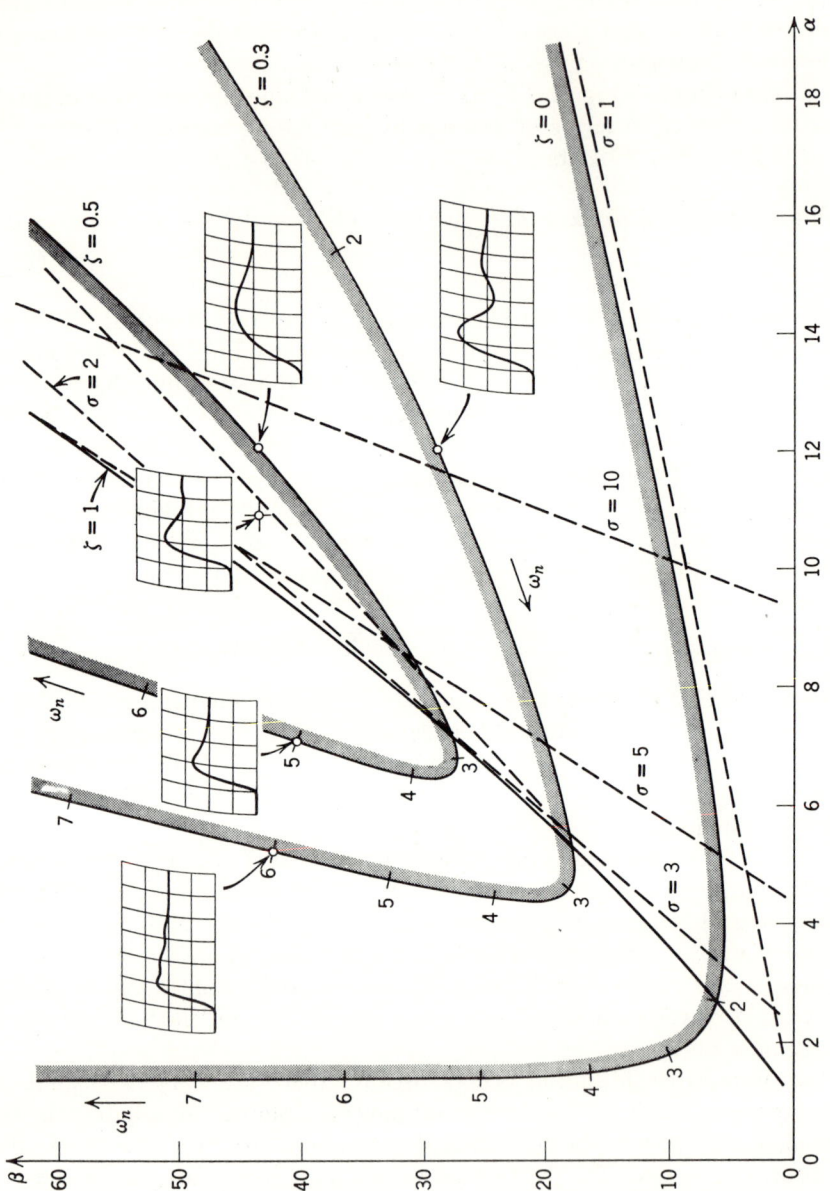

Figure 2.11 Parameter plane diagram with unit-step responses for $K_v = 10$.

usually lead to practically meaningless parameter values. Additional equations are composed directly from the system specifications that are automatically met when the equations are solved, thus yielding practical results. The additional relations associated with two-parameter plane equations are called the *constraint equations and inequalities.* The constraint equations are determined from the characteristic root specifications, dominant-root parameter sensitivity, and error-constant specifications, and are incorporated into the design analytically, whereas the relative stability and dominancy inequality constraints are considered graphically either by direct parameter plane mapping (parameter plane) or the inverse parameter mapping (complex plane). The constrained equations and inequalities are handled so that the multiparameter graphical presentation in a parameter space is avoided, because the graphical presentation in more than two dimensions is difficult to interpret. The preliminary considerations, however, can be carried out in the parameter plane as well as the final step of the design when all the specifications are incorporated in the previous steps and the design is reduced to the two-parameter problem. Along the steps in the design, other methods of linear system analysis can be helpful and should be introduced whenever useful results can be expected. The design techniques to be presented are computer oriented so that readily applied programs can be used in a systematic procedure to achieve an acceptable system performance.

In addition, approximate methods whereby the structure and parameter of a system may be selected to satisfy relative stability and dominancy specifications are also considered. This may be adequate for the design of lower-order systems and may be useful as preliminary investigations in the application of other multiparameter methods for high-order systems.

Root Specifications: The characteristic equation is written as

$$F(s) \equiv \sum_{k=0}^{n} a_k s^k = 0, \tag{2.71}$$

where s is the complex variable,

$$s = \sigma + j\omega, \tag{2.72}$$

and the coefficients a_k, $(k = 0, 1, \ldots, n)$, are functions of the adjustable parameters p_i, $(i = 1, 2, \ldots, r)$,

$$a_k = a_k(p_1, p_2, \ldots, p_r). \tag{2.73}$$

By using the functions $X_k = X_k(\sigma, \omega)$, $Y_k = Y_k(\sigma, \omega)$, (2.71) may be conveniently separated into two equations by equating the real part and the imaginary part to zero independently. Thus

$$\sum_{k=0}^{n} a_k X_k = 0$$
$$\sum_{k=0}^{n} a_k Y_k = 0. \qquad (2.74)$$

For the case in which $s = \sigma$ is real, (2.71) yields

$$\sum_{k=0}^{n} a_k \sigma^k = 0. \qquad (2.75)$$

Equations 2.74 and 2.75 can be used as constraint equations relating the system parameters p_i ($i = 1, 2, \ldots, r$), when complex characteristic roots ($\sigma = \sigma_0$, $\omega = \omega_0$) and real roots ($\sigma = \sigma_0$) respectively, are specified in advance.

Root Sensitivity Specifications: If the sensitivity of a pair of complex characteristic roots is specified, as well as the pair of the roots itself, four constraint equations result. Consider that the sensitivity of the real and imaginary parts of the roots $s_{j,j+1}$ of (2.71) is defined as

$$S_j^\sigma = \frac{\Delta \sigma_j / \sigma_j}{\sum_{i=1}^{r} \Delta p_i / p_i}$$
$$S_j^\omega = \frac{\Delta \omega_j / \omega_j}{\sum_{i=1}^{r} \Delta p_i / p_i}. \qquad (2.76)$$

When σ_j, ω_j, S_j^σ, S_j^ω, p_i and Δp_i, ($i = 1, 2, \ldots, r$) are specified, the constraint equations are given by (2.74) and the following equations:

$$\sum_{k=0}^{n} a_k' X_k(\sigma_j + \Delta \sigma_j, \omega_j + \Delta \omega_j) = 0$$
$$\sum_{k=0}^{n} a_k' Y_k(\sigma_j + \Delta \sigma_j, \omega_j + \Delta \omega_j) = 0, \qquad (2.77)$$

where $a_k' = a_k(p_1 + \Delta p_1, p_2 + \Delta p_2, \ldots, p_r + \Delta p_r)$ and

$$\Delta \sigma_j = S_j^\sigma \sigma_j \sum_{i=1}^{r} \frac{\Delta p_i}{p_i}$$
$$\Delta \omega_j = S_j^\omega \omega_j \sum_{i=1}^{r} \frac{\Delta p_i}{p_i}. \qquad (2.78)$$

Equations 2.77 are readily derived from (2.74).

Similarly, if the sensitivity of a real characteristic root σ_j is specified, as well as the root σ_j itself, two constraint equations result, that is, (2.71) when $s = \sigma_j$, and

$$\sum_{k=0}^{n} a'_k (\sigma_j + \Delta\sigma_j)^k = 0. \tag{2.79}$$

In the above calculations large parameter sensitivities are considered. When small parameter variations are specified, the root sensitivity equations derived in the preceding chapter should be used. These result from the above equations when $\Delta p_i/p_i$ is made sufficiently small.

Error-Constant Specifications: The constraint equation relating the system parameters for the case in which an error constant is specified is readily derived from the appropriate error constant definition; that is,

$$\begin{aligned} K_p &= \lim_{s \to 0} G(s) \\ K_v &= \lim_{s \to 0} sG(s) \\ K_a &= \lim_{s \to 0} s^2 G(s), \end{aligned} \tag{2.80}$$

where K_p, K_v, and K_a are the position, velocity, and acceleration error constants, respectively, and $G(s)$ is the open-loop transfer function.

To the above specifications it is possible to add also bandwidth specifications as presented in the previous section as well as the integral and mean-squared error (see Appendix E), the Q factor, some sensitivity indices [2.13, 14] and, in general, any performance characteristic that can be expressed as a function of system parameters.

To illustrate the application of the constraint equations to linear system design, consider the same system of Figure 2.8 with the same transfer functions

$$G_1(s) = K_1 \frac{s + Z}{s + 0.2}, \qquad G_2(s) = \frac{K_2}{s(s^2 + 0.8s + 1)} \tag{2.81}$$

$$G_{-1}(s) = K_{-1} s, \qquad G_{-2}(s) = K_{-2} s^2$$

except that in $G_1(s)$ the zero Z is assumed to be variable.

The system characteristic equation is

$$s^4 + (1 + p_1)s^3 + (0.2p_1 + p_2 + 1.16)s^2 + (0.2p_2 + p_3 + 0.2)s + p_4 = 0, \tag{2.82}$$

where $p_1 = K_2 K_{-2}$, $p_2 = K_2 K_{-1}$, $p_3 = K_1 K_2$, and $p_4 = K_1 K_2 Z$. The error constant is

$$K_v = \frac{p_4}{0.2(1 + p_2)}. \tag{2.83}$$

92   Linear Systems

The design problem is to choose the parameters p_1, p_2, p_3, and p_4 so that $K_v = 10$ and there is a complex dominant pair of roots $s_{1,2}$ located so that $\zeta = 0.5$, $\omega_n = 2.5$.

By specifying K_v, as well as a pair of complex roots, three equations exist relating the four adjustable parameters, namely (2.83) and (2.74), which become

$$2(1 + p_2) - p_4 = 0$$

$$\sum_{k=0}^{4} a_k X_k = 0 \qquad (2.84)$$

$$\sum_{k=0}^{4} a_k Y_k = 0$$

with

$$\begin{aligned}
a_0 &= p_4 & X_k &= X_k\,(\zeta = 0.5,\ \omega_n = 2.5) \\
a_1 &= 0.2 p_2 + p_3 + 0.2 & Y_k &= Y_k\,(\zeta = 0.5,\ \omega_n = 2.5). \\
a_2 &= 0.2 p_1 + p_2 + 1.16 & & \\
a_3 &= 1 + p_1 & & \\
a_4 &= 1 & &
\end{aligned} \qquad (2.85)$$

To solve the three equations 2.84 for four parameters, one may assume $p_3 = p_4$ and obtain $p_1 = 9.5$, $p_2 = 34$, $p_3 = p_4 = 70$. The remaining two roots of (2.82) are easily calculated by using the specified root values, calculated parameter values, and the long division method of Appendix A to obtain $s_3 = -1.5$, $s_4 = -2.1$.

Difficulties in the above procedure may arise because the remaining roots are automatically given once the constrained equations are specified. These roots may not be outside the dominancy region of the specified roots and their locations may not be acceptable. The design procedure should then be repeated for some other relation between the parameters p_3 and p_4 other than $p_3 = p_4$. This will make possible the use of an additional degree of freedom by adjusting the zero Z of the compensating network $G_1(s)$ of (2.81), since $p_3 = K_1 K_2$, $p_4 = K_1 K_2 Z$.

From (2.82) it is seen that by specifying four roots, four parameter plane equations may be written (two for each complex pair and one for each real root specified). These four equations may then be solved for the four parameters p_1, p_2, p_3, and p_4. For example, specifying $\zeta = 0.3$, $\omega_n = 1.25$, $\sigma_3 = -1.7$, $\sigma_4 = -7.4$, the parameters are $p_1 = 9$, $p_2 = 18.5$, $p_3 = p_4 = 20$. Some difficulties may arise because the error constant is given automatically by (2.83) as $K_v = 5.12$, which in some applications of the system may be considered undesirably small. Moreover, once all the roots are specified, the parameters are calculated as fixed values that may be unrealizable in a physical system. The parameter plane equations to be solved should be linear in terms of the system parameters, which is not always the case.

By specifying not only one pair of complex roots but also their sensitivity to parameter variations, four equations may be written and solved for four parameters (see equations 2.77). For example, when ζ and ω_n of a pair of complex roots are specified as $\zeta = 0.3$, $\omega_n = 1.25$, then the two-parameter plane equations 2.74 may be written relating the parameters for an operating point M_1. Furthermore, if $\zeta = 0.5$ and $\omega_n = 2.25$ when the parameter p_3 is 3.6 times its previous value, then two further parameter plane equations relating the system parameters for point M_2 can be written. Solving the four-parameter plane equations gives

$$M_1: \zeta = 0.3, \omega_n = 1.25, \sigma_3 = -1.7, \sigma_4 = -7.4, K_v = 5.3$$
$$p_1 = 9, p_2 = 18.5, p_3 = p_4 = 20$$
$$M_2: \zeta = 0.5, \omega_n = 2.25, \sigma_3 = -1.5, \sigma_4 = -2.1, K_v = 10$$
$$p_1 = 9.5, p_2 = 34.5, p_3 = p_4 = 72.$$

As demonstrated in the above example, the simultaneous solution of independent constraint equations derived from the specifications may facilitate a multiparameter design. The number of such independent constraint equations that are available and that can be conveniently solved gives the number of adjustable parameters that can be considered. Usually only two, three, four, or possibly five parameters can be considered in this way. In order to apply a straightforward general computer program, the equations must be linear or contain only one product term. For the more complex nonlinear case, even if the equation solution is convenient, the interpretation of the results may be difficult.

The introduction of constraint inequalities into a system design gives more freedom in adapting the system to the application. Requirements on a system transient response may usually be translated into constraints on the system transfer function pole-zero configuration. These may be considered graphically on both the complex plane and the parameter plane, or analytically by the truncation procedure as shown in the following section.

2.5 Design. Compensation and Parameter Adjustment

In general, the problem of linear system design is to select the system structure and adjustable parameters and to achieve desired system performance characteristics. Numerous methods and techniques have been developed for the solution of this problem. Each method has certain advantages and inherent limitations, and the designer normally has to use more than one technique to obtain the required specifications in the actual design. There is a large variety of design aspects and performance characteristics, so that there is no unique and general approach to linear system design. In this section the parameter plane concept is applied to the problem of linear design in which it may prove useful either as a complete design tool or in connection with some of the well-known methods.

In many design problems the variation of parameters is not sufficient to meet all the specifications, and it is necessary to introduce additional blocks and connections into the system and change its structure. The additional blocks, which contain a certain number of adjustable parameters, are called the *compensators*. Compensators have been classified as integral, differential, feedback compensators, and so on, according to their effect and function when introduced into the original system. In classical methods of the root-locus and frequency domain, the compensators are introduced to reshape the corresponding graphs of the original system in order to achieve desired system response characteristics. Except in special cases, reshaping the parameter plane curves is rather difficult since their sketching cannot be performed in a straightforward manner. In the parameter plane approach the direct effects of the additional blocks and connections on the characteristic equation are studied. The compensators are introduced so that the additional adjustable parameters are conveniently distributed among the coefficients of the characteristic equation. This enables the designer to have control over the characteristic roots by varying the introduced parameters [2.15].

The selection of a system structure and its parameters may be based upon the application of the following truncation technique. For high-order multi-loop systems it can be used as the preliminary investigation to a more detailed multiparameter analysis.

Consider the characteristic equation

$$F(s) \equiv \sum_{k=0}^{n} a_k s^k = 0, \qquad (2.26)$$

which is truncated into the two equations

$$\sum_{k=0}^{m} a_k s^k = 0 \qquad (2.86a)$$

$$\sum_{k=m}^{n} a_k s^k = 0. \qquad (2.86b)$$

For the case in which the magnitude of the roots of (2.86a) are much smaller than those of (2.86b), the "dominant" roots of the characteristic equation 2.26 are approximately given by the roots of (2.86a), and the approximate values of the remaining roots of (2.26) are indicated by the values of the roots of (2.86b). The important fact about this statement is that the adjustable parameters and their values appearing in the coefficients a_k are chosen so that this convenient separation of the roots takes place. Moreover, this separation not only ensures the success of the truncation, but is a desirable situation for achieving an appropriate system response.

Therefore, the new structure of the system is generally chosen so that the additional adjustable parameters appear in both equations 2.86. Then the parameters appearing in the coefficients a_k, $(k = m, m+1, \ldots, n)$, of (2.86b) are chosen so that the corresponding roots are far enough to the left of the dominant roots. Then the chosen values of these parameters are introduced in (2.86a) and the remaining adjustable parameters in a_k, $(k = 0, 1, \ldots, m)$, are selected so that the dominant roots are placed at desired locations. With such chosen values of the adjustable parameters, the original characteristic equation may not have a desired root configuration since it is not identical to the one calculated approximately by the truncation. The process can then be repeated, or else the parameter plane technique can be used for final adjustment of the two parameters considered as most powerful in controlling the significant part of the root configuration.

To illustrate the truncation technique, consider the simple control system of Figure 2.12a. The system should have two dominant complex closed-loop poles so that the system transient response is sufficiently close to the response $d = 0.75$ of Figure 2.13. (This figure is Figure C.1d of Appendix C.) The steady-state response is specified by the error constant $K_v \geq 6$.

The characteristic equation of the system is

$$s^3 + 3s^2 + 2s + K_1 = 0 \qquad (2.87)$$

with only the adjustable parameter K_1. Since $K_v = K_1/2$, the system gain K_1 should be $K_1 \geq 12$. With $K_1 = 12$, (2.87) becomes

$$s^3 + 3s^2 + 2s + 12 = 0, \qquad (2.88)$$

which should be solved for the roots to see if the transient response specifications are satisfied.

Along these lines, it is of interest to note that the third-degree algebraic equations can be graphically solved by using charts that can be plotted once and for all [2.16]. In general, the third-degree equation has the form

$$a_3 s^3 + a_2 s^2 + a_1 s + a_0 = 0. \qquad (2.89)$$

By introducing the substitution

$$s = \frac{a_2}{a_3} S, \qquad (2.90)$$

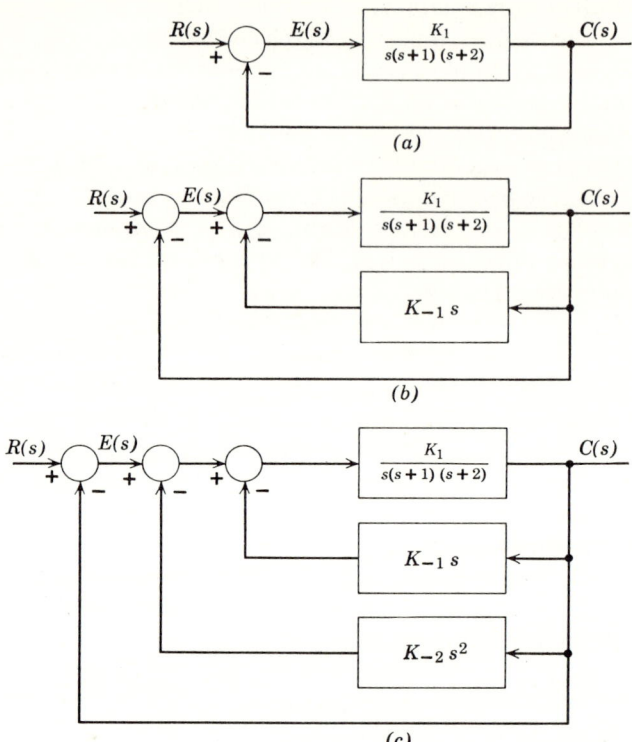

Figure 2.12 System block diagrams.

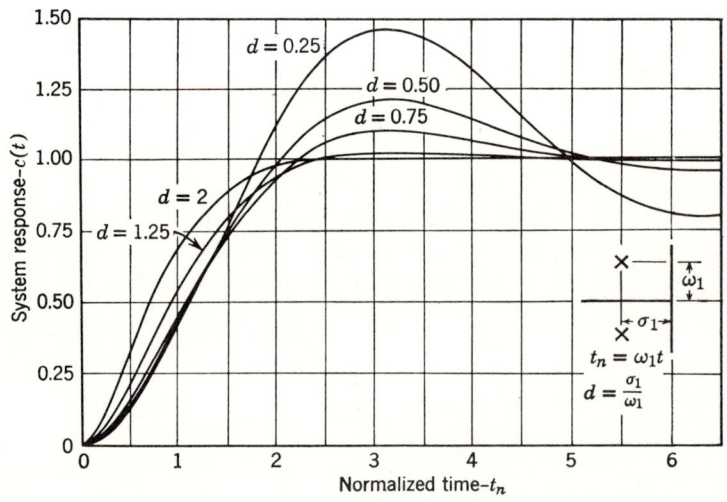

Figure 2.13 Second-order system responses.

(2.89) can be rewritten as

$$S^3 + S^2 + A_1 S + A_0 = 0, \qquad (2.91)$$

where $A_1 = a_1 a_3^2/a_2^2$, $A_0 = a_0 a_3^2/a_2^3$.

If the coefficients A_1 and A_0 are considered as variables α and β, the charts of ζ, ω_{nt}, and σ_t curves can be plotted as in Figures 2.14 and 2.15. The diagrams are used in the normal fashion by considering the position of the point $M(A_1; A_0)$ in relation to the characteristic curves except that the values of ω_n and σ of (2.89) are obtained from the transformed values ω_{nt} and σ_t as

$$\omega_n = \frac{a_2}{a_3}\omega_{nt}, \qquad \sigma = \frac{a_2}{a_3}\sigma_t. \qquad (2.92)$$

It is important to note that ζ remains unchanged in the applied transformation.

From (2.88), the point $M_1(A_1 = 0.222; A_0 = 0.445)$ is in the unstable region as indicated in Figure 2.14. Thus the structure of the system should be changed and some additional adjustable parameter introduced in the design. If a velocity feedback compensation (sometimes called tachometer feedback compensation) is added as shown in Figure 2.12b, the characteristic equation becomes

$$s^3 + 3s^2 + (2 + K_1 K_{-1})s + K_1 = 0. \qquad (2.93)$$

The introduced parameter K_{-1} is not sufficient for two reasons.

First, the error constant K_v is now

$$K_v = \frac{K_1}{2 + K_1 K_{-1}} \geq 6, \qquad (2.94)$$

which requires that the transformed coefficients of (2.93) for the limiting case ($K_v = 6$) satisfy the relationship

$$A_0 = 2A_1, \qquad (2.95)$$

where $A_0 = (2 + K_1 K_{-1})/9$, $A_1 = K_1/27$. Equation 2.95 represents a straight line that lies entirely in the unstable region as shown in Figure 2.14.

Second, even if K_v is allowed to decrease, the parameter K_{-1} is located in the coefficient a_1 only, so that it has a powerful control only on the complex roots, approximately determined by the truncated equation

$$3s^2 + (2 + K_1 K_{-1})s + K_1 = 0. \qquad (2.96)$$

The remaining real root is approximately determined by the rest of (2.93),

$$s + 3 = 0, \qquad (2.97)$$

which is independent of the adjustable parameters. For example, if the point

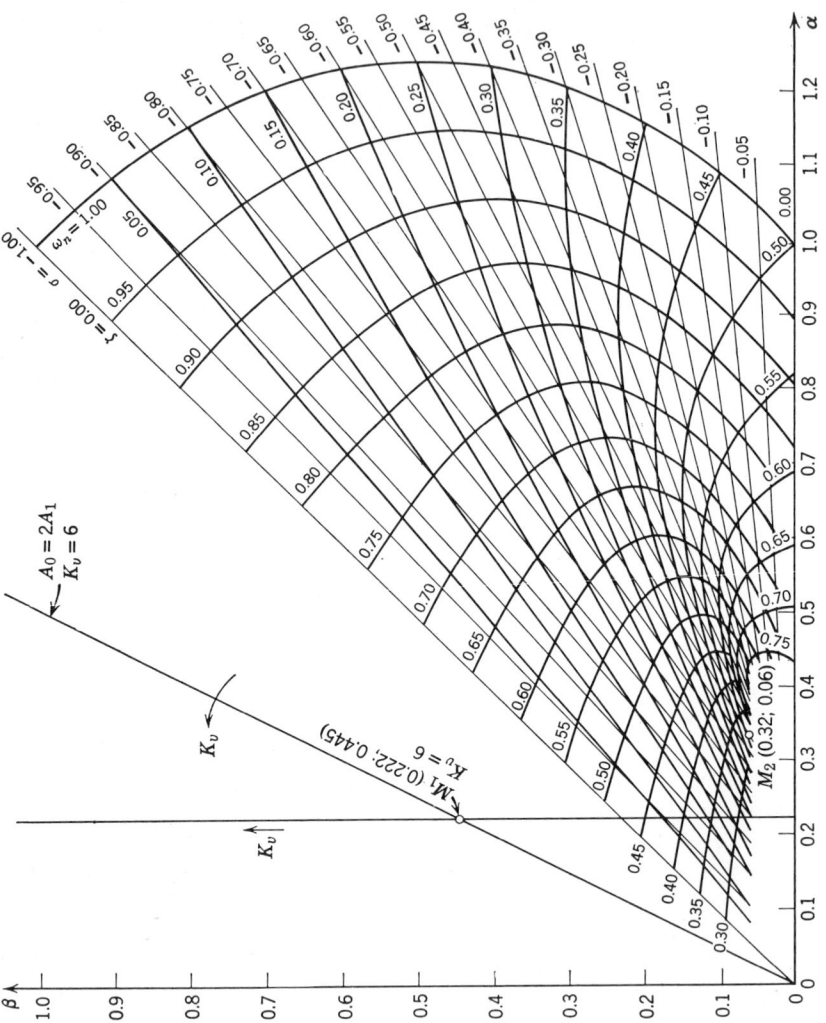

Figure 2.14 Charts for third-degree algebraic equations.

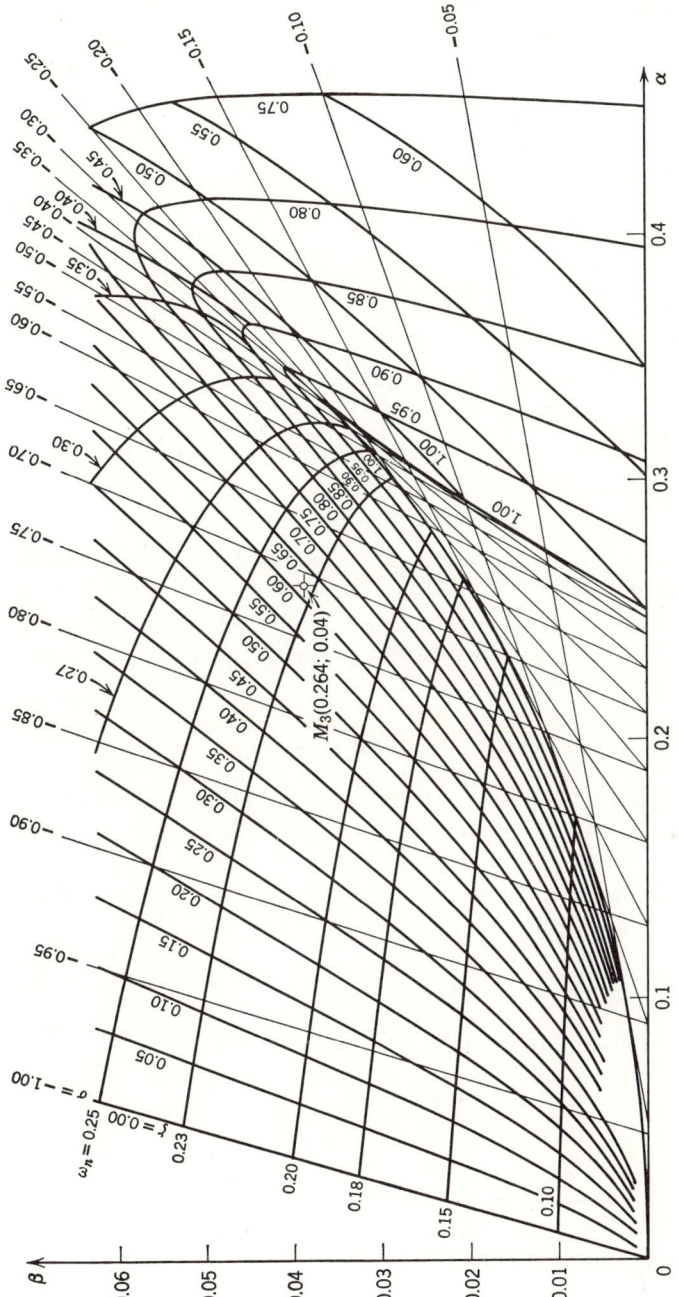

Figure 2.15 Enlarged region around the origin of Figure 2.14.

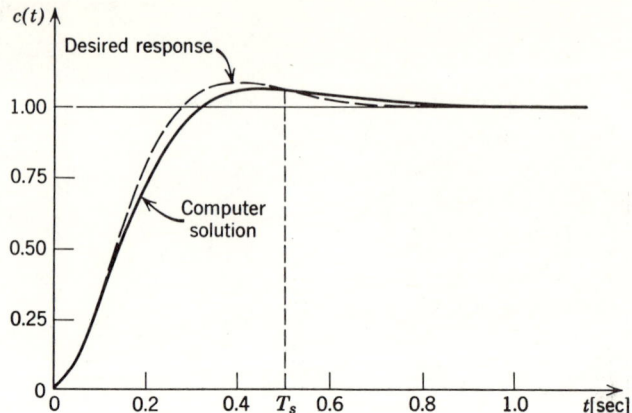

Figure 2.16 System transient response to unit-step function.

$M_2(0.32; 0.06)$ is chosen to have $\zeta = 0.6$, $\omega_n = 1$ ($\omega_{nt} = 0.33$), which yield the response $d = 0.75$ of Figure 2.13, the real root is $\sigma = -1.9$ ($\sigma_t = -0.63$) and sufficient dominancy is not achieved.

To have complete control over all three roots of the characteristic equation, acceleration feedback compensation is added, as shown in Figure 2.12c. The characteristic equation is

$$s^3 + (3 + K_1K_{-2})s^2 + (2 + K_1K_{-1})s + K_1 = 0, \qquad (2.98)$$

whereas the velocity constant K_v remains the same as in (2.94).

To achieve the desired system response with second-order dominancy, let us choose the complex pair of roots determined by $\zeta = 0.7$ and $\omega_n = 10$.[10] To ensure dominancy, choose the parameters $K_1K_{-2} = 47$ so that the real root σ, determined approximately by

$$s + 3 + K_1K_{-2} = 0, \qquad (2.99)$$

is $\sigma \simeq -50$. The dominant roots of (2.98) are given by

$$s^2 + \frac{2 + K_1K_{-1}}{3 + K_1K_{-2}} s + \frac{K_1}{3 + K_1K_{-2}} = 0. \qquad (2.100)$$

---

[10] The values $\zeta = 0.7$ and $\omega_n = 10$ are calculated from the desired system response, that is, $d = \zeta/\sqrt{1 - \zeta^2} = 0.7$, which yields $\zeta \geq 0.6$; and the settling time T_s required for the system response to start lying within 5% of its steady-state value is approximately determined for second-order dominancy by $T_s \simeq 3/\zeta\omega_n$, which is $T_s \simeq 0.45$ (Figure 2.13) and $\omega_n = 10$.

Since $3 + K_1 K_{-2} = 50$ and

$$2\zeta\omega_n = \frac{2 + K_1 K_{-1}}{3 + K_1 K_{-2}}, \qquad \omega_n^2 = \frac{K_1}{3 + K_1 K_{-2}}, \qquad (2.101)$$

where $\zeta = 0.7$, $\omega_n = 10$, we obtain the parameter values as $K_1 = 5000$, $K_{-1} = 0.14$, $K_{-2} = 0.0094$. The compensated characteristic equation is

$$s^3 + 50s^2 + 700s + 5000 = 0. \qquad (2.102)$$

To find the "exact" roots of this equation, note that it corresponds to the point M_3 (0.264; 0.04) on the diagram of Figure 2.15 and that the roots are determined by $\zeta \simeq 0.625$, $\omega_n \simeq 12$ ($\omega_{nt} \simeq 0.24$), $\sigma \simeq -35.4$ ($\sigma_t \simeq -0.705$). The corresponding velocity error constant is from (2.94), obtained as $K_v = 7.15$, and all the necessary specifications are achieved.

The desired system response and the actual response obtained by a computer simulation are shown in Figure 2.16 for comparison.

In concluding the description of truncation, it should be noted that normally the characteristic equation is truncated down to a low-degree equation that can be solved by elementary methods (up to the third degree). The truncation is not necessarily only in two parts and can be extended to truncation in several equations. This, however, is of small practical interest.

A somewhat more complicated situation arises in connection with the system of Figure 2.17. Two inner feedback paths with compensators have been introduced that have three adjustable parameters, p_1, p_2, and p_3. The parameters appear conveniently distributed in the coefficients of the corresponding characteristic equation

$$(19.2p_3)s^5 + (19.2 + 74.5p_3)s^4 + (74.5 + 75.7p_3 + 1440p_1)s^3$$
$$+ (75.7 + 23.6p_3 + 792p_1 + 1228p_2p_3)s^2 \qquad (2.103)$$
$$+ (23.6 + 461p_3 + 36p_1 + 1228p_2)s + 460 = 0.$$

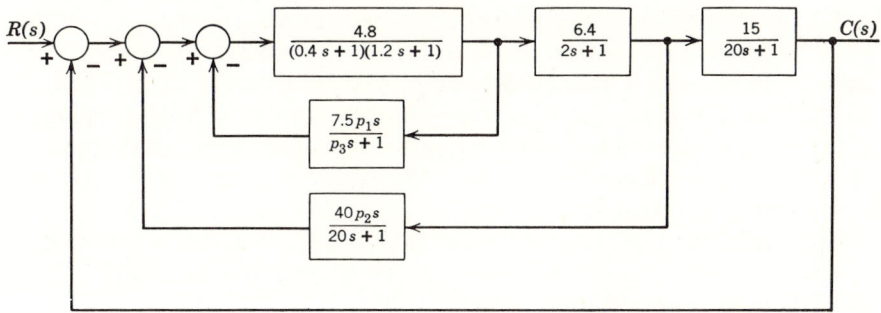

Figure 2.17 Control system block diagram.

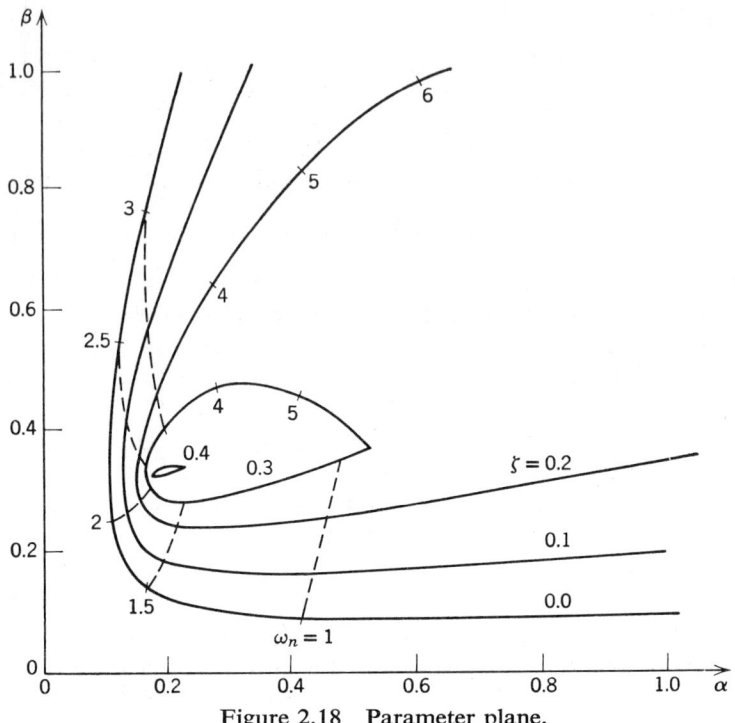

Figure 2.18 Parameter plane.

Equation 2.103 can be appropriately truncated into two low-degree equations:

$$(19.2p_3)s^2 + (19.2 + 74.5p_3)s + (74.5 + 75.7p_3 + 1440p_1) = 0 \quad (2.104a)$$
$$(74.5 + 75.7p_3 + 1440p_1)s^3 + (75.7 + 23.6p_3 + 792p_1 + 1228p_2p_3)s^2$$
$$+ (23.6 + 461p_3 + 36p_1 + 1228p_2)s + 460 = 0. \quad (2.104b)$$

After several trials and applications of the third-degree charts of Figures 2.14 and 2.15, the parameters p_1 and p_3 are chosen to be $p_1 = 0.2$, $p_3 = 1.25$ to yield the roots of (2.104a) as $s_{1,2} = (\omega_n \simeq 6.1, \zeta \simeq 0.39)$. These values of p_1 and p_3 are introduced in (2.104b), and $p_2 = 0.2$ is determined so that $s_{3,4} = (\omega_n \simeq 0.90, \zeta \simeq 0.35)$ and $s_5 = \sigma \simeq -0.4$. As noted, a sufficient separation of the roots is achieved (ω_n of $s_{1,2}$ is approximately six times greater than that of $s_{3,4}$), and the obtained root values are expected to be close to the exact root locations of (2.103).

To estimate the related system transient response to a step input function, it is necessary to include the zero -0.8 of the closed-loop transfer function which is then approximately given as

$$\frac{C}{R}(s) = \frac{460(1.25s + 1)}{(s^2 + 4.75s + 37.2)(s^2 + 0.63s + 0.81)(s + 0.4)}. \quad (2.105)$$

Before calculating the transient response on the basis of (2.105), several questions should be answered: (1) Is the accuracy of the truncation used sufficient for (2.105) to represent the closed-loop of the actual system? (2) If so, does the related transient response have the desired characteristics or can it be improved by changing the values of the adjustable parameters? (3) What are the other possible values of the parameter for the chosen structure? These questions can be considered by applying the parameter plane techniques around the parameter values estimated in the preliminary approximate analysis. Then a computer simulation can be used to check the transient response for a limited range of parameter values that are candidates for giving a satisfactory time-domain behavior.

Parameter plane contours for (2.103) and $p_1 = \alpha$, $p_2 = \beta$, $p_3 = 1.25$ are given on Figures 2.18 and 2.19, which display two sets of nonintersecting contours. It is seen that if a point is taken within the $\zeta = 0.4$ contour and the remaining roots are determined, the relative stability of the system is achieved for $\zeta \simeq 0.4$ and all corresponding values of p_1 and p_2. In fact, the problem of "maximizing the system relative stability" by adjusting the two parameters is

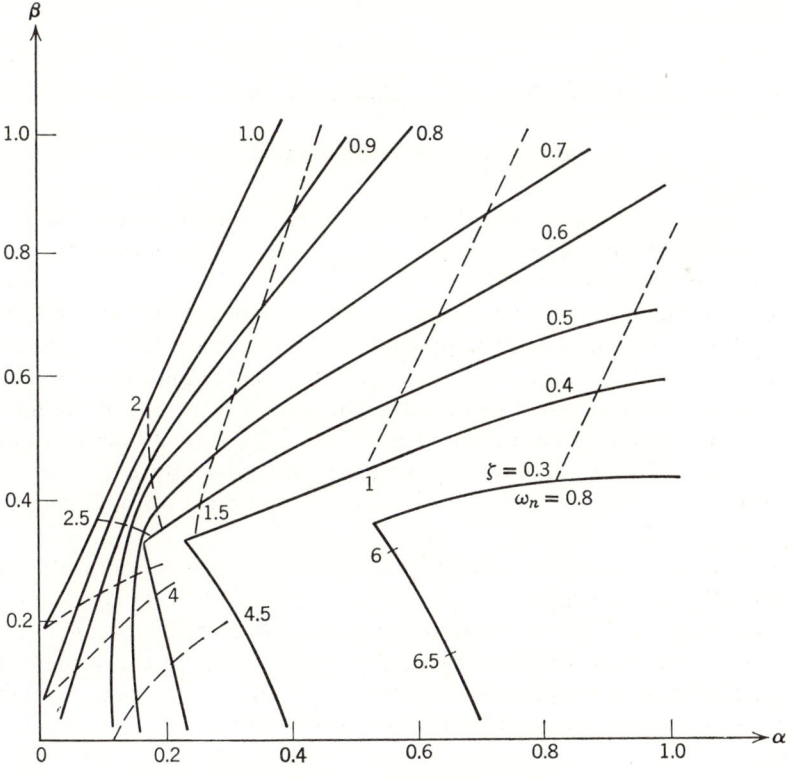

Figure 2.19 Parameter plane.

104   Linear Systems

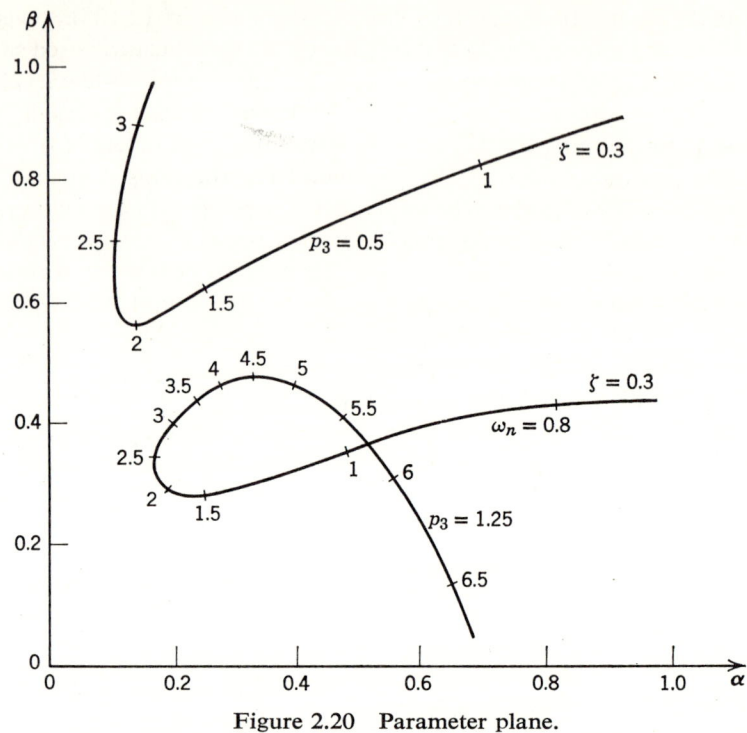

Figure 2.20   Parameter plane.

solved using Figures 2.18 and 2.19. However, in order to determine all the root locations throughout the parameter plane, other contours for interpolation purposes must be added to both diagrams. This can be avoided if attention is focused on the contours that are of interest in the actual design. Thus, by the inverse parameter mapping of the $\zeta = 0.3$ curves of Figure 2.20, plotted for two different values of $p_3 = 1.25, 0.5$, the complex plane diagram is constructed as shown in Figure 2.21. The values of p_1 and p_2 along the complex plane loci are best presented as in Fig. 2.20. It is of interest to note that the approximate root values obtained previously by truncation are sufficiently close to the exact roots obtained on the diagram of Figure 2.20 for $p_1 = p_2 = 0.2, p_3 = 1.25$.

Now the system transient responses to the unit-step function are plotted for points along the straight line $\zeta = 0.3$ of Figure 2.21. They are shown in Figure 2.22 with the corresponding values of the parameters p_1, p_2, p_3 and the values of the undamped natural frequency indicated along the straight line $\zeta = 0.3$. The designer may choose the appropriate transient response and then realize the compensators as RC networks with given values of the adjustable parameters.

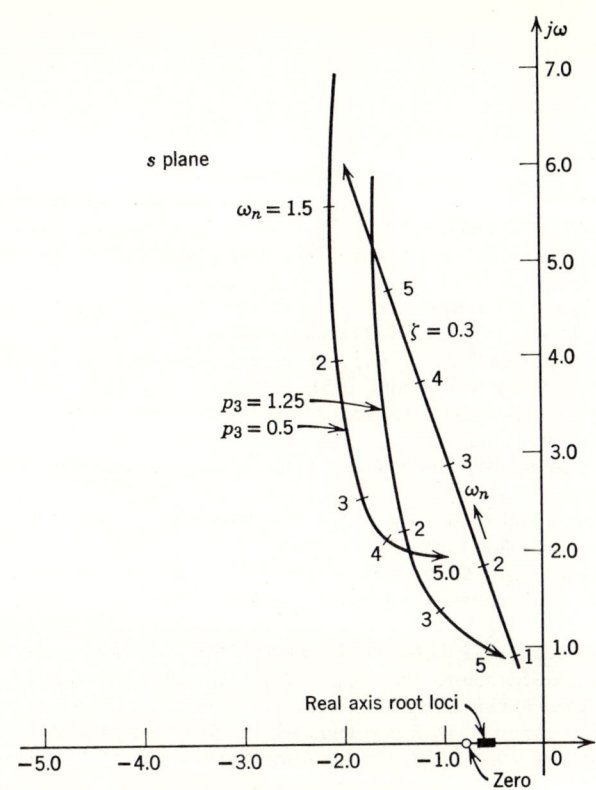

Figure 2.21 Complex plane.

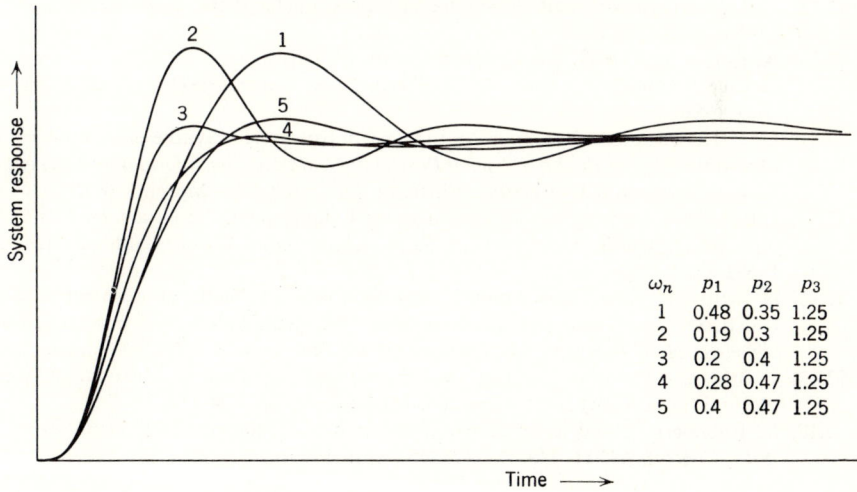

ω_n	p_1	p_2	p_3
1	0.48	0.35	1.25
2	0.19	0.3	1.25
3	0.2	0.4	1.25
4	0.28	0.47	1.25
5	0.4	0.47	1.25

Figure 2.22 Transient responses.

References

[2.1] D. D. Šiljak, Analysis and Synthesis of Feedback Control Systems in the Parameter Plane, Pt. I—Linear Continuous Systems, *IEEE Trans.*, *(Applications and Industry)*, **83**, 449–458 (November 1964).

[2.2] D. D. Šiljak, An Extension of the Parameter Plane Method, *Automatika*, **1**, No. 2, 59–64 (1965).

[2.3] D. D. Šiljak, Generalization of the Parameter Plane Method, *IEEE Trans. on Auto. Control*, **AC-11**, No. 1, 63–70 (January 1966).

[2.4] G. Doetsch, *Handbook of the Laplace Transformation* (in German), Vols. 1–3, Birkhauser, Basel, Switzerland, 1950.

[2.5] W. Kaplan, *Operational Methods for Linear Systems*, Addison-Wesley, Reading, Massachusetts, 1962.

[2.6] P. De Russo, R. Roy, and R. Close, *State Variables for Engineers*, Wiley, New York, 1965.

[2.7] L. A. Zadeh and C. A. Desoer, *Linear System Theory: The State Space Approach*, McGraw-Hill, New York, 1963.

[2.8] L. S. Pontryagin, *Ordinary Differential Equations* (in Russian), FIZMATGIZ, Moscow, 1961 (English translation, Addison-Wesley, Reading, Massachusetts, 1964).

[2.9] F. H. Holister and G. J. Thaler, Symmetrical Parallel-tee Network-Parameter Plane Analysis and Synthesis, *Proc. of the Third Annual Allerton Conf. on Network and System Theory*, University of Illinois, Monticelo, Ill., 430–438 (October 1965).

[2.10] A. C. Barker and A. B. Rosenstein, S-Plane Synthesis of the Symmetrical Twin-T Network, *IEEE Trans.*, Pt. II *(Applications and Industry)*, **83**, 382–389 (November 1964).

[2.11] T. J. Lazear and A. B. Rosenstein, Pole-Zero Synthesis and the General Twin-T, *IEEE Trans.* Pt. II *(Applications and Industry)*, **83**, 389–393 (November 1964).

[2.12] J. B. Moore, Steady-state Response in the Parameter Plane, *Automatika*, **1**, No. 2, 55–58 (1965).

[2.13] A. Burzio and D. D. Šiljak, Minimization of Sensitivity with Stability Constraints in Linear Control Systems, *IEEE Trans. Auto. Control*, **AC-11**, No. 3, 567–569 (July 1966).

[2.14] G. J. Thaler and A. B. Lemanski, Linear Control System Synthesis Using Algebraic Methods, *Proc. of the First Annual Princeton Conference on Information Sciences and Systems*, Princeton University, Princeton, N.J., 268–271 (March 1967).

[2.15] J. B. Moore, Control System Design Using Extensions of the Parameter Plane Concept, Ph.D. Thesis, University of Santa Clara, Santa Clara, California (January 1967).

[2.16] D. Mitrović, Graphical Analysis and Synthesis of Feedback Control Systems: I—Theory and Analysis; II—Synthesis; III—Sampled-data Feedback Control Systems, *AIEE Trans.*, Pt. II *(Applications and Industry)*, **77**, 476–503 (January 1959).

[2.17] T. Brkić and P. Kokotović, Design of the Voltage Regulator by Mitrović's Algebraic Method (in Serbian) *Elektroprivreda*, **15**, No. 6–7, 277–286 (1962).

[2.18] L. Eisenberg, Stability of Linear Systems with Transport Lag, *IEEE Trans. on Auto. Control*, **AC-11**, No. 2, 247–255 (April 1966).

CHAPTER THREE
Introduction to Approximate Analysis of Nonlinear Systems

3.1 Introduction

Nonlinear systems with either inherent nonlinear characteristics or non-linearities deliberately introduced into the system to improve their dynamic characteristics have found wide application in the most diverse fields of engineering. The principal task of nonlinear system analysis is to obtain a comprehensive picture, quantitative if possible, but at least qualitative, of what happens in the system if the variables are allowed, or forced, to move far away from the operating points. This is called the *global*, or in-the-large, behavior. *Local*, or in-the-small, behavior of the system (which means that the system variables are perturbed only slightly from their operating points) can be analyzed on a linearized model of the system. Therefore, the local behavior can be investigated by rather general and efficient linear methods that are based upon the powerful superposition and homogeneity principles (see Section 2.2). If linear methods are extended to the investigation of the global behavior of a nonlinear system, the results can be erroneous both quantitatively and qualitatively since the nonlinear characteristics may be essential but the the linear methods may fail to reveal it. Therefore, there is a strong emphasis on the development of methods and techniques for the analysis and design of nonlinear systems.

The development of nonlinear methods faces real difficulties for various reasons. There are no universal mathematical methods for the solution of non-linear differential equations which are the mathematical models of nonlinear systems. The methods deal with specific classes of nonlinear equations and therefore have only limited applicability to system analysis. The classification of a given system and the choice of an appropriate method of analysis are not at all an easy task. Furthermore, even in simple nonlinear problems, there are numerous new phenomena qualitatively different from those expected in linear system behavior, and it is impossible to encompass all these phenomena in a single and unique method of analysis.

Although there is no universal approach to the analysis of nonlinear systems, by excluding specific techniques we can still conclude that the nonlinear methods generally fall under one of the three following approaches: the phase-space topological techniques, stability analysis method, and the approximate methods of nonlinear analysis. This classification of the nonlinear methods is rather subjective but can be useful in systematization of their review.

The *phase-space*, or more specifically the *phase-plane*, approach has been used for solving problems in mathematics and physics at least since Poincaré. The approach gives both the local and the global behavior of the nonlinear system and provides an exact topological account of all possible system motions under various operating conditions. It is convenient, however, only in the case of second-order equations, and for high-order cases the phase-space approach is cumbersome to use.[1] Nevertheless, it is a powerful concept underlying the entire theory of ordinary differential equations (linear or nonlinear, time varying or time invariant). It can be extended to the study of high-order differential equations in those cases where a reasonable approximation can be made to find an equivalent second-order equation. However, this may lead to either erroneous conclusions about the essential system behavior, such as stability and instability, or various practical difficulties such as time scaling.

The *stability analysis* of nonlinear systems, which is heavily based on the work of Liapunov, is a powerful approach to the qualitative study of the system global behavior.[2] By this approach, the global behavior of the system is investigated utilizing the given form of the nonlinear differential equations but without explicit knowledge of their solutions. Stability is an inherent feature of wide classes of systems, thus system theory is largely devoted to the stability concept and related methods of analysis. Stability analysis, however, does not constitute a complete satisfactory theory for the design of nonlinear systems. The stability conditions, which are often hard to determine, are sufficient but usually not necessary. This comes from the fact that the given equations are reformulated for the application of the stability analysis. In that reformulation certain information about the specific system characteristics is lost and, unfortunately, the amount of information that is lost cannot be estimated. For example, if a nonlinear system is found to be stable for a

---

[1] The basic techniques of the phase plane approach are presented in the first six problems for Chapter 8.
[2] Stability analysis by Liapunov's methods is outlined in Chapter 8.

certain range of parameter values, it is not possible to predict how far from that range the parameter value can be chosen without affecting the system stability.[3] Furthermore the system can be unstable and still be satisfactory for practical applications. For example, a system can exhibit stable periodic oscillations and therefore be unstable. However, in the application of the system, these oscillations may not be observed because their amplitude is sufficiently small and the perturbations permanently acting on the system are large enough to drive the system far from the periodic oscillations.

Approximate methods for solving problems in mathematical physics, which were developed at the beginning of this century, have been received with much interest by engineers and have promptly obtained wide diffusion in diverse fields of system engineering. The basic merit of approximate methods consists in their being direct and efficient, and they permit a simple evaluation of the solution for a wide class of problems arising in the analysis of nonlinear oscillations.[4]

The application of computer techniques and system simulations has given strong emphasis to those approximate methods which employ rather straightforward and realizable solution procedures and calculations. These methods enable a simple estimation of how different system structures and parameters influence the salient system dynamic characteristics. The application of a computer simulation can then provide the actual solution of the design problem. If the system behavior is not satisfactory, or if the computer solution does not agree with the predicted characteristics, the approximate methods can again be applied to guide the next step in the system simulation and also achieve a better solution of the analysis problem. If we interchange these two steps—that is, apply the approximate methods and then the computer simulation—the design converges eventually to a final satisfactory solution. This philosophy in the analysis of nonlinear systems can give improved results not only in a specific system but also in the related class of systems, and thus has an important generality in system theory and application.

---

[3] In certain cases it has been found that the forward gain of a nonlinear system can be as much as a hundred times greater than the limiting value calculated in a stability analysis and the system still will not become unstable. For example, see A. G. Dewey and E. I. Jury, A Stability Inequality for a Class of Nonlinear Feedback Systems, *IEEE Trans. Auto. Control*, **AC-11**, No. 11, 54–62 (January, 1966).

[4] It is of interest to note that only the Krylov-Bogoliubov asymptotical method [3.1] is used here (see Section 3.5), because these methods form the basis of the harmonic linearization and describing function technique (see Section 3.4) employed in the parameter plane analysis of nonlinear systems. For other approximate methods of higher mathematical analysis used in engineering problems, the following reference is recommended: L. V. Kantorovich and V. I. Krylov, *Approximate Methods of Higher Analysis*, Interscience, New York, 1958. An engineering treatment of the methods is given by Bulgakov [3.2] and Hale [3.15].

It is of particular significance to classify the nonlinear problem before a specific technique is applied to its solution. Thus it is necessary to evaluate the potential of both the exact and the approximate methods before they are tested on the actual problem. This involves engineering experience and ingenuity in choosing the appropriate design technique and procedure. If an exact method is to be applied, we should be aware of the fact that it may require that a sequence of simplifications be introduced in the original problem. In the simplifications, certain vital characteristics of the original problem can be lost—for example, the reduction of the order of a differential equation through neglect of one of the system parameters. Then the approximate solution of the original problem may represent more appropriately the actual situation and be of more use in the design. In addition, the approximate methods normally yield more information about the possible performance criteria trade-offs or the structural and parameter changes that might enhance the overall system characteristics. On the other hand, the exact methods can reveal various subtle phenomena in nonlinear system behavior that cannot be discovered by the approximate methods. It can be concluded that in a majority of practical problems both the exact and the approximate methods should be applied to obtain a satisfactory solution of the nonlinear system design problem, and the versatility of the designer in various solution procedures is a prerequisite for a successful system analysis and design.

In the application of the approximate methods, a significant problem is the estimation of their accuracy. A certain degree of accuracy is necessary to guarantee the applicability of the method involved, and to ensure the validity of both the qualitative conclusions and the quantitative results obtained by the approximate analysis. The accuracy problem, however, involves various mathematical difficulties, and the designer is forced to use simple and practical approximate methods despite some pessimism about the validity of the methods. Recently, promising results have been obtained in solving the accuracy problem; such results will be considered in the following sections and in the appendices.

Among the approximate methods used for the analysis of nonlinear oscillations, the Krylov-Bogoliubov [3.1] *asymptotical method* stands out because of its usefulness in system engineering problems. The original method not only enables the determination of steady-state periodic oscillations, but also gives in evidence the transient process corresponding to small amplitude perturbations of the oscillations. The latter is of particular interest in system design, where the transient process is often the ultimate goal. However, the method is applicable to systems described by second-order nonlinear differential equations.

The approximate method to be used in the analysis of nonlinear systems along with the parameter plane concept is the *harmonic linearization method*, often called the *describing function method* or the *method of harmonic balance*. The harmonic linearization is heavily based on the Krylov-Bogoliubov approach, and will be applied to nonlinear systems described by high-order differential equations. This chapter is aimed at developing the basis of the harmonic linearization method. After the mathematical description of nonlinear systems in Section 3.2, we outline in Section 3.3 the concept of the method. The fundamental consideration of the accuracy problem is included in Section 3.4 following the Popov-Palitov approach [3.8]. The Krylov-Bogoliubov method is outlined briefly in Section 3.5 and then compared with the describing function technique in Section 3.6 as proposed in [3.8].

Further consideration of the accuracy problem is carried out in Appendix G on the basis of the work of Rozenvasser and Garber. This is considered sufficient to establish the validity of the describing function technique as applied to specific problems considered in the subsequent chapters.

Appendix F contains the describing function expression, tables, and diagrams for common nonlinearities encountered in system engineering applications.

3.2 Mathematical Description of Nonlinear Systems

In general a nonlinear system consists of *linear* and *nonlinear elements.* The linear elements are described by linear differential equations, as shown in preceding sections. The nonlinear elements, which are normally very limited in number, are described by a nonlinear function or differential equation relating the input and output of the element. The nonlinear input-output relationship can have a rather arbitrary form. The parameter plane analysis to be presented is restricted to a certain class of these relationships, which will be, in part, discussed in this section. More detailed consideration is given in Appendix F.

In treating a real system as linear, we assume that the system is linear in a certain range of operation. The signals appearing in various points of the system are such that the superposition principle is justified. However, if signals in the system go beyond the range of linear operation and, for example, become either very large or very small, the characteristics of the system elements can be essentially different from the linearized characteristics and the system must be treated as nonlinear. Such cases are illustrated graphically by the characteristics shown in Figures 3.1a, 3.1b, and 3.1c, where x denotes the input to the element and the output is given by the value of the function $F(x)$. If the output of the element is denoted by y, the input-output relationship can be written analytically as

$$y = F(x). \tag{3.1}$$

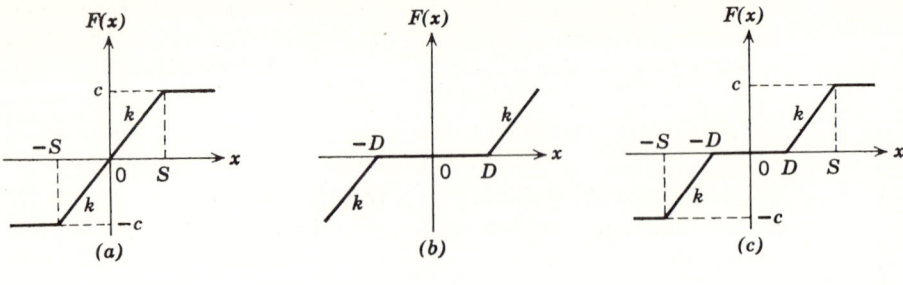

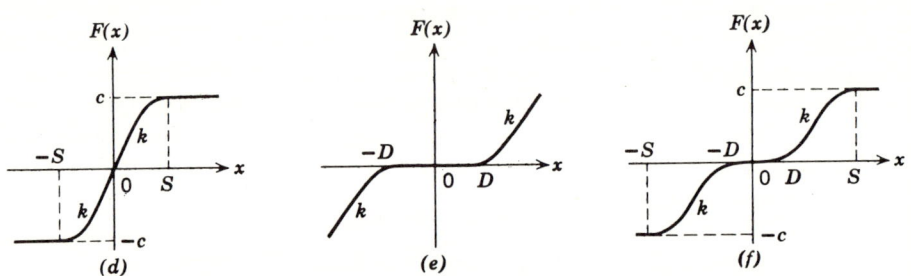

Figure 3.1 Single-valued nonlinear characteristics.

Certain nonlinear characteristics can be given in analytical form. For example, the characteristic of Figure 3.1a can be analytically described by

$$F(x) = \begin{cases} kx, & |x| \leq S \\ c \operatorname{sign} x, & |x| \geq S. \end{cases} \quad (3.2)$$

The characteristic is linear with slope $k = c/S$ for inputs less than S, and it exhibits saturation for input magnitudes greater than S. Similarly, the characteristic of Figure 3.1b can be written as

$$F(x) = \begin{cases} 0, & |x| \leq D \\ k(x \pm D), & |x| \geq D, \end{cases} \quad (3.3)$$

where D represents a dead zone. The characteristic of Figure 3.1c is a combination of the two preceding characteristics.

The more realistic versions of these characteristics are shown in Figures 3.1d, 3.1e, and 3.1f. The characteristic of Figure 3.1d may be given as

$$F(x) = \begin{cases} \dfrac{3c}{2S}\left(1 - \dfrac{1}{3S^2} x^2\right) x, & |x| \leq S \\ c \operatorname{sign} x, & |x| \geq S. \end{cases} \quad (3.4)$$

The characteristics of Figure 3.1 are *continuous characteristics*.

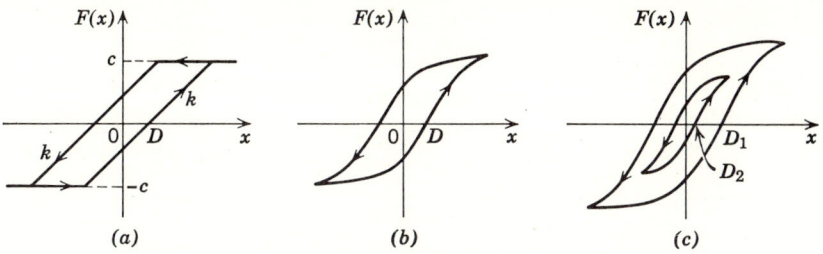

Figure 3.2 Nonlinear characteristics with hysteresis.

In various practical applications the nonlinear characteristic is obtained experimentally, and an adequate analytical expression cannot be justified. On the other hand, some characteristics are conveniently expressed analytically, whereas a graphical interpretation is not possible. This latter case can be illustrated by a nonlinear element that exhibits a *transport lag* T_t with the output y given by

$$y = e^{-T_t s} F(x), \tag{3.5}$$

where $F(x)$ may have any of the characteristics shown in Figure 3.1. Since the static characteristics of nonlinear elements do not have a time coordinate, the function y of (3.5) cannot be displayed graphically as can the characteristics of Figure 3.1.

So far only single-valued nonlinear characteristics have been discussed; namely, in the characteristics of Figure 3.1, to each value of the input x there is one and only one value of the output $y = F(x)$. The characteristics in Figures 3.2a, 3.2b, and 3.2c, which have a hysteresis loop, are multi-valued nonlinear characteristics. The hysteresis property can be such that the loop dimensions depend on the magnitude of the input signal (Figure 3.2c). It is also to be noted that hysteresis type nonlinear characteristics cannot be completely described by the function $y = F(x)$ since the output y inherently depends on the direction of change in the magnitude of the input x. If the rate of change in x is greater than zero, the right-hand side of the loop represents the nonlinear characteristic, and vice versa. Thus the adequate description of the hysteresis type of nonlinearities should be expressed as

$$y = F(x, \text{sign } sx), \qquad s \equiv \frac{d}{dt} \tag{3.6}$$

rather than as $y = F(x)$.

The relay characteristics shown in Figure 3.3 can be related to the above class of nonlinearities. Relay characteristics have a general property in that they are *discontinuous* without any linear portion whatsoever. The

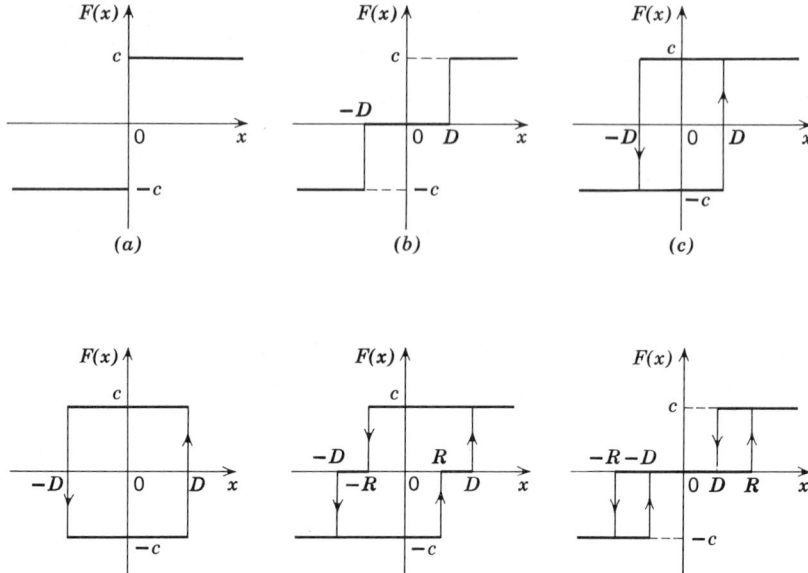

Figure 3.3 Nonlinear relay characteristics.

ideal relay characteristic of Figure 3.3a displays this property, and the nonlinear function is $F(x) = \text{sign } x$. For $x = 0$, the value $F(0)$ is either $+c$ or $-c$, unless a special arrangement is made to have $-c \leq F(0) \leq +c$. Therefore the part of the $F(x)$ axis between $+c$ and $-c$ is represented by a thin line indicating that it does not belong to the function $F(x)$.[5]

The output y of a nonlinear element may depend entirely on the rate sx of its input x; that is,

$$y = F(sx). \tag{3.7}$$

Such is the case with the nonlinear friction characteristics shown in Figure 3.4. The graphical expression of coulomb friction given in Figure 3.4a is the same as that of an ideal relay. It should be noted, however, that in case of coulomb friction, for $sx = 0$, the friction force $F(0)$ may have any value in the interval between $+c$ and $-c$, $c \leq F(0) \leq -c$. In Figure 3.4b a viscous friction characteristic is shown that has a parabolic variation in $F(sx)$ when $sx \neq 0$. The analytic expression for the quadratic friction shown in Figure 3.4c is given as

$$F(sx) = ksx + c(sx)^2 \text{ sign } (sx). \tag{3.8}$$

---

[5] This description of the relay characteristic is an idealization. In practice there are physical imperfections that give rise to certain phenomena (chattering or sliding motions) that are treated in Section H.4 of Appendix H.

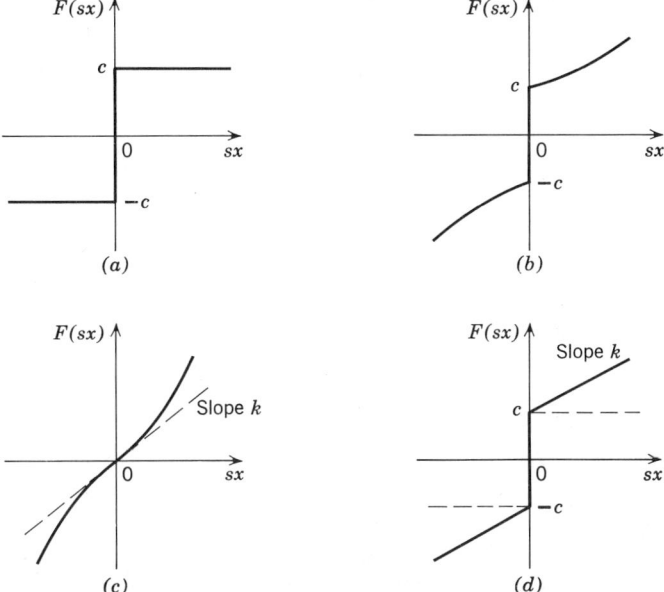

Figure 3.4 Nonlinear friction characteristics.

In Figure 3.4d a combination of linear viscous and coulomb friction is shown. The nonlinear friction characteristics are discussed in Appendix F in more detail.

The nonlinear characteristics discussed so far are all symmetrical about the origin. This is not necessarily the case, and the nonlinear characteristics shown in Figure 3.5 often appear in system design.

Furthermore, some nonlinear characteristics do not relate two signals of the system, but give a relationship between a system parameter and an input signal to a system element. In a differential equation,

$$Tsy + y = Kx, \qquad (3.9)$$

describing an element of the system, both parameters T and K can be functions of x, y, or both. Then (3.9) may have one of the following forms depending on the actual situation:

$$Tsy + y = F(x, sx) \qquad (3.10a)$$

$$F(x, y)sy + y = Kx \qquad (3.10b)$$

$$F(x)sy + y = Kx. \qquad (3.10c)$$

Thus the corresponding element is described by a nonlinear differential equation rather than by a single function $F(x, sx)$.

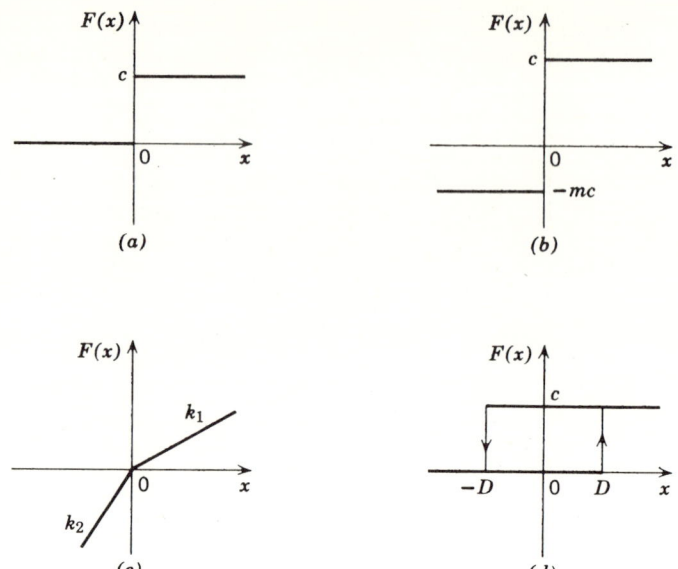

Figure 3.5 Asymmetrical nonlinear characteristics.

The characteristics of nonlinear elements may include also higher derivatives of the input signal. In the following analysis (Chapters 3–7), however, we consider only those elements with characteristics that can be described by a relationship involving a nonlinear function of signals related to the element and their first derivatives. Nonlinear characteristics that belong to this class of relationships are discussed in detail in Appendix F. In the appendix other characteristics that have not been shown above, such as backlash, polynomial characteristics, and so on are also discussed. It is of importance to note that the time varying nonlinear characteristics such as $F(x, t)$ or $F(x, sx, t)$ are not considered.

As shown previously, a linear system can be described, in general, by a set of differential equations

$$D_{i1}x_1 + D_{i2}x_2 + \cdots + D_{il}x_l = f_i, \qquad (i = 1, 2, \ldots, l), \qquad (3.11)$$

where the operators $D_{ij} = D_{ij}(s)$ have the form

$$D_{ij} \equiv u_{ij}s^2 + v_{ij}s + w_{ij}, \qquad (3.12)$$

$x_j = x_j(t)$, $(j = 1, 2, \ldots, l)$, are generalized coordinates of the system, and $f_i = f_i(t)$, $(i = 1, 2, \ldots, l)$, are the external signals (or forces) applied to the system. In the following nonlinear analysis the system description is based upon the above notation.

Mathematical Description of Nonlinear Systems

A nonlinear system with one nonlinear element defined by a function $F(x_k, sx_k)$ can be described by the system of equations:

$$D_{11}x_1 + \cdots + D_{1k}x_k \qquad + \cdots + D_{1l}x_l = f_1$$
$$\vdots$$
$$D_{i1}x_1 + \cdots + D_{ik}x_k + F(x_k, sx_k) + \cdots + D_{il}x_l = f_i \qquad (3.13)$$
$$\vdots$$
$$D_{j1}x_1 + \cdots + D_{jk}x_k \qquad + \cdots + D_{jl}x_l = f_j$$
$$\vdots$$
$$D_{l1}x_1 + \cdots + D_{lk}x_k \qquad + \cdots + D_{ll}x_l = f_l.$$

If, for example, all external forces are identically zero except f_j, equations 3.13 may be rewritten as a single equation

$$B(s)x + C(s)F(x, sx) = H(s)f, \qquad s \equiv \frac{d}{dt}, \qquad (3.14)$$

where the notation $x_k = x$ and $f_j = f$ is used. In (3.14), $B(s)$ is a determinant

$$B(s) = \begin{vmatrix} D_{11} & \cdots & D_{1k} & \cdots & D_{1l} \\ \vdots & & \vdots & & \vdots \\ D_{i1} & \cdots & D_{ik} & \cdots & D_{il} \\ \vdots & & \vdots & & \vdots \\ D_{j1} & \cdots & D_{jk} & \cdots & D_{jl} \\ \vdots & & \vdots & & \vdots \\ D_{l1} & \cdots & D_{lk} & \cdots & D_{ll} \end{vmatrix} \qquad (3.15)$$

and $C(s)$ is the corresponding cofactor of the (i, k)th element. Similarly, $H(s)$ is the cofactor of the (j, k)th element of the determinant (3.15).

If another nonlinear element exists, which is described by a nonlinear function $F_2(x_k, sx_k)$ placed in the jth row of (3.13), and if the external forces f_1 and f_l are applied to the system, then equations 3.11 can be rewritten as

$$D(s)x + C(s)F_1(x, sx) + B(s)F_2(x, sx) = H_1(s)f_1 + H_2(s)f_2, \quad (3.16)$$

where the notation is $x = x_k$, $F_1(x, sx) = F(x, sx)$, and $f_2 = f_l$. Now in (3.16), $D(s)$ is the determinant of (3.13) and $C(s)$ and $B(s)$ are its cofactors of the (i, k)th and (j, k)th elements, respectively. The $H_1(s)$ and $H_2(s)$ are the cofactors of the $(1, k)$th and (l, k)th elements of the determinant (3.13), respectively.

The same description can be used for nonlinear systems having a nonlinear element whose characteristic depends on two or more signals of the system. Moreover, several such elements may be present in the system.

The analysis of nonlinear systems in the parameter plane will be based on single equations 3.14 or 3.16 rather than on the system of equations such as (3.13). It will be assumed that the above procedure, in going from (3.13) to (3.14) or (3.16), is already performed.

Besides the analytical description of nonlinear elements and systems, it is essential to consider the structure of the system, which is usually given in familiar block diagram or signal flow graph form. The structure of the system displays certain inherent features of nonlinear systems that are not apparent in the analytical description.

The basic nonlinear system with one nonlinear element n is shown in Figure 3.6. It should be noted that the function $F(x, sx)$ associated with n does not necessarily represent the nonlinear element as described by a nonlinear differential equation such as (3.10). To make the analysis easier, the nonlinear function $F(x, sx)$ may be isolated in the system, while all the linear relations are joined in the block $G(s)$. For example, if the nonlinear element n is described by (3.10a), it can be split into two equations:

$$(Ts + 1)y = z, \quad (3.17a)$$

$$z = F(x, sx). \quad (3.17b)$$

Then (3.17a) is associated with the other linear elements of the system, and the function $F(x, sx)$ is isolated in the block n. Naturally, the function $F(x, sx)$ does not represent the nonlinear element n and therefore will be called the *nonlinearity*.

The linear elements may be coupled in an arbitrary way to make the equivalent transfer function $G(s)$, whose order is not theoretically limited as far as the parameter plane analysis is concerned. However, certain restrictions on the nature of the function $G(s)$ are imposed in order to justify the application of the approximate analysis. These restrictions are discussed in the following developments. It should be noted that, according to (3.14) and the block

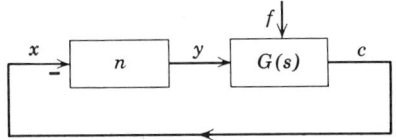

Figure 3.6 Basic nonlinear system.

diagram of Figure 3.6, the transfer function $G(s)$ is

$$G(s) = \frac{C(s)}{B(s)}. \qquad (3.18)$$

The function $f = f(t)$, which may be either a desired input signal or an undesired perturbation, is applied somewhere in the linear part of the system.

The block diagram of Figure 3.6 may represent a nonlinear system having two nonlinear elements connected in cascade, providing it is possible to isolate the two related nonlinearities and join them in one equivalent block. The remaining linear relations can be associated with the linear part of the system.

Sometimes it is impossible to isolate the nonlinearity in a separate block as shown in Figure 3.6. The nonlinear element may be described by (3.10b), in which the nonlinear function $F(x, y)$ is a function of two different signals in the system. Nevertheless, an analytical description of the form of (3.14) may be straightforward. Of course, the functions can be $F(x, sy)$, $F(sx, y)$, or $F(sx, xy)$ as discussed in Appendix F.

Equation 3.16 may describe a nonlinear system of Figure 3.7 that basically has two nonlinear elements, n_1 and n_2, separated by two linear blocks in two different feedback loops. The two nonlinearities have a common input x. The external forces are applied at two points in the linear parts of the system.

A more complicated situation is shown in Figure 3.8, in which two nonlinear elements are separated by a linear part so that the two inputs at the nonlinear elements are related by a linear differential equation. The two nonlinear elements may sometimes be replaced by their related nonlinearities, which can facilitate the analysis.

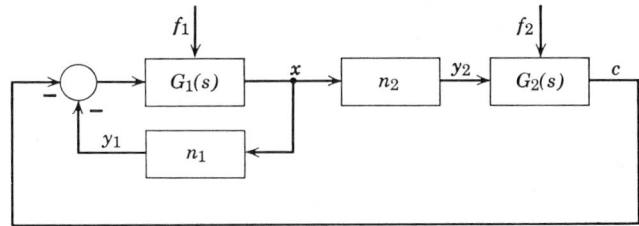

Figure 3.7 System with two nonlinear elements; common input signal.

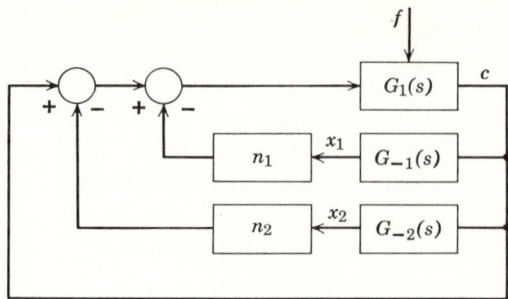

Figure 3.8 System with two nonlinear elements; inputs related by a linear differential equation.

The two nonlinearities, n_1 and n_2, may be placed in the system so that the corresponding inputs are related by a nonlinear differential equation. This situation is illustrated in Figure 3.9.

Besides the analytical and structural representations of nonlinear control systems, discussed above, there are other quite distinct situations.

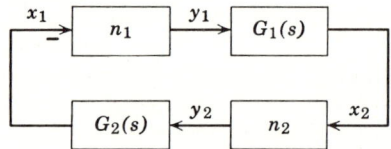

Figure 3.9 System with two nonlinear elements; inputs related by a nonlinear differential equation.

For example, a block diagram of Figure 3.10 illustrates a linear system with a variable structure. The system is actually nonlinear, since the structure is changed depending on the magnitude of the output signal. Instead of the simple relay, the structure may be changed according to a certain nonlinear function of the output or error signal.

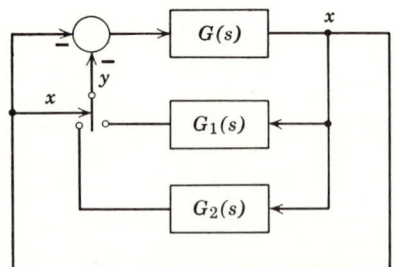

Figure 3.10 System with variable structure.

In practice there are many other system configurations with nonlinear characteristics that can be, by suitable operations, reduced to the ones discussed above, and the analysis in the parameter plane may be conveniently applied. However, whether an actual system can be analyzed by the parameter plane method depends also on various theoretical and practical conditions imposed by the method itself. These conditions are discussed in the following developments.

3.3 Describing Function

Among the methods used for stability analysis and investigation of sustained nonlinear oscillations, sometimes called a limit cycle, the describing function generally stands out because of its usefulness in engineering problems of control system analysis. The describing function technique can be successfully applied to systems other than control whenever the sustained oscillations, which are based on some nonlinear phenomena, represent possible operating conditions.

The theoretical basis of the describing function analysis lies in the van der Pol [3.3] method of slowly varying coefficients as well as in the methods of harmonic balance and equivalent linearization proposed by Krylov and Bogoliubov [3.1] for solving certain problems of nonlinear mechanics. The analysis has been further developed in the work of Goldfarb [3.4] with the emphasis on nonlinear phenomena in feedback systems. Various applications and interpretations of essentially the same method have been presented independently by Kochenburger [3.5], Oppelt [3.6], and Tustin [3.7]. Significant results in the system analysis were obtained by extended describing function techniques proposed by Popov and Palitov [3.8]. These extensions will be used in the parameter analysis outlined in Chapter 4-7.

For presenting the concept of describing function method, an autonomous nonlinear feedback system with a block diagram of Figure 3.6 is considered. The block n represents the isolated nonlinearity described by a given function, $F(x, sx)$. The linear part of the system is presented by a known transfer function $G(s) = C(s)/B(s)$. The external forcing function $f = f(t)$ is identically zero for all values of time t. Thus the *free oscillations* in the system are determined by a nonlinear homogeneous differential equation

$$B(s)x + C(s)F(x, sx) = 0, \qquad s \equiv \frac{d}{dt}. \tag{3.19}$$

In the system analysis it is of basic interest to determine whether the system can exhibit periodic oscillations, often called a limit cycle.[6] The problem

---

[6] The term *limit cycle* has been introduced in the phase $x\dot{x}$ plane analysis of nonlinear systems where it is an isolated path corresponding to a periodic solution.

122 Introduction to Approximate Analysis of Nonlinear Systems

may be reformulated so that it is required to determine the conditions that should be satisfied by the polynomials $B(s)$, $C(s)$ and the nonlinearity $F(x, sx)$ for the periodic oscillations to exist. Then it is desired to determine the characteristics of the oscillations. In mathematical terms this means that it is necessary to investigate the possibility of (3.19) having an isolated periodic solution[7] $x = x(t)$ of frequency Ω, so that $x(t + T) = x(t)$, for all time t, where $T = 2\pi/\Omega$ is a constant and represents the period of oscillation. If possible, it may be required to determine the stability of the solution and the solution $x(t)$ itself. Of particular interest are stable periodic solutions when the system exhibits sustained oscillations[8] without external forcing signals.

To solve the above formulated problem by the describing function approach, it is necessary to make the basic assumption that the variable $x = x(t)$, appearing in the nonlinear function $F(x, sx)$ is sufficiently close to a sinusoidal oscillation; that is,

$$x \simeq A \sin \Omega t, \qquad (3.20)$$

where the amplitude A and the frequency[9] Ω of the oscillation are constant.[10] *Describing function analysis, therefore, belongs to those methods of solving nonlinear differential equations which are based upon an assumed solution.* As such, it requires that conditions for the assumed solution exist. Such an assumption is quite realistic since the nonlinear system may exhibit periodic oscillations arbitrarily close to a pure sinusoid. Furthermore, it is sometimes desired that a nonlinear system have sinusoidal oscillations of certain amplitude and frequency as a normal operating characteristic. Then it is required to determine the conditions under which the desired oscillations occur.

---

[7] The existence, uniqueness, and continuity of solutions of differential equations, as well as other general properties of the solutions, are discussed in Section 8.2 of Chapter 8.

The term *isolated* is used to indicate the fact that if we have a periodic solution of a linear differential equation, then it cannot be isolated since any constant multiple of a solution is also a solution. In nonlinear differential equations the superposition principle is not valid, and they can exhibit isolated periodic solutions except, of course, for a phase shift. In (3.20), for example, the sinusoidal oscillations can be expressed by $x = A \sin \phi$, $\phi = \Omega t + \phi_0$ where ϕ_0 is the phase shift. Since ϕ_0 merely corresponds to a translation of the periodic solution in time, we can put $\phi_0 = 0$ and use (3.20).

[8] *Sustained* oscillations are sometimes referred to as *self-excited* or *self-sustained* oscillations and were first described by Lord Rayleigh. In connection with his work on the theory of sound [3.9], he discovered a nonconservative second-order nonlinear differential equation that could sustain an oscillation without external disturbances.

[9] The frequency is $\Omega = 2\pi/T$, where T is the period that approximately represents the period T' ($T \simeq T'$) of the actual solution $x(t)$.

[10] In the following developments, the notation $\simeq$ for approximate equivalence used in (3.20), is replaced by the equality sign $=$.

Before the conditions that justify the above assumption are discussed, the describing function concept will be developed. If the variable x in the nonlinear function $F(x, sx)$ has the sinusoidal form of (3.20), then the variable $y = F(x, sx)$ is generally complex, but is also a periodic function of time. As such, it can be developed in a Fourier series[11]

$$y = F^0 + N_1 x + \frac{N_2}{\Omega} sx + \cdots. \tag{3.21}$$

Only the first three terms are considered, and the corresponding coefficients are

$$F^0 = \frac{1}{2\pi} \int_0^{2\pi} F(A \sin \phi, A\Omega \cos \phi) \, d\phi \tag{3.22a}$$

$$N_1 = \frac{1}{\pi A} \int_0^{2\pi} F(A \sin \phi, A\Omega \cos \phi) \sin \phi \, d\phi \tag{3.22b}$$

$$N_2 = \frac{1}{\pi A} \int_0^{2\pi} F(A \sin \phi, A\Omega \cos \phi) \cos \phi \, d\phi \tag{3.22c}$$

and $\phi = \Omega t$.

First, a case will be discussed in which the nonlinear function $F(x, sx)$ is symmetrical about the origin and, therefore, the constant term F^0 in the Fourier series (3.21) is $F^0 \equiv 0$. The quantities N_1 and N_2, defined in (3.22b) and (3.22c),

---

[11] The Fourier series for a function $y(t) = F(x, sx)$ where $x = A \sin \phi$ and $\phi = \Omega t$ is

$$y(t) = g_0 + \sum_{k=1}^{\infty} (g_k \sin k\phi + h_k \cos k\phi),$$

where

$$g_0 = \frac{1}{2\pi} \int_0^{2\pi} F(A \sin \phi, A\Omega \cos \phi) \, d\phi$$

$$\left. \begin{array}{l} g_k = \dfrac{1}{\pi} \int_0^{2\pi} F(A \sin \phi, A\Omega \cos \phi) \sin k\phi \, d\phi \\[6pt] h_k = \dfrac{1}{\pi} \int_0^{2\pi} F(A \sin \phi, A\Omega \cos \phi) \cos k\phi \, d\phi. \end{array} \right\} (k \neq 0)$$

The series converges to $y(t)$ for values of t at which the function is continuous. At a finite discontinuity the series converges to the average of the values of $y(t)$ on either side of the discontinuity. The Fourier series converges uniformly for all t, if $y(t)$ is a single-valued periodic function that has in any one period only a finite number of maxima, minima, and discontinuities, and in addition remains finite. If only a finite number of terms are taken, the corresponding Fourier series approximates $y(t)$ with the least mean-squared error. For further study of the properties and applications of Fourier series, the following books can be recommended; G. Sansone, *Orthogonal Functions*, Interscience, New York, 1959; L. V. Kantorovich and V. I. Krylov, *Approximate Methods of Higher Analysis*, Interscience, New York, 1958.

are the coefficients of the describing function $N = N_1 + jN_2$, for which the operator s is interpreted in the frequency domain as $s = j\omega = j\Omega$.[12] To discuss the physical meaning of the describing function N defined above, suppose that the nonlinearity n of the basic nonlinear system, shown in Figure 3.6, is $F(x) = c \text{ sign } x$, which represents the pure relay characteristic of Figure 3.4a. Then, if the input x of the nonlinearity is a sinusoidal oscillation, the output y is a square-wave oscillation as illustrated in Figure 3.11. If the output signal y is developed in a Fourier series, the first harmonic y_1 will be

$$y_1 = A_1 \sin \Omega t, \tag{3.23}$$

where

$$A_1 = \frac{1}{\pi} \int_0^{2\pi} F(A \sin \phi) \sin \phi \, d\phi \tag{3.24}$$

is the amplitude of the harmonic y_1. If only the first harmonic is considered, the describing function $N = N_1$ is defined as the ratio between the amplitude A of the input signal x and the amplitude A_1 of the first harmonic y_1 contained

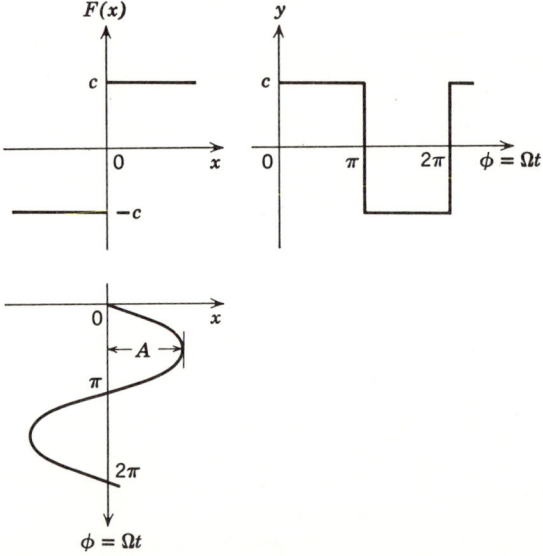

Figure 3.11 Relay response to a sinusoidal input.

---

[12] The notation Ω is used to designate the frequency of the periodic solution for which Ω is constant but not given beforehand. In the following developments, ω is used to designate a frequency when it is a function of the amplitude or time, and Ω is reserved for the periodic solution $x = A \sin \Omega t$ and is always constant.

in the output y; that is,

$$N_1 = \frac{A_1}{A}. \tag{3.25}$$

According to (3.24), the coefficient N_1 has the form

$$N_1 = \frac{1}{\pi A} \int_0^{2\pi} F(A \sin \phi) \sin \phi \, d\phi. \tag{3.26}$$

If the integral on the right side of (3.26) is calculated, we obtain the coefficient $N_1 = N_1(A)$ as

$$N_1 = \frac{4c}{\pi A}. \tag{3.27}$$

It is important to note that the coefficient N_1 is related only to the nonlinearity $F(x)$ and not to the linear part $G(s)$ of the system. This makes it possible to calculate the function $N_1(A)$ once and for all and to use it in the analysis instead of the corresponding nonlinearity. It should be noted, however, that such a result is achieved only under the condition that it is not necessary to consider the higher harmonics of the signal y. For this to be valid, certain conditions should be satisfied by the linear part of the system. These conditions are discussed in the next section.

In the case of a symmetric nonlinear characteristic with a hysteresis loop, when the input signal is a sinusoid $x = A \sin \Omega t$, the first harmonic component y_1 of the output $y = F(x)$ is

$$y_1 = A_1 \sin \Omega t + B_1 \cos \Omega t, \tag{3.28}$$

where

$$\begin{aligned} A_1 &= \frac{1}{\pi} \int_0^{2\pi} F(A \sin \phi) \sin \phi \, d\phi \\ B_1 &= \frac{1}{\pi} \int_0^{2\pi} F(A \sin \phi) \cos \phi \, d\phi \end{aligned} \tag{3.29}$$

and

$$sx = A\Omega \cos \Omega t, \qquad s \equiv \frac{d}{dt}. \tag{3.30}$$

According to the definition of the describing function formulated in (3.21) and (3.22), linearization by use of the first terms of the Fourier series is given by

$$y = N_1 x + \frac{N_2}{\Omega} sx, \tag{3.31}$$

where

$$N_1 = \frac{1}{\pi A} \int_0^{2\pi} F(A \sin \Omega t) \sin \phi \, d\phi$$
$$N_2 = \frac{1}{\pi A} \int_0^{2\pi} F(A \sin \Omega t) \cos \phi \, d\phi. \qquad (3.32)$$

In particular, when a nonlinearity is of the form shown in Figure 3.3c, the coefficients of the describing function $N_1 = N_1(A)$ and $N_2 = N_2(A)$ are obtained (see Appendix F) as

$$N_1 = \frac{4c}{\pi A}\left(1 - \frac{D^2}{A^2}\right)^{1/2}$$
$$N_2 = -\frac{4cD}{\pi A^2}, \qquad (3.33)$$

which are valid for $A \geq D$. Again, the coefficients N_1 and N_2 can be calculated once and for all and put in a diagram or in table form.

In general, when the output of the nonlinear element or nonlinearity is related to the input x by a function $y = F(x, sx)$, as a result of the describing function linearization, the output $y = F(x, sx)$ is expressed by

$$F(x, sx) = N_1 x + \frac{N_2}{\Omega} sx, \qquad (3.34)$$

where, according to (3.22b) and (3.22c), the coefficients N_1 and N_2 are functions of both the amplitude A and the frequency Ω as

$$N_1 = N_1(A, \Omega)$$
$$N_2 = N_2(A, \Omega). \qquad (3.35)$$

In certain cases, either of the coefficients may become identically zero.

For nonlinearities usually encountered in feedback systems, the describing function coefficients are calculated and presented by diagrams in Appendix F.

By using the harmonic linearization of the function $F(x, sx)$ as given by (3.34), the nonlinear differential equation 3.19 can be rewritten as a linear differential equation with constant coefficients

$$\left[B(s) + C(s)\left(N_1 + \frac{N_2}{\Omega}s\right)\right]x = 0. \qquad (3.36)$$

Although the amplitude A and the frequency Ω in the coefficients $N_1(A, \Omega)$ and $N_2(A, \Omega)$ are not given beforehand, they are constant once the sustained oscillations are established. Thus the periodic solutions of the linearized equation 3.36 can be determined by considering the corresponding characteristic equation

$$B(s) + C(s)\left(N_1 + \frac{N_2}{\Omega}s\right) = 0. \tag{3.37}$$

The existence of a periodic solution of (3.19), which is close to

$$x = A \sin \Omega t, \tag{3.20}$$

is reduced to the existence of pure imaginary roots in (3.37). Furthermore, the algebraic equation 3.37 can be used to evaluate approximately the amplitude A and the frequency Ω of the possible periodic solutions and, in addition, determine their stability. The conditions under which the above linearization yields the correct results are presented in the following section. Then, in Chapter 4, the parameter plane method is applied to the algebraic problem of evaluating periodic solutions in high-order nonlinear differential equations.

3.4 Applicability Conditions

Describing function analysis is based upon the assumption that the variable x present in $F(x, sx)$ is $x = A \sin \Omega t$. Physically this means that the input x to the nonlinearity $F(x, sx)$ is a sinusoidal oscillation. No assumptions or restrictions are made for other signals in the system.

In applying an approximate analysis such as the describing function, it is important to have an a priori knowledge about the validity of the above assumption that will justify its application. Unfortunately there are no precise quantitative criteria to establish the validity of approximation before the analysis is actually performed. Then the problem of justification is reduced to the problem of accuracy of the obtained result, which is discussed in Appendix G.

There are, however, rather qualitative measures of the degree of accuracy used in the approximation. In the majority of practical cases these measures are sufficiently precise in estimating the validity of the method. Therefore they can be successfully used as indices guiding the computer study of the system in the design process. In certain cases the qualitative measures are valuable to show how the system can be adjusted to be eligible for the approximate analysis. Then the method is applied to achieve the desired performance characteristics if they do not contradict the applicability conditions of the approximation method used.

The conditions of applicability will be discussed on the basis of the equation

$$B(s)x + C(s)F(x, sx) = 0, \quad s \equiv \frac{d}{dt}, \tag{3.19}$$

where $B(s)$ and $C(s)$ are real polynomials in s, and the degree n of the polynomial $B(s)$ is higher than the degree m of the polynomial $C(s)$. The function $F(x, sx)$ represents an essential nonlinearity.

In order to apply the describing function method and solve (3.19), it is assumed that the related solution $x = x(t)$ is sufficiently close to the solution

$$x = A \sin \Omega t \tag{3.20}$$

of the corresponding linear differential equation

$$\left[B(s) + C(s)\left(N_1 + \frac{N_2}{\Omega} s\right) \right] x = 0, \tag{3.36}$$

obtained from (3.19) by performing the linearization

$$F(x, sx) = N_1 x + \frac{N_2}{\Omega} sx. \tag{3.34}$$

If a solution $x = A \sin \Omega t$ exists, then in the solution process (3.36) is a linear differential equation with constant coefficients. Therefore, the periodic solution $x = A \sin \Omega t$ of (3.19) can be found by establishing the conditions under which the corresponding characteristic equation

$$B(s) + C(s)\left(N_1 + \frac{N_2}{\Omega} s\right) = 0 \tag{3.37}$$

of (3.36) has a pure imaginary root $s = j\Omega$.[13] The problem of finding the periodic solution of (3.19) is thereby reduced to an algebraic problem and the parameter plane method yields an effective solution procedure. In Chapter 4, the procedure is presented for various cases and forms of (3.37). Of interest here, however, are the conditions under which such a procedure is appropriate and can give valid results. These conditions are called the *applicability conditions of the describing function*. They are the conditions that should be satisfied by the polynomials $B(s)$, $C(s)$, and the nonlinear function $F(x, sx)$ so that in the solution process (3.19) can be replaced by (3.36) as long as periodic solutions are being investigated.[14]

---

[13] Note that for $s = j\Omega$, the describing function is $N = N_1 + jN_2$ and (3.37) represents an algebraic equation with complex coefficients. Thus complex roots do not necessarily appear in conjugate pairs (see Section 1.4).

[14] It is not correct to say that (3.19) and (3.36) are "sufficiently close" to each other so that we can use (3.36) to find the periodic solutions of (3.19). Note that if two equations are "near" or "close" to each other it does not mean, in general, that their solutions are also correspondingly "close." For example, the two linear equations $s^2 x + x = 0$ and $s^2 x + \varepsilon sx + x = 0$ can be made as close as desired by choosing a sufficiently small $\varepsilon \neq 0$. Their solutions, however, are essentially different as $t \to \infty$. Therefore, additional conditions have to be satisfied for (3.19) and (3.36) to have sufficiently close respective solutions.

Applicability Conditions

The applicability conditions are obtained in the Popov-Palitov approach [3.8] by considering the solution of (3.19) in the form

$$x = x_1 + \varepsilon\chi, \tag{3.38}$$

where

$$x_1 = A_1 \sin \Omega_1 t, \tag{3.39}$$

$\chi = \chi(t)$ is an arbitrary finite function of time, and ε ($\varepsilon \ll 1$) is a small parameter.[15] Note that the equivalence of the solutions of (3.19) and (3.36) is not assumed here, this will be done in the derivation that follows.

In case a periodic solution (3.38) of (3.19) exists and x_1 is its first harmonic, then the contribution of higher harmonics in the solution is $\varepsilon\chi$ where

$$\chi = \sum_{k=2}^{\infty} A_k \sin (k\Omega_1 t + \theta_k). \tag{3.40}$$

Now let us find the Fourier-series expansion of the function $F(x, sx)$. According to (3.38), we have

$$F(x, sx) = F(x_1 + \varepsilon\chi, sx_1 + \varepsilon s\chi) \tag{3.41}$$

or

$$F(x, sx) = F(x_1, sx_1) + [F(x_1 + \varepsilon\chi, sx_1 + \varepsilon s\chi) - F(x_1, sx_1)]. \tag{3.42}$$

The Fourier-series expansion of the first term $F(x_1, sx_1)$ is

$$F(x_1, sx_1) = F^0 + \left(L_1 + \frac{L_2}{\Omega_1} s\right) \sin \Omega_1 t + G, \tag{3.43}$$

where

$$F^0 = \frac{1}{2\pi} \int_0^{2\pi} F(A_1 \sin \phi, A_1\Omega_1 \cos \phi) \, d\phi \tag{3.44a}$$

$$L_1 = \frac{1}{\pi} \int_0^{2\pi} F(A_1 \sin \phi, A_1\Omega_1 \cos \phi) \sin \phi \, d\phi \tag{3.44b}$$

$$L_2 = \frac{1}{\pi} \int_0^{2\pi} F(A_1 \sin \phi, A_1\Omega_1 \cos \phi) \cos \phi \, d\phi \tag{3.44c}$$

$$G = \sum_{k=2}^{\infty} G_k \sin (k\Omega_1 t + \xi_k). \tag{3.44d}$$

The G_k's in (3.44d) represent the amplitudes of the higher harmonics in (3.43), and no restriction on their magnitudes compared to the magnitude of the first harmonic $(L_1^2 + L_2^2)^{1/2}$ is given. As mentioned before, no restrictions

---

[15] The meaning of the small parameter ε is discussed in more detail in Section 3.5, where the Krylov-Bogoliubov asymptotical methods are presented.

have been imposed on other variables besides x. However, it is still required that $G_k \to 0$ ($k = 2, 3, \ldots$), for $k \to \infty$.

Before the Fourier-series expansion of the second term in square brackets of (3.42) is given, a discussion of the Taylor-series expansion of the same expression is necessary. Thus

$$[F(x_1 + \varepsilon\chi, sx_1 + \varepsilon s\chi) - F(x_1, sx_1)]$$
$$= \varepsilon\left[\frac{\partial}{\partial x} F(x_1, sx_1)\chi + \frac{\partial}{\partial sx} F(x_1, sx_1)s\chi\right] + \varepsilon^2 \cdots. \quad (3.45)$$

From (3.45) it follows that the second term of (3.42) is small with the degree ε, as long as the partial derivatives of $F(x, sx)$ are finite. This is true even for the discontinuous characteristics of relay nonlinearities where the derivatives of $F(x, sx)$ become delta functions.[16]

The Fourier-series expansion of the second term in the square brackets of (3.42) is now

$$[F(x_1 + \varepsilon\chi, sx_1 + \varepsilon s\chi) - F(x_1, sx_1)] = \varepsilon H, \quad (3.46)$$

where

$$H = \sum_{k=0}^{\infty} H_k \sin(k\Omega_1 t + \eta_k). \quad (3.47)$$

Now, according to (3.38), (3.42), (3.43), and (3.46), the nonlinear equation 3.19 can be rewritten as

$$B(s)x_1 + B(s)\varepsilon\chi + C(s)\left(L_1 + \frac{L_2}{\Omega_1} s\right) \sin \Omega_1 t$$
$$+ C(s)F^0 + C(s)G + C(s)\varepsilon H = 0. \quad (3.48)$$

In order to obtain equality with zero in the above equation, it is necessary to equate to zero separately all harmonics of the same order. Therefore, with the help of (3.39) and (3.40), we get

$$F^0 + \varepsilon H_0 = 0 \quad (3.49a)$$

$$B(s)A_1 \sin \Omega_1 t + C(s)\left(L_1 + \frac{L_2}{\Omega_1} s\right) \sin \Omega_1 t$$
$$+ C(s)\varepsilon H_1 \sin(\Omega_1 t + \eta_1) = 0 \quad (3.49b)$$

$$B(s)\varepsilon A_k \sin(k\Omega_1 t + \theta_k) + C(s)G_k \sin(k\Omega_1 t + \xi_k)$$
$$+ C(s)\varepsilon H_k \sin(k\Omega_1 t + \eta_k) = 0, \quad (k = 2, 3, \ldots). \quad (3.49c)$$

Equation 3.49a corresponds to the zeroth-order harmonics and, with the accuracy of degree ε, it follows that $F^0 = 0$. This is equivalent to the requirement that the function $F(x, sx)$ be a symmetric function of the arguments x

---

[16] The delta functions are briefly discussed in Section G.4 of Appendix G, where corresponding references are recommended for further reading.

Applicability Conditions 131

and sx. Consequently, the integral of (3.44a) representing the constant term in the Fourier series is zero.[17]

In the case of the first harmonic component, from (3.49b), we get

$$A_1 \sin \Omega_1 t = -\left|\frac{C(j\Omega_1)}{B(j\Omega_1)}\right| \sqrt{L_1^2 + L_2^2} \sin (\Omega_1 t + \mu + \nu)$$

$$- \varepsilon \left|\frac{C(j\Omega_1)}{B(j\Omega_1)}\right| H_1 \sin (\Omega_1 t + \eta_1 + \nu), \quad (3.50)$$

where

$$\mu = \arctan \frac{L_2}{L_1}$$

$$\nu = \arg \frac{C(j\Omega_1)}{B(j\Omega_1)} \quad (3.51)$$

and it is assumed that $s = j\Omega_1$ is not a zero of the polynomial $B(s)$.

With accuracy of degree ε, (3.50) determines the first approximation A and Ω for the amplitude A_1 and frequency Ω_1 of the first harmonic in the periodic solution by the equations

$$A = \left|\frac{C(j\Omega)}{B(j\Omega)}\right| \sqrt{L_1^2 + L_2^2} \quad (3.52)$$

$$\mu + \nu(\Omega) = \pi.$$

For the relationships of (3.52) to be valid, the restriction about the polynomial $B(s)$ is extended to include the condition that $B(s)$ have no pure imaginary zeros, $s = j\Omega$.

Under the condition that the second term in (3.42) is small, or at least that the corresponding first harmonic H_1 is small with degree ε, the solution of the linear differential equation 3.36 is the first approximation of the first harmonic in the periodic solution of the nonlinear differential equation 3.19 with accuracy of the same degree ε. This may be concluded directly by observing that for $A = A_1$ and $\Omega = \Omega_1$, we have $L_1 = AN_1$, $L_2 = AN_2$, and $x = A \sin \Omega t$, which is the solution of (3.36). It was assumed earlier that a periodic solution of (3.19) exists.

Finally, the conditions (3.49c) give

$$\varepsilon A_k \sin (k\Omega_1 t + \theta_k) = -\left|\frac{C(jk\Omega_1)}{B(jk\Omega_1)}\right| G_k \sin (k\Omega_1 t + \xi_k + \nu_k)$$

$$- \varepsilon \left|\frac{C(jk\Omega_1)}{B(jk\Omega_1)}\right| H_k \sin (k\Omega_1 t + \eta_k + \nu_k),$$

$$(k = 2, 3, \ldots), \quad (3.53)$$

---

[17] This restriction, that $F^0 = 0$ and that $F(x, sx)$ is a symmetric function, is waived in Chapter 5, where asymmetrical nonlinear oscillations are considered.

132   Introduction to Approximate Analysis of Nonlinear Systems

where

$$v_k = \arg \frac{C(jk\Omega_1)}{B(jk\Omega_1)}. \qquad (3.54)$$

In order to have the solution $x = x_1$ with accuracy of degree ε, it is necessary that in $x = x_1 + \varepsilon\chi$ the term

$$\varepsilon\chi = \varepsilon A_k \sin(k\Omega_1 t + \theta_k), \qquad (k = 2, 3, \ldots) \qquad (3.55)$$

be small with the same degree ε. Therefore from (3.53) it follows that

$$\left| \frac{C(jk\Omega_1)}{B(jk\Omega_1)} \right| G_k, \qquad (k = 2, 3, \ldots) \qquad (3.56)$$

has to be small with degree ε, compared to the amplitude A_1 of the first harmonic $x_1 = A_1 \sin \Omega_1 t$. According to (3.52), this means that we should have

$$\left| \frac{C(jk\Omega)}{B(jk\Omega)} \right| G_k \ll \left| \frac{C(j\Omega)}{B(j\Omega)} \right| \sqrt{L_1^2 + L_2^2}, \qquad (k = 2, 3, \ldots). \qquad (3.57)$$

Since G_k cannot be considered small in comparison to $(L_1^2 + L_2^2)^{1/2}$, at least for low values of k, it follows that

$$\left| \frac{C(jk\Omega)}{B(jk\Omega)} \right| \ll \left| \frac{C(j\Omega)}{B(j\Omega)} \right|, \qquad (k = 2, 3, \ldots). \qquad (3.58)$$

Furthermore, it was noted that the degree n of the polynomial $B(s)$ is higher than the degree m of the polynomial $C(s)$, and hence

$$\left| \frac{C(jk\Omega)}{B(jk\Omega)} \right|_{k \to \infty} \to 0. \qquad (3.59)$$

The conditions (3.58) and (3.59) imposed by higher harmonics are vital for the application of the describing function linearization. It is observed that they are rather qualitative criteria for justification of the linearization. Physically, they can be interpreted by considering their relationship to the frequency response of the linear part

$$G(s) = \frac{C(s)}{B(s)} \qquad (3.18)$$

of the corresponding system shown in Figure 3.6. The amplitude characteristic $|G(j\Omega)|$ has to be of a low-pass nature so that it introduces significant attenuation at the frequencies of the higher harmonics. This is illustrated in Figure 3.12a. Since the existence of zeros at the origin of $B(s)$ is not excluded, the characteristic $|G(j\Omega)|$ may have a form as shown in Figure 3.12b. In special cases the characteristic $|G(j\Omega)|$ may have the form of Figure 3.12c, which

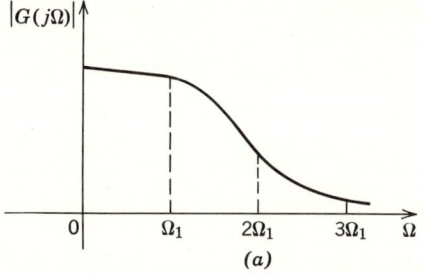

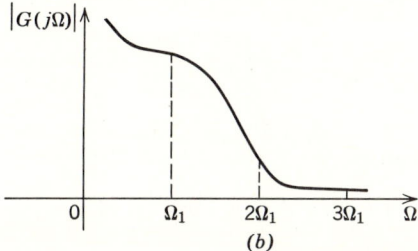

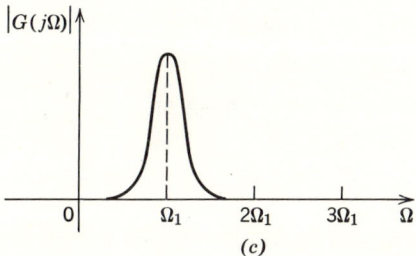

Figure 3.12 Frequency characteristics.

obviously satisfies the requirements.[18] In other words conditions (3.58) and (3.59) represent a generalization of the low-pass filter property for all systems that can be described by (3.19).

The generalization consists of the requirement that the linear part $G(s)$, which contains both the actual linear part of the system and the linear part

---

[18] This case is often referred to as "auto-resonance" [3.10]. This occurs when the nonlinear characteristic is nearly linear—for example, $F(x) = kx + \varepsilon f(x)$, where ε is a small number ($\varepsilon \ll 1$). In addition, the amplitude characteristic $|G(j\Omega)|$ of the linear part of the system has a relatively high and sharp peak. Then we can postulate beforehand that the frequency of the harmonic oscillation close to the actual solution is that corresponding to the peak of the amplitude characteristic.

134   Introduction to Approximate Analysis of Nonlinear Systems

of the eventual nonlinear element, introduce a significant attenuation at higher harmonic frequencies. In the special case when, instead of the nonlinear element, there is only an isolated nonlinearity, the filter property applies to the actual linear part of the system.

Now let us summarize the applicability conditions of describing function linearization, which should be satisfied by the polynomials $B(s)$, $C(s)$ and the nonlinear function $F(x, sx)$, so that the nonlinear differential equation

$$B(s)x + C(s)F(x, sx) = 0, \qquad s \equiv \frac{d}{dt} \qquad (3.19)$$

can be replaced by the corresponding linear differential equation

$$\left[B(s) + C(s)\left(N_1 + \frac{N_2}{\Omega} s\right) \right] x = 0 \qquad (3.36)$$

in the solution process whereby the periodic solutions of (3.19) are investigated. They are the following:

1. The polynomials $B(s)$ and $C(s)$ have to satisfy the following relations:

$$\left| \frac{C(jk\Omega)}{B(jk\Omega)} \right| \ll \left| \frac{C(j\Omega)}{B(j\Omega)} \right|; \qquad (k = 2, 3, \ldots) \qquad (3.58)$$

$$\left| \frac{C(jk\Omega)}{B(jk\Omega)} \right|_{k \to \infty} \to 0, \qquad (3.59)$$

which are interpreted as the generalized filter property of systems described by nonlinear equation 3.19.

2. The polynomial $B(s)$ must not have pure imaginary zeros $s = \pm jk\Omega$, $(k = 1, 2, \ldots)$.

3. The function $F(x, sx)$ should have finite partial derivatives with respect to x and sx, and not be an explicit function of time.

In control system analysis the first condition of the generalized filter property assures, by the low-pass nature of the linear part, that the higher harmonics at the output of the nonlinearity are greatly attenuated in traversing the control loop, and their contribution to the solution is negligible. This condition is commonly satisfied in control systems and, even more so, for higher-order systems. This is one of the advantages of the describing function technique in comparison to other methods of nonlinear analysis.

The second condition, that $B(s)$ must not have pure imaginary zeros $s = jk\Omega$, means that the nonresonant case is considered for which $B(jk\Omega) \neq 0$ for all integers k. Moreover, this requirement is strengthened by the condition that $B(s) = 0$ should have no roots in the right half of the s plane or the "open-loop system is stable." On the other hand, the zero roots of $B(s) = 0$

are allowed and are welcome for they normally improve the approximations involved.[19]

The third condition requiring finite values of the partial derivatives of the nonlinear function $F(x, sx)$ assures that the expression $[F(x_1 + \varepsilon\chi, sx_1 + \varepsilon s\chi) - F(x_1, sx_1)]$ in (3.42) is sufficiently small. This condition is satisfied by a majority of real nonlinear systems, even those employing relay nonlinear characteristics, since the partial derivatives are interpreted as delta functions. This condition may be expressed in terms of the describing function coefficients N_1 and N_2 by requiring that they be monotonic functions of the amplitude A in the neighborhood of the value of A that corresponds to the actual periodic solution.

The above conditions apply also to the case in which two nonlinear elements are present in the system. For the autonomous system of Figure 3.7, which is described by the nonlinear differential equation

$$D(s)x + C(s)F_1(x, sx) + B(s)F_2(x, sx) = 0, \qquad s \equiv \frac{d}{dt}, \qquad (3.60)$$

the applicability conditions for the linear parts are that the polynomial $D(s)$ have no pure imaginary zeros, and that the generalized filter property now given as conditions

$$\left|\frac{C(jk\Omega)}{D(jk\Omega)}\right| \ll \left|\frac{C(j\Omega)}{D(j\Omega)}\right|, \qquad \left|\frac{B(jk\Omega)}{D(jk\Omega)}\right| \ll \left|\frac{B(j\Omega)}{D(j\Omega)}\right|, \qquad (k = 2, 3, \ldots) \tag{3.61}$$

and

$$\left|\frac{C(jk\Omega)}{D(jk\Omega)}\right|_{k\to\infty} \to 0, \qquad \left|\frac{B(jk\Omega)}{D(jk\Omega)}\right|_{k\to\infty} \to 0, \tag{3.62}$$

be satisfied.

The above results apply also to the system of Figure 3.9. The transfer functions $G_1(s)$ and $G_2(s)$ must satisfy the general filter property as well as other applicability conditions. It should be noted, however, that when one of the nonlinearities, n_1 or n_2, does not generate higher harmonics of significant amplitude, the corresponding linear part need only satisfy the condition (3.57), which is weaker than the condition (3.58). This weaker condition is sufficient because the amplitudes N_k of the higher harmonics are small in

---

[19] The generalized filter property can only be realized if the linear part of the system is stable. If a harmonic input is applied to an unstable stage of the linear part, which is not embraced by feedback, no harmonic oscillation appears at its output, since the harmonics of the free oscillations are superimposed. Therefore the filter property has no physical meaning, and so the proposition that the oscillations at the input of the nonlinearity are close to harmonic oscillations is not valid.

comparison to the amplitude $A_1 = \sqrt{L_1{}^2 + L_2{}^2}$ of the first harmonic at the input of the nonlinearity. In certain cases harmonic linearization can be applied even if one of the linear parts, $G_1(s)$ or $G_2(s)$, does not satisfy the filter property.

Although the applicability conditions are given in a rather qualitative form, they are useful adjuncts in computer study of control systems when the region of applicability is desired in the preliminary analysis. Then a computer simulation of the system, and/or a direct experiment, may give justification to the use of describing function linearization. More precise and quantitative applicability conditions are also possible and are discussed in Appendix G.

If the applicability conditions are not satisfied, the describing function analysis may predict sustained oscillations that do not exist, or may fail to predict sustained oscillations that do exist. This later case is illustrated by an example in Section 8.15 of Chapter 8. In such situations the describing function is not an adequate technique and other methods of analysis should be tried.

3.5 Krylov-Bogoliubov Asymptotical Method

Since the parameter plane analysis of nonlinear oscillations is based upon the concept and results of the Krylov-Bogoliubov asymptotical method [3.1], it is of interest to outline, at least briefly, the fundamental aspects of the method. Then the derivations involved in further extensions and applications of the method can be more easily followed. Furthermore, the method is highly applicable to practical problems of nonlinear oscillations and represents a basis for other approximate methods in nonlinear analysis, particularly the describing function technique.

The basis of approximate analysis of nonlinear oscillations is the *small parameter method*[20] introduced in connection with the three-body problem of celestial mechanics. The fundamental concept and certain solution procedures have been postulated in a general form by Poincaré. In this method a second-order nonlinear differential equation describing the oscillations has been formulated so that it incorporates a small parameter ε. The parameter ε is small in the sense that it represents a number of sufficiently small absolute value. For a zero value of ε the nonlinear equation reduces to a linear equation, the solution of which is a harmonic oscillation. The solution of the linear equation is called the *generating solution*. The essential idea of the method is to

---

[20] A rigorous and detailed treatment of the small parameter method has been given by Bulgakov [3.2] and later by Malkin [3.11]. For an outline of the method in English, the references of Minorsky [3.12] and, with an engineering aspect, of Tsien [3.13] and Gibson [3.14] are recommended. An up-to-date treatment of the topic is given by Hale [3.15].

assume the solution of the nonlinear differential equation in the form of an infinite power series in ε. Then, by substituting the solution into the original differential equation, a recursive system of linear nonhomogeneous differential equations with constant coefficients is obtained. Based upon the generating solution, the recursive system can be solved by elementary calculations up to a desired degree of accuracy. The small parameter method has proved useful for solving numerous problems in physics and the technical sciences.

By considering certain nonlinear phenomena in electron tube oscillators, van der Pol [3.3] proposed the method of slowly varying coefficients for evaluation of the related periodic oscillations. The van der Pol method is a variant of the small parameter method, which is heavily based upon the consideration of the first harmonic in the Fourier-series expansion of the nonlinear function, this being the keystone in the describing function analysis. Furthermore, not only is the method convenient for the identification of periodic solutions of second-order nonlinear differential equations, but it also places in evidence the manner in which the possible periodic solutions are established, after small amplitude perturbations, around the solution. The method, however, has been based on a rather intuitive approach and only the first approximation has been considered. From the approach it is not clear how the higher approximations can be made.

The concept of the van der Pol approach has been postulated with all the necessary mathematical rigorousness in the asymptotical method of Krylov and Bogoliubov. Although in a majority of practical situations the theory of the first approximation gives a satisfactory degree of accuracy, the method yields the possibility of further refinement of approximate solutions if greater accuracy should be required. The Krylov and Bogoliubov method has been further extended by E. P. Popov [3.8] to the analysis of transient nonlinear oscillations for large amplitude perturbations.

Based on the Popov extension, the Krylov-Bogoliubov method is used in Chapter 6 for the parameter plane analysis of transient nonlinear oscillations in high-order systems.

The Krylov-Bogoliubov asymptotical method considers a second-order nonlinear differential equation of the form

$$\ddot{x} + F(x, \dot{x}) = 0, \tag{3.63}$$

where $\dot{x} \equiv dx/dt$, $\ddot{x} \equiv d^2x/dt^2$. Suppose that (3.63) can be rewritten as

$$\ddot{x} + \omega_0^2 x = \varepsilon f(x, \dot{x}), \tag{3.64}$$

where $\varepsilon (\varepsilon \ll 1)$ is the small parameter, and

$$\varepsilon f(x, \dot{x}) = \omega_0^2 x - F(x, \dot{x}). \tag{3.65}$$

Therefore, in evaluating the solution of (3.63), it is assumed that (3.63) is close to the linear differential equation

$$\ddot{x} + \omega_0^2 x = 0, \tag{3.66}$$

which is obtained from (3.64) for $\varepsilon = 0$. Equation 3.66 has the solution

$$x_0 = A \sin \phi_0, \tag{3.67}$$

which represents harmonic oscillations with a constant amplitude A and monotonically increasing phase $\phi_0 = \omega_0 t + \theta$. The amplitude A and the constant phase shift θ depend on the initial conditions. The solution x_0 is called the *generating solution*, and (3.66) is called the *generating equation* (sometimes called the *simplified equation* [3.2,8]).

On a purely formal basis, justified later, it can be assumed that for small values of $\varepsilon \neq 0$, the solution $x = x(t)$ of (3.64) can be adequately expressed as a power series

$$x = a \sin \phi + \varepsilon u_1(a, \phi) + \varepsilon^2 u_2(a, \phi) + \cdots + \varepsilon^m u_m(a, \phi) + \cdots, \tag{3.68}$$

where $u_1(a, \phi), u_2(a, \phi), \ldots$, are periodic functions of the angle ϕ with a period 2π.[21] In (3.68), $a = a(t)$ and $\phi = \phi(t)$ are functions of time determined by[22]

$$\begin{aligned} \dot{a} &= \varepsilon P_1(a) + \varepsilon^2 P_2(a) + \cdots + \varepsilon^m P_m(a) + \cdots \\ \dot{\phi} &= \omega_0 + \varepsilon Q_1(a) + \varepsilon^2 Q_2(a) + \cdots + \varepsilon^m Q_m(a) + \cdots. \end{aligned} \tag{3.69}$$

The solution problem now consists of determining the functions $u_i = u_i(a, \phi)$, $P_i = P_i(a)$, and $Q_i = Q_i(a)$ so that their substitution in (3.64) satisfies this differential equation with prescribed accuracy. In other words, the aim of the method is to determine certain periodic functions $u_i(a, \phi)$ that yield the solution (3.68) under the conditions (3.69). Therefore the solution of nonlinear differential equation 3.64 is reduced to the integration of (3.69) with separated variables, which can be solved by elementary calculations.

When the functions $u_i(a, \phi)$ are determined up to $i = m$, we obtain from (3.68) an approximate solution of (3.64) with an accuracy up to the order ε^{m+1}; that is, if m in (3.68) is fixed and all higher terms are deleted, the approximation error tends to zero at a rate proportional to ε^{m+1}. It is also

---

[21] In the original method of Krylov and Bogoliubov, the first approximation has been assumed as $a \cos \phi$ rather than $a \sin \phi$. This, however, is of only formal consequence in the following developments.

[22] The notation a is used to designate the amplitude which is a function of time. For a periodic solution, $a = A =$ const. This is similar to the previous notation for frequency where ω designates a variable frequency that, for a periodic solution, becomes $\omega = \Omega =$ const.

understood that the above series do not converge for $m \to \infty$. This is, however, of no essential consequence since the evaluation of higher terms in (3.68) is difficult and, in practical problems, is normally avoided. The method is called "asymptotic" in the sense that the solution (3.68) for a fixed m has asymptotical properties when $\varepsilon \to 0$.

It can be easily shown that there is an arbitrariness in the definition of the functions $u_i(a, \phi)$ since there are no restrictions in choosing the functions $P_i(a)$ and $Q_i(a)$ that generate functions $u_i(a, \phi)$.[23] To remove this arbitrariness at the start, the additional conditions imposed are

$$\int_0^{2\pi} u_i(a, \phi) \sin \phi \, d\phi = 0, \qquad \int_0^{2\pi} u_i(a, \phi) \cos \phi \, d\phi = 0,$$
$$(i = 1, 2, \ldots). \quad (3.70)$$

These additional conditions, that there be no first harmonic in $u_i(a, \phi)$, reveal intuitively the idea that we wish to have the solution in the form of a Fourier series.

Now the problem of finding the functions $u_i(a, \phi), P_i(a), Q_i(a)$ is attempted. By differentiating (3.68), we obtain

$$\begin{aligned}
\dot{x} &= \dot{a}(\sin \phi + \varepsilon u_1^{(a)} + \varepsilon^2 u_2^{(a)} + \cdots) \\
&+ \dot{\phi}(a \cos \phi + \varepsilon u_1^{(\phi)} + \varepsilon^2 u_2^{(\phi)} + \cdots) \\
\ddot{x} &= \ddot{a}(\sin \phi + \varepsilon u_1^{(a)} + \varepsilon^2 u_2^{(a)} + \cdots) \\
&+ \ddot{\phi}(a \cos \phi + \varepsilon u_1^{(\phi)} + \varepsilon^2 u_2^{(\phi)} + \cdots) \\
&+ \dot{a}^2(\varepsilon u_1^{(aa)} + \varepsilon^2 u_2^{(aa)} + \cdots) \\
&+ 2\dot{a}\dot{\phi}(\cos \phi + \varepsilon u_1^{(a\phi)} + \varepsilon^2 u_2^{(a\phi)} + \cdots) \\
&+ \dot{\phi}^2(-a \sin \phi + \varepsilon u_1^{(\phi\phi)} + \varepsilon^2 u_2^{(\phi\phi)} + \cdots),
\end{aligned} \quad (3.71)$$

---

[23] Suppose that we start with some arbitrary functions $\alpha_1(a), \alpha_2(a), \ldots, \beta_1(a), \beta_2(a), \ldots$, respectively, for $P_1(a), P_2(a), \ldots, Q_1(a), Q_2(a), \ldots$, and in (3.68) and (3.69) replace a and ϕ by the expressions

$$a = b + \varepsilon \alpha_1(b) + \varepsilon^2 \alpha_2(b) + \cdots, \qquad \phi = \psi + \varepsilon \beta_1(b) + \varepsilon^2 \beta_2(b) + \cdots.$$

We obtain then, instead of (3.68) and (3.69), similar equations but with different coefficients:

$$x = b \sin \psi + \varepsilon[\alpha_1(b) \sin \psi + b\beta_1(b) \cos \psi + u_1(b, \psi)] + \varepsilon^2 \cdots,$$

$$\dot{b} = \varepsilon P_1(b) + \varepsilon^2 \left[\frac{dP_1(b)}{db} \alpha_1(b) - \frac{d\alpha_1(b)}{db} P_1(b) + P_2(b) \right] + \varepsilon^2 \cdots,$$

$$\dot{\psi} = \omega_0 + \varepsilon Q_1(b) + \varepsilon^2 \left[\frac{dQ_1(b)}{db} \beta_1(b) - \frac{d\beta_1(b)}{db} P_1(b) + Q_2(b) \right] + \varepsilon^2 \cdots.$$

It is evident that additional conditions (3.70) should be imposed in order to avoid this arbitrariness at the start.

where the partial differentiations are designated by superscripts; thus $u_i^{(a)} = \partial u_i/\partial a$, $u_i^{(aa)} = \partial^2 u_i/\partial a^2$, and so on.

From equations 3.69, we obtain also expressions

$$\ddot{a} = (\varepsilon P_1^{(a)} + \varepsilon^2 P_2^{(a)} + \cdots)(\varepsilon P_1 + \varepsilon^2 P_2 + \cdots)$$
$$= \varepsilon^2 P_1 P_1^{(a)} + \varepsilon^3 (P_2 P_1^{(a)} + P_1 P_2^{(a)}) + \cdots$$
$$\ddot{\phi} = (\varepsilon Q_1^{(a)} + \varepsilon^2 Q_2^{(a)} + \cdots)(\varepsilon P_1 + \varepsilon^2 P_2 + \cdots)$$
$$= \varepsilon^2 P_1 Q_1^{(a)} + \varepsilon^3 (P_1 Q_2^{(a)} + P_2 Q_1^{(a)}) + \cdots \quad (3.72)$$

and likewise, for $\dot{a}^2$, $\dot{a}\dot{\phi}$, and $\dot{\phi}^2$.

By substituting these various expressions into (3.71) and then x, $\ddot{x}$ into the left-hand side of (3.64), it becomes

$$\ddot{x} + \omega_0^2 x = \varepsilon(2\omega_0 P_1 \cos\phi - 2\omega_0 a Q_1 \sin\phi + \omega_0^2 u_1^{(\phi\phi)} + \omega_0^2 u_1)$$
$$+ \varepsilon^2 [(P_1 P_1^{(a)} - aQ_1^2 - 2\omega_0 a Q_2) \sin\phi$$
$$+ (2\omega_0 P_2 + 2P_1 Q_1 + P_1 Q_1^{(a)}) \cos\phi + 2\omega_0 P_1 u_1^{(a\phi)}$$
$$+ 2\omega_0 Q_1 u_1^{(\phi\phi)} + \omega_0^2 u_2 + \omega_0^2 u_2^{(\phi\phi)}] + \varepsilon^3 \cdots. \quad (3.73)$$

The right side of the same equation 3.64 can be rewritten as

$$\varepsilon f(x, \dot{x}) = \varepsilon f(a \sin\phi, a\omega_0 \cos\phi) + \varepsilon^2 [u_1 f^{(x)}(a \sin\phi, a\omega_0 \cos\phi)$$
$$+ (P_1 \sin\phi + aQ_1 \cos\phi + \omega_0 u_1^{(\phi)})$$
$$\times f^{(\dot{x})}(a \sin\phi, a\omega_0 \cos\phi)] + \varepsilon^3 \cdots. \quad (3.74)$$

By equating the terms associated with the like powers of ε on both sides, we obtain a recursive system

$$\omega_0^2(u_1^{(\phi\phi)} + u_1) = f_0(a, \phi) - 2\omega_0 P_1 \cos\phi + 2\omega_0 a Q_1 \sin\phi$$
$$\omega_0^2(u_2^{(\phi\phi)} + u_2) = f_1(a, \phi) - 2\omega_0 P_2 \cos\phi + 2\omega_0 a Q_2 \sin\phi$$
$$\cdots \quad (3.75)$$
$$\omega_0^2(u_m^{(\phi\phi)} + u_m) = f_{m-1}(a, \phi) - 2\omega_0 P_m \cos\phi + 2\omega_0 a Q_m \sin\phi,$$

where

$$f_0(a, \phi) = f(a \sin\phi, a\omega_0 \cos\phi)$$
$$f_1(a, \phi) = u_1 f^{(x)}(a \sin\phi, a\omega_0 \cos\phi) + (P_1 \sin\phi + aQ_1 \cos\phi + \omega_0 u_1^{(\phi)})$$
$$\times f^{(\dot{x})}(a \sin\phi, a\omega_0 \cos\phi) - (P_1 P_1^{(a)} - aQ_1^2) \sin\phi$$
$$- (2P_1 Q_1 + P_1 Q_1^{(a)}) \cos\phi - 2\omega_0 P_1 u_1^{(a\phi)} - 2\omega_0 Q_1 u_1^{(\phi\phi)}$$
$$\cdots \quad (3.76)$$

It is clear that $f_k(a, \phi)$ is a periodic function of the variable ϕ with a period 2π depending also on the amplitude a; its explicit expression is determined as soon as we determine $u_i(a, \phi)$, $P_i(a)$, and $Q_i(a)$ to the kth order.

Consider, first, the functions $f_0(a, \phi)$ and $u_1(a, \phi)$; their Fourier series are

$$f_0(a, \phi) = g_0(a) + \sum_{n=1}^{\infty} [g_n(a) \sin n\phi + h_n(a) \cos n\phi] \tag{3.77}$$

$$u_1(a, \phi) = v_0(a) + \sum_{n=1}^{\infty} [v_n(a) \sin n\phi + w_n(a) \cos n\phi].$$

Equating coefficients of harmonics of the same order, we get

$$g_1(a) + 2\omega_0 a Q_1 = 0, \qquad h_1(a) - 2\omega_0 P_1 = 0, \qquad v_0(a) = \frac{g_0(a)}{\omega_0^2} \tag{3.78}$$

$$v_n(a) = \frac{g_n(a)}{\omega_0^2(1 - n^2)}, \qquad w_n(a) = \frac{h_n(a)}{\omega_0^2(1 - n^2)}, \qquad (n = 2, 3, \ldots).$$

Therefore the functions $P_1(a)$ and $Q_1(a)$ are determined as well as all harmonic components of the function $u_1(a, \phi)$ except the first harmonics $v_1(a)$ and $w_1(a)$. However, in view of the additional requirements of (3.70), we have $v_1(a) = w_1(a) = 0$. Thus

$$u_1(a, \phi) = \frac{g_0(a)}{\omega_0^2} + \frac{1}{\omega_0^2} \sum_{n=2}^{\infty} \frac{g_n(a) \sin n\phi + h_n(a) \cos n\phi}{1 - n^2}. \tag{3.79}$$

Since $u_1(a, \phi)$, $P_1(a)$, and $Q_1(a)$ are evaluated, we know $f_1(a, \phi)$ from (3.76); its Fourier series is

$$f_1(a, \phi) = g_0'(a) + \sum_{n=1}^{\infty} [g_n'(a) \sin n\phi + h_n'(a) \sin n\phi]. \tag{3.80}$$

By using the second equation 3.75 and conditions (3.70), we find

$$g_1'(a) + 2\omega_0 a Q_2 = 0, \qquad h_1'(a) - 2\omega_0 P_2 = 0 \tag{3.81}$$

and

$$u_2(a, \phi) = \frac{g_0'(a)}{\omega_0^2} + \frac{1}{\omega_0^2} \sum_{n=2}^{\infty} \frac{g_n'(a) \sin n\phi + h_n'(a) \cos n\phi}{1 - n^2}. \tag{3.82}$$

The evaluation of the higher approximations is thus sufficiently clear.

The evident complexity of these calculations results from a somewhat complicated substitution of the series development and subsequent differentiations. However, once all this is completed, the rest merely amounts to the consideration of the first, and possibly the second, approximation. In the developments that follow, which are related to the analysis of nonlinear oscillations in the parameter plane, only the first approximation is used since it is sufficient in a majority of applied problems.

Consider more closely the first approximation

$$x = a \sin \phi, \qquad \dot{a} = \varepsilon P_1(a), \qquad \dot{\phi} = \omega_0 + \varepsilon Q_1(a), \tag{3.83}$$

where, according to the above general formulas, we have

$$P_1(a) = \frac{1}{2\pi\omega_0} \int_0^{2\pi} f(a \sin \phi, a\omega_0 \cos \phi) \cos \phi \, d\phi$$

$$Q_1(a) = \frac{1}{2\pi\omega_0 a} \int_0^{2\pi} f(a \sin \phi, a\omega_0 \cos \phi) \sin \phi \, d\phi. \tag{3.84}$$

It has been shown by Krylov and Bogoliubov [3.1] that the first approximation of their asymptotic method is equivalent to the van der Pol method of slowly varying coefficients [3.3], as well as the method of energy balance proposed by Theodorchik [3.16], and the harmonic balance method utilized by Goldfarb [3.4]. It has also been shown that (3.83) and (3.84) can be interpreted in a different form called the *equivalent linearization*. This form represents the basis of the harmonic linearization used for the analysis in the parameter plane. The equivalent linearization will be discussed in Section 3.6. Here the Krylov-Bogoliubov method is illustrated by the following example.

The van der Pol equation is

$$\ddot{x} + x = \varepsilon(1 - x^2)\dot{x}, \tag{3.85}$$

where $f(x, \dot{x}) = (1 - x^2)\dot{x}$, and $\omega_0 = 1$. It is of interest to determine the first approximation of the corresponding solution that is expressed as

$$x = a \sin \phi. \tag{3.86}$$

To determine the amplitude a, (3.83) and (3.84) are used to obtain

$$\dot{a} = \frac{\varepsilon}{2\pi\omega_0} \int_0^{2\pi} f(a \sin \phi, a\omega_0 \cos \phi) \cos \phi \, d\phi$$

$$= \frac{\varepsilon}{2\pi}\left[a \int_0^{2\pi} \cos^2 \phi \, d\phi - a^3 \int_0^{2\pi} \sin^2 \phi \cos^2 \phi \, d\phi \right] = \frac{\varepsilon a}{2}\left(1 - \frac{a^2}{4}\right). \tag{3.87}$$

From (3.87) we have

$$\frac{da^2}{dt} = \varepsilon a^2 \left(1 - \frac{a^2}{4}\right) \tag{3.88}$$

and

$$\frac{da^2}{a^2\left(1 - \dfrac{a^2}{4}\right)} = d\left(\log \frac{a^2}{4 - a^2}\right) = \varepsilon \, dt. \tag{3.89}$$

Upon integration, we obtain

$$a = \frac{a_0 e^{\varepsilon t/2}}{[1 + \tfrac{1}{4}a_0^2(e^{\varepsilon t} - 1)]^{1/2}}, \tag{3.90}$$

which determines the amplitude in (3.86).

Forming the expression for Q_1 given in the second equation 3.84, we find that $Q_1 = 0$. From this, by the last equation 3.83, $\dot\phi = \omega_0 = 1$; that is, $\phi = t + \theta$, where θ is an arbitrary constant.

The first approximation (3.86) of the solution of van der Pol's equation is

$$x = \frac{a_0 e^{\varepsilon t/2}}{[1 + \tfrac{1}{4}a_0^2(e^{\varepsilon t} - 1)]^{1/2}} \sin(t + \phi_0). \tag{3.91}$$

For $t \to \infty$, (3.91) becomes

$$x = 2\sin(t + \phi_0), \tag{3.92}$$

which represents the isolated periodic solution of van der Pol's equation.

As can be seen from (3.91), if $a_0 = 0$, then the amplitude remains equal to zero for all time t and we have $x \equiv 0$, that is, the trivial solution of the van der Pol equation. By using the same equation (3.91), we can readily show that the static regime $x \equiv 0$ is unstable.[24] Actually, no matter how small a_0 is chosen, the amplitude will monotonically increase, leaving the region determined by C_1 in the $x\dot x$ plane of Figure 3.13.

On the other hand, the term $\varepsilon(1 - x^2)\dot x$ in (3.85) may be considered as a friction force that for large values of x has damping effect. As a result of this, we can show that the solutions of the van der Pol equation are bounded and that there actually is a curve C_2 in the $x\dot x$ plane across which the solutions move from the outside inward (see Figure 3.13). Since there are no equilibrium positions in the region between C_1 and C_2, intuition leads to the conjecture that there must be a closed curve C in this region that represents the periodic solution (3.92). This is actually the case and is a consequence of a nontrivial result obtained at the beginning of the century by Bendixson.[25] By a more

---

[24] The van der Pol equation 3.85 can be rewritten to obtain the state variable form

$$\dot x_1 = x_2$$
$$\dot x_2 = -x_1 + \varepsilon(1 - x_1^2)x_2, \tag{i}$$

where $x_1 = x$, $x_2 = \dot x$. To conclude instability of the trivial solution $x \equiv 0$, we observe first that it is the equilibrium point $x_1 = x_2 = 0$ of (i). Then, to study the behavior of nearby solutions, it is possible to use linearization about the equilibrium (see Section 8.2) The linearized equations are

$$\dot x_1 = x_2$$
$$\dot x_2 = -x_1 + \varepsilon x_2 \tag{ii}$$

with $\varepsilon > 0$. Since (ii) represents an unstable system, all solutions of the linearized system (ii) leave the equilibrium point as t increases.

[25] I. Bendixson, Sur les courbes définies par les équations différentialles, *Acta Math.*, **24** (1901). The related theorem is sometimes called the Poincaré-Bendixson theorem, since it was initiated by Poincaré in his paper: Sur les courbes définies par une équation différentielle, *J. Math.*, **3** (1881).

Unfortunately, Bendixson's results cannot be generalized in a simple manner for differential equations of orders higher than two.

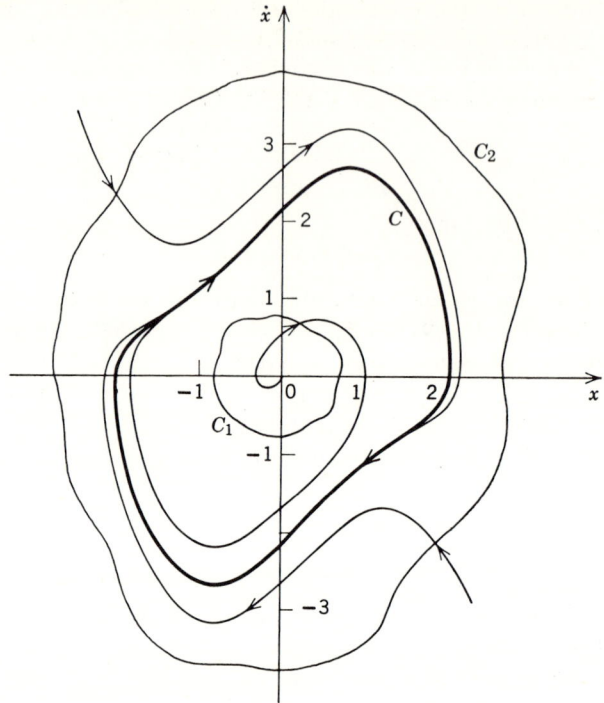

Figure 3.13 Computer solution of the van der Pol equation ($\varepsilon = 1$).

detailed argument, we can actually show that there is a unique closed solution curve of the van der Pol equation, and all other solutions (except the trivial one $x \equiv 0$) approach this curve as time t increases indefinitely. In Figure 3.13 a computer solution for $\varepsilon = 1$ is shown to illustrate this result (see also Problem 3.9).

A detailed study of the existence of limit cycles in high-order nonlinear systems using the harmonic linearization based upon the Krylov-Bogoliubov method is outlined in Chapter 4. If limit cycles exist, the proposed techniques give their amplitudes and frequencies.

It is important to note that the first approximation (3.91) of the solution of van der Pol's equation obtained by the Krylov-Bogoliubov method not only gives in evidence the periodic solution (3.92), but also reveals how this periodic solution is established starting from a certain initial amplitude $a_0 \simeq 2$. This property of the Krylov-Bogoliubov method is used later in Chapter 6 to predict the transient processes in high-order nonlinear systems.

3.6 A Comparison of Describing Function Analysis and the Krylov-Bogoliubov Asymptotical Method

Krylov and Bogoliubov proposed another variant of asymptotic methods for treating directly the second-order nonlinear differential equations in the form

$$\ddot{x} + F(x, \dot{x}) = 0. \tag{3.63}$$

Determine first the expression for ω^2 from the first approximation (3.83) as

$$\omega^2 = \omega_0^2 + 2\omega_0 \varepsilon Q_1(a) + \varepsilon^2 Q_1^2(a). \tag{3.93}$$

If only the first approximation is considered and the term associated with ε^2 is neglected, then using the second equation 3.84, we obtain

$$\omega^2 = \omega_0^2 + \frac{1}{\pi a} \int_0^{2\pi} \varepsilon f(a \sin \phi, a\omega_0 \cos \phi) \sin \phi \, d\phi. \tag{3.94}$$

From (3.65) we have for $x = a \sin \phi$ the relationship

$$\int_0^{2\pi} F(a \sin \phi, a\omega_0 \cos \phi) \sin \phi \, d\phi$$

$$= \omega_0^2 \pi a - \int_0^{2\pi} \varepsilon f(a \sin \phi, a\omega_0 \cos \phi) \sin \phi \, d\phi. \tag{3.95}$$

Thus

$$\omega^2 = \frac{1}{\pi a} \int_0^{2\pi} F(a \sin \phi, a\omega_0 \cos \phi) \sin \phi \, d\phi. \tag{3.96}$$

From (3.65) we also have for $x = a \sin \phi$ the relationship

$$\int_0^{2\pi} F(a \sin \phi, a\omega_0 \cos \phi) \cos \phi \, d\phi$$

$$= -\int_0^{2\pi} \varepsilon f(a \sin \phi, a\omega_0 \cos \phi) \cos \phi \, d\phi. \tag{3.97}$$

According to the approximation of (3.83) and the relationship just derived, we have

$$\dot{a} = -\frac{1}{2\pi\omega_0} \int_0^{2\pi} F(a \sin \phi, a\omega_0 \cos \phi) \cos \phi \, d\phi. \tag{3.98}$$

For a periodic solution $x = A \sin \Omega t$, $(a = A = \text{const}, \omega = \Omega = \text{const})$, (3.95) and (3.97) yield

$$\int_0^{2\pi} F(A \sin \phi, A\omega_0 \cos \phi) \cos \phi \, d\phi = 0$$

$$\Omega^2 = \frac{1}{\pi A} \int_0^{2\pi} F(A \sin \phi, A\omega_0 \cos \phi) \sin \phi \, d\phi. \tag{3.99}$$

These conditions have been derived by Krylov and Bogoliubov for evaluation of the periodic solution of (3.63). It is of interest now to compare these results and the describing function method outlined in the preceding section.

If the periodic solution of nonlinear differential equations is attempted by the describing function method, the corresponding linearized equation corresponding to (3.63) is

$$\ddot{x} + \frac{N_2(A, \Omega)}{\Omega} \dot{x} + N_1(A, \Omega)x = 0, \tag{3.100}$$

in which the linearization $F(x, sx) = N_1(A, \Omega)x + [N_2(A, \Omega)/\Omega]\dot{x}$ is used. The characteristic equation of (3.100) is

$$s^2 + \frac{N_2(A, \Omega)}{\Omega} s + N_1(A, \Omega) = 0. \tag{3.101}$$

For a periodic solution to exist, it is necessary that the algebraic equation 3.101 have a pair of pure imaginary roots $s_{1,2} = \pm j\Omega$. This leads to the condition $N_2(A, \Omega) = 0$, or

$$N_2(A, \Omega) = \frac{1}{\pi A} \int_0^{2\pi} F(A \sin \phi, A\Omega \cos \phi) \cos \phi \, d\phi = 0 \tag{3.102}$$

as defined previously in (3.22c). Then the frequency of the periodic solution from (3.101) is given by $\Omega^2 = N_1(A, \Omega)$, or

$$\Omega^2 = \frac{1}{\pi A} \int_0^{2\pi} F(A \sin \phi, A\Omega \cos \phi) \sin \phi \, d\phi \tag{3.103}$$

as derived from (3.22b).

It can be observed that the conditions (3.102) and (3.103) just derived are identical to those in (3.99), obtained by the Krylov-Bogoliubov approach, except that the frequency Ω of the periodic solution is substituted for the frequency ω_0 of the generating solution. In general, when a nonlinear differential equation is given in the form of (3.63) without the frequency ω_0 of the harmonic generating solution explicitly given, then in applying the Krylov-Bogoliubov asymptotic method the frequency ω_0 should be chosen to be equal to the frequency Ω of the periodic solution. The frequency Ω is computed from the characteristic equation of the linearized equation by requiring that it have a pair of pure imaginary roots. The same facts have been proved to be valid for high-order nonlinear differential equations in the work of Bulgakov [3.2], in which the small parameter method has been applied to the evaluation of limit cycles in high-order automatic control systems. Therefore as far as the first approximation is concerned, the describing function analysis yields the same results as the asymptotical methods in nonlinear system analysis. A comparison of the describing function method with the first approximation of the periodic solution of a second-order differential equation using the Krylov-Bogoliubov method is presented next.

The first approximation by the Krylov-Bogoliubov asymptotic method of the nonlinear differential equation

$$\ddot{x} + \omega_0^2 x = \varepsilon f(x, \dot{x}) \tag{3.64}$$

is

$$x = a \sin \phi, \tag{3.86}$$

where the amplitude a and phase ϕ are determined by

$$\dot{a} = \frac{\varepsilon}{2\pi\omega_0} \int_0^{2\pi} f(a \sin \phi, a\omega_0 \cos \phi) \cos \phi \, d\phi$$

$$\dot{\phi} = \omega(a) \tag{3.104}$$

$$\omega^2(a) = \omega_0^2 + \frac{\varepsilon}{\pi a} \int_0^{2\pi} f(a \sin \phi, a\omega_0 \cos \phi) \sin \phi \, d\phi.$$

This is derived from (3.83) and (3.84).

If we introduce two functions, $k_e(a)$ and $\lambda_e(a)$, of the amplitude a, which are defined as

$$k_e(a) = \omega_0^2 + \frac{\varepsilon}{\pi a} \int_0^{2\pi} f(a \sin \phi, a\omega_0 \cos \phi) \sin \phi \, d\phi$$

$$\lambda_e(a) = \frac{\varepsilon}{2\pi a \omega_0} \int_0^{2\pi} f(a \sin \phi, a\omega_0 \cos \phi) \cos \phi \, d\phi, \tag{3.105}$$

then the equations of the first approximation can be written as

$$\begin{aligned} \dot{a} &= \lambda_e(a) a \\ \dot{\phi} &= \omega(a), \quad \omega^2(a) = k_e(a). \end{aligned} \tag{3.106}$$

By differentiating (3.86) and taking into account (3.106), we have

$$\dot{x} = a\omega(a) \cos \phi + \lambda_e(a) a \sin \phi \tag{3.107}$$

$$\ddot{x} = -a\omega^2(a) \sin \phi + 2\lambda_e(a)\omega(a) a \cos \phi + \lambda_e^2(a) a \sin \phi$$

$$+ \lambda_e(a) \frac{d\omega(a)}{da} a^2 \cos \phi - \frac{d\lambda_e(a)}{da} \lambda_e(a) a^2 \sin \phi$$

$$= -k_e(a) x + 2\lambda_e(a) \dot{x} - \lambda_e^2(a) x + \lambda_e(a) \frac{d\omega(a)}{da} a^2 \cos \phi$$

$$- \frac{d\lambda_e(a)}{da} \lambda_e(a) a^2 \sin \phi. \tag{3.108}$$

On the basis of the definitions (3.105) of the functions $k_e(a)$ and $\lambda_e(a)$, we can rewrite equation 3.108 as

$$\ddot{x} - 2\lambda_e(a)\dot{x} + k_e(a)x = 0(\varepsilon^2), \quad (3.109)$$

where $0(\varepsilon^2)$ is a small number of the order ε^2. Thus, the first approximation (3.86) in the Krylov-Bogoliubov method satisfies the linear differential equation

$$\ddot{x} - 2\lambda_e(a)\dot{x} + k_e(a)x = 0 \quad (3.110)$$

with an accuracy of the order ε^2. Therefore, in the first approximation, the investigation of the nonlinear differential equation 3.64 can be performed by considering the equivalent linear differential equation 3.110 with coefficients dependent on the amplitude of the corresponding solution. This form of analysis is called the *equivalent linearization*.

If the nonlinear differential equation 3.64 is given in the form

$$\ddot{x} + F(x, \dot{x}) = 0, \quad (3.63)$$

the equivalent linearization yields (3.100) as

$$\ddot{x} + \frac{N_2(a, \omega)}{\omega}\dot{x} + N_1(a, \omega)x = 0 \quad (3.111)$$

or, more frequently,

$$\ddot{x} + \frac{N_2(a)}{\omega}\dot{x} + N_1(a)x = 0. \quad (3.112)$$

By comparing (3.112) with (3.100), we observe that (3.108) can be used for the investigation of transient conditions in the establishment of the limit cycles (sustained oscillations) for small deviations from the corresponding periodic solution. In this investigation the above equations can be regarded as linear[26]

$$\ddot{x} - 2\sigma\dot{x} + \omega^2 x = 0. \quad (3.113)$$

However, σ and ω are not constants σ_0 and ω_0 as in the purely linear case

$$\ddot{x} - 2\sigma_0\dot{x} + \omega_0^2 x = 0, \quad (3.114)$$

but rather functions $\sigma = \sigma(a)$, $\omega = \omega(a)$ given by

$$\sigma(a) = -\frac{N_2(a)}{2\omega}, \quad \omega^2(a) = N_1(a). \quad (3.115)$$

Thus in the case of (3.113) the transient oscillatory process should not

---

[26] Note that in (3.113), σ is very small and therefore $\omega_n \simeq \omega$.

be written as $x = a_0 e^{\eta t} \sin \omega t$ in analogy to the linear case where $x = A_0 e^{\sigma_0 t} \sin \omega_0 t$. In accordance with the first approximation,

$$x = a \sin \phi, \qquad \dot{a} = \varepsilon P_1(a), \qquad \dot{\phi} = \omega_0 + \varepsilon Q(a) \qquad (3.83)$$

the transient process about the periodic solution $x = A \sin \Omega t$ should be written in a differential form

$$x = a \sin \phi, \qquad \dot{a} = a\sigma(a), \qquad \dot{\phi} = \omega(a), \qquad (3.116)$$

where $\sigma(a)$ and $\omega(a)$ are determined by (3.115). Note that σ and N_2 are small in magnitude. This can be concluded from (3.97) and the definition of the coefficient N_2. This means that only the transient processes close to the periodic solutions can be investigated by using the linearized equations 3.111 or 3.112. For the periodic solution $x = A \sin \Omega t$, $\sigma = N_2 = 0$, $a = A = $ const, and $\omega = \Omega = $ const.

The stability of the periodic solution can be investigated by considering the sign of $\sigma(a)$, which is in analogy with the linear concept of stability.[27] Here, however, $\sigma(a)$ is a function of the amplitude a, and it is necessary to modify slightly the approach to the question of stability; namely, the stability of the periodic solution $x = A \sin \Omega t$, $(\sigma = 0)$, is determined by the change of $\sigma(a)$ caused by a change in the amplitude a at the value $\sigma(a) = 0$. If, for an increase in the amplitude a from its value A of the periodic solution, the function $\sigma(a)$ changes sign from $+$ to $-$, the periodic solution is stable and vice versa. Thus the stability of the periodic solution is determined by the sign of the ratio $\Delta\sigma/\Delta a$, rather than the sign of $\sigma(a)$. In further analogy to linear analysis, the magnitude of the ratio $\Delta\sigma/\Delta a$ indicates the nature of the transient conditions that arise because of small amplitude perturbations of the periodic solution (sustained oscillations). This postulates the notion of the *relative stability of sustained nonlinear oscillations.*

The above discussion is applied in Chapter 4 to the nonlinear differential equation

$$B(s)x + C(s)F(x, sx) = 0, \qquad s \equiv \frac{d}{dt} \qquad (3.19)$$

and thus the determination of related periodic solutions and their stability are considered in the parameter plane. The analysis is based upon the linearized equation

$$\left[B(s) + C(s)\left(N_1 + \frac{N_2}{\Omega} s\right) \right] x = 0, \qquad (3.36)$$

which has constant coefficients as long as the periodic solutions of the nonlinear equation 3.19 are considered. This stems from the fact that for the periodic solution $x = A \sin \Omega t$, the amplitude A and the frequency Ω, which are

---

[27] In the original Krylov-Bogoliubov method [3.1] the value of $\Phi(a)/a$ is used instead of $\sigma(a)$.

arguments of the coefficients $N_1 = N_1(A, \Omega)$ and $N_2 = N_2(A, \Omega)$, are constant. It should be noted that A and Ω are not known beforehand and are evaluated in the solution procedure. They are variables only in the solution process and therefore (3.36) has coefficients which are adjustable but which are constant in time.

The harmonic linearization is valid not only for the periodic solution but also in the neighborhood of that solution. Therefore (3.36) represents the nonlinear equation 3.19 around the periodic solution and yields information about the transient process during the establishment of the corresponding sustained oscillations. Then the describing function coefficients are

$$N_1 = N_1(a, \omega)$$
$$N_2 = N_1(a, \omega)$$
(3.117)

as used in (3.111) and (3.112). This means that (3.36) becomes an equation with slowly varying coefficients because of the slowly varying amplitude a and frequency ω. Thus the coefficients are not explicit functions of time but rather through the solution itself. As a consequence, (3.36) gives a family of solutions that are slightly damped or undamped exponential functions of time, which correspond to different initial conditions chosen sufficiently close to the periodic solution. This corresponds to the similar conclusions obtained by Bogoliubov [3.1] in connection with the strong stability of a two-parameter family of particular solutions. Important results concerning solutions of high-order nonlinear differential equations with slowly varying coefficients have been obtained by Mitropolsky [3.1].

The above conclusion about the nature of the transient process in the neighborhood of the periodic solution will be used in the stability analysis of the periodic solution. The perturbed solution will be investigated to find whether it is a damped or undamped exponential function. This analysis is presented in the following chapter and is further extended to large amplitude variations and transient nonlinear oscillations in Chapter 6.

References

[3.1] N. N. Bogoliubov and J. A. Mitropolsky, *Asymptotical Methods in the Theory of Nonlinear Oscillations* (in Russian), State Press for Physics and Mathematical Literature, Moscow, 1963. (English translation: Hindustan Publishing Co. Delhi, India; U.S.A. Distr.: Gordon and Breach, New York).

[3.2] B. V. Bulgakov, *Oscillations* (in Russian), GOSTEHIZDAT, Moscow, 1954.

[3.3] B. van der Pol, Forced Oscillations in a Circuit with Non-Linear Resistance (Reception with Reactive Triode), *The London, Edinburgh, and Dublin Philosophical Magazine and Journal of Science*, 3, 65–80 (1927). See also R. Bellman and R. Kalaba, *Selected Papers on Mathematical Trends in Control Theory*, Dover, New York, 1964, pp. 124–141.

References

[3.4] L. C. Goldfarb, On Some Nonlinear Phenomena in Regulatory Systems (in Russian), *Avtomatika i Telemekhanika*, **8,** No. 5, 349–383 (1947). (English translation: National Bureau of Standards, Report 1691, May 29, 1952.)

[3.5] R. J. Kochenburger, A Frequency Response Method for Analyzing and Synthesizing Contactor Servomechanisms, *AIEE Trans.* Pt. I (*Power Apparatus and Systems*), **69,** 270–284 (1950).

[3.6] W. Oppelt, Locus Curve Method for Regulators with Friction, *J. Inst. Elec. Eng.* (*London*), Pt. IIA, **94,** Nos. 1 and 2 (May, 1947).

[3.7] A. Tustin, The Effects of Backlash and of Speed-development Friction on the Stability of Closed-cycle Control Systems, *J. Inst. Elec. Eng.* (*London*), Pt. IIA, **94,** No. 1 (May, 1947).

[3.8] E. P. Popov and I. P. Palitov, *Approximate Methods for Analysis of Nonlinear Automatic Systems* (in Russian), State Press for Physics and Mathematical Literature, Moscow, 1960. (English translation: Foreign Technical Division, AFSC, Wright-Patterson AFB, Ohio, Report FTD-TT-62-910.)

[3.9] J. W. S. Rayleigh, *The Theory of Sound*, Vol. 1, Dover, New York, 1945.

[3.10] M. A. Aizerman, *Lectures on the Theory of Automatic Control* (in Russian) State Press for Physics and Mathematical Literature, Moscow, 1958. (English translation Pergamon and Addison-Wesley, New York, 1963.)

[3.11] I. G. Malkin, *Some Problems of Nonlinear Oscillations* (in Russian), GOSTEHIZDAT Moscow, 1956.

[3.12] N. Minorsky, *Nonlinear Oscillations*, Van Nostrand, New York, 1962.

[3.13] H. S. Tsien, The Poincaré-Lighthill-Kuo Method, in *Advances in Applied Mechanics*, Vol. IV, Academic, New York, 1956, pp. 281–349.

[3.14] J. E. Gibson, *Nonlinear Automatic Control*, McGraw-Hill, New York, 1963.

[3.15] J. K. Hale, *Oscillations in Nonlinear Systems*, McGraw-Hill, New York, 1963.

[3.16] K. F. Theodorchik, *Self-Oscillating Systems* (in Russian), 3d. ed., GOSTEHIZDAT, Moscow, 1952.

CHAPTER FOUR

Symmetrical Oscillations

4.1 Introduction

Nonlinear systems with symmetrical characteristics of the nonlinear elements and free from external disturbances can exhibit the symmetrical self-excited oscillations. These oscillations are represented by periodic solutions of the corresponding nonlinear differential equation describing the system, which are symmetrical with respect to the time axis. The purpose of this chapter is to present the parameter plane analysis [4.1–6] of the periodic solutions in a class of high-order nonlinear differential equations. Techniques and procedures will be shown for the determination of existence, stability, and actual values of the amplitude and frequency of the periodic solutions in the parameter plane.

The existence of self-excited oscillations in nonlinear systems is important for their application. There are systems in which the sensitivity is increased by introducing appropriate self-excited oscillations that eliminate the effects of friction. On the other hand, there are cases in which the self-excited oscillations are not allowed since they deteriorate the system performance, and means should be found to remove them from the system operation. However, a nonlinear system can be such that the self-excited oscillations cannot be avoided and the analysis task is to change the system so that the oscillations are not present for a sufficiently large range of system operation.

The existence of self-excited oscillations can be investigated in the parameter plane as a function of the system parameters and the initial conditions. In the following section a graphical procedure is developed to discover the oscillations and determine the corresponding amplitude and frequency. Because of the nonlinear element, the location of a set of the M points, called the M locus, has to be investigated with respect to the characteristic Σ or ζ

curves each time a parameter along one of the parameter plane axes is changed. The self-excited oscillations are indicated at the intersections of the M locus and the characteristic curve representing the stability boundary. If there is no intersection of the two graphs, the conclusion can be made that there are no self-excited oscillations and the system is either stable or unstable in the usual sense determined by linear system analysis. The validity of such a conclusion depends inherently upon the accuracy of the harmonic linearization employed in the solution procedure. This was discussed in the previous chapter.

Once the intersection of the two mentioned graphs in the parameter plane is found and the self-excited oscillations are indicated, the stability of the oscillation can be determined. In Section 4.3 it is shown how the parameter plane technique can be used to answer the question of the stability of self-excited oscillations. In addition, the sensitivity of the oscillations caused by amplitude perturbations is also discussed and the correlation with stability is shown.

In Section 4.4 the concept of sensitivity is extended to the analysis of amplitude and frequency changes caused by small parameter perturbations. This may be important in the design of nonlinear systems for which the self-excited oscillations represent one of their normal operating characteristics (for example, limit-cycling control systems, plant-adaptive systems, and so on).

System stability is again discussed in Section 4.5. From the relative position of the M locus and the characteristic curves that determine the stability regions in the parameter plane, certain information can be obtained about the global behavior of the system, with due caution concerning the approximation involved. Thus the information is useful when supplemented with the computer study of the systems, which can remove doubt about the validity of the approximation.

Systems with two nonlinearities are considered in Section 4.6. The two nonlinearities must be single valued in order for the procedure to be applied. Multivalued nonlinearities are treated in Section 4.7. Frequency-dependent and variable-pole describing functions representing the nonlinear characteristics are discussed in Section 4.8. In these cases the parameter plane analysis is more convenient to apply than other analysis methods.

4.2 Solution Procedure in the Parameter Plane

Basic solution procedures for determining the sustained oscillations of nonlinear control systems will be presented by considering the nonlinear differential equation

$$B(s)x + C(s)F(x, sx) = 0, \qquad s \equiv \frac{d}{dt}, \tag{4.1}$$

where $B(s)$ and $C(s)$ are polynomials in s and the degree n of the polynomial $B(s)$ is higher than the degree m of the polynomial $C(s)$. The function $F(x, sx)$ that represents the nonlinearity is a symmetric function with respect to the origin.

If a periodic solution $x = x(t)$ of (4.1) exists and is close to

$$x = A \sin \Omega t \tag{4.2}$$

and the applicability conditions of the describing function analysis are satisfied, the solution $x(t)$ can be found by considering the linearized differential equation

$$\left[B(s) + C(s)\left(N_1 + \frac{N_2}{\Omega}s\right) \right] x = 0. \tag{4.3}$$

Equation 4.3 is obtained from (4.1) by performing the harmonic linearization of the nonlinear function $F(x, sx)$ as

$$F(x, sx) = N_1 x + \frac{N_2}{\Omega} sx, \tag{4.4}$$

where $N_1 = N_1(A, \Omega)$, $N_2 = N_2(A, \Omega)$ are coefficients of the describing function $N = N_1 + jN_2$ for which $s = j\Omega$ and

$$\begin{aligned} N_1 &= \frac{1}{\pi A} \int_0^{2\pi} F(A \sin \phi, A\Omega \cos \phi) \sin \phi \, d\phi \\ N_2 &= \frac{1}{\pi A} \int_0^{2\pi} F(A \sin \phi, A\Omega \cos \phi) \cos \phi \, d\phi \end{aligned} \tag{4.5}$$

with $\phi = \Omega t$.

Under the above conditions, for the periodic solution $x(t) = A \sin \Omega t$, the linearized differential equation has constant coefficients and the corresponding characteristic equation is

$$B(s) + C(s)\left(N_1 + \frac{N_2}{\Omega}s\right) = 0. \tag{4.6}$$

Now the periodic solution $x = A \sin \Omega t$ of the linearized equation 4.3 is identified by a pure imaginary root $s = j\Omega$ of the algebraic equation 4.6. Thus the determination of periodic solutions of nonlinear differential equation 4.1 is reduced to an algebraic problem that can be solved in the parameter plane.

The algebraic problem may be considered in two parts. First, for given polynomials $B(s)$ and $C(s)$, it is necessary to determine the values of $N_1(A, \Omega)$ and $N_2(A, \Omega)$ that result in (4.6) having a pure imaginary root $s_1 = j\Omega$. Then the corresponding values of A and Ω determine the desired solution $x = A \sin \Omega t$. The second part of the problem consists in determining if the resulting solution $x = A \sin \Omega t$ represents stable or unstable oscillations. The stability is in the sense that the perturbed oscillations in the neighborhood of the periodic solution should converge to that solution, and it will be specified in more detail later. In algebraic terms, to assure stability of the periodic solution, it is necessary that all the remaining $n - 1$ roots, $s_2, s_3, \ldots, s_n$, of (4.6) have negative real parts for those values of N_1 and N_2 that result in the periodic solution $x = A \sin \Omega t$. Furthermore, for a small increase in amplitude A around the periodic solution, all the roots should have negative real parts for the amplitude to decrease to the nominal value of the solution. Both parts of the algebraic problem are interpreted and solved in the parameter plane as shown in the following developments.

To determine the condition under which (4.6) has a pure imaginary root $s_1 = j\Omega$, note that (4.6) can be written in the form

$$\sum_{k=0}^{n} a_k s^k = 0, \tag{4.7}$$

where the coefficients a_k are linear functions of $N_1(A, \Omega)$ and $N_2(A, \Omega)$. Thus N_1 and N_2 may be considered as parameters α and β

$$\begin{aligned} \alpha &= N_1(A, \Omega) \\ \beta &= N_2(A, \Omega). \end{aligned} \tag{4.8}$$

Equations 4.8 determine the positions of point $M(\alpha; \beta)$ for various values of A and Ω. The corresponding loci of M points in the $\alpha\beta$ plane are called the M locus.

By substituting $s = j\Omega$ in (4.7), the $\Sigma = 0$ curve is determined by equations

$$\begin{aligned} \alpha &= \alpha(\Omega) \\ \beta &= \beta(\Omega) \end{aligned} \tag{4.9}$$

obtained from (4.7) in the usual fashion. As known, the $\Sigma = 0$ curve represents the loci of points in the $\alpha\beta$ plane, which correspond to pure imaginary roots of the algebraic equation 4.7. If both the M locus and the $\Sigma = 0$ curve are plotted in the $\alpha\beta$ plane, the conditions for which (4.7) has a pure imaginary root are satisfied at their intersections. After the intersections are found, the stability of the corresponding periodic solutions is determined by inspection from the parameter plane plot.

The solution procedure will be illustrated by a specific example in which a nonlinear system with one nonlinearity characterized by a single-valued function $F(x)$ is investigated. Other cases will be considered in the following sections.

Consider a control system described by the nonlinear differential equation

$$B(s)x + C(s)F(x) = 0, \tag{4.10}$$

in which $F(x)$ is a symmetric and single-valued nonlinear function of the variable x.

The harmonic linearization of the function $F(x)$ yields

$$F(x) = N_1 x, \qquad (4.11)$$

where $N_1 = N_1(A)$ is a real describing function determined by

$$N_1 = \frac{1}{\pi A} \int_0^{2\pi} F(A \sin \phi) \sin \phi \, d\phi, \qquad \phi = \Omega t. \qquad (4.12)$$

The corresponding linearized differential equation has a characteristic equation

$$B(s) + C(s) N_1 = 0, \qquad (4.13)$$

which can be written as

$$\sum_{k=0}^{n} a_k s^k = 0. \qquad (4.7)$$

The coefficients a_k can be reduced to the linear form $a_k = b_k \alpha + c_k \beta + d_k$, where α represents a linear system parameter and $\beta = N_1(A)$ is the describing function.

A possible periodic solution of (4.10), which is close to

$$x = A \sin \Omega t, \qquad (4.2)$$

is determined at the intersections of the $\Sigma = 0$ curve and the $M[\alpha; \beta = N_1(A)]$ locus.

Let us investigate a nonlinear control system in Figure 4.1 that has the transfer functions

$$G_1(s) = \frac{300(s + 2)}{s(s + 0.536)(s + 5)(s + 7.464)}, \qquad G_{-1}(s) = K_{-1} s, \qquad (4.14)$$

and the nonlinearity n with a characteristic given in the upper left corner of Figure 4.2. The corresponding nonlinear differential equation is

$$[s^4 + 13s^3 + 44s^2 + (20 + K_{-1})s]x + 300(s + 2)F(x) = 0. \qquad (4.15)$$

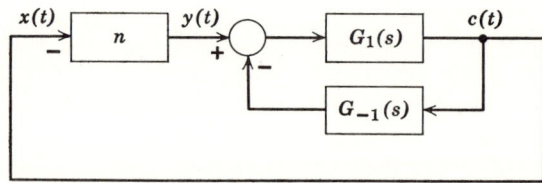

Figure 4.1 Nonlinear system block diagram.

Solution Procedure in the Parameter Plane

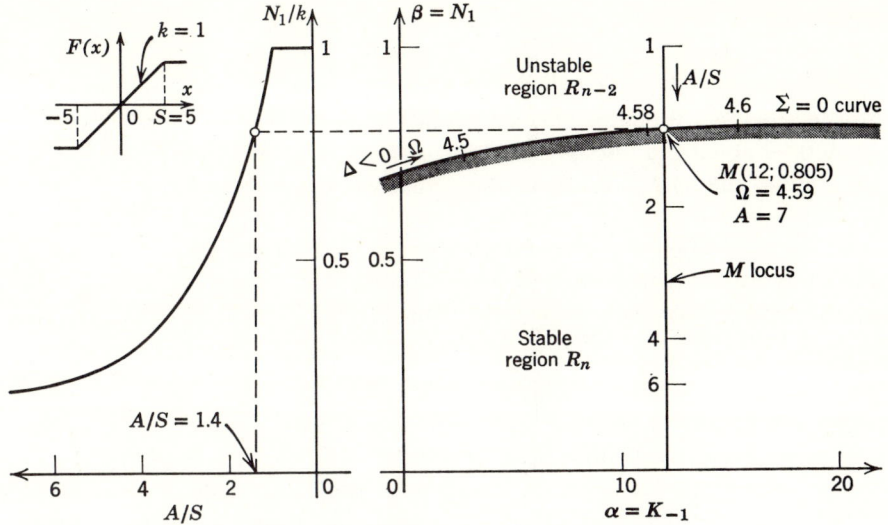

Figure 4.2 Parameter plane diagram.

After the harmonic linearization $F(x) = N_1 x$, the characteristic equation of the corresponding linearized differential quation is

$$s^4 + 13s^3 + 44s^2 + (20 + K_{-1} + 300N_1)s + 600N_1 = 0. \quad (4.16)$$

In order to determine a pair of pure imaginary roots of (4.16) under various conditions, denote

$$\alpha = K_{-1}, \qquad \beta = N_1(A) \quad (4.17)$$

and plot in Figure 4.2 the $\Sigma = 0$ curve determined by

$$\alpha = 0.5\Omega^4 - 9\Omega^2 - 20$$
$$\beta = \frac{1}{600}(44\Omega^2 - \Omega^4). \quad (4.18)$$

The $\Sigma = 0$ curve is shaded according to $\Delta < 0$. Then the stable region, for which all the roots of (4.16) have negative real parts, is determined in the usual fashion.

For a value of $\alpha = 12$, the M locus corresponding to the variation $\beta = N_1(A)$ is plotted in the parameter plane according to the function

$$N_1(A) = \frac{2k}{\pi}\left[\arcsin\frac{S}{A} + \frac{S}{A}\left(1 - \frac{S^2}{A^2}\right)^{1/2}\right], \quad A \geq S, \quad (4.19)$$

or the diagram of the describing function shown in Figure 4.2. Both the expression (4.19) and the diagram of the describing function are obtained by applying the definition

$$N_1(A) = \frac{1}{\pi A} \int_0^{2\pi} F(A \sin \phi) \sin \phi \, d\phi, \qquad \phi = \Omega t \qquad (4.12)$$

to the nonlinear function $F(x)$ shown in Figure 4.2 (see also Appendix F). It should be noted that along the M locus the ratio A/S is interpolated, which determines the amplitude A for the actual value $S = 5$.

At the intersection $M(12; 0.805)$ of the $\Sigma = 0$ curve and the M locus, pure imaginary roots $s_{1,2} = \pm j4.59$ are indicated from the $\Sigma = 0$ curve on which the values of Ω are interpolated. This pair of pure imaginary roots is obtained for an amplitude $A = 7$, calculated from the M locus. Thus a periodic solution of (4.10) is indicated and should be close to

$$x = 7 \sin 4.59t. \qquad (4.20)$$

A stability analysis should now reveal whether the periodic solution (4.20) represents an unstable limit cycle or a stable limit cycle, in the latter case the system of Figure 4.1 will exhibit sustained oscillations. Before the stability problem is considered, it should be noted that from the parameter plane diagram of Figure 4.2 it is easy to study the amplitude and frequency of the periodic solution as functions of the parameter $\alpha = K_{-1}$. This is accomplished by simply shifting the M locus along the α axis, depending on the actual value of $\alpha = K_{-1}$.

The stability problem of a periodic solution will be discussed here in order to present the basic idea and to give a rather physical consideration of the problem. The following section will be completely devoted to the subject.

The first condition for a periodic solution to be stable is that, besides the pair of complex roots on the imaginary axis, all the other roots of the corresponding characteristic equation must be to the left of the imaginary axis. This necessary but not sufficient condition is satisfied if the intersection of the $\Sigma = 0$ curve and the M locus lies on the boundary of the absolute stability region, R_n, in the $\alpha\beta$ plane. (Note that not the whole $\Sigma = 0$ curve represents a boundary of the absolute stability for which all the roots have $\sigma < 0$.) The necessity of the first condition is obvious since the presence of a root with real part greater than zero causes the linearized equation to have a solution that increases without bound. Of course, the linearized equation is a good approximation only in the neighborhood of the periodic solution, and a conclusion that a solution of the corresponding nonlinear differential equation increases without bound cannot be made—the solution may converge to a limit cycle with a larger amplitude than that of the periodic solution under investigation. However, within the accuracy of the analysis by describing function technique, it can be concluded that the investigated periodic solution cannot represent sustained oscillations. Note that the point $M(12; 0.805)$ in Figure 4.2 satisfies this first condition.

The second condition of the stability problem is concerned with the nature of the small amplitude perturbation of the periodic solution. As can be seen from Figure 4.2, when the amplitude changes, the $M(\alpha; \beta)$ point moves along the M locus. If, for an increase of the amplitude, point M moves into the stable region R_n from the $\Sigma = 0$ curve, the linearized equation would represent a stable system and the damped oscillation will decrease in amplitude until the M point reaches the $\Sigma = 0$ curve again. On the other hand, if the amplitude decreases and the M point moves inside the unstable region, the linearized equation corresponds to an unstable system and the oscillations will build up to the amplitude A, which locates the M point on the $\Sigma = 0$ curve. In this case, which is illustrated in Figure 4.2, the periodic solution is stable, and sustained oscillations exist in the system. The second condition is checked in the parameter plane simply by indicating that for an increase in amplitude the point M is moved from the $\Sigma = 0$ curve to the shaded side of the curve. By the same reasoning it can be concluded that a periodic solution is unstable if, for an increase in the amplitude, point M moves into the unstable region (that is, from the $\Sigma = 0$ curve it goes to the unshaded side of the curve) and for a decrease in amplitude, the point M enters the stable region (from the $\Sigma = 0$ curve it goes to the shaded side of the curve).

It should be noted, however, that the above reasoning about the behavior of the system near the periodic oscillations is approximately valid since the equation

$$[B(s) + C(s)N_1]x = 0 \tag{4.21}$$

is a linear differential equation with constant coefficients only for the periodic solution. In the neighborhood of the solution, it may be rewritten as a linear differential equation with a periodic coefficient

$$B(s)\Delta x + C(s)\frac{dF}{dx^*}\Delta x = 0. \tag{4.22}$$

The periodic coefficient is $dF/dx^* = (dF/dx)\big|_{x=x^*}$, $x^* = A \sin \Omega t$. (The starred notation indicates that x^* is the periodic solution for which A and Ω are known constants.) Equation 4.22 is obtained from (4.10) by substituting $x = x^* + \Delta x$, $x^* = A \sin \Omega t$. Thus the stability problem of the periodic solution $x^* = A \sin \Omega t$ is reduced to the stability problem of the corresponding linear differential equation 4.22, which has periodic coefficients and is written for a small variation Δx around the solution $x^* = A \sin \Omega t$. A precise definition of the stability of periodic oscillations is sometimes referred to as the *orbital stability* (see Section 8.4 of Chapter 8). In the analysis here, however, interest is focused on the approximate study of periodic solutions. It will be shown in a following section that the linearized differential equation 4.21 contains sufficient information about the stability of the periodic solution,

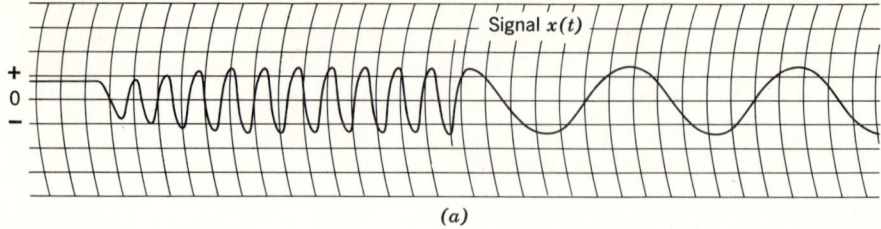

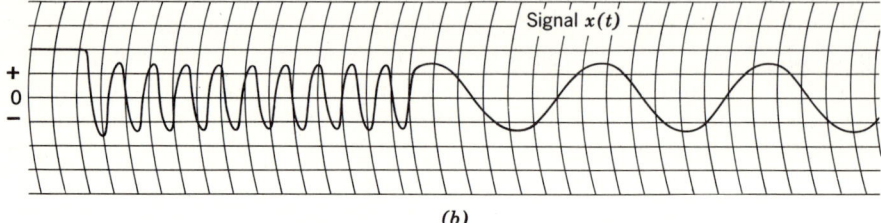

Figure 4.3 Computer solution: (a) $x_0 = 4$; (b) $x_0 = 10$. Amplitude scale: 1 unit/line. Time scale: 1 sec/division and 0.2 sec/division.

provided the applicability conditions of the describing function analysis are satisfied.

The analysis results obtained for the specific system of Figure 4.1 can be checked by computer simulation. In Figure 4.3 two computer solutions are shown, one for an initial condition, $x_0 = 4$, the other for $x_0 = 10$. As can be seen from Figure 4.3, in both cases the solution ends in sustained oscillations with the amplitude and frequency equal to that calculated graphically from Figure 4.2. The phase-plane portrait of the system is shown in Figure 4.4. It illustrates the stability of the limit cycle with the same initial conditions $x_0 = 4$ and 10.

4.3 Stability and Sensitivity Analysis

When a periodic solution of the nonlinear differential equation

$$B(s)x + C(s)F(x, sx) = 0 \tag{4.1}$$

is found, the stability problem arises to determine whether the periodic solution represents stable or unstable periodic oscillations. This problem has been long under consideration and significant results have been obtained by investigating the behavior of the periodic solution subject to small amplitude perturbations. Perturbation techniques based upon the work of Poincaré, van der Pol, Liapunov, Krylov, and Bogoliubov can be applied to this problem.

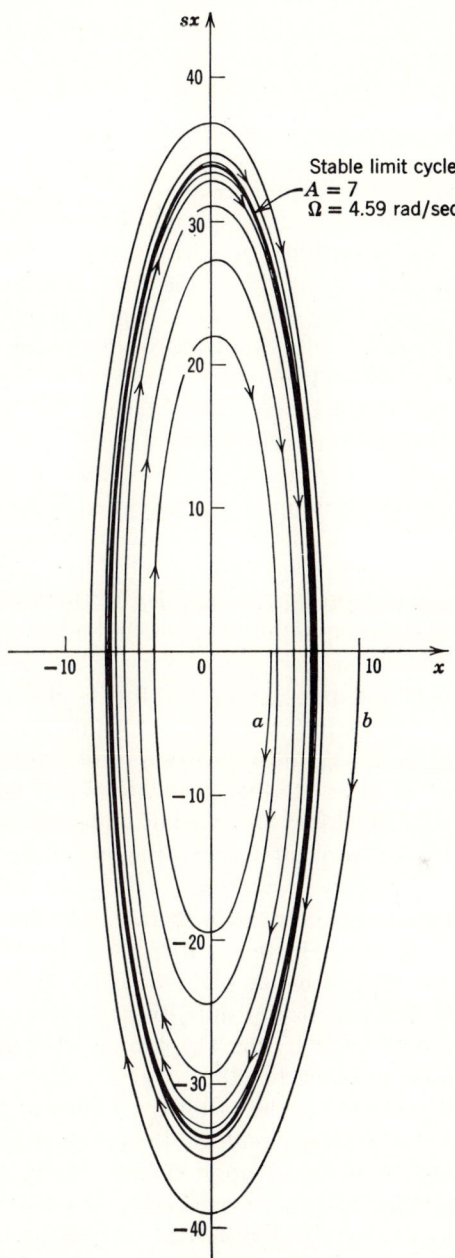

Figure 4.4 Phase-plane portrait: (*a*) initial condition $x_0 = 4$; (*b*) initial condition $x_0 = 10$.

Of particular and practical value are the asymptotical methods of Krylov and Bogoliubov, since they give in evidence the transient process of sustained oscillations resulting from small amplitude perturbations.[1] It has been shown by Popov [4.7] that if only the first approximation in the linearization of a nonlinear system is considered, the describing function technique can be advantageously applied to obtain similar results as achieved by the Krylov-Bogoliubov methods, but with ease, especially in higher-order systems. Popov investigated the relative stability of periodic oscillations by evaluating the effects of small amplitude variations on the imaginary roots of the characteristic equation of the linearized system. The degree of relative stability, which indicates how rapidly the stable oscillations are established, is measured by the real part of the perturbed imaginary roots. This idea will be exploited in the following developments, and some extensions will be made to consider the related problem of sensitivity analysis [4.2].

By using the incremental input describing function proposed by West [4.8] and the type of incremental input signal suggested by Cosgriff [4.9], Lonenn [4.10] presented a graphical procedure for determining the relative stability of sustained oscillations. The procedure, which is based upon the applications of Nyquist diagrams, is conveniently applied to one-loop nonlinear control systems for which the corresponding incremental input describing function is readily available.

Tsypkin [4.11] proposed an exact method for stability analysis of sustained oscillations based upon Liapunov's work. A significant contribution in the Tsypkin method is that the stability problem of periodic oscillations in relay systems is reduced to the stability problem of the corresponding linear sampled-data systems, which can be solved by known methods. The approach, however, may become complicated when the sustained oscillations are not simple, and it does not apply to continuous nonlinearities.

Neither the incremental input describing function concept nor the Tsypkin approach is discussed here. However, there is a possibility of using these methods along with the parameter plane analysis to achieve some further extensions with new and practical results. In particular, a study of sustained oscillations may be performed by the Tsypkin method and the sampled-data analysis in the parameter plane [4.12].

The approximate method for the stability analysis of periodic oscillations is closely related to the previous results in the parameter plane consideration of nonlinear systems. It is based upon the sensitivity analysis outlined in Section 1.7 and reference 4.2, and it represents an analytical approach to the problem of stability of perturbed sustained oscillations. In this method the perturbed motion $\Delta x(t)$ is not considered separately, but as a part of the whole

---

[1] This property of the Krylov-Bogoliubov method was discussed in Section 3.6.

Stability and Sensitivity Analysis 163

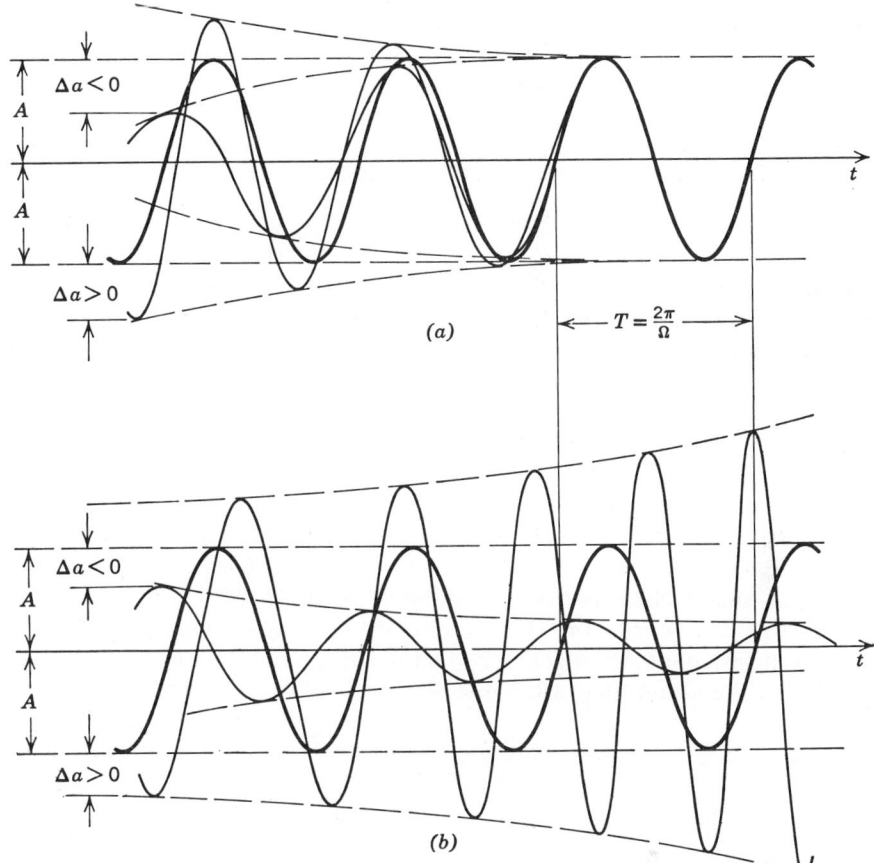

Figure 4.5 Perturbed sustained oscillations: (a) stable; (b) unstable.

solution $x = x^* + \Delta x$, $x^* = A \sin \Omega t$.[2] Thus the solution $x = x(t)$ represents nonstationary oscillations that are close to the sinusoidal oscillations for small amplitude perturbation Δa of the periodic solution x^*. Either the nonstationary oscillations are damped and converge to the periodic solution x^* (Figure 4.5a), or they are undamped and diverge from the solution x^* (Figure 4.5b) for both positive and negative values of Δa. The stability problem is then to determine whether or not the perturbed oscillations converge to the periodic solution x^*.

---

[2] The starred notation x^* for the periodic solution is used since $x = x(t)$ represents the complete transient process that arises after small amplitude perturbations of the periodic oscillations x^*.

Since the perturbed oscillations related to the nonlinear differential equation

$$B(s)x + C(s)F(x, sx) = 0 \qquad (4.1)$$

are close to the corresponding periodic solution $x^* = A \sin \Omega t$, they are properly described by the linearized differential equation

$$\left[B(s) + C(s)\left(N_1 + \frac{N_2}{\omega} s \right) \right] x = 0, \qquad (4.3)$$

provided that the corresponding linear part is stable, that the describing function coefficients N_1 and N_2 and their first derivatives are monotonic functions of a and ω, and that all the roots of the corresponding characteristic equation

$$B(s) + C(s)\left(N_1 + \frac{N_2}{\omega} s \right) = 0, \qquad (4.6)$$

besides the one (or two) on the imaginary axis, are located inside the left half of the s plane. The perturbed solution may then be considered as $x = a \sin \omega t$, where $a = a(t)$ and $\omega = \omega(t)$ are variable in time sufficiently close to the values of A and Ω of the periodic solution $x^* = A \sin \Omega t$.[3]

Under the above conditions, the stability problem of sustained oscillations may be transferred into the algebraic domain and solved by the root sensitivity analysis. To show this, (4.6) is rewritten as

$$\sum_{k=0}^{n} a_k s^k = 0, \qquad (4.7)$$

where $a_k = a_k(a, \omega)$ are

$$a_k = b_k + c_k N_1(a, \omega) + jc_k N_2(a, \omega), \qquad (4.23)$$

provided that $\omega \neq 0$. For $a = A$ and $\omega = \Omega$, (4.7) has a pure imaginary root $s = j\Omega$, and all the other $n - 1$ roots, $s_2, s_3, \ldots, s_n$, are in the left half of the s plane. If for a positive increment Δa of the amplitude a, the pure imaginary

---

[3] Note that algebraic equation 4.6 has, in general, complex coefficients, and the complex roots are not necessarily in conjugate pairs. When the nonlinearity is described by a function $F(x)$, the coefficient $N_2 \equiv 0$ and the coefficients of (4.6) are real. Then the complex roots appear in conjugate pairs only.

In (4.3) and (4.6) the coefficients N_1 and N_2 are considered as functions of a and ω instead of A and Ω. This indicates that the validity of the mentioned equations is assumed in the neighborhood of the periodic solution $A \sin \Omega t$ when the amplitude and frequency become slowly varying functions of time and, therefore, they are denoted by a and ω. This is in accordance with the conclusions derived in Section 3.6, where it is indicated that the functions

$$N_1(a, \omega) = N_1(A, \Omega)\Big|_{\substack{A=a \\ \Omega=\omega}} \qquad N_2(a, \omega) = N_2(A, \Omega)\Big|_{\substack{A=a \\ \Omega=\omega}}$$

and, therefore, the standard expressions of $N_1(A, \Omega)$, $N_2(A, \Omega)$ can be used for $N_1(a, \omega)$, $N_2(a, \omega)$ as derived in Appendix F.

root $j\Omega$ moves into the left half of the s plane to become $\Delta\sigma + j(\Omega + \Delta\omega)$ ($\Delta\sigma < 0$), and the rest of the $n - 1$ roots remain in the left half of the s plane, the linearized differential equation is stable and the corresponding perturbed solution x would converge to the periodic solution x^* (Figure 4.5a, $\Delta a > 0$). If for a negative increment Δa, the root $j\Omega$ moves into the right half of the s plane ($\Delta\sigma > 0$), the linearized equation is unstable and the amplitude a will increase until $\Delta a = 0$ and reaches the value A of the periodic solution x^* (Figure 4.5a, $\Delta a < 0$).

It may now be concluded that if the sensitivity $S_a{}^\sigma$, defined by

$$S_a{}^\sigma \equiv \frac{\partial \sigma}{\partial a}, \tag{4.24}$$

is less than zero for the values $\sigma = 0$, $\omega = \Omega$, and $a = A$, then the periodic solution $x^* = A \sin \Omega t$ represents sustained oscillations. By the same reasoning as given above, it can be shown that a positive value of $S_a{}^\sigma$ indicates unstable periodic oscillations (Figure 4.5b).

The change in the frequency ω because of the small change in the amplitude a is measured by the sensitivity $S_a{}^\omega$ defined as

$$S_a{}^\omega \equiv \frac{\partial \omega}{\partial a}. \tag{4.25}$$

The stability analysis of periodic oscillations may now be performed by determining the sensitivity $S_a{}^\sigma$. The procedure is similar to that of Section 1.7.

By substituting $s = \sigma + j\omega$ into (4.7), it can be rewritten as two equations:

$$\begin{aligned} \sum_{k=0}^{n} a_k X_k &= 0 \\ \sum_{k=0}^{n} a_k Y_k &= 0, \end{aligned} \tag{4.26}$$

where $a_k = a_k(a, \omega)$, $X_k = X_k(\sigma, \omega)$, and $Y_k = Y_k(\sigma, \omega)$. By considering a as an independent variable and differentiating (4.26) with respect to a, we obtain

$$\begin{aligned} P_1 S_a{}^\sigma + Q_1 S_a{}^\omega + R_1 &= 0 \\ P_2 S_a{}^\sigma + Q_2 S_a{}^\omega + R_2 &= 0, \end{aligned} \tag{4.27}$$

where

$$\begin{aligned} P_1 &= \sum_{k=0}^{n} a_k \frac{\partial X_k}{\partial \sigma}, & Q_1 &= \sum_{k=0}^{n} \left(\frac{\partial a_k}{\partial \omega} X_k + a_k \frac{\partial X_k}{\partial \omega} \right) \\ P_2 &= \sum_{k=0}^{n} a_k \frac{\partial Y_k}{\partial \sigma}, & Q_2 &= \sum_{k=0}^{n} \left(\frac{\partial a_k}{\partial \omega} Y_k + a_k \frac{\partial Y_k}{\partial \omega} \right) \\ R_1 &= \sum_{k=0}^{n} \frac{\partial a_k}{\partial a} X_k, & R_2 &= \sum_{k=0}^{n} \frac{\partial a_k}{\partial a} Y_k. \end{aligned} \tag{4.28}$$

166 Symmetrical Oscillations

Equations 4.27 may be solved for $S_a{}^\sigma$ and $S_a{}^\omega$ when $\sigma = 0$ and the values of A, Ω are specified. The sign of $S_a{}^\sigma$ indicates the stability of the periodic solution $x^* = A \sin \Omega t$. The numerical values of $S_a{}^\sigma$ and $S_a{}^\omega$ give further information about the degree of relative stability of the sustained oscillations under investigation. The sensitivity $S_a{}^\sigma$ indicates the rate of decay of small amplitude perturbations applied to the sustained oscillations. The greater the value of $S_a{}^\sigma$, the better the relative stability of the oscillations. The sensitivity $S_a{}^\omega$ determines the corresponding change in the frequency of oscillations. Therefore, in nonlinear system analysis when the stable sustained oscillations represent a desired system characteristic (limit-cycling control systems and oscillators), the values of $S_a{}^\sigma$ and $S_a{}^\omega$ may be interpolated along the $\Sigma = 0$ curve. The operating point on the $\Sigma = 0$ curve may therefore be chosen according to the relative stability of the desired stable limit cycle.

It should be noted that the sensitivity analysis may be performed by considering the sensitivities

$$S_a^\zeta \equiv \frac{\partial \zeta}{\partial a} \quad \text{and} \quad S_a^{\omega_n} \equiv \frac{\partial \omega_n}{\partial a} \tag{4.29}$$

as shown in reference 4.2.

To illustrate the sensitivity analysis, consider a class of nonlinear control systems described by

$$B(s)x + C(s)F(x) = 0, \quad s \equiv \frac{d}{dt}, \tag{4.10}$$

where

$$B(s) = \sum_{k=0}^{n} b_k s^k \quad \text{and} \quad C(s) = \sum_{k=0}^{m} c_k s^k,$$

$n > m$, and $F(x)$ is a single-valued nonlinear function of x. By the harmonic linearization $F(x) = N_1 x$, the characteristic equation of the corresponding linearized differential equation is

$$B(s) + C(s)N_1 = 0. \tag{4.13}$$

By substituting $s = \sigma + j\omega$, (4.13) may be rewritten as two equations

$$\begin{aligned} B_1 + C_1 N_1 &= 0 \\ B_2 + C_2 N_1 &= 0 \end{aligned} \tag{4.30}$$

in which

$$\begin{aligned} B_1 &= \sum_{k=0}^{n} b_k X_k, & C_1 &= \sum_{k=0}^{m} c_k X_k \\ B_2 &= \sum_{k=0}^{n} b_k Y_k, & C_2 &= \sum_{k=0}^{m} c_k Y_k \end{aligned} \tag{4.31}$$

Stability and Sensitivity Analysis   167

and $X_k = X_k(\sigma, \omega)$, $Y_k = Y_k(\sigma, \omega)$, $N_1 = N_1(A)$. By differentiating (4.30) with respect to the amplitude a, which is considered as an independent variable, we obtain

$$S_A^\sigma \left(\frac{\partial B_1}{\partial \sigma} + \frac{\partial C_1}{\partial \sigma} N_1 \right) + S_A^\omega \left(\frac{\partial B_1}{\partial \omega} + \frac{\partial C_1}{\partial \omega} N_1 \right) + C_1 \frac{\partial N_1}{\partial a} = 0$$
$$S_A^\sigma \left(\frac{\partial B_2}{\partial \sigma} + \frac{\partial C_2}{\partial \sigma} N_1 \right) + S_A^\omega \left(\frac{\partial B_2}{\partial \omega} + \frac{\partial C_2}{\partial \omega} N_1 \right) + C_2 \frac{\partial N_1}{\partial a} = 0. \quad (4.32)$$

If the Cauchy-Riemann conditions are applied to (4.32), it can be shown that

$$\frac{\partial B_1}{\partial \sigma} + \frac{\partial C_1}{\partial \sigma} N_1 = \frac{\partial B_2}{\partial \omega} + \frac{\partial C_2}{\partial \omega} N_1$$
$$\frac{\partial B_2}{\partial \sigma} + \frac{\partial C_2}{\partial \sigma} N_1 = -\frac{\partial B_1}{\partial \omega} - \frac{\partial C_1}{\partial \omega} N_1, \quad (4.33)$$

which simplifies the analysis and yields

$$P_1 S_a^\sigma - P_2 S_a^\omega + R_1 = 0$$
$$P_2 S_a^\sigma + P_1 S_a^\omega + R_2 = 0, \quad (4.34)$$

where

$$P_1 = \sum_{k=1}^{n} k(b_k + N_1 c_k) X_{k-1}, \quad R_1 = \frac{\partial N_1}{\partial a} \sum_{k=0}^{m} c_k X_k$$
$$P_2 = \sum_{k=1}^{n} k(b_k + N_1 c_k) Y_{k-1}, \quad R_2 = \frac{\partial N_1}{\partial a} \sum_{k=0}^{m} c_k Y_k. \quad (4.35)$$

In deriving (4.34) and (4.35), the relationship

$$\frac{\partial X_k}{\partial \sigma} = k X_{k-1}, \quad \frac{\partial Y_k}{\partial \sigma} = k Y_{k-1} \quad (4.36)$$

has been used, as proven in Appendix A.

Equations 4.34 represent two equations in two unknowns, S_a^σ and S_a^ω, which may be solved to obtain

$$S_a^\sigma = \frac{-P_2 R_2 - P_1 R_1}{P_1^2 + P_2^2}$$
$$S_a^\omega = \frac{P_2 R_1 - P_1 R_2}{P_1^2 + P_2^2}. \quad (4.37)$$

If a periodic solution $x^* = A \sin \Omega t$ is found previously, the sign of S_a^σ, which is calculated for $\sigma = 0$, $\omega = \Omega$, $a = A$, indicates the stability of the solution. The numerical procedure is illustrated by the following example. It should be noted that in the example and the following developments, the

Symmetrical Oscillations

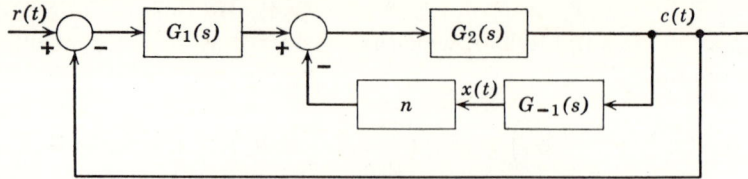

Figure 4.6 System block diagram.

periodic solutions are not given with starred notations unless omitting the star may lead to confusion.

Consider a control system with the block diagram shown in Figure 4.6 with the following transfer functions

$$G_1(s) = K_1, \qquad G_2(s) = \frac{K_2}{s(s+1)(s+2)(s+4)}, \qquad G_{-1}(s) = K_{-1}s \quad (4.38)$$

and the nonlinearity described by a function $F(x)$ given in Figure 4.7. The corresponding linearized system has the characteristic equation

$$s^4 + 7s^3 + 14s^2 + (K_2 K_{-1} N_1 + 8)s + K_1 K_2 = 0. \quad (4.39)$$

If the substitution

$$K_2 K_{-1} N_1 = \alpha$$
$$K_1 K_2 = \beta \quad (4.40)$$

is used, the $\Sigma = 0$ curve is

$$\alpha = 7\Omega^2 - 8 \quad (4.41a)$$
$$\beta = 14\Omega^2 - \Omega^4. \quad (4.41b)$$

The M locus is determined by (4.40) where, in accordance with Figure 4.7, the function $N_1 = N_1(a)$ can be calculated from

$$N_1 = \frac{2k}{\pi}\left[\arcsin\frac{S}{a} + \frac{S}{a}\left(1 - \frac{S^2}{a^2}\right)^{1/2}\right] \quad (4.42)$$

with $k = 1$ and $S = 3$.

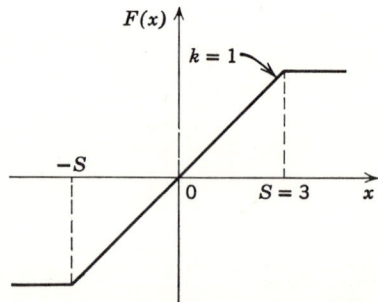

Figure 4.7 Nonlinear characteristic.

Stability and Sensitivity Analysis 169

If the linear parameters K_1, K_2, K_{-1} are specified as $K_1 = 1.5$, $K_2 = 20$, $K_{-1} = 6$, then $\beta = 30$. The M locus is then a straight line $\beta = 30$ in the $\alpha\beta$ plane, and its intersections with the $\Sigma = 0$ curve may be found analytically if $\beta = 30$ is substituted in (4.41b) to obtain

$$\Omega^4 - 14\Omega^2 + 30 = 0. \tag{4.43}$$

There are two intersections corresponding to the two pairs of roots $\Omega_{1,2} = \pm 1.625$ and $\Omega_{3,4} = \pm 3.37$. For these values of Ω, there are two values of α calculated from (4.41a) as $\alpha_1 = 10.5$ and $\alpha_2 = 71.5$. These two values yield two values for N_1 from (4.42), which are 0.0875 and 0.595, respectively. From the diagram of the function $N_1(a)$ that is given in Appendix F (or equation 4.42) for $a = A$, the two corresponding amplitudes are $A_1 = 45$ and $A_2 = 5.3$. It may be concluded that there are two periodic solutions $x_1 = 45 \sin 1.625t$ and $x_2 = 5.3 \sin 3.37t$ that correspond to the two intersections $M_1(10.48; 30)$ and $M_2(71.5; 30)$ of the M locus (equations 4.40) and the $\Sigma = 0$ curve (equations 4.41).

To apply the sensitivity analysis and determine the stability of the periodic solutions x_1 and x_2, it is necessary to check that in both cases all the roots, besides the pure imaginary ones, have negative real parts. Since the imaginary roots are known and the corresponding parameter values are given numerically, by long division process (see Th. A.2 of Appendix A) the characteristic equation can be reduced to a second-degree equation and readily solved for the roots. In doing so, it can be shown that for both points M_1 and M_2 these two remaining roots have negative real parts.

Perform first the sensitivity analysis of the periodic solution $x_1 = 45 \sin 1.625t$. According to (4.39) and (4.42) for $\sigma = 0$, $\omega = \Omega_1 = 1.625$, $a = A_1 = 45$, the coefficients P_1, P_2, R_1, R_2 of (4.34) and (4.35) are calculated as

$$\begin{aligned} P_1 &= \sum_{k=1}^{4} k(b_k + N_1 c_k) X_{k-1} = -47.5 \\ P_2 &= \sum_{k=1}^{4} k(b_k + N_1 c_k) Y_{k-1} = 62.7 \\ R_1 &= C_1 \frac{\partial N_1}{\partial a} = 0 \\ R_2 &= C_2 \frac{\partial N_1}{\partial a} = 0.03. \end{aligned} \tag{4.44}$$

By solving (4.34) for these values of P_1, P_2, R_1, R_2, the sensitivities are

$$S_a^\sigma = 0.31 \times 10^{-3}, \qquad S_a^\omega = -0.235 \times 10^{-3}. \tag{4.45}$$

Since the sensitivity $S_a^\sigma > 0$, the periodic solution $x_1 = 45 \sin 1.625t$ represents unstable periodic oscillations. This is demonstrated by the computer

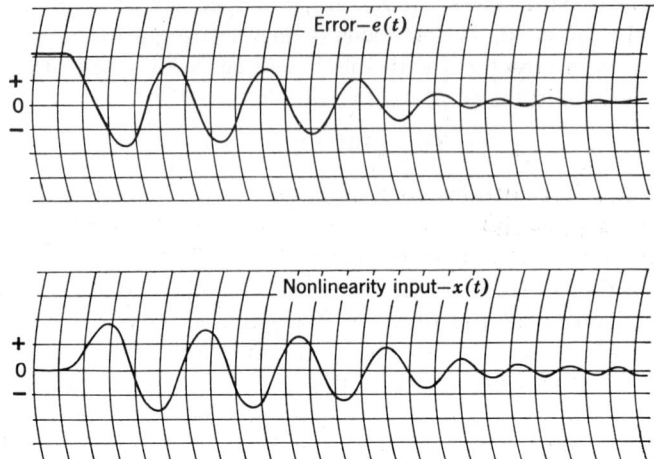

Figure 4.8 Error signal e and nonlinearity input x for 5.8 step input r. Scales: e is 2.5 units/line; x is 25 units/line; time is 1 sec/division.

simulation on Figure 4.8, where the nonlinearity input x and the error signal e are shown for a step input of 5.8 units. Although the amplitude of the signal x reaches the value $A_1 = 45$ of the periodic solution $45 \sin 1.625t$, it is damped down to a lower-level limit cycle. For a step input of 6.22 units, the amplitude of the signal x exceeds the value $A = 45$ and infinitely increases as shown in Figure 4.9. Therefore the periodic solution $x_1 = 45 \sin 1.625t$ is unstable.

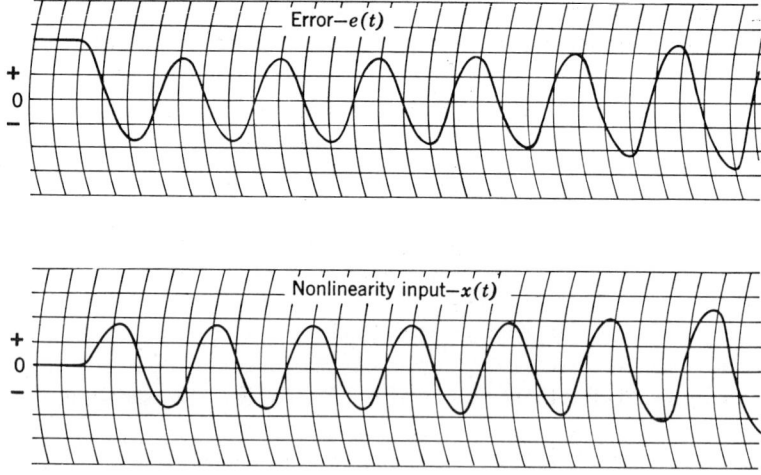

Figure 4.9 Error signal e and nonlinearity input x for 6.22 step input r. Scales: e is 2.5 units/line; x is 25 units/line; time is 1 sec/division.

Stability and Sensitivity Analysis   171

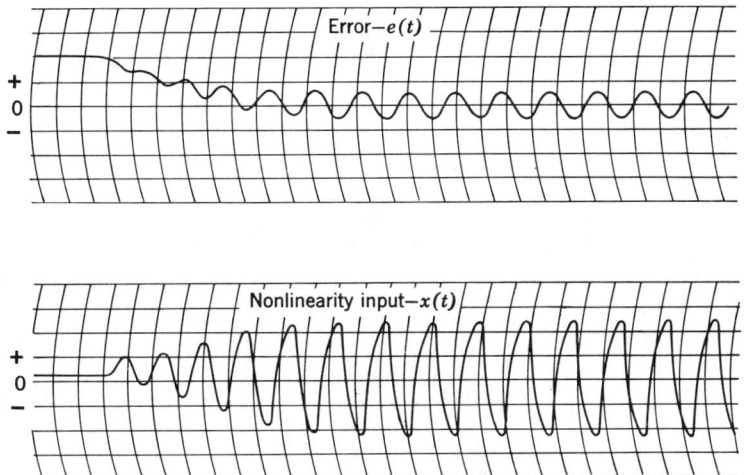

Figure 4.10   Error signal e and nonlinearity input x for unit step input r. Scales: e is 0.5 units/line; x is 2.5 units/line; time is 1 sec/division.

If the sensitivity analysis is repeated for the second periodic solution $x_2 = 5.3 \sin 3.37t$, the sensitivities S_a^σ and S_a^ω are obtained as

$$S_a^\sigma = -0.025, \qquad S_a^\omega = -0.235. \tag{4.46}$$

The sensitivity $S_a^\sigma < 0$ indicates that the periodic solution $x_2 = 5.3 \sin 3.37t$ is a stable limit cycle, and this fact can be checked by computer simulation. For amplitudes of the nonlinearity input higher than $A_2 = 5.3$, the signal x converges to the periodic solution, as already shown in Figure 4.8. For a unit-step input the nonlinearity input x and the error signal e are recorded as shown in Figure 4.10. This recorder trace indicates that for amplitudes smaller than $A_2 = 5.3$ the nonlinearity input tends to the periodic solution $x_2 = 5.3 \sin 3.37t$. Thus the periodic solution x_2 is stable.

The outlined analysis may be illustrated by the phase-plane portrait corresponding to the error signal e. A computer solution is shown in Figure 4.11. Such diagrams are convenient in system stability analysis since they not only give information about the stability of the limit cycles, but also give in evidence the applicability of the system itself.

Before concluding this section, it should be noted that the shading of the $\Sigma = 0$ curve indicates the stability of the periodic oscillations just as the S_a^σ does, because in the nonlinear analysis they have essentially the same meaning. However, the notion of S_a^σ enables the investigation of the sustained oscillations by an analytical procedure, as shown in the previous example.

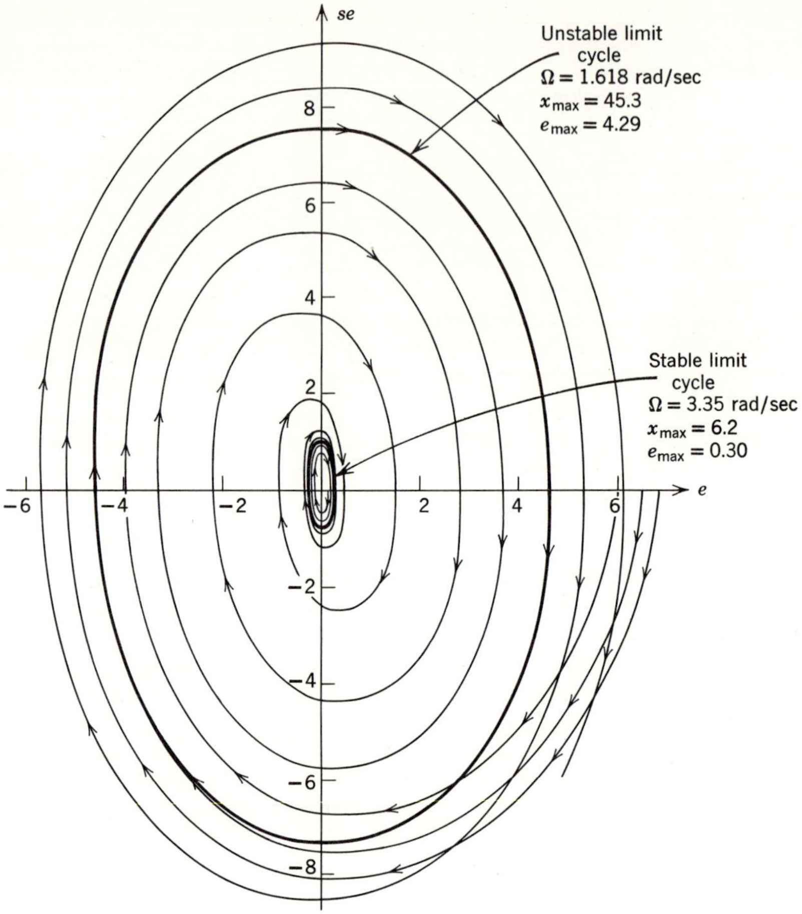

Figure 4.11 Phase-plane portrait corresponding to the error signal e.

4.4 Sensitivity Analysis of Sustained Oscillations for Small Parameter Variations

The previous section considered the stability of periodic oscillations subject to small amplitude perturbations. This part is devoted to the study of the changes in steady-state oscillations, namely, the amplitude and frequency of the oscillations, caused by small changes of system parameters [4.2].

If sustained oscillations can take place in a nonlinear system described by the differential equation

$$B(s)x + C(s)F(x, sx) = 0, \qquad s \equiv \frac{d}{dt}, \qquad (4.1)$$

they can be detected from the characteristic equation

$$B(s) + C(s)\left(N_1 + \frac{N_2}{\Omega}s\right) = 0, \quad (4.6)$$

of the corresponding linearized equation. Since, for sustained oscillations $\sigma = 0$, $\omega = \Omega$, (4.6) may be rewritten as a matrix equation

$$\mathbf{B} + \mathbf{CN} = \mathbf{0}, \quad (4.47)$$

where

$$\mathbf{B} = \begin{bmatrix} B_1 \\ B_2 \end{bmatrix}, \quad \mathbf{C} = \begin{bmatrix} C_1 & -C_2 \\ C_2 & C_1 \end{bmatrix}, \quad \mathbf{N} = \begin{bmatrix} N_1 \\ N_2 \end{bmatrix}, \quad \mathbf{0} = \begin{bmatrix} 0 \\ 0 \end{bmatrix}. \quad (4.48)$$

In (4.48) the corresponding elements are

$$B_1 = \sum_{k=0}^{n} b_k \Omega^k \cos\left(k\frac{\pi}{2}\right), \quad B_2 = \sum_{k=0}^{n} b_k \Omega^k \sin\left(k\frac{\pi}{2}\right)$$

$$C_1 = \sum_{k=0}^{m} c_k \Omega^k \cos\left(k\frac{\pi}{2}\right), \quad C_2 = \sum_{k=0}^{m} c_k \Omega^k \sin\left(k\frac{\pi}{2}\right), \quad (4.49)$$

where, for $\sigma = 0$, $X_k = \Omega^k \cos(k\pi/2)$ and $Y_k = \Omega^k \sin(k\pi/2)$. The trigonometric functions are used to denote the proper sign for Ω^k. In (4.48), $N_1 = N_1(A, \Omega)$ and $N_2 = N_2(A, \Omega)$ are coefficients of the describing function $N = N_1 + jN_2$ given by (4.5) and calculated for various nonlinear functions $F(x, sx)$ in Appendix F.

In general the coefficients b_k and c_k of the polynomials $B(s)$ and $C(s)$ are functions of the system parameters p_i ($i = 1, 2, \ldots, \mu$), that appear in the linear portion of the system and may be referred to as *linear parameters*. Thus

$$\begin{aligned} b_k &= b_k(p_1, p_2, \ldots, p_\mu), \quad (k = 0, 1, \ldots, n) \\ c_k &= c_k(p_1, p_2, \ldots, p_\mu), \quad (k = 0, 1, \ldots, m). \end{aligned} \quad (4.50)$$

The coefficients N_1 and N_2 are considered as functions

$$\begin{aligned} N_1 &= N_1(A, \Omega, q_1, q_2, \ldots, q_\nu) \\ N_2 &= N_2(A, \Omega, q_1, q_2, \ldots, q_\nu), \end{aligned} \quad (4.51)$$

where q_j ($j = 1, 2, \ldots, \nu$) are parameters of the nonlinear device called the *nonlinear parameters*.

Once the sustained oscillations are reached and the corresponding amplitude and frequency have achieved their steady-state values, the sensitivity problem may arise: how do small changes in the system parameters p_i and

q_j affect the steady-state values of the amplitude and frequency of sustained oscillations? Thus it is necessary to define the following sensitivities:[4]

$$S_{p_i}{}^A \equiv \frac{\partial A}{\partial p_i}, \qquad S_{p_i}{}^\Omega \equiv \frac{\partial \Omega}{\partial p_i},$$
$$S_{q_j}{}^A \equiv \frac{\partial A}{\partial q_j}, \qquad S_{q_j}{}^\Omega \equiv \frac{\partial \Omega}{\partial q_j}. \qquad (4.52)$$

In order to evaluate the sensitivities $S_{p_i}{}^A$ and $S_{p_i}{}^\Omega$, which are the amplitude and frequency sensitivities to small variations of linear parameters, it is necessary to differentiate (4.47) with respect to the parameter p_i. Thus

$$S_{p_i}{}^A \left(\mathbf{C} \frac{\partial \mathbf{N}}{\partial A^0} \right) + S_{p_i}{}^\Omega \left(\frac{\partial \mathbf{B}}{\partial \Omega^0} + \frac{\partial \mathbf{C}}{\partial \Omega^0} \mathbf{N} + \mathbf{C} \frac{\partial \mathbf{N}}{\partial \Omega^0} \right) + \frac{\partial \mathbf{B}}{\partial p_i{}^0} + \frac{\partial \mathbf{C}}{\partial p_i{}^0} \mathbf{N} = 0, \qquad (4.53)$$

where the derivatives of $\mathbf{B}$ and $\mathbf{C}$ with respect to Ω are computed readily by the help of (4.49) for the given value of Ω^0. The derivatives $\partial \mathbf{B}/\partial p_i{}^0$ and $\partial \mathbf{C}/\partial p_i{}^0$ are calculated from the same equation for that value of Ω^0 and the given sets of system parameters $p_i{}^0$ ($i = 1, 2, \ldots, \mu$), $q_j{}^0$ ($j = 1, 2, \ldots, \nu$). The derivatives $\partial \mathbf{N}/\partial A^0$ and $\partial \mathbf{N}/\partial \Omega^0$ are evaluated from (4.5) for the given amplitude A^0 and frequency Ω^0 of the sustained oscillations under investigation. The matrix equation 4.53 represents two scalar equations in two unknowns, $S_{p_i}{}^A$ and $S_{p_i}{}^\Omega$, which may be solved for $S_{p_i}{}^A$ and $S_{p_i}{}^\Omega$ numerically.

By differentiating (4.47) with respect to a nonlinear parameter q_j, the following equations result:

$$S_{q_j}{}^A \left(\mathbf{C} \frac{\partial \mathbf{N}}{\partial A} \right) + S_{q_j}{}^\Omega \left(\frac{\partial \mathbf{B}}{\partial \Omega} + \frac{\partial \mathbf{C}}{\partial \Omega} \mathbf{N} + \mathbf{C} \frac{\partial \mathbf{N}}{\partial \Omega} \right) + \mathbf{C} \frac{\partial \mathbf{N}}{\partial q_j} = 0. \qquad (4.54)$$

The values of $S_{q_j}{}^A$ and $S_{q_j}{}^\Omega$ may be calculated similarly from the foregoing matrix equation, as $S_{p_i}{}^A$ and $S_{p_i}{}^\Omega$ can be computed from (4.53), except that the independent term is changed to $\mathbf{C}(\partial \mathbf{N}/\partial q_j)$. This term can be evaluated when the describing function $\mathbf{N} = \mathbf{N}_1 + j\mathbf{N}_2$ is determined.

Before illustrating the procedure for investigating the influence of small parameter variations on sustained oscillations, it should be noted that it is possible to derive a relationship between the sensitivities $S_a{}^\sigma$, $S_a{}^\omega$ on the one hand, and the sensitivities $S_{p_i}{}^A$, $S_{p_i}{}^\Omega$, $S_{q_j}{}^A$, $S_{q_j}{}^\Omega$ on the other. This means that by changing the system parameters, the amplitude and frequency of the sustained oscillations are changed. This in turn introduces changes in the sensitivities $S_a{}^\sigma$ and $S_a{}^\omega$ that are, in general, functions of both the amplitude a and the frequency ω.

---

[4] In actual analysis it may be advisable to use normalized definitions; for example, $S_{p_i}{}^A \equiv \partial \ln A / \partial \ln p_i$. The reader, however, may easily translate the various results given in this text by referring to the normalized definition.

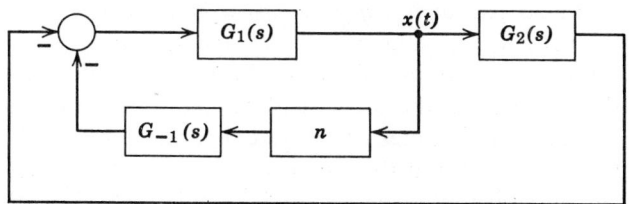

Figure 4.12 System block diagram.

In order to illustrate the sensitivity analysis of small parameter variations, consider the system with a block diagram of Figure 4.12 with the transfer functions

$$G_1(s) = \frac{2}{0.2s^2 + 0.8s + 1}, \quad G_2(s) = \frac{0.5(s+1)}{0.2s + 1} \quad (4.55)$$

$$G_{-1}(s) = \frac{10K_{-1}}{T_{-1}s + 1}.$$

The nonlinearity n has a characteristic shown in Figure 4.13. The corresponding characteristic equation after linearization is

$$0.04T_{-1}s^4 + (0.36T_{-1} + 0.04)s^3 + (2T_{-1} + 0.36)s^2 + (2T_{-1} + 2)s$$
$$+ 2 + (0.4s + 2)K_{-1}N_1 = 0. \quad (4.56)$$

In this equation the coefficients b_k and c_k are functions of two linear parameters $p_1 = T_{-1}$ and $p_2 = K_{-1}$. According to Figure 4.13, the describing function coefficient N_1 is

$$N_1 = k\left[1 - \frac{2}{\pi}\left\{\arcsin\left(\frac{D}{A}\right) + \frac{D}{A}\left[1 - \left(\frac{D}{A}\right)^2\right]^{1/2}\right\}\right], \quad D \le A \le S \quad (4.57a)$$

$$N_1 = k\frac{2}{\pi}\left\{\arcsin\left(\frac{S}{A}\right) - \arcsin\left(\frac{D}{A}\right) + \frac{S}{A}\left[1 - \left(\frac{S}{A}\right)^2\right]^{1/2} - \frac{D}{A}\left[1 - \left(\frac{D}{A}\right)^2\right]^{1/2}\right\},$$
$$A \ge S \quad (4.57b)$$

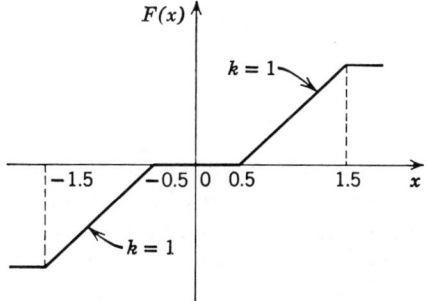

Figure 4.13 Nonlinear characteristic.

where the nonlinear parameters $q_1 = k$, $q_2 = D$, $q_3 = S$. For the parameter values $T_{-1} = 0.16$, $K_{-1} = 7.99$, $k = 1$, $D = 0.5$, and $S = 1.5$, the system exhibits sustained oscillations of an amplitude $A = 2.25$ and a frequency $\Omega = 7.5$ rad/sec. By using the previously defined sensitivities, it is possible to investigate how small changes in the system parameters T_{-1}, K_{-1}, k, D, and S affect the amplitude A and the frequency Ω of the sustained oscillations.

To evaluate, for example, the sensitivities $S_{T_{-1}}^A$ and $S_{T_{-1}}^\Omega$, the following terms of (4.54) should be calculated by applying (4.49) and (4.57):

$$B_1 = -16.2, \quad B_2 = 23.7, \quad C_1 = 32.1, \quad C_2 = 48.2$$

$$\frac{\partial B_1}{\partial \Omega} = 0.56, \quad \frac{\partial B_2}{\partial \Omega} = 14.1, \quad \frac{\partial C_1}{\partial \Omega} = \frac{\partial C_2}{\partial \Omega} = 0$$

$$\frac{\partial B_1}{\partial T_{-1}} = 14.1, \quad \frac{\partial B_2}{\partial T_{-1}} = -137.2, \quad \frac{\partial C_1}{\partial T_{-1}} = \frac{\partial C_2}{\partial T_{-1}} = 0 \qquad (4.58)$$

$$N_1 = 0.5, \quad \frac{\partial N_1}{\partial A} = -0.75.$$

Substituting these numerical values into (4.48) and using (4.54), we obtain

$$S_{T_{-1}}^A = 0.82 \quad \text{and} \quad S_{T_{-1}}^\Omega = 9.6. \qquad (4.59)$$

If the procedure is repeated for the linear parameter K_{-1}, the values

$$S_{K_{-1}}^A = 0.0418 \quad \text{and} \quad S_{K_{-1}}^\Omega = 0 \qquad (4.60)$$

are evaluated. Comparing, then, the obtained results, it may be concluded that small changes in the parameter T_{-1} affect the amplitude and the frequency of the sustained oscillations whereas the changes in the parameter K_{-1} do not influence the frequency at all. Such conclusions are important in the analysis and design of nonlinear systems in which a stable limit cycle is one of the desirable operating conditions.

If a similar procedure as previously outlined is applied to (4.56) and the following derivatives $\partial N_1/\partial k = 0.5$, $\partial N_1/\partial D = -1.27$, $\partial N_1/\partial S = 0.94$ are evaluated, we may obtain

$$S_k^A = -1.68, \quad S_D^A = 1.4, \quad S_S^A = 0.042, \quad S_k^\Omega = S_D^\Omega = S_S^\Omega = 0. \qquad (4.61)$$

These numerical values may be interpreted in a similar manner as the values of sensitivities to linear parameter variations. As can be seen, small variations of the nonlinear parameters do not affect the frequency of the sustained oscillations. This is because the describing function is real and independent of frequency. Thus, for a specified amplitude and frequency of sustained oscillations, any variation in the describing function caused by nonlinear parameter variations can be interpreted as a variation in the gain K_{-1}, which, as has been shown, does not affect the frequency of the oscillations.

4.5 System Stability

Stability of periodic oscillations has been investigated on the basis of small amplitude perturbations. The validity of the linearized equation has been assumed not only at the corresponding periodic solution, but also in the neighborhood of the solution. Then the sensitivity analysis applied to the roots of the corresponding characteristic equation reveals the stability of the related sustained oscillations. It is now of utmost interest to extend the results to the analysis of the behavior of the system and to determine the system stability as a function of both the initial conditions and the system parameters. Before the analysis procedure is presented, some preliminary considerations are necessary.

In approximate analysis by the describing function concept there is no precise formulation of system stability. The stability conditions are, therefore, usually postulated in a descriptive manner to emphasize the practical aspect of the problem. For example, a system may have a small amplitude limit cycle and be considered satisfactorily stable since the external perturbations are sufficiently large so that the system never exhibits the limit cycle. In certain practical applications such a system may be considered satisfactory and a stable limit cycle may even be a required operating characteristic. In another application a stable limit cycle may not be a desirable characteristic of the system. Since the practical aspect is different in various applications, the stability considerations may become ambiguous.

The describing function method, being a linearization technique, suggests an attempt to interpret the stability as in linear system analysis. Because of essential differences between the linear and nonlinear systems, this leads immediately to difficulties. First of all, the stability of a linear system is an inherent feature of the system itself and is not affected by the input signals or initial conditions. The linear system[5] has a unique equilibrium position, and if it is stable the perturbed system returns to that position after an infinite time no matter how large the temporary perturbation is. A nonlinear system may have both stable and unstable equilibrium points and hence may exhibit stable behavior for some values of initial conditions and unstable behavior for others. A second-order nonlinear system with an unstable limit cycle containing an equilibrium state may be considered as stable but, of course, only for the initial conditions chosen in the interior of the limit cycle.

---

[5] The cases with free integrators in the system and the undamped linear oscillator are excluded.

It may be concluded that our intuitive ideas of stability and the notion of stability in linear system analysis are not sufficient to characterize the stability of nonlinear systems. Thus precise stability definitions for nonlinear motion are desirable. Such definitions were introduced at the end of the last century by Liapunov in his fundamental work on stability of motion. Liapunov's stability definitions and related exact methods for nonlinear analysis are considered in Chapter 8.

In a large number of practical cases, however, the mentioned ambiguity about the stability formulation does not seriously affect the application of approximate methods. The interpretation of the stability is based upon the practical results obtained by the analysis, and the methods are only tools for achieving desired system performance characteristics. Thus the success of a method in predicting the system behavior under different operating conditions is the most important factor in nonlinear system design.

In a wide class of nonlinear systems it is desired that the periodic solutions for a certain (or arbitrary) set of initial conditions should not exist, provided that the equilibrium position of the system is stable. This class of nonlinear systems can be analyzed by the proposed method under the condition that, besides the sustained oscillations and equilibrium positions, they have no other characteristics such as separatrix, and so on. It is also assumed that there is only one equilibrium position or zone (for example, when the nonlinearity has a dead zone) of the system, which may be stable or unstable. Two related cases are considered; namely, (a) when the equilibrium position is inside the zone of periodic solutions; (b) when there are no periodic solutions that can be detected by the linearized equation describing the system, and the equilibrium position is unique[6] for the chosen set of system parameters.

In the following stability analysis, attention is focused on the initial amplitudes only. We should be aware of the fact that this may be insufficient for a particular system in question, and that the initial conditions of interest may be given in the phase or state space having a number of dimensions equal to the order of the system. However, in a majority of practical problems the consideration of the stability as a function of the initial amplitudes is satisfactory, at least in the first step of the analysis. This is particularly true when the periodic solutions are present, as was proven by Bogoliubov [4.13] in connection with the conditions of so-called *strong stability*.

---

[6] In fact, if a nonlinear system contains more than one equilibrium position and if these positions are separated by certain corresponding initial conditions and deviations of the related system variables, they can also be separated in the analysis. The stability analysis is applied to each of the equilibrium positions one at a time; however, the stability (or instability) is ensured only for the limited set of initial conditions and deviations of the corresponding system variables. Thus the analysis can be reduced to the case where the nonlinear system has a unique equilibrium position.

It should again be noted that the method used for stability analysis is an approximate method and the results are valid as long as the linearized equations represent adequately the system under investigation. Of course, the analysis may be limited to a certain range of initial conditions and system parameters for which the approximation is satisfactory. The validity of the method and the approximation can be examined a priori by the applicability conditions of the describing function analysis outlined in an earlier section.

When the applicability conditions are not satisfied, the harmonic linearization method may predict stability that does not exist. This is illustrated by an example at the end of Chapter 8, where the method predicts stability in an unstable system exhibiting limit cycles on fundamental and subharmonic frequencies.

We must emphasize that in many cases stability is not sufficient to guarantee satisfactory operation of nonlinear systems for various driving and disturbing influences. Therefore it is important to guarantee a certain degree of stability that can ensure a desired nature of the transient process caused by the external perturbations. This fact leads to the notion of *relative stability* exploited in connection with linear systems (see Chapter 2). The relative stability aspect is considered in Chapter 6 as related to transient nonlinear oscillations.

In the parameter plane the stability of nonlinear systems can be investigated simultaneously with respect to initial amplitudes and system parameters. Therefore it is possible to choose system parameters so that the stability of the system is ensured for a certain set of initial conditions. It may also be possible to choose system parameters so that the system is stable for arbitrary

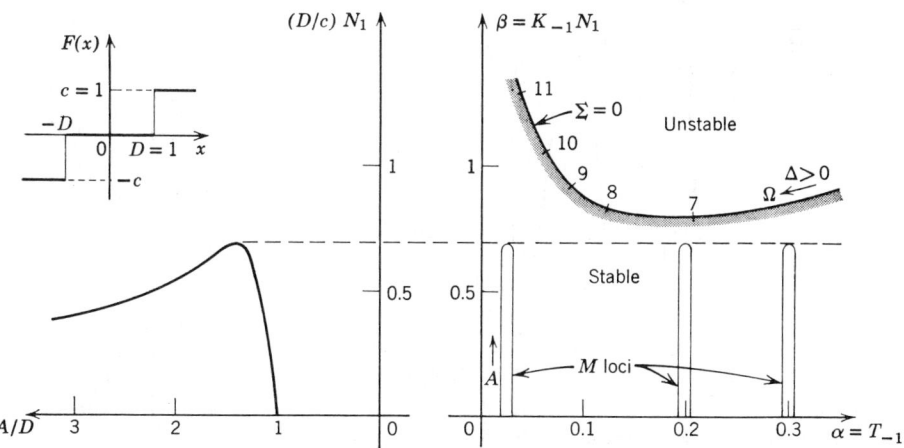

Figure 4.14   Parameter plane diagram for stability analysis.

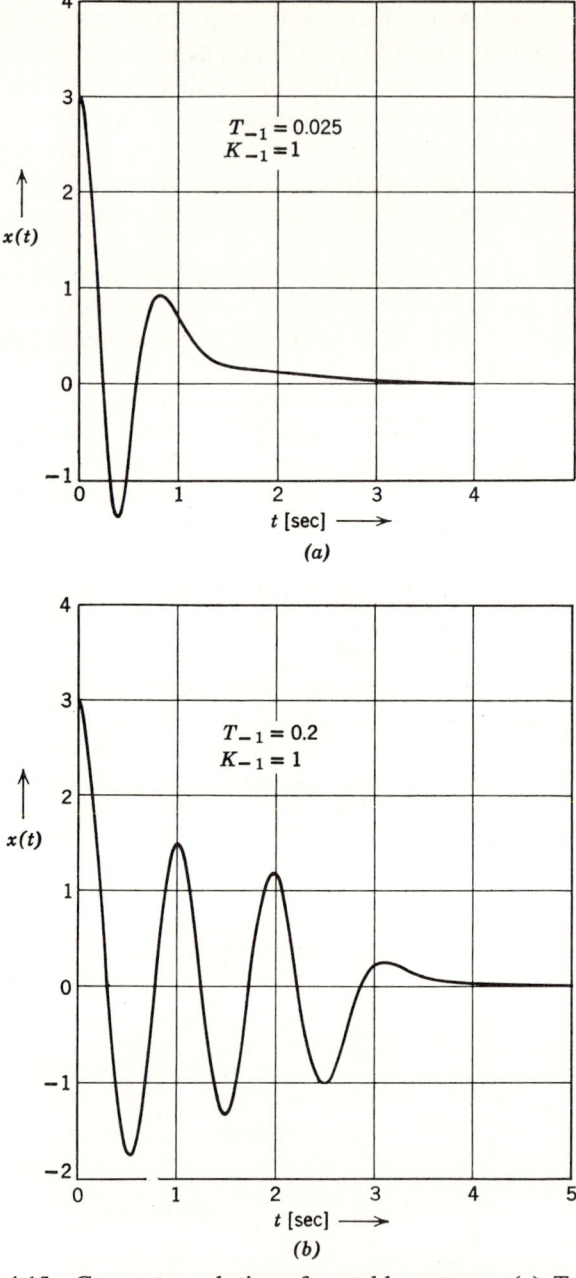

Figure 4.15 Computer solutions for stable systems: (a) $T_{-1} = 0.025$; (b) $T_{-1} = 0.2$; (c) $T_{-1} = 0.3$.

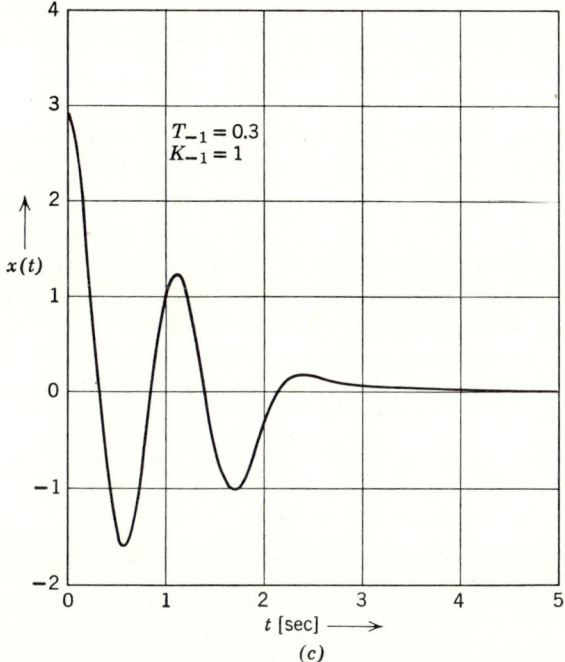

(c)

initial amplitudes provided that the linearization is correct under such conditions.

Consider the first case, in which it is desired to determine the system parameters so that the system is stable for arbitrary initial amplitudes. In terms of parameter plane analysis this problem is readily interpreted as choosing system parameters so that the entire M locus is situated in the stable region R_n. Thus for any initial amplitude there is no intersection with the $\Sigma = 0$ curve, and no periodic solutions are possible. This situation may be illustrated by the system used earlier in connection with the block diagram of Figure 4.12 and the transfer functions of (4.55). The nonlinearity n, in this case, is a relay characteristic with a dead zone shown in Figure 4.14. The characteristic equation of the linearized system and the parameter plane equations are

$$(0.04s^4 + 0.36s^3 + 2s^2 + 2)T_{-1} + (4s + 20)K_{-1}N_1$$
$$+ 0.04s^3 + 0.36s^2 + 2s + 2 = 0 \quad (4.62a)$$

$$T_{-1} = \alpha, \quad K_{-1}N_1 = \beta \quad (4.62b)$$

$$\alpha = \frac{0.064\Omega^2 + 3.2}{0.016\Omega^4 - 0.08\Omega^2 - 4}$$

$$\beta = \frac{0.0016\Omega^6 - 0.003\Omega^4 + 0.256\Omega^2 + 0.4}{0.016\Omega^4 - 0.08\Omega^2 - 4}. \quad (4.62c)$$

182 Symmetrical Oscillations

The variation of $N_1 = N_1(A)$ and the parameter plane diagram are shown in Figure 4.14. As can be seen from Figure 4.14, the entire variation of $N_1(A)$ lies inside the stable region; therefore the system is stable for arbitrary initial conditions. Three particular cases for three different values of system parameters, $K_{-1} = 1$; $\alpha = T_{-1} = 0.025, 0.2$, and 0.3, are simulated on the computer for the initial condition $A_0 = 3$. The results obtained are shown in Figure 4.15, which indicates the stability of the system for the chosen values of system parameters.[7]

The second case refers to the situation where the equilibrium position is stable only for a limited set of initial conditions because of the presence of an unstable limit cycle. This situation was illustrated by Figure 4.5b and is discussed again in the following example.

The system block diagram is shown in Figure 4.16. The relevant transfer functions are

$$G_1(s) = \frac{K_1}{s(s+1)(s+2)(s+4)}, \qquad G_{-1}(s) = K_{-1}s^2, \qquad (4.63)$$

and the nonlinearity n is given in Figure 4.17. Thus the characteristic equation of the linearized differential equation describing the system is

$$s^4 + 7s^3 + (14 + K_1 K_{-1} N_1)s^2 + 8s + K_1 = 0. \qquad (4.64)$$

To determine the $\Sigma = 0$ curve and the stable region R_4 in the parameter plane, substitute $s = j\Omega$, $K_1 K_{-1} N_1 = \alpha$, and $K_1 = \beta$ into (4.64) to obtain

$$\Omega^4 - (14 + \alpha)\Omega^2 + \beta = 0 \qquad (4.65a)$$

$$-7\Omega^3 + 8\Omega = 0. \qquad (4.65b)$$

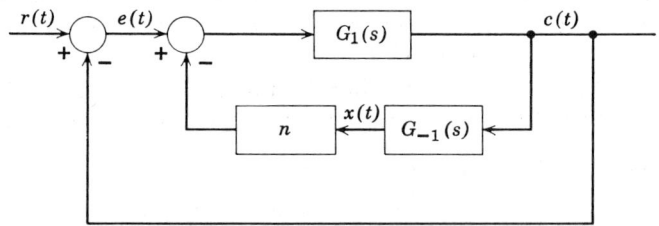

Figure 4.16 System block diagram.

---

[7] These results will be discussed again in Chapter 6, where the relative stability and transient processes are considered.

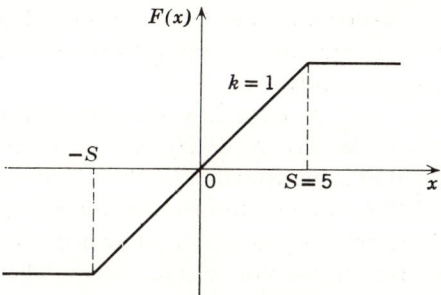

Figure 4.17 Nonlinear characteristic.

It can be observed that equations 4.55 relate to the singular cases in the parameter plane discussed in Derivation A.4 of Appendix A. The $\Sigma = 0$ curve is simultaneously the $\Omega = 1.069$ curve since the values $j\Omega_{1,2} = \pm j1.069$ are roots of (4.65b). Therefore both the $\Sigma = 0$ and the $\Omega = 1.069$ curve are represented by a single straight line in the $\alpha\beta$ plane, determined by (4.65a) for $\Omega = 1.069$. Thus the straight line is

$$-1.14\alpha + \beta - 14.66 = 0, \qquad (4.66)$$

which is the complex root boundary plotted in Figure 4.18. The real root boundary is

$$\beta = 0, \qquad (4.67)$$

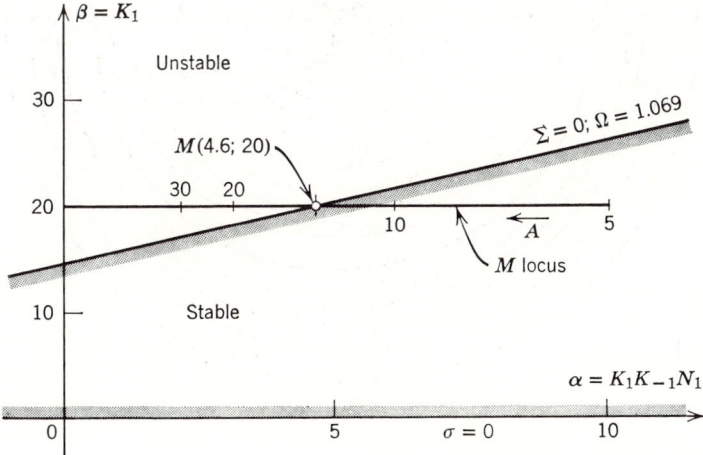

Figure 4.18 Parameter plane diagram.

184 Symmetrical Oscillations

obtained from the characteristic equation by substituting $s = 0$ and $K = \beta$. The boundaries are shaded and the stable region is determined in the usual fashion. The M locus is plotted according to the function $N_1(A)$ associated with the nonlinearity of Figure 4.17.

The intersection $M(4.6; 20)$ of the $\Sigma = 0$ curve and the M locus correspond to an unstable limit cycle. For $K_1 = 20$, $K_{-1} = 0.5$ the limit cycle has an amplitude $A = 10.4$ that is interpolated on the M locus. The frequency of the limit cycle is $\Omega = 1.069$. A computer simulation presented as a phase-plane portrait related to the error signal $e(t)$ is shown in Figure 4.19. The phase-plane portrait illustrates the stability of the system. The equilibrium position

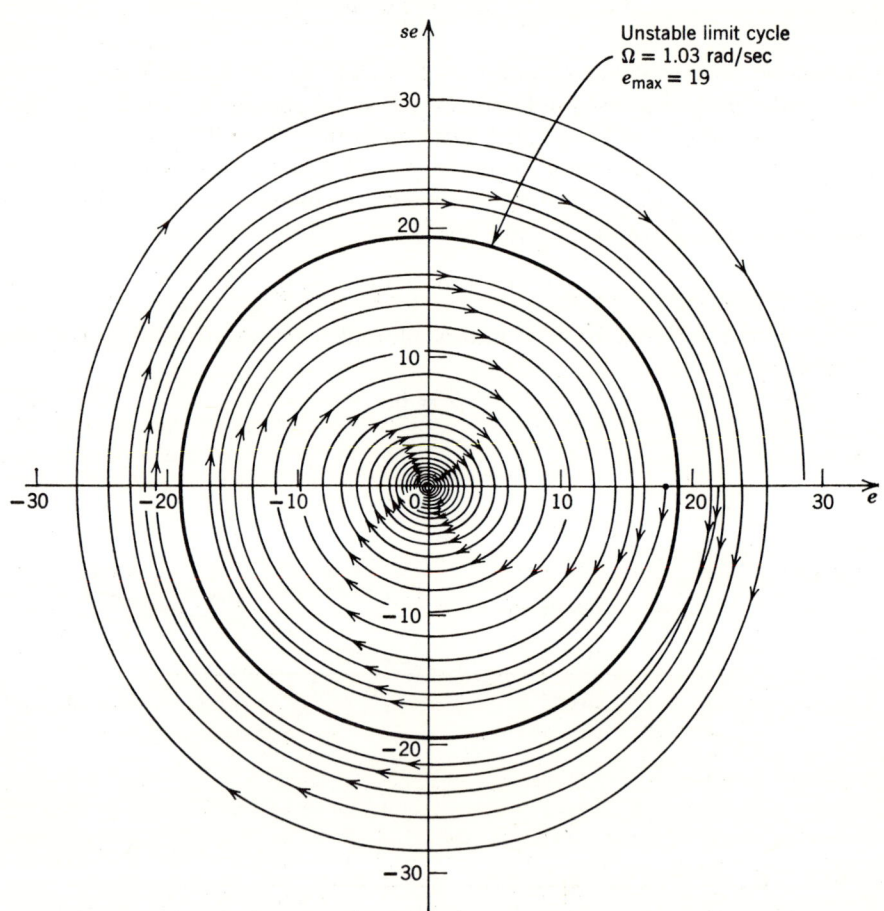

Figure 4.19 Phase-plane portrait of the error signal.

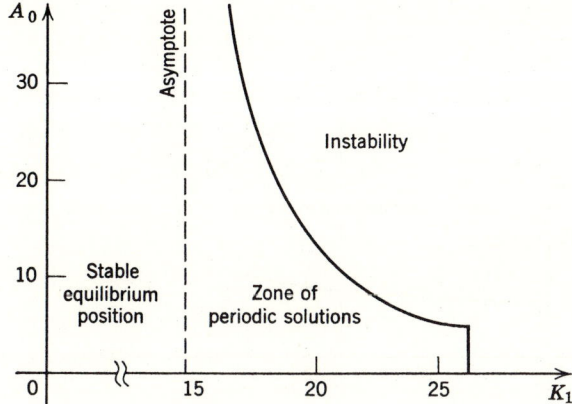

Figure 4.20 Initial amplitude versus system parameter.

is stable and is reached whenever the initial amplitude of $x(t)$, or the corresponding deviation, does not exceed the value of 4.6. In terms of the error signal, the equilibrium is reached under the condition that the error does not exceed the value of 19. This value can be read from the phase-plane portrait of Figure 4.19, in which the traces of the two 18 and 22 step inputs are given.

From the parameter plane diagram it is possible to obtain information about system stability with respect to both the initial conditions and the values of system parameters. By translating the M locus of Figure 4.18 along the β axis and simultaneously reading the amplitude of the unstable limit cycle, which determines the range of the initial conditions for a stable equilibrium, it is possible to obtain the diagram of Figure 4.20. When a value for the system parameter $K_1 = \beta$ is specified, the system is stable whenever an initial amplitude is chosen under the curve plotted in Figure 2.20. Since the curve has an asymptote for $K_1 = 14.66$ along which it goes to infinity, the system is stable for arbitrary initial conditions provided $K_1 \leq 14.66$. If an initial condition is less than $A_0 = 5$, the outlined stability analysis is no longer valid. The excitation is below the saturation limit and the operating region is on the linear part of the nonlinear characteristic of Figure 4.17.[8] Diagrams similar to that of Figure 4.20 can be plotted when the system exhibits several limit cycles. Therefore, in the case of the system analyzed in connection with

---

[8] It may be concluded, however, that the system is stable for $A_0 \leq 5$ if the parameter $\alpha = 20K_{-1}$, ($N_1 = 1$), is chosen so that the point $M(\alpha; 20)$ lies inside the stable region. This can be found by applying the parameter plane analysis to the system regarded as a linear system with $N_1 = 1$.

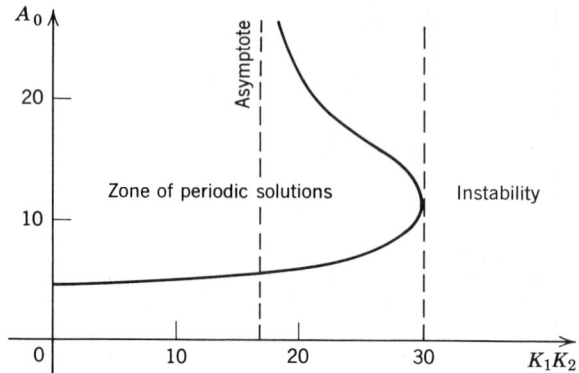

Figure 4.21 Initial conditions versus system parameter.

Figure 4.6, a diagram of initial conditions versus the parameter $\beta = K_1 K_2$ is sketched in Figure 4.21.

In certain cases, such as that illustrated in Figure 4.11, there is no stable equilibrium position because of the stable limit cycle around it. However, the particular system may be considered as satisfactory since the limit cycle has a rather small amplitude (Figure 4.8). In the application, perturbations may be such that the system cannot exhibit the stable limit cycle, and the system can be considered practically stable. Similar reasoning can be applied to the case of Figure 4.4.

4.6 Two Nonlinearities

In Section 3.2 the systems with two nonlinear elements have been conveniently classified into three distinct cases according to the relation between the input signals to the two elements. They are the following:

Case 1: Common input signal to both nonlinear elements (Figure 3.7).

Case 2: Inputs to nonlinear elements related by a linear differential equation (Figure 3.8).

Case 3: Inputs to nonlinear elements related by a nonlinear differential equation (Figure 3.9).

In this section all three cases are considered separately to determine the stability and limit cycles. The analysis is presented by examples.

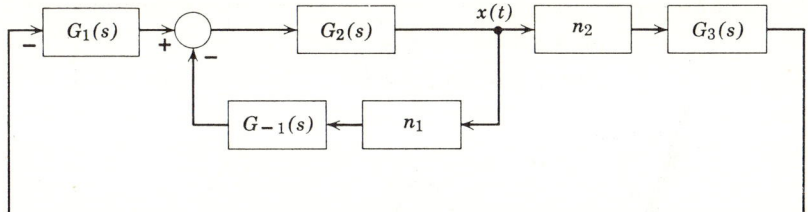

Figure 4.22 System block diagram.

Consider the system of Figure 4.22, which belongs to the first case. The related transfer functions are

$$G_1(s) = \frac{K_1}{0.2s + 1}, \quad G_3(s) = \frac{K_3}{s^2(2s + 1)}$$
$$G_2(s) = \frac{K_2 s}{s + 1}, \quad G_{-1}(s) = \frac{K_{-1}}{0.5s + 1}. \quad (4.68)$$

The two nonlinearities, n_1 and n_2, have piecewise linear characteristics with a dead zone D, saturation S, and slope k of the linear part (No. 4 of Table F.1 in Appendix F).

By denoting the describing functions of the nonlinearities n_1 and n_2 by N_{11} and N_{12}, respectively, and considering N_{11} and N_{12} as variables α and β, the characteristic equation of the linearized system is

$$s^2(0.2s + 1)(2s + 1)K_2 K_{-1}\alpha + (0.5s + 1)K_1 K_2 K_3 \beta$$
$$+ s(0.2s + 1)(0.5s + 1)(s + 1)(2s + 1) = 0. \quad (4.69)$$

The $\zeta = 0$ curve is given by

$$\alpha = \frac{1}{K_2 K_{-1}} \frac{-0.2\Omega^4 - 1.17\Omega^2 + 0.5}{0.1\Omega^4 + 0.85\Omega^2}$$
$$\beta = \frac{1}{K_1 K_2 K_3} \frac{-0.04\Omega^6 - 0.38\Omega^4 + 1.76\Omega^2 + 0.5}{0.1\Omega^2 + 0.85} \quad (4.70)$$

and plotted in Figure 4.23. The stable region is determined in the usual manner. Two M loci are constructed for different values of the ratio S/D, indicated in the upper left corner of Figure 4.23. Two limit cycles are found at the points M_a ($A = 5.2D$; $\Omega = 0.396$ rad/sec) and M_b ($A = 2.92D$; $\Omega = 0.281$ rad/sec).

Another system that belongs to the first case will be considered in Chapter 6, Section 6.5, where the transient oscillations are analyzed. The results will be checked by computer simulation.

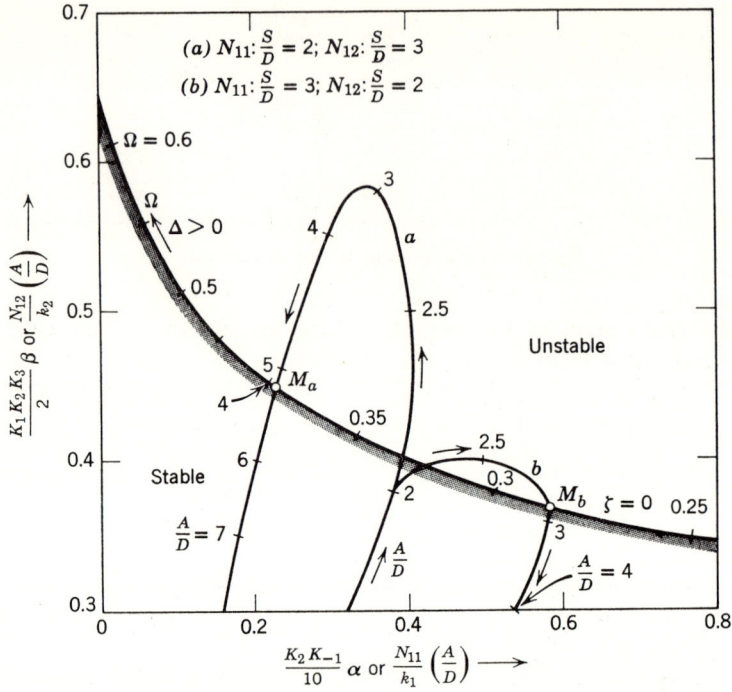

Figure 4.23 Parameter plane diagram for two purely real describing functions.

In Figure 4.24 the block diagram is shown of a nonlinear system with two nonlinearities, n_1 and n_2, and the transfer functions

$$G_1(s) = \frac{20}{s(s+1)(s+2)(s+4)}, \qquad G_{-1}(s) = s. \tag{4.71}$$

The nonlinearities n_1 and n_2 have the characteristics as shown in Figure 4.25.

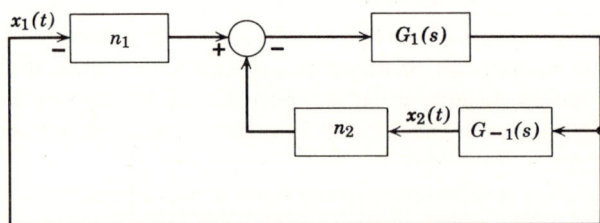

Figure 4.24 System block diagram.

The system belongs to the second case since the two inputs $x_1(t)$ and $x_2(t)$ are related by a linear differential equation, namely,

$$x_2 = -sx_1. \qquad (4.72)$$

Then if the input signal x_1 is assumed as a sinusoid,

$$x_1 = A \sin \Omega t, \qquad (4.73)$$

the signal x_2 is

$$x_2 = A\omega \sin (\Omega t + \pi), \qquad (4.74)$$

where π represents the phase shift which is irrelevant to the analysis. Thus the two describing functions N_{11} and N_{12} corresponding to the two nonlinearities n_1 and n_2 should be calculated for two different amplitudes related by the value of the frequency ω, that is, N_{11} for A and N_{12} for $A\omega$. This affects the plotting of the M locus since the point $M(\alpha; \beta)$, or $M(N_{11}; N_{12})$, is indirectly dependent on both the amplitude A and the frequency ω. Such a dependence for the system under inves-

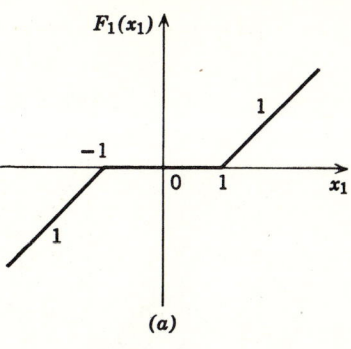
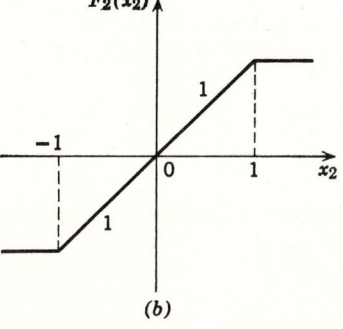

Figure 4.25 Nonlinear characteristics: (a) element n_1; (b) element n_2.

tigation is shown in Figure 4.26. The grid of Figure 4.26 is calculated by using expressions

$$N_{11} = 1 - \frac{2}{\pi}\left\{\arcsin \frac{1}{A} + \frac{1}{A}\left[1 - \left(\frac{1}{A}\right)^2\right]^{1/2}\right\} \qquad (4.75a)$$

$$N_{12} = \frac{2}{\pi}\left\{\arcsin \frac{1}{A\Omega} + \frac{1}{A\Omega}\left[1 - \left(\frac{1}{A\Omega}\right)^2\right]^{1/2}\right\}. \qquad (4.75b)$$

Note that the nonlinearity n_2 has the single-valued characteristic $F(x)$ of Figure 4.25 and the corresponding describing function N_{12} is a function of the input amplitude only. In (4.75b) the frequency Ω appears because the analysis is reduced to the single input $x_1(t)$ of (4.73), and the input amplitude of the signal $x_2(t)$ is $A\Omega$ as indicated in (4.74).

If the grid of Figure 4.26 is superimposed on the parameter plane diagram of Figure 4.27, the M locus is drawn as shown in Figure 4.27. The points of

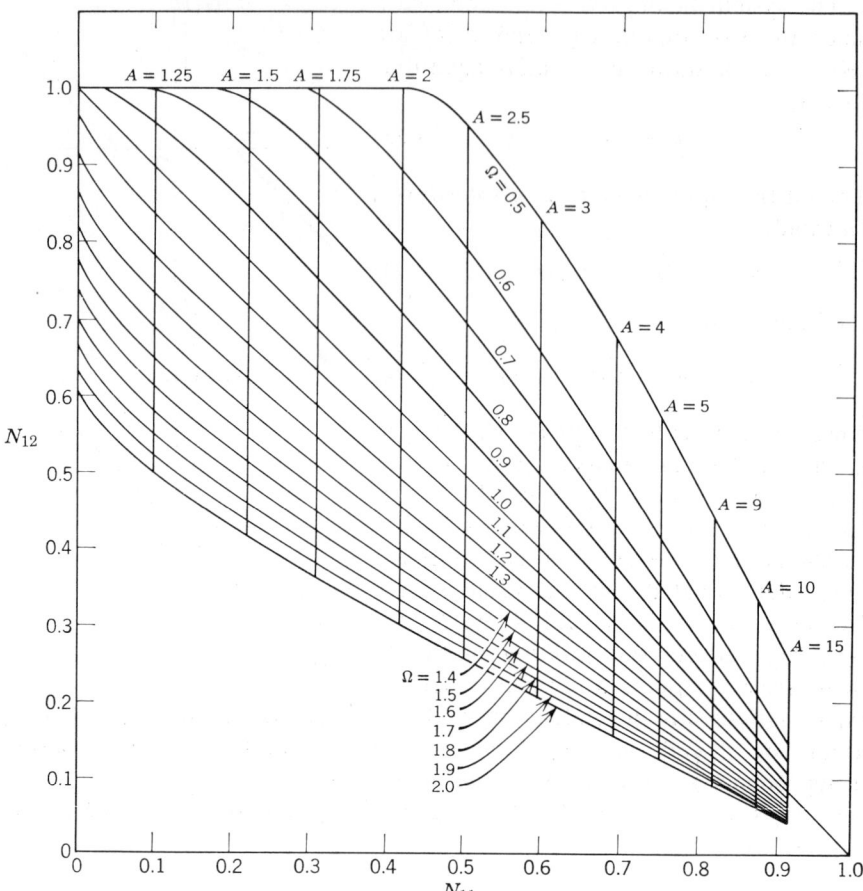

Figure 4.26 The grid for the M locus.

the M locus are chosen so that for a given amplitude A the frequency $\omega = \Omega$ coincides on both the grid and the parameter plane diagram. Of course the analysis is valid only at the intersection of the M locus and the $\zeta = 0$ curve where the periodic solution with amplitude $A = 11.3$ and $\Omega = 1.18$ rad/sec is indicated. This limit cycle is unstable and, for deviations of $x_1(t)$ less than 11.3, the system is stable. This will be checked by computer simulation in Section 6.5, where the diagrams of Figures 4.26 and 4.27 are used to predict the transient process. For that reason, the diagrams are plotted in detail. Otherwise, if only the stability and limit cycle analysis is desired, the diagrams can be largely simplified and the region close to the $\zeta = 0$ curve can be examined.

Two Nonlinearities 191

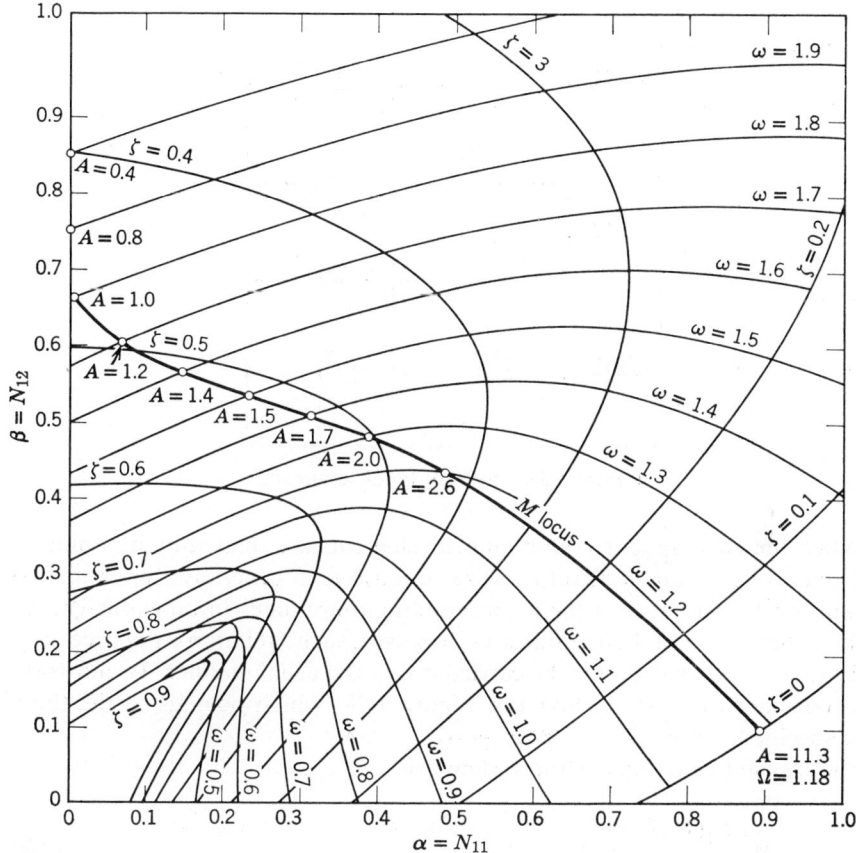

Figure 4.27 Parameter plane diagram.

The procedure can be directly extended to the general situation in which the inputs of the two nonlinearities are

$$x_1 = A_1 \sin \Omega t$$
$$x_2 = A_2 \sin (\Omega t + \psi), \qquad (4.76)$$

and they are related by

$$X_2(s) = G(s) X_1(s). \qquad (4.77)$$

Then the corresponding describing functions $N_{11}(A_1)$ and $N_{12}(A_2)$ are correlated by

$$A_2 = |G(j\Omega)| A_1 \qquad (4.78)$$

and

$$\psi = \arg G(j\Omega). \qquad (4.79)$$

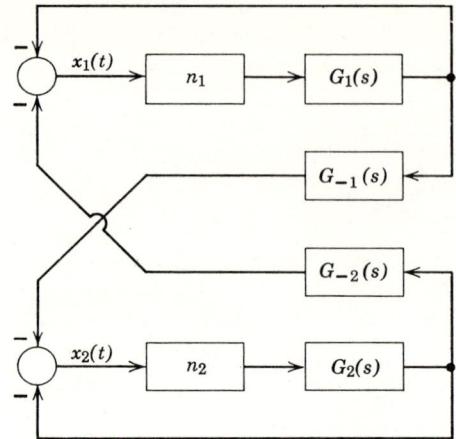

Figure 4.28 System block diagram.

When the two inputs to the nonlinear elements are related by a nonlinear differential equation, difficulties arise because it is generally impossible to obtain a relation between the corresponding amplitudes in an explicit form such as that of (4.78). In certain cases, however, such a relation is not necessary and the stability analysis can be carried out in the usual manner. To illustrate this, consider the system shown in Figure 4.28, which belongs to the third case specified earlier.

The transfer functions of the system are

$$G_1(s) = \frac{K_1}{s(0.01s + 1)(0.05s + 1)}, \quad G_2(s) = \frac{K_2}{s} \quad (4.80)$$
$$G_{-1}(s) = 0.5, \quad G_{-2}(s) = 0.4,$$

and the nonlinear elements n_1 and n_2 have the characteristic of that shown in Figure 4.29a. The characteristic equation of the linearized system is

$$0.0005s^4 + (0.0005\alpha + 0.06)s^3 + (0.06\alpha + 1)s^2 + (\alpha + \beta)s$$
$$+ 0.8\alpha\beta = 0, \quad (4.81)$$

where

$$\alpha = K_2 N_{12}$$
$$\beta = K_1 N_{11}. \quad (4.82)$$

The corresponding $\zeta = 0$ curve is plotted in Figure 4.29b and the stable region is determined in the usual fashion. Since the parameters α and β appear in (4.81) in a product form $\alpha\beta$, the curve $\zeta = 0$ is calculated by using the procedure outlined in Derivation A.2 of Appendix A.

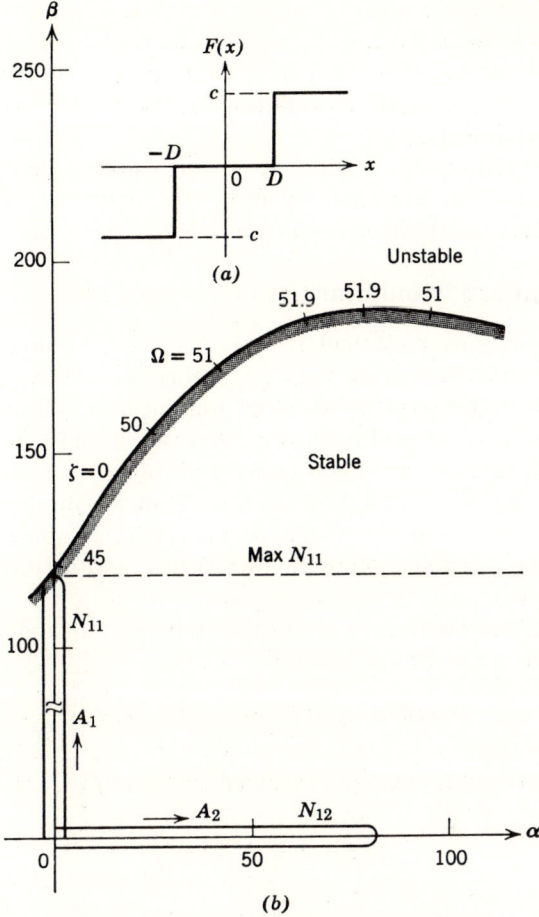

Figure 4.29 Parameter plane diagram.

Since the two signals $x_1(t)$ and $x_2(t)$ are related by a rather complicated nonlinear relation, it is difficult to determine the correlation of the corresponding amplitudes A_1 and A_2 and to plot the locus of the point M [$\alpha = N_{12}(A_2)$; $\beta = N_{11}(A_1)$]. The existence of limit cycles can be predicted, however, without actually plotting the exact M locus. The variations of $N_{11}(A_1)$ and $N_{12}(A_2)$ can be plotted along the corresponding axes β and α independently, as shown in Figure 4.29b. Whenever the system parameters K_1, D_1, and C_1 are chosen so that the entire variation of the describing function $N_{11}(A_1)$ is located under the $\zeta = 0$ curve in the stable region, it is impossible for the M locus to intersect the $\zeta = 0$ curve, whatever the correlation is between A_1 and A_2. Thus there is no limit cycle and the system is stable. Cases may exist, however, in which the M locus exceeds the dotted line max $N_{11}(A_1)$ and the system is still stable since the M locus again lies entirely in the stable region. The analysis of such cases requires a plotting of the M locus as a function of the amplitudes A_1 and A_2.

Before concluding this section, it is worth noting that in Chapter 6 the transient oscillations are analyzed in systems with two nonlinearities. The results obtained prove the analysis presented in this section.

4.7 Multivalued Nonlinearities

When the nonlinear element is described by a nonlinear function $F(x)$, or $F(x, sx)$, which has more than one value for one specified value of its argument, then the term multivalued nonlinearity is used to denote such elements. The class of multivalued nonlinearities that can be considered by the proposed method are continuous and relay characteristics with hysteresis and backlash. The method can be extended to more complex situations such as hysteresis loops for magnetic materials, combinations of backlash and dead zone, certain friction characteristics, and so on. (See Section F.4 of Appendix F.)

If a multivalued nonlinearity is present in the system and it can be described by a nonlinear differential equation

$$B(s)x + C(s)F(x, sx) = 0, \qquad s \equiv \frac{d}{dt}, \qquad (4.1)$$

then the linearization of the multivalued function $F(x, sx)$

$$F(x, sx) = N_1 x + \frac{N_2}{\Omega} sx \qquad (4.4)$$

yields the characteristic equation of the linearized system

$$B(s) + C(s)\left(N_1 + \frac{N_2}{\Omega} s\right) = 0 \qquad (4.6)$$

in the form

$$\sum_{k=0}^{n} a_k s^k = 0. \qquad (4.7)$$

For periodic solutions $x = x(t)$ of (4.1),

$$x = A \sin \Omega t, \qquad (4.2)$$

and thus $s = j\Omega$, the coefficients a_k in (4.7) are expressed as

$$a_k = b_k + c_k N_1 + jc_k N_2. \qquad (4.83)$$

Therefore (4.7) represents an algebraic equation with complex coefficients and, in analyzing (4.7) by the parameter plane method, the results obtained in Section 1.6 can be applied.

If N_1 and N_2 in (4.83) are considered as α and β and the substitution $s = j\Omega$ is used, (4.6) or (4.7) can be rewritten as two equations

$$C_1\alpha - C_2\beta + B_1 = 0$$
$$C_2\alpha + C_1\beta + B_2 = 0, \quad (4.84)$$

where $C(j\Omega) = C_1(\Omega) + jC_2(\Omega)$, $B(j\Omega) = B_1(\Omega) + jB_2(\Omega)$. Solving (4.84) for α and β, we obtain $\alpha = \alpha(\Omega)$, $\beta = \beta(\Omega)$ as

$$\alpha = -\frac{C_1 B_1 + C_2 B_2}{C_1^2 + C_2^2}$$
$$\beta = \frac{C_2 B_1 - C_1 B_2}{C_1^2 + C_2^2}. \quad (4.85)$$

Of certain interest is the fact that the determinant $\Delta = C_1^2 + C_2^2$ of (4.84) is always positive, and the $\Sigma = 0$ curve should be shaded on the left side facing the direction in which Ω increases. Moreover, it can be shown that the $\zeta = 0$ curve is always symmetric with respect to the α axis and $\beta(\Omega) = -\beta(-\Omega)$.

To illustrate the analysis procedure, consider the system of Figure 4.30a

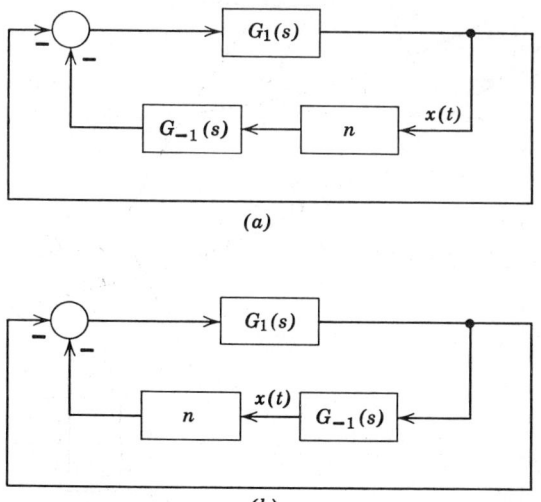

Figure 4.30 System block diagram with different distribution of elements: (a) original system; (b) system with interchanged elements.

196 Symmetrical Oscillations

with the transfer functions

$$G_1(s) = \frac{2}{s(0.4s + 1)(s + 1)}, \quad G_{-1}(s) = \frac{K_{-1}s}{0.5s + 1} \quad (4.86)$$

and the nonlinearity n of Figure 4.31a.

The describing function $N = N_1(A) + jN_2(A)$ of the given nonlinear characteristic is given in Table F.1 of Appendix F. The corresponding characteristic equation for $s = j\Omega$ has the form

$$s(0.4s + 1)(0.5s + 1)(s + 1) + 2(0.5s + 1) + 2(N_1 + jN_2)K_{-1}s = 0. \quad (4.87)$$

The $\zeta = 0$ curve is then

$$\alpha = \frac{1}{2K_{-1}}(1.1\Omega^2 - 2)$$

$$\beta = \frac{1}{2K_{-1}\Omega}(0.2\Omega^4 - 1.9\Omega^2 + 2), \quad (4.88)$$

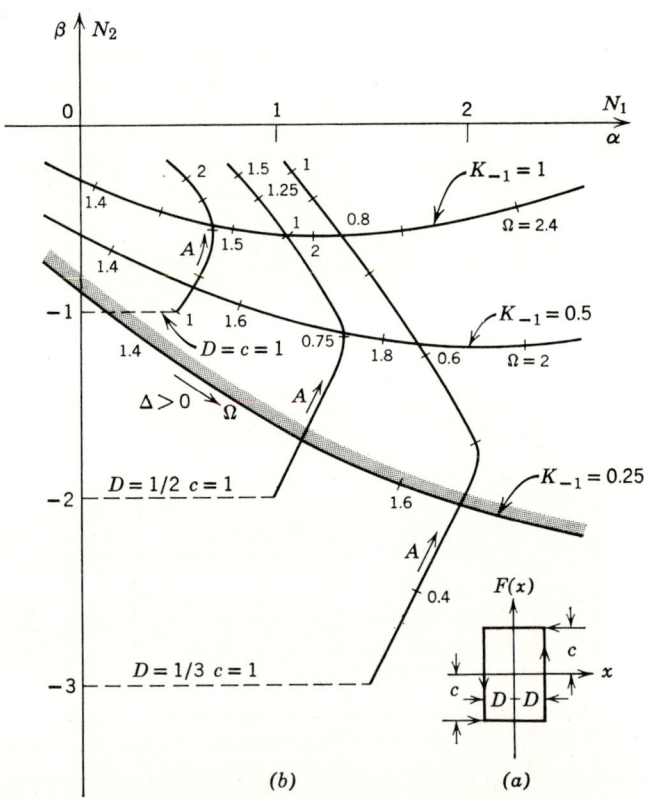

Figure 4.31 Parameter plane diagram for a complex describing function.

which is plotted in Figure 4.31b for various values of the linear parameter $K_{-1} = 0.25, 0.5, 1$. The stable region is above the $\zeta = 0$ curve. The M loci are drawn for different values of the nonlinear parameter $D = \frac{1}{3}, \frac{1}{2}, 1$. The intersections of the M loci and the $\zeta = 0$ curves represent stable limit cycles. The amplitudes of these limit cycles are read on the M locus, and the frequency is to be interpolated on the $\zeta = 0$ curve.

For example, when $K_{-1} = D = c = 1$, the limit cycle is $x = 1.2 \sin 1.52t$. This is simulated on the computer, and the solution obtained is shown in Figure 4.32.

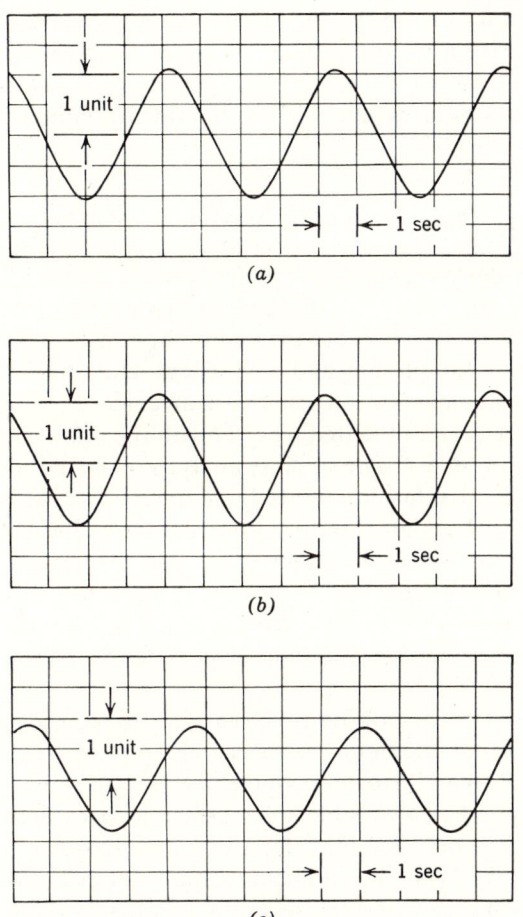

Figure 4.32 Computer solutions: (a) nonlinearity input of the original system; (b) nonlinearity input of the system with interchanged blocks; (c) output of the system with interchanged blocks.

It is of interest to note that if the blocks n and $G_{-1}(s)$ of Figure 4.30a are interchanged as shown in Figure 4.30b, the characteristic equation 4.87 remains the same except that the signal $x(t)$, being the input to the nonlinear element, is carried over to the new position along with the element n. Since the equations are written with respect to the signal $x(t)$, which is the argument of the function $F(x, sx)$, that signal should have approximately the same form for both systems of Figure 4.30. This is demonstrated by the computer solution of Figure 4.32b, which refers to the system of Figure 4.30b. The signals at other points of the system, however, are definitely changed. For example, for the system (Figure 4.30a) the signal $x(t)$ is also the output signal of the system. The output of the system (Figure 4.30b) is related to the input $x(t)$ of the nonlinearity n by the transfer function $G_{-1}(s)$ and, therefore, has a different form as shown in Figure 4.32c.

Another example of a system with multivalued nonlinearity is analyzed in Section 5.5.

4.8 Frequency-Dependent and Variable-Pole Describing Function

The outlined design procedure applies directly to the nonlinearities with both amplitude- and frequency-dependent describing functions. The operations are slightly altered, because the describing function is a function of two variables—that is, amplitude and frequency—and the M loci are usually represented by a family of curves. For convenience in manipulations, the family parameter is often the frequency Ω with the points marked for various values of the amplitude A. The stability analysis in the parameter plane is then performed in the usual fashion by studying the relative location of the M loci and the stability boundary $\zeta = 0$. However, the sustained oscillations are indicated at the intersections of the M loci and the curve $\zeta = 0$ only at the points where the frequency parameter along the curve $\zeta = 0$ has the same value as the frequency for the intersecting M locus. The stability of sustained oscillations is determined in the same manner as described in the preceding sections.

In cases where the describing function may be considered as a variable gain factor of the corresponding transfer function, the frequency dependence gives little or no trouble. However, there are various situations in which the nonlinearity appears as a variable pole of the relevant transfer functions, and the describing function is both amplitude and frequency dependent. The conventional methods are then difficult to apply in any but the simplest design problems. In such cases the analysis in the parameter plane is quite straightforward and it is a powerful tool for stability investigations.

Consider a control system with the block diagram in Figure 4.33. The nonlinearity n is located in the forward path of the inner control loop and is

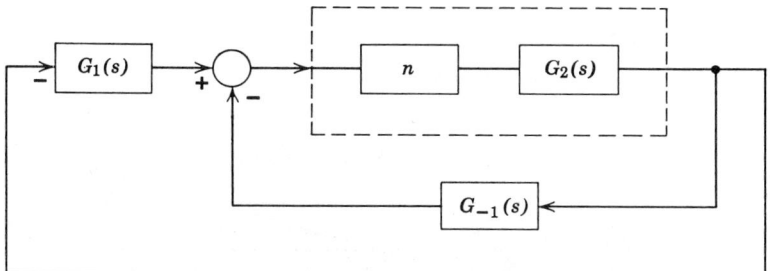

Figure 4.33 System block diagram with friction nonlinearity.

described by the function

$$F(sx) = k_1(sx)^2 \text{ sign } (sx). \tag{4.89}$$

This is a friction characteristic that can be linearized as

$$F(sx) = \frac{8k_1}{3\pi} A\Omega sx. \tag{4.90}$$

Then the transfer function of the linearized forward path in the inner control loop is

$$G_2(s, N_1) = \frac{K_2}{s(T_1s + 1)(T_2s + 1 + N_1)}, \tag{4.91}$$

where

$$N_1 = \left(\frac{8k}{3\Omega}\right) A\Omega. \tag{4.92}$$

This linearization procedure is discussed in detail in Section F.4 of Appendix F.

In applying conventional design techniques difficulties arise because the overall transfer function is $G_2(s, N_1)$, rather than the simple product $N_1 G_2(s)$. In the parameter plane analysis this entails no real difficulty since N_1 can be considered as the variable α or β in the usual manner.

Let the transfer functions of Figure 4.33 be

$$G_1(s) = 0.128$$

$$G_2(s, N_1) = \frac{37.5}{s(0.2s + 1)(2s + 1 + N_1)} \tag{4.93}$$

$$G_{-1}(s) = \frac{K_{-1}s}{s + 1}.$$

The corresponding characteristic equation has the form

$$0.4s^4 + (0.2N_1 + 2.6)s^3 + (1.2N_1 + 3.2)s^2 + (N_1 + 37.5K_{-1} + 5.8)s + 4.8 = 0. \tag{4.94}$$

200 Symmetrical Oscillations

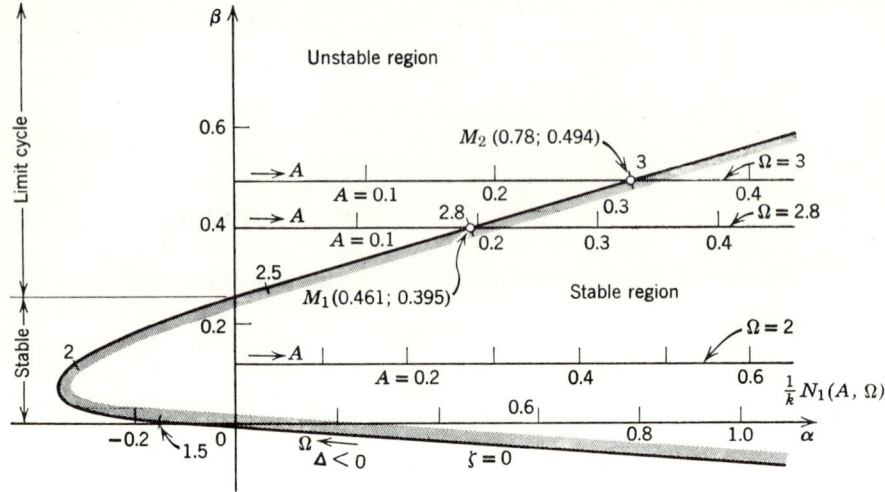

Figure 4.34 Parameter plane diagram for a variable-pole and frequency-dependent describing function.

If $N_1 = \alpha$, $K_{-1} = \beta$, and $s = j\Omega$ are substituted in (4.94), the stability boundary $\zeta = 0$ is obtained as shown in Figure 4.34.

As seen in Figure 4.34, a stable system is obtained whenever the parameter K_{-1} is chosen so that $0 < K_{-1} < 0.256$. When the mentioned parameter exceeds 0.256, however, the stable sustained oscillations occur. It is important to note that, for a given value of the parameter K_{-1}, these oscillations may have only one value of the frequency Ω, which is read on the curve $\zeta = 0$ at the intersection of the curve and the corresponding M locus. Thus the M loci must be plotted so that the family parameter Ω corresponds to the value of Ω at the mentioned intersection of the curve $\zeta = 0$, as shown in Figure 4.34. In doing so two limit cycles are indicated at points M_1 and M_2 with the amplitudes $A_1 = 0.306$ and $A_2 = 0.124$ and the frequencies $\Omega_1 = 3$ radians per second and $\Omega_2 = 2.8$ radians per second, respectively.

References

[4.1] D. D. Šiljak, Analysis and Synthesis of Feedback Control Systems in the Parameter Plane, Part III—Nonlinear Systems, *IEEE Trans.* Pt. II (*Applications and Industry*), **83**, 466–473 (November 1964).
[4.2] D. D. Šiljak and M. R. Stojić, Sensitivity Analysis of Self-Excited Nonlinear Oscillations, *IEEE Trans. Auto. Control*, **AC-10**, No. 4, 413–420 (October 1965).
[4.3] D. D. Šiljak, Generalization of the Parameter Plane Method, *IEEE Trans. Auto. Control*, **AC-11**, No. 1, 63–70 (January 1966).

[4.4] C. A. Pelegrini, An Investigation of Nonlinear System Performance on the Parameter Plane, M.S. Thesis, U.S. Naval Postgraduate School, Monterey, California, 1966.

[4.5] L. Eisenberg, Design of Feedback Control Systems with Transport Lag by Parameter Plane Techniques, Ph.D. Thesis, Newark College of Engineering, Newark, New Jersey, 1966.

[4.6] J. B. Moore, Control System Design Using Extensions of the Parameter Plane Concept, Ph.D. Thesis, University of Santa Clara, Santa Clara, California, 1967.

[4.7] E. P. Popov and I. P. Palitov, *Approximate Methods for Analysis of Nonlinear Automatic Systems* (in Russian), State Press for Physics and Mathematical Literature, Moscow, 1960. (English translation: Foreign Technical Division, AFSC, Wright-Patterson AFB, Ohio, Report FTD-TT-62-910.)

[4.8] U. C. West, *Analytical Techniques for Nonlinear Control Systems*, English University Press, London, 1960.

[4.9] R. L. Cosgriff, *Nonlinear Control Systems*, McGraw-Hill, New York, 1958.

[4.10] Z. Bonenn, Relative Stability of Oscillations in Nonlinear Control Systems, presented at the Second Conference of the International Federation of Automatic Control, Basel, Switzerland, 1963 (Paper No. 214).

[4.11] Ya. Z. Tsypkin, *Theory of Relay Control Systems* (in Russian), State Press for Technical and Theoretical Literature, Moscow, 1955.

[4.12] D. D. Šiljak, Analysis and Synthesis of Feedback Control Systems in the Parameter Plane, Part II—Sampled-Data Systems, *IEEE Trans*. Pt. II (*Applications and Industry*), **83**, 458–466 (November 1964).

[4.13] N. N. Bogoliubov and Yu. A. Mitropolsky, *Asymptotical Methods in the Theory of Nonlinear Oscillations* (in Russian), State Press for Physics and Mathematical Literature, Moscow, 1963. (English translation: Hindustan Publishing Company; Delhi, India; U.S. distr.: Gordon and Breach, New York.)

CHAPTER FIVE

Asymmetrical Oscillations

5.1 Introduction

In certain classes of nonlinear systems the oscillations consist of a limit cycle superimposed on a constant or slowly varying signal. These oscillations are referred to as the *asymmetrical oscillations* because the center of the limit cycle is shifted according to the corresponding value of the constant or slowly varying signal. In general, asymmetrical oscillations may occur when the input-output characteristic of the nonlinearity in the system is not symmetrical about the origin, or when the system is subjected to forcing signals. When the nonlinear characteristic is asymmetric, the output of the nonlinearity may contain a constant term even though the corresponding input is a single sinusoidal wave. If the nonlinear characteristic is symmetric, asymmetrical oscillations can arise whenever the system is subject to forcing input signals. Evidently these oscillations may take place at certain points of the system if both conditions are present. Before the analysis of asymmetrical oscillations in the parameter plane is presented, the previous work and the results in considering these oscillations and related problems are reviewed.

It has been shown first by MacColl [5.1] that the introduction of an external sinusoidal signal at the input to an on-off servomechanism yields a system that behaves like a linear one for small inputs superimposed on the sinusoidal signal. This phenomenon has been later investigated under various names, such as "dither effect," "signal stabilization," and so on. In the following, the term "asymmetrical nonlinear oscillations" is used as the most appropriate term for such phenomena.

In analyzing a carrier-controlled relay servo, Lozier [5.2] has introduced an idea for accomplishing the linearization of the relay by a limit cycle existing in the system and without an external signal. This idea has been further developed by several authors [5.3–9], and a detailed treatment of the problem has been given by Popov and Palitov [5.8]. On the other hand, the external signal application has been developed by Loeb [5.9] and by Oldenburger and his associates [5.10–12]. Oldenburger introduced the name "signal stabilization" to indicate that the nonlinear system is stabilized in the state of sustained oscillations with sufficiently high frequency. The stabilization is actually a consequence of the linearizing effect discovered by MacColl. The concept of signal stabilization has been extended by Sridhar [5.13–14] to the case of a nonlinear system that has one single-valued nonlinearity in the loop, and the stabilizing signal is a stationary random process with a Gaussian distribution and obeys the ergodic hypothesis.

The problem defined above can be treated by dual-input describing functions as proposed by West [5.15]. This approach has been significantly simplified by Boyer [5.16] as outlined by Gibson [5.17]. A similar approach is used by Gelb and Van der Velde [5.18], and significant results have been obtained by Atherton and others [5.19-20], who made a comparison of the utilized concept with the Tsypkin method [5.21].

The study of asymmetrical nonlinear oscillations has been extensively performed in the analysis and design of a large class of plant-adaptive control systems. This class of system is sometimes called *limit-cycling adaptive* because the existing limit cycle is used as an identification signal. Some of the references on this subject are listed here [5.22–26]. A majority of the authors proposed an external sinusoidal signal for identification. More recently, Gelb and Van der Velde [5.18] have examined to a limited extent and in a quantitative manner the properties of self-oscillating adaptive systems that have several advantages over the external adaptation, such as simplicity, cost, reliability, and others. The following analysis of asymmetrical nonlinear oscillations in the parameter plane can be applied directly to self-oscillating adaptive systems.

In the following developments the asymmetrical nonlinear oscillations are analyzed in the parameter plane [5.27]. The control systems with asymmetrical nonlinear characteristics are considered to determine stability and sustained oscillations. The same type of oscillation is investigated in nonlinear control systems subject to constant reference and perturbing input signals. The procedure is further extended to the analysis of the systems with slowly varying input signals. In this case it is shown how a nonlinear characteristic can be modified for the slowly varying signal. The analysis presented is performed with respect to both input signals and the values of adjustable system parameters. The analysis procedure is illustrated by examples in which multiloop feedback structures with several adjustable parameters are considered. In addition, various nonlinear characteristics are used in either the forward or the feedback path. The results obtained are checked by computer simulations that indicate a sufficient accuracy of the presented procedure.

5.2 Basic Developments

Consider a system described by the nonlinear differential equation

$$B(s)x + C(s)F(x, sx) = H(s)f, \quad s \equiv \frac{d}{dt}, \qquad (5.1)$$

where $B(s)$, $C(s)$, and $H(s)$ are polynomials in s and the degree of the polynomial $B(s)$ is greater than the degree of the polynomial $C(s)$ and greater or at least equal to the degree of the polynomial $H(s)$. The function $F(x, sx)$ describes the nonlinearity. Function $f = f(t)$ is a forcing signal, which may be either a reference input or a perturbing signal, and it is assumed to be a constant or a slowly varying function of time.

As a first approximation the steady-state solution $x = x(t)$ of (5.1), which represents the input to the nonlinearity, is assumed to be

$$x = x^0 + x^*, \qquad (5.2)$$

where $x^0 = x^0(t)$ is either a slowly varying function of time or is constant, and x^*, which is

$$x^* = A \sin \phi, \quad \phi = \Omega t + \theta, \qquad (5.3)$$

represents the periodic component of the solution $x(t)$. Since θ in (5.3) merely corresponds to a shift in t, we can put $\theta = 0$ and use $x^* = A \sin \Omega t$.

The forcing function $f(t)$ is considered as a slowly varying function of time if it can be assumed approximately as constant over any cycle of the periodic component x^*; that is,

$$|f(t + T) - f(t)| \ll |f(t)|, \qquad (5.4)$$

where the period $T = 2\pi/\Omega$. In the frequency domain (5.4) means that the frequency Ω of the periodic component x^* is much greater (practically ten times or more) than the highest frequency of the slowly varying component x^0. In this case no harmonic relation between the components x^0 and x^* of the solution x needs to be considered. In that respect the nonlinear system behaves like a linear system and certain phenomena, which may be inherent features of a nonlinear system subject to forcing signals, such as jump resonance, generation of subharmonics, and so on, cannot take place. The forced nonlinear oscillations for which the condition (5.4) is not necessarily satisfied are considered in Chapter 7.

Under the condition (5.4), the values of x^0, A, and Ω, which appear in the solution $x = x^0 + A \sin \Omega t$, are slowly varying quantities in time. This enables the extension of the conventional harmonic linearization in which the describing function is defined for the signal $x = x^0 + x^*$ as an input to the nonlinear element. Thus the nonlinear function $F(x, sx)$ is approximately expressed by the first terms of the Fourier series as

$$F(x, sx) = F^0 + N_1 x^* + \frac{N_2}{\Omega} sx^*, \tag{5.5}$$

where

$$F^0 = \frac{1}{2\pi} \int_0^{2\pi} F(x^0 + A \sin \phi, A\Omega \cos \phi) \, d\phi \tag{5.6a}$$

$$N_1 = \frac{1}{\pi A} \int_0^{2\pi} F(x^0 + A \sin \phi, A\Omega \cos \phi) \sin \phi \, d\phi \tag{5.6b}$$

$$N_2 = \frac{1}{\pi A} \int_0^{2\pi} F(x^0 + A \sin \phi, A\Omega \cos \phi) \cos \phi \, d\phi \tag{5.6c}$$

and $\phi = \Omega t$.

As can be seen from (5.5) and (5.6a), the component F^0 of the output of the nonlinearity $F(x, sx)$ is not considered zero, as was the case in the analysis of symmetrical nonlinear oscillations presented in the previous chapter. This results from the fact that either the nonlinear function $F(x, sx)$ is not symmetric or the system is subject to an external input signal, or that both are present in the system.

According to (5.6) all coefficients F^0, N_1, and N_2 are generally functions of x^0, A, and Ω; that is,

$$F^0 = F^0(x^0, A, \Omega), \quad N_1 = N_1(x^0, A, \Omega), \quad N_2 = N_2(x^0, A, \Omega). \tag{5.7}$$

For a majority of the nonlinear functions $F(x, sx)$ encountered in practical applications, the above functions (5.7) are obtained once and for all and given in Table F.2 of Appendix F.

By applying the linearization of the function $F(x, sx)$ given in (5.5), the solution $x = x^0 + x^*$ of (5.1) can be obtained by considering the following linearized differential equation:

$$B(s)(x^0 + x^*) + C(s)\left(F^0 + N_1 x^* + \frac{N_2}{\Omega} sx^*\right) = H(s)f \tag{5.8}$$

instead of (5.1). If x^0, A, and Ω are slowly varying functions of time as a consequence of the same property associated with the forcing function f, (5.8) can be rewritten as two simultaneous equations corresponding to the slowly varying signal x^0 and the periodic signal x^* as follows:

$$B(s)x^0 + C(s)F^0 = H(s)f \tag{5.9a}$$

$$B(s)x^* + C(s)\left(N_1 x^* + \frac{N_2}{\Omega} sx^*\right) = 0. \tag{5.9b}$$

Equations 5.9, however, cannot be solved independently since they are related to each other by the nonlinear equations 5.7. This fact indicates that the applied linearization preserves the essential feature of nonlinear systems, and the superposition principle from linear analysis is not valid.

An analytical solution of (5.9) is difficult to obtain since F^0 in (5.9a) is usually a transcendental function with respect to x^0. A graphical procedure is presented for solving (5.9) in the parameter plane. A necessary condition for (5.1) to have a solution $x(t)$ close to (5.2) is that the characteristic equation

$$B(s) + C(s)\left(N_1 + \frac{N_2}{\Omega}s\right) = 0, \tag{5.10}$$

corresponding to the linearized differential equation 5.9b, have a pure imaginary root $s = j\Omega$. This is similar to the case of (4.6), considered in the previous chapter, where the symmetrical oscillations were analyzed, except that in (5.10) the describing functions N_1 and N_2 are functions not only of A and Ω but also of x^0, as indicated in (5.7) and (5.8).

By using the parameter plane approach, (5.10) can be solved for α and β as

$$\alpha = \alpha(\Omega)$$
$$\beta = \beta(\Omega), \tag{5.11}$$

where α and β are N_1, N_2 or some of the adjustable system parameters. Equations 5.11 represent the $\Sigma = 0$ (or $\zeta = 0$) curve for which $s = j\Omega$. The $\Sigma = 0$ curve determines the stable region in the $\alpha\beta$ plane in the usual manner. After the stable region is found, the loci of points $M(\alpha; \beta)$ are plotted according to the variations of α and/or β representing N_1 and/or N_2. The M loci incorporates the additional variable x^0, and a family of the loci should be constructed for different values of x^0. Then the stability of the nonlinear system is determined by the relative location of the Σ curve and the M loci, and the limit cycles are found at their intersections. The stability of the limit cycles is determined in the usual manner. This part of the solution process will be best described by the examples that follow.

The presence of a limit cycle in the system can modify the nonlinear characteristic for the slowly varying input signal. In order to determine the modified characteristic, the intersections of the $\Sigma = 0$ curve and the M loci are considered to evaluate the amplitude A and the frequency Ω of the limit cycle as functions of the slowly varying component x^0; that is,

$$A = A(x^0), \qquad \Omega = \Omega(x^0). \tag{5.12}$$

These functions, when substituted into the function $F^0(x^0, A, \Omega)$, yield the modified nonlinear characteristic for the slowly varying signal

$$F^0 = \psi(x^0). \tag{5.13}$$

The function $\psi(x^0)$ is continuous in a limited range of x^0, which indicates the smoothing effect caused by the presence of the limit cycle.

Substitution of (5.13) into (5.9a) gives

$$B(s)x^0 + C(s)\psi(x^0) = H(s)f. \tag{5.14}$$

Equation 5.14 is a nonlinear differential equation in x^0, which can be solved graphically for x^0 after the function $\psi(x^0)$ is obtained. This, in turn, yields the related values of the functions $A(x^0)$ and $\Omega(x^0)$ of (5.12), and the solution $x = x^0 + A \sin \Omega t$ is thereby determined.

The function $\psi(x^0)$ is a continuous function of x^0 and it can be assumed approximately linear for small variations of x^0. Then the stability problem related to (5.14) can be solved by known linear methods. If it is regarded as a nonlinear function of x^0, it can be linearized by harmonic linearization and the results of the previous chapter can be applied.

It should be noted here that the same parameter plane procedure can be used when the right side of (5.1) has more than one forcing function; that is, the right-hand side is expressed by $\sum_{i=1}^{r} H_i(s)f_i$. The solution x, however, must be found by considering all existing inputs simultaneously, since the superposition principle of linear analysis is not valid. Furthermore, if the polynomial $H(s)$ of (5.1) can be factored in the form $sH_1(s)$, the procedure applies to the case in which the rate sf of the function f is considered as a slowly varying signal; that is, $|sf(t + T) - sf(t)| \ll |sf(t)|$.

The graphical procedure presented here can be extended to nonlinear control systems with two nonlinear functions $F_1(x)$ and $F_2(x)$, whereby the following nonlinear differential equation is investigated:

$$B(s)x + C(s)F_1(x) + D(s)F_2(x) = H(s)f. \tag{5.15}$$

In this case a procedure similar to that given in Section 4.6 can be extended to determine the solution $x = x^0 + x^*$.

The general procedure outlined in this section is modified depending on the actual problem involved. These problems may be divided into three major groups: asymmetrical nonlinearities, constant forcing signals, and slowly varying signals. In the following sections each group is considered separately.

5.3 Asymmetrical Nonlinearities

In an autonomous nonlinear system, which is described by the differential equation 5.1 and in which $f \equiv 0$, the asymmetrical oscillations may occur whenever the function $F(x, sx)$ is not symmetric to the origin. Then, under the conditions discussed in the previous section, the system may be described

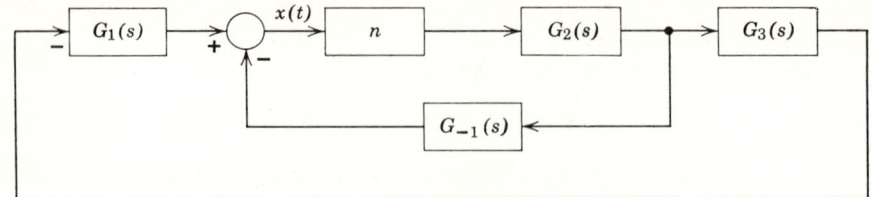

Figure 5.1 System block diagram.

by (5.9) in the form

$$B(0)x^0 + C(0)F^0 = 0 \tag{5.16a}$$

$$\left[B(s) + C(s)\left(N_1 + \frac{N_2}{\Omega}s\right)\right]x^* = 0. \tag{5.16b}$$

In (5.16a), which corresponds to (5.9a), there is no slowly varying forcing function ($f \equiv 0$), and in the steady-state solution $x = x^0 + x^*$ the x^0 is constant and hence s is replaced by zero in $B(s)$ and $C(s)$.

In practical situations $B(0)$ or $C(0)$ can be zero. Also, the nonlinearity in the system is often described by a single-valued function $F(x)$ and $N_2 \equiv 0$. Thus an adjustable parameter appearing in $B(s)$ or $C(s)$ can be chosen as one of the axes in the parameter $\alpha\beta$ plane, while the other axis is related to the describing function coefficient N_1. Some of these situations are discussed in the following examples.

Consider a feedback control system with the block diagram of Figure 5.1 in which the transfer functions are

$$G_1(s) = K_1, \quad G_2(s) = \frac{K_2}{s(s+1)}, \quad G_3(s) = \frac{K_3}{s+2}, \quad G_{-1}(s) = K_{-1}s. \tag{5.17}$$

The nonlinearity n has the form shown in the upper left corner of Figure 5.2. Equations 5.16, for the system under investigation, have the form

$$F^0 = 0 \tag{5.18a}$$

$$\{s(s+1)(s+2) + [K_2K_{-1}s(s+2) + K_1K_2K_3]N_1\}x^* = 0, \tag{5.18b}$$

where, according to the function $F(x)$ of Figure 5.2 and (5.6), we have

$$F^0 = \frac{(1-m)c}{2} + \frac{(1+m)c}{\pi}\arcsin\frac{x^0}{A} \tag{5.19a}$$

$$N_1 = \frac{2(1-m)c}{\pi A}\left[1 - \left(\frac{x^0}{A}\right)^2\right]^{1/2} \tag{5.19b}$$

$$N_2 \equiv 0, \tag{5.19c}$$

Asymmetrical Nonlinearities

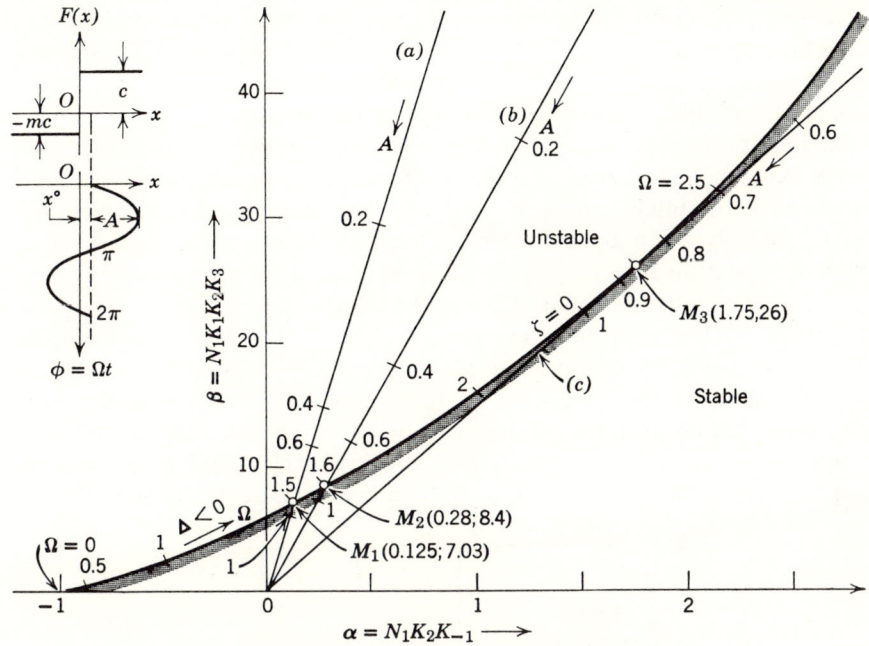

Figure 5.2 Parameter plane diagram.

and $x = x(t)$ is the input signal to the nonlinearity n as indicated in Figure 5.1.

The characteristic equation of (5.18b) is

$$s(s+1)(s+2) + [K_2 K_{-1} s(s+2) + K_1 K_2 K_3]N_1 = 0. \quad (5.20)$$

By denoting $K_2 K_{-1} N_1 = \alpha$ and $K_1 K_2 K_3 N_1 = \beta$, the $\zeta = 0$ curve is obtained as

$$\begin{aligned} \alpha &= \tfrac{1}{2}(\Omega^2 - 2) \\ \beta &= \tfrac{1}{2}\Omega^2(\Omega^2 + 4), \end{aligned} \quad (5.21)$$

and the stable region is determined in the $\alpha\beta$ plane in the usual fashion, as shown in Figure 5.2.

From (5.18a) and (5.19a), we obtain

$$x^0 = A \cos \frac{\pi}{1+m} \quad (5.22)$$

and N_1 of (5.19b) becomes

$$N_1 = \frac{2(1+m)c}{\pi A} \sin \frac{\pi}{1+m}. \quad (5.23)$$

210 Asymmetrical Oscillations

By using (5.23) and the expressions $\alpha = K_2 K_{-1} N_1$, $\beta = K_1 K_2 K_3 N_1$, three M loci, (a), (b), and (c), are drawn in Figure 5.2. They correspond to the parameter values $m = 0.5$, $c = 1$, $K_2 = 1$, and (a) $K_1 K_3 = 7.03$, $K_{-1} = 0.125$; (b) $K_1 K_3 = 8.39$, $K_{-1} = 0.28$; (c) $K_1 K_3 = 26$, $K_{-1} = 1.75$. The stable asymmetrical oscillations are found at the point M_1 and M_2 where the M loci (a) and (b) intersect the $\zeta = 0$ curve. The amplitudes of the oscillations are approximately $A_1 = 0.85$ and $A_2 = 0.8$, which is read from the M loci (a) and (b) at the intersections M_1 and M_2. The corresponding frequencies $\Omega_1 = 1.5$ and $\Omega_2 = 1.6$ are indicated on the $\zeta = 0$ curve. The related value of x^0 in the solution $x = x^0 + A \sin \Omega t$ is calculated for each point M_1 and M_2 using (5.22); thus we obtain $x_1{}^0 = -0.42$ and $x_2{}^0 = -0.39$.

In Figure 5.3 the solution $x_1 = -0.42 + 0.85 \sin 1.5t$ for the case (a) is shown as obtained by a computer simulation. The calculated results are sufficiently close to that obtained by the simulation. From Figure 5.3 it can be seen that an initial condition $x_1(0) = 4.3$ is chosen and the variable $x_1(t)$ approaches a stable limit cycle. That the limit cycle is stable and will be reached by $x_1(t)$ starting from $x_1(0) = 4.3$ can be concluded from the relative

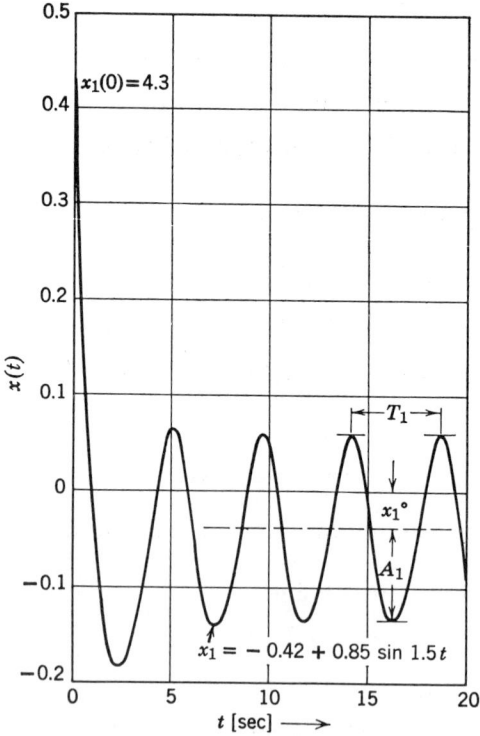

Figure 5.3 Computer solution in case (a).

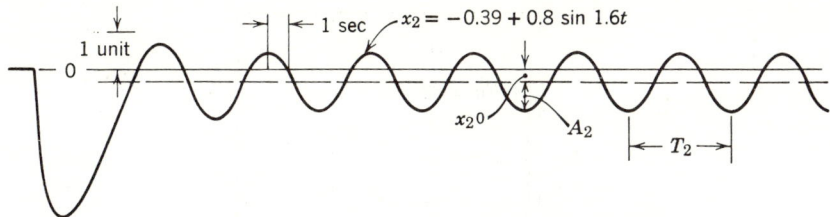

Figure 5.4 Computer solution in case (b).

location of the $\zeta = 0$ curve and the M locus (a), as explained in the preceding chapter on the symmetrical oscillations. The additional component x^0 of the solution $x(t)$ does not alter the stability analysis of the oscillations.

A computer simulation of the case (b) gives the solution $x_2 = -0.39 + 0.8 \sin 1.6t$, as shown in Figure 5.4. A sufficient accuracy is indicated. The initial condition $x_2(0) = 0$ and $x_2(t)$ reaches a limit cycle. This could be concluded from Figure 5.2, as previously noted.

It is of particular interest to consider the case (c) of Figure 5.2. The M locus (c) is tangent to the $\zeta = 0$ curve and corresponds to the ratio $\alpha/\beta = K_1 K_3/K_{-1} = 14.8$. If this ratio is higher than 14.8, there is a limit cycle as shown by cases (a) and (b). On the other hand, if this ratio is less than 14.8, the entire M locus is situated in the stable region and the corresponding system is always stable. The tangent case (c)—$m = 0.5$, $c = 1$, $K_2 = 1$, $K_1 K_3 = 26$, $K_{-1} = 1.75$—is simulated on a computer and the obtained solution $x_3(t)$ is shown in Figure 5.5, which indicates that the system is stable.

In the above example it was possible to express explicitly x^0 as a function of A as shown in (5.23). The solution procedure is thereby reduced to that described in the preceding chapter. In general, however, this is not possible and a different graphical procedure should be used.

5.4 Constant Forcing Signals

When the forcing signal at certain points of a nonlinear system is constant, the solution $x = x^0 + A \sin \Omega t$ (if it exists) will have x^0, A, and Ω as constant values. To determine these values, note that the equations to be solved in the presence of a constant forcing signal f^0 have the form

$$B(0)x^0 + C(0)F^0 = H(0)f^0 \qquad (5.24a)$$

$$\left[B(s) + C(s)\left(N_1 + \frac{N_2}{\Omega} s\right) \right] x^* = 0. \qquad (5.24b)$$

In general, $B(0)$, $C(0)$, and $H(0)$ are constants different from zero, and the solution procedure is somewhat more complicated to perform than in the previous section, where the right side of (5.24a) was zero.

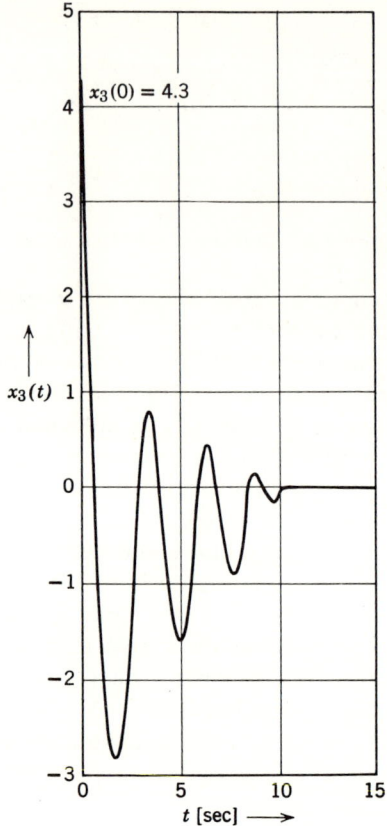

Figure 5.5 Computer solution in case (c).

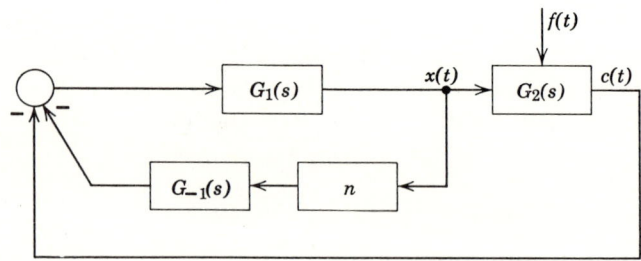

Figure 5.6 System block diagram.

To illustrate the solution procedure, consider a nonlinear feedback system, with the block diagram of Figure 5.6 and the transfer functions

$$G_1(s) = \frac{2}{0.2s^2 + 0.8s + 1}, \quad G_2(s) = \frac{0.5(s+1)}{0.2s+1}, \quad G_{-1}(s) = \frac{K_{-1}}{T_{-1}s+1}. \tag{5.25}$$

The nonlinearity n is given in Figure 5.7a. The input to the system is a perturbation signal $f = f(t)$ that is related to the signal $x = x(t)$ and $c = c(t)$ of Figure 5.6 as

$$(0.2s+1)c = 0.5(s+1)x - f. \tag{5.26}$$

If the perturbation signal is $f(t) = f^0 = \text{const}$, equations 5.24 have the form

$$x^0 + K_{-1}F^0 = f^0 \tag{5.27a}$$

$$(0.04s^4 + 0.36s^3 + 2s^2 + 2s)T_{-1} + (0.4s+2)K_{-1}N_1$$
$$+ 0.04s^3 + 0.36s^2 + 2s + 2 = 0, \tag{5.27b}$$

where (5.27b) represents the characteristic equation of equation 5.24b.

By substituting $T_{-1} = \alpha$ and $K_{-1}N_1 = \beta$, the parameter plane diagram is plotted in Figure 5.7b according to the parameter plane equations

$$\alpha = \frac{0.064\Omega^2 + 3.2}{0.016\Omega^4 - 0.08\Omega^2 - 4}$$

$$\beta = \frac{0.016\Omega^6 - 0.03\Omega^4 + 2.56\Omega^2 + 4}{0.016\Omega^4 - 0.08\Omega^2 - 4}. \tag{5.28}$$

The variation of the M point due to the function $N_1 = N_1(x^0, A)$ given as

$$N_1 = k - \frac{k}{\pi}\left\{\arcsin\frac{D-x^0}{A} + \arcsin\frac{D+x^0}{A}\right.$$
$$\left. + \frac{D-x^0}{A}\left[1 - \left(\frac{D-x^0}{A}\right)^2\right]^{1/2} + \frac{D+x^0}{A}\left[1 - \left(\frac{D+x^0}{A}\right)^2\right]^{1/2}\right\},$$
$$A \geq D + |x^0| \tag{5.29}$$

is plotted in Figure 5.7c. The expression (5.29), which corresponds to the nonlinearity of Figure 5.7, is calculated in Appendix F and given in Table F.2. In order to find a solution $x = x^0 + x^*$ of (5.27), the function $F^0(x^0, A)$ is plotted in Figure 5.7d by using

$$F^0 = \frac{kA}{\pi}\left\{\left[1 - \left(\frac{D-x^0}{A}\right)^2\right]^{1/2} - \left[1 - \left(\frac{D+x^0}{A}\right)^2\right]^{1/2}\right\} + kx^0$$
$$+ \frac{k}{\pi}\left[D\left(\arcsin\frac{D-x^0}{A} - \arcsin\frac{D+x^0}{A}\right)\right.$$
$$\left. - x^0\left(\arcsin\frac{D-x^0}{A} + \arcsin\frac{D+x^0}{A}\right)\right], \quad A \geq D + |x^0|. \tag{5.30}$$

It is assumed that $k = 1$.

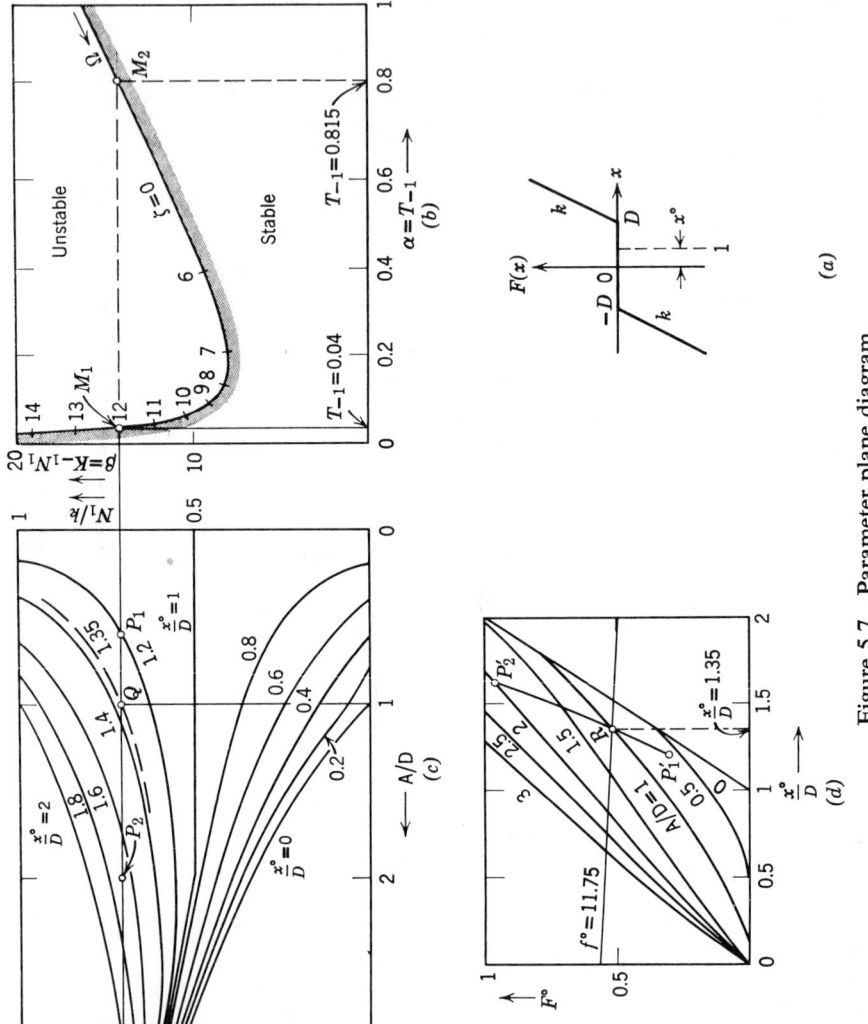

Figure 5.7 Parameter plane diagram.

Constant Forcing Signals

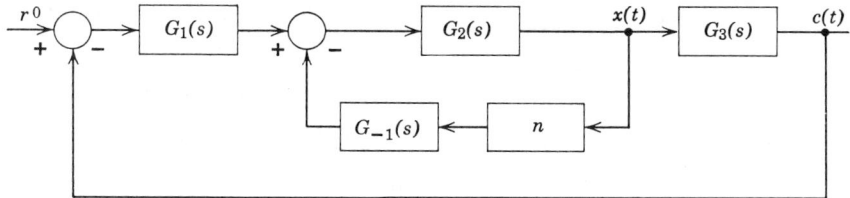

Figure 5.8 System block diagram.

For $T_{-1} = 0.04$, the point $M_1(0.04; 14.3)$ corresponds to a solution $x = x^0 + x^*$ that will have $\Omega = 12$ rad/sec as indicated on the curve $\zeta = 0$. If $K_{-1} = 20$, from M_1 it follows that $N_1 = \beta/K_{-1} = 0.715$. This value of N_1 determines the relationship between the values of x^0 and A for a possible solution x. This relationship, expressed as a function $A = A(x^0)$, can be graphically obtained from the diagram $N_1 = N_1(x^0, A)$ by plotting the straight line P_1P_2 corresponding to the value $N_1 = 0.715$.

The function $A = A(x^0)$ represents the solution of (5.27b) only. The pair of values (x^0, A) which enter into the actual solution of (5.27), is replotted on the diagram $F^0 = F^0(x^0, A)$ of Figure 5.7d into the curve $P_1'P_2'$. Suppose that the constant perturbing signal has a value $f^0 = 11.75$; then (5.27a) determines the straight line $f^0 = 11.75$ plotted in the diagram $F^0 = F^0(x^0, A)$ of Figure 5.7d. The intersection R of that straight line and the curve $P_1'P_2'$ gives the pair (x^0, A) of the solution $x(t)$ that satisfies both parts of (5.27) simultaneously. At this point R the values are $x^0/D = 1.35$ and $A/D = 1$. The same values are obtained at the point Q on the diagram $N_1 = N_1(x^0, A)$ and the solution $x = x^0 + A \sin \Omega t$ of (5.27) is found. If $D = 1$, it is $x = 1.35 + \sin 12t$. Note that the same solution is obtained if the point M_2 of Figure 5.7b is considered, except that the frequency Ω is lower (approximately $\Omega = 5.5$ rad/sec).

Simpler situations may occur if one of the values $B(0)$ or $C(0)$ is zero. To illustrate, consider the nonlinear system of Figure 5.8. The transfer functions are

$$G_1(s) = K_1, \quad G_2(s) = \frac{K_2}{s(s+1)}, \quad G_3(s) = \frac{K_3}{s+2},$$

$$G_{-1}(s) = K_{-1}s \quad (5.31)$$

and the nonlinearity n in the system is given by the function $F(x)$ of Figure 5.9. The input to the system is the reference constant input signal $r(t) = r^0$.

The nonlinear differential equation describing the above system is

$$[s(s+1)(s+2) + K_1K_2K_3]x + K_2K_{-1}s(s+2)F(x) = K_1K_2(s+2)r^0, \quad (5.32)$$

which may be rewritten according to (5.24) as

$$K_1K_2K_3x^0 = 2r^0 \quad (5.33a)$$

$$[s(s+1)(s+2) + K_1K_2K_3 + K_2K_{-1}s(s+2)N_1]x^* = 0. \quad (5.33b)$$

216 Asymmetrical Oscillations

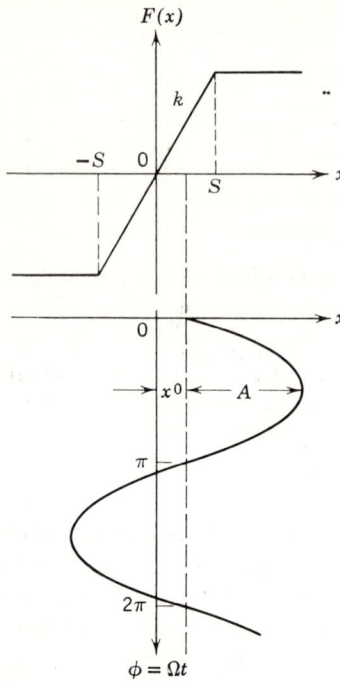

Figure 5.9 Nonlinear characteristic.

The characteristic equation of (5.33b) is evidently

$$s(s+1)(s+2) + K_1 K_2 K_3 + K_2 K_{-1} s(s+2) N_1 = 0. \quad (5.34)$$

By denoting

$$\begin{aligned}\alpha &= N_1 \\ \beta &= K_3,\end{aligned} \quad (5.35)$$

the parameter plane diagram is plotted in Figure 5.10 in the usual fashion. The function $N_1 = N_1(A, x^0)$, which appears as a variation of α in the point $M(\alpha; \beta)$, is plotted in Figure 5.11 by using general formula (5.6b).

From (5.33a), we can derive the following relationship among the input r^0, the constant term x^0, and the parameter $\beta = K_3$:

$$S\beta = \frac{2r^0}{x^0/S}, \quad (5.36)$$

where S is the parameter of the nonlinearity $F(x)$ of Figure 5.9. The function $S\beta$ given in (5.36) is plotted in Figure 5.10.

Now, by using Figures 5.10 and 5.11, it is possible to determine the sustained oscillations and their stability for various values of system parameters K_1, K_2, K_3, K_{-1}, S, k, and the input r^0. For example, if $K_1 = 1$, $K_2 = 10$, $K_3 = 1.75$, $K_{-1} = 1$, $S = 1$, $k = 1$, and $r^0 = 1.1$, then by the values $x^0 = 1.2$, $A = 0.3$, and $\Omega = 2.1$ rad/sec the solution of (5.33) is determined to be approximately

$$x = 1.2 + 0.3 \sin 2.1t. \quad (5.37)$$

For a given value of $\beta = K_3 = 1.75$, $r^0 = 1.1$, and $S = 1$, the value of $x^0 = 1.2$ is read from the left part of Figure 5.10. Then the value of $K_1 K_2 \beta = 17.5$ determines the point $M(1.2; 17.5)$ on the $\zeta = 0$ curve where $\Omega = 2.1$ rad/sec. At this point, $K_2 K_{-1} \alpha = 1.2$, which gives $N_1 = \alpha = 0.12$. Figure 5.11 is used to evaluate the amplitude $A = 0.3$ from the curve $x^0/S = 1.2$. The value $A = 0.3$ is read directly from the diagram $N_1(A, x^0)$ of Figure 5.11, since $K = S = 1$ are the parameters of the given nonlinearity in Figure 5.9.

The solution (5.37) is stable because an increase in the amplitude A causes the point M to move into the stable region and a decrease in the amplitude

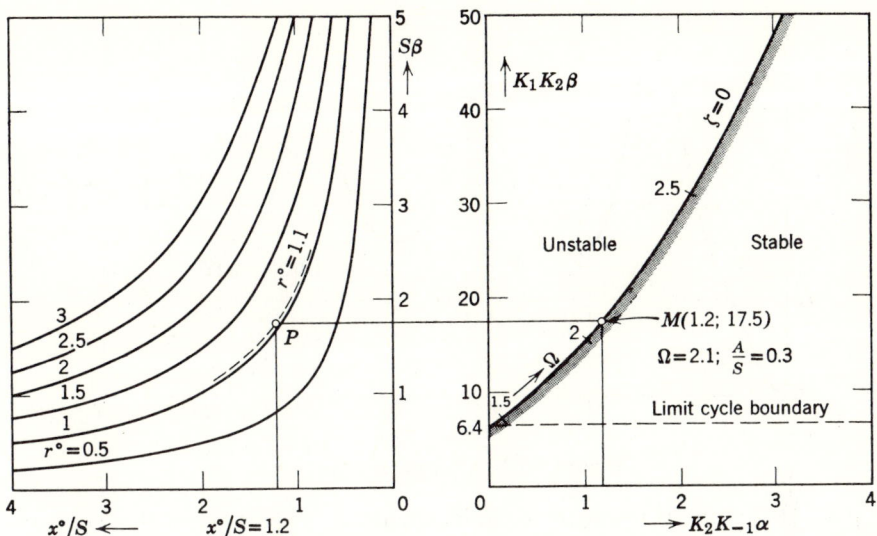

Figure 5.10 Parameter plane diagram.

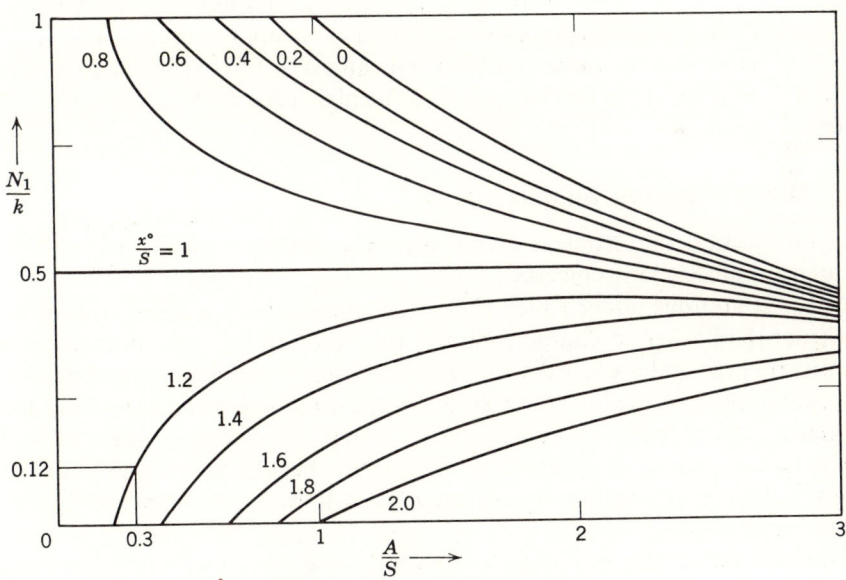

Figure 5.11 Function $N_1(A, x^\circ)$.

217

218 Asymmetrical Oscillations

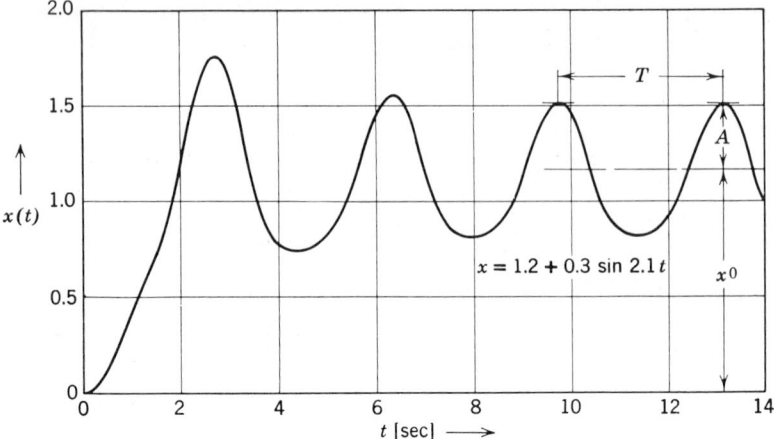

Figure 5.12 Computer solution.

A places the point M inside the unstable region of the parameter plane (Figure 5.10). It is of interest to note that if the product $K_1 K_2 \beta$ where $\beta = K_3$ is such that it is less than 6.4, the system is always stable, because there are no intersections of the variation of the M point and the $\zeta = 0$ curve.

The above solution (5.37) is checked by computer simulation to obtain the curve in Figure 5.12. The accuracy of the calculated solution is sufficiently high and, for calculated values of x^0, A, and Ω, is approximately 10%. On the other hand, the computer solution indicates a distortion of the assumed solution $x = x^0 + A \sin \Omega t$ that is due to the higher harmonics present in the actual solution.

5.5 Slowly Varying Signals

In this section the problem of linearizing a nonlinear system by a high-frequency limit cycle is considered in more detail. The objective is to determine the conditions under which such a linearization is possible and then to construct the linearized characteristic of the nonlinearity. This linearization has several practical aspects discussed in Section 5.1, which are based upon a general property of the linearized system that, for a limited magnitude of the reference signal, behaves like a linear system. Therefore results of the nonlinearities, such as dead-zone, hysteresis, backlash, and so on, are eliminated. The procedure to achieve this will be best illustrated in the following examples.

Consider the system in Figure 5.13 with the transfer functions

$$G_1(s) = K, \quad G_2(s) = \frac{K_2}{s^2 + 0.8s + 1}, \quad G_3(s) = \frac{K_3}{s(s+1)}, \quad G_{-1}(s) = K_{-1}$$

(5.38)

Slowly Varying Signals 219

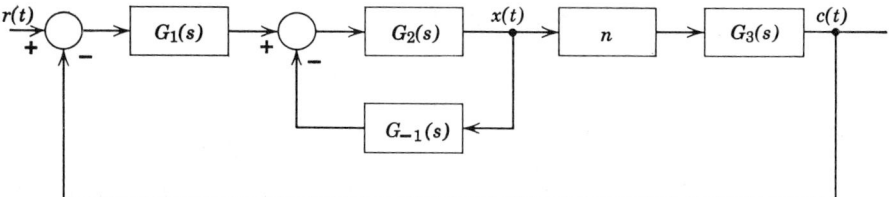

Figure 5.13 System block diagram.

and the nonlinearity n as shown in Figure 5.14. The input to the system is a slowly varying reference signal $r = r(t)$.

The equation that describes the system is

$$[s(s+1)(s^2 + 0.8s + 1) + K_2 K_{-1} s(s+1)]x + K_1 K_2 K_3 F(x) = K_1 K_2 s(s+1)r \tag{5.39}$$

where the signal $x = x(t)$ is the input to the nonlinearity. Equation 5.39 can be rewritten in terms of (5.9) as

$$[s(s+1)(s^2 + 0.8s + 1) + K_2 K_{-1} s(s+1)]x^0 + K_1 K_2 K_3 F^0 = K_1 K_2 s(s+1)r$$

$$\{[s(s+1)(s^2 + 0.8s + 1) + K_2 K_{-1} s(s+1)] + K_1 K_2 K_3 N_1\}x^* = 0. \tag{5.40}$$

The characteristic equation of the second equation 5.40 is

$$s(s+1)(s^2 + 0.8s + 1) + K_2 K_{-1} s(s+1) + K_1 K_2 K_3 N_1 = 0. \tag{5.41}$$

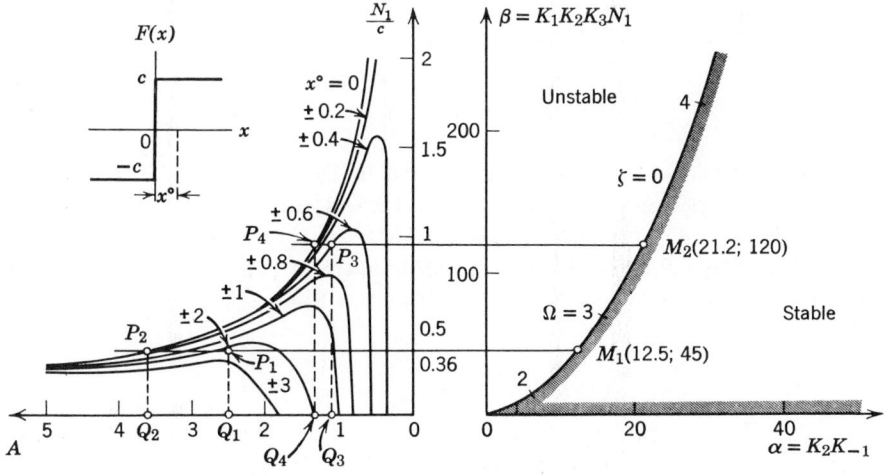

Figure 5.14 Parameter plane diagram.

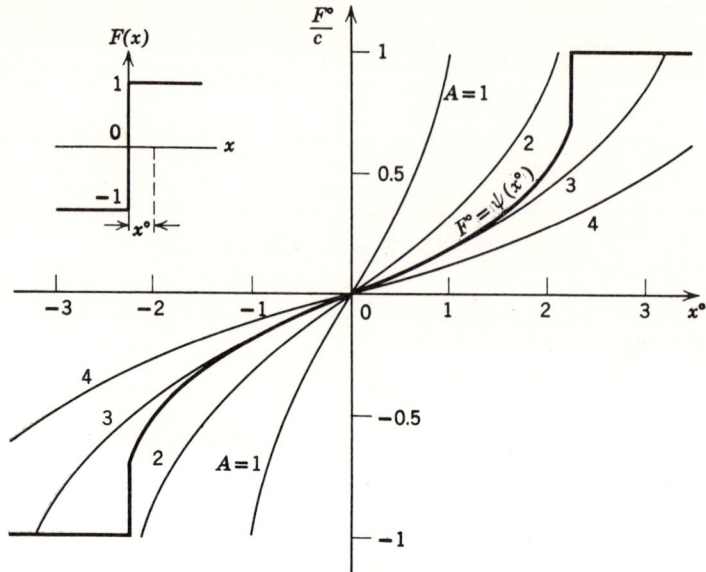

Figure 5.15 Function $\psi(x^0)$.

Substituting $K_2 K_{-1} = \alpha$, $K_1 K_2 K_3 N_1 = \beta$, and $s = j\Omega$ into (5.41), we obtain the parameter plane equations of the $\zeta = 0$ curve as

$$\alpha = 1.8\Omega^2 - 1$$
$$\beta = 0.8\Omega(\Omega + 1). \tag{5.42}$$

The curve $\zeta = 0$ is plotted in Figure 5.14. The variations of the M point are also plotted in Figure 5.14 according to

$$N_1 = \frac{4c}{\pi A}\left[1 - \left(\frac{x^0}{A}\right)^2\right]^{1/2}, \quad A \geq |x^0|, \tag{5.43}$$

which is derived in Appendix F and listed in Table F.2.

The system parameters $K_1 = 1$, $K_2 = 12.5$, $K_3 = 10$, $K_{-1} = 1$ result in the point $M_1(12.5; 45)$. If $c = 1$, this point M_1 gives $N_1 = \beta/K_1 K_2 K_3 = 0.36$, and the straight line $P_1 P_2$ is plotted on the diagram of function $N_1 = N_1(x^0, A)$. After the diagram $F^0 = F^0(x^0, A)$ is plotted in Figure 5.15 using

$$F^0 = \frac{2c}{\pi} \arcsin \frac{x^0}{A}, \quad A \geq |x^0| \tag{5.44}$$

of Table F.2 in Appendix F, the replotting of the straight line $P_1 P_2$ on the diagram $F^0(x^0, A)$ yields the function $\psi(x^0)$ of Figure 5.15. The replotting

Slowly Varying Signals 221

procedure is the same as that used in the previous section; that is, for each pair of values (x^0, A) read on the straight line P_1P_2, the corresponding pair exists in the diagram $F^0(x^0, A)$, which determines one point on the curve $\psi(x^0)$.

Function $\psi(x^0)$ of Figure 5.15 is smooth and represents the nonlinearity for the slowly varying signal x^0. The shape of $\psi(x^0)$ explains the smoothing effect of the high-frequency limit cycle that has a slowly varying amplitude, the value of which is located between the points Q_1 and Q_2 on the A axis of Figure 5.14. The frequency Ω is approximately constant and has the value $\Omega = 2.7$ rad/sec. According to $\psi(x^0)$, the smoothing effect of the limit cycle is present under the condition that $|x^0| \leq 2.25$. For small values of x^0 it is possible to consider $\psi(x^0) = Kx^0$, where $K = $ const. Then the stability of the system with respect to slowly varying signals may be investigated by well-known linear methods outlined in Chapter 2. In the specific example the equation of interest is

$$s(s + 1)(s^2 + 0.8s + 1) + K_2K_{-1}s(s + 1) + KK_1K_2K_3 = 0. \quad (5.45)$$

Finally, it is to be noted that for the smoothing effect to take place, the amplitude A should be $A \geq |x^0|$, as stated in (5.43) and (5.44).

The results of the above analysis are checked by simulating the above system on a computer. Three cases are considered. In Figure 5.16 the input to the nonlinearity $x = x^0 + A \sin \Omega t$ and the system output $x = x(t)$ are

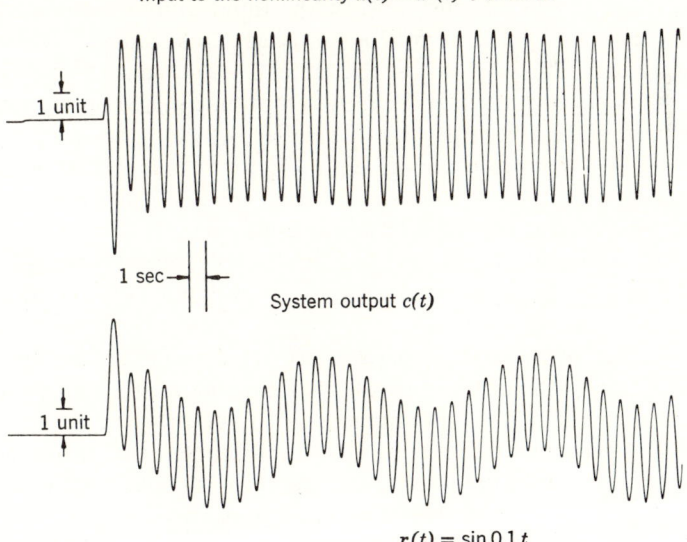

Figure 5.16 Computer solution.

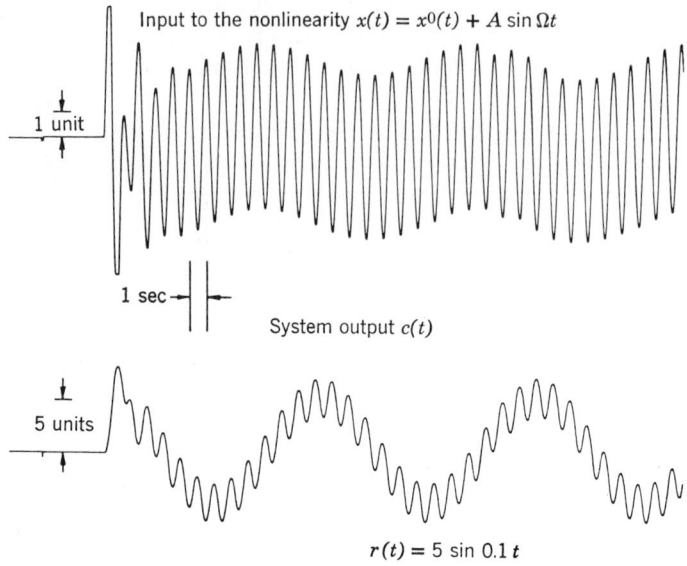

Figure 5.17 Computer solution.

shown when the input signal is $r = \sin 0.1t$. The obtained computer solution agrees with the prediction. The output $c(t)$ exhibits a smaller amplitude limit cycle with the same frequency. When the input amplitude is increased five times, the diagram of Figure 5.17 is obtained. This change increased x^0, but the amplitude A remained almost the same. The frequency Ω did not change. Similar results occurred when the input amplitude increased ten times except that the amplitude A became slightly smaller, which agrees with the diagram of Figure 5.14. The third case is given in Figure 5.18. It should be noted from these computer solutions that the output signal $c(t)$ represents the input signal $r(t)$ except for the superimposed limit cycle. It can be eliminated by introducing sufficient filtering in the block $G_3(s)$ of the system of Figure 5.13 or by readjusting the system parameters to obtain a higher frequency limit cycle.

If the values of the system parameters are chosen so that the operating point is $M_2(21.2; 120)$ of Figure 5.14, the frequency of the limit cycle becomes higher. However, the corresponding range of variations of x^0 is decreased to $|x^0| < 0.7$, together with the range of the amplitude A that is between Q_3 and Q_4. This indicates that the presented procedure is convenient to apply when the system parameters and operating conditions are changed.

If the nonlinearity n is changed in the system of Figure 5.13 by introducing a considerable dead zone D, the diagram of Figure 5.19 is obtained. The

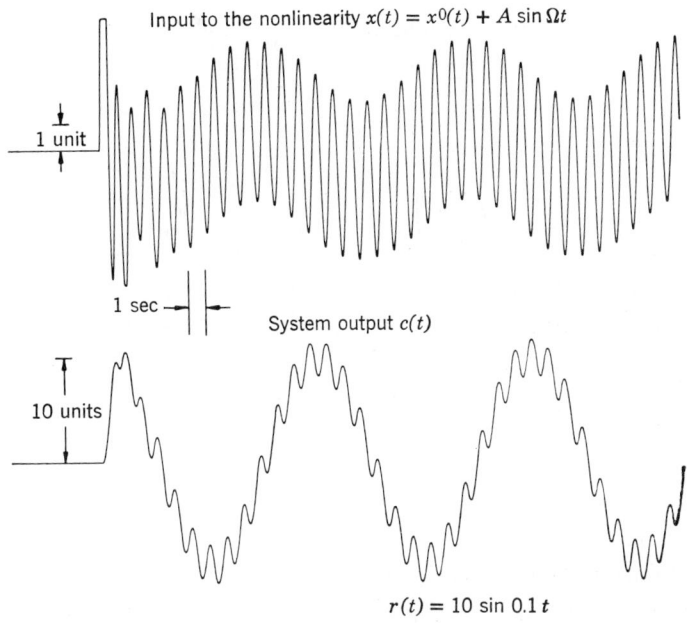

Figure 5.18 Computer solution.

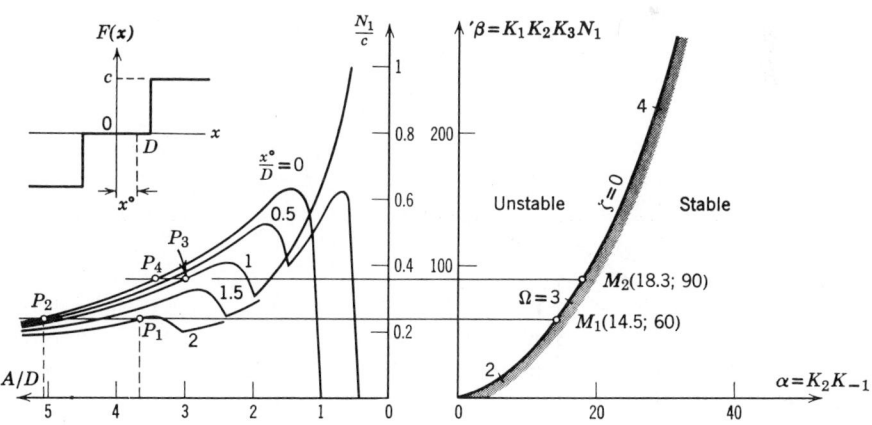

Figure 5.19 Parameter plane diagram.

224 Asymmetrical Oscillations

variation of the M point is calculated by using (5.6b) for the given nonlinearity of Figure 5.19. Two cases should be considered separately:

$$N_1 = \frac{2c}{\pi A}\left\{\left[1 - \left(\frac{x^0 + D}{A}\right)^2\right]^{1/2} + \left[1 - \left(\frac{x^0 - D}{A}\right)^2\right]^{1/2}\right\}, \quad A \geq |x^0| + D \tag{5.46a}$$

$$N_1 = \frac{2c}{\pi A}\left[1 - \left(\frac{x^0 - D}{A}\right)^2\right]^{1/2}, \quad |x^0| - D \leq A \leq |x^0| + D \tag{5.46b}$$

and the diagram $N_1(x^0, A)$ is shown in Figure 5.19. By using (5.6a), the corresponding diagram $F(x^0, A)$ of Figure 5.20 is plotted according to

$$F^0 = \frac{c}{\pi}\left(\arcsin\frac{x^0 + D}{A} + \arcsin\frac{x^0 - D}{A}\right), \quad A \geq |x^0| + D \tag{5.47a}$$

$$F^0 = \frac{c}{\pi}\left(\frac{\pi}{2} + \arcsin\frac{|x^0| - D}{A}\right)\operatorname{sign} x^0, \quad |x^0| - D \leq A \leq |x^0| + D. \tag{5.47b}$$

If the points M_1 and M_2 are chosen in Figure 5.19 as operating points, the replotting of the straight lines P_1P_2 and P_3P_4 results in the two linearized characteristics (a) and (b) of Figure 5.20, respectively. They are constructed for the values of nonlinear parameters $c = D = 1$. As can be seen from

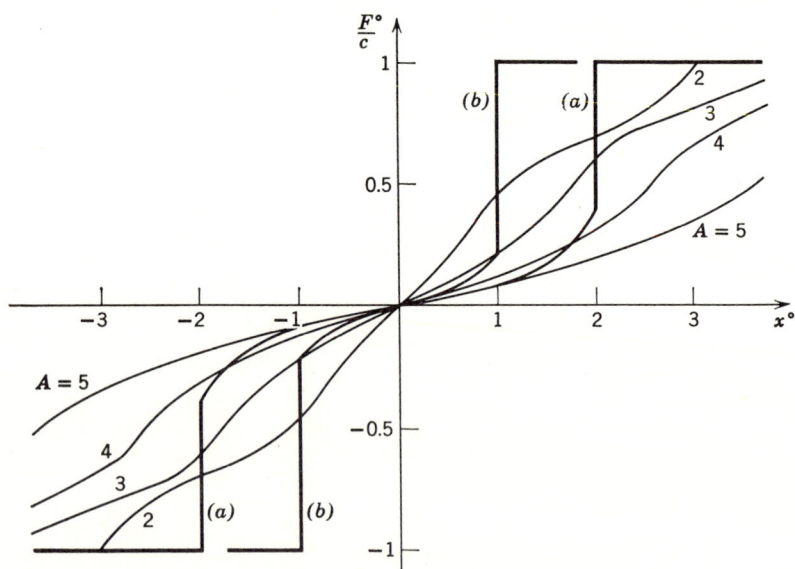

Figure 5.20 Linearized characteristic.

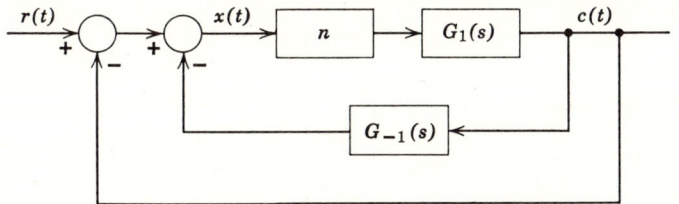

Figure 5.21 System block diagram.

Figure 5.20, the dead zone is eliminated as far as the slowly varying signals are concerned. For this to take place, it is necessary to choose operating conditions so that (5.47a) is valid. This means that the amplitude A of the limit cycle must be greater than $|x^0| + D$. Otherwise the linearized characteristic $\psi(x^0)$ does not go to zero when $x^0 = 0$, since F^0 does not go to zero for $x^0 = 0$. This is indicated in Figure 5.20, where $F^0 = 0$ for $x^0 = 0$ and the dead zone is eliminated.

By the outlined technique it is possible to eliminate the hysteresis and backlash in systems with multivalued nonlinearities. The linearization yields a single-valued function $\psi(x^0)$ that is linear in a certain limited range of values of the variable x^0 about the origin. To illustrate this, consider a nonlinear system with the block diagram of Figure 5.21 and the transfer functions

$$G_1(s) = \frac{K_1}{s(s+1)(s+2)}, \quad G_{-1}(s) = K_{-1}s. \tag{5.48}$$

The nonlinear function $F(x)$ of the nonlinearity n is given in Figure 5.22.

The equation describing the system is

$$s(s+1)(s+2)x + (K_{-1}s + 1)K_1 F(x) = 0. \tag{5.49}$$

After harmonic linearization of (5.49), the corresponding characteristic equation is

$$s(s+1)(s+2) + K_1(K_{-1}s + 1)\left(N_1 + \frac{N_2}{\Omega}s\right) = 0. \tag{5.50}$$

If $K_1 = 50$, $K_{-1} = 1$,

$$\begin{aligned}\alpha &= N_1 \\ \beta &= N_2,\end{aligned} \tag{5.51}$$

and $s = j\Omega$, we obtain the $\zeta = 0$ curve as

$$\begin{aligned}\alpha &= \tfrac{1}{50}\Omega^2 \\ \beta &= \tfrac{1}{25}\Omega.\end{aligned} \tag{5.52}$$

226 Asymmetrical Oscillations

The curve is plotted in Figure 5.22. On the same plot the variation of the point $M(N_1; N_2)$ is constructed according to

$$N_1 = \frac{2c}{\pi A}\left\{\left[1 - \left(\frac{D - x^0}{A}\right)^2\right]^{1/2} + \left[1 - \left(\frac{D + x^0}{A}\right)^2\right]^{1/2}\right\},$$

$$N_2 = -\frac{4cD}{\pi A^2}, \qquad A \geq D + |x^0| \tag{5.53}$$

and the nonlinearity $F(x)$ of Figure 5.22 for which $c = D = 1$. From the intersections of the $\zeta = 0$ curve and the variation of the M point, we can determine the amplitude A and the frequency Ω as functions of x^0; that is,

$$A = A(x^0)$$
$$\Omega = \Omega(x^0). \tag{5.54}$$

Then, by using the expression

$$F^0 = \frac{c}{\pi}\left(\arcsin\frac{D + x^0}{A} - \arcsin\frac{D - x^0}{A}\right), \qquad A \geq D + |x^0| \tag{5.55}$$

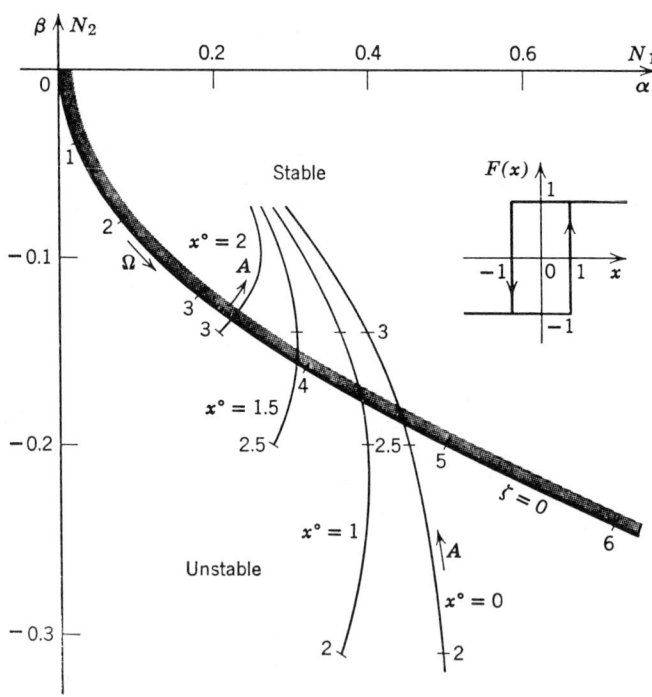

Figure 5.22 Parameter plane diagram.

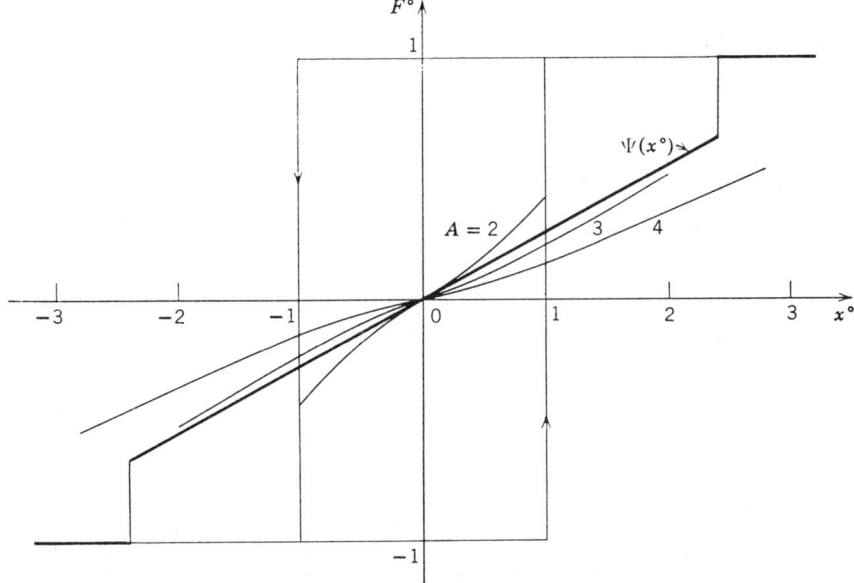

Figure 5.23 Function $\psi(x^\circ)$.

for $c = D = 1$, a family of curves with constant amplitude A is plotted in Figure 5.23. If the first equation 5.54 is mapped onto the family of constant amplitude A, the function $\psi(x^0)$ is obtained as shown in Figure 5.23. The function $\psi(x^0)$ is a single-valued function of x^0, which is linear in the range $0 \leq |x^0| \leq 2.4$.

For an input $r(t) = 5 \sin 0.5t$, the computer solution is shown in Figure 5.24. The amplitude A and the frequency Ω of the limit cycle are slowly varying quantities according to (5.54) and the slowly varying variable x^0. Their average values, however, are close to that which can be predicted from the parameter plane diagram of Figure 5.22, that is, $A = 2.8$ and $\Omega = 4.5$ rad/sec. This can be concluded from the diagram (a) of Figure 5.24. The diagram (b) shows the output signal $c(t)$ whereby the limit cycle is largely attenuated by the block $G_1(s)$ of Figure 5.21. The low-frequency component in the signal $c(t)$ represents the input $r(t) = 5 \sin 0.5t$ at the output of the system.

Of course if the input $r(t)$ is not present, the system will exhibit a limit cycle that can be determined from the intersection of the M locus $x^0 = 0$ and the $\zeta = 0$ curve in Figure 5.22 as $x = A \sin \Omega t$, $A = 2.6$, $\Omega = 4.8$. This is checked by a computer simulation and the resulting solution is shown in Figure 5.25. This consideration of limit cycle without external input to the system is related to Section 4.7.

228   Asymmetrical Oscillations

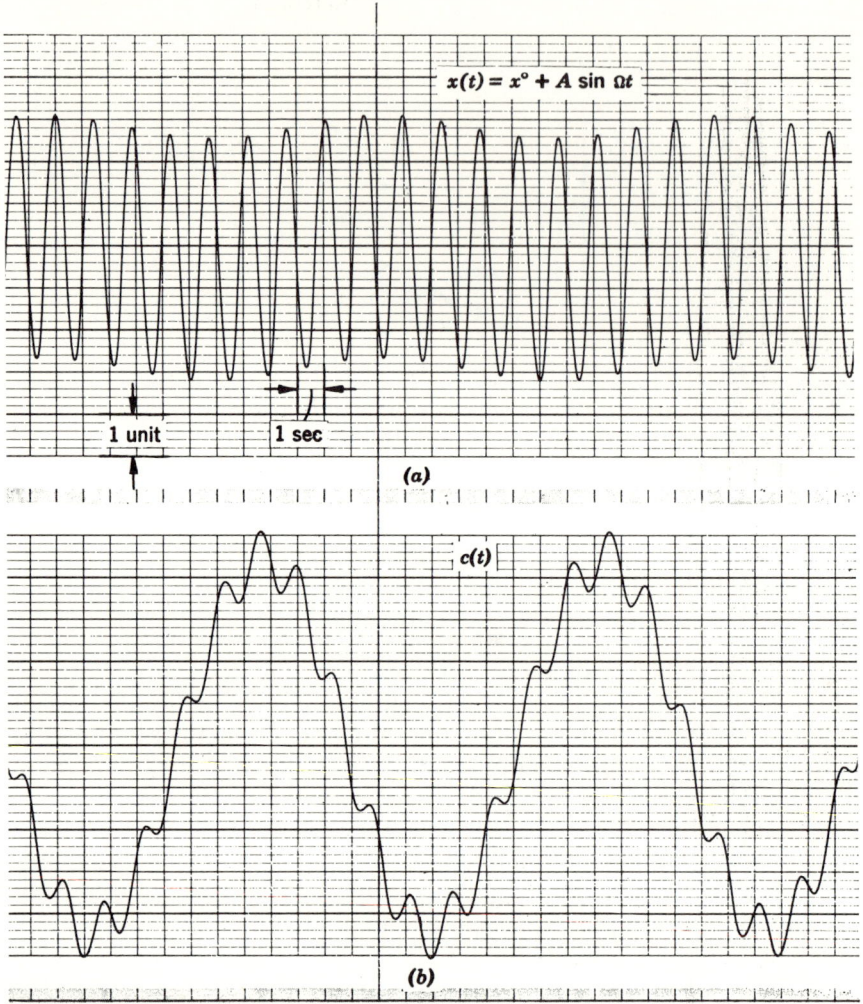

Figure 5.24   Computer solution.

5.6 Conclusion

The parameter plane method has been used to indicate the existence of asymmetrical oscillations in nonlinear control systems. A procedure has been developed to determine the oscillations for different values of system parameters and input signals. It has been shown how a limit cycle can modify the nonlinear characteristic for slowly varying signals. This modification may be of importance when a high-accuracy control system has to be designed in the

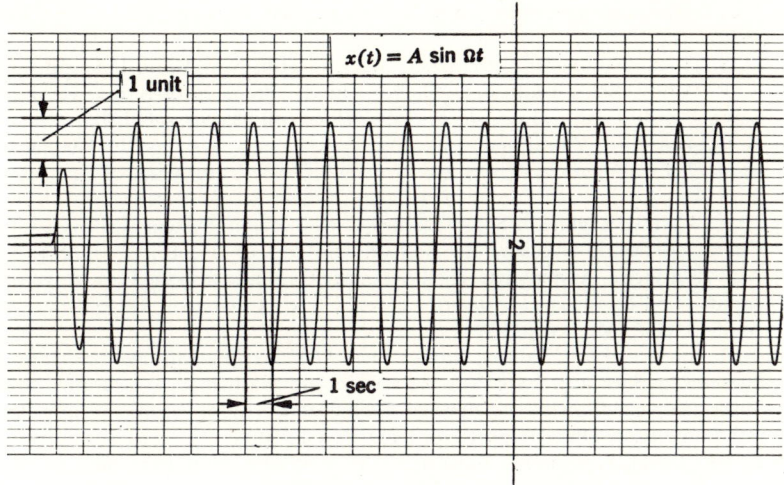

Figure 5.25 Computer solution for $x = 2.6 \sin 4.8t$.

presence of nonlinearities with excessive dead zone, hysteresis, backlash, and so on. The design technique can be directly applied to a large class of plant-adaptive control systems where a sinusoidal signal is used as an identification signal.

In a future study the technique may be extended to the investigation of transient asymmetrical oscillations. Thus, to investigate how these oscillations are established after certain amplitude perturbation, this study should be largely based upon the material presented in the following chapter.

It may also be shown [5.16, 17] that the presented analysis can be extended not only to the case where the signal superimposed on a sinusoid is a constant or slowly varying sinusoid, but also where the additional signal is described as a Gaussian process, provided that the amplitude or standard deviation of the additional signal is small relative to the amplitude of the sinusoid. The generalization that we can with some justification draw from this is that the corresponding describing function coefficients can be calculated as proposed, and that the bandwidth of the small additional signal is of no consequence in the analysis. This further supports the idea of applying the dual-input describing function [5.15, 17] along with the parameter plane method to investigate the case in which the input to a nonlinearity of the system is a combination of two similar sinusoidal signals.

References

[5.1] L. A. MacColl, *Fundamental Theory of Servomechanisms*, Van Nostrand, Princeton, New Jersey, 1945, pp. 78–87.

[5.2] J. C. Lozier, Carrier-Controlled Relay Servos, *Elec. Eng.*, **69**, 1052–1056 (December 1950).

[5.3] H. S. Tsien, *Engineering Cybernetics*, McGraw-Hill, New York, 1954, pp. 73–82.

[5.4] E. P. Popov, Self-Sustained Oscillations in the Presence of Slowly-varying Input Signals (in Russian), *Trans. Acad. Sci. USSR*, Vol. 98, No. 4 (1954).

[5.5] G. S. Pospelov, Vibrational Linearization of Relay Systems (in Russian), *Proc. of the 2nd National Conference on Automatic Control*, ed. Acad. of Science, USSR, I, 1955.

[5.6] B. A. Riabov, Regime of Self-Sustained Oscillations in Systems with Asymmetrical Characteristics (in Russian), *Avtomatika*, ed. Acad. of Science, USSR, No. 2, 1956.

[5.7] M. V. Starikova, Asymmetrical Self-Sustained Oscillations in the Presence of External Inputs (in Russian), *Automatic Control and Computer Techniques*, 2d ed., MASCHGIZ, Moscow, 1959.

[5.8] E. P. Popov and I. P. Palitov, *Approximate Methods for Analyzing Nonlinear Control Systems* (in Russian), State Press for Physics and Mathematical Literature, Moscow, 1960. (English translation: Foreign Technical Division, AFSC, Wright-Patterson Air Force Base, Ohio, Report No. FTD-TT-62-910.)

[5.9] J. M. Loeb, A General Linearizing Process for Nonlinear Control Systems, *Automatic and Manual Control*, A. Tustin, ed., Academic, New York, 1952, pp. 275–283.

[5.10] R. Oldenburger, Signal Stabilization of a Control System, *Trans. Am. Soc. Mech. Engrs.*, **79**, No. 8, 1869–1872 (November 1957).

[5.11] R. Oldenburger and C. C. Liu, Signal Stabilization of a Control System, *AIEE Trans.*, Pt. II (*Applications and Industry*), **78**, 96–100 (May 1959).

[5.12] R. Oldenburger and T. Nakada, Signal Stabilization of Self-Oscillating Systems *IRE Trans. Auto. Control*, **AC-6**, 319–325 (September 1961).

[5.13] R. Sridhar, Signal Stabilization of a Control System with Random Inputs, Ph.D. Thesis, Purdue University, Lafayette, Indiana, January, 1960.

[5.14] R. Oldenburger and R. Shridar, Signal Stabilization of a Control System with Random Inputs, *AIEE Trans.* Pt. II (*Applications and Industry*), **80**, 260–267 (November 1961).

[5.15] J. C. West, *Analytical Technique for Nonlinear Control Systems*, English University Press, London, 1960.

[5.16] R. C. Boyer, Sinusoidal Signal Stabilization, M.S. Thesis, Purdue University, Lafayette, Indiana, January, 1960.

[5.17] J. E. Gibson, *Nonlinear Automatic Control*, McGraw-Hill, New York, 1963.

[5.18] A. Gelb and W. E. Van der Velde, On Limit Cycling Control Systems, *IRE Trans. Auto. Control*, **AC-8**, 142–157 (April 1963).

[5.19] D. P. Atherton and G. F. Turnbull, Response of Nonlinear Characteristics to Several Inputs and the Use of the Modified Nonlinearity Concept in Control Systems, *Proc. IEE*, **111**, 157–164 (January 1964).

[5.20] G. F. Turnbull, D. P. Atherton, and J. M. Townsend, Method for the Theoretical Analysis of Relay Feedback Systems, *Proc. IEE*, **112**, No. 5, 1039–1055 (May 1965).

[5.21] Ya. Z. Tsypkin, *Theory of Relay Control Systems* (in Russian), GOSTEHIZDAT, Moscow, 1955.

[5.22] R. J. McGrath and V. C. Rideout, A Simulator Study of a Two-Parameter Adaptive System, *IRE Trans. Auto. Control*, **AC-6**, 35–42 (February 1961).

[5.23] V. Rajaramau, Theory of a Two-Parameter Adaptive Control System, *IRE Trans. Auto. Control*, **AC-7**, 20–26 (July 1962).

[5.24] Kelvin C. Smith, Adaptive Control Through Sinusoidal Response, *IRE Trans. Auto. Control*, **AC-7**, 129–139 (March 1962).

[5.25] R. K. Smyth and M. E. Nahi, Phase and Amplitude Sinusoidal Dither Adaptive Control System, *Proceedings of the Joint Automatic Control Conference*, A.I.Ch.E., New York, 1963, pp. 302–312.

[5.26] I. M. Horowitz, Comparison of Linear Feedback Systems with Self-Oscillating Adaptive Systems, *IEEE Trans. Auto. Control*, **AC-9**, No. 4, 386–393 (October 1964).

[5.27] D. D. Šiljak, Analysis of Asymmetrical Nonlinear Oscillations in the Parameter Plane, *IEEE Trans. Auto. Control*, **AC-11**, No. 2, 82–87 (April 1966).

CHAPTER SIX

Transient Oscillations

6.1 Introduction

The objective of this chapter is to consider the transient oscillations in nonlinear systems on the basis of the harmonic linearization and describing function technique. In Chapter 4 the same tools were used for analyzing the free steady-state oscillations, and the problem of the transient process in the establishment of the periodic oscillations was only partially discussed. The nature of the transient process in the neighborhood of the periodic oscillations indicated their stability. It has been tacitly assumed that the perturbations that caused the transient process to take place are sufficiently small and close to the periodic self-excited oscillations. This small perturbation analysis of the steady-state oscillations has been based on the Krylov-Bogoliubov asymptotical methods [6.1]. In Section 4.3 the sensitivity indices were established to investigate the stability and the rate of the amplitude change during the corresponding transient oscillation.

The Krylov-Bogoliubov approach linearized the nonlinear system with the assumption that, in the neighborhood of the steady-state oscillations, the transient process is close to the transient process of an equivalent linear system with a slowly varying damping factor and frequency. The rates of change of these two characteristics have been neglected. On the basis of rather intuitive ideas, Grensted [6.2, 3] proposed an approach to transient oscillations analysis that includes the previously missing rates of change of damping and frequency. In addition, the damping is not necessarily small, and large amplitude perturbations about either the periodic oscillations or the equilibrium positions can be considered. The Grensted approach is restricted to second-order nonlinear differential equations, and no indication is given for a possible extension to higher-order equations.

The Grensted results were obtained by E. P. Popov [6.4] on the basis of extending the original Krylov-Bogoliubov method. The extension is derived rather rigorously and is applied to the transient oscillations caused by large amplitude perturbations. The rates of change in damping and frequency of the linearized equations are still lacking, and in that respect the Popov extension of the Krylov-Bogoliubov method is less accurate than the Grensted analysis. Popov, however, indicates the possibility of treating high-order nonlinear differential equations when the damping and frequency can be explicitly determined from the given equation. This is possible in a certain class of third-order differential equations to which the method has been applied.

The parameter plane method permits the generalization of the Popov and Grensted results so that high-order nonlinear differential equations can be analyzed for transient oscillations. A first attempt has been made in reference 6.5, where the relative stability of nonlinear systems has been analyzed in the parameter plane by utilizing the Popov extension. If the whole M locus lies inside the region R_n related to the condition $\zeta \geq \zeta_0$, then something can be concluded about the transient process on the basis of the specified number ζ_0. This allows a rather qualitative analysis, and no quantitative conclusions can be drawn from the relative stability specifications. Some quantitative results have been obtained in reference 6.6, where the sensitivity of self-excited oscillations to small amplitude perturbations has been investigated.

In this chapter the parameter plane analysis of the transient process in high-order nonlinear differential equations is presented. Both small and large amplitudes are considered and an approximate procedure is proposed for the evaluation of the transient oscillations from the parameter plane diagram. The procedure takes into account the rates of change of the damping and frequency of the equivalent linearized equation.

In general there are two aspects of the transient oscillation analysis: (1) to get an actual transient solution for a given set of system parameters and initial conditions; (2) to obtain a bird's-eye view of the totality of transient oscillations under all admissible conditions—a view distant enough so that it will not be confused by irrelevant details yet sharp enough so that it can recognize factors of importance.

At the present time the first of these tasks is no more difficult since modern methods of machine computing, simulations, and so on have made the evaluation of specific solutions of differential equations a routine task. The parameter plane attempts the second task, giving in evidence the nature of the transient oscillations as a function of various system parameter values and different initial conditions. Thus in this chapter a computer-oriented method is proposed for the analysis of transient nonlinear oscillations that can be advantageously used as a guide to computer simulation in the design of nonlinear systems.

6.2 Extension of the Krylov-Bogoliubov Method

The original Krylov-Bogoliubov method was presented in Chapter 3, where the second-order nonlinear differential equation

$$\ddot{x} + F(x, \dot{x}) = 0 \tag{6.1}$$

was considered. It has been rewritten in terms of the small parameter as

$$\ddot{x} + \omega_0^2 x = \varepsilon f(x, \dot{x}). \tag{6.2}$$

The corresponding solution $x = x(t)$ has been expressed in a series

$$x = a \sin \phi + \varepsilon u_1(a, \phi) + \varepsilon^2 u_2(a, \phi) + \cdots + \varepsilon^m u_m(a, \phi) + \cdots, \tag{6.3}$$

where

$$\begin{aligned} \dot{a} &= \varepsilon P_1(a) + \varepsilon^2 P_2(a) + \cdots + \varepsilon^m P_m(a) + \cdots \\ \dot{\phi} &= \omega_0 + \varepsilon Q_1(a) + \varepsilon^2 Q_2(a) + \cdots + \varepsilon^m Q_m(a) + \cdots. \end{aligned} \tag{6.4}$$

In the first approximation we have

$$x = a \sin \phi, \quad \dot{a} = \varepsilon P_1(a), \quad \dot{\phi} = \omega_0 + \varepsilon Q_1(a). \tag{6.5}$$

From this form of solution it is apparent that both the amplitude a and the frequency ω are slowly varying functions of time. As indicated in Chapter 3, that fact permits an analysis of the slightly damped, or undamped, transient processes close to the ordinary harmonic oscillations

$$x = a_0 \sin(\omega_0 t + \phi_0) \tag{6.6}$$

but with a slowly varying amplitude a and frequency ω around the values a_0 and ω_0. In (6.6), ϕ_0 denotes the phase shift. This was explicitly expressed by the equivalent linearization discussed in the Chapter 3.

The objective of Popov's extension of the original method is to consider transient processes in nonlinear systems that are close to the well-damped oscillations

$$x = a_0 e^{\sigma_0 t} \sin(\omega_0 t + \phi_0), \tag{6.7}$$

with σ and ω as slowly varying functions of time around the values σ_0 and ω_0 on a certain limited interval of time. The solution (6.7) corresponds to the second-order linear differential equation

$$\ddot{x} + 2p\dot{x} + q^2 x = 0, \tag{6.8}$$

where

$$\sigma_0 = -p, \quad \omega_0 = (q^2 - p^2)^{1/2}. \tag{6.9}$$

Thus the Popov extension is to consider a nonlinear differential equation of the form

$$\ddot{x} + 2p\dot{x} + q^2 x = \varepsilon f(x, \dot{x}), \tag{6.10}$$

where p and q are real constants, $f(x, \dot{x})$ is a known nonlinear function, and ε is the small parameter.

The solution of (6.10) is assumed in the form

$$x = a \sin \phi + \varepsilon U_1(a, \phi) + \varepsilon^2 U_2(a, \phi) + \cdots + \varepsilon^m U_m(a, \phi) + \cdots, \tag{6.11}$$

where

$$\begin{aligned} \dot{a} &= -pa + \varepsilon R_1(a) + \varepsilon^2 R_2(a) + \cdots + \varepsilon^m R_m(a) + \cdots \\ \dot{\phi} &= \omega_0 + \varepsilon Q_1(a) + \varepsilon^2 Q_2(a) + \cdots + \varepsilon^m Q_m(a) + \cdots, \end{aligned} \tag{6.12}$$

In the first approximation, which is discussed here, the solution is represented by

$$x = a \sin \phi, \quad \dot{a} = -pa + \varepsilon R_1(a), \quad \dot{\phi} = \omega_0 + \varepsilon Q_1(a), \tag{6.13}$$

where ω_0 is determined from (6.9).

Popov's extension is essentially different from the original approach given in (6.5) since, in the expression for the derivative a, there is a finite and not necessarily small term $-pa$. Therefore the extension considers nonlinear oscillations that are close to rapidly damped linear oscillations with a finite value σ_0 obtained for $\varepsilon = 0$. In this case the slowly varying quantities are not a and ϕ but the values of $\sigma = \sigma(a)$ and $\omega = \omega(a)$ given as

$$\begin{aligned} \sigma(a) &= -p + \frac{\varepsilon}{a} R_1(a) \\ \omega(a) &= \omega_0 + \varepsilon Q_1(a). \end{aligned} \tag{6.14}$$

The amplitude a can vary at a finite rate $\dot{a}$, approximately determined by the constant p.

Again, the approximate solution is expressed in the form of a series and the problem consists in determining the functions $U_i(a, \phi)$, $R_i(a)$, $Q_i(a)$, $(i = 1, 2, \ldots, m)$, so that the approximate solution, given in (6.11) and (6.12), satisfies the given nonlinear differential equation with an accuracy up to the order ε^{m+1}.

To simplify the following derivations, introduce a new variable y so that

$$x = y e^{-pt}. \tag{6.15}$$

Then

$$\dot{x} = e^{-pt}(\dot{y} - py) \tag{6.16}$$

and the nonlinear differential equation 6.10 becomes

$$e^{-pt}(\ddot{y} + \omega_0^2 y) = \varepsilon f(x, \dot{x}) \tag{6.17}$$

236   Transient Oscillations

with $\omega_0^2 = q^2 - p^2$ as given in (6.9). Equations 6.11 and 6.12 are now in the form

$$y = a_* \sin \phi + \varepsilon u_1(a, \phi) + \varepsilon^2 u_2(a, \phi) + \cdots + \varepsilon^m u_m(a, \phi) \quad (6.18)$$

and

$$\begin{aligned} \dot{a}_* &= \varepsilon P_1(a) + \varepsilon^2 P_2(a) + \cdots + \varepsilon^m P_m(a) \\ \dot{\phi} &= \omega_0 + \varepsilon Q_1(a) + \varepsilon^2 Q_2(a) + \cdots + \varepsilon^m Q_m(a), \end{aligned} \quad (6.19)$$

where

$$\begin{aligned} a &= a_* e^{-pt}, & \dot{a} &= -pa + \dot{a}_* e^{-pt} \\ U_i &= u_i e^{-pt}, & R_i &= P_i e^{-pt}, & (i = 1, 2, \ldots, m). \end{aligned} \quad (6.20)$$

By differentiating (6.18) twice and using (6.19) and (6.20), we have

$$\begin{aligned} \dot{y} = {}& a_* \omega_0 \cos \phi + \varepsilon(P_1 \sin \phi + a_* Q_1 \cos \phi + \omega_0 u_1^{(\phi)} - pa u_1^{(a)}) \\ &+ \varepsilon^2(P_2 \sin \phi + a_* Q_2 \cos \phi + Q_1 u_1^{(\phi)} + \omega_0 u_2^{(\phi)} \\ &+ e^{-pt} P_1 u_1^{(a)} - pa u_2^{(a)}) + \varepsilon^3 \cdots \end{aligned} \quad (6.21)$$

and

$$\begin{aligned} \ddot{y} = {}& -a_* \omega_0^2 \sin \phi + \varepsilon(-2\omega_0 Q_1 a_* \sin \phi - pa P_1' \sin \phi + 2\omega_0 P_1 \cos \phi \\ &- pa Q_1' a_* \cos \phi + \omega_0^2 u_1^{(\phi\phi)} - 2pa\omega_0 u_1^{(a\phi)} + p^2 a^2 u_1^{(aa)} + p^2 a u_1^{(a)}) \\ &+ \varepsilon^2(-Q_1^2 a_* \sin \phi - 2\omega_0 Q_2 a_* \sin \phi - pa P_2' \sin \phi + 2\omega_0 P_2 \cos \phi \\ &+ 2P_1 Q_1 \cos \phi + P_1 P_1' e^{-pt} \sin \phi + P_1 Q_1' e^{-pt} a_* \cos \phi - pa Q_2' a_* \cos \phi \\ &+ 2\omega_0 Q_1 u_1^{(\phi\phi)} + \omega_0^2 u_2^{(\phi\phi)} - pa Q_1' u_1^{(\phi)} + 2\omega_0 P_1 e^{-pt} u_1^{(a\phi)} - 2pa Q_1 u_1^{(a\phi)} \\ &- 2pa\omega_0 u_2^{(a\phi)} - 2p P_1 e^{-pt} u_1^{(a)} - pa P_1' e^{-pt} u_1^{(a)} + p^2 a u_2^{(a)} - 2pa e^{-pt} P_1 u_1^{(aa)} \\ &+ p^2 a^2 u_2^{(aa)}) + \varepsilon^3 \cdots, \end{aligned} \quad (6.22)$$

where the partial differentiation are designated by superscripts as $u_i^{(a)} = \partial u_i/\partial a$, $u_i^{(\phi)} = \partial u_i/\partial \phi$, and so on, and $u_i^{(a\phi)} = u_i^{(\phi a)}$. The differentiation with respect to a is denoted by $P_i' = dP_i/da$, $Q_i' = dQ_i/da$.

According to (6.16), $\dot{x}$ can be expressed as

$$\dot{x} = a\omega_0 \cos \phi - pa \sin \phi \\ + \varepsilon e^{-pt}(P_1 \sin \phi + a_* Q_1 \cos \phi - pu_1 - pa u_1^{(a)} + \omega_0 u_1^{(\phi)}) + \varepsilon^2 \cdots, \quad (6.23)$$

and the Taylor series of the function $\varepsilon f(x, \dot{x})$ is

$$\begin{aligned} \varepsilon f(x, \dot{x}) = {}& \varepsilon f(a \sin \phi, a\omega_0 \cos \phi - pa \sin \phi) \\ &+ \varepsilon^2 e^{-pt}[u_1 f^{(x)}(a \sin \phi, a\omega_0 \cos \phi - pa \sin \phi) \\ &+ f^{(\dot{x})}(a \sin \phi, a\omega_0 \cos \phi - pa \sin \phi) \\ &\times (P_1 \sin \phi + Q_1 a_* \cos \phi - pu_1 - pa u_1^{(a)} + \omega_0 u_1^{(\phi)})] + \varepsilon^3 \cdots. \end{aligned} \quad (6.24)$$

By substituting these various expressions into (6.17) and equating the terms associated with the like powers of ε on both sides, we obtain a recursive system:

$$-2\omega_0 Q_1 a \sin \phi - pa R_1' \sin \phi + 2\omega_0 R_1 \cos \phi - pa^2 Q_1' \cos \phi + \omega_0^2 U_1$$
$$+ \omega_0^2 U_1^{(\phi\phi)} - 2pa\omega_0 U_1^{(a\phi)} + p^2 a^2 U_1^{(aa)} + p^2 a^2 U_1^{(a)}$$
$$= f(a \sin \phi, a\omega_0 \cos \phi - pa \sin \phi) \quad (6.25)$$

$$-Q_1^2 a \sin \phi - 2\omega_0 Q_2 a \sin \phi - pa R_2' \sin \phi + 2\omega_0 R_2 \cos \phi + 2R_1 Q_1 \cos \phi$$
$$+ R_1 R_1' \sin \phi + a R_1 Q_1' \cos \phi - pa^2 Q_2' \cos \phi + \omega_0^2 U_2 + 2\omega_0 Q_1 U_1^{(\phi\phi)}$$
$$+ \omega_0^2 U_2^{(\phi\phi)} - pa Q_1' U_1^{(\phi)} + 2\omega_0 R_1 U_1^{(a\phi)} - 2pa Q_1 U_1^{(a\phi)} - 2pa\omega_0 U_2^{(a\phi)}$$
$$- 2pR_1 U_1^{(a)} - pa R_1' U_1^{(a)} + p^2 a U_2^{(a)} - 2pa R_1 U_1^{(aa)} + p^2 a^2 U_2^{(aa)}$$
$$= U_1 f^{(x)}(a \sin \phi, a\omega_0 \cos \phi - pa \sin \phi) + (R_1 \sin \phi + aQ_1 \cos \phi$$
$$- pU_1 - pa U_1^{(a)} + \omega_0 U_1^{(\phi)}) f^{(\dot{x})}(a \sin \phi, a\omega_0 \cos \phi - pa \sin \phi). \quad (6.26)$$

$$\cdot$$
$$\cdot$$
$$\cdot$$

It is assumed that $U_1(a, \phi)$ does not incorporate the terms $\sin \phi$ and $\cos \phi$ since they are completely represented by the first part $a \sin \phi$ of the solution x given in (6.11). Now the procedure follows the same pattern as in the original Krylov-Bogoliubov method and develops the function $f(a \sin \phi, a\omega_0 \cos \phi - pa \sin \phi)$ into the Fourier series. This, however, requires a special consideration, which will be given later. Therefore, by developing the mentioned function in the trigonometric series and equating the corresponding terms in (6.25), we obtain

$$2\omega_0 R_1 - pa^2 Q_1' = h_1$$
$$-2a\omega_0 Q_1 - pa R_1' = g_1 \quad (6.27)$$

and

$$pa^2 U_1^{(aa)} + p^2 a^2 U_1^{(a)} + \omega_0^2 U_1^{(\phi\phi)} - 2pa\omega_0 U_1^{(a\phi)} + \omega^2 U_1$$
$$= g_0 + \sum_{n=2}^{\infty} [g_n \sin n\phi + h_n \cos n\phi], \quad (6.28)$$

where $g_0 = g_0(a)$, $g_n = g_n(a)$, and $h_n = h_n(a)$ are given by

$$g_0(a) = \frac{1}{2\pi} \int_0^{2\pi} f(a \sin \phi, a\omega_0 \cos \phi - pa \sin \phi) \, d\phi$$

$$g_n(a) = \frac{1}{\pi} \int_0^{2\pi} f(a \sin \phi, a\omega_0 \cos \phi - pa \sin \phi) \sin n\phi \, d\phi$$

$$h_n(a) = \frac{1}{\pi} \int_0^{2\pi} f(a \sin \phi, a\omega_0 \cos \phi - pa \sin \phi) \cos n\phi \, d\phi,$$

$$(n = 1, 2, \ldots, m, \ldots). \quad (6.29)$$

238   Transient Oscillations

From (6.27) it is possible to determine the functions $R_1(a)$ and $Q_1(a)$ and thereby the first approximation (6.13). Then (6.28) enables the calculation of the function $U_1(a, \phi)$ in the form of a series

$$U_1(a, \phi) = v_0(a) + \sum_{n=2}^{\infty} [v_n(a) \sin n\phi + w_n(a) \cos n\phi], \quad (6.30)$$

the coefficients of which are not yet known. By substituting (6.30) into (6.28), we can evaluate the unknown coefficients v_0, v_n, w_n in terms of the known coefficients g_0, g_n, h_n.

After the functions $R_1(a)$, $Q_1(a)$, and $U_1(a, \phi)$ are determined, (6.26) is used to calculate functions $R_2(a)$, $Q_2(a)$, which, in turn, determine the second approximation and the function $U_2(a, \phi)$. The evaluation of the higher approximations is thus sufficiently clear.

Of particular interest in the following is the first approximation, since it is a sufficient representation of the solution in a majority of practical problems. This statement is enforced by the fact that the evaluation of higher approximations is rather complex. Thus the solution of (6.10) is considered in the form

$$x = a \sin \phi, \quad (6.31)$$

where

$$\dot{a} = a\sigma(a), \qquad \dot{\phi} = \omega(a) \quad (6.32)$$

and

$$\sigma(a) = -p + \frac{\varepsilon}{a} R_1(a), \qquad \omega(a) = \omega_0 + \varepsilon Q_1(a). \quad (6.14)$$

Equations 6.27 may now be rewritten as

$$\begin{aligned} 2\omega_0 R_1(a) &= \eta(a) + pa^2 Q_1'(a) \\ 2\omega_0 Q_1(a) &= \xi(a) - p R_1'(a), \end{aligned} \quad (6.33)$$

where $R'(a) = dR_1(a)/da$, $Q_1'(a) = dQ_1(a)/da$, and

$$\xi(a) = -\frac{g_1(a)}{a} = -\frac{1}{a\pi} \int_0^{2\pi} f(a \sin \phi, a\omega_0 \cos \phi - pa \sin \phi) \sin \phi \, d\phi \quad (6.34)$$

$$\eta(a) = h_1(a) = \frac{1}{\pi} \int_0^{2\pi} f(a \sin \phi, a\omega_0 \cos \phi - pa \sin \phi) \cos \phi \, d\phi.$$

It is now assumed that the functions $\xi(a)$ and $\eta(a)$ are with small curvatures. Thus their corresponding second and higher derivatives, if compared to the values of $\xi(a)$ and $\eta(a)$, are with a smallness of the order ε and can be neglected in the first approximation. Then the particular solutions of simultaneous equations 6.33 can be written as

$$2\omega_0 R_1(a) = \eta(a) + \frac{p\omega_0}{2\omega_0^2 + p^2} a^2 \xi'(a)$$
$$2\omega_0 Q_1(a) = \xi(a) - \frac{p}{2\omega_0} \eta'(a) - \frac{p^2}{2\omega_0^2 + p^2} a\xi'(a),$$
(6.35)

where $\xi'(a) = d\xi(a)/da$ and $\eta'(a) = d\eta(a)/da$. Equations 6.35 are obtained by sequential approximations and can be verified by applying (6.33).

In general, by using (6.14), we can derive

$$\sigma = -p + \frac{pa}{2\omega_0} \omega' + \frac{\varepsilon}{2\omega_0} \frac{\eta(a)}{a}$$
$$\omega^2 = \omega_0^2 - p^2 - p\sigma - pa\sigma' + \varepsilon\xi(a),$$
(6.36)

or

$$\sigma = -p - \frac{1}{2\omega_0} \dot{\omega} + \frac{\varepsilon}{2\omega_0} \frac{\eta(a)}{a}$$
$$\omega^2 = \omega_0^2 - p^2 - p\sigma + \dot{\sigma} + \varepsilon\xi(a),$$
(6.37)

where the prime sign designates the derivation with respect to a and the dot means the derivation with respect to time, as used before.

The same relationship (6.37) has been obtained by Grensted [6.2, 3] by a rather different approach without introducing the small parameter. This relationship is in the differential form and can be rewritten as

$$\sigma = -p + \frac{\varepsilon}{2\omega_0} \frac{\eta(a)}{a} + \frac{\varepsilon p}{2(2\omega_0^2 + p^2)} a\xi'(a)$$
$$\omega^2 = \omega_0^2 + \varepsilon\xi(a) - \varepsilon\frac{p}{2\omega_0} \eta'(a) - \frac{\varepsilon p^2}{2\omega_0^2 + p^2} a\xi'(a).$$
(6.38)

Finally, in case $\xi(a)$ and $\eta(a)$ are slowly varying functions with respect to the amplitude a, and if their derivatives $\xi'(a)$ and $\eta'(a)$ can be considered as small of the order ε if compared to the actual values of $\xi(a)$ and $\eta(a)$, (6.38) can be rewritten as

$$\sigma = -p + \frac{\varepsilon}{2\omega_0} \frac{\eta(a)}{a}$$
$$\omega^2 = \omega_0^2 + \varepsilon\xi(a),$$
(6.39)

where, according to (6.9), we have $\omega_0 = (q^2 - p^2)^{1/2}$.

Therefore, by Popov's extension of the Krylov-Bogoliubov method, the first approximation of the solution of second-order nonlinear differential equation 6.2 is given as

$$x = a \sin \phi \qquad (6.31)$$

and
$$\dot{a} = a\sigma, \quad \dot{\phi} = \omega, \tag{6.32}$$

where σ and ω are unknown and should be determined from (6.39). As indicated before, the quantities σ and ω are supposed to be slowly varying functions of time. This, however, enables a consideration of well-damped or undamped oscillations, because the damping depends on the actual value of the quantity σ at a given time and not on its rate of change. Thus limitations are not imposed on the amplitude variation in time as was the case in the original Krylov-Bogoliubov method, but on the variation of σ and ω only. Furthermore, the actual values of σ are also not limited and the solution may represent slightly damped (small σ) or well-damped (large σ) transient processes. Of course, if σ is so small that it is almost zero, it is not necessary to use the above extension, and the original method of Krylov-Bogoliubov outlined in Section 3.5 is more convenient to apply.

It is now necessary to discuss the development of the function $f(a \sin \phi, a\omega_0 \cos \phi - pa \sin \phi)$ in the Fourier series in which the amplitude a and ϕ are variables in time close to $a_0 e^{-pt}$ and $\omega_0 t$, respectively. This is an essential difference from the earlier developments in the Fourier series with respect to ϕ for each point of the curve $\phi(t)$ determined on the ϕt plane. Such a development on a purely formal basis does not ensure all the necessary characteristics of the Fourier series. Moreover, for an arbitrary variation of $a(t)$, the periodic terms may arise in the series and the development may have no practical value.

In the outlined extension, however, the variation of $a(t)$ is not at all an arbitrary process since the essential requirement is that, at least on a limited interval of time, the amplitude $a(t)$ variations are close to the function $a_0 e^{-pt}$. Similarly, the curve $\phi(t)$, along which the Fourier series are calculated on the ϕt plane, is close to the linear function $\omega_0 t$ of time t. Therefore, instead of developing *nonlinear periodic oscillations* in a Fourier series having harmonics with constant amplitudes and frequencies, the *nonlinear damped oscillations* are developed in a Fourier series with "damped harmonics" having slowly varying frequencies, the ratios of which are still expressible by integer values. The same conclusion has been obtained by Grensted [6.2, 3], who pointed out that the higher harmonics generated by the nonlinear function whose argument varies as a damped or undamped sinusoid may be neglected despite the fact that the corresponding frequency is no longer constant. It may therefore be concluded that the development of nonstationary nonlinear oscillations in the Fourier series has not only a formal basis, but a strong physical motivation for the consideration of the first harmonic in that development as the only representative of the actual oscillations in their approximate analysis.

The above statement enables an equivalent linearization to be applied to the equation

$$\ddot{x} + 2p\dot{x} + q^2 x = \varepsilon f(x, \dot{x}) \tag{6.10}$$

as derived for the equation

$$\ddot{x} + \omega_0^2 x = \varepsilon f(x, \dot{x}) \tag{6.2}$$

in Section 3.5, where the Krylov-Bogoliubov method was presented. Thus, by using the assumption (6.31) and expanding the function $\varepsilon f(x, \dot{x})$ into its Fourier series, we have

$$\varepsilon f(x, \dot{x}) = \varepsilon[g_1(a) \sin \phi + h_1(a) \cos \phi] \tag{6.40}$$

if restricted to the first term. From (6.31), it follows that

$$\sin \phi = \frac{x}{a}, \quad \cos \phi = \frac{1}{a\omega_0}(\dot{x} + px). \tag{6.41}$$

Then, according to expressions (6.34) for $\xi(a)$ and $\eta(a)$, (6.10) can be rewritten as

$$\ddot{x} + \left[2p - \frac{\varepsilon}{\omega_0} \frac{\eta(a)}{a}\right] \dot{x} + \left[q^2 + \varepsilon \xi(a) - \frac{\varepsilon p}{\omega_0} \frac{\eta(a)}{a}\right] x = 0, \tag{6.42}$$

which is a quasilinear differential equation with the corresponding characteristic equation

$$s^2 + \left[2p - \frac{\varepsilon}{\omega_0} \frac{\eta(a)}{a}\right] s + q^2 + \varepsilon \xi(a) - \frac{\varepsilon p}{\omega_0} \frac{\eta(a)}{a} = 0. \tag{6.43}$$

The roots of (6.43) are $s_{1,2} = \sigma \pm j\omega$, where σ and ω are exactly the same as those obtained in (6.39) by the rigorous approach. Therefore the "equivalent" linear differential equation corresponding to (6.10) is

$$\ddot{x} - 2\sigma \dot{x} + (\sigma^2 + \omega^2) x = 0, \tag{6.44}$$

where

$$\sigma = -p + \frac{\varepsilon}{2\omega_0} \frac{\eta(a)}{a}$$
$$\omega^2 = \omega_0^2 + \varepsilon \xi(a), \tag{6.39}$$

and $\omega_0^2 = q^2 - p^2$. Thus the expressions for σ and ω, obtained by the equivalent linearization, are the same as those determined rigorously by using the asymptotical method approach. This result may now be transferred to the harmonic linearization that can treat the nonlinear equation

$$\ddot{x} + F(x, \dot{x}) = 0 \tag{6.1}$$

without introducing the small parameter.

Transient Oscillations

The solution $x = x(t)$ of (6.1) is assumed to be

$$x = a \sin \phi, \tag{6.31}$$

where

$$\dot{a} = a\sigma, \qquad \dot{\phi} = \omega. \tag{6.32}$$

Solution (6.31) is supposed to be close to the solution of a linear differential equation

$$\ddot{x} - 2\sigma_0 \dot{x} + (\sigma_0^2 + \omega_0^2)x = 0, \tag{6.45}$$

which is

$$x = a_0 e^{\sigma_0 t} \sin(\omega_0 t + \phi_0), \tag{6.46}$$

where a_0 is the constant initial amplitude and ϕ_0 is the constant initial phase. The term "close" means that at a given limited interval of time the values of σ and ω may be considered as slowly varying quantities around the values σ_0 and ω_0. Therefore, from (6.31) and (6.32), the solution $x(t)$ can be expressed in an integral form

$$x = a_0 \exp\left(\int_0^t \sigma \, dt\right) \sin\left(\int_0^t \omega \, dt + \phi_0\right), \tag{6.47}$$

which indicates that the basic unknown variables are $\sigma = \sigma(t)$ and $\omega = \omega(t)$. These quantities are considered as variables in time and, although they have slowly varying values, they are essentially different from the previous constant values σ_0 and ω_0. The rate of the variation of σ and ω determines the validity of the whole approach and is imposed by the original nonlinear differential equation itself. A significant fact, however, is that the rate of change of the amplitude $a(t) = a_0 \exp\left(\int_0^t \sigma \, dt\right)$ is not limited and (6.47) may represent slightly damped as well as well-damped oscillations.

In the harmonic linearization that follows, the notation for amplitude and frequency is a and ω instead of the previous A and Ω. This indicates that the amplitude and frequency are considered variable in time. Furthermore, if the value of σ is not small, by differentiating (6.31), we obtain

$$sx = a\omega \cos \phi + a\sigma \sin \phi. \tag{6.48}$$

From (6.48) and (6.31)

$$\sin \phi = \frac{x}{a}, \qquad \cos \phi = \frac{s - \sigma}{a\omega} x. \tag{6.49}$$

Therefore the first "damped harmonic" of the nonlinear function $y = F(x, \dot{x})$ when $x = a(t) \sin \phi(t)$ is

$$F(x, \dot{x}) = N_1 x + N_2 \frac{s - \sigma}{\omega} x = \left(N_1 - \frac{\sigma}{\omega} N_2\right)x + \frac{N_2}{\omega} sx, \tag{6.50}$$

where the coefficients N_1 and N_2, being analogous to the coefficients $\xi(a)$ and $\eta(a)$ discussed above, are given by

$$N_1 = \frac{1}{\pi a} \int_0^{2\pi} F(a \sin \phi, a\omega \cos \phi + a\sigma \sin \phi) \sin \phi \, d\phi$$

$$N_2 = \frac{1}{\pi a} \int_0^{2\pi} F(a \sin \phi, a\omega \cos \phi + a\sigma \sin \phi) \cos \phi \, d\phi. \tag{6.51}$$

For $s = \sigma + j\omega$, they are coefficients of the describing function $N = N_1 + jN_2$, which are functions of a, ω, and σ; that is,

$$N_1 = N_1(a, \omega, \sigma), \qquad N_2 = N_2(a, \omega, \sigma). \tag{6.52}$$

In case of a nonlinear function $F(x)$, the coefficients N_1 and N_2 are

$$N_1 = \frac{1}{\pi a} \int_0^{2\pi} F(a \sin \phi) \sin \phi \, d\phi$$

$$N_2 = \frac{1}{\pi a} \int_0^{2\pi} F(a \sin \phi) \cos \phi \, d\phi \tag{6.53}$$

and therefore they are functions of the amplitude a only as

$$N_1 = N_1(a), \qquad N_2 = N_2(a). \tag{6.54}$$

For a given nonlinear function $F(x)$, the calculations of $N_1(a)$ and $N_2(a)$ can be performed by using the results of Appendix F without any further considerations.[1]

---

[1] When the nonlinear function is $F(x, \dot{x})$, the harmonic linearization of (6.50) has an additional term $(\sigma/\omega)N_2$ that did not exist in the linearization corresponding to sustained oscillations as outlined in Section 3.3. Thus the harmonic linearization for transient analysis, which is rather complex, should be used when the quantity $|\sigma/\omega|$ is not small. Otherwise the terms in (6.48)–(6.51) associated with σ/ω can be neglected and the coefficients N_1 and N_2 are obtained by

$$N_1 = \frac{1}{\pi a} \int_0^{2\pi} F(a \sin \phi, a\omega \cos \phi) \sin \phi \, d\phi$$

$$N_2 = \frac{1}{\pi a} \int_0^{2\pi} F(a \sin \phi, a\omega \cos \phi) \cos \phi \, d\phi,$$

which are equivalent to those used in Chapter 3, except that A and Ω are substituted by a and ω, the quantities a and ω being slowly varying functions of time. Therefore, in connection with the analysis of slowly varying signals presented in Chapter 3, the same expressions of N_1 and N_2 have been used as employed in the consideration of periodic oscillations, where the amplitude A and the frequency Ω were constant.

Transient Oscillations

In order to illustrate the presented harmonic linearization, we apply Popov's extension to a variant of van der Pol's equation, which is

$$\ddot{x} - p(1 - x^2)\dot{x} + x = 0, \tag{6.55}$$

where p is not considered small. Equation 6.55 may be rewritten as

$$(s^2 - ps + 1)x + px^2sx = 0, \qquad s \equiv \frac{d}{dt}. \tag{6.56}$$

For the nonlinear function

$$F(x, sx) = px^2sx, \tag{6.57}$$

the describing function coefficients are

$$\begin{aligned}
N_1 &= \frac{1}{\pi a} \int_0^{2\pi} pa^2 \sin^2 \phi (a\omega \cos \phi + a\sigma \sin \phi) \sin \phi \, d\phi = \frac{3p}{4} a^2 \sigma \\
N_2 &= \frac{1}{\pi a} \int_0^{2\pi} pa^2 \sin^2 \phi (a\omega \cos \phi + a\sigma \sin \phi) \cos \phi \, d\phi = \frac{p}{4} a^2 \omega.
\end{aligned} \tag{6.58}$$

The harmonic linearization of the function given in (6.57) is obtained from (6.50) by using the expressions of N_1 and N_2 calculated in (6.58); that is,

$$px^2sx = \frac{p}{2} a^2 \sigma x + \frac{p}{4} a^2 sx. \tag{6.59}$$

The equivalent linear differential equation is now

$$[s^2 - p(1 - \tfrac{1}{4}a^2)s + 1 + \tfrac{1}{2}pa^2\sigma]x = 0. \tag{6.60}$$

By solving the corresponding characteristic equation, we obtain the roots $s_{1,2} = \sigma \pm j\omega$ with

$$\begin{aligned}
\sigma &= \frac{p}{2} (1 - \tfrac{1}{4}a^2) \\
\omega^2 &= 1 - \frac{p^2}{4} (1 - \tfrac{3}{2}a^2 + \tfrac{5}{16}a^4).
\end{aligned} \tag{6.61}$$

From (6.61) it follows that for $\sigma = 0$, $a = A = 2$ and $\omega = \Omega = 1$, which determine the periodic solution $x = 2 \sin (t + \phi_0)$. The solution is the same as that obtained by the original Krylov-Bogoliubov method in Section 3.5. Furthermore, the solution is stable since for $a > 2$, $\sigma < 0$, and for $a < 2$, $\sigma > 0$.

By using (6.31) and (6.32) of the first approximation, we can obtain the amplitude function $a = a(t)$ by

$$t = \int_{a_0}^{a} \frac{da}{a\sigma(a)} = \frac{1}{p} \left(\ln \frac{a^2}{|4 - a^2|} - \ln \frac{a_0^2}{|4 - a^2|} \right), \tag{6.62}$$

from which

$$a = \frac{a_0 e^{(p/2)t}}{[1 + \tfrac{1}{4}a_0^2(e^{pt} - 1)]^{1/2}}. \tag{6.63}$$

This equation is identical to (3.90) of Section 3.5.

The frequency ω as given in the second equation 6.61 is not the same as the frequency obtained by the Krylov-Bogoliubov method, where it was independent of the amplitude a and equal to one. In Popov's extension, however, the constant p has a finite value. If $p = \varepsilon$, where ε is the small parameter, then the term in (6.61) associated with p^2 can be neglected and we obtain the value $\omega = 1$.

6.3 High-Order Systems. Parameter Plane Analysis

A majority of practical nonlinear systems cannot be adequately described by a second-order nonlinear differential equation. Therefore, in order to study the transient oscillations in high-order nonlinear systems, it is necessary to generalize the approach outlined in the previous section and apply it to a nonlinear differential equation of the form

$$B(s)x + C(s)F(x, sx) = 0, \qquad s \equiv \frac{d}{dt}, \tag{6.64}$$

where $B(s)$ and $C(s)$ are polynomials in s with the degree of the polynomial $B(s)$ higher than the degree of the polynomial $C(s)$, and the function $F(x, sx)$ represents the nonlinearity. The generalization of the transient response analysis based on the Krylov-Bogoliubov method consists in determining the conditions under which (6.64) has a solution $x = x(t)$ close to

$$x = a \sin \phi \tag{6.31}$$

with $a = a(t)$, $\phi = \phi(t)$, given as

$$\dot{a} = \sigma a, \qquad \dot{\phi} = \omega. \tag{6.32}$$

This means that the solution $x(t)$ for limited intervals of time and certain regions of initial conditions is close to a damped sinusoid $a_0 e^{-\sigma_0 t} \sin \omega_0 t$ with fixed values σ_0 and ω_0. By the concept of equivalent linearization the above can also be interpreted as a problem of finding a second-order linear differential equation

$$\ddot{x} - 2\sigma\dot{x} + (\sigma^2 + \omega^2)x = 0, \tag{6.44}$$

which is equivalent to (6.64) in the sense that their approximate solutions are sufficiently close to each other and may be expressed as

$$x = a_0 \exp\left(\int_0^t \sigma \, dt\right) \sin\left(\int_0^t \omega \, dt + \phi_0\right). \tag{6.47}$$

246 Transient Oscillations

Equation 6.47 is obtained by integrating (6.32) and substituting the derived functions $a = a(t)$ and $\omega = \omega(t)$ into (6.31). By this equivalent linearization, the problem of finding the amplitude a and the frequency ω as functions of time, $a(t)$ and $\omega(t)$ can be reduced to a determination of the quantities σ and ω as functions of the amplitude

$$\sigma = \sigma(a), \qquad \omega = \omega(a). \tag{6.65}$$

Therefore the equivalent equation 6.44 is a second-order linear differential equation with slowly varying coefficients expressed by the functions $\sigma(a)$ and $\omega(a)$. The quantities σ and ω are slowly varying functions of time that are not explicitly dependent on time but rather through the amplitude $a = a(t)$. As a result of this fact, the rate of change of the amplitude $a(t)$ with respect to time is not limited, and the assumed solution x of (6.47) may represent both slightly damped and well-damped oscillations. The condition of slow variations in time is imposed on the quantities σ and ω, and whether it is satisfied or not depends inherently upon the original equation 6.64.

By the above approach two problems should be solved. First, we must obtain the equivalent second-order differential equation from the given high-order nonlinear differential equation, which reduces to the evaluation of the corresponding functions $\sigma(a)$ and $\omega(a)$. The second problem is to determine the amplitude $a = a(t)$ and the phase $\phi = \phi(t)$ and, thereby, the first approximation of the solution $x = a \sin \phi$ by using (6.47) and the derived functions $\sigma(a)$ and $\omega(a)$. It should be noted that in a majority of practical situations the second problem is not necessarily considered since the functions $\sigma(a)$ and $\omega(a)$ yield a satisfactory description of the required solution $x = x(t)$. Nevertheless, a graphical solution of the second problem will be presented after the first problem has been considered.

To perform the harmonic linearization of the nonlinear function $F(x, sx)$ when the variable x is determined by (6.31) and (6.32), it is necessary to express $\dot{x} = sx = a\omega \cos \phi + a\sigma \sin \phi$. Then, using the expressions for x and sx, we derive

$$\sin \phi = \frac{x}{a}, \qquad \cos \phi = \frac{s - \sigma}{a\omega} x. \tag{6.49}$$

The first "damped harmonic" of the nonlinear function $y = F(x, sx)$ is

$$F(x, sx) = \left(N_1 - \frac{\sigma}{\omega} N_2\right) x + \frac{N_2}{\omega} sx, \tag{6.50}$$

where the coefficients $N_1 = N_1(a, \sigma, \omega)$, $N_2 = N_2(a, \sigma, \omega)$ of the describing function $N = N_1 + jN_2$ are given by

$$N_1 = \frac{1}{\pi a} \int_0^{2\pi} F(a \sin \phi, a\omega \cos \phi + a\sigma \sin \phi) \sin \phi \, d\phi$$

$$N_2 = \frac{1}{\pi a} \int_0^{2\pi} F(a \sin \phi, a\omega \cos \phi + a\sigma \sin \phi) \cos \phi \, d\phi, \tag{6.51}$$

High-Order Systems. Parameter Plane Analysis 247

as derived in the previous section. Often the nonlinear function is $F(x)$ and the coefficients N_1 and N_2 of the approximate expression (6.50) are

$$N_1 = \frac{1}{\pi a} \int_0^{2\pi} F(a \sin \phi) \sin \phi \, d\phi$$
$$N_2 = \frac{1}{\pi a} \int_0^{2\pi} F(a \sin \phi) \cos \phi \, d\phi \qquad (6.53)$$

and the results of Appendix F can be used directly for calculation of $N_1 = N_1(a)$ and $N_2 = N_2(a)$, except that A is replaced by a.

In expansion (6.50) of the function $F(x, sx)$, the higher "damped harmonics" are neglected for exactly the same reasons as given by the applicability conditions outlined in Section 3.4, where the periodic solutions were considered. The fact that σ and ω are no longer constant does not alter the argument justifying this step, and the same order of accuracy is possible and can be expected for transient solutions obtained by using this approximation as was found for steady-state periodic solutions.

With the harmonic linearization of the function $F(x, sx)$ introduced in (6.50), the nonlinear differential equation can be rewritten as a quasilinear differential equation

$$\left[B(s) + C(s)\left(N_1 + \frac{s - \sigma}{\omega} N_2\right) \right] x = 0 \qquad (6.66)$$

with the corresponding characteristic equation

$$B(s) + C(s)\left(N_1 + \frac{s - \sigma}{\omega} N_2\right) = 0, \qquad (6.67)$$

where $N_1 = N_1(a, \sigma, \omega)$, $N_2 = N_2(a, \sigma, \omega)$. Equation 6.66 is called quasilinear in the sense that its coefficients, namely N_1 and N_2, are time varying quantities that are not explicitly dependent on time, but rather through the amplitude $a = a(t)$ and the frequency $\omega = \omega(t)$ of the assumed solution $x = a \sin \phi$.

Now high-order quasilinear differential equation 6.66 can be approximately represented by the second-order quasilinear differential equation

$$s^2 x - 2\sigma s x + (\sigma^2 + \omega^2) x = 0, \qquad s \equiv \frac{d}{dt}, \qquad (6.44)$$

where

$$\sigma = \sigma(a), \qquad \omega = \omega(a), \qquad (6.65)$$

if the characteristic equation

$$s^2 - 2\sigma s + \sigma^2 + \omega^2 = 0 \qquad (6.68)$$

of (6.44) is an adequate substitute for the characteristic equation 6.67 corresponding to the high-order quasilinear equation 6.66. Thus the problem of equivalence of the (6.66) and (6.44) is reduced to the problem of root dominancy discussed in Appendix C in connection with time solutions of linear differential equations with constant coefficients. In other words, (6.66) and (6.44) are equivalent if the two roots $s_{1,2} = \sigma \pm j\omega$ of (6.68) are dominant roots[2] of the characteristic equation 6.67. Since the quantities σ and ω determining the roots $s_{1,2}$ are functions of the amplitude, the variation of $\sigma(a)$ and $\omega(a)$ should be such that the dominancy condition for roots $s_{1,2}$ is satisfied for all possible values of the amplitude a. This dominancy condition not only enables a simple analysis of transient solutions, but is also a desired property of the high-order differential equations representing systems in a majority of practical situations.

The determination of the actual functions $\sigma(a)$, $\omega(a)$, checking that the dominancy condition is being satisfied, can be performed in the parameter plane by a variant of the inverse mapping. In general, when the describing function coefficients N_1 and N_2 are functions of a, σ, and ω, then the position of the point $M(\alpha = N_1; \beta = N_2)$ is specified by the functions

$$\alpha = N_1(a, \sigma, \omega), \qquad \beta = N_2(a, \sigma, \omega). \tag{6.69}$$

On the other hand, the characteristic curves are determined in the usual fashion by the functions

$$\alpha = \alpha(\sigma, \omega), \qquad \beta = \beta(\sigma, \omega). \tag{6.70}$$

If (6.69) and (6.70) are plotted in the parameter $\alpha\beta$ plane, a graphical solution of the functions

$$\sigma = \sigma(a), \qquad \omega = \omega(a) \tag{6.65}$$

is possible. Equations 6.69 and 6.70 represent two simultaneous equations in three unknowns, a, σ, and ω, which are solved for σ and ω as indicated in (6.65). After the characteristic Σ and ω curves are plotted by (6.70) and the M loci are constructed using (6.69), the competent points from which the

---

[2] As discussed in Appendix C, the contribution of large-magnitude roots of the characteristic equation to a solution of the corresponding linear differential equation is small, in comparison to the contribution of the dominant roots lying close to the origin of the s plane. It is irrelevant whether such large-magnitude roots are actually real or complex as long as they are far from the $j\omega$ axis of the s plane. The dominant roots are normally a pair of complex conjugate roots called "the control roots" (in certain cases a dominant root may be a real root, either alone or compounded with two closer complex conjugate roots). The criterion commonly adopted in design is that any roots that are to contribute negligibly to the actual solution should be placed at least six times as far from the $j\omega$ axis as the dominant roots governing the solution of the corresponding linear differential equation.

functions $\sigma(a)$ an d$\omega(a)$ are determined are found from the intersections of the two families of curves at which the values of σ and ω are identical. The related value of the amplitude a is read from the M loci. This procedure of determining the curves $\sigma(a)$ and $\omega(a)$ will be illustrated later in a specific example. It may be noted now that the described procedure is a variant of the inverse parameter plane mapping in which the parameters α and β are composite functions of the amplitude a. The results of the mapping are therefore the functions $\sigma(a)$ and $\omega(a)$ rather than the functions $\sigma(\alpha, \beta)$ and $\omega(\alpha, \beta)$ as previously defined. Nevertheless, the results obtained in Section 1.6, where the inverse parameter mapping was considered, are extensively used in deriving the functions $\sigma(a)$ and $\omega(a)$. Finally, a significant fact in the whole approach is that by the parameter plane analysis and the inverse mapping the regions of the parameter values and the initial conditions, for which the dominancy conditions is satisfied, can be found in a straightforward manner. In this respect the material outlined in Section 1.6 can be used to obtain the mentioned regions in which the above approach is justified, and an equivalent quasilinear differential equation of the second-order can be an adequate approximation of the corresponding high-order nonlinear differential equation.

Once the functions $\sigma(a)$ and $\omega(a)$ are calculated, and by solving the two first-order differential equations

$$\dot{a} = \sigma a, \qquad \dot{\phi} = \omega, \tag{6.32}$$

it is possible to find the first approximation

$$x = a \sin \phi \tag{6.31}$$

to the solution of the nonlinear differential equation

$$B(s)x + C(s)F(x, sx) = 0. \tag{6.64}$$

The integrals of (6.32) for the given initial conditions ($t = 0$)

$$a = a_0, \qquad \phi = \phi_0 \tag{6.71}$$

have the form

$$\int_{a_0}^{a} \frac{da}{a\sigma(a)} = t, \qquad \phi = \int_0^t \omega(a) \, dt + \phi_0, \tag{6.72}$$

which corresponds to the solution form of (6.47). From the first equation (6.72) the function $a(t)$ can be found. The second equation then is used with the function $a(t)$ to determine the function $\phi(t)$. Thus the approximate solution $x(t) = a(t) \sin \phi(t)$ given in (6.31) is obtained.

The solution process of (6.32) is performed graphically. In a majority of practical cases the process may be omitted since the functions $\sigma(a)$ and $\omega(a)$ yield sufficient information about the solution. This process, however, may be important and the corresponding procedure will be outlined in the following section.

The parameter plane concept offers the possibility for a more accurate analysis of the transient oscillations and further generalization of the outlined approach. Consider again the differential equation

$$s^2 x - 2\sigma s x + (\sigma^2 + \omega^2)x = 0, \tag{6.44}$$

which represents (6.64) for the assumed solution $x = x(t)$ given as

$$x(t) = a_0 \exp\left(\int_0^t \sigma_1 \, dt\right) \sin\left(\int_0^t \omega_1 \, dt + \phi_0\right). \tag{6.73}$$

Note that in this equation ω_1 and σ_1 are used to denote ω and σ of (6.47). Now the conditions should be found on σ_1 and ω_1 for $x(t)$ to be the solution of (6.44).

If (6.73) is substituted in (6.44), we obtain the required conditions in the form

$$\dot{\sigma}_1 + \sigma_1^2 - 2\sigma\sigma_1 - \omega_1^2 + \sigma^2 + \omega^2 = 0$$
$$2\sigma_1\omega_1 + \dot{\omega}_1 - 2\sigma\omega_1 = 0. \tag{6.74}$$

If we consider that $\sigma_1 = \sigma$ and $\omega_1 = \omega$, then the conditions (6.74) yield $\dot{\sigma} = \dot{\omega} = 0$, which is in agreement with the previously used assumption that σ and ω are slowly varying functions of time. In other words, the rate of change in time of the values σ and ω as the amplitude a is varied is sufficiently small and can be neglected. Conditions (6.74) take into account the rate of change in σ and ω, but as a result the functions $\sigma(a)$ and $\omega(a)$ cannot be used directly to obtain the solution $x(t)$ as given in (6.47). The solution $x(t)$ should then be calculated from (6.73), where σ_1 and ω_1 are obtained using conditions (6.74). This improved approach may be considered as a generalization of the Grensted [6.3] results obtained for second-order equations.

To find the values σ_1 and ω_1 from the given functions $\sigma(a)$, $\omega(a)$, and conditions (6.74), we can solve (6.74) for σ_1 and ω_1^2 as

$$\sigma_1 = \sigma - \frac{\dot{\omega}_1}{2\omega_1}$$
$$\omega_1^2 = \dot{\sigma} + \omega^2 - \frac{\ddot{\omega}_1}{2\omega_1} + \frac{3\dot{\omega}_1^2}{4\omega_1^2}. \tag{6.75}$$

By neglecting the higher-order terms ($\ddot{\omega}_1, \dot{\omega}_1^2, \ddot{\sigma} \simeq 0$), we may derive from (6.75) the relationship

$$\sigma_1 = \sigma - \frac{\omega\dot{\omega}}{2(\dot{\sigma} + \omega^2)}$$
$$\omega_1 = (\dot{\sigma} + \omega^2)^{1/2}. \tag{6.76}$$

Thus, once the functions $\sigma = \sigma(a)$ and $\omega = \omega(a)$ are calculated from the parameter plane diagram, the quantities σ_1 and ω_1 can be evaluated from (6.76) and used in (6.73) to obtain the solution $x(t)$. The numerical procedure is the same as that when the functions $\sigma(a)$ and $\omega(a)$ are used, and is described in the following section. The comparison between the first approximation (σ, ω) and the second approximation (σ_1, ω_1) is suggested in Problem 6.1. This problem, however, should be attempted only after the material in the next section is studied. It should also be noted that the use of the values σ_1 and ω_1 in (6.73) requires computer application with the result that a better approximation is normally achieved.

6.4 Solution Procedure and Design

As outlined previously, the parameter plane analysis of transient nonlinear oscillations has two basic steps. First, the functions $\sigma(a)$ and $\omega(a)$ are determined from the $\alpha\beta$ plane where the Σ and ω curves as well as the M loci are plotted. The second step is the evaluation of the transient oscillations $x(t)$ from the obtained diagrams of $\sigma(a)$, $\omega(a)$ and the value of the initial condition x_0. Both steps in the solution procedure will be best illustrated in the following example.

Consider a nonlinear feedback system shown in Figure 6.1 with the transfer functions

$$G_1(s) = \frac{2}{0.2s^2 + 0.8s + 1}, \quad G_2(s) = \frac{0.5(s + 1)}{0.2s + 1},$$

$$G_{-1}(s) = \frac{10K_{-1}}{T_{-1}s + 1}, \quad (6.77)$$

and the nonlinearity n shown in Figure 6.2. The corresponding characteristic equation is

$$0.04T_{-1}s^4 + (0.36T_{-1} + 0.04)s^3 + (2T_{-1} + 0.36)s^2$$
$$+ (2T_{-1} + 4K_{-1}N_1 + 2)s + 20K_{-1}N_1 + 2 = 0, \quad (6.78)$$

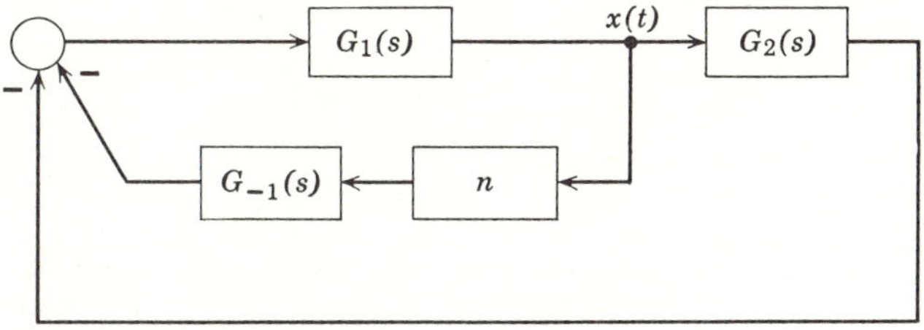

Figure 6.1 System block diagram.

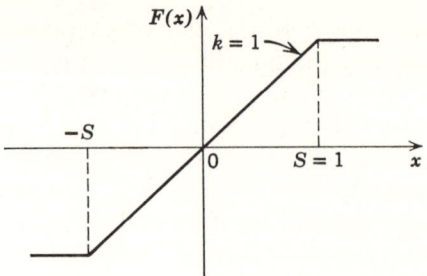

Figure 6.2 Nonlinear characteristic.

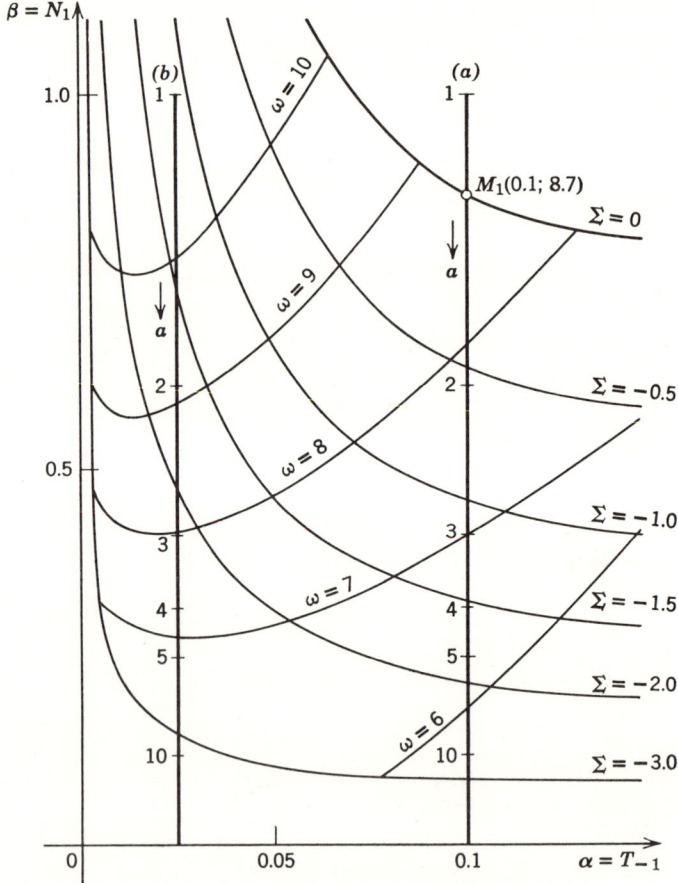

Figure 6.3 Parameter plane diagram.

where N_1 is the describing function coefficient representing the nonlinearity. If the linear parameter K_{-1} is equal to one, by substituting

$$\alpha = T_{-1}, \qquad \beta = N_1 \tag{6.79}$$

into (6.78), we obtain the Σ and ω curves in the usual manner as shown in Figure 6.3. From the net of Σ and ω curves, it is possible to evaluate the two complex conjugate roots $s_{1,2} = \sigma \pm j\omega$ for various values of α and β. This was shown once in Section 1.6, where root evaluation and inverse parameter mapping were considered. Thus a diagram of Figure 6.4 can be plotted by evaluating σ and ω along the M locus (a) in Figure 6.3. The M locus represents the variation of the point $M(\alpha; \beta)$ caused by the variation of the amplitude a in the describing function $N_1 = N_1(a) = \beta$. The variation of the M point that corresponds to the function $F(x)$ of Figure 6.2 can be obtained from diagram No. 2 of Table F.1 in Appendix F, or analytically by

$$N_1 = \left[\arcsin \frac{1}{a} + \frac{1}{a}\left(1 - \frac{1}{a^2}\right)^{1/2} \right], \qquad a \geq 1. \tag{6.80}$$

The position (a) of the M locus corresponds to $\alpha = T_{-1} = 0.1$.

Before the diagrams $\sigma(a)$ and $\omega(a)$ are utilized in the second step of the solution procedure in which the solution $x = x(t)$ is to be obtained, it is of importance to note that the dominancy of the roots $s_{1,2} = \sigma \pm j\omega$ should be checked. Therefore, two remaining roots are to be evaluated in the usual fashion by plotting either the Σ and ω curves or the σ curves that correspond to the case where the remaining roots are real. It can be observed that the system under investigation has the characteristic equation 6.78, which has

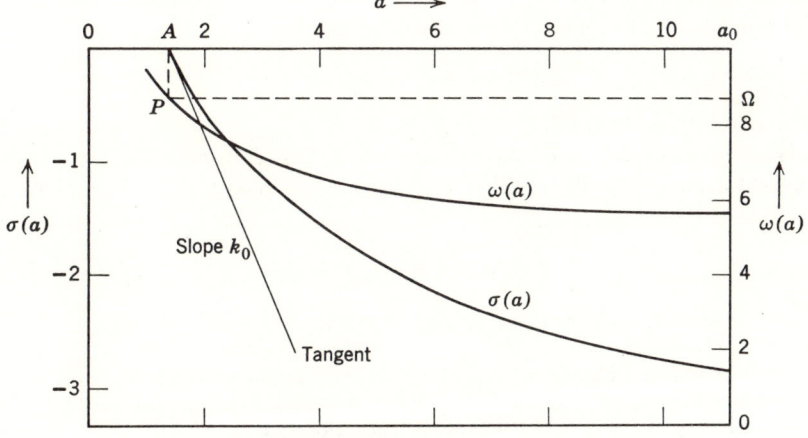

Figure 6.4 Diagram of functions $\sigma(a)$ and $\omega(a)$.

254   Transient Oscillations

already been considered in Figures 1.25 and 1.27. From these figures it is easy to conclude that the variation of the M point along the locus (a) of Figure 6.3 lies completely to the left convex side of the $\zeta = 1$ curve and, consequently, two tangents can always be drawn to the $\zeta = 1$ curve from the M locus (a). The two remaining roots, $s_{3,4}$, are therefore real and will not contribute to the oscillatory nature of the solution $x(t)$. On the other hand, they will cause a certain amplitude change in the solution $x(t)$, which cannot be detected by the following approximate procedure since it considers only the roots $s_{1,2} = \sigma \pm j\omega$ assuming their dominancy.

Consider now the diagram of functions $\sigma(a)$ and $\omega(a)$ in Figure 6.4. It may be concluded by inspection that whatever the initial condition is, the oscillations will end up in a stable limit cycle since the curve $\sigma(a)$ intersects the a axis at which $\sigma = 0$. At this intersection the tangent has a negative slope k_0 and $S_a{}^\sigma < 0$ which ensures stability of the limit cycle. The amplitude A of the limit cycle is read on the a axis to be approximately 1.33. The corresponding value of the frequency $\Omega = 8.7$ rad/sec is obtained as indicated by the point P and the dashed lines on the diagram $\omega(a)$.

Before an attempt is made to calculate the solution

$$x = a \sin \phi \qquad (6.31)$$

of the differential equation

$$B(s)x + C(s)F(x, sx) = 0, \qquad s \equiv \frac{d}{dt} \qquad (6.64)$$

as the solution of the differential equation

$$s^2 x - 2\sigma s x + (\sigma^2 + \omega^2)x = 0 \qquad (6.44)$$

from the functions $\sigma = \sigma(a)$ and $\omega = \omega(a)$, it is necessary to specify the initial conditions. For $t = 0$, it is assumed that[3]

$$x(0) = x_0, \qquad \dot{x}(0) = 0 \qquad (6.81)$$

as discussed in Section 4.5.

It is also assumed that at the instant $t = 0$, the solution $x = a \sin \phi$ starts as the solution of the linear differential equation 6.44, for which $\sigma = \sigma_0 = \sigma(a_0)$, $\omega = \omega_0 = \omega(a_0)$, and

$$x = a_0 e^{\sigma_0 t} \sin(\omega_0 t + \phi_0) \qquad (6.46)$$

$$\dot{x} = a_0 \sigma_0 e^{\sigma_0 t} \sin(\omega_0 t + \phi_0) + a_0 \omega_0 e^{\sigma_0 t} \cos(\omega_0 t + \phi_0). \qquad (6.82)$$

---

[3] For the second-order differential equation 6.44 that represents (6.64), it is sufficient to assume that $x(0) = x_0$ and $\dot{x}(0) = 0$. For (6.64), however it is assumed that $x(0) = x_0$, and $\dot{x}(0) = \ddot{x}(0) = \cdots = 0$, which is in accordance with the notion of the "strong stability" introduced by Bogoliubov [6.1].

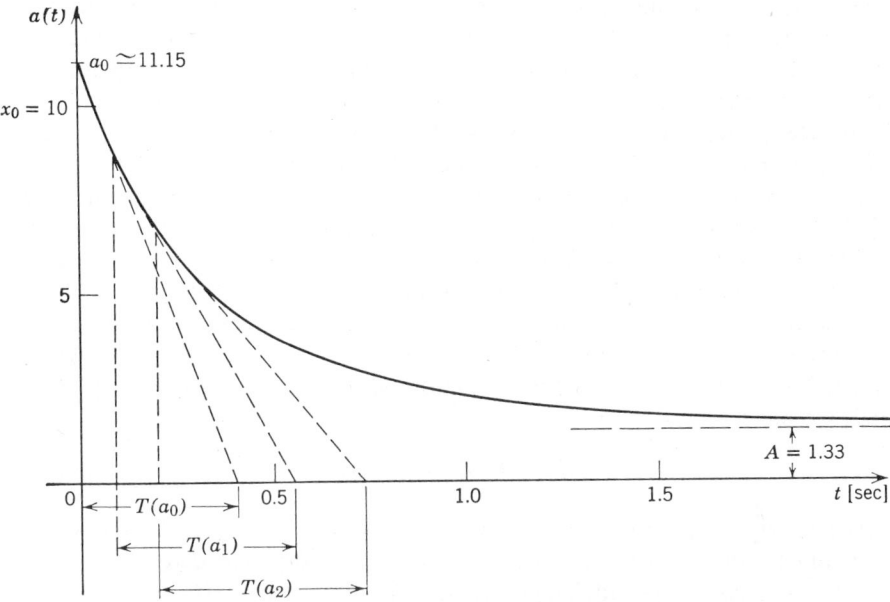

Figure 6.5 The envelope $a(t)$.

By using (6.81) we obtain from (6.46) and (6.82) the relationship

$$x_0 = a_0 \sin \phi_0, \quad \phi_0 = \arctan\left(-\frac{\omega_0}{\sigma_0}\right) \qquad (6.83)$$

among the starting values x_0, a_0, ϕ_0. If, in the example under consideration, the initial value $x_0 = 10$ is specified, the values $a_0 \simeq 11.2$ and $\phi_0 = 1.1$ rad are approximately determined from (6.83). (In cases where the dominant complex pair of roots is sufficiently close to the imaginary axis and $\sigma_0 \ll \omega_0$, the phase shift ϕ_0 may be considered as close to $\pi/2$ and, consequently, $a_0 \simeq x_0$.)

To construct the transient oscillation $x(t)$ from the functions $\sigma(a)$ and $\omega(a)$, a variable time constant $T(a)$ is introduced as

$$T(a) = -\frac{1}{\sigma(a)}. \qquad (6.84)$$

Then the first equation of

$$\dot{a} = \sigma a, \quad \dot{\phi} = \omega \qquad (6.32)$$

becomes

$$\dot{a} = -\frac{a}{T(a)}. \qquad (6.85)$$

If an initial condition x_0 is given and a_0 and ϕ_0 are evaluated from (6.83), then a distance $T(a_0)$ can be determined on the t axis as shown in Figure 6.5.

The straight line connecting the point a_0 and the point $T(a_0)$ is a tangent to the desired curve $a(t)$ at the point where $t = 0$ and $a = a_0$. It is assumed further that this tangent approximates the curve $a(t)$ on a certain limited interval of time up to a value $a = a_1$. Then a new value $T(a_1)$ is calculated from (6.80) and another tangent is plotted as shown in Figure 6.5. By repeating this process, the curve $a(t)$ is constructed in a few steps. The values $T(a_i)$ can be chosen at will, and usually few such values are enough since the whole procedure is approximate, having the goal to sketch the transient oscillation. Moreover, it is known that the function $a(t)$ tends to the value $A = 1.33$, which is the amplitude of the limit cycle.

Once the function $a(t)$ is constructed, it is easy to recalculate the curve $\omega(a)$ into $\omega(t)$ by using the diagram $a(t)$. The curve $\omega(t)$, which is shown in Figure 6.6, is used to obtain the function $\phi(t)$. From the second equation 6.32, we have

$$\phi(t) = \int_0^t \omega(t)\, dt + \phi_0. \tag{6.86}$$

According to (6.86), the function $\phi(t)$ should be determined as the area under the curve $\omega(t)$ given in Figure 6.6, provided that $\phi_0 = \arctan(-\omega_0/\sigma_0) = 1.1$ rad. Therefore the instant t_0—when $x(t) = a(t)$ occurs for $\omega t + \phi_0 = \pi/2$ and it is determined from the diagram of the function $\omega(t)$ as the area under the curve—becomes $\pi/2 - \phi_0$. The first crossover point t_1 of the solution $x(t)$ with the time axis is determined at the time at which the area under $\omega(t)$ at the interval (t_0, t_1) is equal to $\pi/2$. The instant

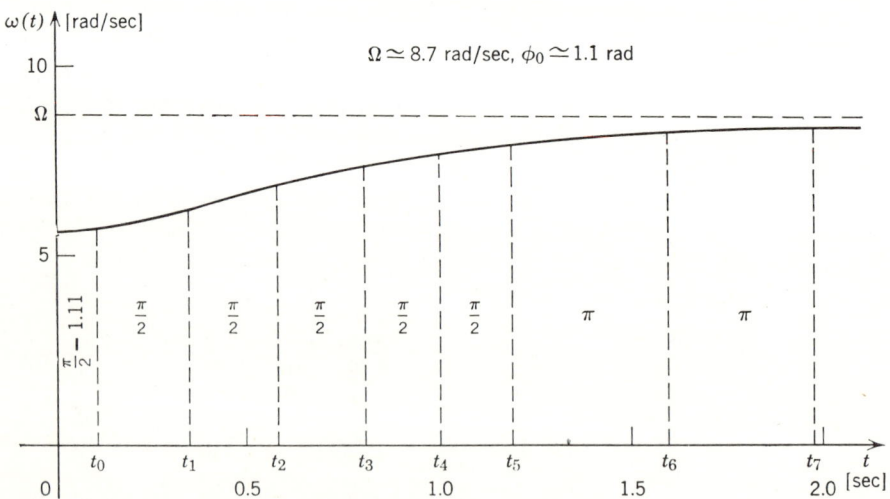

Figure 6.6 The function $\omega(t)$.

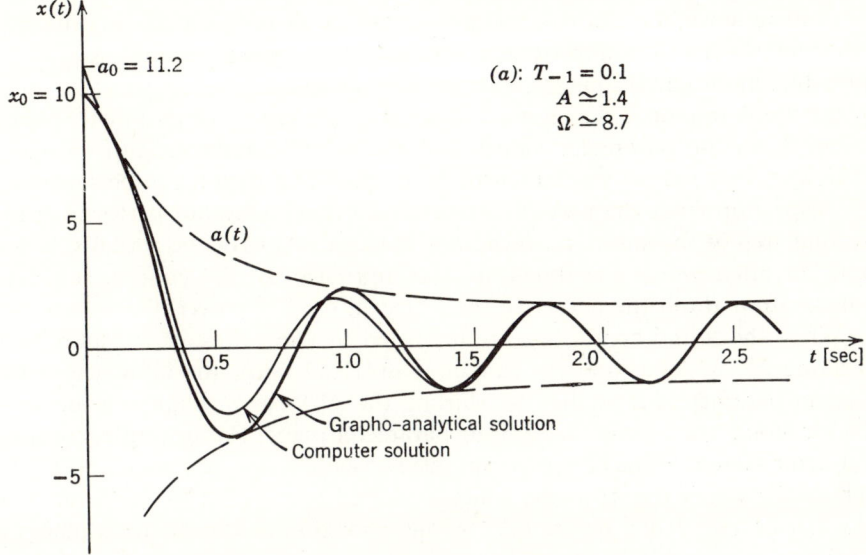

Figure 6.7 Transient oscillation $x(t)$.

of time t_2 is determined in the same way, fulfilling the area equal to $\pi/2$ at the interval (t_1, t_2), and so on. After a sufficient number of steps, the curve $\omega(t)$ approaches the asymptote $\Omega = 8.7$ rad/sec at the frequency of the limit cycle, and the instants denoting maxima, minima, and crossover points of the oscillation $x(t)$ are uniformly distributed along the t axis.

After the envelope $a(t)$ and the characteristic time intervals of $\phi(t)$ are determined, the oscillation $x = a \sin \phi$ can be sketched as shown in Figure 6.7. It should be noted that the sketching procedure can be readily programmed for computer calculation so that the curves $\sigma(a)$ and $\omega(a)$ are used to supply the input data.[4]

To check the accuracy of the obtained grapho-analytical result, a computer solution obtained by simulating the system is plotted on the same diagram. As can be seen, the computer solution is different from the one obtained by

---

[4] Once the values a_0 and ϕ_0 are calculated from (6.83) for a given x_0, the subsequent values of $x(t)$ at discrete instants of time $t_i = t_{i-1} + \Delta t$ can be calculated according to

$$x_i = a_i \sin \phi_i,$$

where a_i and ϕ_i are determined by

$$a_i = a_{i-1} \exp\left[\sigma(a_{i-1})\right]$$
$$\phi_i = \omega(a_{i-1}) \Delta t + \phi_{i-1}.$$

This algorithm is in accordance with the assumption that the representative equation 6.44 is linear on the interval (t_{i-1}, t_i) with constant values of $\sigma = \sigma(a_{i-1})$ and $\omega = \omega(a_{i-1})$.

the proposed approximation. In system design this is not a significant disadvantage since the above analysis is used as a guide in the preliminary considerations of the appropriate values of the system parameters and the possible initial conditions. The accurate solutions can be obtained by computer simulation once the region of interest is focused in the parameter plane, where both the parameter values and the initial conditions can be easily reviewed. In a majority of practical cases the useful region can be separated by inspection from the plotted characteristic curves without performing the second step of the solution procedure—the evaluation of the actual solution $x(t)$. In other words the functions $\sigma(a)$ and $\omega(a)$ usually contain sufficient information about the nature of the transient oscillation $x(t)$.

While the initial conditions are indicated by the values of the amplitude a along the M locus and the functions $\sigma(a)$ and $\omega(a)$, the influence of the system parameters is studied by shifting the M locus or changing the scale factor along one or both axes of the parameter plane. For example, consider the same system of the previous example but with the value $\alpha = T_{-1} = 0.025$. This affects only the M locus, which is shifted to position (b) in Figure 6.3. As can be seen from Figure 6.3, the new position of the M locus places it entirely in the stable region and there is no intersection with the $\Sigma = 0$ curve as previously (the point M_1). Thus the oscillations will eventually end up in the equilibrium position and die out completely. If the initial condition is again $x_0 = 10$, the outlined procedure will yield the curve $a(t)$ given in Figure 6.8, which proves the previous statement. By repeating the whole procedure, the solution $x(t)$ is obtained as shown in Figure 6.8. The computer solution again validates the grapho-analytical result.

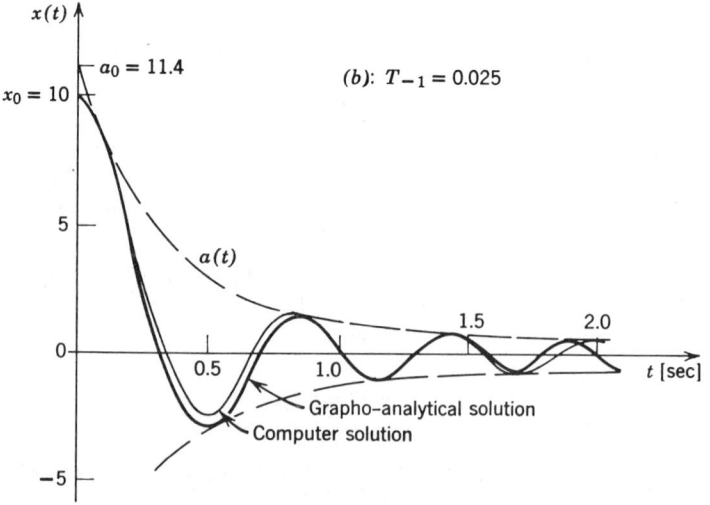

Figure 6.8 Transient oscillation $x(t)$.

Solution Procedure and Design 259

The transient response analysis can be performed by using the ζ and ω_n characteristic curves. The functions of interest are $\zeta = \zeta(a)$ and $\omega_n = \omega_n(a)$ instead of $\sigma = \sigma(a)$ and $\omega = \omega(a)$. This introduces no essential difference, since the two pairs of quantities are simply related by

$$\sigma = -\omega_n \zeta, \qquad \omega = \omega_n \sqrt{1 - \zeta^2}. \tag{6.87}$$

On the other hand, the information contained in the function $\zeta(a)$ and $\omega_n(a)$ may appear more useful than that of the functions $\sigma(a)$ and $\omega(a)$, especially if the relative stability and the actual overshoot of the solution $x(t)$ are of interest. The analysis procedure in this case is left to the reader, as stated in Problem 6.3.

In the design process a quick estimation of the transient response characteristics, such as the settling time T_s, overshoot x_m, and the number of oscillations λ during the time T_s, is desirable to perform directly from the curves $\sigma(a)$ and $\omega(a)$. Thus the plotting of the whole transient solution $x(t)$ is avoided until the above characteristics have satisfactory values. These values are rough approximations of the actual ones that will be shown later, but often enough give insight into the system behavior in the preliminary stage of the design. Once a region is found in the parameter plane that may give promising results, the best combination of the system parameters can be found for a given set of initial conditions by focusing the region and applying the outlined detailed procedure along with computer simulation. The rough estimation of the transient solution is performed according to the following considerations.

The settling time T_s of the transient $x(t)$ can be defined as the time required for the amplitude a to change from an initial value a_0 to a given value a_s. From the first equation of (6.72), T_s is given as

$$T_s = \int_{a_0}^{a_s} \frac{da}{a\sigma(a)}. \tag{6.88}$$

For a rough estimation, T_s may then be calculated by

$$T_s \simeq \frac{1}{\sigma_{av}} \ln \frac{a_s}{a_0}, \tag{6.89}$$

where σ_{av} is the average value of σ in the interval (a_0, a_s) determined from the function $\sigma(a)$. A better approximation can be obtained by dividing the interval (a_0, a_s) into several subintervals and using the formula

$$T_s = \sum_{i=1}^{n} \frac{1}{\sigma_i} \ln \frac{a_{si}}{a_{0i}}, \tag{6.90}$$

where n is the number of subintervals (a_{0i}, a_{si}) in which σ_i represents the average value of σ.

The overshoot x_m is determined as the value of the amplitude a for $\phi = 3\pi/2$. Thus, from the second equation of (6.32), it is

$$\phi = \int_{a_0}^{a} \frac{\omega(a)}{a\sigma(a)} \, da + \phi_0. \tag{6.91}$$

For $\phi = 3\pi/2$, and assuming roughly that $\phi_0 = \pi/2$, we obtain

$$\int_{a_0}^{a} \frac{\omega(a)}{a\sigma(a)} \, da = \pi. \tag{6.92}$$

From (6.92) the overshoot x_m is roughly given by

$$\left| \frac{x_m}{x_0} \right| \simeq \exp\left[\pi(\sigma_{av}/\omega_{av})\right], \tag{6.93}$$

where $\sigma_{av}(\sigma_{av} < 0)$ and ω_{av} are average values of σ and ω in the interval (a_0, a_s) and $|x_0| = a_0$.

For a better approximation, the interval (a_0, a_s) should be divided into several subintervals and expressed as

$$\phi - \phi_0 = \sum_{i=1}^{n} \frac{\omega_i}{\sigma_i} \ln \frac{a_{si}}{a_{0i}}, \tag{6.94}$$

where ω_i and σ_i are average values at the corresponding subintervals (a_{0i}, a_{si}). The computation by (6.94) should be performed starting from the value a_0 up to the point when $\phi - \phi_0 = \pi$. To that point corresponds $|x_m| = a_{si}$.

The number of oscillations λ in the interval (a_0, a_s) is given by

$$\lambda = \frac{\phi_s - \phi_0}{2\pi}, \tag{6.95}$$

where ϕ_s is the area under the curve $\omega(t)$ for $0 \leq t \leq T_s$, and $\phi_0 = \pi/2$. The number λ can therefore be roughly calculated from (6.94) as

$$\lambda = \frac{1}{2\pi} \sum_{i=1}^{n} \frac{\omega_i}{\sigma_i} \ln \frac{a_{si}}{a_{0i}}. \tag{6.96}$$

A simpler formula can be derived as

$$\lambda \simeq \frac{\omega_{av}}{2\pi\sigma_{av}} \ln \frac{a_s}{a_0}, \tag{6.97}$$

where ω_{av} and σ_{av} are average values of ω and σ in the whole interval (a_0, a_s). The application of the outlined rough approximation is left to the reader, as stated in Problem 6.4.

Solution Procedure and Design 261

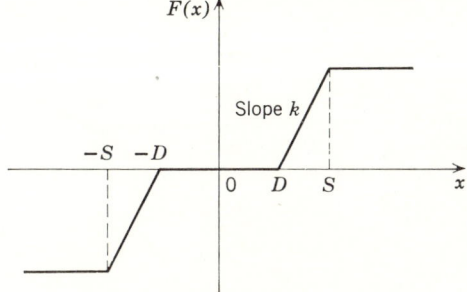

Figure 6.9 Nonlinear characteristic.

More can be learned about the transient process if a relative stability idea is introduced in the system analysis [6.5–7]. In terms of the parameter plane analysis the notion of the relative stability is readily interpreted as the requirement that the entire M locus is situated inside the relative stability region specified by either $\zeta \geq \zeta_0$ or $\sigma \leq \sigma_0$. As an illustration of such an analysis, consider again the control system of Figure 6.1 with the same transfer functions $G_1(s)$, $G_2(s)$, $G_{-1}(s)$; the nonlinearity n is that of Figure 6.9, with unspecified parameters k, D, and S. It is necessary to study the relative stability of the system, which is specified by $\zeta_0 = 0.5$, for various values of system parameters K_{-1} and T_{-1}, k, D, and S.

With the above specifications, the parameter plane diagram of the curve $\zeta = 0.5$ is plotted in Figure 6.10, where several M loci are drawn for different values of the linear system parameter $\alpha = T_{-1}$ and the ratio S/D of the nonlinear system parameters D and S. Figure 6.10 indicates that the locus (a) is unsatisfactory since a portion of the locus is outside the region $\zeta > 0.5$. However, if the initial value of the amplitude A is chosen so that it lies on the left side of the M locus and under the curve $\zeta = 0.5$, the relative stability requirement is fulfilled. The relative stability requirement is also satisfied by the locus (c) (which is the same as locus (a), but shifted along the α axis), independent of the signal level $x(t)$. The loci (b) and (d), which are plotted for different values of the ratio S/D and $T_{-1} = \alpha$, also satisfy the prescribed degree of relative stability.

The relative stability analysis is effective for a quick estimate of the transient process. The analysis can be improved if the $\zeta = 1$ curve is plotted on the same parameter plane diagram as shown in Figure 6.10. In normal control system design it is desired that the characteristic equation of the corresponding linearized system have only one pair of complex control roots, and that the other roots be real for certain variations of the signal $x(t)$. If we refer to Figure 6.10, it is easy to conclude that the mentioned root

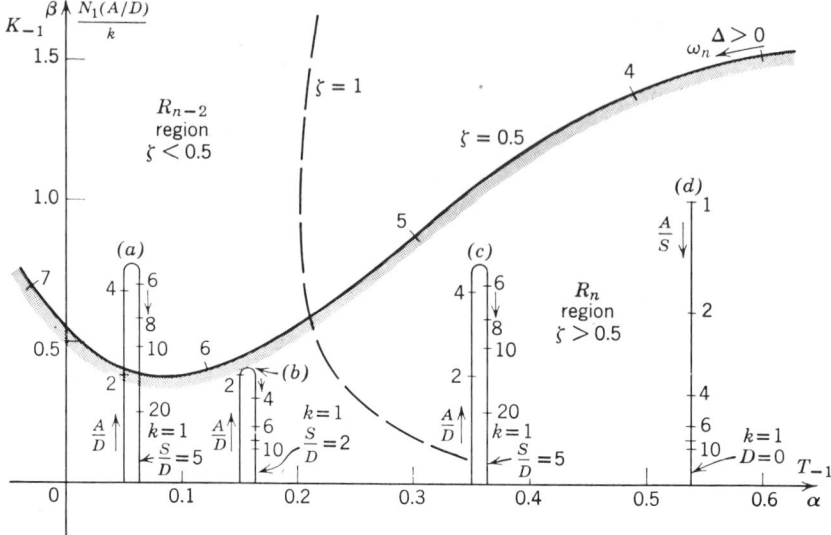

Figure 6.10 Parameter plane diagram for relative stability analysis.

configuration is obtained with the M locus (*b*) since two tangents can be drawn to the curve $\zeta = 1$ for the whole variation of the amplitude A. As already known, these tangents determine the real roots of the characteristic equation. The remaining two roots are the complex control pair related to the ζ curve.

Relative stability may be useful in a quick separation of the regions of parameter values and initial conditions that can result in satisfactory transient processes. If the obtained information is not sufficient, a detailed study of the transient process can be carried out by the procedure outlined previously.

6.5 Quasioptimization

In a wide class of control system applications, the minimum settling time is an essential performance requirement. If a system is considered as linear, the problem is to choose an appropriate location for dominant roots so that the corresponding damping factor and natural frequency result in a short duration of the transient process with a satisfactory overshoot and rise-time values. Thus a compromise should be found between the two second-order system error signals *a* and *b* in Figure 6.11. When the damping factor is large (curve *a*), there is no overshoot, but the rise time and settling time are unacceptable in most cases. If a compromise is found between the two curves

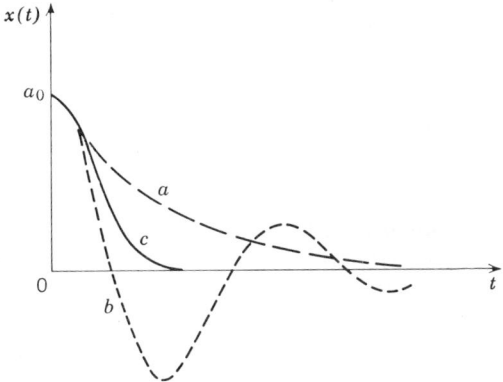

Figure 6.11 Transient processes of linear (dashed lines) and nonlinear (solid line) system.

on Figure 6.11, their good features deteriorate. This comes from the fact that in linear time invariant systems, the damping factor has a constant value.[5] For small values of the damping factor (curve b), the rise time of the transient process is short but the overshoot is excessive and so is the settling time.

To improve the system performance a variable damping factor may be introduced. Quantitative rules for the nature of the damping factor variation are easily evolved from physical reasoning. When the error is large, the damping should be small to permit short rise time; as the error becomes small, the damping should be high to prevent the overshoot and yield a short settling time. This would give a response similar to curve c of Figure 6.11. The approximate minimization of the duration of the transient process by introducing a variable damping is called the quasioptimization.[6]

The variable damping can be accomplished in a nonlinear system as shown previously. Thus in essentially linear systems a certain nonlinear element can be deliberately introduced to yield a desired effect. Then the analysis in the

---

[5] That a purely linear system is subject to limitations is readily apparent from a simple physical reasoning. The linear operation permits applications of maximum power to a driving element of the system through only a short portion of the operating period, and this maximum power may represent only a fraction of the peak power capability of the system to suppress the existing error. Further limitations are evident in the fact that the disturbances are corrected irrespective of their magnitude. In addition, the transient process of a linear system persists for infinite time; that of a nonlinear system may be of finite duration theoretically as well as practically.

[6] For exact treatment of time-optimal control systems, see M. Athans and P. L. Falb, *Optimal Control: An Introduction to the Theory and Its Applications*, McGraw-Hill, New York, 1966.

264 Transient Oscillations

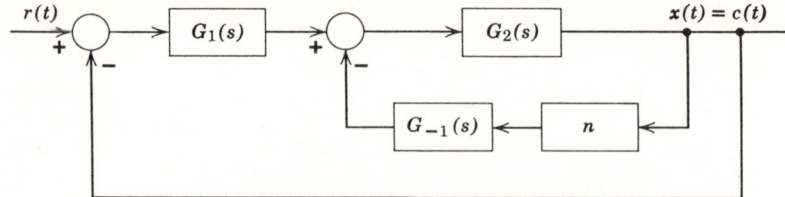

Figure 6.12 System block diagram.

parameter plane can give the quantitative results—what kind of nonlinear element should be used, where it should be placed in the system, what linear compensator should be added, and so on. The design procedure is illustrated by the following example.

Given the system of Figure 6.12 in which

$$G_1(s) = \frac{20}{s + 20}, \qquad G_2(s) = \frac{15K_2}{s(s + 1)(s + 15)}, \qquad G_{-1}(s) = K_{-1}s. \qquad (6.98)$$

The parameters K_2, K_{-1} and the characteristic of nonlinear element n should be chosen so that the transient oscillations will have a less than 15% overshoot for initial values a_0 from $a_0 = 0$ to $a_0 = 10$; and the rise time and settling time are to be as short as possible.

The design philosophy is to increase the amount of the inner-loop feedback as time progresses, and to move the initial dominant roots of the linearized system, from $\zeta = 0.4$ to a $\zeta > 1$ in the final state. The corresponding characteristic equation of the linearized system is

$$s^4 + 36s^3 + 335s^2 + (300 + 15K_2K_{-1}N_1)s + 300K_2 = 0. \qquad (6.99)$$

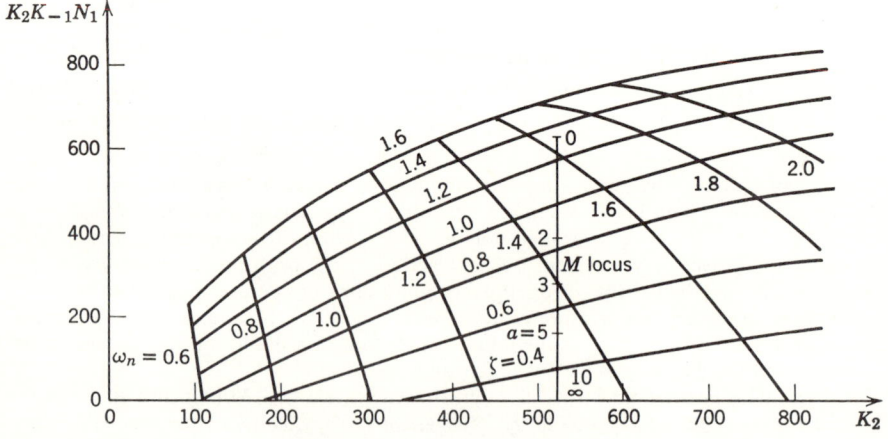

Figure 6.13 Parameter plane.

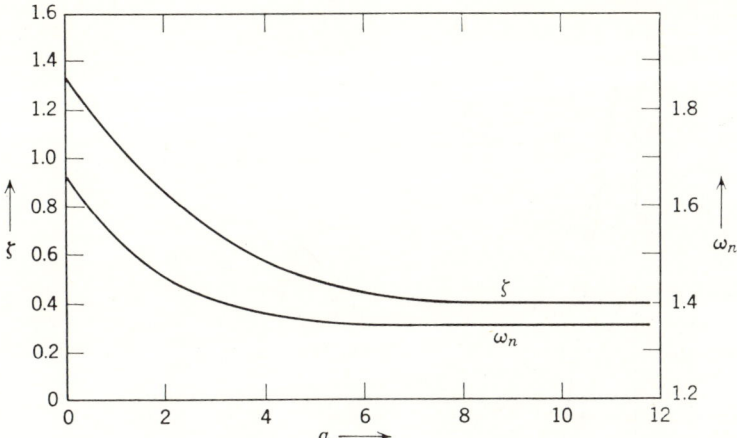

Figure 6.14 $\zeta(a)$ and $\omega_n(a)$ curves.

If $K_2 K_{-1} N_1$ and K_2 are considered as variable parameters, the parameter plane diagram is plotted in Figure 6.13. From the diagram it can be concluded that the desired variation of damping can be accomplished by a nonlinear element having a linear gain with saturation. The corresponding M locus is placed at the value $\alpha = K_2 = 525$. The value of the parameter K_{-1} is 1.2.

Figure 6.14 gives the $\zeta(a)$ and $\omega_n(a)$ plots determined in the usual manner from the intersections of the M locus and the characteristic ζ and ω_n curves. The transient oscillations for various initial conditions are shown in Figure 6.15 and may be compared to the response of the linear system having dominant roots at $\omega_n = 1.35$ and $\zeta = 0.4$ or $\zeta = 0.7$.

6.6 Two Nonlinearities

The procedure outlined in the preceding sections can be directly extended to systems with two nonlinear elements. Two classes defined in Section 4.6 are considered; namely, the class in which the two nonlinear elements have a common input and the class where the inputs are related by a linear differential equation. The application of the procedure is best shown by examples.

Consider the system shown in Figure 6.16 with the transfer functions

$$G_1(s) = \frac{10}{s^2}, \quad G_{-1}(s) = \frac{s}{s+10} \quad (6.100)$$

and the nonlinear elements n_1 and n_2 having the piecewise linear characteristics as shown in Figure 6.17 with dead zones denoted by D_1 and D_2, respectively.

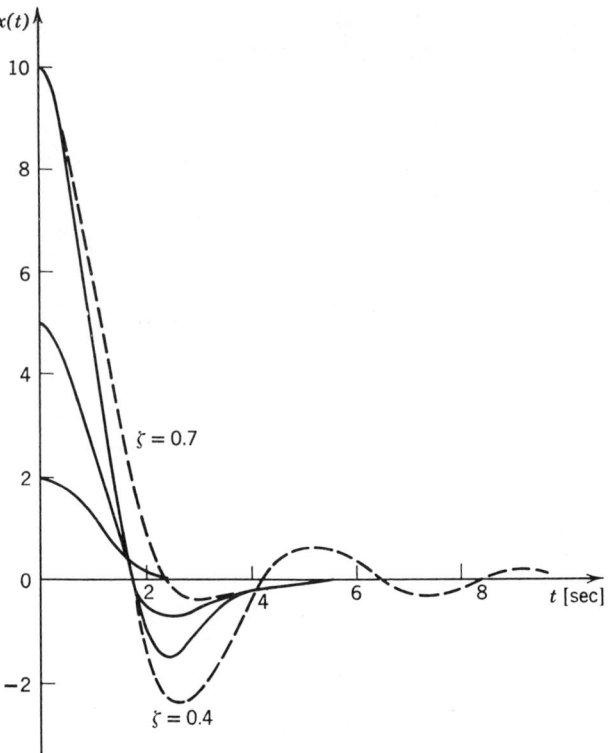

Figure 6.15 Transient oscillations.

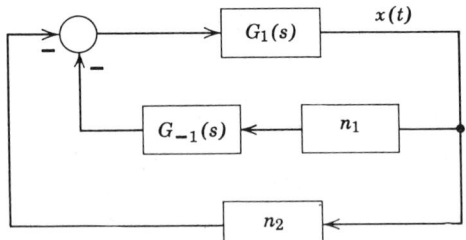

Figure 6.16 System block diagram.

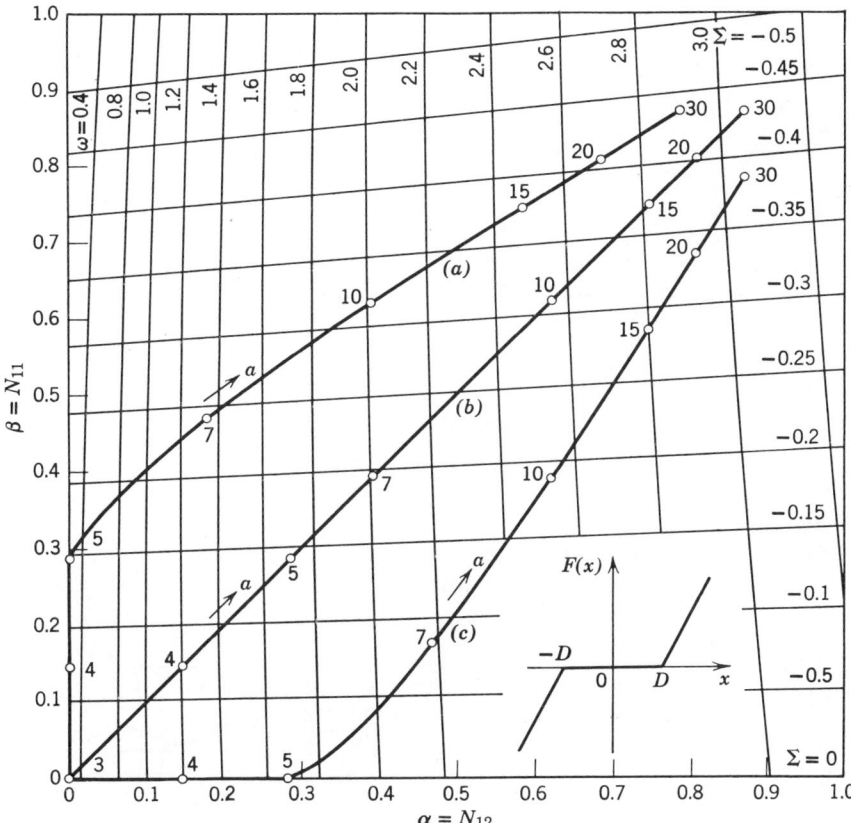

Figure 6.17 Parameter plane diagram.

The system belongs to the first class of systems with two nonlinear elements since the elements have a common input $x = x(t)$ assumed as

$$x = a \sin \phi, \quad \phi = \omega t. \tag{6.101}$$

Thus, according to Table F.1 of Appendix F, the describing functions N_{11} and N_{12} of the elements n_1 and n_2 are given as

$$N_{11} = 1 - \frac{2}{\pi}\left\{\arcsin\frac{D_1}{a} + \frac{D_1}{a}\left[1 - \left(\frac{D_1}{a}\right)^2\right]^{1/2}\right\}$$

$$N_{12} = 1 - \frac{2}{\pi}\left\{\arcsin\frac{D_2}{a} + \frac{D_2}{a}\left[1 - \left(\frac{D_2}{a}\right)^2\right]^{1/2}\right\}, \tag{6.102}$$

where it is assumed that both nonlinear elements have the slope k of the linear portion $k = 1$.

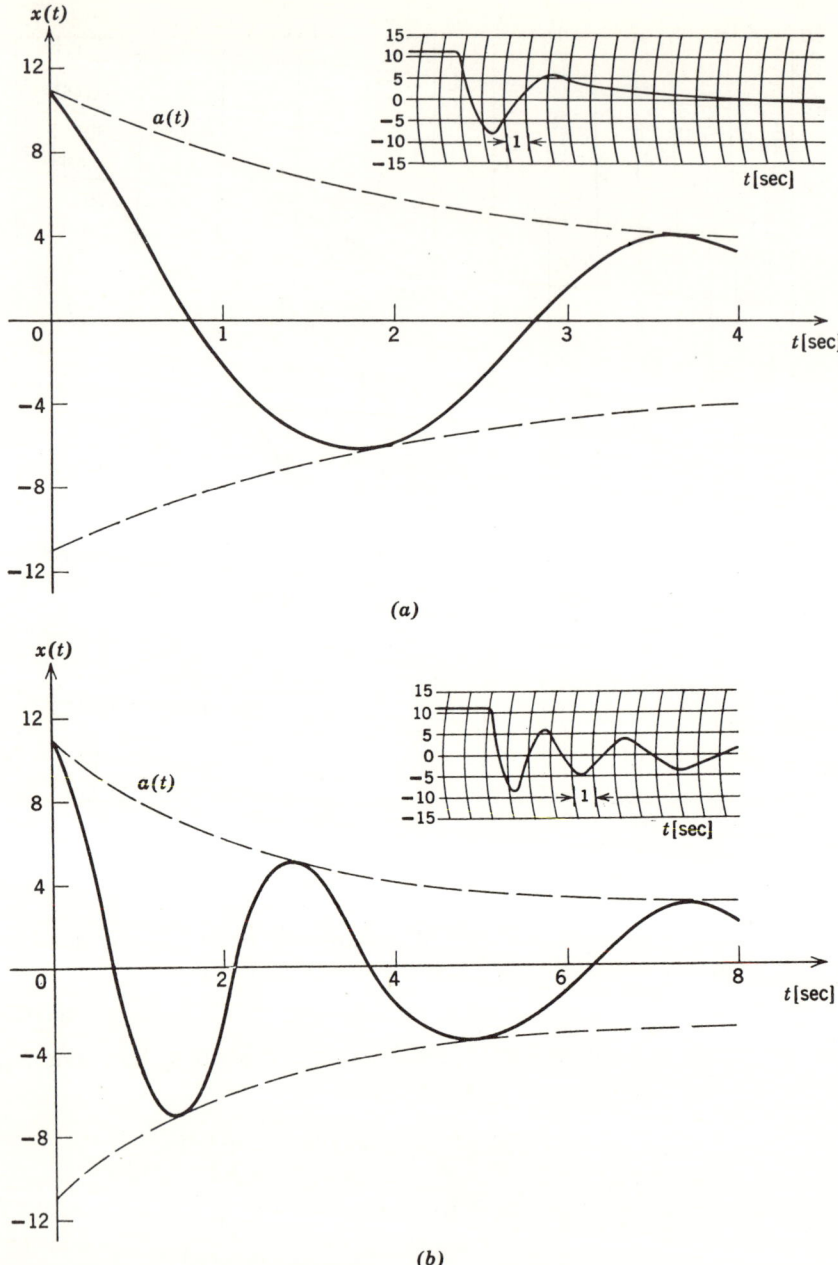

Figure 6.18 Transient solutions for the three different combinations of nonlinear elements: (a) $D_1 = 3$, $D_2 = 5$; (b) $D_1 = 3$, $D_2 = 3$; (c) $D_1 = 5$, $D_2 = 3$.

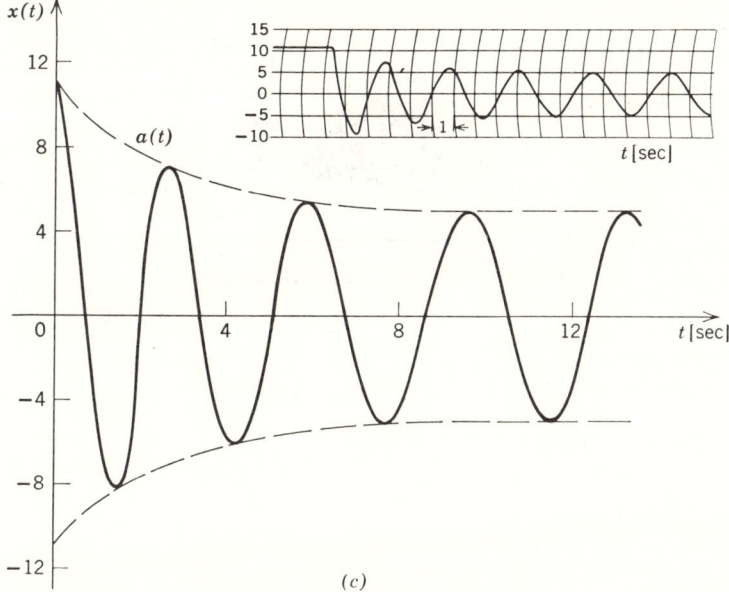

Figure 6.18 (*Continued*)

The effect of the nonlinear parameters on the transient process can easily be studied since only the M locus has to be replotted when a parameter is changed. The characteristic parameter plane curves are invariant to the variation in nonlinear parameters and are affected only by the changes in the linear ones.

The parameter plane diagram is plotted in Figure 6.17 and the three M loci are constructed, namely, (*a*) $D_1 = 3$, $D_2 = 5$; (*b*) $D_1 = D_2 = 3$; (*c*) $D_1 = 5$, $D_2 = 3$. Since the nonlinear elements have common inputs, the M loci are plotted in a straightforward manner without taking into account the frequency ω. Thus, directly using equations 6.102, the Σ and ω curves are determined in the usual manner from the characteristic equation of the linearized system, which is

$$s^3 + 10s^2 + (10\alpha + 10\beta)s + 100\alpha = 0, \qquad (6.103)$$

where $\alpha = N_{12}$ and $\beta = N_{11}$.

The three transient solutions are given in Figures 6.18*a*, 6.18*b*, and 6.18*c*. They are all calculated for the same initial condition $a_0 = 11$. The nature of the three solutions is quite different and can be roughly estimated by studying the motion of the M point along the M loci with respect to the Σ and ω curves of Figure 6.17. The respective computer solutions are also given for comparison.

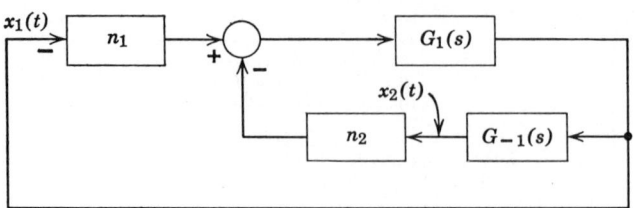

Figure 6.19 System block diagram.

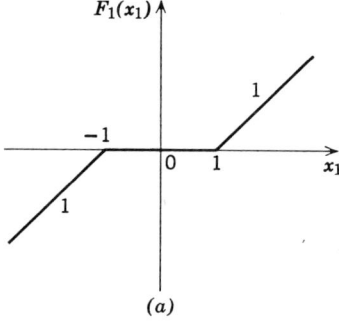

(a)

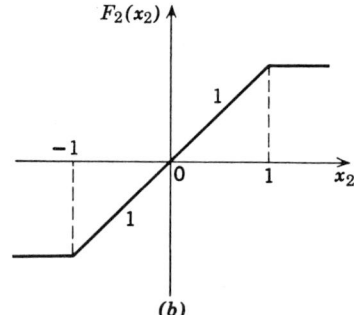

(b)

Figure 6.20 Nonlinear characteristics: (a) for element n_1; (b) for element n_2.

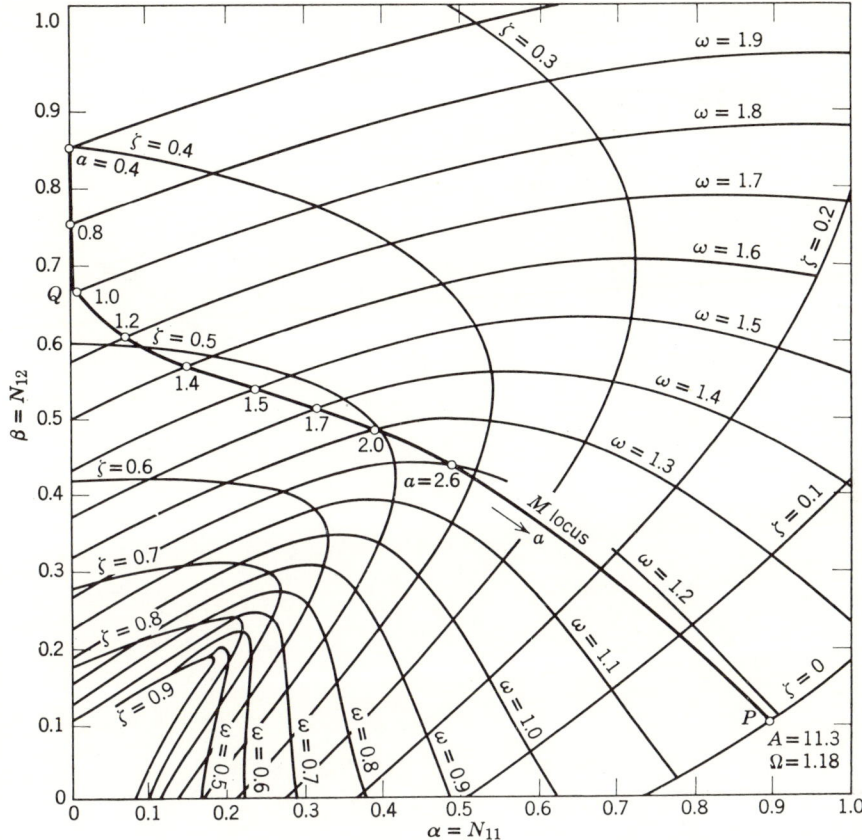

Figure 6.21 Parameter plane diagram.

Finally, consider again the system of Figure 6.19, which has already been considered in Section 4.6 in connection with Figures 4.27–4.30. The system has been examined for stability and limit cycles. It is now of interest to consider the transient process under different initial conditions.

The system of Figure 6.19 has the transfer functions

$$G_1(s) = \frac{20}{s(s+1)(s+2)(s+4)}, \qquad G_{-1}(s) = s \qquad (6.104)$$

and the nonlinear elements n_1 and n_2 with the characteristics given in Figures 6.20a and 6.20b, respectively. Figure 6.21, which is the same as Figure 4.30, represents the parameter plane diagram and is obtained as

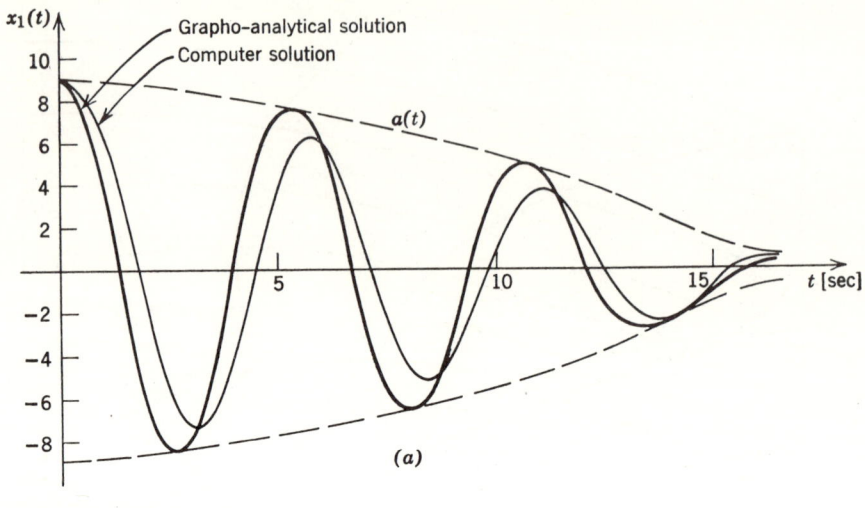
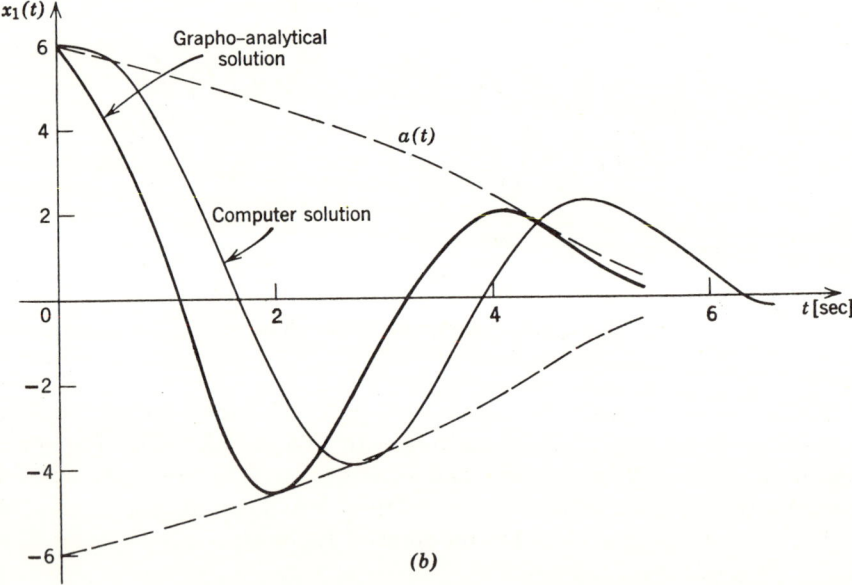

Figure 6.22 Transient solutions: (a) for $a_0 = 9$; (b) for $a_0 = 6$; (c) for $a_0 = 3$.

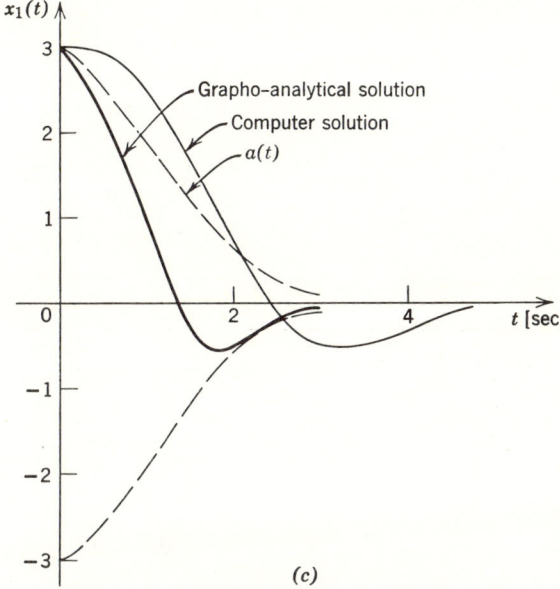

Figure 6.22 (*Continued*)

explained in Section 4.6 except that the ζ curves are used instead of the σ curves.[7]

On the basis of Figure 6.21, we can conclude that there is an unstable limit cycle with $A = 11.3$ and $\Omega = 1.18$, which is read at the intersection of the $\zeta = 0$ curve and the M locus. If an initial condition a_0 is chosen in the neighborhood of the limit cycle as $a_0 < A$, then, because of the small ζ, the response will begin as a slightly damped oscillation. In that region the nature of the ζ curves is such that a large variation of a does not increase the damping ratio ζ significantly and the whole response will continue to be a slightly damped oscillation. This fact is demonstrated in Figure 6.22a, where an initial condition $a_0 = 9$ is chosen.

If the initial condition is selected to be $a_0 = 6$, better damping is obtained than previously, as shown in Figure 6.22b. Lowering the initial amplitude to $a_0 = 3$, the transient solution becomes almost aperiodic (Figure 6.22c). This comes from the fact that for $a_0 = 3$ the M point starts to move from higher values of ζ and, for small changes of the amplitude a, reaches the domains of Figure 6.21, where ζ is sufficiently high to cause the oscillations to die out.

In Figures 6.22a, 6.22b, and 6.22c, the computer solutions are superimposed

---

[7] This system belongs to the second class, in which the two inputs to the nonlinearities x_1 and x_2 are related by a linear differential equation. The way of plotting the corresponding motion of the M point is explained in Section 4.6 and Problem 6.5.

274   Transient Oscillations

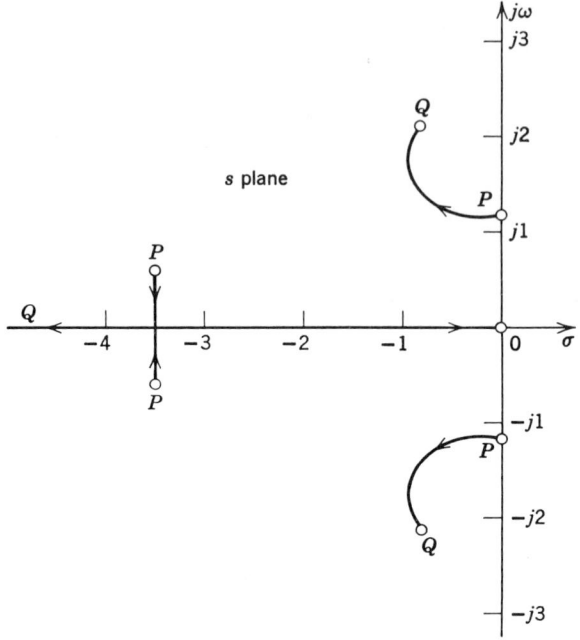

Figure 6.23 Inverse parameter plane root locus.

for comparison. As can be seen from these diagrams, the approximation of the calculated solution is getting poorer as the initial amplitude decreases. The delay in the computer solutions is proportionally increased. This phenomena occurs because, in plotting the approximate solutions, it is assumed that $\phi_0 = \pi/2$ and, consequently, $a_0 = x_0$. The approximation could be improved if the correct values of a_0 and ϕ_0 had been computed from (6.23). To check for dominancy of the complex conjugate pair of roots, $s_{1,2} = \sigma \pm j\omega$, the corresponding inverse parameter plane diagram is shown in Figure 6.23. It can be concluded that a better dominancy is present for higher values of initial conditions when the two complex roots are closer to the imaginary axis.

References

[6.1] N. Bogoliubov and Yu. Mitropolsky, *Asymptotical Methods in the Theory of Nonlinear Oscillations* (in Russian), State Press for Physics and Mathematical Literature, Moscow, 1963. (English translation: Hindustan Publishing Company, Delhi, India; U.S. dist.: Gordon and Breach, New York.)

[6.2] P. E. W. Grensted, The Frequency-response Analysis of Nonlinear Systems, *Trans. Inst. Elec. Engrs.*, Monograph No. 126, 244–253 (April 1955).

[6.3] P. E. W. Grensted, Analysis of the Transient Response of Nonlinear Control Systems, *Trans. ASME*, **80**, 427–432 (February 1958).

[6.4] E. P. Popov and I. P. Palitov, *Approximate Methods for Analysis of Nonlinear Automatic Systems* (in Russian), State Press for Physics and Mathematical Literature, Moscow, 1960. (English translation: Foreign Technical Division, AFSC, Wright-Patterson AFB, Ohio, Report FTD-TT-62-910.)

[6.5] D. Šiljak, Analysis and Synthesis of Feedback Control Systems in the Parameter Plane, Pt. III—Nonlinear Systems, *IEEE Trans.*, Pt. II (*Applications and Industry*), **83**, 466–473 (November 1964).

[6.6] D. Šiljak and M. Stojić, Sensitivity Analysis of Self-excited Nonlinear Oscillations, *IEEE Trans. Auto. Control*, **AC-10**, No. 4, 413–420 (October 1965).

[6.7] D. Šiljak, Generalization of the Parameter Plane Method, *IEEE Trans. Auto. Control*, **AC-11**, No. 1, 63–71 (January 1966).

[6.8] C. A. Pelegrini, An Investigation of Nonlinear System Performance on the Parameter Plane, M.S Thesis, U.S. Naval Postgraduate School, Monterey, California, 1966.

[6.9] R. L. Huston and L. F. Doty, Note on the Krylov-Bogoliubov Method Applied to Linear Differential Equations, *J. SIAM Appl. Math.*, **14**, No. 3, 424–428 (May 1966).

[6.10] R. Bittel, Parameter Analysis of Linear Time-Varying Systems, M.S. Thesis, University of Santa Clara, Santa Clara, California, May, 1967.

[6.11] E. A. Freeman, Characterization of Nonlinearity for Transient Process: Its Evaluation and Application, *IEEE Trans. Auto. Control*, **AC-12**, No. 5, 491–501 (October 1967).

CHAPTER SEVEN

Forced Oscillations

7.1 Introduction

Mostly free nonlinear oscillations have been considered so far, and the right side of differential equations describing the systems has been taken to be identically equal to zero. The only cases in which an external forcing signal was present were analyzed in Chapter 5 in connection with asymmetrical oscillations. The analysis, however, was performed under the assumption that the external forcing function is either constant or a slowly-varying function of time with respect to the corresponding periodic solution. In other words, the frequency of the external signal has been sufficiently lower than the frequency of the existing limit cycle. This assumption greatly simplified the analysis and gave rise to useful practical applications.

When no restrictions are imposed on the frequency of the external signal, the oscillations can become complex even in case of second-order systems [7.1–3]. Several new phenomena can occur that could not take place in a free nonlinear system. In this chapter, however, attention will be focused on the *jump resonance* that is of importance in a number of technical problems.

The jump resonance phenomenon in high-order nonlinear systems has been long under consideration in connection with the evaluation of the closed-loop frequency response of feedback control systems containing a nonlinearity [7.4–18]. The methods exclusively apply the describing function technique. This application of the describing function to nonautonomous systems has been based on intuition and has remained suspect until recently, when the more exact conditions under which such application is correct have been developed (see Appendix G).

The methods of Levinson [7.4] and Prince [7.5] involve the definition of an equivalent gain that approximates the given nonlinear characteristic. The methods are convenient for certain single-valued limiting nonlinearities, but can hardly be adapted for more complicated nonlinear characteristics. This concept, however, has been successfully extended by Booton [7.6, 16] to nonlinear systems with stochastic signals. West and others [7.7, 8, 16] applied the dual-input describing function to determine the conditions of the jump resonance to take place in systems with a polynomial nonlinearity. The extension to other kinds of nonlinearities is limited by the labor involved in calculating the corresponding dual-input describing function. Ogata's analytic techniques [7.9, 10] are limited to relay characteristics or, perhaps, saturating nonlinearities, since the output wave shape of the nonlinearity should be assumed. Considerable work is required to make a proper assumption in cases other than the ideal relay characteristic. Stein and Thaler [7.11] proposed a trial-and-error procedure to calculate the closed-loop frequency response by using the Nichols chart. A similar concept has been used with some modifications in reference 7.14.

The idea of finding conditions under which the jump resonance occurs and then designing the linear part of the system to avoid them (an idea proposed by West and others [7.7]) has been successfully applied along with the common describing function in references 7.8, 13, and 17. The methods, however, cannot give an insight into the frequency response of the closed-loop systems. Moreover, the procedures are extremely cumbersome when applied to the simplest multivalued nonlinearities or to nonlinearities with the frequency-dependent describing functions.

The frequency response can be analyzed by a different approach proposed by Gibson [7.16], which has some apparent advantages over the previously presented procedures. It enables readily obtained information about the effects on the frequency response of varying the nonlinear parameters. In addition, it does not require the analytical expression of the describing function and can thus be applied to experimentally obtained data (see Section F.7 of Appendix F). The procedure, however, has to be repeated each time the amplitude of the forcing periodic signal is changed. This can be circumvented by the technique of Popov and Palitov [7.12]; however, the nonlinearity cannot be varied without recalculating all the necessary curves.

In this chapter the separate ideas of the references mentioned in the preceding paragraph will be employed using the parameter plane concept [7.18]. The proposed analysis is directed toward the consideration of jump phenomena in high-order nonlinear systems.

By using the dual-input describing function proposed by West [7.8] and others [7.7], the proposed procedures can be readily extended to the analysis of *subharmonic oscillations*. Difficulties arise, however, because of computations involved in calculating the dual-input describing function. Nevertheless, the dual-describing function throws some light on several aspects of nonlinear systems—the stability of forced oscillations, jump phenomena, hyperharmonic and subharmonic resonance. In the analysis of subharmonic oscillations it seems to be the only method with some successful results [7.16].

7.2 Periodic Solutions. Jump Resonance

Consider the nonlinear differential equation

$$B(s)x + C(s)F(x, sx) = H(s)f, \qquad s \equiv \frac{d}{dt}, \qquad (7.1)$$

where, at the right side, the forcing function $f = f(t)$ is

$$f = A_f \sin(\Omega_f t - \varphi) \qquad (7.2)$$

and $B(s)$, $C(s)$, $F(x, sx)$ satisfy the conditions listed previously (see Chapter 3), and $H(s)$ is a polynomial in s with a degree less than or equal to that of the polynomial $B(s)$. The basic problem is to determine the conditions under which (7.1) has a periodic solution $x = x(t)$ sufficiently close to

$$x = A \sin \Omega_f t, \qquad (7.3)$$

where the frequency Ω_f is the known frequency of the forcing function $f(t)$ in (7.2), and then to evaluate the unknown values of the amplitude A and the phase shift φ.

By a suitable transformation the above problem can be reduced to that of free symmetrical oscillations considered in Chapter 4. To bring this about, note that the function f can be related to x by deriving

$$f = A_f \cos \varphi \sin \Omega_f t - A_f \sin \varphi \cos \Omega_f t. \qquad (7.4)$$

From (7.3), and

$$sx = A\Omega_f \cos \Omega_f t, \qquad (7.5)$$

we have

$$f = \frac{A_f}{A}\left(\cos \varphi - \frac{\sin \varphi}{\Omega_f} s\right) x. \qquad (7.6)$$

By substituting this expression of f into (7.1), we obtain

$$\left[B(s) - H(s)\frac{A_f}{A}\left(\cos \varphi - \frac{\sin \varphi}{\Omega_f} s\right)\right] x + C(s)F(x, sx) = 0. \qquad (7.7)$$

Therefore the nonhomogeneous differential equation 7.1 can be reduced to a homogeneous one, (7.7), if we know the forcing function $f = f(t)$ and assume the form of the solution $x = x(t)$. To determine the amplitude A and the phase shift φ of the solution $x(t)$, the methods for the analysis of symmetrical self-excited oscillations as outlined in Chapter 4 can be used with minor

Periodic Solutions. Jump Resonance

modifications. The applicability conditions of these methods derived in Chapter 3 for the equation

$$B(s)x + C(s) F(x, sx) = 0 \tag{7.8}$$

are the same except that the polynomial $B(s)$ is represented by the expression

$$B(s) - H(s)\frac{A_f}{A}\left(\cos \varphi - \frac{\sin \varphi}{\Omega_f} s\right), \tag{7.9}$$

which is given within square brackets in (7.7).

If the applicability conditions for symmetrical oscillations are satisfied, the nonlinearity $F(x, sx)$ can be harmonically linearized as

$$F(x, sx) = N_1 x + \frac{N_2}{\Omega_f} sx, \tag{7.10}$$

where the coefficients $N_1 = N_1(A, \Omega_f)$ and $N_2 = N_2(A, \Omega_f)$ are

$$N_1 = \frac{1}{\pi A}\int_0^{2\pi} F(A \sin \phi, A\Omega_f \cos \phi) \sin \phi \, d\phi$$

$$N_2 = \frac{1}{\pi A}\int_0^{2\pi} F(A \sin \phi, A\Omega_f \cos \phi) \cos \phi \, d\phi \tag{7.11}$$

and

$$\phi = \Omega_f t. \tag{7.12}$$

The linearized differential equation has the form

$$\left[B(s) - H(s)\frac{A_f}{A}\left(\cos \varphi - \frac{\sin \varphi}{\Omega_f} s\right) + C(s)\left(N_1 + \frac{N_2}{\Omega_f} s\right)\right] x = 0 \tag{7.13}$$

and the corresponding characteristic equation is

$$B(s) - H(s)\frac{A_f}{A}\left(\cos \varphi - \frac{\sin \varphi}{\Omega_f} s\right) + C(s)\left(N_1 + \frac{N_2}{\Omega_f} s\right) = 0. \tag{7.14}$$

The periodic solution $x = A \sin \Omega_f t$ with the frequency Ω_f can be determined from (7.14) by substituting $s = j\Omega_f$ and using the condition that the summation of reals and imaginaries must go to zero independently. Thus, by certain simple algebraic manipulations, we obtain

$$B_1 A - H_1 A_f \cos \varphi - H_2 A_f \sin \varphi + C_1 A N_1 - C_2 A N_2 = 0$$

$$B_2 A + H_1 A_f \sin \varphi - H_2 A_f \cos \varphi + C_1 A N_2 + C_2 A N_1 = 0, \tag{7.15}$$

where $B_1 = B_1(\Omega_f)$, $B_2 = B_2(\Omega_f)$, and so on, are given as

$$B_1 = \sum_{k=0}^{n} b_k X_k(0, \Omega_f), \qquad B_2 = \sum_{k=0}^{n} b_k Y_k(0, \Omega_f), \tag{7.16}$$

280 Forced Oscillations

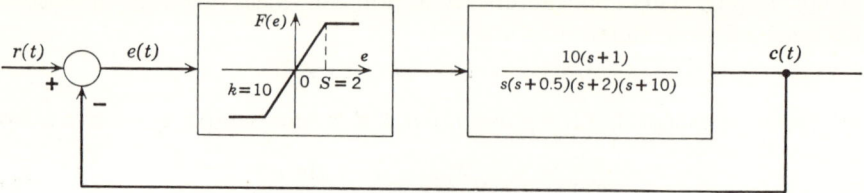

Figure 7.1 System block diagram.

and so on. By denoting

$$\alpha = A_f \cos \varphi$$
$$\beta = A_f \sin \varphi, \quad (7.17)$$

equations 7.15 may represent two equations in two unknowns, α and β, that can be solved for α and β as

$$\alpha = -A \frac{H_1(B_1 + C_1N_1 - C_2N_2) + H_2(B_2 + C_1N_2 + C_2N_1)}{H_1^2 + H_2^2}$$

$$\beta = A \frac{H_1(B_2 + C_1N_2 + C_2N_1) - H_2(B_1 + C_1N_1 - C_2N_2)}{H_1^2 + H_2^2}. \quad (7.18)$$

In the $\alpha\beta$ plane equations 7.18 may be interpreted as equations of the $\zeta = 0$ curve. The solution procedure starts with the plotting of a family of $\zeta = 0$ curves for different values of the frequency Ω_f appearing in the coefficients B_1, B_2, C_1, C_2, H_1, H_2, N_1, N_2 of (7.18). The unknown amplitude A, which enters both explicitly and as an argument of N_1 and N_2 in (7.18), is interpolated along the $\zeta = 0$ curves. The loci of the point $M(\alpha = A_f \cos \varphi; \beta = A_f \sin \varphi)$ are concentric circles with a radius A_f and the phase shift φ interpolated along the circles. Once the ζ and M-point loci are plotted in the $\alpha\beta$ plane, the amplitude A, frequency Ω_f, and the phase shift φ of the possible periodic solution $x = A \sin \Omega_f t$ are determined at their intersections. The stability of the periodic solutions is determined graphically from the obtained plot in a straightforward manner, as shown in the following example.[1]

Consider a control system with the block diagram shown in Figure 7.1. The related differential equation has the form

$$s(s + 0.5)(s + 2)(s + 10)x + 10(s + 1)F(x) = s(s + 0.5)(s + 2)(s + 10)f, \quad (7.19)$$

where the variable $x = x(t)$ represents the error signal $e(t)$ and the forcing function $f = f(t)$ is the input $r(t)$.

---

[1] This example has been treated analytically in reference 7.11, where for every set of values (A_f, Ω_f), at least one transcendental equation has to be solved.

For the sinusoidal input function $r(t) = A_f \sin(\Omega_f t - \varphi)$, the characteristic equation of the corresponding linearized system is

$$s(s + 0.5)(s + 2)(s + 10)\left[A - A_f\left(\cos\varphi - \frac{\sin\varphi}{\Omega_f}s\right)\right]$$
$$+ 10(s + 1)AN_1(A) = 0, \quad (7.20)$$

where the solution $e(t)$ of (7.19) is assumed as $e(t) = A \sin \Omega_f t$.

After the substitution of

$$\alpha = A_f \cos\varphi, \qquad \beta = A_f \sin\varphi, \quad (7.17)$$

and $s = j\Omega_f$ into (7.20), we obtain the equations 7.18 of the $\zeta = 0$ curve as

$$\alpha = \alpha(A, \Omega_f)$$
$$\beta = \beta(A, \Omega_f). \quad (7.18)$$

For the specific equation 7.20, (7.17) of the M-point loci and (7.18) of the $\zeta = 0$ curve are plotted in Figure 7.2 for different values of the amplitude A_f and the frequency Ω_f, respectively, as family parameters.

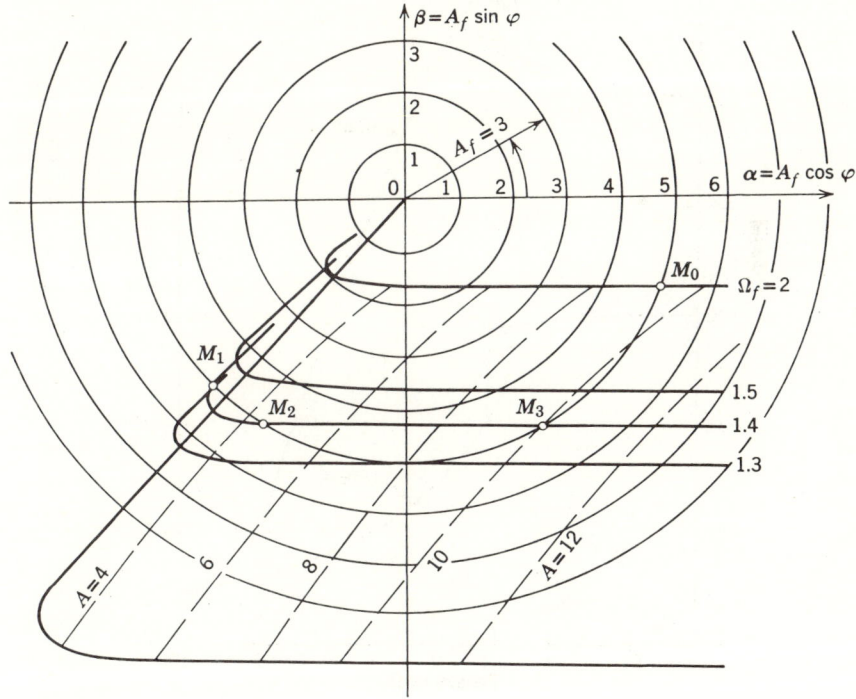

Figure 7.2 Parameter plane diagram for forced nonlinear oscillations.

282 Forced Oscillations

The diagram of Figure 7.2 is interpreted in the usual manner. For example, if the amplitude and frequency of the forcing signal $f(t)$ are $A_f = 5$ and $\Omega_f = 2$, then the related point is M_0, for which the periodic solution $x(t)$ has the amplitude $A \simeq 9.3$ and the phase shift $\varphi \simeq -19°$.

It is of particular significance to note that if the frequency Ω_f is decreased to $\Omega_f = 1.4$ rad/sec, and by keeping $A_f = 5$, the same M locus intersects the $\zeta = 0$ curve at three points, M_1, M_2, and M_3, which correspond to three different periodic solutions with the same frequency $\Omega_f = 1.4$ rad/sec. This is due to the so-called *jump resonance*, which can be better understood by plotting the frequency response characteristics from the diagram of Figure 7.2.

In Figure 7.3 the closed-loop frequency response relating the input and output of the system under investigation is plotted by an analog computer for the input amplitude $A_f = 5$. The jump resonance is indicated between the

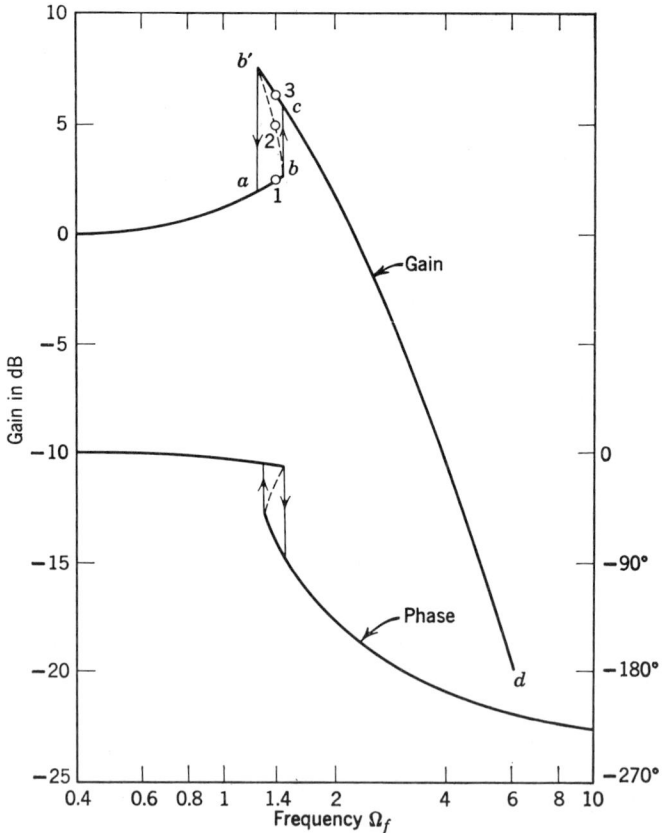

Figure 7.3 Closed-loop frequency response for the input amplitude $A_f = 5$.

frequencies $\Omega_f = 1$ and $\Omega_f = 2$. When the frequency Ω_f of the input signal is gradually increased, the first part $0ab$ of the gain curve $[C/R](j\Omega_f)$ is obtained in Figure 7.3. When the frequency Ω_f is increased at the point b, the system output changes abruptly, reaching the point c. With a further increase in the input frequency, the continuous part cd is obtained. If the procedure is reversed and the input frequency is decreased gradually from a higher value, the part dcb' of the gain curve is plotted. When the point b' is reached, a decrease in Ω_f causes the output amplitude to drop discontinuously to the point a and then follows the part $a0$ as the frequency Ω_f decreases to zero. The corresponding phase characteristic follows the discontinuous jumps in the amplitude, as shown in Figure 7.3.

The three points 1, 2, 3 on the gain characteristic of Figure 7.3 correspond to the points M_1, M_2, M_3 in the diagram of Figure 7.2. The lower point 1 and the upper point 3 are related to the stable periodic solution, whereas the middle point 2 is unstable and is not observed in the experiments. The jump resonance obtained by the computer checks the results available in the diagram of Figure 7.2. For example, the $\zeta = 0$ curve for $\Omega_f = 1.3$ is tangent to the M locus plotted for $A_f = 5$. This corresponds to the points a and b'. By interpolating between the $\zeta = 0$ curves plotted for $\Omega_f = 1.4$ and $\Omega_f = 1.5$, the frequency of the jump bc can be evaluated. (Note that the diagram of Figure 7.2 relates the input $r(t)$ and the error signal $e(t)$, while the plot of Figure 7.3 relates $r(t)$ and the output $c(t)$. This is of no essential importance since a simple relationship $e = r - c$ holds.)

Another observation of the jump resonance can be made if the input frequency is held constant but the input amplitude is varied so that A_f varies while Ω_f is constant. By using again the diagram of Figure 7.2, it is not difficult to show that if Ω_f is chosen to be 1.4 and A_f is varied, a diagram of Figure 7.4 can be calculated from that of Figure 7.2. Once again there is a range of values of A_f for which three values of A are possible. For $A_f = 5$, there are three values, A_1, A_2, A_3, of the amplitude A that correspond to the three points M_1, M_2, M_3 of Figure 7.2. The lower A_1 and upper A_3 correspond to stable solutions, whereas A_2 is unstable and cannot be obtained experimentally. The corresponding phase diagram is shown in Figure 7.5.

It is of interest to note that all the diagrams—Figures 7.3, 7.4, and 7.5—are obtained for specific values of the amplitude A_f (Figure 7.3, $A_f = 5$) and the frequency Ω_f (Figures 7.4 and 7.5, $\Omega_f = 1.4$). If these values are changed, all the diagrams have to be replotted again since they cannot be normalized with respect to the input as is possible in linear systems. The effects on (A, φ) of changing (A_f, Ω_f), however, can be studied directly from the diagram of Figure 7.2. Furthermore, the presented procedure can be applied equivalently to single-valued and common multivalued nonlinearities as well as to nonlinearities with the frequency-dependent describing functions. In

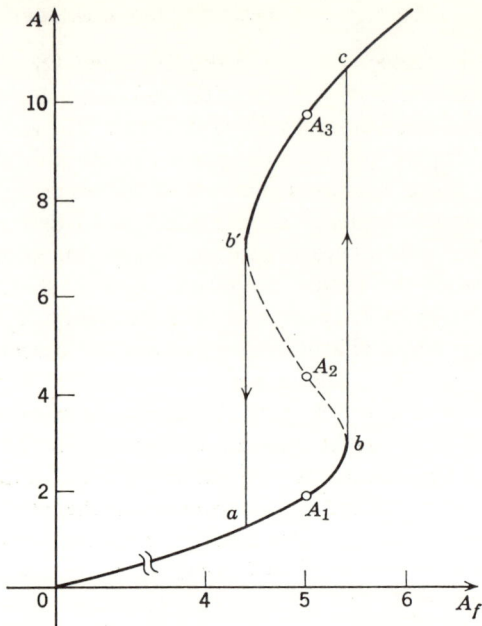

Figure 7.4 Amplitude diagram for $\Omega_f = 1.4$.

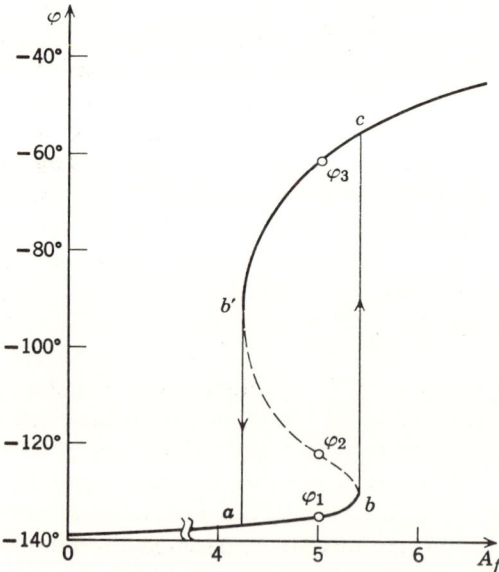

Figure 7.5 Phase diagram for $\Omega_f = 1.4$.

certain cases the procedure can be extended to the analysis of the jump phenomena in nonlinear systems with two nonlinearities, provided that the applicability conditions of the describing function are satisfied.

7.3 Stability and Sensitivity of the Periodic Solutions

In general the jump phenomenon can be more complex than that examined in the previous section. It may contain more than one curl, and thus more than three periodic solutions are possible. After these solutions are determined, the stability problem arises to separate the stable from the unstable solutions.

The unstable solutions are related to the case where an increase in the input amplitude A_f results in a decrease of the amplitude A of the corresponding periodic solution. By examining Figure 7.4, we can conclude that the unstable solution amplitudes are located along the part of the curve for which the slope is negative. The stability can now be checked from the parameter plane diagram. Thus, for an increase in the amplitude A along the curve $\Omega_f = 1.4$ at the point M_1, there is an increase in the amplitude A_f, and the related solution is stable. The same reasoning reveals that the point M_2 is unstable but M_3 is again stable.

The stability of periodic solutions can be checked analytically by determining the sign of the derivative $\partial A/\partial A_f$. This derivative is calculated from (7.15), in which the frequency Ω_f is considered constant, and the amplitude A and phase φ are assumed as functions of the amplitude A_f. This is similar to the procedure for the sensitivity analysis of self-excited oscillations presented in Section 4.3. Thus it is left to the reader to derive the derivatives $\partial A/\partial A_f$ and $\partial \varphi/\partial A_f$ from (7.15).

Likewise, the sensitivity analysis related to the stability problem, the concept of sensitivity analysis of small parameter variations in self-excited oscillations developed in Section 4.4, can be extended to the forced oscillations. The coefficients of the polynomials $B(s)$, $C(s)$, and $H(s)$ in the characteristic equation

$$B(s) - H(s)\frac{A_f}{A}\left(\cos\varphi - \frac{\sin\varphi}{\Omega_f}s\right) + C(s)\left(N_1 + \frac{N_2}{\Omega_f}s\right) = 0 \quad (7.14)$$

can be considered as functions of μ linear parameters p_i ($i = 1, 2, \ldots, \mu$). The describing function $N = N_1 + jN_2$ can be considered as a function of A, Ω_f, and ν nonlinear parameters q_j ($j = 1, 2, \ldots, \nu$). Then the sensitivities

$$\begin{aligned} S_{p_i}^A &\equiv \frac{\partial \ln A}{\partial \ln p_i}, & S_{p_i}^\varphi &\equiv \frac{\partial \ln \varphi}{\partial \ln p_i} \\ S_{q_j}^A &\equiv \frac{\partial \ln A}{\partial \ln q_j}, & S_{q_j}^\varphi &\equiv \frac{\partial \ln \varphi}{\partial \ln q_j} \end{aligned} \quad (7.19)$$

286   Forced Oscillations

can be calculated by differentiating (7.15) with respect to either p_i or q_j. The obtained sensitivity values indicate how the steady-state values of the amplitude A and phase φ are affected by the small parameter variations. Again, since the procedure is the same as that of Section 4.4, it is left to the reader to perform the necessary derivations and apply them to a specific example.

It is of interest to note that the possibility of the jump resonance to occur can be indicated by the condition that the derivative $\partial A/\partial A_f = \infty$; that is, $\partial A_f/\partial A = 0$. This condition has been examined entirely in references 7.13 and 7.17, where a graphical procedure has been proposed to check the condition being satisfied. The procedure of reference 7.17 is convenient for adjusting the linear part of the system so that jump resonance is avoided either entirely or for some range of input amplitudes.

7.4 Variable Nonlinear Characteristic

In the case of single-valued nonlinear characteristics when the nonlinear differential equation has the form

$$B(s)x + C(s)F(x) = H(s)f \qquad (7.20)$$

and the forcing function $f = f(t)$ is

$$f(t) = A_f \sin(\Omega_f t - \varphi), \qquad (7.2)$$

the analysis of the periodic solution $x = x(t)$

$$x(t) = A \sin \Omega_f t \qquad (7.3)$$

can be performed by a different approach, proposed by Gibson [7.16]. This approach is convenient for the analysis of the effects on jump resonance of varying the nonlinear characteristic, that is, the nonlinear parameters. Moreover, the approach can be advantageous in cases where the nonlinear single-valued characteristic is found experimentally and the describing function is determined graphically as shown in Section F.7 of Appendix F. Of course, it applies to situations in which the describing function of the nonlinearity is plotted experimentally or by using computers. The approach will be generalized by the parameter plane concept.

To describe the approach, note that the same harmonic linearization presented in section 7.3, if applied to (7.20), yields

$$\left[B(s) - H(s)\frac{A_f}{A}\left(\cos\varphi - \frac{\sin\varphi}{\Omega_f}s\right) + C(s)N_1\right]x = 0, \qquad (7.21)$$

where $N_1 = N_1(A)$. Equation 7.21 can be obtained from (7.13) by assuming $N_2 \equiv 0$, which is true for a single-valued nonlinearity $F(x)$. The corresponding characteristic equation can be written as

$$B(s)A + C(s)AN_1 - H(s)A_f\left(\cos \varphi - \frac{\sin \varphi}{\Omega_f}s\right) = 0. \quad (7.22)$$

After substituting $s = j\Omega_f$ into (7.22), we obtain two equations in two unknowns, α and β. Thus

$$\begin{aligned} B_1\alpha + C_1\alpha\beta - H_1 A_f \cos \varphi - H_2 A_f \sin \varphi = 0 \\ B_2\alpha + C_2\alpha\beta - H_1 A_f \sin \varphi - H_2 A_f \cos \varphi = 0, \end{aligned} \quad (7.23)$$

where

$$\begin{aligned} \alpha &= A \\ \beta &= N_1(A). \end{aligned} \quad (7.24)$$

Equations 7.23 can be solved for α and β to obtain

$$\begin{aligned} \frac{\alpha}{A_f} &= \frac{(C_1 H_2 - C_2 H_1)\cos \varphi - (C_2 H_2 + C_1 H_1)\sin \varphi}{B_2 C_1 - B_1 C_2} \\ \beta &= \frac{(B_1 H_2 - B_2 H_1)\cos \varphi - (B_2 H_2 + B_1 H_1)\sin \varphi}{(C_2 H_1 - C_1 H_2)\cos \varphi + (C_2 H_2 + C_1 H_1)\sin \varphi}. \end{aligned} \quad (7.25)$$

Now, for a given value of the forcing input frequency Ω_f, the coefficients B_1, B_2, C_1, C_2, H_1, and H_2, which are functions of Ω_f only, can be evaluated numerically. For different values of the phase shift φ, the corresponding values of α and β are calculated from (7.25). Thus a locus with constant frequency Ω_f can be plotted in the $\alpha\beta$ plane, where the scale factor along the α axis is given by A_f. On the other hand, the M locus is simply the describing function curve itself. Any change in the nonlinear parameters affects only the M locus, and any change in the A_f input amplitude simply means a scale change along the α axis of the $\alpha\beta$ plane for the M locus only.

To illustrate the analysis procedure, consider the same example as in the previous section. Applying the outlined procedure we obtain the parameter plane diagram shown in Figure 7.6. The diagram is plotted for $A_f = 5$, and therefore it can be compared with the computer simulation diagram of Figure 7.3 to indicate the accuracy. The points 1, 2, 3 of the intersections of the M locus with the $\Omega_f = 1.4$ correspond to the points 1, 2, 3 of Figure 7.3. The corresponding values of the phase shift φ are read on the Ω_f curves.

Any change in the amplitude A_f can be interpreted merely as a shift of the M locus relative to the Ω_f-constant curves of Figure 7.6. Thus a slight modification of the M locus should be made each time the amplitude A_f is changed.

288 Forced Oscillations

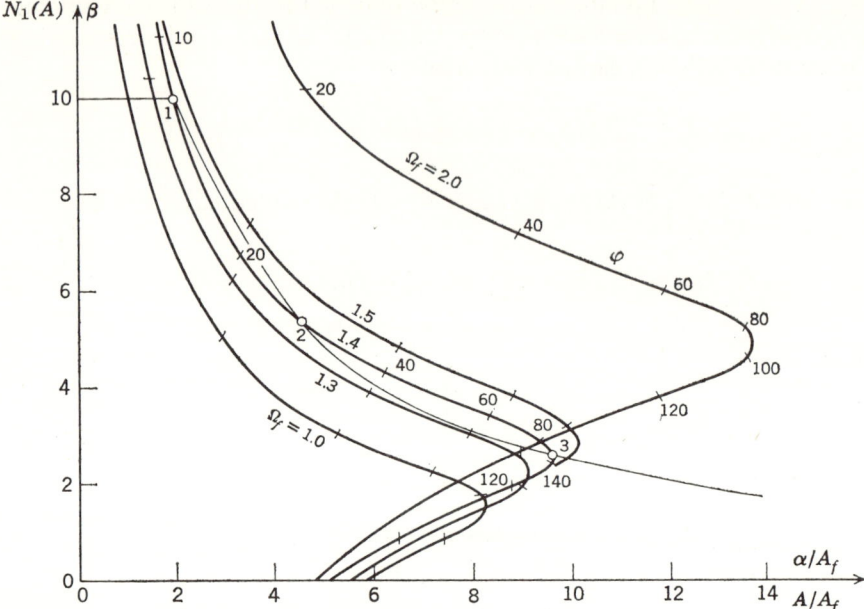

Figure 7.6 Parameter plane diagram for the "variable nonlinearity" analysis.

References

[7.1] N. N. Bogoliubov and Yu. A. Mitropolsky, *Asymptotical Methods in the Theory of Nonlinear Oscillations* (in Russian), FIZMATGIZ, Moscow, 1963. (English translation: Delhi, India; Hindustan Publishing Company, U.S.A. distributors: Gordon and Breach, New York.)

[7.2] W. J. Cunningham, *Introduction to Nonlinear Analysis*, McGraw-Hill, New York, 1958.

[7.3] Ch. Hayashi, *Oscillations in Physical Systems*, McGraw-Hill, New York, 1964.

[7.4] E. Levinson, Some Saturation Phenomena in Servomechanisms with Emphasis on the Tachometer Stabilized Systems, *AIEE Trans.*, Pt. II (*Applications and Industry*), **72**, 1–9 (March 1953).

[7.5] L. T. Prince, Jr., A Generalized Method for Finding the Closed-Loop Frequency Response of Nonlinear Systems, *AIEE Trans.*, Pt. II (*Applications and Industry*), **73**, 217–224 (September 1954).

[7.6] R. C. Booton, Jr., The Analysis of Nonlinear Control with Random Inputs, *Proceedings of the Symposium on Nonlinear Circuit Analysis*, Vol. 2, Polytechnic Institute of Brooklyn, Brooklyn, New York, 1953, pp. 369–391.

[7.7] J. C. West, J. L. Douce, and R. K. Livesley, The Dual-Input Describing Function and Its Use in the Analysis of Nonlinear Feedback Systems, *IEE Proc.*, **103B**, 463–474 (1956).

[7.8] J. C. West, *Analytical Techniques for Nonlinear Control Systems*, Van Nostrand, Princeton, New Jersey, 1960.

[7.9] K. Ogata, An Analytic Method for Finding the Closed-Loop Frequency Response of Nonlinear Feedback-Control Systems, *AIEE Trans.*, Pt. II. (*Applications and Industry*), **76**, 277–285 (November 1957).

[7.10] A. M. Hopkin and K. Ogata, An Analytic Frequency Response Solution for a Higher Order Servomechanism with a Nonlinear Control Element, *J. Basic Eng.*, *ASME Trans.*, Series D, **81**, 41–45 (March 1959).

[7.11] W. A. Stein and G. J. Thaler, Obtaining the Frequency Response Characteristic of a Nonlinear Servomechanism for an Amplitude- and Frequency-Sensitive Describing Function, *AIEE Trans.*, Pt. II (*Applications and Industry*), **77**, 91–96 (May 1958).

[7.12] E. P. Popov and I. P. Palitov, *Approximate Methods for Analyzing Nonlinear Automatic Systems* (in Russian), FITZMATGIZ, Moscow, 1960. (English translation: Foreign Technical Division, AFSC, Wright-Patterson Air Force Base, Ohio, Report FTD-TT-62-910.)

[7.13] J. C. Gille and P. Decaulne, On the Forcing Oscillations of Nonlinear Servosystems (in French), *Automatic and Remote Control*, Vol. 1, Butterworth, London, 1961, pp. 205–210.

[7.14] A. S. McAllister, A Graphical Method for Finding the Frequency Response of Nonlinear Closed-Loop Systems, *AIEE Trans.*, Pt. II (*Applications and Industry*), **80**, 268–277 (November 1961).

[7.15] H. Hatanaka, The Frequency Response and Jump-Resonance Phenomena of Nonlinear Feedback Control Systems, *J. Basic Eng.*, *ASME Trans.*, Series D, **85**, 236–242 (June 1963).

[7.16] J. E. Gibson, *Nonlinear Automatic Control*, McGraw-Hill, New York, 1963.

[7.17] A. Fukuma and M. Matsubara, Jump Resonance Criteria of Nonlinear Control Systems, *IEEE Trans.*, **AC-11**, No. 4, 699–707 (October 1966).

[7.18] J. Moore and D. D. Šiljak, Parameter Plane Analysis of Forced Nonlinear Oscillations, *Proceedings of the Fourth Annual Allerton Conference on Circuit and System Theory*, Monticello, Illinois, 887–903 (October 1966).

CHAPTER EIGHT
Stability Analysis

8.1 Introduction

Stability as an essential characteristic of the motion of dynamic systems has long been under consideration. Lagrange formulated the first theorems for the stability of equilibrium positions. Maxwell postulated the stability problem in linear systems that was solved later by Routh, Hurwitz, Nyquist, and others. Numerous important results and ideas connected with stability considerations of differential equations were developed by Poincaré. The general theory of stability with most useful aspects in applications to problems of physics and engineering, however, was founded by Liapunov.

After formulating a precise definition of stability, Liapunov proposed two methods for the stability analysis. While the "first method" did not find a wide application to stability problems, the "second method" of Liapunov offered much promise for further advances in the stability theory of non-linear systems. The second method attempts to determine the stability solely from equations describing the system, thus avoiding the need for the explicit solution. To achieve this objective, it is necessary to find from a given form of differential equations an appropriate Liapunov function that plays the role of energy. Then intuition leads to a conjecture that if the rate of change of the function is negative for every possible state of the system except for a single equilibrium position, where the function has its minimum value, the function will constantly decrease until the states of the system reach the equilibrium position—this proves stability.

The major difficulty in applying the Liapunov stability concept lies in the fact that there is no explicit systematic method in general for finding appropriate Liapunov functions. Therefore a strong effort in the theory of differential equations and system analysis has been directed toward achieving efficient procedures for constructing Liapunov functions and solving related problems in various classes of differential equations and systems. Numerous important results are outlined in the listed references.

A significant problem in the Liapunov stability theory for a wide class of nonlinear systems was proposed and partially solved by Lur'e. He formulated the *absolute stability in the sense of Liapunov* for systems with a linear part and an essential nonlinearity satisfying certain general conditions given in advance. Lur'e then suggested a systematic method for constructing the Liapunov functions of the type "a quadratic form plus an integral of the nonlinearity" that assures the absolute stability of the mentioned systems. Extended versions of this "Lur'e problem" have been solved by a large number of authors, as outlined in the survey section of Appendix H.

As distinct from the other solutions to the Lur'e problem, V. M. Popov expressed the sufficient conditions for absolute stability in terms of the *frequency characteristic* of the linear part of the system. This resulted in a simple and convenient graphical interpretation common in linear system analysis. The Popov solution has been further extended by many authors to various problems of nonlinear system analysis.

It is of interest to note that Popov obtained his results without use of Liapunov functions. The relationship between the Popov absolute stability criterion and the Liapunov functions was first established by Yakubovich and later refined by Kalman and Meyer. The relationship is given in the form of a lemma which shows that for the existence of the Lur'e type of Liapunov function the Popov conditions are both necessary and sufficient. The lemma is known as the Yakubovich-Kalman lemma.

The objective of this chapter is to postulate precisely the absolute stability concept of Lur'e in the context of Liapunov stability theory, and then by the Yakubovich-Kalman lemma to establish the Popov stability criterion with the ultimate goal of interpreting the absolute stability conditions in the *parameter space*. This self-contained exposition will allow the study of the influence of system parameters on the absolute stability of nonlinear systems.

In achieving the above objective, after a brief discussion of the state-space description of dynamic systems, the various definitions of stability are given and the basic Liapunov stability theorem is outlined. This enables an adequate basis for the formulation of the "Lur'e problem" and the introduction of the absolute stability notion. The Popov solution of the "Lur'e problem" in the frequency domain is presented by several stability theorems. These theorems are extended versions of the original Popov work. They were formulated and then proved by Yakubovich on the basis of his lemma, referred to as the Yakubovich-Kalman lemma. The stability theorems treat various types of systems and related problems. Finally, the derived Popov stability conditions are interpreted in the parameter space. The boundary of these stability regions is readily found as an envelope of surfaces determined from the Popov criterion. The calculation of the envelope can be programmed conveniently for computer solution.

Although the "Lur'e problem" seems a rather particular problem in the stability theory of nonlinear systems, it should be noted that numerous useful models of physical systems can be cast in that of Lur'e. Moreover, most of the results outlined here for systems with a single nonlinear element can and have been extended to systems with multiple nonlinearities and time varying gains. By studying the rather self-contained material of this chapter, the reader will be in a position to turn to the technical literature for these more general results.

It is of interest to see how this chapter is related to the previous ones. It contains an exact stability concept that can be contrasted to that used in the approximate system analysis. In most cases the stability conditions are only sufficient, and to extend the regions of stability toward the necessary conditions an additional analysis is desired. This analysis, however, can be advantageously carried out by the approximate analysis. On the other hand, the exact analysis is extremely important to verify some of the conclusions drawn from the approximate analysis. The relationship between the two approaches can be partially made by the Aizerman conjecture outlined in the last section of this chapter. This conjecture suggests a validation of the linearization techniques and can only be examined by studies in the exact analysis context.

An extensive literature exists on the Liapunov stability theory and application. Some of the important ones are listed here. Besides the original Liapunov monograph [8.1], there are the well-known books by Malkin [8.2], Letov [8.3], Tsypkin [8.4], Lur'e [8.14], Nemytskii and Stepanov [8.9], the widely known introductory monograph of LaSalle and Lefschetz [8.11], the book by Zadeh and Desoer [8.12], the expository papers of Antosiewicz [8.6] and Kalman and Bertram [8.10], as well as the survey books of Hahn [8.7] and Cesari [8.8] and papers of Massera [8.5] and a recent one by Brockett [8.13]. They provide excellent foundation and guidance in an extensive study of Liapunov stability theory.

With respect to the references on the "Lur'e problem" of absolute stability, there are the original book of Lur'e [8.14] and the well-known monographs by Aizerman and Gantmacher [8.15] and Lefschetz [8.17], and survey papers by Gantmacher and Yakubovich [8.16] and Brockett [8.13], which can be used for further studies of the Lur'e problem. In addition, the books of Gibson [8.18] and DeRusso, Roy, and Close [8.19] should be considered since they have chapters on stability theory with a well-presented introduction to the Lur'e problem.

In this chapter, the introduction to stability theory follows the references of Kalman and Bertram [8.10], and Malkin [8.2]. The "Lur'e problem" and the corresponding Popov solution are presented on the basis of the exposition by Aizerman and Gantmacher [8.15]. The absolute stability theorems and their proofs are outlined by using the original references of Yakubovich [8.20–23], with the exception of the Lefschetz [8.17] version of the Yakubovich-Kalman lemma. The parameter space interpretation of the Popov stability conditions is based on references [8.24] and [8.25].

Finally, it should be mentioned that the structure of the exposition is the same as in the previous chapters. That is, the theorems and statements outlined in this chapter are proved in the associated appendix—Appendix H.

In this chapter and the associated appendix, the matrix notation is used. Lowercase boldface letters denote vectors—for example, **x**, **r**. Capital boldface letters denote matrices—**P**, **H** (an exception is the transition vector function **X**, or sometimes **Y**). The unit matrix is denoted by **I**. The transpose of a vector or matrix is designated by T—for example, $\mathbf{x}^T$, $\mathbf{P}^T$. If $\mathbf{P} = (p_{ij})$ is a complex matrix, $\mathbf{P}^* = (\bar{p}_{ji})$ denotes a transpose conjugate. If the quadratic form $\mathbf{x}^T \mathbf{H} \mathbf{x}$ is >0 (<0) for all $\mathbf{x} \neq \mathbf{0}$, we write $\mathbf{H} > 0$ (< 0).

8.2 Motion of Nonlinear Systems in the State Space

In stability analysis of nonlinear systems, the description of the system in the state space by first-order vector differential equations is essential. This is particularly true in defining the system motion and postulating precise stability definitions. Later, however, some of the important results are interpreted by means of the operational form of differential equations that was extensively used in the preceding parts of the book.[1]

Generally a nonlinear system can be described by an nth order scalar differential equation written in the operational form as

$$s^n x + F(x, sx, \ldots, s^{n-1}x, t) = 0, \quad s \equiv \frac{d}{dt}, \tag{8.1}$$

or by a vector differential equation of the first order,

$$\dot{\mathbf{x}} = \mathbf{X}(\mathbf{x}, t), \tag{8.2}$$

which is equivalent to the set of n scalar differential equations

$$\dot{x}_i = X_i(x_1, x_2, \ldots, x_n, t), \quad (i = 1, 2, \ldots, n). \tag{8.3}$$

---

[1] A brief review of the state space description of linear systems is given in Section C.3 of Appendix C. In addition, a transition from the operational form to state space description of linear systems is considered.

The vector $\mathbf{x} = \{x_1, x_2, \ldots, x_n\}$ in (8.2) is the *state* of the system, its components x_i, ($i = 1, 2, \ldots, n$), are *state variables*.[2] The system is specified by the vector-valued function $\mathbf{X}(\mathbf{x}, t)$. The integer n is the *order* of the system.

The operational form (8.1) can be readily reduced to the state space description (8.2) or (8.3) by substitutions

$$x_1 = x, \quad x_2 = sx, \ldots, \quad x_n = s^{n-1}x, \tag{8.4}$$

which yield

$$\begin{aligned}\dot{x}_1 &= x_2 \\ \dot{x}_2 &= x_3 \\ &\vdots \\ \dot{x}_{n-1} &= x_n \\ \dot{x}_n &= -F(x_1, x_2, \ldots, x_n, t).\end{aligned} \tag{8.5}$$

If the differential equation 8.2 represents a *continuous-time dynamic system*, the function $\mathbf{X}(\mathbf{x}, t)$ is sufficiently smooth so that the equation has a unique solution starting at initial state $\mathbf{x}_0$ and at initial time t_0. More precisely, we always assume that in an n-dimensional vector space $\mathscr{X}$—referred to as the *state space*, there exists a function $\mathbf{x}(t, \mathbf{x}_0, t_0)$ so that for any admissible fixed $\mathbf{x}_0, t_0$:

(a) $\mathbf{x}(t_0, \mathbf{x}_0, t_0) = \mathbf{x}_0$;

(b) $\dfrac{d}{dt} \mathbf{x}(t, \mathbf{x}_0, t_0) \equiv \mathbf{X}[\mathbf{x}(t, \mathbf{x}_0, t_0), t]$;

(c) $\mathbf{x}(t, \mathbf{x}_0, t_0)$ is defined for all t; \hfill (8.6)

(d) $\mathbf{x}(t, \mathbf{x}_0, t_0)$ is unique, that is, $\mathbf{x}[t_2, \mathbf{x}(t_1, \mathbf{x}_0, t_0), t_1] = \mathbf{x}(t_2, \mathbf{x}_0, t_0)$ for all t_1, t_2; and

(e) $\mathbf{x}(t, \mathbf{x}_0, t_0)$ is continuous in all arguments.

The vector function $\mathbf{x}(t, \mathbf{x}_0, t_0)$ is called the *solution* of $\dot{\mathbf{x}} = \mathbf{X}(\mathbf{x}, t)$.

Sufficient Cauchy-Lipschitz conditions for validity of the last two assumptions in (8.6) are formulated in terms of the function $\mathbf{X}(\mathbf{x}, t)$ by the classical existence, uniqueness, and continuity theorem.[3] Under these conditions, the

---

[2] To save space, a vector is written using the curly braces to identify it as a column matrix.

[3] The theorem states: Let $\mathbf{X}(\mathbf{x}, t)$ be a continuous function in t and satisfy a Lipschitz condition in some region $R(\mathbf{x}_0, t_0)$ about any $\mathbf{x}_0, t_0$:

$$R(\mathbf{x}_0, t_0) = \begin{cases} |t - t_0| \leq a(t_0) \\ \|\mathbf{x} - \mathbf{x}_0\| \leq b(\mathbf{x}_0), \end{cases}$$

where a and b are positive constants which depend only on t_0 and $\mathbf{x}_0$, respectively, that is, there is a positive constant k that depends only on a and b, so that for every $\mathbf{x}_1$ and $\mathbf{x}_2$ in

$R(\mathbf{x}_0, t_0)$ the following inequality holds:

$$\|\mathbf{X}(\mathbf{x}_1, t) - \mathbf{X}(\mathbf{x}_2, t)\| \leq k \|\mathbf{x}_1 - \mathbf{x}_2\|,$$

and let

$$M(\mathbf{x}_0, t_0) = \max \|\mathbf{X}(\mathbf{x}, t)\|$$

in the closed, bounded set $R(\mathbf{x}_0, t_0)$. Then (a) there exists a unique solution $\mathbf{x} = \boldsymbol{\phi}(t, \mathbf{x}_0, t_0)$ of equation $\dot{\mathbf{x}} = \mathbf{X}(\mathbf{x}, t)$ starting at $\mathbf{x}_0, t_0$, for all $|t - t_0| \leq c(t_0)$ where

$$c(t_0) \leq \min \{a(t_0), b(\mathbf{x}_0)/M(\mathbf{x}_0, t_0)\}$$

and (b) in a sufficiently small neighborhood of $\mathbf{x}_0, t_0$, the solution is a continuous function of its arguments.

A proof of this theorem is given in reference 8.9.

It is important to note that only the local Lipschitz condition is required by the above theorem, which assures the desired properties of the solution $\boldsymbol{\phi}(t, \mathbf{x}_0, t_0)$ only in the immediate neighborhood of $\mathbf{x}_0$ and t_0. Thus, to eliminate a possibility of phenomena such as finite escape time (see the following example in the text), which is required by property (8.6c), the Lipschitz conditions on $\mathbf{X}(\mathbf{x}, t)$ should hold *everywhere*. Although the global Lipschitz conditions rule out the possibility of the finite escape time, they may sometimes be too restrictive.

Observe that if a function $\mathbf{X}(\mathbf{x}, t)$ satisfies the Lipschitz condition then $\mathbf{X}$ is uniformly continuous in $\mathbf{x}$ for each fixed t, although nothing is implied concerning the continuity of $\mathbf{X}$ with respect to t. An application of the mean-value theorem of differential calculus shows that the boundedness of the partial derivatives of $\mathbf{X}$ with respect to $\mathbf{x}$ on a region R (convex in $\mathbf{x}$), implies the Lipschitz condition. In fact, by the law of the mean in the scalar case, $|X(x_1, t) - X(x_2, t)| = |X_x[x_2 + \theta(x_1 - x_2), t]| \, |x_1 - x_2| \leq k |x_1 - x_2|$, where k is the least upper bound of $|X_x| = |\partial X/\partial x|$ in R. This extends directly to the vector-valued function $\mathbf{X}(\mathbf{x}, t)$.

It should be noted that for existence of the solution, the Lipschitz condition is too strong and it is sufficient that the function $\mathbf{X}(\mathbf{x}, t)$ is continuous in $\mathbf{x}$ and t. Without the additional requirement that the function satisfies a Lipschitz condition, however, the uniqueness property of solutions need not follow. For example, in equation

$$\dot{x} = x^{1/3}$$

the right-hand side is continuous at the point $(x, t) = (0, 0)$ but there are two solutions passing through that point, namely,

(a) $\qquad\qquad\qquad\qquad x \equiv 0$

(b) $\qquad\qquad\qquad\qquad \begin{cases} x = (\tfrac{2}{3}x)^{3/2}, & x \geq 0 \\ x = 0, & x \leq 0 \end{cases}$

Apparently, $x^{1/3}$ does not satisfy the Lipschitz condition at $x = 0$, since if $x_1 = \delta$ and $x_2 = -\delta$, then $|X(x_1) - X(x_2)|/|x_1 - x_2| = 1/\delta^{2/3}$ which is unbounded for δ arbitrary small.

Note that in formulating the above theorem the explicit notation $\boldsymbol{\phi}(t, \mathbf{x}_0, t_0)$ is used instead of $\mathbf{x}(t, \mathbf{x}_0, t_0)$. In mathematical literature the explicit notations are required because of clarity in the relevant theorems and definitions. In engineering literature, however, the notation $\mathbf{x}(t, \mathbf{x}_0, t_0)$ is used and thus is accepted in the following developments.

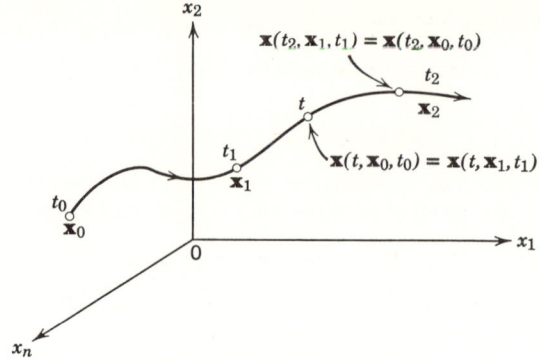

Figure 8.1 State space trajectory.

function $X(x, t)$ can be interpreted as the infinitesimal transition of the system state $x \to x + dx$ that takes place during the infinitesimal change in time $t \to t + dt$. With this interpretation, the solution $x(t, x_0, t_0)$, which passes through the state x_0 at time t_0 and is observed at time t, may be regarded as a *transition function* that specifies how the state $x(t_0)$ is transformed into $x(t)$. Therefore the solution $x(t, x_0, t_0)$ describes the motion of the system from the state $x(t_0)$ to some state $x(t)$. Hypotheses (8.6) about the solution $x(t, x_0, t_0)$ may be used to define axiomatically a continuous-time dynamic system by its motion [8.9].

A solution $x(t, x_0, t_0)$ determines all the states x_i as functions of time; that is, it yields the functions

$$x_i = x_i(t), \qquad (i = 1, 2, \ldots, n). \tag{8.7}$$

In n-dimensional state space $\mathscr{X}$ (Figure 8.1), equations 8.7 define a curve on which the time t appears as a parameter. This curve is referred to as the *trajectory* in the state space $\mathscr{X}$. Besides being defined and continuous for any x_0, t_0, t, the trajectory has the uniqueness property of (8.6d), illustrated in Figure 8.1.

Now it is of interest to discuss different classes of systems in terms of the function $X(x, t)$.[4]

If it is necessary to distinguish between *forced* and *free* (*unforced*) systems, the state equation can be written as

$$\dot{x} = X(x, f, t), \tag{8.8}$$

where the vector $f = f(t)$ is the external *forcing function* or *input*. If $f(t) \equiv 0$

---

[4] We may compare the discussion that follows with the one outlined in Section 2.2 using the operational form of differential equations.

for all t, the system is free and can be described by

$$\dot{\mathbf{x}} = \mathbf{X}(\mathbf{x}, t). \tag{8.2}$$

Then the system is *time varying*. If the system is both unforced and time invariant, it can be specified by a function $\mathbf{X}$ that depends on $\mathbf{x}$ alone and not on time t, that is,

$$\dot{\mathbf{x}} = \mathbf{X}(\mathbf{x}). \tag{8.9}$$

A system of this nature is called *autonomous*. In general, whenever the stated equations do not incorporate time explicitly but only implicitly in terms of the derivatives with respect to time, the system is autonomous. Otherwise, it is *nonautonomous*.

The motion of autonomous systems has an important additional property in that it is invariant under translation in time; that is,

$$\mathbf{x}(t, \mathbf{x}_0, t_0) = \mathbf{x}(t + \tau, \mathbf{x}_0, t_0 + \tau), \quad \tau = \text{const} \tag{8.10}$$

for any admissible $\mathbf{x}_0, t_0, \tau$.

In dealing with stability analysis we are constantly concerned with the *equilibrium states* (*or points*) defined as constant solutions $\mathbf{x}_e$ of a free dynamic system for which

$$\mathbf{X}(\mathbf{x}_e, t) = \mathbf{0}, \quad \text{for all } t \tag{8.11}$$

or, equivalently,

$$\mathbf{x}(t, \mathbf{x}_e, t_0) = \mathbf{x}_e, \quad \text{for all } t. \tag{8.12}$$

Therefore, a motion passing through an equilibrium point at any time remains at the same point for all times. Note that when we replace $\mathbf{x} - \mathbf{x}_e$ by $\mathbf{y}$, then again $\mathbf{y}$ by $\mathbf{x}$, the system is still of type (8.2) but with $\mathbf{X}(\mathbf{0}, t) = \mathbf{0}$ for all t: the equilibrium state is the origin $\mathbf{x}_e = \mathbf{0}$ (it is sometimes called the null solution). This is the general assumption in most of the following developments.

If the superposition principle is valid for a dynamic system, the system is linear and the function $\mathbf{X}(\mathbf{x}, \mathbf{f}, t)$ is a linear function of $\mathbf{x}$ and $\mathbf{f}$. Assuming for simplicity that the origin of the state space $\mathscr{X}$ is an equilibrium point and $\mathbf{X}(\mathbf{0}, \mathbf{0}, t) \equiv \mathbf{0}$, the system can be described by

$$\dot{\mathbf{x}} = \mathbf{P}(t)\mathbf{x} + \mathbf{Q}(t)\mathbf{f}, \tag{8.13}$$

where $\mathbf{P}(t) = (\partial X_i/\partial x_j)$ and $\mathbf{Q}(t) = (\partial X_i/\partial f_j)$ are $n \times n$ and $n \times m$ matrices, respectively, which depend on time. When the system is time invariant, the matrices $\mathbf{P}$ and $\mathbf{Q}$ are constant.

For the linear case (8.13) the general solution is available and has the form [8.12]

$$\mathbf{x}(t, \mathbf{x}_0, t_0) = \mathbf{\Phi}(t, t_0)\mathbf{x}_0 + \int_{t_0}^{t} \mathbf{\Phi}(t, \tau)\mathbf{Q}(\tau)\mathbf{f}(\tau) \, d\tau, \qquad t \geq t_0 \qquad (8.14)$$

where $\mathbf{\Phi}(t, t_0)$ is called the *state transition matrix*.[5]

When the linear system is autonomous and is described by

$$\dot{\mathbf{x}} = \mathbf{P}\mathbf{x}, \qquad (8.15)$$

where $\mathbf{P}$ is a square $n \times n$ constant matrix $\mathbf{P} = (p_{ij})$, with (8.15) there is associated the characteristic equation

$$|\mathbf{P} - s\mathbf{I}| = 0 \qquad (8.16)$$

or, equivalently,

$$\begin{vmatrix} p_{11} - s & p_{12} & \cdots & p_{1n} \\ p_{21} & p_{22} - s & \cdots & p_{2n} \\ \vdots & \vdots & & \vdots \\ p_{n1} & p_{n2} & \cdots & p_{nn} - s \end{vmatrix} = 0. \qquad (8.17)$$

The rules for expansion of a determinant show then that the characteristic equation can be rewritten as

$$\sum_{k=0}^{n} a_k s^k = 0. \qquad (8.18)$$

---

[5] The elements of this matrix can be regarded as the *impulse responses* so that $\phi_{ij}(t, t_0)$ is the response of the ith state variable to a forcing function $f_j(t) = \delta(t - t_0)$, where $\delta(t)$ is the *Dirac delta function* (for a definition of this function, see Section G.4). The rest of the forcing functions $f_k(t) \equiv 0$, $k \neq j$, $\mathbf{x}_0 = \mathbf{0}$ and $\mathbf{Q}(t) \equiv \mathbf{I}$, where $\mathbf{I}$ is the identity matrix.

As a consequence of linearity, axioms (8.6a) and (8.6b) imply that $\mathbf{\Phi}(t_0, t_0) = \mathbf{I}$ for all t_0, and $\mathbf{\Phi}(t_2, t_0) = \mathbf{\Phi}(t_2, t_1)\mathbf{\Phi}(t_1, t_0)$ for all t_0, t_1, t_2. The later group property of the transition matrix can be shown from the relations $\mathbf{x}(t_2) = \mathbf{\Phi}(t_2, t_1)\mathbf{x}(t_1) = \mathbf{\Phi}(t_2, t_0)\mathbf{x}(t_0)$ and $\mathbf{x}(t_1) = \mathbf{\Phi}(t_1, t_0)\mathbf{x}(t_0)$. An immediate consequence of this is also that $\mathbf{\Phi}^{-1}(t_1, t_0) = \mathbf{\Phi}(t_0, t_1)$ for all t, t_0; and it can be finally shown that the transition matrix satisfies its own differential equation, that is, $d\mathbf{\Phi}(t, t_0)/dt = \mathbf{P}(t)\mathbf{\Phi}(t, t_0)$ for all t, t_0.

In time invariant systems, $\mathbf{\Phi}(t, t_0)$ depends only on the difference $t - t_0$. Then it can be shown [8.12] that it is

$$\mathbf{\Phi}(t, t_0) = \exp\left[(t - t_0)\mathbf{P}\right] = \sum_{k=0}^{\infty} [(t - t_0)\mathbf{P}]^k/k!.$$

A procedure for calculation of the coefficients a_k from the given elements p_{ij} of the matrix **P** is outlined in Section C.3 of Appendix C. The characteristic equation plays an important role in the stability analysis.

It is a well-known fact that a nonlinear system in the sufficiently small neighborhood about its equilibrium points may behave somewhat similarly to a linear system. It is then often possible to utilize this property to determine the stability of the equilibrium states of a nonlinear system by applying known linear methods.

Consider the autonomous case when the equations of motion are

$$\dot{\mathbf{x}} = \mathbf{X}(\mathbf{x}), \tag{8.9}$$

where the function $\mathbf{X}(\mathbf{x})$ satisfies the Cauchy-Lipschitz conditions. The equilibrium points given by $\dot{\mathbf{x}} = \mathbf{0}$ are determined from

$$\mathbf{X}(\mathbf{x}) = \mathbf{0}. \tag{8.19}$$

Unlike linear systems, nonlinear systems can have more than one equilibrium point (it is not unique), as is clear from (8.19), which can have more than one solution. Without loss in generality, consider the case where the origin of the state space is an equilibrium point ($\mathbf{x}_e = \mathbf{0}$). Then, expanding each component of $\mathbf{X}(\mathbf{x})$ in a Taylor series about the origin and considering only the linear terms, we get the so-called *variational equation*

$$\dot{\mathbf{x}} = \mathbf{J}(0)\mathbf{x}, \tag{8.20}$$

where $\mathbf{J}(\mathbf{x})$ is the Jacobian matrix evaluated at the equilibrium point $\mathbf{x} = \mathbf{0}$; that is,

$$\mathbf{J}(0) = \begin{bmatrix} \dfrac{\partial X_1}{\partial x_1} & \dfrac{\partial X_1}{\partial x_2} & \cdots & \dfrac{\partial X_1}{\partial x_n} \\ \dfrac{\partial X_2}{\partial x_1} & \dfrac{\partial X_2}{\partial x_2} & \cdots & \dfrac{\partial X_2}{\partial x_n} \\ \cdot & \cdot & & \cdot \\ \cdot & \cdot & & \\ \cdot & \cdot & & \\ \dfrac{\partial X_n}{\partial x_1} & \dfrac{\partial X_n}{\partial x_2} & \cdots & \dfrac{\partial X_n}{\partial x_n} \end{bmatrix}_{\mathbf{x}=0} \tag{8.21}$$

The variational equation is a linear homogeneous differential equation that, for the autonomous case under consideration, has constant Jacobian matrix $\mathbf{J}$.[6] The stability of the linearized equation is then determined by the roots of the characteristic equation

$$|\mathbf{J}(0) - s\mathbf{I}| = 0. \tag{8.22}$$

The question then arises of how the behavior of the nonlinear system (8.9) is represented by the linearized model (8.20). In particular, it is of interest to see how the stability of the linearized model is related to the stability of the nonlinear system.

To illustrate some of the above statements, consider a first-order nonlinear differential equation[7]

$$\dot{x}_1 = -x_1 + x_1^2, \qquad x_{10} = x_1(t_0). \tag{8.23}$$

The general solution is

$$x_1 = \frac{Ce^{-t}}{1 - C + Ce^{-t}}, \tag{8.24}$$

where C is the integration constant to be determined from the initial value x_{10}. If it is assumed that the initial time is $t_0 = 0$, the state space $\mathscr{X}$ (in this case one-dimensional space) and the solutions as functions of time for different initial conditions are shown in Figure 8.2. There are two equilibrium points at $x_1 = 0$ and $x_1 = 1$ that are the solutions of the equation $-x_1 + x_1^2 = 0$. The trajectories in the state space all lie on the x_1 axis. The time as a parameter can be interpolated on the trajectories by means of the state-time diagram.

If attention is focused on the equilibrium point $x_1 = 0$, the Jacobian $J(0) = -1$ and the linearized model (8.20) can be written as

$$\dot{x}_1 = -x_1, \tag{8.25}$$

---

[6] In the nonautonomous case where $\dot{\mathbf{x}} = \mathbf{X}(\mathbf{x}, t)$ and $\mathbf{x}_e$ is the equilibrium state, let $\mathbf{y} = \mathbf{x} - \mathbf{x}_e$ and assume that $\mathbf{X}$ is analytic in a neighborhood of $\mathbf{x}_e$, and in that neighborhood $\mathbf{X}$ can be developed in a Taylor series

$$\dot{\mathbf{y}} = \mathbf{X}(\mathbf{x}, t) = \mathbf{X}(\mathbf{x}_e, t) + [\mathbf{J}(t) + \mathbf{K}(\mathbf{y}, t)](\mathbf{x} - \mathbf{x}_e) = \mathbf{J}(t)\mathbf{y} + \mathbf{K}(\mathbf{y}, t)\mathbf{y},$$

where $\mathbf{J}(t)$ is the Jacobian matrix of $\mathbf{X}$ evaluated at $\mathbf{x}_e, t$ (see equation 8.21); and it is assumed that $\mathbf{K}(\mathbf{y}, t)$ representing the higher-order term in the Taylor expansion is such that $\|\mathbf{K}\mathbf{y}\|$ is small relative to $\|\mathbf{y}\|$ when the latter is small. In other words, it means that

$$\|\mathbf{K}(\mathbf{y}, t)\mathbf{y}\|/\|\mathbf{y}\| \to 0 \quad \text{with} \quad \|\mathbf{y}\| \to 0 \quad \text{for} \quad t \geq 0.$$

Under this condition it is appropriate to linearize the nonlinear system about the equilibrium state $\mathbf{x}_e$ and consider only small deviations from $\mathbf{x}_e$. It is possible to give a rigorous proof of the legitimacy of this procedure by means of the second method of Liapunov outlined in the following section (for a proof see reference 8.10).

[7] This example was used by J. K. Hale and J. P. LaSalle in their tutorial paper, Differential Equations: Linearity vs. Nonlinearity, *SIAM Review*, 5, No. 3, 249–272 (July, 1963). The paper gives an excellent introduction to nonlinear phenomena.

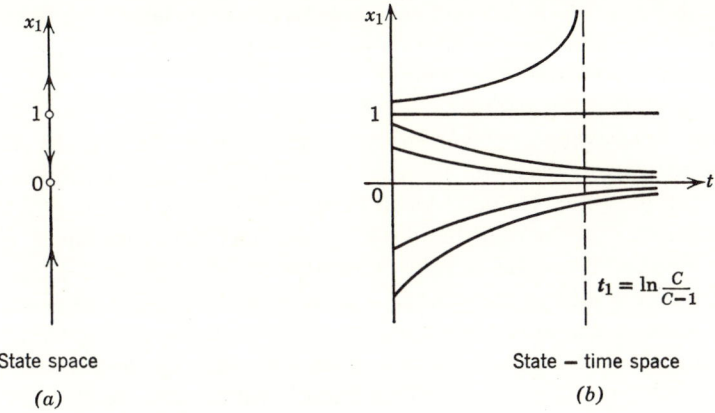

Figure 8.2 Solutions of equation $\dot{x}_1 = -x_1 + x_1^2$.

which has the general solution

$$x_1 = Ce^{-t}. \tag{8.26}$$

This reveals that the equilibrium $x_1 = 0$ is stable. An inspection of the curves in Figure 8.2 supports this statement. It is, however, a *local* property and cannot be extended for all initial values. If the transformation of coordinate $x_1 = 1 + y_1$ is used to consider the equilibrium at $x_1 = 1$ as equilibrium at $y_1 = 0$, the new equation is $\dot{y}_1 = y_1 + y_1^2$. Near $y_1 = 0$ the linearization yields $\dot{y}_1 = y_1$ and the solution of the linearized model is $y_1 = Ce^t$, where C is the initial value $y_{10} = y_1(0)$. The solution $y_1(t)$ shows that the solution $x_1(t)$ of the original equation will diverge from $x_1 = 1$, which is actually the case. The linearization apparently cannot predict *global* behavior of the solutions for all initial values and determine, for example, the region of attraction of the equilibrium point at $x_1 = 0$.

Even local stability (or instability) of equilibrium points cannot always be entirely predicted by the linearization. It is possible to conclude that if all the characteristic roots of the linearized equation have negative (nonzero) real parts, then the corresponding equilibrium point is stable. If some of the roots have positive real parts, the equilibrium state is unstable. The linearization fails to indicate stability (or instability) if there are pure imaginary or zero roots and the rest have negative real parts (see Problem 8.10).

A detailed study of the local stability of nonlinear systems has been presented by Liapunov. The method that Liapunov calls "first method" investigates stability on the basis of the variational equation. More particularly, it considers the solutions of the variational equation by means of successive approximations. The Liapunov "first method" did not find wide application in the stability analysis of nonlinear systems, since it deals with questions of stability by an explicit representation of the solutions of a differential equation describing the considered system. It will not be considered in the following stability analysis.

Another interesting observation can be made in Figure 8.2. As seen from the state-time space diagram, there is a possibility for the solution to become infinite in finite time. Namely, when time t approaches $t_1 = \ln [C/(C - 1)]$, the denominator of the right-hand side in (8.24) tends to zero and x_1 increases without bound. Thus the solution that started at some $x_{10} > 1$ cannot be extended over t_1 unless the state $x_1(t_1)$ "jumps" from $+\infty$ to $-\infty$. In this case we say that the solution of the differential equation "escapes" to infinity in finite time or has *finite escape time*. This phenomenon cannot take place in physical systems and is ruled out by explicit assumption to the contrary as provided by condition (c) of the motion properties (8.6).[8]

8.3 Liapunov's Stability Concept

Stability is an essential property of motion. Under certain general conditions systems exhibit a *fixed motion* and the question of stability arises when the conditions are changed: Does the perturbed motion stay always in the neighborhood of the original motion or not? Thus the stability analysis is concerned with qualitative investigation of the change in time of the "distance" between the two motions.

In mathematical terms it means that a system is described by

$$\dot{\mathbf{x}} = \mathbf{X}(\mathbf{x}, \mathbf{f}, t) \tag{8.8}$$

and it is of interest to analyze qualitatively the deviations about some fixed trajectory $\mathbf{x}^0(t)$ corresponding to a specific forcing function $\mathbf{f}(t)$. We want to find out whether the trajectories $\mathbf{x}(t)$ that start close to $\mathbf{x}^0(t)$ stay close as time progresses, or they diverge from it indefinetely. In the former case we have stability, in the latter case we have instability. Consequently, the distance between the two motions $\mathbf{x}^0(t)$ and $\mathbf{x}(t)$ as a function of time is essential (Figure 8.3).

The "deviation" of the motion $\mathbf{x}(t)$ from $\mathbf{x}^0(t)$ is a vector $\mathbf{y} = \mathbf{y}(t)$ defined by

$$\mathbf{y} = \mathbf{x} - \mathbf{x}^0. \tag{8.27}$$

To measure the deviation $\mathbf{y}$ it is necessary to introduce the notion of the

---

[8] If the global Lipschitz condition holds for the infinitesimal transition function $\mathbf{X}(\mathbf{x}, t)$ there can be no finite escape time. [8.10]

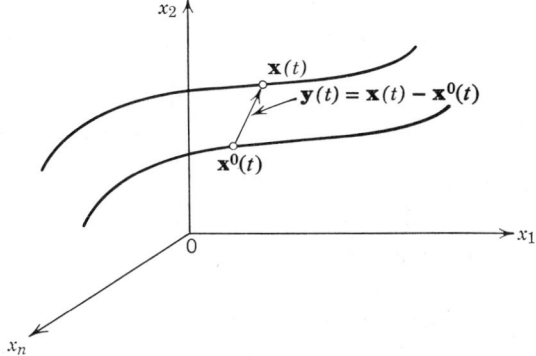

Figure 8.3 Fixed and perturbed motions.

norm[9] $\|\mathbf{y}\|$, which is a generalization of the idea of length. There are several ways a norm may be defined. In the analysis that follows, however, it will suffice to use the Euclidean distance as a norm that is defined on $\mathscr{X}$ as

$$\|\mathbf{x}\| = (\mathbf{x}^T\mathbf{x})^{1/2} = \left(\sum_{i=1}^{n} x_i^2\right)^{1/2}. \tag{8.28}$$

It is important to note that the stability analysis of motions can be reduced to a stability analysis of the equilibrium state or equivalently to the distance between the instantaneous state $\mathbf{x}(t)$ and the equilibrium state $\mathbf{x}_e$, which is a point in the state space. To show this, let $\mathbf{x}^0(t)$ be a fixed motion of (8.8) for some specified $\mathbf{f}(t)$. Then the deviation $\mathbf{y}(t)$ between the fixed motion $\mathbf{x}^0(t)$ and the perturbed motion $\mathbf{x}(t)$ is given by (8.27). From (8.8) and (8.27) we have

$$\begin{aligned}\dot{\mathbf{y}} &= \dot{\mathbf{x}} - \dot{\mathbf{x}}^0 \\ &= \mathbf{X}(\mathbf{x}^0 + \mathbf{y}, \mathbf{f}, t) - \mathbf{X}(\mathbf{x}^0, \mathbf{f}, t) \\ &= \mathbf{Y}(\mathbf{y}, t).\end{aligned} \tag{8.29}$$

Since $\mathbf{x}^0$ is a solution of (8.8), we have

$$\dot{\mathbf{y}} = \mathbf{Y}(\mathbf{y}, t), \quad \mathbf{Y}(\mathbf{0}, t) = \mathbf{0}, \quad \text{for all } t. \tag{8.30}$$

---

[9] The norm $\|\mathbf{x}\|$ of a vector $\mathbf{x}$ is a real number defined on a vector space $\mathscr{X}$, which must satisfy the following axioms:

(a) $\|\mathbf{x}\| \geq 0$ for all $\mathbf{x}$;
(b) $\|\mathbf{x}\| = 0$ if and only if $\mathbf{x} = \mathbf{0}$;
(c) $\|a\mathbf{x}\| = |a| \|\mathbf{x}\|$ for all numbers a and vectors $\mathbf{x}$; and
(d) $\|\mathbf{x}_1 + \mathbf{x}_2\| \leq \|\mathbf{x}_1\| + \|\mathbf{x}_2\|$ for all pairs $\mathbf{x}_1, \mathbf{x}_2$ (triangle inequality).

Then, $\mathscr{X}$ is said to be a *normed space*. When the norm is the Euclidean distance (8.28), $\mathscr{X}$ is an *Euclidean space*.

Hence, for specified $\mathbf{f} = \mathbf{f}(t)$, deviations $\mathbf{y}(t)$ from the fixed motion $\mathbf{x}^0(t)$ are represented by deviations from an equilibrium point in a free dynamic system.

Although the intuitive idea of stability seems to be straightforward, by a closer investigation it becomes rather complex. Little help is offered by the notion of stability in linear systems as was illustrated by the last example in Section 8.2.[10] Therefore, numerous precise definitions of stability have been proposed for nonlinear systems. They are merely extensions and modifications of the original Liapunov ideas and frequently their slight differences may lead to confusion. In this section, the basic definitions will be presented and their relationships to one another will be discussed.

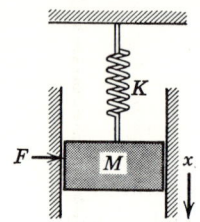

Figure 8.4 Mass, damper, and spring system.

Besides precisely formulating the stability definitions for nonlinear systems, Liapunov proposed the "second method," or "direct method," which answers the questions of stability without solving the related differential equations. The name is somewhat misleading with respect to the contents it represents, since the "second method" offers a powerful general approach rather than a systematic method of analysis.

The idea underlying the second method may be conveniently introduced by a simple physical reasoning based upon the energy concept. In general, however, significant difficulties are involved in expressing the energy of a system when the equations of motion are written in a purely mathematical form. Therefore, Liapunov in his approach departs from the concept of energy by introducing a function which takes the role of energy but can be constructed in different forms to provide information about system stability. This function is known as the *Liapunov function* and will be precisely defined later. Here, it is of interest to illustrate by a simple example the idea of Liapunov's approach and emphasize its most attractive property—the geometric insight into stability.

Consider a free linear mass, damper, and spring system in Figure 8.4. In the conservative or nondissipative case ($F \equiv 0$), after normalization of the angular frequency $\omega = \sqrt{K/M}$ to 1, the system is described by the differential equation

$$\ddot{x} + x = 0. \qquad (8.31)$$

By denoting $x = x_1$, $\dot{x} = x_2$, (8.31) can be rewritten in the state variable form

$$\begin{aligned} \dot{x}_1 &= x_2 \\ \dot{x}_2 &= -x_1. \end{aligned} \qquad (8.32)$$

---

[10] At the beginning of Section 4.5 this fact is discussed in more detail.

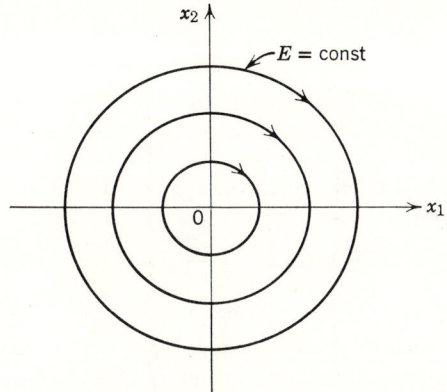

Figure 8.5 State space trajectories of harmonic oscillator.

The trajectories in the state space $\mathscr{X}$ are determined by the equation

$$\frac{dx_2}{dx_1} = -\frac{x_1}{x_2}, \tag{8.33}$$

which is obtained from (8.32). Equation 8.33 can be integrated by separation of the variables to get

$$\frac{x_1^2}{2} + \frac{x_2^2}{2} = C^2, \tag{8.34}$$

where C is a constant. Thus the trajectories are concentric circles about the origin, which is the equilibrium state, as shown in Figure 8.5 for various values of C corresponding to various initial conditions (x_{10}, x_{20}).

The total energy $E(\mathbf{x})$ of this conservative system is the potential energy $x_1^2/2$, plus the kinetic energy $x_2^2/2$. Then, from (8.34) it is clear that the total energy

$$E(\mathbf{x}) = \tfrac{1}{2}x_1^2 + \tfrac{1}{2}x_2^2 \tag{8.35}$$

is constant and that its derivative

$$\dot{E}(\mathbf{x}) = x_1\dot{x}_1 + x_2\dot{x}_2 \tag{8.36}$$

by (8.32) is zero along any trajectory in Figure 8.5. Therefore, for sufficiently small perturbations of any of the trajectories, the perturbed trajectory will stay near the original trajectory for all time.

Now introduce damping so that the system is described by

$$\ddot{x} + \dot{x} + x = 0 \tag{8.37}$$

(for simplicity, it is assumed that $M = F = K = 1$). The state space equations are

$$\dot{x}_1 = x_2$$
$$\dot{x}_2 = -x_1 - x_2. \tag{8.38}$$

The expression (8.34) for the total energy $E(\mathbf{x})$ remains the same, but its derivative $\dot{E}(\mathbf{x})$ becomes

$$\dot{E}(\mathbf{x}) = -x_2^2. \tag{8.39}$$

Thus the motion of the system is from one contour of constant energy to a contour of lesser constant energy, except on the x_1 axis ($x_2 = 0$), where $\dot{E}(\mathbf{x}) = 0$. This requires additional comment. The motion cannot terminate on the x_1 axis since there are no equilibrium positions except at $x_1 = 0$. Therefore the system approaches asymptotically the equilibrium state at the origin, as shown in Figure 8.6.

In the damped case it is actually possible to improve the analysis by using a Liapunov function $V(\mathbf{x})$ of the form

$$V(\mathbf{x}) = \tfrac{3}{2} x_1^2 + x_1 x_2 + x_2^2. \tag{8.40}$$

Its derivative

$$\dot{V}(\mathbf{x}) = -(x_1^2 + x_2^2) \tag{8.41}$$

is negative unless both x_1 and x_2 are zero. This shows that V is constantly decreasing along any trajectory in the state space, and no additional comment

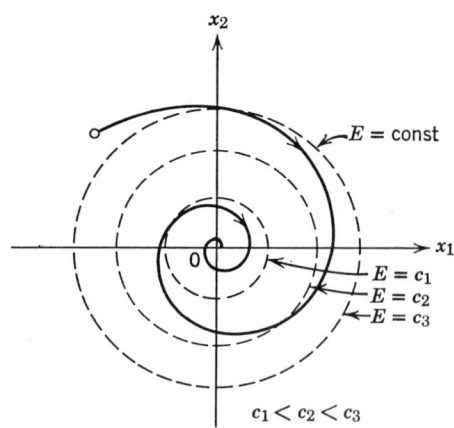

Figure 8.6 State space trajectory and constant energy contours of harmonic oscillator with damping.

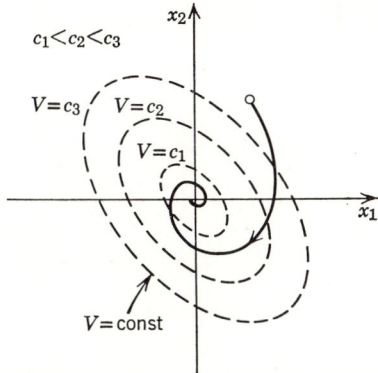

Figure 8.7 Liapunov function contours for damped system.

is necessary as it was when the energy concept was used. As seen in Figure 8.7, the $V = \text{const}$ contours form a family of concentric ellipses surrounding the origin, which is clear from (8.40). It follows that the trajectories cross the boundary of every region $V(\mathbf{x}) \leq \text{const}$ from the outside toward the inside and *asymptotically* approach the equilibrium state at the origin.

The Liapunov function may be geometrically interpreted as a measure of the distance between the instantaneous state $\mathbf{x}(t)$ and the origin with the property that $V(\mathbf{x}) > 0$ when $\mathbf{x} \neq \mathbf{0}$ and $V(\mathbf{0}) = 0$. If $\dot{V}[\mathbf{x}(t)] < 0$ for $\mathbf{x} \neq \mathbf{0}$ and the distance of the state $\mathbf{x}(t)$ from the origin is continually decreasing as $t \to \infty$, then $\mathbf{x}(t) \to \mathbf{0}$.

An immediate advantage of the Liapunov function over the energy concept is that it is not unique and may be constructed in various ways to answer questions of stability even in those cases where the energy fails to provide the necessary information. It should be noted, however, that in general there are no systematic methods for choosing an appropriate Liapunov function that will measure the distance of the system state from the origin at every instant of time and indicate stability. Moreover, even if a certain Liapunov function fails to reveal the stability, the system is not necessarily unstable. In other words, the existence of a Liapunov function gives only sufficient conditions for stability, and this is often a significant drawback in an actual system analysis.

8.4 Stability Definitions

As was pointed out in the previous section, in stability analysis of deviations about some fixed motion, it suffices to analyze deviations from an equilibrium state $\mathbf{x}_e$ of a free dynamic system

$$\dot{\mathbf{x}} = \mathbf{X}(\mathbf{x}, t). \tag{8.2}$$

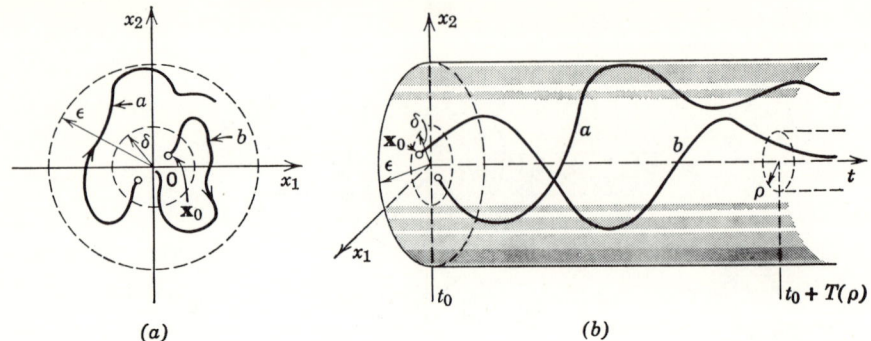

Figure 8.8 The concept of stability in the sense of Liapunov: (*a*) state space; (*b*) state-time space.

Therefore the stability definitions *in the sense of Liapunov* will consider the equilibrium state of (8.2) without loss in generality.

The classical Liapunov stability definition is as follows.

Definition 8.1. Stability. An equilibrium state $\mathbf{x}_e$ of a free dynamic system is stable if for every real number $\varepsilon > 0$ there exists a real number $\delta(\varepsilon, t_0) > 0$ so that $\|\mathbf{x}_0 - \mathbf{x}_e\| \leq \delta$ implies

$$\|\mathbf{x}(t, \mathbf{x}_0, t_0) - \mathbf{x}_e\| \leq \varepsilon, \quad \text{for all } t \geq t_0.$$

If $\mathbf{x}_e$ is not stable, it is said to be unstable.

A geometric interpretation of this definition is shown in Figure 8.8 for a second-order system. If $\mathbf{x}_0$ is chosen in the interior of the circle with radius δ, then the trajectory $\mathbf{x}(t, \mathbf{x}_0, t_0)$ remains for all future times inside the cylinder of radius ε.

It is important to note that the above concept of stability in the sense of Liapunov is a *local* one. It is often referred to as the *stability in the small* since we do not know a priori how small it may be necessary to choose δ.[11]

If for every $\varepsilon > 0$ there is a $\delta > 0$ independent of t_0 so that the above inequality holds, then the equilibrium state $\mathbf{x}_e$ is said to be *uniformly stable in the sense of Liapunov*.[12]

---

[11] For example, a stable equilibrium point could be surrounded in the phase plane by an unstable limit cycle and if δ is chosen too large, some solutions would diverge to infinity.

[12] It is of interest to note that if there is stability for some initial time t_0, there is stability for any other initial time t_1, provided that all motions are continuous in the initial state $\mathbf{x}_0$. It has been shown by Kalman and Bertram [8.10] that by using axiom (8.6e), which requires continuity of $\mathbf{x}(t, \mathbf{x}_0, t_0)$ with respect to all arguments, we can show that the stated proposition is true.

In most practical situations Definition 8.1 is not satisfactory, for it is not sufficient that the solution stay near the equilibrium state (curve *a* of Figure 8.8). It is often necessary that after perturbation the motion return to equilibrium position (curve *b* of Figure 8.8). This means that we prefer a damped harmonic oscillator (Figure 8.6) to the undamped one (Figure 8.5). This leads to the notion of the *asymptotic stability*.

Definition 8.2. Asymptotic Stability. An equilibrium state $\mathbf{x}_e$ of a free dynamic system is asymptotically stable if:
 (a) it is stable; and
 (b) there is some real constant $v(t_0) > 0$ and to every real number $\rho > 0$ there corresponds a real number $T(\rho, \mathbf{x}_0, t_0)$ so that $\|\mathbf{x}_0 - \mathbf{x}_e\| \leq v(t_0)$ implies
$$\|\mathbf{x}(t, \mathbf{x}_0, t_0) - \mathbf{x}_e\| \leq \rho, \quad \text{for all} \quad t \geq t_0 + T$$
(in other words, every motion that starts sufficiently close to $\mathbf{x}_e$ converges to $\mathbf{x}_e$ as $t \to \infty$).

A geometric interpretation of asymptotic stability is given in Figure 8.8. It is important to note that the statement (a) in Definition 8.2 is necessary and it is not implied in Definition 8.2(b) (see Problem 8.9).

Asymptotic stability is also a local concept. It is not clear a priori how small $v(t_0)$ may have to be. For example, consider (8.23) in the form

$$\dot{x}_1 = -ax_1 + x_1^2 \tag{8.42}$$

where a is a positive constant. Then, as is clear from Figure 8.2, the upper limit of the domain of attraction[13] for the equilibrium state $x_1 = 0$ can be made arbitrarily small by a suitable choice of a. In other words, equilibrium state $x_1 = 0$ is asymptotically stable, and yet there are motions close to it that diverge from it. Therefore, in various applications, we are more interested in assuring a large region of attraction or, if possible, *global stability*, which is often referred to as the *stability in the large*.

Definition 8.3. Asymptotic Stability in the Large. An equilibrium state $\mathbf{x}_e$ of a free dynamic system is asymptotically stable in the large (or completely stable) if:
 (a) it is stable; and
 (b) every motion converges to $\mathbf{x}_e$ as $t \to \infty$ (in other words, the region of attraction for the equilibrium state $\mathbf{x}_e$ is the whole state space $\mathscr{X}$).

In the following discussion this kind of stability is often called for short ASIL (asymptotic stability in the large).

---
[13] Finite domains of stability are discussed in Section H.7 of Appendix H.

Evidently ASIL implies that the state x_e is the only equilibrium state of the system. It is also clear that if a linear system is asymptotically stable it is also asymptotically stable in the large; that is, the stability is independent of the distance of the initial state x_0 from the equilibrium x_e. By Definition 8.3 we require that a nonlinear system has this important property of linear systems.

Further refinements of the concept of stability to include the uniformity property with respect to initial conditions are of interest in the following developments. If, besides being stable, x_e has a property that every motion starting sufficiently close to x_e converges to x_e *uniformly* in x_0, in other words, T in Definition 8.2 is of the form $T[\rho, v(t_0), t_0]$, then x_e is said to be *equi-asymptotically stable*.[14] If, in addition δ, v, and T are independent of t_0, we have *uniform asymptotic stability*.[15]

For example [8.7], the first-order equation

$$\dot{x}_1 = -\frac{x_1}{1+t} \qquad (8.43)$$

has the general solution

$$x_1(t, x_{10}, t_0) = x_{10}\frac{1+t_0}{1+t}, \qquad (8.44)$$

which does not tend to zero uniformly in t_0, that is, a system described by (8.43) is not uniformly asymptotically stable.[16]

This example motivates the following definition of stability.

Definition 8.4. **Uniform Asymptotic Stability in the Large.** An equilibrium state x_e of a free dynamic system is uniformly asymptotically stable in the large if:

(a) it is uniformly stable;

(b) every motion is uniformly bounded, that is, given any $v > 0$, there is some $N(v)$ such that $\|x_0 - x_e\| \leq v$ implies $\|x(t, x_0, t_0) - x_e\| \leq N$ for all $t \geq t_0$; and

(c) every motion converges to x_e as $t \to \infty$ uniformly in t_0 and $\|x_0\| \leq v$, where v is fixed but arbitrarily large (in other words, given any $v > 0$ and

---

[14] If all motions are continuous in x_0, it can be shown [8.10] that an equiasymptotic version of Definition 8.2-b implies stability Definition 8.1.

[15] It is clear that uniform asymptotic stability involves the explicit assumption of uniform stability. It can also be shown [8.10] that asymptotic and equiasymptotic stability are equivalent if they are uniform in t_0. This is based upon the continuity of motion in the initial state x_0.

[16] In (8.44) it suffices to show that for a given ρ in Definition 8.2 a T cannot be found independent of t_0, and there is no uniform stability.

$\rho > 0$, there is some $T(\rho, \nu)$ such that $\|\mathbf{x}_0 - \mathbf{x}_e\| \leq \nu$ implies

$$\|\mathbf{x}(t, \mathbf{x}_0, t_0) - \mathbf{x}_e\| \leq \rho$$

for all $t \geq t_0 + T$).[17]

In the case of autonomous systems uniform asymptotic stability is implied by asymptotic stability. As illustrated by (8.42), this conclusion is not valid even for linear nonautonomous systems. Because certain classes of nonautonomous system are considered in the following sections it is necessary to formulate precisely the uniform stability definition. Then, in the next section and Section H.2, a Liapunov stability theorem will be proved that assures this kind of stability.

In most engineering applications it is not sufficient to know that the system is asymptotically stable and that all the solutions tend uniformly towards the equilibrium state $\mathbf{x}_e$ as $t \to \infty$. There is a need to estimate the rate of convergence of the solutions and have information about rapidity of the transient process due to interminent perturbations, parameter variations, and so on.

A common comparison function for the estimation of the decay rate of transient processes is the exponential function. This may be related to the fact that autonomous linear systems have exponential behavior; that is, if they are stable, the transient process decays faster than an exponential (if they are unstable, they are exponentially unstable). This can be concluded from the derivation of the relative stability conditions in Section 2.3 of Chapter 2.

Therefore the following stronger stability definition is desired.

Definition 8.5 *Exponential Stability.* An equilibrium state $\mathbf{x}_e$ of a free dynamic system described by the differential equation $\dot{\mathbf{x}} = \mathbf{X}(\mathbf{x}, t)$ is exponentially stable if there exist two positive constants M and μ that are independent of initial values $(\mathbf{x}_0, t_0)$, such that for sufficiently small initial values the inequality

$$\|\mathbf{x}(t, \mathbf{x}_0, t_0)\| \leq M \|\mathbf{x}_0\| e^{-\mu(t-t_0)}$$

is satisfied.

Apparently the solution (8.44) does not tend exponentially toward zero, hence the corresponding linear system (8.43) is not exponentially stable. In the following sections a useful and broad class of nonlinear systems is considered and conditions for their exponential stability are derived. The local property of the above definition is extended to include the entire state space and thus guarantee global exponential stability, that is, exponential stability in the large.

---

[17] Similarly, the state $\mathbf{x}_e$ is equiasymptotically stable in the large if for any fixed ν however large, $\|\mathbf{x}_0\| \leq \nu$ implies that every motion $\mathbf{x}(t, \mathbf{x}_0, t_0)$ converges to $\mathbf{x}_e$ as $t \to \infty$, uniformly in $\mathbf{x}_0$.

From Definition 8.5 it is clear that exponential stability implies uniform asymptotic stability.

In a large number of practical situations the concept of Liapunov stability may be too restrictive. For instance [8.2], consider the first-order differential equation

$$\dot{x}_1 = x_1(a^2 - x_1^2), \qquad x_1(t_0) = x_{10} \tag{8.45}$$

where $a \neq 0$ is a constant. There are three equilibrium points; at $x_1 = 0$ and $x_1 = \pm a$. The general solution of (8.45) can be expressed as

$$\ln \left| \frac{x_1 \sqrt{a^2 - x_{10}^2}}{x_{10} \sqrt{a^2 - x_1^2}} \right| = a^2(t - t_0). \tag{8.46}$$

Equation 8.46 yields real values for the solution $x_1 = x_1(t, x_{10}, t_0)$ when either $|x_{10}| < a$ or $|x_{10}| > a$. In both these cases

$$\lim_{t \to \infty} x_1 = \pm a,$$

which means that the equilibrium at the origin is unstable, whereas the other two at $\pm a$ are stable as shown in Figure 8.9. The stability of the equilibrium points can be concluded using linearization, as in (8.23).

Despite the fact that the equilibrium point $x_1 = 0$ is unstable, the system described by (8.45) can be useful. If a is sufficiently small so that the deviations from the origin which are equal to (or less) than a are tolerated, then the system can be considered "practically stable." Moreover, the stability is "global and asymptotic."[18]

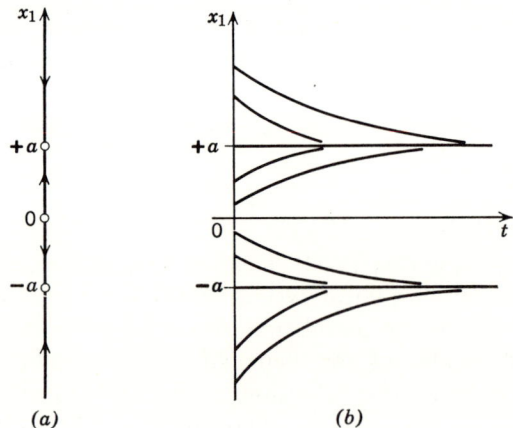

Figure 8.9 Solutions of equation $\dot{x}_1 = x_1(a^2 - x_1^2)$: (a) state space; (b) state-time space.

---

[18] Similar situations may arise in cases when a system has a stable limit cycle with a sufficiently small amplitude.

From the foregoing it may be concluded that in certain situations it is satisfactory to guarantee a boundedness of all the motions so that they may deviate from a certain equilibrium point but never go too far from it.[19]

The boundedness of motion is based on the general idea of the boundedness of solutions in differential equations theory, and when interpreted as a stability property it is usually called the *stability in the sense of Lagrange*.[20]

A type of Lagrange stability that has found important applications is the *bounded input-bounded output stability* usually called *BIBO stability*. BIBO stability places emphasis on forced systems in cases when external forces or inputs are not specified completely but belong to a class of functions with assigned boundedness property.

The definition of the BIBO stability is based on the system description

$$\dot{\mathbf{x}} = \mathbf{X}(\mathbf{x}, \mathbf{f}, t), \qquad (8.8)$$

in which the external force $\mathbf{f} = \mathbf{f}(t)$ enters explicitly into the infinitesimal transition function $\mathbf{X}$.

Definition 8.6. *Bounded Input–Bounded output Stability.* A dynamic system described by a vector differential equation $\dot{\mathbf{x}} = \mathbf{X}(\mathbf{x}, \mathbf{f}, t)$ is BIBO stable (any bounded input $\mathbf{f}$ produces a bounded output $\mathbf{x}$) if for all $\mathbf{x}_0$ and t_0, and for all bounded inputs $\mathbf{f}(t)$, $t_0 \leq t < \infty$, which satisfy the inequality $\|\mathbf{f}\| < M$ for all $t \geq t_0$ where $M < \infty$ is a constant there exists a finite number $N(\mathbf{x}_0, M)$ independent of t_0, such that

$$\|\mathbf{x}(t, \mathbf{x}_0, t_0)\| < N, \qquad \text{for all} \quad t \geq t_0.$$

The output motion $\mathbf{x}(t, \mathbf{x}_0, t_0)$ is bounded.

If the system is not BIBO stable it is said to be BIBO unstable.

It may be concluded that the Liapunov stability concept is concerned with the *internal dynamics* of nonlinear systems, whereas the BIBO stability reflects their *external behavior*.

In general, Liapunov stability does not guarantee BIBO stability, and vice versa. It is a well-known fact that if a linear, time-invariant, differential (lumped) system is asymptotically stable (in the large) in the Liapunov sense, it is also BIBO stable; that is, any bounded input will produce a bounded output (see for example reference [8.12]). The converse is also true,

---

[19] Motion is said to be *bounded* for every $\mathbf{x}_0$, t_0 if there is some constant $N(\mathbf{x}_0, t_0)$ such that $\|\mathbf{x}(t, \mathbf{x}_0, t_0)\| \leq N$ for all $t \geq t_0$. The motion is *equibounded* if $N(\mathbf{x}_0, t_0) \leq N(\nu, t_0)$ for all $\|\mathbf{x}_0\| \leq \nu$, $t \geq t_0$. If, in addition, N is independent of t_0, the motion is *uniformly* bounded.

[20] Certain important conclusions can be drawn about the Lagrangian stability using the Liapunov concept of stability. The boundedness of motions and the correlation between the two kinds of stability have been systematically analyzed by Yoschizawa as reported by Hahn [8.7].

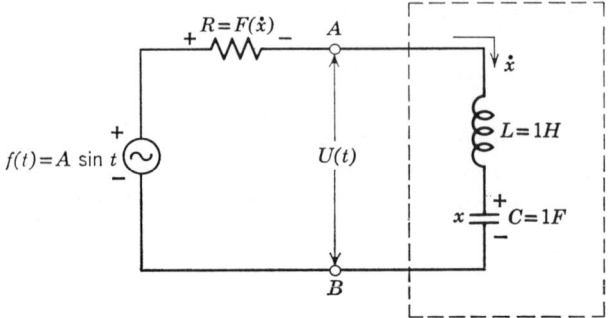

Figure 8.10 Nonlinear RLC circuit.

provided the system is completely observable and controllable.[21] If the system is allowed to be time-varying, the implication is false, as shown by a simple example from Kalman and Bertram [8.10] (see Problem 8.18). It has been shown by Desoer and Liu [8.27] that the implication is also not true for a nonlinear, time-invariant, differential system. Desoer and Liu used the following example.

Consider the RLC electric circuit analog (Figure 8.10) of the mass, spring, and damper system used previously (Figure 8.4). It is assumed that the resistance effect is nonlinear, as represented by the nonlinear function $R = F(\dot{x})$ in the differential equation

$$\ddot{x} + x + F(\dot{x}) = f \tag{8.47}$$

describing the system. Equation 8.47 is obtained by applying Kirchhoff's voltage law. The forcing function $f = f(t)$ is considered to be the input and $x = x(t)$ the output. Suppose that F satisfies the Lipschitz condition and the conditions

$$\begin{aligned} yF(y) &> 0, \quad \text{for all } y \neq 0 \\ F(0) &= 0, \quad \text{and} \quad |F(y)| \leq 1, \quad -\infty < y < \infty. \end{aligned} \tag{8.48}$$

It will be shown that the system is ASIL, and yet for the bounded input

$$f(t) = A \sin t, \quad A > 4/\pi \tag{8.49}$$

it has an unbounded zero-state response.[22]

---

[21] For otherwise there might be unstable modes in the interior dynamics of the system which are uncoupled to the input or output of the system. For notions of observability and controllability see Section C.3.

[22] The zero-state response of a system to a forcing function $f(t)$ that is defined for $t \geq t_0$ is the response of the system to $f(t)$ when it is initially at its zero-state, that is, $x_0 = 0$.

To show that the system is asymptotically stable in the large, let $f(t) \equiv 0$, $x = x_1$, and $\dot{x} = x_2$ in (8.47). Then the system is described by

$$\begin{aligned} \dot{x}_1 &= x_2, \\ \dot{x}_2 &= -x_1 - F(x_2). \end{aligned} \qquad (8.50)$$

From (8.48) it can be concluded that the system (8.50) has a unique equilibrium state $\mathbf{x}_e = \mathbf{0}$. To prove that it is asymptotically stable in the large, we can use again the function

$$V(\mathbf{x}) = \tfrac{1}{2}x_1^2 + \tfrac{1}{2}x_2^2, \qquad (8.35)$$

which is the total energy of the system $V(\mathbf{x}) = E(\mathbf{x})$. The rate of change of energy is $\dot{V}(\mathbf{x}) = -x_2 F(x_2)$. By (8.48), $\dot{V}$ is always negative and the energy is decreasing along any solution of (8.50), except on the line $x_2 = 0$. On this line, for $x_1 \neq 0$, there are no equilibrium positions, and the system cannot stop on the line unless $x_1 = 0$, which corresponds to the equilibrium point $\mathbf{x}_e = \mathbf{0}$. Thus the system is ASIL.

Referring to Figure 8.10, we can see that the state $x_1 = x$ is the charge on the 1F capacitor; $x_2 = \dot{x}$ is then the current in the loop. Using (8.14), the zero-state response of this circuit to $f(t) = A \sin t$ applied at $t_0 = 0$ is obtained by inspection as

$$x_2 = \int_0^t \cos(t - \tau)[A \sin \tau - \mathscr{F}(\tau)]\, d\tau, \qquad (8.51)$$

where $\mathscr{F}(t) = F[x_2(t)]$ is the voltage across the nonlinear resistance at time t, and at the same time $U(t) = A \sin t - \mathscr{F}(t)$ is the voltage between the points A and B. Then, by a simple calculation, we get

$$x_2 = \frac{A}{2} t \sin t - \int_0^t \cos(t - \tau)\mathscr{F}(\tau)\, d\tau. \qquad (8.52)$$

Differentiating, we obtain

$$\dot{x}_2 + \mathscr{F}(t) = \frac{A}{2} t \cos t + \frac{A}{2} \sin t + \int_0^t \sin(t - \tau)\mathscr{F}(\tau)\, d\tau. \qquad (8.53)$$

Comparing this equation with (8.47), and having in mind that $f(t) = A \sin t$, $x_1 = x$, and $x_2 = \dot{x}$, we have

$$x(t) = \frac{A}{2} \sin t - \frac{A}{2} t \cos t - \int_0^t \sin(t - \tau)\mathscr{F}(\tau)\, d\tau. \qquad (8.54)$$

From (8.48), $|\mathscr{F}(\tau)| \leq 1$ for all τ and the integral in (8.54) is bounded as

$$\left| \int_0^t \sin(t - \tau)\mathscr{F}(t)\, d\tau \right| \leq \int_0^t |\sin(t - \tau)|\, d\tau. \qquad (8.55)$$

316 Stability Analysis

Given any $\varepsilon > 0$, there is a lower bound on t—for example, $t_1(\varepsilon)$—so that for all $t > t_1(\varepsilon)$

$$\int_0^t |\sin(t-\tau)|\, d\tau < \frac{2}{\pi}(1+\varepsilon)t. \tag{8.56}$$

By substituting (8.56) into (8.54), we obtain

$$x(t) > \frac{A}{2}\sin t - t\left[\frac{A}{2}\cos t + \frac{2}{\pi}(1+\varepsilon)\right]. \tag{8.57}$$

Selecting the instants of time $t_n = (2n+1)\pi$, and choosing a sufficiently large n, we have

$$x(t_n) > (2n+1)\pi\left[\frac{A}{2} - \frac{2}{\pi}(1+\varepsilon)\right].$$

Consequently, when $A/2 > 2/\pi$, an appropriate value for $\varepsilon > 0$ exists so that for sufficiently large n the sequence $\{x(t_n)\} \to \infty$. We have shown that for $A > 4/\pi$ there is a bounded input that produces an unbounded zero-state output. The system is BIBO unstable.

In the following, however, cases are considered in which nonlinear systems are both ASIL and BIBO stable.

Before concluding this section, it is of interest to mention that in this context it is convenient to formulate precisely the stability of periodic oscillations that was extensively considered in the previous chapters. The definition of orbital stability[23] [8.28] provides the desired formulation.

8.5 Liapunov Stability Theorem. Liapunov Function

The Liapunov stability theorems are based on the notion of sign-definite functions of state variables defined in some region around the origin of the state space. In that region these functions have always the same sign and go to zero only at the origin. In addition the sign-definite functions to be used

---

[23] The orbital stability is a precise stability concept to investigate stability of limit cycles. In Section 4.3 the stability of periodic solutions was investigated by an approximate method. In this context it is convenient to give a rigorous definition of the orbital stability and possibly improve the discussion in Section 4.3. Let $\mathbf{x}^0(t)$ be a nonconstant periodic solution of equation $\dot{\mathbf{x}} = \mathbf{X}(\mathbf{x})$, C be the closed curve defined by $\mathbf{x}^0(t)$ in $\mathscr{X}$, and $d(\mathbf{x}, C) = \inf \|\mathbf{x} - \mathbf{y}\|$ for $\mathbf{y}$ on C. The solution $\mathbf{x}^0(t)$ is said to be *orbitally stable* (to the right) if, for every $\varepsilon > 0$, there is a $\delta > 0$ such that for every solution $\mathbf{x}(t)$ of $\dot{\mathbf{x}} = \mathbf{X}(\mathbf{x})$ with $d[\mathbf{x}(t_0), C] < \delta$, we have $d[\mathbf{x}(t), C] < \varepsilon$ for all $t \geq t_0$. If $\mathbf{x}^0$ is orbitally stable and, in addition, $d[\mathbf{x}(t_0), C] < \delta$ implies $d[\mathbf{x}(t), C] \to 0$ as $t \to \infty$, then we say $\mathbf{x}^0$ is *asymptotically orbitally stable*. If $\mathbf{x}^0$ is asymptotically orbitally stable and for every solution $\mathbf{x}(t)$ with $d[\mathbf{x}(t_0), C] < \delta$ there exists a constant φ such that $\|\mathbf{x}(t) - \mathbf{x}^0(t+\varphi)\| \to 0$ as $t \to \infty$, then $\mathbf{x}^0$ is called *asymptotically orbitally stable with asymptotic phase*. When $\mathbf{x}^0$ is not orbitally stable it is *unstable*.

have continuous partial derivatives in the mentioned region. This is more precisely stated in the following definition.

Definition 8.7. Sign-Definite Function $V(\mathbf{x})$. A function $V(\mathbf{x})$ is said to be positive-definite (negative-definite) if, in the neighborhood of the origin of the vector space $\mathscr{X}$ in some region $R = \{\|\mathbf{x}\| < h\}$, where $h > 0$ is a constant, it satisfies the following conditions:
(a) $V(\mathbf{x})$ has continuous first partial derivatives with respect to $\mathbf{x}$;
(b) $V(\mathbf{0}) = 0$; and
(c) $V(\mathbf{x}) > 0$, $\mathbf{x} \neq \mathbf{0}$ (positive-definite)
   $V(\mathbf{x}) < 0$, $\mathbf{x} \neq \mathbf{0}$ (negative-definite).

In the following the fact that $V(\mathbf{x})$ is positive-definite will often be written $V > 0$, with the understanding that this inequality is preserved everywhere except at the origin, where $\mathbf{x} = \mathbf{0}$. Analogously, we use the notation $V < 0$ to indicate that V is negative-definite.

If in (c) for $\mathbf{x} \neq \mathbf{0}$, $V(\mathbf{x}) \geq 0$ or $V(\mathbf{x}) \leq 0$, $V(\mathbf{x})$ is said to be *positive-semidefinite* or *negative-semidefinite*, respectively.

Let us give a geometric interpretation of the sign-definite function. For simplicity, consider a positive-definite function $V(\mathbf{x})$ for $n = 2$ when $\mathbf{x} = \{x_1, x_2\}$. Let us consider the surface

$$z = V(x_1, x_2) \tag{8.58}$$

in $x_1 x_2 z$ space. It is easy to conclude that this is a concave cup as shown in Figure 8.11a. If the $x_1 x_2$ plane is considered, another interpretation is possible. In the $x_1 x_2$ plane the level curves

$$V(x_1, x_2) = c, \tag{8.59}$$

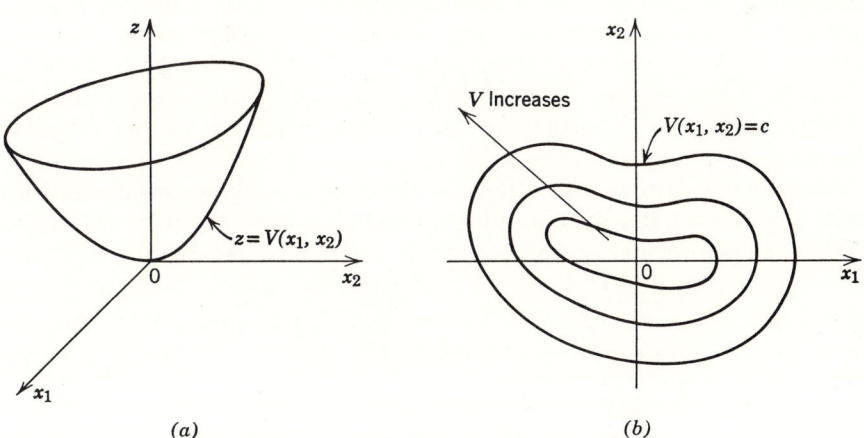

Figure 8.11 Positive-definite function.

where c is a positive constant, are represented by a family of closed but not intersecting curves which surround the origin as shown in Figure 8.11b. These curves can be obtained by cutting the cup of Figure 8.11a by horizontal planes and then projecting the sections on the $x_1 x_2$ plane. Although it is intuitively clear that the curves (8.59) have the form of Figure 8.11b, it is a consequence of a nontrivial rigorous proof [8.2]. In a multidimensional case ($n > 2$), the situation is analogous and we have closed nonintersecting surfaces surrounding the origin of the state space. Note also that the geometric properties of the described curves (surfaces) are generally of a limited extent being restricted to the region R defined above.

We can now state the Liapunov stability theorem for autonomous systems described by equation

$$\dot{\mathbf{x}} = \mathbf{X}(\mathbf{x}). \tag{8.9}$$

Theorem 8.1. Liapunov Stability Theorem. In order for the equilibrium state $\mathbf{x}_e = \mathbf{0}$ of the system (8.9) to be asymptotically stable in the large, it is sufficient that there exists a scalar function $V(\mathbf{x})$ with continuous first partial derivatives with respect to $\mathbf{x}$, so that $V(\mathbf{0}) = 0$ and:

(a) $V(\mathbf{x})$ is a positive-definite function, that is, $V(\mathbf{x}) > 0$ for all $\mathbf{x} \neq \mathbf{0}$;

(b) The total derivative $\dot{V}(\mathbf{x})$ of $V(\mathbf{x})$ along the motions of system (8.9) is $\dot{V}(\mathbf{x}) \equiv dV(\mathbf{x})/dt = (\partial V/\partial \mathbf{x})\mathbf{X}(\mathbf{x}) < 0$, for all $\mathbf{x} \neq \mathbf{0}$;[24] and

(c) $V(\mathbf{x}) \to \infty$ with $\|\mathbf{x}\| \to \infty$. \hfill (8.60)

The function $V(\mathbf{x})$ is called a Liapunov function of the system (8.9).

This is not the most general theorem of Liapunov but it is sufficient for most of the following developments. Moreover, it allows a convenient geometric proof that will be given here. The more general theorem is outlined in Section H.1 of the appendix, where it is shown that the above theorem is a corollary of the more general version. This general version of the Liapunov theorem provides sufficient conditions for uniform stability in the large. Then the other, weaker types of stability are obtained in Appendix H as another corollary by weakening the various requirements on Liapunov functions.

The geometric proof is illustrated in the easily visualized two dimensions, using Figure 8.12, but the reasoning is valid for any finite number of dimensions. We first prove under the assumptions of the theorem that for any given $\varepsilon > 0$ there is a $\delta > 0$ so that $\|\mathbf{x}_0\| < \delta$ implies $\|\mathbf{x}(t)\| < \varepsilon$ for all $t \geq t_0$, where $\|\mathbf{x}\| = \sqrt{x_1^2 + x_2^2}$. This means stability (Definition 8.1). Then, in addition, we wish to show that actually $\|\mathbf{x}\| \to 0$ as $t \to \infty$, which is asymptotic stability (Definition 8.2).

---

[24] Condition (b) can also be formulated as: $\dot{V}(\mathbf{x}) \leq 0$, for all $\mathbf{x}$; and $\dot{V}[\mathbf{x}(t, \mathbf{x}_0, t_0)]$ does not vanish identically in $t \geq t_0$ for any t_0 and $\mathbf{x}_0 \neq \mathbf{0}$.

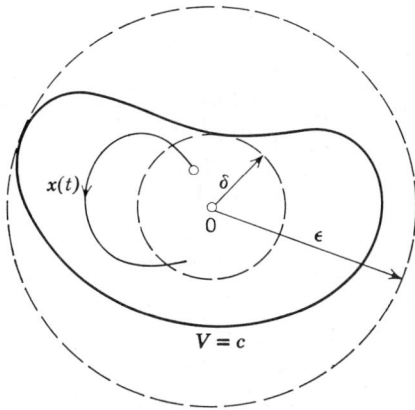

Figure 8.12 Geometric interpretation of stability.

To show stability, let us consider any solution of (8.9) that starts at t_0 from an arbitrary point in the neighborhood of the origin where V is positive-definite. If only $\dot{V} \leq 0$, the solution cannot cross any closed curve $V = c$ from inside to outside because V cannot increase with time ($\dot{V} \leq 0$) along any solution of (8.9) that starts sufficiently close to the origin. In other words, if a solution starts inside a curve $V = c$, it will stay inside the curve for all future time.

Let us now consider Figure 8.12. Because of the positive-definite property of V, it must have a minimum $V = c$ on the circle $\|\mathbf{x}\| = \varepsilon$. On the basis of the same property, a δ can be chosen such that for all $\mathbf{x}$ inside the circle $\|\mathbf{x}\| = \delta$ we have $V(\mathbf{x}) < c$. Since $\dot{V}$ is never positive, any motion that originates inside the circle $\|\mathbf{x}\| = \delta$ must remain inside $V = c$, that is inside the circle $\|\mathbf{x}\| = \varepsilon$. Therefore the origin is stable.

It is now easy to show that in the stronger condition (b) Theorem 8.1 $\dot{V} < 0$ implies asymptotic stability. Function V has to decrease continually along any solution of (8.9). Since $\dot{V}$ vanishes only at the origin, the solution cannot be stalled anywhere and must continually tend to the origin. This is asymptotic stability. Condition (c) of the theorem assures that the properties of the Liapunov function V are preserved in the entire state space and imply asymptotic stability in the large.

It should be noted that under the conditions of Theorem 8.1 the stability is equiasymptotic and the motions of (8.9) will converge to the equilibrium at the origin uniformly in the initial states; that is, the stability is independent of the distance of the initial state from the equilibrium point (the property common in linear systems). This property of the system (8.9) under the conditions stated in Theorem 8.1 is proved in Section H.2 of Appendix H.

There it is shown that the generalized Liapunov function $V(\mathbf{x}, t)$, which is an explicit function of time, may guarantee other properties of systems, such as absence of the finite-escape time, stability of the system under constantly acting perturbations or parameter disturbances, and so on.

It is of interest to note also that when there is a function $V(\mathbf{x})$ defined in a neighborhood R of the origin where $V(0) = 0$, and it can take positive values arbitrarily close to the origin, then, if $\dot{V}$ is positive-definite on R, the origin is *unstable*. This is the so-called *first Liapunov theorem on instability* (see Problem 8.10), which was generalized by Chetaev [8.7]. A geometric interpretation of instability is based on the reasoning used to prove stability [8.11] and can easily be worked out by the reader.

Finally, from the above Liapunov stability theorem, it is clear that the key problem in the stability analysis is the existence of a Liapunov function. It has been already pointed out that in general there is no systematic procedure for finding such a function and it is still a matter of the intuition and experience of the user of the method. However, in several useful system configurations we are about to consider there is a systematic way of constructing suitable Liapunov functions and performing the stability analysis in the sense of Liapunov.

Another apparent drawback of the outlined results is that the existence of an appropriate Liapunov function is only a sufficient condition for stability. Necessary conditions are not provided, and if a function cannot be found this does not necessarily indicate instability. Moreover, in system design we may ask that the system fulfill sufficient conditions that are too demanding. This, however, is all we can do at the present time, unless use of approximate methods outlined in Chapters 3 to 7 is indicated.

8.6 Absolute Stability. The Lur'e Problem

The Lur'e problem of absolute stability can be generally formulated for the class of nonlinear systems that are described by the differential equation

$$\dot{\mathbf{x}} = \mathbf{P}\mathbf{x} + \mathbf{q}F(y) + \mathbf{f}(t), \qquad y = \mathbf{r}^T\mathbf{x}, \tag{8.61}$$

where $\mathbf{P}$ is an $n \times n$ constant matrix, $\mathbf{q}$ and $\mathbf{r}$ are constant vectors of order $n \times 1$, and the function $F(y)$ is generally an arbitrary, single-valued, and discontinuous real function, defined for all real values of y and satisfying the condition $F(0) = 0$ and an additional inequality specified later. The n vector function $\mathbf{f}(t)$ is an external forcing function bounded for all t.

When the nonlinear function $F(y)$ is continuous, the system has an infinitesimal transition function

$$\mathbf{X} = \mathbf{P}\mathbf{x} + \mathbf{q}F(\mathbf{r}^T\mathbf{x}) + \mathbf{f}(t) \tag{8.62}$$

that satisfies the usual conditions for existence, uniqueness, and continuity of the solution $\mathbf{x}(t, \mathbf{x}_0, t_0)$ for all $t, \mathbf{x}_0, t_0$.

If the function $F(y)$ is discontinuous, a more general definition of the solution is necessary—the solution *in the Filippov sense* as specified later.[25]

Equation 8.61 actually describes a forced linear system with an essential nonlinearity described by an additively entering function $F(y)$. Indeed, when $y \equiv 0$ and $\mathbf{f}(t) \equiv \mathbf{0}$, from (8.61) we obtain a linear differential equation

$$\dot{\mathbf{x}} = \mathbf{P}\mathbf{x} \tag{8.63}$$

that describes the linear part of the system. It is now of interest to consider in more detail the linear and nonlinear parts of the system and to define the different subclasses of systems to be analyzed.

With the linear part (8.63) there is associated the characteristic equation

$$|\mathbf{P} - s\mathbf{I}| = 0. \tag{8.16}$$

According to the distribution of the characteristic roots of (8.16) (eigenvalues of $\mathbf{P}$) with respect to the imaginary axis, we shall distinguish the following cases.

Definition 8.8. Principal Case. A system (8.61) is said to be the principal case if all n characteristic roots $s_k = \sigma_k + j\omega_k$ lie in the left half of the s plane ($\sigma_k < 0$). (In this case it is sometimes said that the matrix $\mathbf{P}$ is a "Hurwitz matrix.")

Definition 8.9. Particular Case. A system (8.61) is said to be a particular case if n_1 characteristic roots $s_i = j\omega_i$ are on the imaginary axis ($\sigma_i = 0$) and the other n_2 roots $s_k = \sigma_k + j\omega_k$ are in the left half of the s plane ($\sigma_k < 0$). Of course, $n_1 + n_2 = n$.

Definition 8.10. Simplest Particular Case. A system (8.61) is said to be the simplest particular case if there is one distinct characteristic root $s_n = 0$ at the origin of the s plane (the zero root) and the other $n - 1$ roots $s_k = \sigma_k + j\omega_k$ are in the left half of the s plane ($\sigma_k < 0$).

In the Lur'e terminology the principal case refers to the "direct control" and the simplest particular case is called the "indirect control." In the following discussion, however, this terminology is not essential and will not be used.

More specifically, a nonlinear system with the block diagram shown in Figure 8.13 can be identified by (8.61) written in a slightly different form:

$$\dot{\mathbf{x}} = \mathbf{P}\mathbf{x} + \mathbf{q}F(y) + \mathbf{d}f(t), \qquad y = \mathbf{r}^T\mathbf{x}, \tag{8.64}$$

---

[25] See Section H.4 on discontinuous nonlinearities and their effect on the motion of dynamic systems.

Stability Analysis

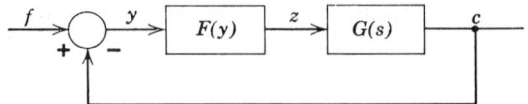

Figure 8.13 System block diagram.

where the forcing function $\mathbf{f}(t)$ is assumed to be $\mathbf{d}f(t)$ and $\mathbf{d}$ is $n \times 1$ constant vector. If we denote $d/dt \equiv s$ in (8.64) and rewrite this equation in the form

$$(s\mathbf{I} - \mathbf{P})\mathbf{x} = \mathbf{q}F(y) + \mathbf{d}f(t), \qquad y = \mathbf{r}^T\mathbf{x}, \qquad (8.65)$$

then it readily follows that the transfer function $G(s)$ of the linear part from the input $z = F(y)$ to the output $c = -y$ (of course, in defining $G(s)$ it is assumed that $\mathbf{f}(t) \equiv \mathbf{0}$) is given as

$$G(s) = \frac{C(s)}{B(s)} = \mathbf{r}^T(\mathbf{P} - s\mathbf{I})^{-1}\mathbf{q}, \qquad (8.66)$$

where

$$B(s) = |\mathbf{P} - s\mathbf{I}| = \sum_{k=0}^{n} b_k s^k, \qquad (8.67)$$

$$C(s) = \mathbf{r}^T [\text{adj } (\mathbf{P} - s\mathbf{I})]\mathbf{q} = \sum_{k=0}^{m} c_k s^k, \qquad (8.68)$$

and $n > m$.[26] Therefore the characteristic equation of the linear part is $B(s) = 0$.

It is easy now to write the accompanying equation of (8.64) using the operational notation. That is,

$$B(s)y + C(s)F(y) = H(s)f(t), \qquad s \equiv d/dt, \qquad (8.69)$$

in which $B(s)$ and $C(s)$ are given by (8.67) and (8.68), and

$$H(s) = \mathbf{r}^T [\text{adj } (\mathbf{P} - s\mathbf{I})]\mathbf{d}. \qquad (8.70)$$

Equation 8.69 was the object of extensive study in the preceding chapters. It is of interest to note that the conditions for absolute stability will again be interpreted in terms of the operational form (8.69) rather than in the state space notation.

---

[26] It will be assumed that the pair $(\mathbf{P}, \mathbf{q})$ is completely controllable and $(\mathbf{P}, \mathbf{r}^T)$ is completely observable. From the controllability theory [8.26] the following equivalent statements are needed: (a) $(\mathbf{P}, \mathbf{q})$ is completely controllable; (b) $\det [\mathbf{q}, \mathbf{Pq}, \ldots, \mathbf{P}^{n-1}\mathbf{q}] \neq 0$; (c) $\mathbf{x}^T [\exp \mathbf{P}t]\mathbf{q} \equiv 0$ for all t implies $\mathbf{x} = \mathbf{0}$. In addition, by the definition, $(\mathbf{P}, \mathbf{r}^T)$ is completely observable if and only if $(\mathbf{P}^T, \mathbf{r})$ is completely controllable.

It has been shown that if the transfer function $G(s)$ is *nondegenerate function* (that is, $B(s)$ and $C(s)$ do not have common factors and $B(s)$ is *of degree not less than n*) then the controllability and observability conditions are satisfied (see Section H.3).

For more detailed discussion of the controllability and observability concepts, see Section C.3.

The nonlinear part is described by a nonlinear function $F(y)$. As mentioned previously, this function satisfies the condition

$$F(0) = 0 \tag{8.71}$$

and is generally discontinuous with only isolated points of discontinuity of the first kind. This means that for any point of discontinuity y_0 the limits

$$F_-(y_0) = \lim_{y \to y_0} F(y), \qquad F_+(y_0) = \overline{\lim_{y \to y_0}} F(y) \tag{8.72}$$

exist and are finite. In case of continuous function $F_-(y_0) = F_+(y_0)$. For example, the common continuous nonlinear functions are shown in Figure 3.1, and the discontinuous functions in Figure 3.3 in Chapter 3.

If $F(y)$ is discontinuous, a sliding motion[27] is possible. To incorporate this motion, we define the solution of the fundamental system (8.61) thus: *a solution of the system* (8.61) *is any function* $\mathbf{x}(t)$ *in the state space* $\mathscr{X}$ *such that for all t and for any function*[28] $\mathscr{F}(t) = F[y(t)]$,

$$\dot{\mathbf{x}} = \mathbf{P}\mathbf{x} + \mathbf{q}\mathscr{F}(t) + \mathbf{f}(t), \tag{8.73}$$

$$F_-[y(t)] \leq \mathscr{F}(t) \leq F_+[y(t)], \qquad y = \mathbf{r}^T\mathbf{x}(t). \tag{8.74}$$

The function $\mathscr{F}(t)$, uniquely determined by the solution $\mathbf{x}(t)$, is called the *complementary function* for $F[y(t)]$.

It is also necessary to assume that the nonlinear characteristic is such that the integral

$$\int_0^y F(y)\,dy \to \infty \qquad \text{when} \quad y \to \pm\infty, \tag{8.75}$$

that is, it diverges.

Besides these general conditions, the function $F(y)$ should satisfy some type of inequality depending on the class of problems under consideration. If the function $F(y)$ is continuous, the simplest inequality is

$$0 \leq yF(y) \leq ky^2, \tag{8.76}$$

where $k \leq +\infty$ is a positive constant or infinity. In the yz plane, (8.76) means that the curve $z = F(y)$ lies within the sector determined by the straight line $z = ky$ and the y axis (Figure 8.14a). Since in (8.76) equality is admissible, curve $z = F(y)$ and sides of the sector may have points in common (the curve may even coincide with one side of the sector).

---

[27] The notion of sliding motion is given in Section H.4 of Appendix H.
[28] In Section H.4 of Appendix H the motion of systems with discontinuous nonlinearities is briefly discussed and the definition of the solution of the corresponding differential equations in the Filippov sense is given.

324   Stability Analysis

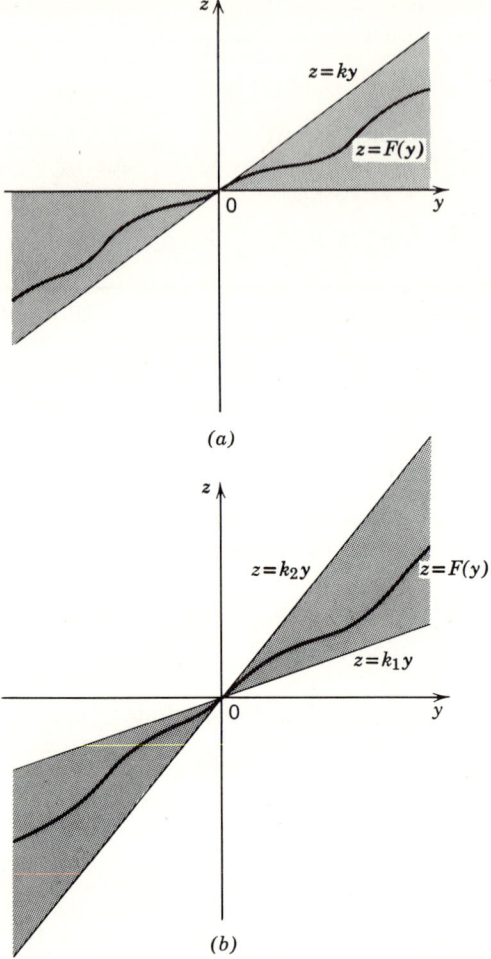

Figure 8.14 Nonlinearity sectors: (a) sector [0, k]; (b) sector [k_1, k_2].

Inequality (8.76) is meaningful only in principal cases. In a particular case when $F(y) \equiv 0$, the inequality is satisfied, but the null solution $\mathbf{x}(t) = \mathbf{0}$ is not asymptotically stable. Hence when we consider the absolute stability in the particular cases it is necessary to use stronger conditions and require the inequality

$$0 < yF(y) \leq ky^2 \tag{8.77}$$

to be satisfied for all $y \neq 0$. Occasionally the requirement is even stronger. For all $y \neq 0$ the following should hold:

$$0 < \varepsilon y^2 \leq yF(y) \leq ky^2 \tag{8.78}$$

for some positive number ε. The latest inequality excludes the nonlinear characteristics that are tangent to the y axis at the point $y = 0$, and also the ones for which the direction of the vector drawn from the origin of the yz plane to the point $[y, F(y)]$ on the curve $F(y)$ becomes arbitrary close to the direction of the y axis as $y \to \infty$.

It is often convenient to write $F(y) \in [0, k]$ and indicate that the "function $F(y)$ belongs to the sector $[0, k]$," which means that $F(y)$ satisfies inequality 8.76. If the sector is open from the left, the notation $F(y) \in (0, k]$ is used that refers to inequality 8.77. For inequality 8.78, the notation is obviously $F(y) \in [\varepsilon, k]$; that is, the interval is closed from both sides.

It may seem that the inequality

$$k_1 y^2 \leq y F(y) \leq k_2 y^2 \qquad (8.79)$$

is more general than that of (8.76), as can be concluded by comparing Figures 8.14a and b. However, the simple transformation $F_{tr}(y) = F(y) - k_1 y$ results in a system of the same type as the original system (8.61), except that the nonlinear characteristic of the transformed system is $F_{tr}(y)$, which, as a consequence of (8.79), satisfies (8.76) for $k = k_2 - k_1$.

In certain cases it will be necessary to specify the variation of the slope of $F(y)$. Then the inequality for continuous functions may be

$$0 \leq F'(y) \leq k. \qquad (8.80)$$

When the function $F(y)$ is discontinuous, it will be required that $F(y)$ satisfies the condition

$$0 \leq \frac{F(y_1) - F(y_2)}{y_1 - y_2} \leq k, \qquad (8.81)$$

where $k \leq +\infty$, y_1, and y_2 are points of continuity of $F(y)$, and $-\infty < y_1 < y_2 < +\infty$. Evidently, if $F(y)$ is continuous, then condition (8.81) is equivalent to (8.80).

Now it is possible to discuss the constraints on the linear part $G(s)$ in the light of the constraints on the nonlinear part $F(y)$. One may gain the impression that the restrictions on $G(s)$ are too strong, since it is required that $G(s)$ have only stable poles, or at most poles on the imaginary axis of the s plane. This is not true. For the absolute stability of the system (8.61), it is necessary that the null solution $\mathbf{x} = \mathbf{0}$ is asymptotically stable for characteristics $z = \varepsilon y$ where ε is an arbitrarily small positive number. For $\varepsilon \to 0$, the linear system must be stable or be on the limit of stability. The former is the principal case, the latter is the particular case. Therefore all cases for which absolute stability is meaningful are covered.

Finally, we can formulate the absolute stability definition for a free dynamic system described by the equation

$$\dot{\mathbf{x}} = \mathbf{P}\mathbf{x} + \mathbf{q}F(y), \qquad y = \mathbf{r}^T\mathbf{x}. \tag{8.82}$$

Definition 8.11. Absolute Stability. A free dynamic system (8.82) is absolutely stable in a sector $[0, k]$ if, for any continuous single-valued nonlinear characteristic $F(y) \in [0, k]$, $F(0) = 0$, the equilibrium position $\mathbf{x} = \mathbf{0}$ of the system is asymptotically stable in the large.

This definition may be modified depending on the sector specification, which, in turn, depends on the case under consideration (principal or particular case) and the nonlinearity in the system. The principal task of the following development is to find the necessary and sufficient conditions, which ensures the absolute stability of nonlinear systems described by (8.82) for any $F(y) \in [0, k]$. This constitutes essentially the *Lur'e problem*. In the following section the solution of this problem devised by V. M. Popov will be outlined.

The absolute stability definition given above can be extended to forced systems (8.61) and systems with discontinuous nonlinear characteristics.

8.7 Popov's Method. Free Dynamic Systems

The basic idea of the Popov solution to the Lur'e problem is to express the sufficient conditions for absolute stability in terms of the frequency characteristic of the linear part which has a simple graphical interpretation. The conditions are given by a certain inequality known as the *Popov inequality*. Whenever this inequality is satisfied by the linear part for a given class of nonlinearities, the system is absolutely stable. To prove that this last statement is true, we have only to show that the Popov condition is necessary and sufficient for the existence of a Liapunov function that ensures absolute stability of the system. The Liapunov function is of the type "a quadratic form plus an integral of the nonlinearity." The necessary and sufficient conditions for the Popov-Liapunov correspondence are supplied by the Yakubovich-Kalman lemma given in Appendix H.

Let us consider a free dynamic system described by the differential equation

$$\dot{\mathbf{x}} = \mathbf{P}\mathbf{x} + \mathbf{q}F(y), \qquad y = \mathbf{r}^T\mathbf{x} \tag{8.82}$$

with the companion equation

$$B(s)y + C(s)F(y) = 0, \qquad s \equiv d/dt. \tag{8.83}$$

We assume that it is a principal case and that the matrix $\mathbf{P}$ is Hurwitz, that

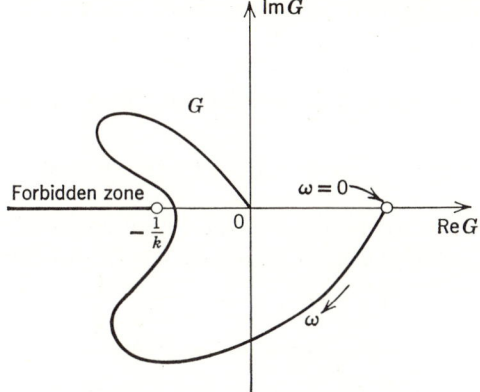

Figure 8.15 Nyquist diagram.

is, the roots of $B(s) = 0$, or poles of the transfer function

$$G(s) = \frac{C(s)}{B(s)} = \mathbf{r}^T(\mathbf{P} - s\mathbf{I})^{-1}\mathbf{q}, \tag{8.66}$$

are all in the left half of the s plane. The nonlinearity $F(y) \in [0, k]$.

As a preliminary step to the statement of the Popov theorem, consider the linear system for which

$$F(y) = Ky. \tag{8.84}$$

Then by the Nyquist stability criterion (see Section D.3 of Appendix D) for the system (8.82) or (8.83) to be stable for any K in the sector $0 < K < k$, it is necessary and sufficient that the locus $G(j\omega)$ in $G(s)$ plane does not intersect the real axis from $-\infty$ to and including the critical point $(-1/k, j0)$.[29] The graphical interpretation of this result is given in Figure 8.15. The conclusion remains valid for $k = \infty$, since the forbidden zone is the entire negative real axis but the origin is excluded. The characteristics $z = Ky$ are considered for any finite K but not for $k = \infty$.

It is clear that if the system is going to be absolutely stable as nonlinear, it must necessarily be stable as linear. Therefore *the necessary condition for absolute stability* of nonlinear system (8.82) is that the frequency characteristic $G(j\omega)$ never intersect the "forbidden zone" (see Figure 8.15) determined by the value $-1/k$. The problem then is to constitute the sufficient conditions that will guarantee the absolute stability of the system (8.82) or (8.83). They are formulated in the following Popov theorem.

---

[29] One may gain an impression that "no intersections" is an overly strict requirement. However, if there are any intersections of the Nyquist plot and the real axis it would always be possible to find a value K_0 in the sector $[0, k]$ so that the system is unstable.

328   Stability Analysis

Theorem 8.2. Popov Theorem. Principal Case.   In order for a free dynamic system (8.82), or equivalently (8.83), to be absolutely stable in the sector $[0, k]$ for $k < \infty$, and principal case, it is sufficient that there exists a real number[30] q so that for all real $\omega \geq 0$ the following inequality is satisfied:

$$\pi(\omega) \equiv \frac{1}{k} + \text{Re}\,(1 + jq\omega)G(j\omega) > 0. \qquad (8.85)$$

To give a graphical interpretation of the Popov inequality, it is necessary to introduce the *modified frequency characteristic* $G^*(\omega)$ defined by

$$\begin{aligned}\text{Re}\,G^*(\omega) &= \text{Re}\,G(j\omega) \\ \text{Im}\,G^*(\omega) &= \omega\,\text{Im}\,G(j\omega).\end{aligned} \qquad (8.86)$$

This means that in plotting the locus $G^*(\omega)$ from the locus $G(j\omega)$ we have to multiply all the ordinates of $G(j\omega)$ by the corresponding value of ω, which will "stretch" or "shrink" the locus $G(j\omega)$ in the direction parallel to the imaginary axis. Since $G^*(-\omega) = \overline{G^*(\omega)}$, only $\omega \geq 0$ need be considered.

If we write

$$G^*(\omega) = R + jI, \qquad (8.87)$$

then

$$\text{Re}\,(1 + jq\omega)G(j\omega) = R - qI \qquad (8.88)$$

and inequality 8.85 becomes

$$R - qI + \frac{1}{k} > 0, \quad \text{for all} \quad \omega \geq 0. \qquad (8.89)$$

Now the equation

$$R - qI + \frac{1}{k} = 0 \qquad (8.90)$$

---

[30] It is of particular interest to discuss the case when q is infinite, that is, $q = \pm\infty$. [8.15]

By letting $q = \mu/\tau$ where μ and τ are real numbers, and multiplying the Popov inequality (8.85) by τ, it assumes the form

$$\frac{\tau}{k} + \text{Re}\,(\tau + j\mu\omega)G(j\omega) > 0. \qquad (i)$$

Thus for $\tau = 0$ the inequality (*i*) becomes

$$\text{Re}\,[\pm j\omega G(j\omega)] > 0 \qquad (ii)$$

or

$$\pm\text{Im}\,\omega G(j\omega) < 0, \qquad (iii)$$

where the sign $\pm$ is the same as that of μ. In this case we say that the Popov inequality holds for infinite values of $q = \pm\infty$ ($q = \mu/\tau$). In geometric terms this means that the Popov line is actually the real axis and that the modified frequency response $G^*(\omega)$ is situated entirely under the axis ($q = +\infty$) or entirely above the axis ($q = -\infty$). The modified frequency characteristic and the Popov line are defined immediately after Theorem 8.2.

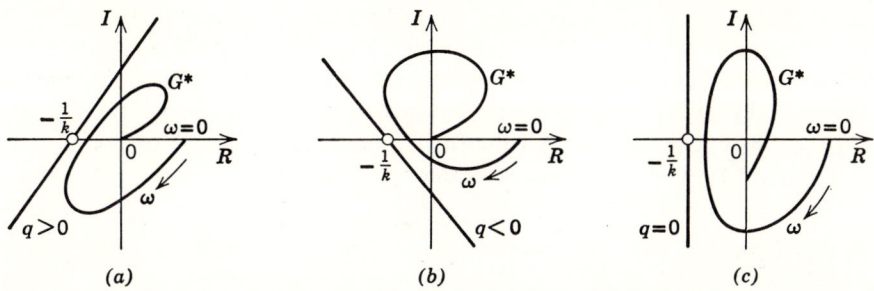

Figure 8.16 Graphical interpretation of the Popov inequality for stable systems.

determines a straight line in the RI plane with slope $1/q$ that intersects the real axis at the point $(-1/k, 0)$. It is usually referred to as the *Popov line*. This line divides the plane into two half-planes and inequality 8.89 represents the half of the plane that is to the right of the Popov line. Therefore the geometric interpretation of inequality 8.85 is that the modified frequency characteristic $G^*(\omega)$ should lie entirely to the right of the Popov line having no points in common with it.[31]

If Figure 8.16 several cases are illustrated in which the Popov condition is satisfied. Figure 8.17 shows the case when the Popov inequality cannot be satisfied but the necessary conditions are fulfilled. The curve G^* (or equivalently the curve G) does not have points in common with the forbidden zone. In this case it is not possible to determine stability (or instability) of

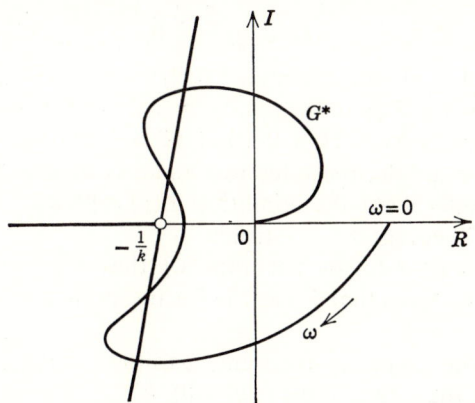

Figure 8.17 Popov inequality cannot be satisfied.

---

[31] The modified characteristic $G^*(\omega)$ is defined only for finite values of ω. Thus the limit point $G^*(\infty)$ does not belong to $G^*(\omega)$. The point $G^*(\infty)$ may be located on the Popov line.

the system since Popov conditions are only sufficient conditions for stability. On the other hand, the Popov theorem always guarantees the satisfaction of the necessary conditions for absolute stability, since the locus G^* must lie to the right of the straight line passing through the point $(-1/k, 0)$ and, consequently, cannot intersect the forbidden zone.

To prove Popov's theorem, consider a positive-definite function
$$V = V(\mathbf{x}, y)$$
in the form
$$V = \mathbf{x}^T \mathbf{H} \mathbf{x} + q \int_0^y F(y)\, dy, \tag{8.91}$$

where q is the parameter in the Popov inequality 8.85, and the expression $\mathbf{x}^T \mathbf{H} \mathbf{x}$ represents a quadratic form with the symmetric matrix $\mathbf{H} = \mathbf{H}^T$.

Differentiating, we obtain
$$-\dot V = [\mathbf{x}^T(\mathbf{G} - \mathbf{g}\mathbf{g}^T)\mathbf{x} + (\sqrt{\gamma}\,\mathscr{F} + \mathbf{g}^T\mathbf{x})^2] + \Omega, \tag{8.92}$$
where
$$-\mathbf{G} = \mathbf{H}\mathbf{P} + \mathbf{P}^T\mathbf{H}, \qquad -\sqrt{\gamma}\,\mathbf{g} = \mathbf{H}\mathbf{q} + \tfrac{1}{2}(q\mathbf{P}^T + \mathbf{I})\mathbf{r},$$
$$-\gamma = q\mathbf{r}^T\mathbf{q} - \frac{1}{k}, \qquad \Omega = \left(y - \frac{\mathscr{F}}{k}\right)\mathscr{F}. \tag{8.93}$$

To the quantity obtained by differentiation we add the null expression $\Omega - (\mathbf{x}^T\mathbf{r} - \mathscr{F}/k)\mathscr{F}$.

In order that the $-\dot V$ of (8.92) be a positive definite function, it is necessary that a matrix $\mathbf{H}$ we are seeking satisfy the quadratic matrix inequality
$$\mathbf{G} - \mathbf{g}\mathbf{g}^T > 0. \tag{8.94}$$

By the Yakubovich-Kalman lemma, for such a matrix $\mathbf{H}$ to exist it is necessary and sufficient that the Popov inequality be satisfied. From (8.94) it follows that $\mathbf{G} > 0$, which implies $\mathbf{H} > 0$ (that $\mathbf{G} > 0$ implies $\mathbf{H} > 0$ when $\mathbf{P}$ is Hurwitz see, for example, the reference 8.11). Note also that $\Omega > 0$ unless $y = 0$, which results from the condition $F(y) \in [0, k]$. Thus the condition (b) of Liapunov Theorem 8.1 is satisfied.

The last requirement (c) in Theorem 8.1 that $V \to \infty$ with $\|\mathbf{x}\| \to \infty$ is also satisfied on the basis that V has the form (8.91) and the condition (8.75) is true.

Therefore, by the Liapunov theorem, it follows that the system is asymptotically stable in the large for all $F(y) \in [0, k]$.

8.8 Particular Cases

Before considering the particular cases of nonlinear systems, assume that a system (8.82) is linear. Suppose that the locus $G(j\omega)$ or (which is the same) the locus $G^*(\omega)$ have no points in common with the forbidden zone. According to the Nyquist criterion, this means that all the characteristic roots stay in the left half of the s plane for all values of K from the sector $0 \leq K \leq k$. For the principal case this means that the characteristic equation

$$B(s) + KC(s) = 0 \qquad (8.95)$$

is Hurwitz for all values of $K \in [0, k]$, since (8.95) is Hurwitz for $K = \varepsilon$ where ε is a small positive number. In the particular cases, an additional requirement is that the corresponding linear system (8.82) in which $F(y) = Ky$ be *stable in the limit*, that is, stable for $K = \varepsilon > 0$. That constitutes necessary conditions for the absolute stability of system (8.82). In other words the particular cases are stable in the limit if for $F = \varepsilon y$, when ε goes from zero to a small positive number, all the characteristic roots that, for $\varepsilon = 0$, were on the imaginary axis move immediately to the left. Note also that, for particular cases, absolute stability in the sector $[0, k]$ is impossible since the systems are not asymptotically stable for $y = 0$. We can consider only the sector $(0, k]$.

Now the sufficient conditions for the absolute stability in the particular cases are expressed by the Popov Theorem.

Theorem 8.3. Particular Cases. For the free dynamic system (8.82) in a particular case to be absolutely stable in the sector $[\varepsilon, k]$ for $k < \infty$ and arbitrarily small positive number ε, it is sufficient that there exist a finite real number q so that for all real $\omega \geq 0$ the following inequality is satisfied:

$$\pi(\omega) \equiv \frac{1}{k} + \mathrm{Re}\,(1 + jq\omega)G(j\omega) > 0 \qquad (8.85)$$

and, in addition, that the conditions for stability in the limit be satisfied.

It is clear that by the conditions of Theorem 8.3 the necessary conditions for absolute stability—that the corresponding linear system be stable—are automatically fulfilled.

Proof of the above theorem can be reduced to that of the principal case by the change of variables [8.33, 34]

$$F(y) = F_{tr}(y) + \varepsilon y. \qquad (8.96)$$

This change of variables transforms the system (8.82) into another system but with different coefficients and with a different characteristic

$$F_{tr}(y) = F(y) - \varepsilon y. \qquad (8.97)$$

If the function $F(y) \in [\varepsilon, k]$, from (8.96) it follows that $F_{tr}(y) \in [0, k - \varepsilon]$.

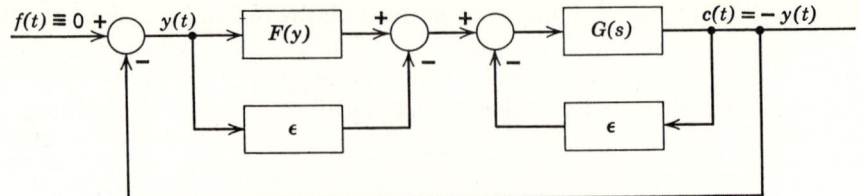

Figure 8.18 Transformed system.

Therefore, if the original system was stable for $F(y) = ky$, the transformed system will be stable for $F_{tr}(y) = 0$; that is, the transformation (8.96) changes a particular case into the principal case.

It is now of interest to find relationship between the frequency characteristic $G(s)$ of the original system and the transformed one $G_{tr}(s)$. Note that

$$-y = G(s)z = G(s)[z_{tr} + \varepsilon y], \qquad (8.98)$$

where $z = F(y)$ and $z_{tr} = F_{tr}(y)$. Therefore

$$-y = \frac{G(s)}{1 + \varepsilon G(s)} z_{tr} \qquad (8.99)$$

and the transformed transfer function $G_{tr}(s)$ is given as

$$G_{tr}(s) = \frac{G(s)}{1 + \varepsilon G(s)}. \qquad (8.100)$$

The transformed system has a block diagram shown in Figure 8.18.

It is now left to show that the inequality

$$\frac{1}{k - \varepsilon} + \mathrm{Re}\,(1 + jq\omega)G_{tr}(j\omega) > 0, \qquad \text{for all real } \omega \geq 0 \qquad (8.101)$$

follows from the Popov inequality 8.85:

$$\frac{1}{k} + \mathrm{Re}\,(1 + jq\omega)G(j\omega) > 0, \qquad \text{for all real } \omega \geq 0. \qquad (8.85)$$

After simple transformation, (8.101) can be represented in the form

$$\frac{1}{k - \varepsilon} + \mathrm{Re}\,(1 + jq\omega)G(j\omega) + \frac{k}{k + \varepsilon}|G(j\omega)|^2 > 0,$$

$$\text{for all real } \omega \geq 0. \qquad (8.102)$$

From this, for $\varepsilon = 0$, we obtain inequality 8.85. It is important to note that inequality 8.85 can be weakened to "$\geq$."[32]

Now, to prove stability in the particular case, we transform the system and use the Popov theorem for the principal case. Then, by (8.101) we show stability of the transformed system in the sector $[0, k - \varepsilon]$ which implies stability of the original system in the sector $[\varepsilon, k]$.

The stability in the limit can be determined directly from the algebraic equation

$$A(s) \equiv B(s) + \varepsilon C(s) = 0, \qquad (8.103)$$

which is obtained from (8.95) for $K = \varepsilon$ where $\varepsilon > 0$ is arbitrarily small. For $\varepsilon = 0$, (8.103) becomes $A(s) \equiv B(s) = 0$ and some zeros of $B(s)$ and consequently of $A(s)$ are on the imaginary axis. If $B(s)$ is given in the factored form, these roots are given explicitly and will be denoted as $s_k = j\omega_k$ ($k = 1, 2, \ldots, n_1$). The other n_2 ($n_1 + n_2 = n$) zeros of $B(s)$ are in the left half of the s plane. Then to determine the stability in the limit, the root sensitivity (Section 1.7) can be applied effectively. Note that it is necessary to calculate Δs_k for $\varepsilon = 0$ where ε is considered as a parameter.

According to equation A.141 in Appendix A, the differential change Δs_k

---

[32] In particular cases inequality 8.85 can be replaced by the weakened Popov condition

$$\frac{1}{k} + \text{Re}\,(1 + jq\omega)G(j\omega) \geq 0, \quad \text{for all real} \quad \omega \geq 0. \qquad (i)$$

This can be shown [8.15] if the transformation $F(y) = ky - F_{\text{tr}}(y)$ is used. Then $G_{\text{tr}}(s) = G(s)/[1 + kG(s)]$, and the following relation holds:

$$\frac{1}{k} + \text{Re}\,(1 - jq\omega)G_{\text{tr}}(j\omega) = \frac{1}{|1 + kG(j\omega)|^2}\left[\frac{1}{k} + \text{Re}\,(1 + jq\omega)G(j\omega)\right]. \qquad (ii)$$

Apparently the equality in (i) takes place for each $\omega = \omega_k$ where $j\omega_k$ is a root of the characteristic equation $B(s) = 0$, since $\lim_{\omega \to \omega_k} 1/|1 + kG(j\omega)|^2 = 0$. Equality in (i) means that $G^*(\omega)$ may have points in common with the Popov line but not the point $(-1/k, 0)$, since we require that the system be stable for $F(y) = ky$.

It is thus necessary that the modified frequency response $G^*(\omega)$, and hence also $G(j\omega)$, do not pass through the point of the real axis whose abscissa is $-1/k$. In this manner it is possible to replace the Popov condition (8.85) by

$$\frac{1}{k} + \text{Re}\,(1 + jq\omega)G(j\omega) \geq 0, \quad \text{for all real} \quad \omega \geq 0 \qquad (iii)$$

$$G(j\omega) \neq -\frac{1}{k}, \qquad (iv)$$

and weaken the strict inequality "$>$" to "$\geq$" for particular cases of system (8.82).

is given as

$$\Delta s_k = \left[-\frac{\nu!}{\dfrac{\partial^\nu A}{\partial s^\nu}(s)} \frac{\partial A}{\partial \varepsilon} \Delta \varepsilon \right]^{1/\nu}_{\substack{\varepsilon=0 \\ s=s_k}}, \qquad (8.104)$$

where ν is the multiplicity of the root s_k.

By a simple argument[33] using the root-locus technique, it can be shown that the stability in the limit cannot be achieved if the multiplicity $\nu > 2$.[34] Thus from (8.103), (8.104), and $\Delta s_k = \Delta \sigma_k + j(\omega_k + \Delta \omega_k)$, we have for $\nu = 1$

$$\Delta \sigma_k = \operatorname{Re} \left[-\frac{C(s_k)}{B'(s_k)} \varepsilon \right] \qquad (8.105)$$

and for $\nu = 2$

$$\Delta \sigma_k = \operatorname{Re} \left[-\frac{2C(s_k)}{B''(s_k)} \varepsilon \right]^{1/2}, \qquad (8.106)$$

where $\Delta \varepsilon$ is replaced by $\varepsilon > 0$ sufficiently small.

The conditions for the stability in the limit are obviously that

$$\Delta \sigma_k < 0, \qquad \text{for } k = 1, 2, \ldots, n_1. \qquad (8.107)$$

(This condition has been obtained in terms of $G^*(\omega)$ in reference 8.15.)

Extensive treatment of various particular cases is given in the work of Yakubovich [8.21]. He shows that the above conclusions are also valid in cases with discontinuous nonlinearities where certain additional conditions are satisfied.

The simplest particular case is considered also along with the Yakubovich-Kalman lemma in Appendix H.

8.9 The Case $k = \infty$

In stating the Popov theorem, it was assumed that k is finite. It is possible to extend the theorem to include the case $k = \infty$ if some additional

---

[33] In this case ν branches of the root locus start at the point $s_k = j\omega_k$ of the s plane. The branches of the root locus divide 2π into ν equal angles. Since $\nu \geq 2$, some of the branches must go to the right of the imaginary axis and for $\varepsilon > 0$ small (8.103) will not be Hurwitz. Therefore for $\nu > 2$, the corresponding particular case is not stable in the limit. [8.15]

[34] It should be noted that in the case $\nu = 2$, $s_k = j\omega_k$, $\omega_k \neq 0$, the Popov inequality cannot be satisfied since the locus $G^*(\omega)$ cannot lie to the right of the Popov line (even if the inequality is weakened to ≥ 0). In this case the Popov line coincides with the real axis, and the locus of the modified frequency characteristic $G^*(\omega)$ must lie either above or below the real axis. It has been shown [8.15], however, that the stability in the limit requires that the locus $G^*(\omega)$ lie to the right of the Popov line.

conditions are satisfied. This case is of particular interest in the analysis of nonlinear systems with discontinuous nonlinearities.

We will first consider the principal case. For $k = \infty$, the Popov line always goes through the origin. Thus, in order to satisfy the Popov inequality, it is necessary to have $G^*(\infty) \neq 0$. This takes place when in $G(s) = C(s)/B(s)$ the difference between the degree of $B(s)$ exceeds that of $C(s)$ exactly by one ($n - m = 1$). Indeed, in the necessary condition for absolute stability we require that the polynomial

$$B(s) + KC(s) = b_n s^n + (b_{n-1} + Kc_m)s^{n-1} + \cdots + b_0 + Kc_0$$

is Hurwitz for all $K \in [0, k]$. This is possible only if $c_m > 0$. Therefore, the limit point $G^*(\infty) = -jc_m$ must lie somewhere on the negative part of the imaginary axis. Consequently, in the Popov inequality for $k = \infty$, it follows that $q \geq 0$. This results in the following theorem [8.15].

Theorem 8.4. Principal Case for $k = \infty$. The system (8.82), or equivalently (8.83), for the principal case is absolutely stable in the sector $(0, \infty)$ if the Popov inequality

$$\operatorname{Re}(1 + jq\omega)G(j\omega) > 0 \qquad (8.108)$$

is satisfied for some $q \geq 0$ and all real $\omega \geq 0$ and in addition

$$\lim_{\omega \to \infty} j\omega G(j\omega) \neq 0. \qquad (8.109)$$

(The condition (8.109) is equivalent to the requirement that $n - m = 1$.)

Since the Popov line passes through the origin but the limit point $G^*(\infty)$ lies elsewhere on the imaginary axis, it is always possible to choose q in such a way that the Popov line has no common points with $G^*(\omega)$ even for $\omega = \infty$. It suffices to rotate the Popov line through a small angle in the proper direction about the origin.

It has been shown by Yakubovich [8.22] that the principal case is absolutely stable for $k = \infty$ if the Popov inequality 8.108 is satisfied, and in addition, whenever

$$\lim_{\omega \to \infty} \pi(\omega) = 0 \qquad (8.110)$$

the inequality

$$\lim_{\omega \to \infty} \pi(\omega)\omega^2 > 0 \qquad (8.111)$$

holds. This is relevant for the case $n - m = 2$, since in this case the limit point $G^*(\infty)$ is at the origin. Condition (8.111) means that the modified frequency characteristic $G^*(\omega)$ is not tangent to the Popov line at the limit point $G^*(\infty)$, that is, at the origin, since $G^*(\infty) = 0$.[35] The preceding reasoning indicates that such a tangency can always be avoided by performing a small rotation of the Popov line about the origin.

For the case $n - m > 2$ and $k = \infty$, the system (8.83) cannot be stable even with linear characteristic $F(y) = Ky$ since for $K > 0$ large $G(j\omega)$ must intersect the real axis. Hence the Popov condition cannot be satisfied because it guarantees the absolute stability for linear characteristics in the required sector. Therefore, for any k (finite or infinite) the Popov condition

$$\pi(\omega) \equiv \frac{1}{k} + \text{Re}\,(1 + jq\omega)G(j\omega) > 0, \quad \text{for all real} \quad \omega \geq 0 \quad (8.85)$$

is sufficient for the absolute stability of the *principal case* of system (8.61).

For the *simplest particular case*, Popov [8.35] established the following theorem:

Theorem 8.5. Simplest Particular Case ($k = \infty$). If for the simplest particular case of (8.82), for some real q and for all real $\omega \geq 0$, the condition

$$\text{Re}\,(1 + jq\omega)G(j\omega) \geq 0 \quad (8.108)$$

is satisfied and, in addition, the condition $\text{Im}\,G^*(0) < 0$ for stability in the limit is satisfied, then this system is absolutely stable in the sector $(0, \infty)$.

An extended version of this theorem was proved by Kalman [8.36] by using his results in the Yakubovich-Kalman Lemma (see Section H.3 of Appendix H).

The Popov theorem was extended to the *general particular case* by Yakubovich [8.21]. The extension includes the case $k = \infty$.

Note that Theorem 8.5 cannot be proved using the results obtained in Theorem 8.3, since the transformation applied in the proof of Theorem 8.3 is meaningless for $k = \infty$.

8.10 Degree of Absolute Stability

The idea of the degree of absolute stability is closely related to the relative stability concept used in the linear system analysis (Chapter 2, Section 2.3).

---

[35] If the function $G(s)$ is expanded in the power series in terms of $1/s$, then it is not difficult to obtain an expansion of $\pi(\omega)$ in the powers of $1/\omega^2$ to get

$$\pi(\omega) = \frac{A_1}{\omega^2} + \frac{A_2}{\omega^4} + \cdots.$$

A condition in the proof of absolute stability [8.15] requires that $A_1 > 0$, which is fulfilled if (8.111) holds. The tangency interpretation of (8.111) is now also clear.

A specific degree of stability guarantees that the motion approaches the equilibrium faster than an exponential. The following theorem of Yakubovich [8.22] gives this idea precisely.

Theorem 8.6. Exponential Stability. In a free dynamic system (8.82) the inequality $0 \leq F'(y) \leq k$ and $F(0) = 0$ are satisfied, where $F'(y)$ is a nondecreasing function of y. Let the roots of the equation $B(s) = \det(\mathbf{P} - s\mathbf{I}) = 0$ be in the domain $\operatorname{Re} s < \sigma \leq 0$. If, for some q and all real $\omega \geq 0$,

$$\pi(\omega) \equiv \frac{1}{k} + \operatorname{Re}(1 + jq\omega)G(\sigma + j\omega) > 0, \quad (8.112)$$

$$\lim_{\omega \to \infty} \pi(\omega)\omega^2 > 0, \quad \text{if} \quad \lim_{\omega \to \infty} \pi(\omega) = 0, \quad (8.113)$$

then there exist constants $M > 0$ and $\mu > 0$ so that, for any solution $\mathbf{x}(t)$ and any $t \geq t_0$, we have

$$\|\mathbf{x}(t)\| \leq M \|\mathbf{x}(t_0)\| \exp[(\sigma - \mu)(t - t_0)]. \quad (8.114)$$

The numbers M and μ do not depend on the function $F(y)$, but they generally depend on the number k.

The proof of this theorem outlined in Appendix H follows that of Yakubovich [8.22], who claims its validity under the weaker condition (8.76). This claim, however, is verified in the proof of Appendix H when either q or σ is zero.

An interpretation of inequality (8.112) in the parameter space is given in Section 8.14.

8.11 Forced Systems

The Popov conditions can be applied to forced systems—this is a significant result achieved by Yakubovich [8.22]. He showed that the exponential stability can be determined for forced oscillations, and thus answered the questions of jump resonance and existence of subharmonic resonance in systems with one essential nonlinearity. This theorem is his major result.

Theorem 8.7. Exponential Stability of Forced Oscillations. Unknown Solution. Given the system

$$\dot{\mathbf{x}} = \mathbf{P}\mathbf{x} + \mathbf{q}F(y) + \mathbf{f}(t), \quad y = \mathbf{r}^T\mathbf{x}, \quad (8.61)$$

we assume that the roots of equation $\det(\mathbf{P} - s\mathbf{I}) = 0$ are in the domain $\operatorname{Re} s < \sigma \leq 0$, and that the nonlinear function $F(y)$ satisfies the condition

$$0 \leq \frac{F(y_1) - F(y_2)}{y_1 - y_2} \leq k, \quad (8.81)$$

where $k \neq +\infty$, y_1, and y_2 are points of continuity of $F(y)$, and $-\infty \leq y_1 \leq y_2 \leq +\infty$. For all $\omega \geq 0$, let

$$\pi(\omega) \equiv \frac{1}{k} + \text{Re } G(\sigma + j\omega) > 0, \qquad (8.115)$$

and if $k = \infty$ then we have also

$$\lim_{\omega \to \infty} \pi(\omega)\omega^2 > 0. \qquad (8.116)$$

In addition, the forcing function $\mathbf{f}(t)$ is bounded in $(-\infty, +\infty)$.

Then the system (8.61) has in the time interval $(-\infty, +\infty)$ a unique solution $\mathbf{x}^0(t)$. This solution is exponentially stable in the large with exponent less than σ; that is, there are numbers $M > 0$ and $\mu > 0$ so that for any $t \geq t_0$ and any solution $\mathbf{x}(t)$ of the system (8.61), we have

$$\|\mathbf{x}(t) - \mathbf{x}^0(t)\| \leq M \|\mathbf{x}(t_0) - \mathbf{x}^0(t_0)\| \exp\left[(\sigma - \mu)(t - t_0)\right]. \qquad (8.117)$$

The numbers M and μ do not depend on the function $F(y)$, but they depend generally on k. If $\mathbf{f}(t)$ is a T-periodic or almost periodic vector function, then the solution $\mathbf{x}^0(t)$ is also respectively a T-periodic or almost periodic vector function.

If the conditions of the above theorem are satisfied, it is clear that jump and subharmonic resonance cannot take place and the system is BIBO stable.[36]

In Chapter 7 methods are given for finding the forced oscillations. The case when the solution $\mathbf{x}^0(t)$ is known is therefore of interest. In this case the number k in the condition (8.115) can generally be somewhat smaller than the following theorem would indicate.

Theorem 8.8. Exponential Stability of Forced Oscillations. Known Solution.
Let $\mathbf{x}^0(t)$ be a solution of (8.61) bounded in $(-\infty, +\infty)$, let the equation $|\mathbf{P} - s\mathbf{I}| = 0$ have all roots in the left half of the s plane with $\sigma < 0$, and let

$$0 \leq \frac{F(y) - \mathscr{F}^0(t)}{y - y^0(t)} \leq k, \qquad (k \leq +\infty), \qquad (8.118)$$

where $y^0(t) = \mathbf{r}^T \mathbf{x}^0$, $\mathscr{F}^0(t) = F[y^0(t)]$, $-\infty < t < +\infty$, $-\infty < y < +\infty$, and y is a point of continuity of the function $F(y)$, $y \neq y^0(t)$. If inequality 8.115 is also satisfied for $k = \infty$—that is, if (8.116) is satisfied—then $\mathbf{x}^0(t)$ is the only bounded solution on $(-\infty, +\infty)$. For any solution $\mathbf{x}(t)$ the

---

[36] From (8.117) it follows that all forced solutions approach each other regardless of their initial state. It is also clear that the system is globally and uniformly BIBO stable.

inequality 8.117 holds with numbers $M > 0$ and $\mu > 0$ independent of the function $F(y)$. If $\mathbf{f}(t)$ is T-periodic or almost periodic, the solution $\mathbf{x}^0(t)$ is also T-periodic or almost periodic, respectively.

When $\mathbf{f} \equiv 0$ and $\mathbf{x}^0 \equiv 0$ is a solution of (8.61), Theorem 8.8 extends Theorem 8.6 for $q = 0$ to discontinuous $F(y)$.

It is important to note that under the conditions of Theorem 8.7 the system will be BIBO stable. This is expressed in the following proposition and its simple proof due to Yakubovich [8.22]. The exposition will illustrate the method of proving the above theorems, which is used in Appendix H.

Given the system (8.61), let $\mathbf{P}$ be a Hurwitz matrix (all roots of equation $|\mathbf{P} - s\mathbf{I}| = 0$ are in the left half of the s plane Re $s < \sigma = 0$), let $F(y)$ satisfy the inequalities 8.81, and let $\mathbf{f}(t)$ be bounded in $(-\infty, +\infty)$. Let the inequality

$$\pi(\omega) \equiv \frac{1}{k} + \text{Re } G(j\omega) > 0, \quad \text{for all real} \quad \omega \geq 0 \quad (8.119)$$

and, in addition, if $k = \infty$, the condition

$$\lim_{\omega \to \infty} \omega^2 G(j\omega) > 0, \quad (8.120)$$

be satisfied.

Then (1) any solution $\mathbf{x}(t)$ of the system (8.61) is bounded in (t_0, ∞); (2) in the state space $\mathscr{X}$ there is a bounded region R so that any solution reaches this region at some time, and for $t \geq t_0$ and $\mathbf{x}(t_0) \in$ R it follows that $\mathbf{x}(t) \in$ R; and (3) there are numbers $M > 0$ and $\mu > 0$ so that, for any two solutions $\mathbf{x}_1(t)$ and $\mathbf{x}_2(t)$ and $t > t_0$, we have

$$\|\mathbf{x}_1(t) - \mathbf{x}_2(t)\| \leq M \|\mathbf{x}_1(t_0) - \mathbf{x}_2(t_0)\| \exp\left[-\mu(t - t_0)\right]. \quad (8.121)$$

The proof is as follows. For a Liapunov function we take the quadratic form $V(\mathbf{x}) = \mathbf{x}^T \mathbf{H} \mathbf{x}$. Differentiating $V(\mathbf{x})$ and using (8.61), we obtain

$$\dot{V} = 2\mathbf{x}^T \mathbf{H} \dot{\mathbf{x}} = \mathbf{x}^T (\mathbf{HP} + \mathbf{P}^T \mathbf{H}) \mathbf{x} + 2\mathbf{x}^T \mathbf{H} \mathbf{q} \mathscr{F} + 2\mathbf{x}^T \mathbf{H} \mathbf{f}, \quad (8.122)$$

where $\mathscr{F}(t) = F[y(t)]$, and $\mathbf{f} = \mathbf{f}(t)$.

We write $F_1(y) = F(y) - F(0)$. From (8.81), we obtain

$$[y - (\mathscr{F}_1/k)]\mathscr{F}_1 \equiv [\mathbf{x}^T \mathbf{r} - (\mathscr{F}_1/k)]\mathscr{F}_1 \geq 0.$$

Hence, by adding the null expression

$$[\mathbf{x}^T \mathbf{r} - (\mathscr{F}_1/k)]\mathscr{F}_1 - [(y - (\mathscr{F}_1/k)]\mathscr{F}_1 = 0$$

to $\dot{V}$, we obtain for $k \neq \infty$,

$$-\dot{V} = [\mathbf{x}^T (\mathbf{G} - \mathbf{g}\mathbf{g}^T)\mathbf{x} + (\sqrt{\gamma} \mathscr{F}_1 + \mathbf{g}^T \mathbf{x})^2] + \Omega_1 + \Omega_2, \quad (8.123)$$

where

$$-\mathbf{G} = \mathbf{HP} + \mathbf{P}^T\mathbf{H}, \qquad -\sqrt{\gamma}\,\mathbf{g} = \mathbf{Hq} + \tfrac{1}{2}\mathbf{r}$$

$$\gamma = \frac{1}{k}, \quad \Omega_1 = \left(y - \frac{\mathscr{F}_1}{k}\right)\mathscr{F}_1, \quad -\Omega_2 = 2\mathbf{x}^T\mathbf{H}[\mathbf{f} + \mathbf{q}F(0)]. \quad (8.124)$$

Stability requires that the matrix $\mathbf{H} = \mathbf{H}^T$ satisfies the quadratic matrix inequality

$$\mathbf{G} - \mathbf{g}\mathbf{g}^T > 0. \qquad (8.94)$$

From the Yakubovich-Kalman lemma of Appendix H, such a matrix exists if and only if the Popov inequality 8.119 is satisfied. In (8.94), we have $\mathbf{G} > 0$ and, since $\mathbf{P}$ is Hurwitz, that implies $\mathbf{H} > 0$ (that $\mathbf{G} = -(\mathbf{HP} - \mathbf{P}^T\mathbf{H}) > 0$ implies $\mathbf{H} > 0$ see, for example, reference 8.11).

Now, let us consider two positive-definite quadratic forms

$$\theta_1(\mathbf{x}) = \mathbf{x}^T(\mathbf{G} - \mathbf{g}\mathbf{g}^T)\mathbf{x} \quad \text{and} \quad \theta_2(\mathbf{x}) = \mathbf{x}^T\mathbf{H}\mathbf{x}.$$

Since we can always find a constant $\mu > 0$ so that $\theta_1(\mathbf{x}) > 2\mu\theta_2(\mathbf{x})$, it follows that

$$\dot{V} \leq -2\mu V + 2\mathbf{x}^T\mathbf{H}[\mathbf{f} + \mathbf{q}F(0)]. \qquad (8.125)$$

Furthermore, $\mathbf{f}(t)$ is bounded and we can choose a constant c so that the last term in (8.125) is smaller than $2cV^{1/2}$, that is

$$\dot{V} \leq -2\mu V + 2cV^{1/2}. \qquad (8.126)$$

Denoting $V^{1/2} = \xi$, we have $\dot{\xi} < -\mu\xi + c$ and

$$V[\mathbf{x}(t)]^{1/2} \equiv \xi(t) \leq \left[\xi(t_0) - \frac{c}{\mu}\right]\exp[-\mu(t - t_0)] + \frac{c}{\mu}. \qquad (8.127)$$

Now, given any constant $c_1 = c/\mu$, a t_1 exists such that for $t \geq t_1$ we have $\xi \leq c_1$; that is, for $t \geq t_1$ all motions that start outside the region $R = \{V \leq c_1^2\}$ will enter the region after t_1. Since on the boundary of the region R we have $\xi = c_1$, $\dot{\xi} \leq -\mu c_1 + c < 0$, and $\dot{V} < 0$, we conclude that the motions which are at some time t_0 inside the region R will never leave it, that is, the region R is *invariant*.

If $k = \infty$, the matrix $\mathbf{H}$ should satisfy the inequality 8.94 for $\mathbf{g} = 0$. From Yakubovich-Kalman lemma, it follows that such a matrix exists if and only if the Popov condition (8.119) supplemented with (8.120) is satisfied. Note also that for $k = \infty$, we have $\gamma F_1 = 0$.

Thus, we conclude that the assertions (1) and (2) of the above proposition are true.

To prove the assertion (3) of the proposition, let us use the transformations $\mathbf{z} = \mathbf{x}_1 - \mathbf{x}_2$, $y_1 = \mathbf{r}^T\mathbf{x}_1$, $y_2 = \mathbf{r}^T\mathbf{x}_2$, $y_0 = y_1 - y_2$, $\mathscr{F}_0 = F[y_1(t)] - F[y_2(t)]$, and inequality (8.81). Then, we obtain

$$-\dot{V} = [\mathbf{z}^T(\mathbf{G} - \mathbf{g}\mathbf{g}^T)\mathbf{z} + (\sqrt{\gamma}\mathscr{F}_0 + \mathbf{g}^T\mathbf{x})^2] + \Omega_0, \qquad (8.128)$$

where

$$\Omega_0 = (y_0 - \mathscr{F}_0/k)\mathscr{F}_0 \geq 0. \qquad (8.129)$$

By the same arguments used above, we find that

$$\dot{V}(\mathbf{z}) \leq -2\mu V(\mathbf{z}), \quad \text{and} \quad V[\mathbf{z}(t)] \leq V[\mathbf{z}(t_0)]e^{-2\mu(t-t_0)}. \qquad (8.130)$$

This relationship yields inequality 8.121, and the assertion (3) is also proved.

It is of interest to note that the invariancy of the region R guarantees existence of a bounded solution $\mathbf{x}^0(t)$ in the time interval $(-\infty, +\infty)$. From inequality 8.121 we conclude that the solution $\mathbf{x}^0(t)$ is unique. Indeed, for $t \to \infty$, we have $\mathbf{x}_1(t) \equiv \mathbf{x}_2(t)$. Furthermore, when the forcing function $\mathbf{f}(t)$ is bounded and T-periodic, then the solution $\mathbf{x}^0(t)$ is bounded, unique, and T-periodic, $\mathbf{x}^0(t + T) \equiv \mathbf{x}^0(t)$. It can also be shown that the assertion (1) above is true even for $\mathbf{f} = \mathbf{f}(\mathbf{x}, t)$ which motivates the discussion in the following section.

8.12 Nonlinear Time Varying System

The Popov criterion was shown to be valid for the principal case of the time varying system by Rozenvasser [8.37]. The development is similar to that of the previous section. The quantity q is again zero. The reason for eliminating the integral term in the Lur'e form of the Liapunov function (equation 8.91) is that the integral term is time varying and only creates difficulties when the time derivative of the function is taken.

Rozenvasser considered the system

$$\dot{\mathbf{x}} = \mathbf{P}\mathbf{x} + \mathbf{q}F(y, t), \quad y = \mathbf{r}^T\mathbf{x}, \qquad (8.131)$$

where $\mathbf{P}$, $\mathbf{q}$, and $\mathbf{r}$ are the same as in the fundamental system (8.61). The continuous nonlinear, time varying function $F(y, t)$ satisfies the conditions[37]

$$F(0, t) = 0, \quad 0 < yF(y, t) < ky^2. \qquad (8.132)$$

To derive the stability condition, use the function

$$V(\mathbf{x}) = \mathbf{x}^T\mathbf{H}\mathbf{x} \qquad (8.133)$$

once again. Taking the time derivative and then adding and subtracting $\Omega(y, t) = (y - \mathscr{F}/k)\mathscr{F} \geq 0$ gives

$$-\dot{V} = [\mathbf{x}^T(\mathbf{G} - \mathbf{g}\mathbf{g}^T)\mathbf{x} + (\sqrt{\gamma}\mathscr{F} + \mathbf{g}^T\mathbf{x})^2] + \Omega, \qquad (8.134)$$

---

[37] The right-hand inequality condition can be weakened to "$\leq$" with some care. See Theorem H.2 outlined in Appendix H.

where

$$\mathbf{G} = \mathbf{HP} + \mathbf{P}^T\mathbf{H}, \qquad -\sqrt{\gamma}\,\mathbf{g} = \mathbf{Hq} + \tfrac{1}{2}\mathbf{r} \qquad (8.135)$$

$$\gamma = \frac{1}{k}, \qquad \Omega = (y - \mathscr{F}/k)\mathscr{F}, \qquad \mathscr{F} = F[y(t), t].$$

We again require that the matrix $\mathbf{H} = \mathbf{H}^T$ that we are seeking satisfy the quadratic inequality

$$\mathbf{G} - \mathbf{gg}^T > 0. \qquad (8.94)$$

By the Yakubovich-Kalman lemma of Appendix H, such a matrix $\mathbf{H}$ exists if and only if the Popov inequality

$$\pi(\omega) \equiv \frac{1}{k} + \operatorname{Re} G(j\omega) > 0 \qquad (8.119)$$

is satisfied for all real $\omega \geq 0$.

It has been shown [8.38] that condition (8.119) also holds in the simplest particular case where the equation $B(s) = |\mathbf{P} - s\mathbf{I}| = 0$ has a zero root and all the other roots have negative real parts (see Problem 8.15).[38]

Along these lines, Yakubovich [8.22] obtained important results. They are outlined in Theorem H.2 and its proof, in Section H.5 of Appendix H.

8.13 Parameter Analysis. Sensitivity

A study of physical systems is usually based on experimental measurements by which the corresponding initial conditions and parameters are determined. The ultimate accuracy of such measurements is limited and the results of the system analysis would be of little practical interest if slight inaccuracies in the data upon which the analysis is based could greatly change the obtained results. Therefore it is of fundamental importance to assure that small errors in the data have a relatively small effect on the system performance characteristics.

Furthermore, when we describe a physical system by the differential equations as a model, a certain idealization is always involved with the intention of isolating the most significant features of the system and thus simplifying the analysis. It is then necessary to guarantee that the solutions of all "neighboring" differential equations with "neighboring" initial

---

[38] It is of interest to mention also the results of the circle criterion obtained by Bongiorno [8.39], Sandberg [8.40], Narendra and Goldwyn [8.41], and Zames [8.42], which represent further extensions of the Popov results.

conditions and parameters are "sufficiently close together" and in fact indistinguishable within the context of the given idealization.

Sensitivity analysis is used to investigate how small (or large) changes in system parameters affect its performance characteristics, and thus to distinguish the problems in system analysis that are well formulated and have practical meaning from those that yield either a badly behaved solution or none at all.[39] The analysis is both qualitative and quantitative and is often expected to give information not only about the existence of a parametric family in a sufficiently small neighborhood of the reference solution but also about how fast the reference solution varies when one or more parameters are given slightly different values. Moreover, the sensitivity analysis may be both a local and global analysis, since it can be used to find the extent of the domain of parameter variations in which the resulting solutions are qualitatively of the same type as the reference solution.[40]

It should also be noted that a study of parameter changes can give information about possible modification and improvement of system performance. If the system parameters are divided into two groups—"controllable" and "uncontrollable" parameters—it may be necessary to estimate on which of the controllable parameters we should act to make system performance invariant despite the changes in the uncontrollable parameters.

For more detailed discussion of the parameter analysis, let us consider the n-vector differential equation

$$\dot{\mathbf{x}} = \mathbf{X}(\mathbf{x}, \mathbf{p}, t), \qquad \mathbf{x}(t_0) = \mathbf{x}_0, \tag{8.136}$$

which contains the l-vector $\mathbf{p}$ of an Euclidean vector space $\mathscr{P}$. The solution $\mathbf{x}(t, \mathbf{p}, \mathbf{x}_0, t_0)$ of (8.136) is a function of both the parameters and the initial conditions. If for certain constant values $\mathbf{x}_0^0$ and $\mathbf{p}^0$ we have a unique solution $\mathbf{x}^0 = \mathbf{x}^0(t, \mathbf{p}^0, \mathbf{x}_0^0, t_0)$, then it is required that the neighboring solutions of $\mathbf{x}^0$, existing for values $\mathbf{x}_0$, $\mathbf{p}$ other than $\mathbf{x}_0^0$, $\mathbf{p}^0$, be qualitatively of the same type as $\mathbf{x}^0$ and differ little from $\mathbf{x}^0$ when $\mathbf{x}_0$, $\mathbf{p}$ differ little from $\mathbf{x}_0^0$, $\mathbf{p}^0$. This requirement amounts to the condition that the solution $\mathbf{x}(t, \mathbf{p}, \mathbf{x}_0, t_0)$ be a continuous function in all of its arguments. Such property of the solution

---

[39] An excellent survey of the sensitivity analysis is outlined by P. V. Kokotović and R. J. Rutman [8.43].

[40] Sensitivity analysis may be regarded as an extension of a relatively old concept introduced in the theory of partial differential equations as the "correctly set problem in the sense of Hadamard." Under this problem it is understood that the solutions exist not only for isolated parameter values, but also in the neighborhood sufficiently close to these values. Being interested in evaluation of the representative solutions, the theory of partial differential equations developed basically the qualitative aspect of the problem. Sensitivity analysis is concerned with effective estimates of the changes in the solutions caused by parameter variations and, therefore, develops extensively the quantitative aspect of the correctly set problem.

is assured by function **X** being continuous and Lipschitz, which is a well-known qualitative result.[41]

The qualitative analysis of the effects of small parameter changes on the system motion is studied by introducing the variations of the parameters into equation of motion (8.136) about a given reference solution $\mathbf{x}^0(t) = \mathbf{x}(t, \mathbf{p}^0, \mathbf{x}_0^0, t_0)$. It is assumed that the function **X** is continuously twice differentiable and admits a unique solution $\mathbf{x}(t, \mathbf{p}, \mathbf{x}_0, t_0)$ continuous with respect to its arguments in a certain neighborhood of the given values $(\mathbf{x}_0^0, \mathbf{p}^0)$.

Let us consider the solution $\mathbf{x}(t, \mathbf{p}, \mathbf{x}_0, t_0)$ as a function of the variables $\mathbf{x}_0, \mathbf{p}$ and let us introduce the variations $\delta \mathbf{x}_0$ and **p** as

$$\delta \mathbf{x}_0 = \mathbf{x}_0 - \mathbf{x}_0^0, \qquad \delta \mathbf{p} = \mathbf{p} - \mathbf{p}^0, \qquad (8.137)$$

which will cause the variation $\delta \mathbf{x}(t)$ of the solution given as

$$\delta \mathbf{x}(t) = \mathbf{x}(t, \mathbf{p}, \mathbf{x}_0, t_0) - \mathbf{x}^0(t). \qquad (8.138)$$

Then the use of Taylor's expansion yields

$$\delta \mathbf{x}(t) = \left[\frac{\partial \mathbf{x}}{\partial \mathbf{x}_0}\right]_0 \delta \mathbf{x}_0 + \left[\frac{\partial \mathbf{x}}{\partial \mathbf{p}}\right]_0 \delta \mathbf{p} + 0(\|\delta \mathbf{x}_0\|^2) + 0(\|\delta \mathbf{p}\|^2), \qquad (8.139)$$

where the subscript "0" of the brackets means that the expressions in the brackets are calculated for $\mathbf{x}_0^0, \mathbf{p}^0$. The existence of the partial derivatives in the brackets is provided by the differentiability of function **X**. The functions $0(\|\delta \mathbf{x}_0\|^2), 0(\|\delta \mathbf{p}\|^2)$ represent the higher-order terms in the expansion (if k is a positive integer, the expression $\mathbf{f}(\mathbf{x}) = 0(\|\mathbf{x}\|^k)$ means that there exist a

---

[41] The precise result is the following theorem:
If (a) The function $\mathbf{X}(\mathbf{x}, \mathbf{p}, t)$ of (8.136) is a real n vector function of the state n vector **x**, parameter l vector **p**, and the scalar t, which is defined and has bounded continuous derivatives with respect to **x**, **p**, and t about any point $\mathbf{x}_0, \mathbf{p}^0, t_0$ on some set S of $(n + l + 1)$-dimensional Euclidean space, and if the set S is the Cartesian product of R and D, that is, $S = R \times D$, where R is a region $\|\mathbf{x} - \mathbf{x}_0\| < a$, $|t - t_0| < b$, D is a domain $\|\mathbf{p} - \mathbf{p}^0\| < c$, and a, b, c are positive constants; and if (b) (Lipschitz condition) the following inequality is satisfied for $(\mathbf{x}_1, \mathbf{p}, t), (\mathbf{x}_2, \mathbf{p}, t)$ on S:

$$\|\mathbf{X}(\mathbf{x}_1, \mathbf{p}, t) - \mathbf{X}(\mathbf{x}_2, \mathbf{p}, t)\| \le k \|\mathbf{x}_1 - \mathbf{x}_2\|,$$

where k is a positive constant independent of **x**, **p**, t.
Then (1) there is a unique solution $\mathbf{x}(t, \mathbf{p}, \mathbf{x}_0, t_0)$ of (8.136) starting at $\mathbf{x}_0, t_0$, for all $|t - t_0| \le d$, $d = \min\{b, a/M\}$, $M = \max \|\mathbf{X}\|$ on S; and (2) the solution $\mathbf{x}(t, \mathbf{p}, \mathbf{x}_0, t_0)$ has continuous derivatives with respect to its arguments on the set $I \times S$, where I is the closed interval determined by $|t - t_0| \le d$ of (1).

For a proof of this result we can use E. A. Coddington and N. Levinson, *Theory of Ordinary Differential Equations*, McGraw-Hill, New York, 1955.

It should also be noted by the Cartesian product $S = R \times D$ we mean the set of all points $(\mathbf{x}, \mathbf{p}, t)$ so that $(\mathbf{x}, t)$ belongs to R and **p** belongs to D.

finite constant M and a constant $\varepsilon > 0$ so that $(\|\mathbf{f}(\mathbf{x})\|/\|\mathbf{x}\|^k) < M$ for all $\|\mathbf{x}\| < \varepsilon$).

The bracketed functions in (8.139) are sometimes called "sensitivity coefficients" [8.43] and they contain the information about the effect of parameter variations. The validity of these functions in calculation of this effect may be "in the large" or only "in the small," depending, of course, on the contribution of the higher-order terms in Taylor's expansion (8.139) for the range of $(\mathbf{x}_0, \mathbf{p})$ considered.[42]

By differentiating (8.136) with respect to $\mathbf{x}_0$, we obtain the sensitivity function $[\partial \mathbf{x}/\partial \mathbf{x}_0]_0$ by solving a *linear* differential equation

$$\frac{d}{dt}\left[\frac{\partial \mathbf{x}}{\partial \mathbf{x}_0}\right]_0 = \left[\frac{\partial \mathbf{X}}{\partial \mathbf{x}}\right]_0 \left[\frac{\partial \mathbf{x}}{\partial \mathbf{x}_0}\right]_0, \qquad \left[\frac{\partial \mathbf{x}}{\partial \mathbf{x}_0}\right]_0 (t_0) = \mathbf{I}, \qquad (8.140)$$

where $[\partial \mathbf{X}/\partial \mathbf{x}]_0$ is to be evaluated for the reference solution $\mathbf{x}^0(t)$ and $\mathbf{p} = \mathbf{p}^0$.

Similarly, the sensitivity function $[\partial \mathbf{x}/\partial \mathbf{p}]_0$ corresponding to parameter variation $\delta \mathbf{p}$ is obtained from the equation

$$\frac{d}{dt}\left[\frac{\partial \mathbf{x}}{\partial \mathbf{p}}\right]_0 = \left[\frac{\partial \mathbf{X}}{\partial \mathbf{x}}\right]_0 \left[\frac{\partial \mathbf{x}}{\partial \mathbf{p}}\right]_0 + \left[\frac{\partial \mathbf{X}}{\partial \mathbf{p}}\right]_0, \qquad \left[\frac{\partial \mathbf{x}}{\partial \mathbf{p}}\right]_0 (t_0) = \mathbf{0}, \qquad (8.141)$$

where again $[\partial \mathbf{X}/\partial \mathbf{p}]_0$ is calculated for the reference solution $\mathbf{x}^0(t)$ and $\mathbf{p} = \mathbf{p}^0$.

If (8.140) is postmultiplied by $\delta \mathbf{x}_0$ and (8.141) by $\delta \mathbf{p}$ and the result is added, (8.139) can be rewritten as $\delta \mathbf{x} = \mathbf{y}(t) + 0(\|\delta \mathbf{x}_0\|^2) + 0(\|\delta \mathbf{p}\|^2)$, where $\mathbf{y}(t)$ is determined from

$$\dot{\mathbf{y}} = \left[\frac{\partial \mathbf{X}}{\partial \mathbf{x}}\right]_0 \mathbf{y} + \left[\frac{\partial \mathbf{X}}{\partial \mathbf{p}}\right]_0 \delta \mathbf{p}, \qquad \mathbf{y}(t_0) = \delta \mathbf{x}_0. \qquad (8.142)$$

Equation 8.142 is called the *variational equation*. The significant characteristic of the variational equation is that it is linear and the effects of small perturbations $\mathbf{x}_0$ and $\mathbf{p}$ in the considered class of nonlinear systems can be evaluated by known methods. Moreover, convenient computer techniques exist for calculating the sensitivity functions appearing in the variational equation 8.142 (see reference 8.43). It should be noted, however, that the reference solution $\mathbf{x}^0(t)$ must be known.

It is also important to note that the variational equation can be used to study small perturbations in the form of forcing functions. If a small scalar forcing function $f = f(t)$ enters the equation of motion as

$$\dot{\mathbf{x}} = \mathbf{X}(\mathbf{x}, f, t), \qquad \mathbf{x}(t_0) = \mathbf{x}_0, \qquad (8.143)$$

---

[42] The domain of validity of the sensitivity functions may be extended if expansions other than Taylor's expansion are used as proposed by Gumowski [8.44].

then by repeating essentially the same procedure as above keeping $\mathbf{x}_0$ constant, we obtain the variational equation

$$\dot{\mathbf{y}} = \left[\frac{\partial \mathbf{X}}{\partial \mathbf{x}}\right]_0 \mathbf{y} + \left[\frac{\partial \mathbf{X}}{\partial f}\right]_0 f, \qquad \mathbf{y}(t_0) = \mathbf{0}, \tag{8.144}$$

where the expressions in the brackets are evaluated for the reference solution $\mathbf{x}^0(t)$ and $f = 0$. Therefore disturbances of the forcing function type can be considered in the context of parameter variation.

From the above discussion it can be concluded that the sensitivity analysis is based on the "imbedding" of the reference solution in a parametric family of solutions. Depending on the nature of the imbedding, the resulting sensitivity functions will be valid "globally" or only "locally." The limits of validity for a particular parametric family are related to *critical* or *bifurcation values* of the parameters. If for some changes of a parameter the solution varies without undergoing any qualitative change, such values of the parameter are called *ordinary* values; otherwise they are bifurcation values.[43]

It is a real situation that the parameters are subject to either desired or undesired perturbations; the designer would like to assure a proper functioning of the system by estimating the domain of the ordinary parameters, and, if necessary, to remove the unwanted bifurcation parameter values. For example, numerous computing and adaptive schemes have been devised that use small parameter variations for adjusting the values of parameters so that the system has a "best" performance in the sense of a prescribed performance criterion [8.43]. To utilize the adaptive schemes properly, the resulting system should be guaranteed stable under all expected parameter variations. Therefore, if a system is described by (8.136), and it is known that for certain parameter values $\mathbf{p} = \mathbf{p}^0$ the system is stable, then the system should also be stable for $\mathbf{p} = \mathbf{p}^0 + \Delta \mathbf{p}$ where $\Delta \mathbf{p}$ represents a small (or large) increment in $\mathbf{p}$. Then the modified differential equation

$$\dot{\mathbf{x}} = \mathbf{X}(\mathbf{x}, \mathbf{p}^0, t) + \mathbf{Y}(\mathbf{x}, \Delta \mathbf{p}, t) \tag{8.145}$$

is to be analyzed and the effect of the additional term $\mathbf{Y}(\mathbf{x}, \Delta \mathbf{p}, t)$ on the stability of the equilibrium of (8.136) must be determined.

In situations when the perturbations are not accurately known and only their estimates are available, that is, we have no information about $\mathbf{Y}$ save that "hopefully" it is not excessively large, the notion of *total stability* is useful. It is an extension of the Liapunov stability concept, and this kind of stability can effectively be guaranteed by the existence of an appropriate

---

[43] For example, in the equation $\ddot{x} + \varepsilon b \dot{x} + ax = 0$, the value $\varepsilon = 0$ is a bifurcation value of the parameter ε since the trajectories in the state space undergo a qualitative change when ε is varied about $\varepsilon = 0$, no matter how small the variations may be.

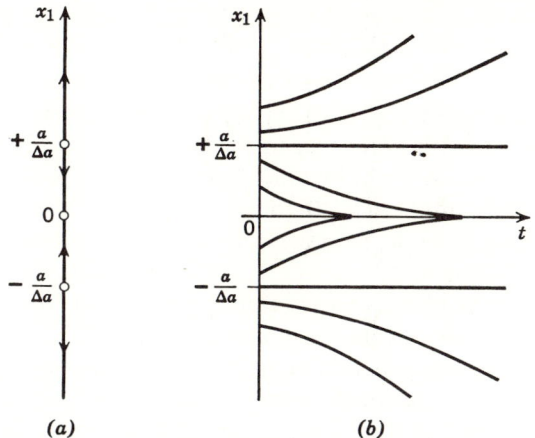

Figure 8.19 Solution of equation $\dot{x}_1 = -a^2 x_1 + (\Delta a)^2 x_1^3$: (a) state space; (b) state-time space.

Liapunov function. It has been shown by Malkin [8.2] that if a dynamic system is uniformly asymptotically stable, it is also totally stable. This important qualitative result and the definition of total stability are given in Section H.2 of the appendix. In the following, however, we are interested in more quantitative results concerning parameter perturbations and their effect on system stability behavior.

In order to illustrate the role of parameter variations in system stability behavior, let us consider a modified version of (8.45):

$$\dot{x}_1 = -a^2 x_1 + (\Delta a)^2 x_1^3. \tag{8.146}$$

The characteristic root of the linearized equation is $-a^2$ and the unperturbed motion is globally asymptotically stable. The trajectories of (8.146) shown in Figure 8.19 indicate, however, that the extent of the stability domain is finite and depends on the perturbed ratio $a/\Delta a$. If $a/\Delta a$ is relatively small, the perturbed system may be considered "practically unstable" despite the fact that the origin is asymptotically stable in the sense of Liapunov. Hence, for proper functioning of the system, the values of a and Δa are essential.

A generalization of the above example is of particular practical interest. The perturbed system is described by the equation

$$\dot{\mathbf{x}} = \mathbf{P}(\mathbf{p}^0)\mathbf{x} + \mathbf{Y}(\mathbf{x}, \Delta \mathbf{p}), \tag{8.147}$$

where $\mathbf{P}(\mathbf{p})\mathbf{x}$ is the linear term and $\mathbf{Y}(\mathbf{x}, \Delta \mathbf{p})$ represent the second or higher-order terms.

348 Stability Analysis

Let us assume that the stability domain in the parameter space is determined for the linearized version of (8.146). In other words, the domain of parameter values is found for which all the characteristic roots of $\mathbf{P(p)}$ have negative real parts (see Chapter 2). Then, if the parameters are chosen from that domain and, in addition, $\mathbf{Y} \equiv \mathbf{0}$, the system (8.147) is globally asymptotically stable. When $\mathbf{Y}$ is not identically zero, the stability will be generally restricted to a finite domain in the state space. The extent of the domain will depend on how close the parameter values are to the boundary of stability for the linearized system.

Let $\mathbf{p}^0$ in (8.147) be chosen on the stability boundary and let $\mathbf{p} = \mathbf{p}^0 + \Delta \mathbf{p}$ be a neighboring point outside the stability domain. The additional term $\mathbf{Y}$ in (8.147) is zero for $\Delta \mathbf{p} = \mathbf{0}$, and because it is continuous with respect to $\Delta \mathbf{p}$ we can use the Malkin result for total stability (see Corollary C in Section H.2) to conclude that if the equilibrium of $\dot{\mathbf{x}} = \mathbf{P(p^0)x}$ is asymptotically stable, the maximum deviation of the motion caused by $\Delta \mathbf{p}$ can be kept arbitrarily small by keeping $\|\Delta \mathbf{p}\|$ sufficiently small. When the unperturbed motion is unstable, initial perturbations, however small, will cause the perturbed motion to diverge from the unperturbed one as the time increases.

The above can be summarized by saying that the unperturbed motion in the neighborhood of the stability boundary may be considered practically stable or unstable depending on whether the motion on the boundary is stable or unstable in the sense of Liapunov. The boundaries on which the unperturbed motion is stable are called "safe" and the boundaries on which the motion is unstable are called "dangerous." Analysis of "dangerous" and "safe" boundaries reduces to the solution of the stability problem in particular cases [8.2].

Let us consider as an illustrative example the following system of differential equations:[44]

$$\ddot{x} + N\dot{x} + Mx = Lz$$
$$\dot{z} = F(y) \qquad (8.148)$$
$$y = -\beta \dot{x} - \alpha x,$$

where

$$F(y) = Ky + \kappa y^3, \qquad (8.149)$$

and all the parameters $L, N, M, \alpha, \beta, K, \kappa$ are positive.

A system block diagram that corresponds to (8.148) is shown in Fig. 8.20.
By introducing the variables

$$z = x_1, \qquad x = x_2, \qquad \dot{x} = x_3, \qquad (8.150)$$

---

[44] Under certain simplifying assumptions, (8.148) describes the motion of an airplane with an auto-pilot [8.2].

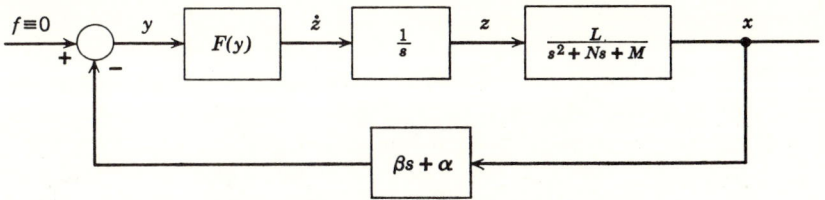

Figure 8.20 System block diagram.

we obtain from (8.148) the state equations in the form

$$\dot{x}_1 = F(y) = Ky + \kappa y^3$$
$$\dot{x}_2 = x_3$$
$$\dot{x}_3 = -Lx_1 - Mx_2 - Nx_3 \quad (8.151)$$
$$y = -\alpha x_2 - \beta x_3.$$

The transfer function of the linear part of the system is

$$G(s) = \frac{L(\beta s + \alpha)}{s(s^2 + Ns + M)}. \quad (8.152)$$

Let us consider first the linearized version of (8.151), that is, $F(y) = Ky$, $\kappa = 0$. The corresponding characteristic equation is

$$s^3 + Ns^2 + (M + KL\beta)s + KL\alpha = 0. \quad (8.153)$$

For the linearized system to be stable it is necessary and sufficient that the Hurwitz inequalities (see Appendix D) are satisfied; that is,

$$N > 0, \quad M + KL\beta > 0, \quad KL\alpha > 0$$
$$\Sigma = N(M + KL\beta) - KL\alpha > 0. \quad (8.154)$$

If α and β are considered as variable parameters, the stability boundary

$$\Sigma = -KL\alpha + KLN\beta + NM = 0 \quad (8.155)$$

is plotted in Figure 8.21.

Now, let us consider the nonlinear system with the characteristic (8.149) when $\kappa \neq 0$. In order to determine stability we should plot the Popov modified frequency characteristic $G^*(\omega)$. For the transfer function $G(s)$ of (8.152), two typical plots of the modified characteristic $G^*(\omega)$ are shown in Figure 8.22. The modified characteristic is always a "stationary" curve so that a Popov line can be drawn through the point $(-1/k, j0)$ tangent to the curve $G^*(\omega)$ and the entire curve (for all $\omega \geq 0$) is situated to the right of the Popov line (Figure 8.22). Cases represented by the plot of $G^*(\omega)$ in Figure 8.17 cannot take place. Therefore, for nonlinear systems with the transfer function $G(s)$ of (8.152) to be absolutely stable in the sector $(0, k)$, it is

350 Stability Analysis

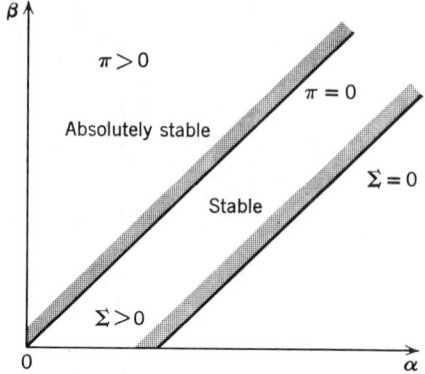

Figure 8.21 Parameter plane.

necessary and sufficient that the characteristic $G^*(\omega)$ does not intersect the forbidden zone $(-\infty, -1/k)$ of the real axis (curve a in Figure 8.22).

In the case under investigation the nonlinearity is (8.149) and the sector is $(0, \infty)$. On the basis of Theorem 8.5 and the above discussion, the system (8.148) is absolutely stable if and only if the curve $G^*(\omega)$ does not intersect the negative part of the real axis allowing k to go to infinity (curve b in Figure 8.22). Because the modified characteristic $G^*(\omega)$ differs from the Nyquist plot $G(j\omega)$ only in scale change along the imaginary axis, the system is absolutely stable if it is stable for all linear gains in the sector $(0, \infty)$.

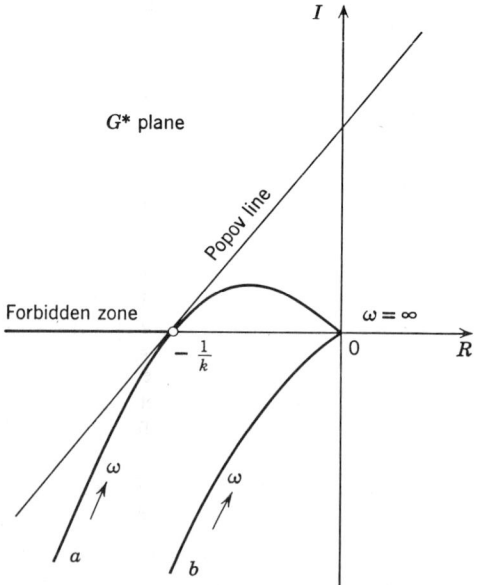

Figure 8.22 Popov diagram.

This takes place when, in the condition
$$\Sigma = NM + KL(N\beta - \alpha) > 0, \tag{8.156}$$
the inequality
$$\pi = N\beta - \alpha > 0 \tag{8.157}$$
is satisfied. The absolute stability region determined by (8.157) is represented by the shaded region $\pi > 0$ in Figure 8.21.

It is of interest to note that in the preceding example the absolute stability analysis was reduced to stability analysis of the linearized system, which can be performed by well-known methods. As a result, the absolute stability conditions were conveniently interpreted in the parameter plane. As is shown in Section 8.15, this simplification is not always possible and is restricted to certain classes of nonlinear systems [8.51]. More general methods for interpretation of absolute stability conditions in the parameter space are given in the next section. The emphasis in the following discussion is on finding regions in the parameter space of ordinary values that do not affect the system stability. In other words, a joint stability and sensitivity analysis is directed toward estimating domains in the parameter space that do not contain bifurcation parameter values with respect to stability.

Before outlining methods for solving the joint stability-sensitivity problem, an important aspect of parameter analysis should be mentioned. In formulating the mathematical model for a given physical system, certain small parasitic parameters are neglected (set to zero). It often happens that they enter as the coefficients attached to the higher derivatives; then, upon setting these parameters equal to zero, we obtain a system of equations of lower order and also frequently leave the higher derivatives unsolved. Generally, under these circumstances, not considering the parameters can lead to an incorrect qualitative description of the physical phenomena. This problem is related to the problem of integral manifolds.[45]

---

[45] A wide class of systems can be described by a differential equation of the form
$$\dot{\mathbf{x}} = \mathbf{X}(\mathbf{x}, t, \varepsilon), \tag{i}$$
in which $\mathbf{x}$, $\mathbf{X}$, are n vectors of an n-dimensional Euclidean space $\mathscr{X}^n$, t stands for time, and $\varepsilon > 0$ is a small parameter. Function $\mathbf{X}$ satisfies certain general conditions for $-\infty < t < +\infty$ and for $\mathbf{x} \in U^n$, an open subset of $\mathscr{X}^n$. Suppose that each t of the interval $(-\infty, +\infty)$ corresponds to some set S^l of points $\mathbf{x}$, which can be represented by an analytical form in terms of equation $\mathbf{x} = \boldsymbol{\phi}(t, C_1, C_2, \ldots, C_l)$, where function $\boldsymbol{\phi}$ satisfies the Lipschitz condition with respect to $C_1, C_2, \ldots, C_l$ in the complete domain of their variation.

An analytical definition of the integral manifold is the following: set S^l is an l-dimensional ($l \leq n$) integral manifold of (i) if, given any solution $\mathbf{x} = \mathbf{x}(t)$ of this equation, the validity of $\mathbf{x}(t) \in S^l$ at any moment $t = t_0$ implies its validity for arbitrary t, $(-\infty < t < +\infty)$.

In geometrical terms an integral manifold is a hypersurface with the property that if any value of the solution of (i) lies on that hypersurface, then the entire solution lies on the hypersurface.

A concise treatment of integral manifolds is found in J. K. Hale, *Oscillations in Nonlinear Systems*, McGraw-Hill, New York, 1963.

8.14 Absolute Stability in the Parameter Space

Stability and sensitivity are two essential properties of dynamic systems. While stability assures a proper functioning of the system, the sensitivity indicates the ability of the system to retain required performance characteristics despite changes in the operating conditions. These changes may occur if the parameters of physical systems deviate from their nominal values either because of inaccuracies in the system components (time invariant case), or because the system parameters vary in time (time varying case). Therefore a simultaneous consideration of stability and sensitivity in system analysis is desired [8.24, 25, 54, 56].

In absolute stability analysis the nonlinear characteristic is not completely specified and should only belong to a certain class of functions defined by sector inequalities (see Section 8.6). On the other hand, the parameters of the linear part are specified numerically. The purpose of this section is to propose an *absolute stability definition that relaxes the conditions on the linear part and allows system parameters to deviate from their nominal values.* Then graphical and analytical procedures are presented to determine regions in the parameter space that correspond to an absolutely stable system. While the graphical technique based upon an envelope construction will provide a largest possible region assumed by the Popov inequality, the analytical technique will be used as an approximation method and will serve actually as an interpretation technique of the absolute stability regions in the parameter space.

Before the two techniques are presented, it is necessary to give an appropriate absolute stability definition that includes the parameter variations.

Let us consider again a system described by the equations

$$\dot{\mathbf{x}} = \mathbf{P}\mathbf{x} + \mathbf{q}F(y), \qquad y = \mathbf{r}^T\mathbf{x}, \tag{8.82}$$

where $\mathbf{x}, \mathbf{q}, \mathbf{r}$ are real n vectors, $\mathbf{P}$ is a real $n \times n$ matrix,[46] and $F(y)$ is generally an arbitrary, real- and single-valued nonlinear function, defined for all real values of y and contained in a certain sector $[0, k]$, that is, $F(y) \in [0, k]$. The linear part of system (8.82) is described by the transfer function $G(s)$ given by

$$G(s) = \mathbf{r}^T(\mathbf{P} - s\mathbf{I})^{-1}\mathbf{q} = \frac{C(s)}{B(s)}, \tag{8.66}$$

---

[46] It is assumed again without loss of generality that the pair $(\mathbf{P}, \mathbf{q})$ is completely controllable and that $(\mathbf{P}, \mathbf{r}^T)$ is completely observable.

where

$$B(s) = |\mathbf{P} - s\mathbf{I}| = \sum_{k=0}^{n} b_k s^k \qquad (8.67)$$

$$C(s) = \mathbf{r}^T [\text{adj} (\mathbf{P} - s\mathbf{I})]\mathbf{q} = \sum_{k=0}^{m} c_k s^k \qquad (8.68)$$

and $n > m$.

Let us now assume that the transfer function $G(s, p_1, p_2, \ldots, p_l)$ is a function of $s = \sigma + j\omega$ and l parameters $(p_1, p_2, \ldots, p_l)$, and let us also suppose that the solution $\mathbf{x}(t, \mathbf{p})$ of (8.82) is well defined for those values of parameters that belong to a certain region R of the l-dimensional Euclidean space $\mathscr{P}$. Then Definition 8.11 of absolute stability can be reformulated to read as follows.

Definition 8.12. Absolute Stability in the Parameter Space. A free dynamic system (8.82) is absolutely stable in a sector $[0, k]$ and a region R if for any (continuous or discontinuous) single-valued nonlinear characteristic $F(y) \in [0, k]$, $F(0) = 0$, and any set of parameters $(p_1, p_2, \ldots, p_l) \in R$, the equilibrium position $\mathbf{x} = \mathbf{0}$ of the system is asymptotically stable in the large.

When system (8.82) is specified, (1) we wish to find the greatest value of k (the largest sector) for a given set of parameter values $(p_1^0, p_2^0, \ldots, p_l^0) \in R$; (2) a value of k is given and the largest region R is to be determined; and (3) the greatest value of k and the largest region R are to be found. In the following developments all three aspects of the problem will be considered.

From Theorem 8.6 of Section 8.10, it follows that the absolute stability of (8.82) can be verified by the Popov inequality[47]

$$\pi(\omega) \equiv \frac{1}{k} + \text{Re}\,(1 + jq\omega)G(\sigma + j\omega) > 0, \qquad \forall \omega \geq 0. \quad (8.112)$$

Function $G(\sigma + j\omega)$ can be written as

$$G = \frac{C_1 + jC_2}{B_1 + jB_2}, \qquad (8.158)$$

where

$$\begin{aligned} B_1 &= \sum_{k=0}^{n} b_k X_k, & B_2 &= \sum_{k=0}^{n} b_k Y_k \\ C_1 &= \sum_{k=0}^{m} c_k X_k, & C_2 &= \sum_{k=0}^{m} c_k Y_k \end{aligned} \qquad (8.159)$$

---

[47] For convenience in further developments, the symbol "$\forall$" is introduced; it means "for all values of." It is also assumed that ω is real.

and $X_k = X_k(\sigma, \omega)$, $Y_k = Y_k(\sigma, \omega)$ are functions defined as[48]

$$X_k = \sum_{v=0}^{k} (-1)^v \binom{k}{2v} \sigma^{k-2v} \omega^{2v}$$

$$Y_k = \sum_{v=1}^{k} (-1)^{v-1} \binom{k}{2v-1} \sigma^{k-2v+1} \omega^{2v-1}.$$

(8.160)

Functions X_k and Y_k can readily be calculated from the recurrence formulas

$$X_{k+1} - 2X_1 X_k + (X_1^2 + Y_1^2) X_{k-1} = 0$$

$$Y_{k+1} - 2X_1 Y_k + (X_1^2 + Y_1^2) Y_{k-1} = 0,$$

(8.161)

where $X_0 \equiv 1$, $X_1 \equiv \sigma$, $Y_0 \equiv 0$, $Y_1 \equiv \omega$.

If σ is specified, inequality 8.112 becomes equivalent to the inequality

$$\Pi(\omega) \equiv S(\omega)\frac{1}{k} + Q(\omega)q + T(\omega) > 0, \qquad \forall \omega \geq 0, \qquad (8.162)$$

where $S = S(\omega)$, $Q = Q(\omega)$, and $T = T(\omega)$ are polynomials in ω,

$$S = B_1^2 + B_2^2, \qquad Q = \omega(B_2 C_1 - B_1 C_2), \qquad T = B_1 C_1 + B_2 C_2.$$

(8.163)

Let us assume that the transfer function $G(s)$ of the linear part is a function of $l - 2$ parameters $p_1, p_2, \ldots, p_{l-2}$. These parameters appear in the coefficients b_k and c_k of (8.159). Thus (8.162) can be finally rewritten as

$$\Pi(\omega, \mathbf{p}) \equiv \sum_{k=0}^{n} a_{2k} \omega^{2k} > 0, \qquad \forall \omega \geq 0, \qquad (8.164)$$

where the coefficients $a_{2k} = a_{2k}(\mathbf{p})$ are functions of the parameter l vector $\mathbf{p} = (p_1, p_2, \ldots, p_l)$. For convenience, $1/k$ and q of (8.162) are considered as parameters.[49]

It may now be concluded that the problem of absolute stability has been reduced to the problem of finding the conditions on the coefficients a_{2k}, or parameters p_i, so that the polynomial $\Pi(\omega, \mathbf{p})$ of (8.164) has no positive real zeros. Two methods will be presented to check these conditions being satisfied: (1) the envelope graphical criterion and (2) the analytical method.

Envelope Criterion: To describe the envelope graphical procedure, let us consider the case where there are two variable parameters, α, β, and where

---

[48] See Derivation A.1 of Appendix A.
[49] Note, however, that q is not a physical parameter and only its existence is required so that $\pi > 0$, $\forall \omega \geq 0$.

(8.164) has a general form

$$\Pi(\omega, \alpha, \beta) > 0, \qquad \forall \omega \geq 0. \tag{8.165}$$

Taking equality in (8.152), we obtain

$$\Pi(\omega, \alpha, \beta) = 0, \tag{8.166}$$

which for a specific value of ω determines a curve C in the parameter $\alpha\beta$ plane. This curve divides the $\alpha\beta$ plane into regions $\Pi > 0$ and $\Pi < 0$. Then, to determine the set of points $\{(\alpha, \beta) \in R \mid \pi > 0, \forall \omega \geq 0\}$ in the parameter plane,[50] it is necessary to construct the *envelope E* of all the curves C obtained for various values of ω.

To derive the equations that determine the envelope E, let us assume that such an envelope exists, and let $M(\alpha; \beta)$ be the point of tangency of E with the one curve C that corresponds to a certain value ω.[51] The quantities $\alpha = \alpha(\omega)$ and $\beta = \beta(\omega)$ are unknown functions of ω that satisfy (8.166). In order to determine these functions, it is necessary to use the fact that the tangents to the two curves C and E coincide for all values of ω. Then a necessary condition for tangency is

$$\frac{d\alpha/d\omega}{\delta\alpha} = \frac{d\beta/d\omega}{\delta\beta}, \tag{8.167}$$

where $d\alpha/d\omega$ and $d\beta/d\omega$ are the derivatives of the unknown functions α and β, and $\delta\alpha$ and $\delta\beta$ are two quantities proportional to the direction cosines of the tangent to the curve C.

Since ω in (8.166) has a constant value for the specific curve C under consideration, we have

$$\frac{\partial \Pi}{\partial \alpha} \delta\alpha + \frac{\partial \Pi}{\partial \beta} \delta\beta = 0, \tag{8.168}$$

which determines the tangent to C. The two unknown functions $\alpha(\omega)$ and $\beta(\omega)$ satisfy also (8.166), where ω now is the independent variable. Therefore

$$\frac{\partial \Pi}{\partial \alpha} \frac{d\alpha}{d\omega} + \frac{\partial \Pi}{\partial \beta} \frac{d\beta}{d\omega} + \frac{\partial \Pi}{\partial \omega} = 0. \tag{8.169}$$

By combining equations (8.166–8.169), we get

$$\Pi(\omega, \alpha, \beta) = 0, \qquad \frac{\partial \Pi}{\partial \omega} = 0, \tag{8.170}$$

---

[50] $\{p \in R \mid P\}$ denotes a set of p's that belong to R with a property P.
[51] An exposition on envelopes can be found in E. Goursat, *A Course in Mathematical Analysis*, Vol. I, Chapters X and XI, Dover Publications, New York, 1959.

and the unknown functions $\alpha(\omega)$ and $\beta(\omega)$ are solutions of equations 8.170. Thus the equations of the envelope, when an envelope exists, should be found by eliminating ω from (8.170).[52]

In case of inequality 8.164, (8.166) is

$$\Pi(\omega, \alpha, \beta) \equiv \sum_{k=0}^{n} a_{2k}\omega^{2k} = 0, \qquad (8.171)$$

where $a_{2k} = a_{2k}(\alpha, \beta)$. For specific values of α and β, (8.171) will generally have $2n$ distinct roots and through the corresponding point $M(\alpha; \beta)$ pass $2n$ different curves of the given family. But if the point M lies on the curve $E(\alpha, \beta) = 0$, equations 8.170 are satisfied simultaneously, and (7.171) for the chosen specific values of α and β has a double root. If only real positive values of ω are considered, as is required in the inequality 8.164, $E(\alpha, \beta) = 0$ represents the double-real-positive-root locus in the $\alpha\beta$ plane. In the parameter analysis this locus is called the $\zeta = 1$ curve and has the important property that the number of tangents from a point $M_1(\alpha_1; \beta_1)$ to the curve is equal to the number of real roots of (8.171) for $\alpha = \alpha_1$ and $\beta = \beta_1$ (see Derivation A.3 of Appendix A). Therefore, when the envelope E is plotted on the $\alpha\beta$ plane for values of $\omega \geq 0$, the region R of absolute stability is determined by inspection as the *convex region* from which no tangents can be drawn to the envelope E.

To illustrate the procedure, let us consider the system (8.82) with the transfer function

$$G(s) = \frac{40}{s(s+1)(s^2 + 0.8s + 16)}. \qquad (8.172)$$

It is required to evaluate the maximum value k of the nonlinearity sector so that the system is exponentially stable with $\sigma = 0$.

After the polynomials B_1, B_2, C_1, and C_2 are determined from (8.172), the two equations 8.170 are obtained in the form

$$\Pi \equiv S\alpha + Q\beta + T = 0$$
$$\frac{\partial \Pi}{\partial \omega} \equiv S'\alpha + Q'\beta + T' = 0, \qquad (8.173)$$

where S, Q, T are those of (8.163), and S', Q', T' denote their derivatives with respect to ω. In (8.173), $\alpha = 1/k$, $\beta = q$.

Equations 8.173 represent two equations in two unknowns α and β, which can be solved for α and β each time ω is changed. That process will yield the points of the envelope $E(\alpha, \beta) = 0$, which is traced in Figure 8.23.

---

[52] The elimination is possible if the corresponding Jacobian is different from zero. See Appendix A.

Absolute Stability in the Parameter Space

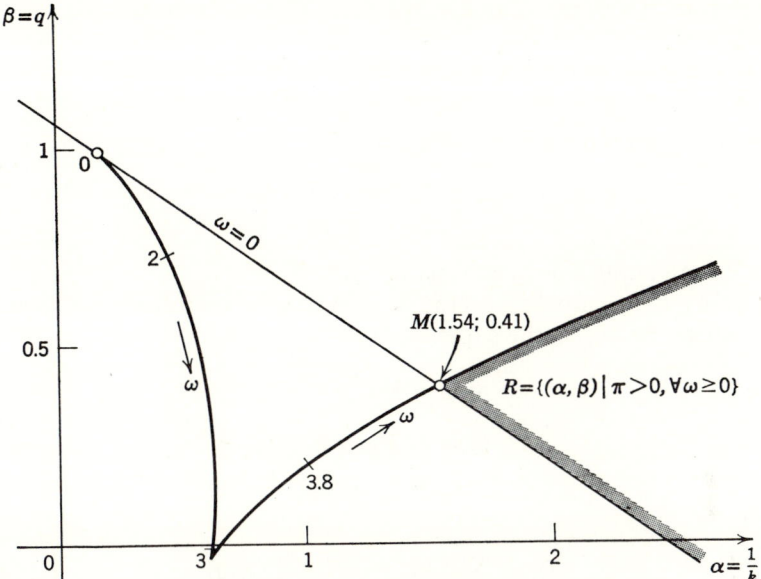

Figure 8.23 Parameter plane diagram.

If $\omega = 0$, the part of the curve E is a locus of singular points that is represented by the straight line $\omega = 0$. This straight line is determined by the first equation 8.173 since, in the second equation for $\omega = 0$, the left-hand side vanishes identically.

The shaded region R in the $\alpha\beta$ plane diagram of Figure 8.23 is readily determined as the region of absolute stability $\{(\alpha, \beta) \mid \pi > 0, \forall \omega \geq 0\}$. It is such that no tangent can be drawn to the plotted envelope. From the diagram the minimum value of $\alpha = 1/k$ that corresponds to maximum value of the sector $(0, k)$ is found by inspection to be at the point $M(1.54; 0.41)$. Thus the maximum value of $k = 0.65$.

A general computer program can be written to plot the envelope E for any given $G(s)$ and specified σ. The program is written to solve equations 8.173 on the basis of recurrence formulas 8.161 and

$$X'_{k+1} - 2X_1 X'_k + (X_1^2 + Y_1^2) X'_{k-1} + 2Y_1 X_{k-1} = 0 \qquad (8.174)$$
$$Y'_{k+1} - 2X_1 Y'_k + (X_1^2 + Y_1^2) Y'_{k+1} + 2Y_1 Y_{k-1} = 0,$$

where X'_k and Y'_k are derivatives of functions X_k and Y_k with respect to ω.

As another example, consider a forced system

$$\dot{\mathbf{x}} = \mathbf{P}\mathbf{x} + \mathbf{q}F(y) + \mathbf{f}(t), \qquad y = \mathbf{r}^T \mathbf{x} \qquad (8.61)$$

Stability Analysis

the exponential absolute stability of which is based on inequality

$$\pi(\omega) \equiv \frac{1}{k} + \operatorname{Re} G(\sigma + j\omega) > 0, \qquad \forall \omega \geq 0 \qquad (8.115)$$

as stated in Theorems 8.7 and 8.8. Let

$$G(s) = \frac{C(s) + \beta D(s)}{B(s)}, \qquad (8.175)$$

where β is an adjustable system parameter.

To verify the inequality 8.115, it is necessary to solve again equations 8.173, where, however, the expressions

$$S = B_1^2 + B_2^2$$
$$Q = B_1 D_1 + B_2 D_2 \qquad (8.176)$$
$$T = B_1 C_1 + B_2 C_2$$

and

$$D_1 = \sum_{k=0}^{m} d_k X_k, \qquad D_2 = \sum_{k=0}^{m} d_k Y_k \qquad (8.177)$$

should be used.

A specific case,

$$G(s) = \frac{s^2 + \beta}{(s+1)(s+2)(s+3)}, \qquad (8.178)$$

may be used for illustration purposes. If $\sigma = 0$, the diagram in Figure 8.24 is obtained. The locus of singular points is again the $\omega = 0$ straight line. As ω increases, the envelope is obtained following the points A, B, C, D; this envelope approaches the two asymptotes as indicated in Figure 8.24. The shaded region $R = \{(\alpha, \beta) \mid \pi > 0, \forall \omega \geq 0\}$ is determined by inspection and the minimum of $1/k$ is on the β axis. Consequently, absolute stability in this case can be assured in the sector consisting of the entire first and third quadrant ($k = \infty$) by properly choosing the value of β.[53]

If an exponential absolute stability with $\sigma = -0.5$ is desired, the parameter plane diagram has the form shown in Figure 8.25. The envelope is similar to that of Figure 8.24 plotted for $\sigma = 0$, but the parameter β cannot be chosen to yield $k = \infty$.

It should be noted that the equations

$$\Pi(\omega, \alpha, \beta) = 0, \qquad \frac{\partial \Pi}{\partial \omega} = 0 \qquad (8.170)$$

---

[53] Of course, to guarantee the absolute stability of the system under investigation, it is also necessary to verify the condition (8.116) associated with inequality 8.115.

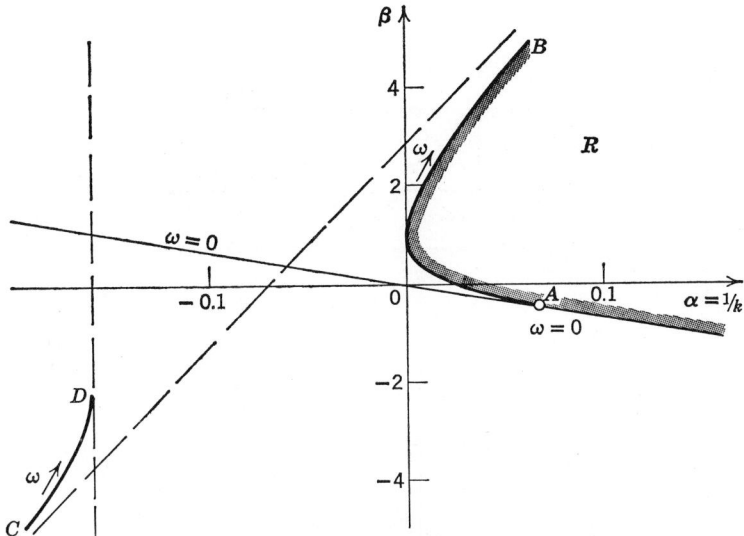

Figure 8.24 Parameter plane diagram for $\sigma = 0$.

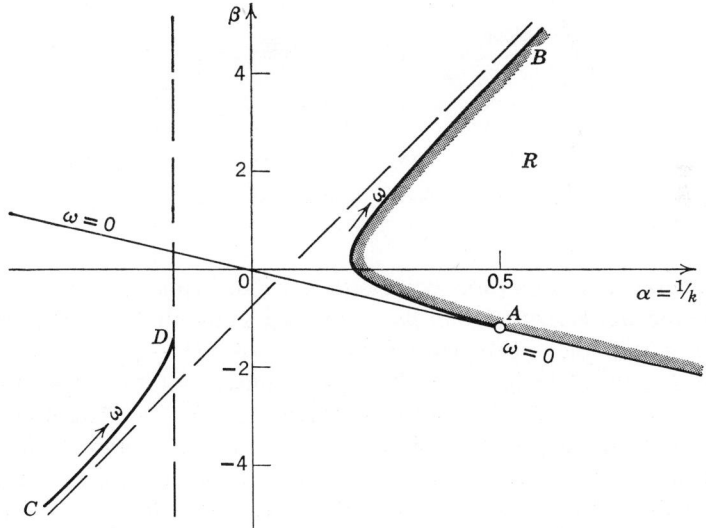

Figure 8.25 Parameter plane diagram for $\sigma = -0.5$.

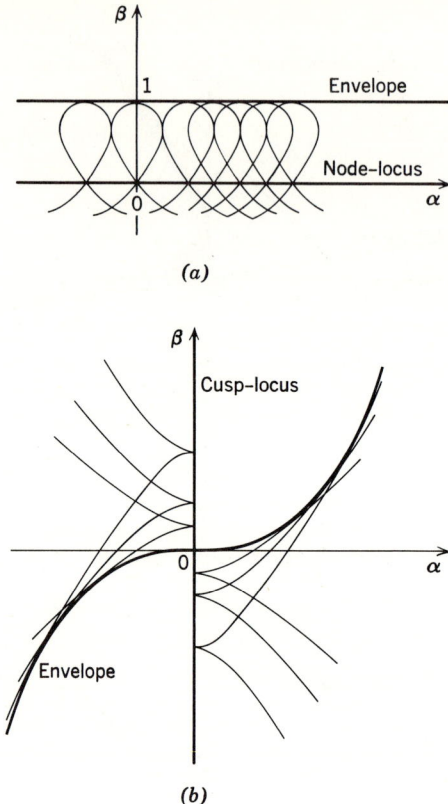

Figure 8.26 Singular loci.

may, in general, yield other loci besides the envelope (if an envelope exists at all). These loci are *loci of singular points* and are of no interest in the stability analysis. For example, when $\Pi(\omega) \equiv (\alpha - \omega)^2 - \beta^2(1 - \beta) = 0$, we obtain the curve $E(\alpha, \beta) \equiv \beta^2(1 - \beta) = 0$, which consists of the true envelope $\beta = 1$ and the node-locus $\beta = 0$ as shown in Figure 8.26a. The nodes may contract into cusps as in the case $\Pi(\omega) \equiv (2\omega + 3\beta)^2 - 4\alpha^3\omega = 0$, which represents a family of semicubical parabolas with their cusps on the axis of β as shown in Figure 8.26b. The curve $E(\alpha, \beta) \equiv \alpha^2(6\beta - \alpha^3) = 0$ consists of the envelope $6\beta = \alpha^3$ and the cusp-locus $\alpha = 0$. Therefore, although equations 8.170 seem to be necessary, they are by no means a sufficient condition to ensure the existence of an envelope. It may now be concluded that an envelope E of the family $\Pi(\omega, \alpha, \beta) = 0$ of curves has the following properties.

Absolute Stability in the Parameter Space

1. Every point of the envelope $E(\alpha, \beta) = 0$ is on some curve of the family $\Pi(\omega, \alpha, \beta) = 0$.
2. The tangents of the envelope E and the particular curve C at their common points coincide.
3. The envelope E and the particular curves C for each value of the variable parameter ω have only isolated points in common.

The above result can be generalized for the case of l parameters ($p_1, p_2, \ldots, p_l$) when the inequality is that of (8.164). Then equation

$$\Pi(\omega, \mathbf{p}) \equiv \sum_{k=0}^{n} a_{2k}(\mathbf{p})\omega^{2k} = 0, \tag{8.179}$$

arising from inequality 8.164, may represent a one-parameter family of hypersurfaces in the l-dimensional Euclidean parameter space $\mathscr{P}$; ω is the parameter of the family.

Let us assume that a surface S of the family (8.179) corresponding to an arbitrary (fixed) value of ω intersects every surface corresponding to a value $\omega + \delta\omega$ ($|\delta\omega|$ is sufficiently small). This intersection C can be represented in the form

$$\Pi(\omega, \mathbf{p}) = 0, \qquad \Pi(\omega + \delta\omega, \mathbf{p}) = 0. \tag{8.180}$$

Since C lies also on the surface $\Pi(\omega + \delta\omega, \mathbf{p}) - \Pi(\omega, \mathbf{p}) = 0$, we may replace the second equation 8.180 by

$$\frac{1}{\delta\omega}\{\Pi(\omega + \delta\omega, \mathbf{p}) - \Pi(\omega, \mathbf{p})\} = 0. \tag{8.181}$$

As the increment $\delta\omega$ tends to zero, we assume that the intersection C tends to a limiting position given by equations

$$\Pi(\omega, \mathbf{p}) = 0, \qquad \frac{\partial}{\partial \omega}\Pi(\omega, \mathbf{p}) = 0. \tag{8.182}$$

This limiting intersection represents the so-called characteristic curve of the family of hypersurfaces (8.179). Geometrically, it is the curve on the hypersurface that contains every point of the set to which the points of intersection given by (8.180) tend when $\delta\omega \to 0$.[54]

If the characteristics of the family (8.179) exist and if their totality obtained by letting ω assume all possible values generates a surface, this surface represents the envelope of the family (8.179). Of course, equations may

---

[54] However, C will contain, in general, other points as well. It may even be that neighboring surfaces of the family do not intersect at all but nevertheless (8.182) determines a characteristic on each of the surfaces.

yield singular solutions together with the envelope as shown above. If the family parameter ω can be eliminated from (8.182), the envelope is represented by $E(\mathbf{p}) = 0$.

In absolute stability analysis the hypersurface $E(\mathbf{p}) = 0$ determines the stability boundary in the parameter space $\mathscr{P}$. In case of several variable parameters the problem of applying the envelope graphical technique is essentially one of interpretation. For this reason, the envelope technique is convenient only for easily visualized two-dimensional space (Figures 8.23 to 8.25). Therefore, for more than two parameters the following analytical technique may be suggested.

Analytical Technique: By this simple technique, a region R is first determined in the parameter space $\mathscr{P}$ in terms of algebraic inequalities involving parameters. Then a rectangular parallelepiped of maximum volume is imbedded in the region to yield a convenient interpretation of the absolute stability region in the parameter space.[55]

When σ is specified in an absolute stability problem and the transfer function $G(s)$ of the linear part of the system is a rational function in $s = \sigma + j\omega$, the problem reduces to that of finding the conditions on the coefficients $a_{2k} = a_{2k}(\mathbf{p})$ ($k = 0, 1, \ldots, n$) under which the following inequality

$$\Pi(\omega, \mathbf{p}) \equiv \sum_{k=0}^{n} a_{2k}\omega^{2k} > 0, \qquad \forall \omega \geq 0 \qquad (8.164)$$

is satisfied. In other words, a system is absolutely stable if the corresponding polynomial Π has no positive real zeros. For this to take place, it is sufficient that the algebraic inequalities

$$a_0 > 0, \, a_{2k} \geq 0, \qquad (k = 1, 2, \ldots, n), \qquad (8.183)$$

are satisfied.

Inequalities 8.183 specify a region $\bar{R}(\bar{R} \subset R)$[56] of absolute stability in the parameter space $\mathscr{P}$ that may appear to be an overly strict region since conditions (8.183) are only sufficient for $\pi > 0$, $\forall \omega \geq 0$. Conditions (8.183), however lead to a convenient interpretation of the stability regions.

For example, if the transfer function of a forced system (8.61) is

$$G(s) = \frac{s^2 + p_2 s + p_3}{p_1(s + 1)(s + 2)(s + 3)}, \qquad (8.184)$$

---

[55] This interpretation technique was proposed by George for linear system analysis and approximation of finite stability regions in the state space for nonlinear systems (for reference, see Section D.5)

[56] The symbol "$\subset$" means in $\bar{R} \subset R$, "$\bar{R}$ is a subset of R."

Absolute Stability in the Parameter Space 363

$k = 1$, and $\sigma = 0$ ($s = j\omega$) are specified, we obtain Π from (8.115) as

$$\Pi(\omega, p_1, p_2, p_3) \equiv p_1\omega^6 + (14p_1 - p_2 + 6)\omega^4 + (49p_1 + 11p_2 - 6p_3 - 6)\omega^2 \\ + 36p_1 + 6p_3. \quad (8.185)$$

Inequalities 8.183 are

$$\begin{aligned} 6p_1 + p_3 &> 0 \\ 49p_1 + 11p_2 - 6p_3 - 6 &\geq 0 \\ 14p_1 - p_2 + 6 &\geq 0 \\ p_1 &\geq 0, \end{aligned} \quad (8.186)$$

which determine the boundaries of $\bar{R}$.

After inequalities 8.186 (or, generally, 8.183) are obtained, the problem of using them in a practical problem consists in interpreting them in the parameter space.

In general, to interpret the absolute stability region, let us imbed a parallelepiped P into the convex region $\bar{R}$ determined by inequalities 8.183 which has sides perpendicular to the coordinate axes of the parameter space $\mathscr{P}$ and center at the known stable point $\bar{M}(\bar{p}_1; \bar{p}_2; \ldots ; \bar{p}_l)$.[57] Let the volume v of P be defined as

$$v = 2^l(p_1 - \bar{p}_1)(p_2 - \bar{p}_2) \cdots (p_l - \bar{p}_l). \quad (8.187)$$

Now the function v should be maximized with respect to each inequality 8.183 separately considered as a constraint. Thus a constraint

$$a_k(\mathbf{p}) = 0, \quad (8.188)$$

arising from (8.183), may be represented as

$$p_1 = p_1(p_2, p_3, \ldots, p_l). \quad (8.189)$$

Substituting (8.189) into (8.187) and extremizing, a necessary condition[58]

---

[57] If the region $\bar{R}$ is not convex, see footnote 5 in Section D.5, Appendix D, in connection with equation D.38.
[58] The following theorem (found in H. Hancock, *Theory of Maxima and Minima*, Dover, New York, 1960) provides the necessary and sufficient conditions for locating the extrema of a function $v(\mathbf{p})$. Let $v(\mathbf{p})$, $a_k(\mathbf{p})$, ($k = 0, 1, \ldots, m$) be continuously differentiable function on R, and $\mathbf{p}$ is an l vector.
Let

$$F(\mathbf{p}) = v(\mathbf{p}) + \sum_{k=0}^{m} \lambda_k a_k(\mathbf{p}), \quad (i)$$

where the λ_k are unspecified constants.

for $(p_2^0, p_3^0, \ldots, p_l^0)$ to yield a maximum of v is that it be a solution of

$$\frac{\partial v}{\partial p_i} = 0, \quad (i = 2, 3, \ldots, l). \tag{8.190}$$

Let the solution $\mathbf{p} = \mathbf{p}^0$ occur at the maximum value of v under the constraints (8.183). Then the desired parallelepiped P is given as

$$P = \left\{ \mathbf{p} \,\middle|\, |p_i - \bar{p}_i| < \min_k |p_i - p_i^0|_k, \right.$$
$$\left. (i = 1, 2, \ldots, l; k = 0, 1, \ldots, n) \right\}. \tag{8.191}$$

Since each vertex point of P is located in R containing the stable point $\bar{M}(\bar{p}_1; \bar{p}_2; \ldots; \bar{p}_l)$, it follows that the parallelepiped P is completely imbedded in $\bar{R}$, that is, $P \subset \bar{R}$.

In case of the above specific example let us choose the stable point $\bar{M}(0.2; 0; 0)$. The volume to be maximized is

$$v = 8(p_1 - 0.2)p_2 p_3. \tag{8.192}$$

Maximization of v with respect to the constraint

$$49p_1 + 11p_2 - 6p_3 - 6 = 0 \tag{8.193}$$

yields $p_1^0 = 0.178$, $p_2^0 = -0.123$, $p_3^0 = 0.226$. According to these values of parameters, the parallelepiped P is determined by

$$|p_1 - 0.2| < 0.022, \quad |p_2| < 0.123, \quad |p_3| < 0.226. \tag{8.194}$$

A *necessary condition* for an extremum of $v(\mathbf{p})$ to exist subject to $a_k(\mathbf{p}) = 0$, $(k = 0, 1, \ldots, m)$ at $\mathbf{p} = \mathbf{p}^0$ is that a solution to the $l + m + 1$ equations

$$\frac{\partial F}{\partial p_i} = 0, \quad (i = 1, 2, \ldots, l) \tag{ii}$$

$$a_k = 0, \quad (k = 0, 1, \ldots, m)$$

exist at $\mathbf{p} = \mathbf{p}^0$.

Let

$$V = \mathbf{p}^T \left(\frac{\partial^2 F}{\partial p_i \partial p_j} \bigg|_{\mathbf{p}=\mathbf{p}^0} \right) \mathbf{p} \tag{iii}$$

and define

$$U = \left\{ \mathbf{p} \,\middle|\, \sum_{i=1}^{l} \left(\frac{\partial a_k}{\partial p_i} \bigg|_{\mathbf{p}=\mathbf{p}^0} \right) p_i = 0, \quad k = 0, 1, \ldots, m \right\}. \tag{iv}$$

A *sufficient condition* for an extremum of $v(\mathbf{p})$ to occur at $\mathbf{p} = \mathbf{p}^0$ is that V is sign-definite when the equation defining U is eliminated from V. If V is negative-definite when the equation defining U is eliminated from V, then $v(\mathbf{p})$ has a maximum at $\mathbf{p} = \mathbf{p}^0$ subject to $a_k(\mathbf{p}) = 0$, $(k = 0, 1, \ldots, m)$; and if V is positive-definite when the equation defining U is eliminated from V, $v(\mathbf{p})$ has a minimum of $\mathbf{p} = \mathbf{p}^0$ subject to $a_k = 0$, $(k = 0, 1, \ldots, m)$.

We can readily check that all the vertex points of P satisfy the rest of the constraints of (8.186).[59] Therefore (8.194) is the solution of the interpretation problem under consideration. It may be concluded that the corresponding system motion is asymptotically stable in the large for all nonlinearities $F(y)$ in the sector [0, 1] and parameters $(p_1, p_2, p_3) \in P$ determined by (8.194). Thus the parallelepiped P represents the solution of both the stability and the sensitivity problems stated at the beginning of this section.

Absolute Stability Test: It is shown above that absolute stability can be concluded without plotting any curves. It is sufficient to verify that the coefficients a_{2k} of the Popov polynomial Π satisfy inequalities 8.183. This test is particularly simple if the coefficients a_{2k} are given numerically, that is, when the sector, system parameters, and number q are specified. However, the conditions 8.183 are only sufficient for the Popov inequality to be satisfied. Hence, it is desirable to have a simple analytic test which will provide both necessary and sufficient conditions for satisfaction of the Popov inequality. It will be shown here how after simple transformation of the Popov polynomial, the straightforward application of the well known Routh test [8.53] may be used to verify the Popov inequality and determine absolute stability.

Note that the Popov inequality is satisfied if and only if the Popov polynomial

$$\Pi(\omega) \equiv \sum_{k=0}^{n} a_{2k}\omega^{2k} \qquad (8.195)$$

has no positive real zeros. Polynomial $\Pi(\omega)$ is an even polynomial with real coefficients and, therefore, has $2n$ zeros symmetrically distributed with respect to both the real and imaginary axis of the ω plane. Now, let us consider the polynomial

$$\Pi(j\omega) \equiv \sum_{k=0}^{n} (-1)^k a_{2k}\omega^{2k}, a_0 > 0 \qquad (8.196)$$

and conclude that the previous symmetry is preserved but real zeros of $\Pi(\omega)$ (if there are any) became pure imaginary zeros of $\Pi(j\omega)$. Consequently, if $\Pi(j\omega)$ has n zeros with positive real parts, there are no positive real zeros of $\Pi(\omega)$ and the Popov inequality is satisfied.

Note that the polynomial $\Pi(j\omega)$ is an even polynomial and the Routh test cannot be applied directly since the second row of the Routh array is identically zero. This is known as a special case of the Routh algorithm and is

---

[59] It should be noted that some of the constraints in (8.183) may not contain all the parameters, as is clear from inequalities 8.186. Then some of the parameters in incomplete inequalities are arbitrary, and to make the maximization of v meaningful, we should consider the arbitrary parameters as constant.

For example, the optimization of v in (8.192) with respect to the constraints $14p_1 - p_2 + 6 \geq 0$ of (8.186) should be performed with $p_3 = c$ ($c \neq 0$). Then the maximization of $v = 8c(p_1 - 0.2)p_2$ gives $p_1^0 = -0.122$, $p_2^0 = 4.292$. In applying (8.191) to determine the parallelepiped P, these values are discarded and P is given by (8.194).

treated in the usual manner [8.53]. It is only necessary to form the second row using the coefficients of the derivative of $\Pi(j\omega)$ and continue the test.

Therefore, the Routh array in this case is

$$\begin{array}{c|cccc} \omega^{2n} & (-1)^n a_{2n} & (-1)^{n-1} a_{2(n-1)} & \cdots & -a_2 \quad a_0 \\ \omega^{2n-1} & (-1)^n 2n a_{2n} & (-1)^{n-1} 2(n-1) a_{2(n-1)} & \cdots & -2a_2 \\ \vdots & \vdots & \vdots & \vdots & \\ \omega^0 & a_0 & & & \end{array} \qquad (8.197)$$

and the following result is relevant.

Theorem 8.9 *Absolute Stability Test.* For the Popov inequality

$$\pi(\omega) \equiv \frac{1}{k} + \mathrm{Re}\,(1 + jq\omega) G(\sigma + j\omega) > 0, \qquad \forall \omega \geq 0 \qquad (8.112)$$

to be satisfied it is necessary and sufficient that the coefficients a_{2k} of the corresponding Popov polynomial

$$\Pi(\omega) \equiv \sum_{k=0}^{n} a_{2k} \omega^{2k} \qquad (8.195)$$

yield the first column of the Routh array (8.197)

$$(-1)^n a_{2n}, \quad (-1)^n 2n a_{2n}, \ldots, a_0$$

with exactly n sign changes and $a_0 > 0$.

As is well known [8.53], the first two rows of the Routh array

$$\begin{array}{cccc} a_n & a_{n-2} & a_{n-4} & \cdots \\ a_{n-1} & a_{n-3} & a_{n-5} & \cdots \\ b_0 & b_1 & b_2 & \cdots \\ c_0 & c_1 & c_2 & \cdots \\ \cdots & \cdots & \cdots & \end{array} \qquad (8.198)$$

are formed by the coefficients of the polynomial $\sum_{k=0}^{n} a_k \omega^k$ to be tested. The next rows are obtained by the recursive algorithm

$$b_0 = \frac{a_{n-1}a_{n-2} - a_n a_{n-3}}{a_{n-1}}, \quad b_1 = \frac{a_{n-1}a_{n-4} - a_n a_{n-5}}{a_{n-1}}, \quad \cdots$$

$$c_0 = \frac{b_0 a_{n-3} - a_{n-1} b_1}{b_0}, \quad c_1 = \frac{b_0 a_{n-5} - a_{n-1} b_2}{b_0}, \quad \cdots \qquad (8.199)$$

which is self-explanatory. Two special cases may occur. If any number in the first column of (8.198) becomes zero the algorithm cannot be continued. To remedy this situation the original polynomial should be multiplied by

($\omega + c$) where $c > 0$ is a constant. This multiplication does not change the number of the zeros with positive real parts, but the Routh criterion can be continued. Secondly, an entire row may become a row of zeros. To overcome this difficulty, it is necessary to replace the row of zeros by coefficients of the first derivative of the polynomial formed by the coefficients of the preceding row as demonstrated in (8.197).

As an example, let us prove that the point $\bar{M}(0.2; 0; 0)$ in the previous section corresponds to a stable system. The corresponding Popov polynomial (8.185) is

$$\Pi(\omega) = 0.2\omega^6 + 8.8\omega^4 + 3.8\omega^2 + 7.2 \tag{8.200}$$

and the Routh array (8.197) is

$$
\begin{array}{c|cccc}
\omega^6 & -0.2 & 8.8 & -3.8 & 7.2 \\
\omega^5 & -1.2 & 35.2 & -7.6 & \\
\omega^4 & 3 & -2.54 & 7.2 & \\
\omega^3 & 34.15 & -4.62 & & \\
\omega^2 & -2.14 & 7.2 & & \\
\omega^1 & 101.38 & & & \\
\omega^0 & 7.2 & & & \\
\end{array}
\tag{8.201}
$$

The sequence of the first column

$$-0.2, -1.2, 3, 34.15, -2.14, 101.38, 7.2$$

has exactly $n = 3$ sign changes and the corresponding system is absolutely stable.

Apparently, the absolute stability test just described requires that the sector and parameter q in the Popov inequality be specified numerically. This is a disadvantage with respect to the graphical solution unless the Popov inequality prescribes the sector and $q = 0$ as is often the case. Then the simple analytical test is far better than the graphical procedure since computer solution is readily available.

8.15 Linearization. Aizerman's Conjecture

In the previous chapters the harmonic linearization (describing function) has been used to study the stability and periodic solutions of nonlinear systems. The linearization essentially implies an approximation of the nonlinear characteristic $F(y)$ by a linear function $N_1 y$ in which the slope N_1 depends on the amplitude A of the signal at the input of the nonlinearity; that is,

$$F = N_1 y, \tag{8.202}$$

where $N_1 = N_1(A)$ is determined by

$$N_1(A) = \frac{1}{\pi A} \int_0^{2\pi} F(A \sin \phi) \sin \phi \, d\phi \qquad (8.203)$$

with assumption that $y = A \sin \phi$, $\phi = \Omega t$.

The stability analysis is then performed by a linear technique verifying that for the entire variation of the coefficient N_1 the system considered as linear remains stable. The entire variation of the slope N_1 of the assumed linear characteristic will define a sector in the yF plane that may be called the *describing function sector*. The sector in which the corresponding linear system with

$$F = Ky \qquad (8.204)$$

is stable for all values of K inside the open interval (k_1, k_2) is called the *Hurwitz sector*. By the rule above, the nonlinear system is supposed to be stable whenever $k_1 < N_1(A) < k_2$ for $0 < A < \infty$; that is, the describing function sector is contained inside the Hurwitz sector.

The harmonic linearization being an approximate approach may fail to indicate instability, as will be shown in the following example. Attempt has been made by Aizerman [8.45] to improve the linearization by introducing the *total gain* concept. He asked whether a nonlinear system is absolutely stable if the nonlinear characteristic is situated in the Hurwitz sector. This is evidently a stronger requirement than that of the describing function because the maximum of N_1 is less or at the most equal to the maximum of the total gain defined as

$$K(F, y) = \frac{F(y)}{y}. \qquad (8.205)$$

This can be observed from the definition (8.203) of N_1, which is an averaging process applied to the total gain $K(F, y)$. Pliss [8.46] however, has shown that this Aizerman conjecture is not true. He found a system that satisfies the condition of the Aizerman conjecture but exhibits periodic oscillations (limit cycle) and is unstable. Dewey and Jury [8.47] gave another specific counterexample to the Aizerman conjecture with numerical values and a given nonlinear characteristic.

Recently Fitts [8.48] discovered counterexamples to Aizerman's conjecture that have arisen from an investigation of the frequency domain stability criteria for nonlinear systems. They are also counterexamples to Kalman's conjecture [8.49], which requires that nonlinear systems have the entire variation of the *incremental gain*

$$K(F, y) = \frac{dF(y)}{dy} \qquad (8.206)$$

inside the Hurwitz sector to be ASIL.

Before presenting the Fitts example, let us note that the Popov stability theorem cannot be used to disprove the Aizerman conjecture because it gives only sufficient and not necessary conditions. It has been found however, that for certain subclasses of nonlinear systems discussed in this chapter the Popov theorem provides also the necessary conditions for ASIL, and the Aizerman conjecture is valid [8.50, 51]. (See Problem 8.19.) In other words, the *Popov sector* in which the nonlinear system is absolutely stable coincides with the Hurwitz sector. If the characteristics that asymptotically approach the straight lines bounding the Hurwitz sector are excluded, it was shown that the two sectors coincide for all second-order systems.

It is then clear that to disprove the Aizerman conjecture it is necessary to find a system that exhibits a periodic solution but satisfies conditions of the conjecture. As a counterexample to the conjecture, Fitts considered the system of Figure 8.13 with

$$G(s) = \frac{s(s + a)}{[(s + b)^2 + 0.9^2][(s + b)^2 + 1.1^2]} \quad (8.207)$$

and

$$F(y) = ky^3. \quad (8.208)$$

Let us assume that in $G(s)$ the parameter $b = 0.01$, and let us analyze the system using a linearization so that $F(y) = Ky$. The characteristic equation is

$$s^4 + 0.04s^3 + (K + 2.02)s^2 + (Ka + 0.04)s + 0.98 = 0. \quad (8.209)$$

Denoting $\alpha = K$, $\beta = a$, we obtain the parameter plane diagram of the $\zeta = 0$ curve as shown in Figure 8.27. Apparently, if the parameter a is set to zero, the system should be stable for all linear gains K inside the interval $[0, \infty]$. This means that the Hurwitz sector consists of the entire first and

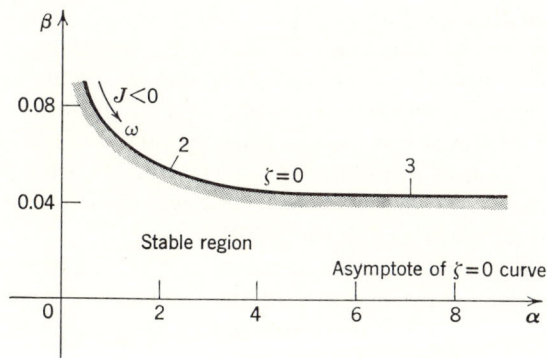

Figure 8.27 Parameter plane diagram.

third quadrants. Thus the conditions of the Aizerman conjecture are satisfied and the system should be ASIL.

By analog computer simulation of the above system Fitts showed that sustained oscillations existed for parameter values given under the three cases shown in Table 8.1. The underlined limits in the table were imposed by the computer and not by the fact that the system failed to oscillate. To start the oscillations, it was necessary to choose rather large initial conditions. However, the shape of the nonlinearity did not affect the existence of the oscillations. Other monotonic nonlinearities were tried, such as the quadratic $F(y) = y|y|$ and a piecewise linear characteristic with the dead-zone $D = 1$ and the slope of the linear part $k = 1$, and the system did not fail to oscillate.

Table 8.1 Parameter Ranges for Oscillation of the Fourth-Order System

Case	k	a	b
1	$0.1 \leq k \leq 1000$	0	0.01
2	10	$0 \leq a \leq 0.02$	0.01
3	10	0	$0.01 \leq b \leq 0.75$

Depending upon the initial conditions and the damping of the poles in the transfer function (8.203), the system exhibited different modes of oscillations as illustrated in Figure 8.28.

All these different fourth-order systems of Table 8.1 represent counterexamples to Aizerman's conjecture, since their Hurwitz sector contains the first and third quadrants (together with the y axis), in which the entire nonlinearity is situated. The systems are also counterexamples to Kalman's conjecture.

Despite the fact that counterexamples to Aizerman's and Kalman's conjectures are found, the mechanism of instability is still unclear. We did not discover the physical and mathematical reasons for the failure of linearization to predict unstable behavior of a nonlinear system. Even the cases where the Popov condition is both necessary and sufficient and where the linearization gives a valid answer to question of stability did not contribute effectively to the general problem of instability [8.50, 51]. Further attempts were made to use the perturbation theory in order to discover the reasons for a possible failure of the linearization techniques [8.52].

With respect to the application of harmonic linearization to the above counterexample, it is of interest to plot the frequency characteristic of the linear part represented by the transfer function $G(s)$ and verify the validity of the harmonic linearization in this case. The characteristic is shown in Figure 8.29 and apparently does not encourage the use of the linearization

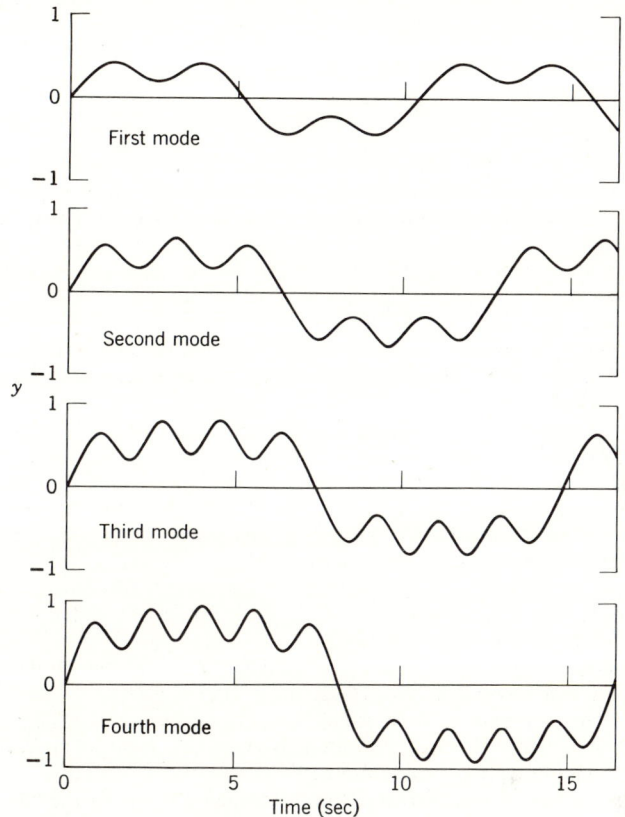

Figure 8.28 The first four oscillation modes in the system with the parameter values in Case 1, Table 8.1, when $k = 10$.

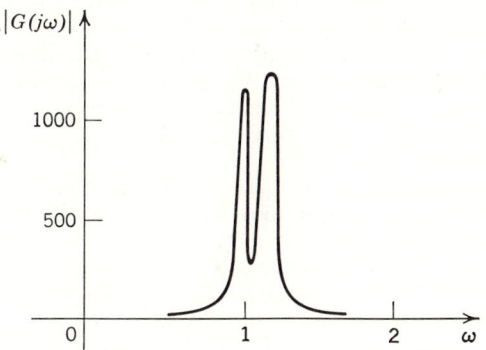

Figure 8.29 The frequency characteristic.

based upon the filter hypothesis that the linear part following the nonlinearity is of a low-pass nature.

References

[8.1] M. A. Liapunov, Problème général de la stabilité du mouvement (in French), *Commun. Soc. Math. Kharkow*, 1893. (American edition of a 1907 French translation: Annals of Mathematics Study, No. 17, Princeton University Press, Princeton, New Jersey, 1949).

[8.2] I. G. Malkin, *Theory of Stability of Motion* (in Russian), GOSTEKHIZDAT, Moscow, 1952. (English translation: AEC Translation 3352, Department of Commerce, Washington, D.C., 1958.)

[8.3] A. M. Letov, *Stability of Nonlinear Regulating Systems* (in Russian), GOSTEKHIZDAT, Moscow, 1955. (English translation: Princeton University Press, Princeton, New Jersey, 1961).

[8.4] Ya. Z. Tsypkin, *Theory of Automatic Regulating Relay Systems* (in Russian), GOSTEKHIZDAT, Moscow, 1955.

[8.5] J. L. Massera, Contributions to Stability Theory, *Ann. Math.*, **64**, 182–206 (1956).

[8.6] H. A. Antosiewicz, A Survey of Liapunov's Second Method, *Ann. Math. Stud.*, Vol. 41. *Contributions to the Theory of Nonlinear Oscillations*, Vol. 4, Princeton University Press, Princeton, New Jersey, 1958.

[8.7] W. Hahn, Theorie und Anwendungen der Direkten Methode von Liapunov, *Eng. der Math.*, No. 22, Springer, Berlin, 1959. (English translation: Prentice-Hall, Englewood Cliffs, New Jersey, 1963.)

[8.8] L. Cesari, *Asymptotic Behavior and Stability Problems in Ordinary Differential Equations, Eng. der Math.*, No. 16, Springer, Berlin, 1959.

[8.9] V. V. Nemytskii and V. V. Stepanov, *Qualitative Theory of Differential Equations*, Princeton University Press, Princeton, New Jersey, 1960. (Translation of a 1951 Russian edition with some additions.)

[8.10] R. E. Kalman and J. E. Bertram, Control System Design via the "Second Method" of Liapunov—Pt. I: Continuous-time Systems, *J. Basic Eng., Trans. ASME*, **82**, 371–393 (June 1960).

[8.11] J. LaSalle and S. Lefschetz, *Stability by Liapunov's Direct Method with Applications*, Academic, New York, 1961.

[8.12] L. Zadeh and C. A. Desoer, *Linear System Theory: The State Space Approach*, McGraw-Hill, New York, 1963.

[8.13] R. W. Brockett, The Status of Stability Theory for Deterministic Systems (a survey paper), *IEEE Trans. Auto. Control*, **AC-11**, 596–607 (July 1966).

[8.14] A. I. Lur'e, Some Nonlinear Problems in the Theory of Automatic Control, (in Russian), GOSTEKHIZDAT, Moscow, 1951. (English translation: Her Majesty's Stationery Office, London, 1957.)

[8.15] M. A. Aizerman and F. R. Gantmacher, *Absolute Stability of Regulator Systems* (in Russian), *Acad. Sci. USSR*, Moscow, 1963. (English translation: Holden-Day, San Francisco, 1964.)

[8.16] F. R. Gantmacher and V. A. Yakubovich, Absolute Stability of Nonlinear Regulator Systems (in Russian), *Proceedings of Trudi II Vsesoyuznogo Sezda po Teoreticheskoi i Prikladnoy Mekhanike*, Nauka, Moscow, January 29–February 5, 1965.

[8.17] S. Lefschetz, *Stability of Nonlinear Control Systems*, Academic, New York, 1965.

[8.18] J. E. Gibson, *Nonlinear Automatic Control*, McGraw-Hill, New York, 1963.

[8.19] P. DeRusso, R. Roy, and R. Close, *State Variables for Engineers*, Wiley, New York, 1965.

[8.20] V. A. Yakubovich, Solution of Certain Matrix Inequalities Encountered in the Theory of Automatic Control (in Russian), *Dokl. Akad. Nauk SSSR*, **143**, No. 6, 1304–1307 (1962).

[8.21] V. A. Yakubovich, Absolute Stability of Nonlinear Control Systems in Critical Cases (in Russian), *Avtomatika i Telemekhanika*, Pt. I: **24**, No. 3, 293–303 (March 1963); Pt. II: **24**, No. 6, 717–731 (June 1963); Pt. III: **25**, No. 5, 601–612 (May 1964).

[8.22] V. A. Yakubovich, The Method of Matrix Inequalities in the Stability Theory of Nonlinear Control Systems, Pt. I—Absolute Stability of Forced Oscillations (in Russian), *Avtomatika i Telemekhanika*, **25**, No. 7, 1017–1029 (July 1964).

[8.23] V. A. Yakubovich, The Method of Matrix Inequalities in the Stability Theory of Nonlinear Control Systems, Pt. II—Absolute Stability in a Class of Nonlinearities with a Condition on the Derivative (in Russian), *Avtomatika i Telemekhanika*, **26**, No. 4, 577–590 (April 1965).

[8.24] D. D. Šiljak, Popov Inequality via Parameter Plane, *Proceedings of the First Princeton Conference on Information Sciences and Systems*, 183–187 (March 1967).

[8.25] D. D. Šiljak, Absolute Stability in the Parameter Space, *Proceedings of the First Asilomar Conference on Circuit and Systems*, Asilomar, California, 624–632 (November 1967).

[8.26] R. E. Kalman, C. Y. Ho, and K. S. Narendra, Controllability of Linear Dynamical Systems, *Contrib. Differential Equations*, **1**, 189–213 (1963).

[8.27] C. A. Desoer and R. Liu, Linearity vs. Nonlinearity and Asymptotic Stability in the Large, *IEEE Trans. Circuit Theory*, **CT-12**, No. 1, 117–118 (March 1965).

[8.28] J. K. Hale, *Oscillations in Nonlinear Systems*, McGraw-Hill, 1963.

[8.29] A. F. Filippov, Differential Equations with Discontinuous Right-Hand Sides, *Mat. Sb.*, Vol. 51(93), No. 1, 1960.

[8.30] A. F. Filippov, Application of the Theory of Differential Equations with Discontinuous Right-Hand Sides to Nonlinear Problems in Automatic Control, *Proc. 1st Inter. Congr. of IFAC*, **1**, 1098–1100 (1960) Moscow (published by Butterworths, London).

[8.31] J. André and P. Seibert, Uber stuckweise lineare Differentalgleichungen die bei Regelungsproblemen auftreten, *I. Arch. Math.*, **7**, 148–156 (1956).

[8.32] V. I. Zubov, *The Methods of A. M. Liapunov and their Applications* (in Russian), Press of Leningrad State University, Leningrad, 1957.

[8.33] Ya. Z. Tsypkin, Absolute Stability of Equilibrium Position and Transients in Nonlinear Sampled-Data Systems (in Russian), *Avtomatika i Telemekhanika*, **24**, No. 12, 1601–1615 (December, 1963).

[8.34] B. N. Naumov and Ya. Z. Tsypkin, A Frequency Criterion for Absolute Stability in Nonlinear Automatic Control Systems (in Russian), *Avtomatika i Telemekhanika*, **25**, No. 6, 852–867 (June 1964).

[8.35] V. M. Popov, Absolute Stability of Nonlinear Systems of Automatic Control (in Russian), *Avtomatika i Telemekhanika*, **22**, No. 8, 961–979 (August 1961).

[8.36] R. E. Kalman, Lyapunov Functions for the Problem of Lur'e in Automatic Control, *Proc. Nat. Acad. Sci. U.S.*, **49**, No. 2, 201–205 (February 1963).

[8.37] E. N. Rozenvasser, The Absolute Stability of Nonlinear Systems (in Russian), *Avtomatika i Telemekhanika*, **24**, No. 3, 304–313 (March 1963).

[8.38] W. T. Higgins, Jr. and D. G. Shultz, The Stability of Certain Nonlinear Time-Varying Systems, Control System Laboratory, Report No. CSL-66-2, University of Arizona, Tucson, Arizona, May 1966.

[8.39] J. J. Bongiorno, Jr., An Extension of the Nyquist-Barkhausen Stability Criterion to Linear Lumped-Parameter Systems with Time-Varying Elements, *IEEE Trans. Auto. Control*, **AC-8**, No. 2, 166–170 (April 1963).

[8.40] I. W. Sandberg, A Frequency-Domain Condition for the Stability of Feedback Systems Containing a Single Time-Varying Nonlinear Element, *Bell System Tech. J.*, **43**, Pt. 2, 1601–1608 (July 1964).

[8.41] K. S. Narendra and R. M. Goldwyn, A Geometrical Criterion for the Stability of Certain Nonlinear Nonautonomous Systems, *IEEE Trans. Circuit Theory*, **CT-11**, No. 3, 406–408 (September 1964).

[8.42] G. Zames, On the Input-Output Stability of Time-Varying Nonlinear Feedback Systems—Part II: Conditions Involving Circles in the Frequency Plane and Sector Nonlinearities, *IEEE Trans. Auto. Control*, **AC-11**, No. 3, 465–476 (July 1966).

[8.43] P. V. Kokotović and R. S. Rutman, Sensitivity of Automatic Control Systems (Survey) (in Russian), *Avtomatika i Telemekhanika*, **26**, No. 4, 730–750 (April 1965).

[8.44] I. Gumowski, Sensitivity Analysis and Lyapunov Stability, *Proceedings of the First Dubrovnik Symposium on Sensitivity Analysis*, Pergamon, 1966, pp. 2–35.

[8.45] M. A. Aizerman, On the Effect of Nonlinear Functions of Several Variables on the Stability of Automatic Control Systems (in Russian), *Avtomatika i Telemekhanika*, **8**, No. 1, 64–72 (January 1947).

[8.46] V. A. Pliss, Certain Problems in the Theory of Stability in the Whole (in Russian), Leningrad University Press, Leningrad, 1958.

[8.47] A. G. Dewey and E. I. Jury, A Note on Aizerman's Conjecture, *IEEE Trans. Auto. Control*, **AC-10**, No. 4, 482–483 (October 1965).

[8.48] R. E. Fitts, Two Counterexamples to Aizerman's Conjecture, *IEEE Trans. Auto. Control*, **AC-11**, No. 3, 553–556 (July 1966).

[8.49] R. E. Kalman, Physical and Mathematical Mechanisms of Instability in Nonlinear Automatic Control Systems, *J. Basic Eng., Trans. ASME*, **79**, 553–556 (April 1957).

[8.50] R. W. Brockett and J. L. Williams, Frequency Domain Stability Criteria—Pts. I and II, *IEEE Trans. Auto. Control*, **AC-10**, No. 3 and 4, 255–261, 407–413 (July and October 1965).

[8.51] G. Schmidt and G. Preusche, Popov Stability Criterion as a Tool in Verification of Aizerman's Conjecture (in German), *Regelungs technik*, **15**, No. 1, 20–25 (1967).

[8.52] J. C. Williams, Perturbation Theory for the Analysis of Instability in Nonlinear Feedback Systems, *Proceedings of the Fourth Allerton Conference on Circuit and System Theory*, University of Illinois, Urbana, Illinois, 836–848 (1966).

[8.53] F. R. Gantmacher, *The Theory of Matrices*, Vol. II, Chelsea Publ. Co., New York, 1960.

[8.54] D. D. Šiljak, On Absolute Stability and Sensitivity, *Second IFAC Symposium on System Sensitivity and Adaptivity*, Dubrovnik, Yugoslavia, August, 1968.

[8.55] D. D. Šiljak and S. Weissenberger, Regions of Exponential Stability for the Problem of Lur'e, *Network Theory Symposium*, Belgrade, Yugoslavia, September, 1968.

[8.56] D. D. Šiljak, Absolute Stability and Parameter Sensitivity, *Int. J. Control* (to be published in 1968).

APPENDIX A

Basic Theorems and Derivations

Introduction

This section gives various theorems and derivations related to algebraic equations and polynomials to which reference will be made in the development of the parameter plane mapping. Moreover, some of the theorems and derivations have practical value in the application of the parameter mapping to actual design problems. For additional study of the algebraic equations, references [A.1–4] may be used.

The mathematical object of investigation is a function $F(s)$ of the form

$$F(s) \equiv \sum_{k=0}^{n} a_k s^k = a_n s^n + a_{n-1} s^{n-1} + \cdots + a_1 s + a_0, \qquad (A.1)$$

where $a_0, a_1, \ldots, a_n$ are real or complex numbers and s is the complex variable. The function $F(s)$ so defined is called a *polynomial in s*. The quantities $a_k (k = 0, 1, \ldots, n)$ are called *coefficients*. If $a_n \neq 0$, the polynomial is of degree n. The monomials $a_n s^n, a_{n-1} s^{n-1}, \ldots, a_1 s, a_0$ are denoted as *terms* of the polynomial $F(s)$. The term $a_n s^n$ is called the leading term and the term a_0, which does not contain s, is called the absolute term. The terms with coefficients equal to zero are normally omitted, and their number is not limited except that a polynomial $F(s)$ must involve the variable s. As a convention, a constant different from zero may be considered as a polynomial of degree zero. A polynomial identically equal to zero (all of whose coefficients are equal to zero) is called an *identically vanishing polynomial* and is replaced by zero. Normally no degree is attributed to such a polynomial. For formal reasons, the degree of identically vanishing polynomials is sometimes set equal to minus infinity. A polynomial is said to be *complete* when it contains all the terms from zero to n, and *incomplete* when it has some coefficients equal to zero in terms of that range. Any value of the variable s that makes the polynomial vanish is called the *zero of the polynomial*. Two polynomials are called *equal* if they are identical term for term.

A polynomial $F(s)$ defined in equation A.1 is a *rational* and *integral* (or *entire*) function of s since the variable s enters the function in a rational and integral form only. The polynomial $F(s)$ is therefore an analytic function of the complex variable s in the entire s plane. Consequently, when the complex variable s is expressed as

$$s = \sigma + j\omega, \tag{A.2}$$

where σ and ω are real numbers and $j^2 = -1$, and the polynomial $F(s)$ is rewritten as

$$F(s) = R + jI, \tag{A.3}$$

the real part R and the imaginary part I satisfy the Cauchy-Riemann conditions

$$\frac{\partial R}{\partial \sigma} = \frac{\partial I}{\partial \omega}, \qquad \frac{\partial R}{\partial \omega} = -\frac{\partial I}{\partial \sigma} \tag{A.4}$$

in the entire s plane.

The relation

$$F(s) \equiv \sum_{k=0}^{n} a_k s^k = 0, \tag{A.5}$$

in which the polynomial is equated to zero, is called the *algebraic equation*, and any value of s that satisfies the equation is called a *root* of the equation. The collection of all possible roots constitutes the *complete solution of the algebraic equation*. The *classical algebraic* problem consists in evaluating the complete solution of the given algebraic equation when its coefficients are given in the numerical form. This is considered in the following theorem.

Theorem A.1. Fundamental Theorem of Algebra. If $F(s)$ is a polynomial of degree n, $(n \geq 1)$, with real or complex coefficients, then the equation $F(s) = 0$ has at least one root.

The proof of this theorem is displayed in almost all mathematical literature concerned with the theory of equations, or functions, of the complex variable; for example, in references A.1–7. However, it has been found unavoidable to present the proof here since the parameter mapping is inherently a mapping through the algebraic equations. Moreover, the idea of the proof will be used to develop the computer scheme for numerical calculations of the root values.

Historically, a proof of the theorem was first given in 1746 by D'Alambert and later improved by Weierstrass. The first complete proof of the fundamental theorem was presented by Gauss at the beginning of the past century.

Fundamental Theorem of Algebra

By its nature the fundamental theorem belongs rather to analysis than to algebra, and the application of Liouville's theorem [A.6, 7] leads to an almost trivial proof. The proof of the fundamental theorem given here, however, follows that shown in reference A.5 (Volume 1, Chapter 3), where nonalgebraic elements are reduced to a minimum.

Let us consider a polynomial $F(s)$ of degree n,

$$F(s) \equiv \sum_{k=0}^{n} a_k s^k, \tag{A.1}$$

where the coefficients $a_k = b_k + jc_k$, $(k = 0, 1, \ldots, n)$, are complex numbers. By substituting $s = \sigma + j\omega$ into (A.1), it can be rewritten as $F(s) = R + jI$, where

$$R \equiv \sum_{k=0}^{n} (b_k X_k - c_k Y_k)$$
$$I \equiv \sum_{k=0}^{n} (c_k X_k + b_k Y_k). \tag{A.6}$$

Functions $X_k = X_k(\sigma, \omega)$, $Y_k = Y_k(\sigma, \omega)$ are defined by the recurrence formulas

$$X_{k+1} - 2\sigma X_k + (\sigma^2 + \omega^2) X_{k-1} = 0$$
$$Y_{k+1} - 2\sigma Y_k + (\sigma^2 + \omega^2) Y_{k-1} = 0, \tag{A.7}$$

where $X_0 \equiv 1$, $X_1 \equiv \sigma$, $Y_0 \equiv 0$, $Y_1 \equiv \omega$. We then have

$$|F(s)| = (R^2 + I^2)^{1/2}. \tag{A.8}$$

First, it is to be shown that $|F(s)|$, or, what amounts to the same thing, $R^2 + I^2$, cannot be at a minimum value for $\sigma = \omega = 0$ unless $b_0 = c_0 = 0$. For this purpose, the substitution $s = \omega_n(\cos\theta + j\sin\theta)$ is used in (A.1), and it is assumed that the first coefficient after a_0 that does not vanish in (A.1) is a_p. Then (A.6) can also be written as

$$R = b_0 + \omega_n^p(b_p \cos p\theta - c_p \sin p\theta) + \cdots$$
$$I = c_0 + \omega_n^p(c_p \cos p\theta + b_p \sin p\theta) + \cdots \tag{A.9}$$

according to the Moivre formula

$$s^k = \omega_n^k(\cos\theta + j\sin\theta)^k = \omega_n^k(\cos k\theta + j\sin k\theta). \tag{A.10}$$

The terms not written out in (A.9) are of higher degree than p with respect to ω_n; then

$$R^2 + I^2 = b_0^2 + c_0^2 + 2\omega_n^p[(b_0 b_p + c_0 c_p)\cos p\theta$$
$$+ (c_0 b_p - b_0 c_p)\sin p\theta] + \cdots. \tag{A.11}$$

However, the equation

$$(b_0 b_p + c_0 c_p)\cos p\theta + (c_0 b_p + b_0 c_p)\sin p\theta = 0 \tag{A.12}$$

gives $\tan p\theta = \kappa$, which determines p straight lines, separated by angles each equal to $2\pi/p$. Therefore it is not possible that $R^2 + I^2$ has a minimum for $\sigma = \omega = 0$ unless the expressions

$$b_0 b_p + c_0 c_p, \quad c_0 b_p - c_p b_0$$

both vanish. However, since $b_p^2 + c_p^2 \neq 0$, this requires that $b_0 = c_0 = 0$, that is, $R = I = 0$ for $\sigma = \omega = 0$. When $|F(s)|$ has a minimum for $\sigma = \sigma_0$, $\omega = \omega_0$, the conclusion may be reduced to the preceding by substituting $s = \sigma_0 + j\omega_0 + s'$. Therefore, $|F(s)|$ cannot be at a minimum unless R and I go to zero independently for $\sigma = \sigma_0$, $\omega = \omega_0$.

The absolute value of $F(s)$ must have a minimum value for at least one value of s, for it increases indefinitely as the absolute value of s increases indefinitely. In other words, we have

$$R^2 + I^2 = (b_n^2 + c_n^2)\omega_n^{2n} + \cdots, \tag{A.13}$$

where the omitted terms are of a degree less than $2n$ in ω_n. Equation A.13 may be rewritten as

$$(R^2 + I^2)^{1/2} = \omega_n^n[(b_n^2 + c_n^2)^{1/2} + \varepsilon], \tag{A.14}$$

where ε approaches zero as ω_n increases indefinitely. Thus a circle may be constructed whose radius is so large that the value $(R^2 + I^2)^{1/2}$ is greater at every point of the circumference than it is at the origin, for example. We conclude that there is at least one point $\sigma = \sigma_0$, $\omega = \omega_0$ inside the circle for which $(R^2 + I^2)^{1/2}$ has a minimum. Furthermore, it follows that the point $\sigma = \sigma_0$, $\omega = \omega_0$ is a point of intersection of the two curves $R = 0$, $I = 0$, which amounts to saying that $s = \sigma_0 + j\omega_0$ is a root of the equation $F(s) = 0$.

In the above proof it was assumed that a function of the two variables, σ and ω, that is continuous in the interior of a limited region assumes a minimum value inside or on the boundary of that region. The proof of this statement is necessary although it is intuitively evident. The proof is omitted here and can be found elsewhere, for example, in reference A.5 (Volume 1, Chapter 6).

After we have proved the fundamental theorem of algebra—that an algebraic equation has at least one root—a problem of the number of roots of an equation arises. This problem can be solved after we prove the following theorem.

Theorem A.2. Remainder Theorem. Synthetic Division. The remainder obtained in dividing $F(s)$ by $s - \lambda$ is the value of the polynomial $F(s)$ for $s = \lambda$ that is, $F(\lambda)$.

To prove this theorem, it is necessary first to consider the division of polynomials. Let

$$F(s) = \sum_{k=0}^{n} a_k s^k \qquad (A.15)$$

$$H(s) = \sum_{k=0}^{m} b_k s^k$$

be two polynomials of degrees n and m, respectively, and $n \geq m$. We can obtain a polynomial

$$F_1(s) = F(s) - \frac{a_n}{b_m} s^{n-m} H(s), \qquad (A.16)$$

which, if not vanishing identically, is of degree v smaller than n. Then setting

$$F_1(s) = \sum_{k=0}^{v} c_k s^k, \qquad (A.17)$$

we form the polynomial

$$F_2(s) = F_1(s) - \frac{c_v}{b_m} s^{v-m} H(s), \qquad (A.18)$$

which, if not vanishing identically, is of a degree smaller than v. Therefore we continue until we reach the polynomial

$$F_i(s) = F_{i-1} - \frac{d_\mu}{a_n} s^{\mu-m} H(s), \qquad (A.19)$$

which, if not vanishing identically, is of a degree smaller than m; and then, adding, we obtain

$$F(s) = \frac{a_n s^{n-m} + c_v s^{v-m} + \cdots + d_\mu s^{\mu-m}}{b_m} H(s) + F_k(s) \qquad (A.20)$$

or

$$F(s) = Q(s)H(s) + R. \qquad (A.21)$$

Such an operation is ordinary division of two polynomials, $F(s)$ and $H(s)$, giving $Q(s)$ as the quotient and R as the remainder.

Now, if the polynomial $H(s) = s - \lambda$, we have from (A.21)

$$F(s) = (s - \lambda)Q(s) + R, \qquad (A.22)$$

where the remainder R, being a function of s of lower degree than $s - \lambda$, is constant. By substituting $s = \lambda$ in (A.22), we obtain

$$F(\lambda) = (\lambda - \lambda)Q(\lambda) + R = R \qquad (A.23)$$

and the remainder theorem is thereby proved.

The quotient $Q(s)$ in the division can be found by a very convenient process known as *synthetic division*. Substituting from

$$Q(s) = \sum_{k=0}^{n-1} c_k s^k \qquad (A.24)$$

with degree one less than the degree of $F(s)$, we have

$$F(s) = \sum_{k=0}^{n-1} c_k s^{k+1} + R - \lambda \sum_{k=0}^{n-1} c_k s^k, \tag{A.25}$$

whence

$$a_k = c_k - \lambda c_{k-1}, \quad (k = 0, 1, \ldots, n) \tag{A.26}$$

and, consequently,

$$c_k = a_k + \lambda c_{k-1}, \quad (k = 0, 1, \ldots, n). \tag{A.27}$$

The calculation of the coefficients c_k of the quotient $Q(s)$ is of a recursive nature and can be arranged more conveniently in a scheme

$$\begin{array}{c} \lambda) \quad a_n \quad a_{n-1} \quad \cdots \quad a_1 \quad a_0 \\ + \quad \quad c_{n-1}\lambda \cdots c_1\lambda \quad c_0\lambda \\ \hline a_n = c_{n-1} \quad c_{n-2} \quad \cdots \quad c_0 \quad R \end{array}, \tag{A.28}$$

which is called the synthetic division scheme. Since the remainder is $F(\lambda)$, the synthetic division provides a convenient means of calculating the value of a polynomial for a given value of the variable $s = \lambda$. If, in turn, the value λ is a root of equation $F(s) = 0$, the synthetic division may be used to factor out the term $s - \lambda$ from the equation $F(s) = 0$. We solve for the remaining roots from the quotient equation $Q(s) = 0$, which has the degree one less than the degree of $F(s) = 0$. Besides this application, *in the parameter plane mapping this division may be used to check the points obtained for plotting the characteristic curves*. If, for certain values of σ and ω, the value of α and β are calculated, then, for these α and β equations, $F(s) = 0$ has a root $s = \sigma + j\omega$ and the remainder R has to vanish.

After the fundamental and remainder theorems are proved, it is almost trivial to derive the following:

$$\begin{aligned} F(s) &= (s - s_1)Q(s) \\ Q(s) &= (s - s_2)Q_1(s) \\ &\quad \vdots \\ Q_{n-2}(s) &= (s - s_n)Q_{n-1}(s), \end{aligned} \tag{A.29}$$

where $s_i (i = 1, 2, \ldots, n)$ are the roots of $F(s) = 0$, and consequently

$$F(s) = a_n(s - s_1)(s - s_2) \cdots (s - s_n), \tag{A.30}$$

which leads directly to the following conclusions:

1. A polynomial $F(s)$ of degree n is the product of n linear factors.
2. An algebraic equation $F(s) = 0$ of degree n has exactly n roots.

This answers the question about the number of roots of an algebraic equation.

In the parameter mapping the mapping functions are given by means of implicit equations. The existence of the solution of these equations is then of interest. Let us consider first the real root case and prove the following theorem [A.5].

Theorem A.3. Fundamental Implicit Theorem. Let $\alpha = \alpha_0$, $\beta = \beta_0$, $\sigma = \sigma_0$ be a set of values that satisfy the equation

$$F(\alpha, \beta, \sigma) = 0 \tag{A.31}$$

and let us suppose that the function F, together with its first derivatives, is continuous in the neighborhood of this set of values. If the derivative $\partial F/\partial \alpha$ does not vanish for these values, there exists one and only one continuous function $\alpha(\beta, \sigma)$ of the independent variables β and σ that satisfied (A.31), and that assumes the values α_0 when β and σ assume the values β_0 and σ_0, respectively.

Since $\partial F/\partial \alpha$ does not vanish for $\alpha = \alpha_0$, $\beta = \beta_0$, $\sigma = \sigma_0$, let us suppose, for definiteness, that it is positive. The function F and the derivatives $\partial F/\partial \alpha$, $\partial F/\partial \beta$, $\partial F/\partial \sigma$, being continuous in the neighborhood, let us choose a positive constant l sufficiently small so that these four functions are continuous for all values α, β, σ that satisfy the inequalities

$$|\alpha - \alpha_0| \leq l, \quad |\beta - \beta_0| \leq l, \quad |\sigma - \sigma_0| \leq l \tag{A.32}$$

and that, for these values of α, β, σ, we have $\partial F/\partial \alpha > P$, where P is a positive constant. Let Q be another positive constant so that $|\partial F/\partial \beta|$, $|\partial F/\partial \sigma| < Q$ in the same region.

Giving α, β, σ values that satisfy the inequalities (A.32), we may write the following relationship:

$$F(\alpha, \beta, \sigma) - F(\alpha_0, \beta_0, \sigma_0) = F(\alpha, \beta, \sigma) - F(\alpha, \beta, \sigma_0) + F(\alpha, \beta, \sigma_0) \\ - F(\alpha, \beta_0, \sigma_0) + F(\alpha, \beta_0, \sigma_0) - F(\alpha_0, \beta_0, \sigma_0). \tag{A.33}$$

Applying the law of mean to each of these differences, and observing that $F(\alpha_0, \beta_0, \sigma_0) = 0$,

$$F(\alpha, \beta, \sigma) = (\sigma - \sigma_0)\frac{\partial F}{\partial \sigma}[\alpha, \beta, \sigma_0 + \theta(\sigma - \sigma_0)] \\ + (\beta - \beta_0)\frac{\partial F}{\partial \beta}[\alpha, \beta_0 + \theta'(\beta - \beta_0), \sigma_0] \\ + (\alpha - \alpha_0)\frac{\partial F}{\partial \alpha}[\alpha_0 + \theta''(\alpha - \alpha_0), \beta_0, \sigma_0] \tag{A.34}$$

or

$$F(\alpha, \beta, \sigma) = A(\alpha, \beta, \sigma)(\sigma - \sigma_0) + B(\alpha, \beta, \sigma)(\beta - \beta_0)$$
$$+ C(\alpha, \beta, \sigma)(\alpha - \alpha_0), \qquad (A.35)$$

where $|A|$, $|B| < Q$, and $|C| > P$ for all sets of values of α, β, σ that satisfy (A.32). Now let ε be a positive number less than l, and η the smaller of the two numbers l and $P\varepsilon/2Q$. Let us suppose that β and σ in (A.31) have certain values that satisfy the conditions $|\beta - \beta_0|, |\sigma - \sigma_0| < \eta$, and that we seek the number of roots of that equation, α being regarded as the unknown, which lies between $\alpha_0 - \varepsilon$ and $\alpha_0 + \varepsilon$. In (A.35) for $F(\alpha, \beta, \sigma)$ the sum of the first two terms is always less than $2Q\eta$ in absolute value, and the absolute value of the third term is greater than $P\varepsilon$ if α is substituted by $\alpha_0 \pm \varepsilon$. According to the way η is chosen, it follows that this last term determines the sign of F. It is evident, then, that $F(\alpha_0 - \varepsilon, \beta, \sigma) < 0$ and $F(\alpha_0 + \varepsilon, \beta, \sigma) > 0$ and (A.31) have at least one root that lies between $\alpha_0 - \varepsilon$ and $\alpha_0 + \varepsilon$. Furthermore, this root is unique since the derivative $\partial F/\partial \alpha$ is positive for all values of α between $\alpha_0 - \varepsilon$ and $\alpha_0 + \varepsilon$. Consequently, (A.31) has one and only one root, and this root approaches α_0 as β and σ approach β_0 and σ_0, respectively. The above theorem is thereby proved.

When the complex roots of an algebraic equation are considered as the functions of the two parameters α and β, then the parameter mapping can be applied, provided that the conditions expressed by the following theorem are satisfied.

Theorem A.4. Parameter Mapping. Let us consider a system of two simultaneous equations

$$R(\alpha, \beta, \sigma, \omega) = 0 \qquad (A.36a)$$

$$I(\alpha, \beta, \sigma, \omega) = 0 \qquad (A.36b)$$

between variables $\alpha, \beta, \sigma, \omega$. Suppose that these equations are satisfied for the values $\alpha = \alpha_0, \beta = \beta_0, \sigma = \sigma_0, \omega = \omega_0$; that the functions R and I are continuous and possess first partial derivatives that are continuous in the neighborhood of this system of values; and, finally, that the determinant does

$$J\left(\frac{R, I}{\alpha, \beta}\right) = \begin{vmatrix} \dfrac{\partial R}{\partial \alpha} & \dfrac{\partial R}{\partial \beta} \\ \dfrac{\partial I}{\partial \alpha} & \dfrac{\partial I}{\partial \beta} \end{vmatrix} = \dfrac{\partial R}{\partial \alpha}\dfrac{\partial I}{\partial \beta} - \dfrac{\partial R}{\partial \beta}\dfrac{\partial I}{\partial \alpha} \qquad (A.37)$$

not vanish for $\alpha = \alpha_0, \beta = \beta_0, \sigma = \sigma_0, \omega = \omega_0$. Under these conditions, there exists one and only one system of continuous functions $\alpha = \alpha(\sigma, \omega), \beta = \beta(\sigma, \omega)$ that satisfies (A.36) and that reduces to α_0, β_0 for σ_0, ω_0.

Since the determinant J of (A.37), which is called the Jacobian determinant, or Jacobian for brevity, does not vanish, then at least one of the derivatives $\partial R/\partial \beta$, $\partial I/\partial \beta$ does not vanish for these same values $\alpha = \alpha_0$, $\beta = \beta_0$, $\sigma = \sigma_0$, $\omega = \omega_0$. Assume, for definiteness, that $\partial R/\partial \beta$ does not vanish. By applying the theorem proved above for the single equation A.31, we conclude that the equation A.36a defines a function

$$\beta = f(\alpha, \sigma, \omega), \tag{A.38}$$

which becomes β_0 for $\alpha = \alpha_0$, $\sigma = \sigma_0$, $\omega = \omega_0$. Substituting β in (A.36b), we obtain

$$H(\alpha, \sigma, \omega) = I[\alpha, f(\alpha, \sigma, \omega), \sigma, \omega] = 0. \tag{A.39}$$

Equation A.39 is satisfied for $\alpha = \alpha_0$, $\sigma = \sigma_0$, $\omega = \omega_0$. It follows that

$$\frac{\partial H}{\partial \alpha} = \frac{\partial I}{\partial \alpha} + \frac{\partial I}{\partial \beta}\frac{\partial f}{\partial \alpha}, \tag{A.40}$$

and from (A.36a) we obtain

$$\frac{\partial R}{\partial \alpha} + \frac{\partial R}{\partial \beta}\frac{\partial f}{\partial \alpha} = 0. \tag{A.41}$$

By combining (A.40) and (A.41), we get

$$\frac{\partial H}{\partial \alpha} = -\frac{J(R, I/\alpha, \beta)}{\partial R/\partial \beta}. \tag{A.42}$$

It is clear that the $\partial H/\partial \alpha$ does not go to zero for α_0, β_0, σ_0, ω_0, and $H = 0$ is satisfied when α is replaced by $\alpha = \alpha(\sigma, \omega)$, which is continuous and equal to α_0 when $\sigma = \sigma_0$, $\omega = \omega_0$. By substituting α by $\alpha(\sigma, \omega)$ in $f(\alpha, \sigma, \omega)$, we obtain also for β a certain continuous function $\beta(\sigma, \omega)$. The above theorem is thereby proved. (This theorem is a special case of the General Implicit Theorem [A.5].)

It should be noted here that the restrictions on the functions R, I, and F imposed by the Implicit Theorem can be readily transferred to the functions $a_k = a_k(\alpha, \beta)$, since the functions R, I, and F are linear functions of the coefficients a_k. Several cases of a_k functions are discussed in Derivation A.2 of this appendix.

Theorem A.5. **Theorem on Inverse Mapping.** Let $\alpha = \alpha(\sigma, \omega)$, $\beta = \beta(\sigma, \omega)$ and define a continuously differentiable transformation for all pairs (σ, ω) in some neighborhood of a point $s_0(\sigma_0, \omega_0)$. Let $\alpha_0 = \alpha(\sigma_0, \omega_0)$, $\beta_0 = \beta(\sigma_0, \omega_0)$, and suppose that Jacobian

$$J_1\left(\begin{matrix}\alpha, \beta \\ \sigma, \omega\end{matrix}\right) = \frac{\partial \alpha}{\partial \sigma}\frac{\partial \beta}{\partial \omega} - \frac{\partial \alpha}{\partial \omega}\frac{\partial \beta}{\partial \sigma} \tag{A.43}$$

is not zero at $s_0(\sigma_0, \omega_0)$. Then there exists one and only one system of continuous functions $\sigma = \sigma(\alpha, \beta)$, $\omega = \omega(\alpha, \beta)$ that corresponds to the functions $\alpha(\sigma, \omega)$, $\beta(\sigma, \omega)$.

The theorem is basically just a special case of Theorem A.3 for the system of equations

$$\alpha(\sigma, \omega) - \alpha = 0, \qquad \beta(\sigma, \omega) - \beta = 0, \qquad (A.44)$$

and there is nothing more to prove.

It should be noted that the inverse parameter mapping is based on the above theorem.

Derivation A.1. X_k and Y_k Functions. Functions $X_k = X_k(\sigma, \omega)$, $Y_k = Y_k(\sigma, \omega)$ as defined in Chapter 1 represent the real and imaginary part of the function s^k; that is,

$$s^k = X_k + jY_k \qquad (A.45)$$

when $s = \sigma + j\omega$. By expressing the function $s^k = (\sigma + \omega e^{j\pi/2})^k$ and using the binomial formula, we can write

$$(\sigma + \omega e^{j\pi/2})^k = \sum_{r=0}^{k} \binom{k}{r} \sigma^{k-r} \omega^r e^{j\pi r/2}. \qquad (A.46)$$

By equating the reals and imaginaries of (A.46), according to (A.45), we obtain

$$X_k = \sum_{\nu=0}^{k} (-1)^\nu \binom{k}{2\nu} \sigma^{k-2\nu} \omega^{2\nu} \qquad (A.47a)$$

$$Y_k = \sum_{\mu=1}^{k} (-1)^{\mu-1} \binom{k}{2\mu - 1} \sigma^{k-2\mu+1} \omega^{2\mu-1}, \qquad (A.47b)$$

which is a definition of X_k and Y_k functions by means of series.

A significant fact about X_k and Y_k functions is that they can be obtained by the recurrence formulas[1]

$$X_{k+1} - 2\sigma X_k + (\sigma^2 + \omega^2) X_{k-1} = 0 \qquad (A.48a)$$

$$Y_{k+1} - 2\sigma Y_k + (\sigma^2 + \omega^2) Y_{k-1} = 0, \qquad (A.48b)$$

where $X_0 \equiv 1$, $X_1 \equiv \sigma$, $Y_0 \equiv 0$, $Y_1 \equiv \omega$.

---

[1] From $s = \sigma + j\omega$ and

we can derive
$$s^2 - 2\sigma s + (\sigma^2 + \omega^2) = 0,$$

$$s^{k-1}[s^2 - 2\sigma s + (\sigma^2 + \omega^2)] = 0$$

or

$$s^{k+1} - 2\sigma s^k + (\sigma^2 + \omega^2) s^{k-1} = 0.$$

From this equation the above recurrence relationships are obtained by substituting $s^k = X_k + jY_k$.

By using either the recurrence formulas (A.48) or the series expressions (A.47), a table of several functions X_k and Y_k can be obtained, such as that of Table A.1.

A simple relationship between the functions X_k and Y_k can be derived if (A.45) is written for $k-1$ as $s^{k-1} = X_{k-1} + jY_{k-1}$. By the definition of X_k and Y_k functions, it follows that

$$s^k = (\sigma + j\omega)s^{k-1} = (\sigma + j\omega)(X_{k-1} + jY_{k-1})$$
$$= (\sigma X_{k-1} - \omega Y_{k-1}) + j(\sigma Y_{k-1} + \omega X_{k-1}). \quad (A.49)$$

From (A.45) and (A.48) we have

$$X_k = \sigma X_{k-1} - \omega Y_{k-1}$$
$$Y_k = \sigma Y_{k-1} + \omega X_{k-1}. \quad (A.50)$$

By using (A.48) and (A.50) we can derive

$$\omega X_k = \sigma Y_k - (\sigma^2 + \omega^2)Y_{k-1}$$
$$\omega Y_k = -\sigma X_k + (\sigma^2 + \omega^2)X_{k-1}. \quad (A.51)$$

When the complex variable $s = \omega_n e^{j\theta}$ is expressed by

$$s = -\omega_n \zeta + j\omega_n \sqrt{1 - \zeta^2}, \quad (A.52)$$

where $\omega_n = \sqrt{\sigma^2 + \omega^2}$ is the undamped natural frequency and ζ is the relative damping coefficient $\zeta = -\cos\theta$, alternate definitions of functions X_k and Y_k can be postulated in terms of ω_n and ζ,

$$X_{k+1} + 2\omega_n \zeta X_k + \omega_n^2 X_{k-1} = 0$$
$$Y_{k+1} + 2\omega_n \zeta Y_k + \omega_n^2 Y_{k-1} = 0, \quad (A.53)$$

in which $X_k = X_k(\sigma, \omega)$, $Y_k = Y_k(\sigma, \omega)$, and $X_0 \equiv 1$, $X_1 \equiv -\omega_n \zeta$, $Y_0 \equiv 0$, $Y_1 \equiv \omega_n \sqrt{1 - \zeta^2}$. The substitution used to obtain this formulation of functions X_k and Y_k from recurrence formulas (A.48) is $\sigma = -\omega_n \zeta$, $\omega = \omega_n \sqrt{1 - \zeta^2}$. Therefore recurrence formulas (A.48) and (A.53) can be rewritten in a general form as

$$X_{k+1} - 2X_1 X_k + (X_1^2 + Y_1^2)X_{k-1} = 0$$
$$Y_{k+1} - 2X_1 Y_k + (X_1^2 + Y_1^2)Y_{k-1} = 0. \quad (A.54)$$

The alternative expression (A.53) of the functions X_k and Y_k leads immediately to the formulation of real and imaginary parts of a complex variable s of the power k in terms of the widely used Chebyshev functions as proposed in reference A.8.

The Chebyshev functions $T_k(\zeta)$ and $U_k(\zeta)$ of an argument ζ, which is $0 \leq |\zeta| \leq 1$, are treated in detail in reference A.9, and only basic properties of these functions are given here. The functions are defined starting from

Table A.1 Functions $X_k(\sigma, \omega)$, $Y_k(\sigma, \omega)$

$X_0 \equiv 1$
$X_1 \equiv \sigma$
$X_2 = \sigma^2 - \omega^2$
$X_3 = \sigma^3 - 3\sigma\omega^2$
$X_4 = \sigma^4 - 6\sigma^2\omega^2 + \omega^4$
$X_5 = \sigma^5 - 10\sigma^3\omega^2 + 5\sigma\omega^4$
$X_6 = \sigma^6 - 15\sigma^4\omega^2 + 15\sigma^2\omega^4 - \omega^6$
$X_7 = \sigma^7 - 21\sigma^5\omega^2 + 35\sigma^3\omega^4 - 7\sigma\omega^6$
$X_8 = \sigma^8 - 28\sigma^6\omega^2 + 70\sigma^4\omega^4 - 28\sigma^2\omega^6 + \omega^8$
$X_9 = \sigma^9 - 36\sigma^7\omega^2 + 126\sigma^5\omega^4 - 84\sigma^3\omega^6 + 9\sigma\omega^8$
$X_{10} = \sigma^{10} - 45\sigma^8\omega^2 + 210\sigma^6\omega^4 - 210\sigma^4\omega^6 + 45\sigma^2\omega^8 - \omega^{10}$

$Y_0 \equiv 0$
$Y_1 \equiv \omega$
$Y_2 = 2\sigma\omega$
$Y_3 = 3\sigma^2\omega - \omega^3$
$Y_4 = 4\sigma^3\omega - 4\sigma\omega^3$
$Y_5 = 5\sigma^4\omega - 10\sigma^2\omega^3 + \omega^5$
$Y_6 = 6\sigma^5\omega - 20\sigma^3\omega^3 + 6\sigma\omega^5$
$Y_7 = 7\sigma^6\omega - 35\sigma^4\omega^3 + 21\sigma^2\omega^5 - \omega^7$
$Y_8 = 8\sigma^7\omega - 56\sigma^5\omega^3 + 56\sigma^3\omega^5 - 8\sigma\omega^7$
$Y_9 = 9\sigma^8\omega - 84\sigma^6\omega^3 + 126\sigma^4\omega^5 - 72\sigma^2\omega^7 + \omega^9$
$Y_{10} = 10\sigma^9\omega - 120\sigma^2\omega^3 + 252\sigma^6\omega^5 - 120\sigma^4\omega^7 + 10\sigma\omega^9$

Moivre's formula
$$(\cos\theta + j\sin\theta)^k = \cos k\theta + j\sin k\theta \quad (A.55)$$
and putting
$$\cos\theta = \zeta \quad (A.56a)$$
$$\sin\theta = \sqrt{1-\zeta^2}. \quad (A.56b)$$
Then
$$\cos k\theta + j\sin k\theta = (\zeta + j\sqrt{1-\zeta^2})^k, \quad (A.57)$$
whence, separating real and imaginary parts,
$$\cos k\theta = \zeta^k - \binom{k}{2}\zeta^{k-2}(1-\zeta^2) + \binom{k}{4}\zeta^{k-4}(1-\zeta^2)^2 - \cdots \quad (A.58)$$
$$\frac{\sin k\theta}{\sin\theta} = \binom{k}{1}\zeta^{k-1} - \binom{k}{3}\zeta^{k-3}(1-\zeta^2) + \binom{k}{5}\zeta^{k-5}(1-\zeta^2)^2 - \cdots. \quad (A.59)$$

The coefficient of ζ^k on the right of (A.58) is
$$1 + \binom{k}{2} + \binom{k}{4} + \cdots = 2^{k-1}. \quad (A.60)$$
The same holds for the coefficient of ζ^{k-1} on the right of (A.59):
$$\binom{k}{1} + \binom{k}{3} + \cdots = 2^{k-1}. \quad (A.61)$$

The polynomials $T_k(\zeta)$ and $U_k(\zeta)$ thus defined in (A.58) and (A.59) are called Chebyshev polynomials of the first and second kind, respectively.

From the above it can be derived that
$$T_k(\zeta) = \cos(k \arccos \zeta)$$
$$U_k(\zeta) = \frac{\sin(k \arccos \zeta)}{(1-\zeta^2)^{1/2}}. \quad (A.62)$$

It can also be shown that Chebyshev's functions can be obtained by applying recurrence formulas
$$T_{k+1}(\zeta) - 2\zeta T_k(\zeta) + T_{k-1}(\zeta) = 0 \quad (A.63a)$$
$$U_{k+1}(\zeta) - 2\zeta U_k(\zeta) + U_{k-1}(\zeta) = 0, \quad (A.63b)$$
where $T_0(\zeta) \equiv 1$, $T_1(\zeta) \equiv \zeta$, $U_0(\zeta) \equiv 0$, $U_1(\zeta) \equiv 1$.

In Table A.2 explicit expressions of the first ten $T_k(\zeta)$ and $U_k(\zeta)$ polynomials are shown. The numerical values of $T_k(\zeta)$ and $U_k(\zeta)$ for pertinent values of ζ are given in Tables A.3 and A.4, respectively.

By using the trigonometric addition formulas translated into the variable ζ, we can show that there is a relationship between the $T_k(\zeta)$ and $U_k(\zeta)$

Table A.2 Chebyshev Functions $T_k(\zeta)$, $U_k(\zeta)$

$T_0 = 1$
$T_1 = \zeta$
$T_2 = 2\zeta^2 - 1$
$T_3 = 4\zeta^3 - 3\zeta$
$T_4 = 8\zeta^4 - 8\zeta^2 + 1$
$T_5 = 16\zeta^5 - 20\zeta^3 + 5\zeta$
$T_6 = 32\zeta^6 - 48\zeta^4 + 18\zeta^2 - 1$
$T_7 = 64\zeta^7 - 112\zeta^5 + 56\zeta^3 - 7\zeta$
$T_8 = 128\zeta^8 - 256\zeta^6 + 160\zeta^4 - 32\zeta^2 + 7$
$T_9 = 256\zeta^9 - 576\zeta^7 + 432\zeta^5 - 120\zeta^3 + 9\zeta$
$T_{10} = 512\zeta^{10} - 1280\zeta^8 + 1120\zeta^6 - 400\zeta^4 + 50\zeta^2 - 1$

$U_0 \equiv 0$
$U_1 \equiv 1$
$U_2 = 2\zeta$
$U_3 = 4\zeta^2 - 1$
$U_4 = 8\zeta^3 - 4\zeta$
$U_5 = 16\zeta^4 - 12\zeta^2 + 1$
$U_6 = 32\zeta^5 - 32\zeta^3 + 6\zeta$
$U_7 = 64\zeta^6 - 80\zeta^4 + 24\zeta^2 - 1$
$U_8 = 128\zeta^7 - 192\zeta^5 + 80\zeta^3 - 8\zeta$
$U_9 = 256\zeta^8 - 448\zeta^6 + 240\zeta^4 - 40\zeta^2 + 1$
$U_{10} = 512\zeta^9 - 1024\zeta^7 + 672\zeta^5 - 160\zeta^3 + 10\zeta$

Table A.3 Functions $T_k(\zeta)$

ζ	T_0	T_1	T_2	T_3	T_4	T_5	T_6	T_5	T_4	T_5	T_{10}
0.00		0.00	−1.000	0.0000	1.00000	0.000000	−1.0000000	0.00000000	1.000000000	0.0000000000	−1.00000000000
0.05		0.05	−0.995	−0.1495	0.98005	0.347505	−0.9452995	−0.44203495	0.901096005	0.5321445505	−0.84788154995
0.10		0.10	−0.980	−0.2960	0.92080	0.480160	−0.8247680	−0.64511360	0.695745280	0.7842626560	−0.53889274880
0.15		0.15	−0.955	−0.4365	0.82405	0.683715	−0.6189355	−0.86939565	0.358116805	0.9768306915	−0.06506759755
0.20		0.20	−0.920	−0.5680	0.69280	0.845120	−0.3547520	−0.98702080	−0.040056320	0.9709982720	0.42845562880
0.25		0.25	−0.875	−0.6875	0.53125	0.953125	−0.0546875	−0.98046875	−0.435546875	0.7626953125	0.81689453125
0.30		0.30	−0.820	−0.7920	0.34480	0.998880	0.2545280	−0.84616320	−0.762225920	0.3888276480	0.99552250880
0.35		0.35	−0.755	−0.8785	0.14005	0.976535	0.5435245	−0.59606785	−0.960771995	−0.0764725465	0.90724121245
0.40		0.40	−0.680	−0.9440	−0.07520	0.883840	0.7822720	−0.25802240	−0.988689920	−0.5329295360	0.46234629120
0.45		0.45	−0.595	−0.9855	−0.29195	0.722745	0.9424205	0.12543345	−0.829530395	−0.8720108055	0.04472067005
0.50	1	0.50	−0.500	−1.0000	−0.50000	0.500000	1.0000000	0.50000000	−0.500000000	−1.0000000000	−0.50000000000
0.55		0.55	−0.395	−0.9845	−0.68795	0.227755	0.9384805	0.80457355	−0.053449595	−0.8633681045	−0.98625531995
0.60		0.60	−0.280	−0.9360	−0.84320	−0.075840	0.7521920	0.97847040	0.421972480	−0.4721034240	−0.98849658880
0.65		0.65	−0.155	−0.8515	−0.95195	−0.386035	0.4501045	0.97117085	0.812407605	0.0849590365	−0.70196085755
0.70		0.70	−0.020	−0.7280	−0.99920	−0.670880	0.0599680	0.75483520	0.996801280	0.6406865920	−0.09984005120
0.75		0.75	0.125	−0.5625	−0.96875	−0.890625	−0.3671875	0.33984375	0.876953125	0.9755859375	0.58642578125
0.80		0.80	0.280	−0.3520	−0.84320	−0.997120	−0.7521920	−0.20638720	0.421972480	0.8815431680	0.98849658880
0.85		0.85	0.445	−0.0935	−0.60395	−0.933215	−0.9825155	−0.73706135	−0.270488795	0.2772303985	0.74178047245
0.90		0.90	0.620	0.2160	−0.23120	−0.632160	−0.9066880	−0.99987840	−0.893093120	−0.6076892160	−0.20074746880
0.95		0.95	0.805	0.5795	0.29605	−0.017005	−0.3283595	−0.60687805	−0.824708795	−0.9600686605	−0.99942165995
1.00		1.00	1.000	1.0000	1.00000	1.000000	1.0000000	1.00000000	1.000000000	1.0000000000	1.00000000000

Table A.4 Functions $U_k(\zeta)$

ζ	U_{-1}	U_0	U_1	U_2	U_3	U_4	U_5	U_6	U_7	U_8	U_9	U_{10}
0.00	−1	0	1	0.0	−1.00	0.000	1.0000	0.00000	−1.000000	0.0000000	1.00000000	0.000000000
0.05				0.1	−0.99	−0.199	0.9701	0.29601	−0.940499	−0.3900599	0.90149301	0.480209201
0.10				0.2	−0.96	−0.392	0.8816	0.56832	−0.767936	−0.7219072	0.62355456	0.846618112
0.15				0.3	−0.91	−0.573	0.7381	0.79443	−0.499771	−0.9443613	0.21646261	1.009300083
0.20				0.4	−0.84	−0.736	0.5456	0.95424	−0.163904	−1.0198016	−0.24401664	0.922194944
0.25				0.5	−0.75	−0.875	0.3125	1.03125	0.203125	−0.9296875	−0.66796875	0.595703125
0.30				0.6	−0.64	−0.984	0.0496	1.01376	0.558656	−0.6785664	−0.96579584	0.099088896
0.35				0.7	−0.51	−1.057	−0.2299	0.89607	0.857149	−0.2960657	−1.06439499	−0.449010793
0.40				0.8	−0.36	−1.088	−0.5104	0.67968	1.054144	0.1636352	−0.92323584	−0.902223872
0.45				0.9	−0.19	−1.071	−0.7739	0.37449	1.110941	0.6253569	−0.54811979	−1.118664711
0.50				1.0	0.00	−1.000	−1.0000	0.00000	1.000000	1.0000000	0.00000000	−1.000000000
0.55				1.1	0.21	−0.869	−1.1659	−0.41349	0.711061	1.1956571	0.60416181	−0.531079109
0.60				1.2	0.44	−0.672	−1.2464	−0.82368	0.257984	1.1332608	1.10192896	0.189053952
0.65				1.3	0.69	−0.403	−1.2139	−1.17507	−0.313691	0.7672717	1.31114421	0.937215773
0.70				1.4	0.96	−0.056	−1.0384	−1.39776	−0.918464	0.1119104	1.07513856	1.393283584
0.75				1.5	1.25	0.375	−0.6875	−1.40625	−1.421875	−0.7265625	0.33203125	1.224609375
0.80				1.6	1.56	0.896	−0.1264	−1.09824	−1.630784	−1.5110144	−0.78683904	0.252071936
0.85				1.7	1.89	1.513	0.6821	−0.35343	−1.282931	−1.8275527	−1.82390859	−1.273091903
0.90				1.8	2.24	2.232	1.7776	0.96768	−0.035776	−1.0320768	−1.82196224	−2.247455232
0.95				1.9	2.61	3.059	3.2021	3.02499	2.545381	1.9112339	0.89596341	−0.108903421
1.00				2.0	3.00	4.000	5.0000	6.00000	7.000000	8.0000000	9.00000000	10.000000000

functions
$$T_k(\zeta) = \zeta U_k(\zeta) - U_{k-1}(\zeta), \tag{A.64}$$
as was the case with functions X_k and Y_k.

In the stability analysis of linear systems, it is postulated by convention that the relative damping coefficient $\zeta > 0$ for the values of s in the left half of the s plane and $\zeta < 0$ in the right half of the s plane. This is contrary to the convention used in defining the Chebyshev polynomials, and the sign of ζ has to be reversed. (Compare equations A.52 and A.56a.) This reversal of sign causes no particular difficulty since the following relationship holds:
$$\begin{aligned} T_k(-\zeta) &= (-1)^k T_k(\zeta) \\ U_k(-\zeta) &= (-1)^{k+1} U_k(\zeta). \end{aligned} \tag{A.65}$$

Therefore, according to (A.52) and (A.65), we can express s^k as
$$s^k = (-1)^k \omega_n^k T_k(\zeta) + j(-1)^{k+1} \omega_n^k \sqrt{1-\zeta^2}\, U_k(\zeta). \tag{A.66}$$

If (A.66) is substituted in the characteristic equation A.1, the latter can be rewritten as
$$\begin{aligned} R &\equiv \sum_{k=0}^{n} (-1)^k a_k \omega_n^k T_k(\zeta) = 0 \\ I &\equiv \sum_{k=0}^{n} (-1)^{k+1} a_k \omega_n^k \sqrt{1-\zeta^2}\, U_k(\zeta) = 0. \end{aligned} \tag{A.67}$$

By using the relationship (A.64), equations A.67 can be rewritten again to yield
$$\begin{aligned} \sum_{k=0}^{n} (-1)^k a_k \omega_n^k U_{k-1}(\zeta) &= 0 \\ \sum_{k=0}^{n} (-1)^k a_k \omega_n^k U_k(\zeta) &= 0, \end{aligned} \tag{A.68}$$

which are the basic equations of the parameter plane mapping given in terms of Chebyshev's polynomials of the second kind. In deriving (A.68), it was assumed that $\zeta \neq 1$. However, when the double-real roots are discussed in a later development, it will be proved that equations A.68 are valid also for the case where $\zeta = 1$.

It is of interest now to write the relationship between the X_k, Y_k functions and the Chebyshev polynomials U_k and T_k. The relationship follows directly by comparing (A.48) and (A.66),
$$\begin{aligned} X_k(\omega_n, \zeta) &= (-1)^k \omega_n^k T_k(\zeta) \\ Y_k(\omega_n, \zeta) &= (-1)^{k+1} \omega_n^k \sqrt{1-\zeta^2}\, U_k(\zeta). \end{aligned} \tag{A.69}$$

On the basis of these relationships, other properties of the functions X_k and Y_k can be derived from the properties of Chebyshev's polynomials that are described in detail in reference A.9.

The starred polynomials Y_k^*, introduced by

$$Y_k = \omega_n \sqrt{1 - \zeta^2}\, Y_k^*, \qquad (A.70)$$

are related to Chebyshev's polynomials $U_k(\zeta)$ as

$$Y_k^* = (-1)^{k+1} \omega_n^{k-1} U_k(\zeta), \qquad (A.71)$$

and they are polynomials in ω_n and ζ, while Y_k are not because of the term $(1 - \zeta^2)^{1/2}$ that appears in all functions of Y_k. Explicit expressions of the first-term Y_k^* polynomials are given in Table A.5.

Because of the recursion formulas for Y_k^* polynomials, they are easily implemented in a digital program, and a general program can be readily devised for plotting the characteristic curves in the parameter plane on the basis of equations

$$\sum_{k=0}^{n} a_k Y_{k-1}^* = 0 \qquad (A.72a)$$

$$\sum_{k=0}^{n} a_k Y_k^* = 0. \qquad (A.72b)$$

For simple problems to be worked longhand, the characteristic curves can be plotted using (A.68) and Table A.4 for computing the polynomials $U_k(\zeta)$.

The argument principle of Cauchy serves as a basis for various stability criteria for linear feedback systems, as will be shown later. It also has significant applications in separation of the roots of algebraic polynomials and proofs of various theorems related to polynomial functions.

Theorem A.6. Cauchy Principle of Argument. Let $F(s)$ be the analytic interior to a single closed Jordan contour C and continuous and different from zero on C. Let K be the curve described in the w plane by the point $w = F(s)$ and let $\Delta_c \arg F(s)$ denote the net change in $\arg F(s)$ as the point s traverses C once over in the counterclockwise direction. Then the number p of zeros of $F(s)$ interior to C, counted with their multiplicities, is

$$p = \frac{1}{2\pi} \Delta_c \arg F(s); \qquad (A.73)$$

that is, it is the net number of times that K encircles the origin of the w plane.

The proof of the above theorem is given only in the case where $F(s)$ is a polynomial [A.4]. If $s_1, s_2, \ldots, s_p$ denote the zeros of $F(s)$ inside C and

Table A.5 Y_k^* Functions

$Y_k^* = Y_k^*(\sigma, \omega)$

$Y_0^* = 0$
$Y_1^* = 1$
$Y_2^* = 2\sigma$
$Y_3^* = 3\sigma^2 - \omega^2$
$Y_4^* = 4\sigma^3 - 4\omega^2$
$Y_5^* = 5\sigma^4 - 10\sigma^2\omega^2 + \omega^4$
$Y_6^* = 6\sigma^5 - 20\sigma^3\omega^2 + 6\sigma\omega^4$
$Y_7^* = 7\sigma^6 - 35\sigma^4\omega^2 + 21\sigma^2\omega^4 - \omega^6$
$Y_8^* = 8\sigma^7 - 56\sigma^5\omega^2 + 56\sigma^3\omega^4 - 8\sigma\omega^6$
$Y_9^* = 9\sigma^8 - 84\sigma^6\omega^2 + 126\sigma^4\omega^4 - 72\sigma^2\omega^6 + \omega^8$
$Y_{10}^* = 10\sigma^9 - 120\sigma^8\omega^2 + 252\sigma^6\omega^4 - 120\sigma^4\omega^6 + 10\sigma\omega^8$

$Y_k^* = Y_k^*(\omega_n, \zeta)$

$Y_0^* = 0$
$Y_1^* = 1$
$Y_2^* = -2\omega_n\zeta$
$Y_3^* = 4\omega_n^2\zeta^2 - \omega_n^2$
$Y_4^* = -8\omega_n^3\zeta^3 + 4\omega_n^3\zeta$
$Y_5^* = 16\omega_n^4\zeta^4 - 12\omega_n^4\zeta^2 + \omega_n^4$
$Y_6^* = -32\omega_n^5\zeta^5 + 32\omega_n^5\zeta^3 - 6\omega_n^5\zeta$
$Y_7^* = 64\omega_n^6\zeta^6 - 80\omega_n^6\zeta^4 + 24\omega_n^6\zeta^2 - \omega_n^6$
$Y_8^* = -124\omega_n^7\zeta^7 + 192\omega_n^7\zeta^5 - 80\omega_n^7\zeta^3 + 8\omega_n^7\zeta$
$Y_9^* = 256\omega_n^8\zeta^8 - 448\omega_n^8\zeta^6 + 240\omega_n^8\zeta^4 - 40\omega_n^8\zeta^2 + \omega_n^8$
$Y_{10}^* = -512\omega_n^9\zeta^9 + 1024\omega_n^9\zeta^7 - 672\omega_n^9\zeta^5 + 160\omega_n^9\zeta^3 - 10\omega_n^9\zeta$

394 Basic Theorems and Derivations

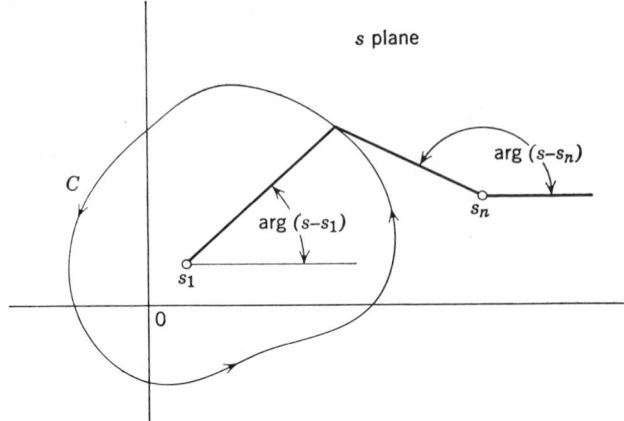

Figure A.1 The s plane contour C.

$s_{p+1}, s_{p+2}, \ldots, s_n$ denote those outside C, then

$$F(s) = a_n \prod_{k=1}^{p}(s - s_k) \cdot \prod_{k=p+1}^{n}(a - a_k), \quad a_n \neq 0 \tag{A.74}$$

$$\arg F(s) = \arg a_n + \sum_{k=0}^{n} \arg(s - s_k) + \sum_{k=p+1}^{n} \arg(s - s_k).$$

As the point s describes the contour C counterclockwise, as indicated in Figure A.1, $\arg(s - s_k)$ increases by 2π when $1 \leq k \leq p$, but has a zero net change when $p < k \leq n$. This has a direct consequence in (A.73).

For the purpose of investigating the number of roots at a particular point of the $\alpha\beta$ plane, the polynomial $F = F(s)$ of

$$F(s) \equiv \sum_{k=0}^{n} a_k s^k = 0 \tag{A.1}$$

can be expressed as

$$F = R + jI, \tag{A.3}$$

where $a_k = b_k + jc_k$ and

$$R \equiv \sum_{k=0}^{n}(b_k X_k - c_k Y_k)$$

$$I \equiv \sum_{k=0}^{n}(c_k X_k + b_k Y_k). \tag{A.6}$$

If the characteristic equation is given and a point $M(\alpha; \beta)$ is specified, then the coefficients a_k are evaluated numerically. In addition, if a closed contour is specified, then it can be mapped by (A.6) onto the $F(s)$ plane with R and I as coordinate axes. By applying the Cauchy argument principle to the obtained plot, we can conclude that the hodograph $F(s)$ will describe an argument $2\pi p$, where p is the number of roots of $F(s) = 0$ inside the specified contour.

To illustrate the above, determine the number of roots at the point $M(0.2; 0.5)$ of Figure 1.22. For $\alpha = 0.2$ and $\beta = 0.5$, the polynomial (1.79) is $F(s) = 0.008s^4 + 0.112s^3 + 0.76s^2 + 2.6s + 3 = 0$. For the s plane contour of Figure 1.21b and $\zeta = 0.5$, equations A.6 are

$$R(\omega_n, 0.5) = 3 - 1.3\omega_n - 0.38\omega_n^2 + 0.112\omega_n^3 - 0.004\omega_n^4$$
$$I(\omega_n, 0.5) = 0.866(2.6\omega_n - 0.76\omega_n^2 + 0.008\omega_n^4). \quad \text{(A.75)}$$

When the tracing point in the s plane moves radially out from the origin along the constant-damping line $\overline{AB}$ and ω_n increases from 0 to ∞, the vector $F(\omega_n, 0.5)$ rotates simultaneously in a counterclockwise direction and describes the hodograph computed by (A.74) and plotted in Figure A.2. As seen from the hodograph, all four roots are located inside the specified contour, since the total argument described by the vector $F(\omega_n, 0.5)$ is $480°$, the angle $\theta = \pi - \arccos 0.5 = 120°$, and consequently $p = n = 4$.

Note that for $\omega_n \to \infty$ (that is, $s \to \infty$) $F(s)$ behaves as $a_n s^n$ and the increment of the argument of $F(s)$ along the circular arc of infinite radius of contour C (Figure 1.21b) will be $2n(\pi - \theta)$. Furthermore, because of symmetry of the contour C, the argument may be investigated only along the straight line $\overline{AB}$, and for all the roots to lie inside C, this argument is

$$\tfrac{1}{2}[2n\pi - 2n(\pi - \theta)] = n\theta.$$

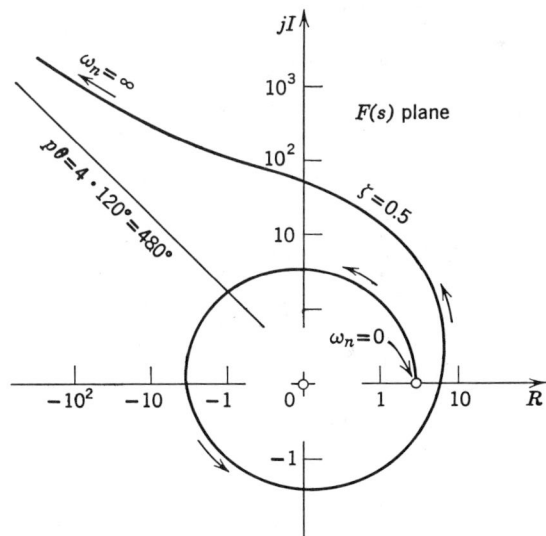

Figure A.2 The $F(\omega_n, 0.5)$ curve.

396   Basic Theorems and Derivations

By using the result of Theorem A.5, the following Rouché Theorem [A.3, 4] is derived. The Rouché theorem can be applied to prove an important proposition that the zeros of a polynomial are continuous functions of the coefficients of the polynomial. If, in addition, the coefficients of a polynomial are continuous functions of some parameters, then the zeros of the polynomial are also continuous functions of the parameters, which is a direct consequence of the previous statement.

Theorem A.7.   Rouché's Theorem.   If $P(s)$ and $Q(s)$ are analytic interior to a simple closed Jordan contour C and if they are continuous on C and

$$|P(s)| < |Q(s)| \tag{A.76}$$

on C, then the function $F(s) = P(s) + Q(s)$ has the same number of zeros interior to C as does $Q(s)$.

To prove the above theorem, express

$$F(s) = wQ(s), \qquad w = 1 + [P(s)/Q(s)]. \tag{A.77}$$

If p denotes the number of zeros of $Q(s)$ inside C, then according to Theorem A.5,

$$\Delta_c \arg Q(s) = 2\pi p. \tag{A.78}$$

Since $|P(s)/Q(s)| < 1$ on C, the point w defined in (A.71) describes a closed contour C_1 that lies inside the circle with a center at $w = 1$ and radius 1, as shown in Figure A.3. Therefore the point w remains always in the right half of the w plane and the net change in arg w on the contour C_1 is zero. According to (A.61) and (A.78), this fact leads at once to the following equation:

$$\Delta_c \arg F(s) = \Delta_c \arg w + \Delta_c \arg Q(s) = 2\pi p. \tag{A.79}$$

As shown in Theorem A.6, (A.79) proves that $F(s)$ has also p zeros in C.

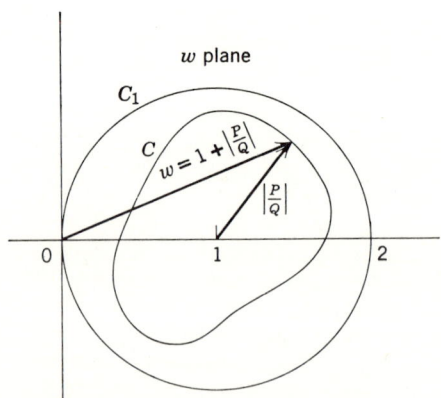

Figure A.3   The w plane contours.

Now Rouché's Theorem will be applied to prove a statement that the zeros of a polynomial are continuous functions of a parameter of the polynomial. The proof will be based upon the proof of a similar theorem given in reference A.4.

Theorem A.8. Continuity of Zeros. Let

$$F(s) \equiv \sum_{k=0}^{n} a_k s^k = a_n \prod_{j=1}^{p} (s - s_j)^{m_j}, \qquad a_n \neq 0, \qquad (A.80)$$

where the coefficients $a_k = a_k(\alpha)$, $k = 0, 1, \ldots, n - 1$, are continuous functions of a real parameter α, and let us give α a definite value, real or complex, for instance, $\alpha = \alpha_0$. (If for $\alpha = \alpha_0$ some or all of the coefficients a_k go to infinity, then α_0 is an exceptional point on the α axis.) Let also

$$f(s) = a_n s^n + \sum_{k=0}^{n} a_k(\alpha_0 + \Delta\alpha) s^k \qquad (A.81)$$

$$0 < r_i < \min |s_i - s_j|, \qquad (j = 1, 2, \ldots, i - 1, k + 1, \ldots, p) \qquad (A.82)$$

for the specified value of $\alpha = \alpha_0$.

There exists a number $\Delta\alpha$ so that $F(s)$ has precisely m_k zeros in the circle C_i with a center at s_i and radius r_i.

To prove the above theorem, note that (A.81) can be rewritten as

$$f(s) = a_n s^n + \sum_{k=0}^{n-1} (a_k + \varepsilon_k) s^k. \qquad (A.83)$$

Now show that there exists a positive number ε so that, if

$$|\varepsilon_k| \leq \varepsilon, \qquad (k = 0, 1, \ldots, n - 1), \qquad (A.84)$$

then $f(s)$ has precisely m_i zeros in the circle C_i with a center at s_i and radius r_i. To prove this proposition it is necessary to note [A.4] that on C_i the polynomial

$$h(s) = \sum_{k=0}^{n} \varepsilon_k s^k \qquad (A.85)$$

has the property

$$|h(s)| \leq \varepsilon M_i, \qquad M_i = \sum_{j=0}^{n-1} (r_i + |s_i|)^j, \qquad (A.86)$$

whereas, on C_i

$$|F(s)| \geq |a_n| r_i^{m_i} \prod_{\substack{j=1 \\ j \neq i}}^{p} (|s_j - s_i| - r_i)^{m_i} = \delta_i > 0. \qquad (A.87)$$

If ε is chosen so that $\varepsilon > \delta_i M_i$, it follows that $|h(s)| < |F(s)|$ on C_i.

398   Basic Theorems and Derivations

According to Rouché's theorem, this means that $f(s)$ of (A.83) has the same number of zeros in C_i as does $F(s)$. Since inequality A.82 insures that the only zero of $F(s)$ in C_i is the one of multiplicity m_i at s_i, it follows that $f(s)$ of (A.83) has precisely m_i zeros in C_k.

Since the coefficients $a_k(\alpha)$ are continuous functions of the parameter α—that is, for each ε there is a number δ' so that

$$|a_k(\alpha_0 + \Delta\alpha) - a_k(\alpha_0)| = |\varepsilon_k| < \varepsilon, \qquad (k = 0, 1, \ldots, n-1) \quad \text{(A.88)}$$

whenever $|\Delta\alpha| < \delta'$. It is always possible, therefore, to determine a number $\Delta\alpha$ so that the inequalities A.82 are satisfied, and Theorem A.8 is thereby proved.

Derivation A.2.   Functions $a_k(\alpha, \beta)$.   In this section the form

$$a_k = e_k\alpha^2 + g_k\beta^2 + 2h_k\alpha\beta + 2b_k\alpha + 2c_k\beta + d_k \quad \text{(A.89)}$$

of the a_k functions will be discussed for solving the basic equations

$$R \equiv \sum_{k=0}^{n} a_k X_k = 0 \quad \text{(A.90a)}$$

$$I \equiv \sum_{k=0}^{n} a_k Y_k = 0 \quad \text{(A.90b)}$$

in terms of α and β as

$$\alpha = \alpha(\sigma, \omega), \qquad \beta = \beta(\sigma, \omega). \quad \text{(A.91)}$$

Form (A.89) is of interest in practical applications of the parameter mapping.

Since the functions a_k are continuous functions of α and β, it may be concluded according to Theorem A.8 that the parameters α and β as given in (A.91) will be continuous functions with respect to the arguments σ and ω. To obtain these functions α and β it is necessary to solve the two simultaneous quadratics

$$R(\alpha, \beta) = E_1\alpha^2 + G_1\beta^2 + 2H_1\alpha\beta + 2B_1\alpha + 2C_1\beta + D_1 = 0 \quad \text{(A.92a)}$$

$$I(\alpha, \beta) = E_2\alpha^2 + G_2\beta^2 + 2H_2\alpha\beta + 2B_2\alpha + 2C_2\beta + D_2 = 0, \quad \text{(A.92b)}$$

which are obtained from (A.90) by substituting a_k from (A.92). The coefficients in (A.92) are given numerically when coefficients on the right side of (A.89) and σ, ω are specified numerically. Then the functions $\alpha(\sigma, \omega)$ and $\beta(\sigma, \omega)$ are obtained by solving (A.92) for different values of the pair (σ, ω), provided that the coefficients a_k are specified as (A.89).

To discuss the solution of (A.92), form an equivalent system of equations

$$\begin{aligned} R'(\alpha, \beta) &= G_2 R(\alpha, \beta) - G_1 I(\alpha, \beta) \\ I'(\alpha, \beta) &= -E_2 R(\alpha, \beta) + E_1 I(\alpha, \beta), \end{aligned} \quad \text{(A.93)}$$

which has the same solutions as (A.92) if

$$E_1G_2 - E_2G_1 \neq 0. \tag{A.94}$$

Equations A.93 become

$$R'(\alpha, \beta) = E_1'\alpha^2 + 2H_1'\alpha\beta + 2B_1'\alpha + 2C_1'\beta + D_1' = 0 \tag{A.95a}$$

$$I'(\alpha, \beta) = G_2'\alpha^2 + 2H_2'\alpha\beta + 2B_2'\alpha + 2C_2'\beta + D_2' = 0, \tag{A.95b}$$

where

$$E_1' = E_1G_2 - E_2G_1, \quad H_1' = H_1G_2 - H_2G_1, \ldots . \tag{A.96}$$

Thus we replaced the system of (A.92) by a simpler equivalent system, provided that $E_1G_2 - E_2G_1 \neq 0$. If, however, $E_1G_2 - E_2G_1 = 0$, then in (A.89b), we have $G_2' = E_1G_2 - E_2G_1 = 0$, and (A.92) can be replaced by an equivalent system

$$E_1\alpha^2 + G_1\beta^2 + 2H_1\alpha\beta + 2B_1\alpha + 2C_1\beta + D_1 = 0 \tag{A.92a}$$

$$2H_2'\alpha\beta + 2G_2'\alpha + 2C_2'\beta + D' = 0. \tag{A.97}$$

By means of (A.97) we can eliminate the $\alpha\beta$ term from (A.92a), giving rise to a new equivalent system

$$\begin{aligned} E_1'\alpha^2 + G_1'\beta^2 + 2B_1'\alpha + 2C_1'\beta + D_1' &= 0 \\ 2H_2'\alpha\beta + 2B_2'\alpha + 2C_2'\beta + D_2' &= 0. \end{aligned} \tag{A.98}$$

In addition, if in (A.95) we have $H_2 = E_1H_2 - E_2H_1 = 0$, equations A.95 can be reduced to a linear form

$$\begin{aligned} B_1'\alpha + C_1'\beta + D_1' &= 0 \\ B_2'\alpha + C_2'\beta + D_2' &= 0. \end{aligned} \tag{A.99}$$

When two simultaneous quadratics are reduced to the form of either (A.95), (A.98), or (A.99), we say that they are *normalized*. The normalized system (A.95) is suitable, for it enables one of the unknowns to be expressed as a rational function of the other.

To discuss the existence of solutions of the two quadratics, assume that $E_1G_2 - E_2G_1 \neq 0$ and normalize the system to (A.95). From (A.92b) we have

$$\alpha = -\frac{G_2'\beta^2 + 2B_2'\beta + D_2'}{2(H_2'\beta + C_2')}, \tag{A.100}$$

so that α is a rational function of β. By substituting this value of α into (A.95a), we obtain

$$E_1'(G_2'\beta^2 + 2C_2'\beta + D_2')^2 - 4H_1'\beta(H_2'\beta + B_2')(G_2'\beta^2 + 2C_2'\beta + D_2') \\ - 4B_1'(H_2'\beta + B_2')(G_2'\beta^2 + 2C_2'\beta + D_2') + 4(H_2'\beta + B_2')(2C_1' + D_1') = 0. \tag{A.101}$$

400   Basic Theorems and Derivations

Equation A.101 is a fourth-degree algebraic equation in β, and therefore has four roots, $\beta_1, \beta_2, \beta_3, \beta_4$. If we substitute these four roots in (A.100), we shall have four corresponding values, $\alpha_1, \alpha_2, \alpha_3, \alpha_4$, for α. Thus the four solutions of the two simultaneous quadratics are found. In the parameter mapping we are interested only in the pairs of solutions (α, β) that are real, since the parameters are real numbers.

Various situations may arise in solving the two simultaneous quadratics, depending on which of the coefficients in (A.92) or (A.95) are equal to zero. Of particular interest is the linear case discussed in Chapter 1, because a majority of practical cases can be reduced to the linear combination of parameters in the coefficients of relevant characteristic equations. Note that the parameters α and β do not necessarily coincide with the physical parameters of the system, and α and β can be a combination of them, as demonstrated in Chapter 2.

Let us discuss the important case where the coefficients are linear functions of the parameters α and β and their product $\alpha\beta$:

$$a_k = h_k \alpha\beta + b_k \alpha + c_k \beta + d_k. \tag{A.102}$$

The corresponding system of two simultaneous quadratics can be written as

$$\begin{aligned} H_1 \alpha\beta + B_1 \alpha + C_1 \beta + D_1 &= 0 \\ H_2 \alpha\beta + B_2 \alpha + C_2 \beta + D_2 &= 0. \end{aligned} \tag{A.103}$$

By using the notation

$$a = B_2 H_1 - B_1 H_2, \quad b = C_2 H_1 - C_1 H_2, \quad c = C_1 D_2 - C_2 D_1$$
$$d = B_1 D_2 - B_2 D_1, \quad e = B_2 C_1 - B_1 C_2 + H_1 D_2 - H_2 D_1$$
$$f = C_2 B_1 - B_2 C_1 + H_1 D_2 - H_2 D_1, \quad J = -a\alpha + b\beta + B_1 C_2 - B_2 C_1 \tag{A.104}$$

and eliminating β from (A.103), we obtain

$$\alpha_{1,2} = \frac{-e \pm (e^2 - 4ac)^{1/2}}{2a}, \quad \beta_{1,2} = -\frac{B_1 \alpha_{1,2} + D_1}{H_1 \alpha_{1,2} + C_1}. \tag{A.105}$$

Elimination of α from (A.103) yields

$$\beta_{1,2} = \frac{-f \pm (f^2 - 4bd)^{1/2}}{2b}, \quad \alpha_{1,2} = -\frac{C_1 \beta_{1,2} + D_1}{H_1 \beta_{1,2} + B_1}. \tag{A.106}$$

As can be seen from (A.105) and (A.106), there are generally two pairs of solutions (α_1, β_1), (α_2, β_2) of (A.103). If $a = 0$, the solutions are given by (A.106), provided that $b \neq 0$. In the special case where $a = b = 0$, the solutions for α and β are

$$\alpha = -\frac{c}{e}, \quad \beta = -\frac{d}{f}. \tag{A.107}$$

If $H_1 = H_2 = 0$, the case of (A.103) reduces to the linear form. Since the parameter values of interest are real, in calculating α and β from (A.105) or (A.106), additional requirements are that

$$e^2 - 4ac \geq 0, \qquad f^2 - 4bd \geq 0. \tag{A.108}$$

Finally, for (A.103) to have any solutions, Jacobian J of (A.104) has to be different from zero.

Theorem A.9. Complex Roots of Equations with Real Coefficients. Symmetry Principle. If an algebraic equation with real coefficients has a complex root $\sigma + j\omega$ of multiplicity ν, it has also the conjugate root $\sigma - j\omega$ of the same multiplicity, or complex roots occur in conjugate pairs.

By applying the symmetry principle, the proof of the above theorem is extremely simple. Consider an s plane region $C = \bar{C}$, in which case C is said to be symmetric with the real axis, and $\bar{C}$ is a region obtained by reflecting C in the real axis of the s plane. Then C is a connected region and it intersects the real axis. If a function $F(s)$ is analytic in C, and if it is real on the intersection of C with the real axis, then by the symmetry principle we have $F(s) = \overline{F(\bar{s})}$. This means that $F(s)$ has a conjugate value at conjugate points. To show that this is true, it is sufficient to note that the function $F(s) - \overline{F(\bar{s})}$ is analytic in C and equal to zero on the real axis. Then it must vanish identically in C and we obtain $F(s) = \overline{F(\bar{s})}$, or $\overline{F(s)} = F(\bar{s})$. Consequently, if $F(s) = R + jI$ is a polynomial that vanishes for some value of $s = \sigma + j\omega$ (that is, $R = I = 0$), then its conjugate $\overline{F(s)} = R - jI$ also vanishes for that value of s. By the symmetry principle the value of $F(\bar{s})$ simultaneously goes to zero and $\bar{s} = \sigma - j\omega$ is also a zero of the polynomial $F(s)$ whenever s is zero of that polynomial. This reasoning directly extends to zeros of multiplicity ν, and the above theorem is thereby proved.

Derivation A.3. The $\zeta = 1$ Curve. Double-Real Roots. It has just been proved that when an equation

$$F(s) \equiv \sum_{k=0}^{n} a_k s^k = 0 \tag{A.1}$$

has a root value

$$s = -\omega_n \zeta + j\omega_n \sqrt{1 - \zeta^2}, \tag{A.52}$$

then it also has a root that is a conjugate value $\bar{s} = -\omega_n \zeta - j\omega_n\sqrt{1 - \zeta^2}$. In special cases where $\zeta = 1$, the two complex conjugate roots become a double-real root $s = -\omega_n = \sigma$. Therefore the $\zeta = 1$ curve represents the loci of points that correspond to the double-real root of (A.1). The first objective to consider in connection with the $\zeta = 1$ curve is that this curve can be obtained from the general equations

$$\sum_{k=0}^{n} a_k Y_{k-1}^* = 0 \tag{A.72a}$$

$$\sum_{k=0}^{n} a_k Y_k^* = 0, \tag{A.72b}$$

where Y_k^* functions are related to Chebyshev's functions $U_k(\zeta)$ as

$$Y_k^* = (-1)^{k+1} \omega_n^{k-1} U_k(\zeta). \tag{A.71}$$

To prove the above statement, note that the polynomial $F(s)$ is a continuous function of s and introduce the derivative $F'(s)$ of $F(s)$. We proved in the section on Theorem A.2 that if a polynomial $F(s)$ has a zero λ of multiplicity ν, then it can be written as

$$F(s) = (s - \lambda)^\nu Q(s), \tag{A.109}$$

where $Q(s)$ is a quotient polynomial in s. By differentiating (A.109) with respect to s, we obtain

$$F'(s) = \nu(s - \lambda)^{\nu-1} Q(s) + (s - \lambda)^\nu Q'(s). \tag{A.110}$$

From (A.109), it follows that λ is a zero of multiplicity $\nu - 1$ of the polynomial $F'(s)$. By repeating the differentiation of (A.109) exactly $\nu - 1$ times, we can see that λ is a zero of multiplicity 1 of the polynomial $F^{(\nu-1)}(s) = 0$.

In the case of $\nu = 2$, in which the polynomial $F(s)$ has a double-real zero $s = \lambda = \sigma$, it follows from (A.1) that

$$F'(\sigma) \equiv \sum_{k=0}^{n} k a_k \sigma^{k-1} = 0. \tag{A.111}$$

It can be shown now that for $\zeta = 1$, (A.72a) is equivalent to (A.111). From the relationship

$$T_k(\zeta) = \cos(k \arccos \zeta) \tag{A.62}$$

and

$$T_k(\zeta) = \zeta U_k(\zeta) - U_{k-1}(\zeta) \tag{A.64}$$

it follows that $U_k(1) = k$. Then, from (A.71) we have $Y_k^*(1) = k\sigma^{k-1}$ (for $\zeta = 1$, $\omega_n = -\sigma$) and (A.72b) becomes

$$\sum_{k=0}^{n} a_k k \sigma^k = 0. \tag{A.112}$$

If (A.112) is multiplied by σ^{-1}, it becomes equivalent to (A.111), provided that $\sigma \neq 0$. Therefore (A.72) can be derived from

$$\sum_{k=0}^{n} a_k X_k = \sum_{k=0}^{n} a_k(-\omega_n \zeta Y_k^* - \omega_n^2 Y_{k-1}^*) = 0 \qquad (A.113)$$

$$\sum_{k=0}^{n} a_k Y_k = \sum_{k=0}^{n} a_k \omega_n \sqrt{1 - \zeta^2} Y_k^* = 0$$

regardless of the fact that ζ may be equal to 1. The above statement is thereby proved. In other words, for $\zeta = 1$, equations A.72 are derived from the condition $F(\sigma) = F'(\sigma) = 0$, rather than from $R(\sigma) = I(\sigma) = 0$.

The fact that along the $\zeta = 1$ curve both $F(\sigma)$ and $F'(\sigma)$ vanish leads to a conclusion that the $\zeta = 1$ curve is an envelope of the σ curves [A.10]; or, for each point on the $\zeta = 1$ curve, there is a fixed σ curve that has at that point a common tangent with the curve $\zeta = 1$. To prove this, consider the equation

$$\sum_{k=0}^{n} a_k s^k = 0, \qquad (A.1)$$

where s is replaced by σ. If α is considered as an independent variable by differentiating (A.1) with respect to α, and remembering that $a_k = a_k(\alpha, \beta)$, we obtain

$$\sum_{k=0}^{n} \left[\frac{\partial a_k}{\partial \alpha} + \frac{\partial a_k}{\partial \beta} \frac{\partial \beta}{\partial \alpha} \right] \sigma^k - F'(\sigma) \frac{\partial \sigma}{\partial \alpha} = 0. \qquad (A.114)$$

Along the σ constant curve, $\partial \sigma / \partial \alpha = 0$, and the slope $\partial \beta / \partial \alpha$ of the curve is obtained from (A.114) as

$$\frac{\partial \beta}{\partial \alpha} = - \frac{\sum_{k=0}^{n} (\partial a_k / \partial \alpha) \sigma^k}{\sum_{k=0}^{n} (\partial a_k / \partial \beta) \sigma^k}. \qquad (A.115)$$

Since along the $\zeta = 1$ curve $F(\sigma) = F'(\sigma) = 0$, from (A.114) it follows that the slope $\partial \beta / \partial \alpha$ of $\zeta = 1$ curve is the same as that of (A.115), and this curve is an envelope of the σ curves.

The envelope property of the $\zeta = 1$ curve is important in the case where the coefficients a_k are linear functions of system parameters and the σ curves are straight lines in the $\alpha\beta$ plane. Then by drawing a tangent to the curve $\zeta = 1$ from a specific point $M_1(\alpha_1; \beta_1)$, it is possible to determine the negative real root σ_1 of the characteristic equation, which corresponds to the parameter values α_1 and β_1. The real roots are read from the points of tangency at the $\zeta = 1$ curve on which the values of ω_n are interpolated. Furthermore, it can be concluded that there are as many negative real roots as there are tangents to the $\zeta = 1$ curve that are drawn from a chosen point $M_1(\alpha_1; \beta_1)$. This fact is illustrated in Figure 1.25 of Section 1.6.

Derivation A.4. Singular Cases. As shown in Section 1.4, the mapping of the points from the real axis of the s plane by an algebraic equation with real coefficients results in an infinite set of points that lie on the corresponding σ curves. The mapping of the real values of s may, therefore, be considered as a singular case in which the Jacobian $J \equiv 0$. There are also other cases in which the images of the points lying elsewhere in the s plane are not distinct points but rather loci of points in the parameter $\alpha\beta$ plane. These cases are treated here in more detail.

Consider a polynomial

$$F(s) \equiv \sum_{k=0}^{n} a_k s^k, \qquad a_n \neq 0 \tag{A.1}$$

in which the coefficients $a_k = b_k \alpha + c_k \beta + d_k$ are linear functions of the parameters α and β. As shown, by substituting $s = \sigma + j\omega$, (A.1) may be rewritten as two linear equations in two unknowns α and β as

$$\begin{aligned} \alpha B_1(\sigma, \omega) + \beta C_1(\sigma, \omega) + D_1(\sigma, \omega) &= 0 \\ B_2(\sigma, \omega) + \beta C_2(\sigma, \omega) + D_2(\sigma, \omega) &= 0. \end{aligned} \tag{A.116}$$

For the values of σ and ω for which the rank of the matrix

$$\begin{bmatrix} B_1(\sigma, \omega) & C_1(\sigma, \omega) & D_1(\sigma, \omega) \\ B_2(\sigma, \omega) & C_2(\sigma, \omega) & D_2(\sigma, \omega) \end{bmatrix} \tag{A.117}$$

is equal to two, (A.116) can be solved for α and β as

$$\begin{aligned} \alpha &= \frac{C_1 D_2 - C_2 D_1}{B_1 C_2 - B_2 C_1} \\ \beta &= \frac{B_2 D_1 - B_1 D_2}{B_1 C_2 - B_2 C_1}. \end{aligned} \tag{A.118}$$

For certain values of $s = \sigma + j\omega$ the rank of matrix (A.117) may be equal to one and the solution is a straight line

$$\{\alpha B(s) + \beta C(s) + D(s)\}_{s=\sigma+j\omega} = 0. \tag{A.119}$$

For the case where the rank of matrix (A.117) is zero, it means that $s = \sigma + j\omega$ is a common zero of the polynomials $B(s)$, $C(s)$, and $D(s)$, and the case requires no particular consideration.

In Section 1.5 the singular case of the σ curves for which $\omega = 0$ was considered. The cases in which the rank of matrix (A.117) is equal to one but $\omega \neq 0$ is discussed here. This is related to the case where equations A.116 have the form

$$\alpha B_1 + \beta C_1 + D_1 = 0 \tag{A.120a}$$

$$D_2 = 0 \tag{A.120b}$$

and the parameter plane curve is a straight line determined by (A.120a) for those values of σ and ω that are the real roots of (A.120b).

If the straight line (A.120a) is considered as a Σ or ω curve, the shading is determined by the sign of the derivatives $\partial \sigma / \partial \alpha$ or $\partial \omega / \partial \alpha$, respectively. To determine these derivatives, differentiate (A.120) with respect to α as an independent variable to obtain

$$\begin{aligned} \frac{\partial R}{\partial \sigma} \frac{\partial \sigma}{\partial \alpha} + \frac{\partial R}{\partial \omega} \frac{\partial \omega}{\partial \alpha} + B_1 &= 0 \\ \frac{\partial I}{\partial \sigma} \frac{\partial \sigma}{\partial \alpha} + \frac{\partial I}{\partial \omega} \frac{\partial \omega}{\partial \alpha} &= 0, \end{aligned} \tag{A.121}$$

where $R = \alpha B_1 + \beta C_1 + D_1 = 0$ and $I = D_2$. From (A.121) we obtain

$$\begin{aligned} \frac{\partial \sigma}{\partial \alpha} &= -\frac{B_1 \, \partial D_2/\partial \omega}{J_1} \\ \frac{\partial \omega}{\partial \alpha} &= -\frac{B_1 \, \partial D_2/\partial \sigma}{J_1}, \end{aligned} \tag{A.122}$$

where J_1 is always positive (see Section 1.5). Thus

$$\begin{aligned} \operatorname{sign} \partial \sigma / \partial \alpha &= \operatorname{sign} (-B_1 \partial D_2/\partial \omega) \\ \operatorname{sign} \partial \omega / \partial \alpha &= \operatorname{sign} (-B_1 \partial D_2/\partial \sigma). \end{aligned} \tag{A.123}$$

Similarly, conditions (A.123) can be derived with respect to the parameter β.

To illustrate the above development, consider the algebraic equation

$$s^4 + s^3 + \alpha s^2 + s + \beta = 0 \tag{A.124}$$

and determine the $\Sigma = 0$ curve. For $s = j\omega$ ($\sigma = 0$), equations A.124 can be rewritten as

$$-\alpha \omega^2 + \beta + \omega^4 = 0 \tag{A.125a}$$

$$-\omega^3 + \omega = 0. \tag{A.125b}$$

The roots of (A.125) are $\omega_1 = 0$, $\omega_{2,3} = \pm 1$. Therefore the $\Sigma = 0$ curve is represented by two straight lines

$$\beta = 0 \tag{A.126a}$$

$$-\alpha + \beta + 1 = 0, \tag{A.126b}$$

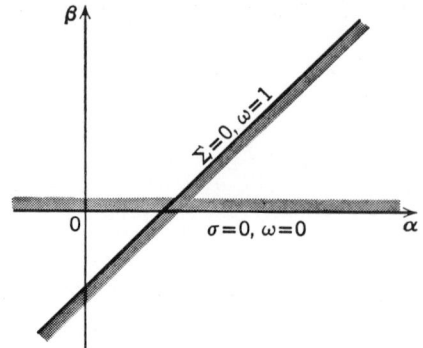

Figure A.4 Parameter plane diagram for the singular case.

which are plotted in Figure A.4. The straight line $\beta = 0$ is a real root boundary of which the line (A.126b) is a complex root boundary. To shade these lines, note that for $\omega_1 = 0$ we have sign $\partial\sigma/\partial\beta = $ sign $(-C_1 \partial D_2/\partial\omega) = $ sign $(3\omega^2 - 1) = -1$, and the upper side of the $\beta = 0$ curve is shaded (since that side corresponds to a decrease in σ). For $\omega_{2,3} = \pm 1$ we get sign $\partial\sigma/\partial\alpha = $ sign $(-B_1 \partial D_2/\partial\omega) = $ sign $(-3\omega^4 + \omega^2) = -1$, and the complex root boundary $\Sigma = 0$ is shaded as shown in Figure A.3.

It is of interest to note that in the singular case discussed above the $\Sigma = 0$ curve is simultaneously the $\omega = 1$ curve. Therefore the complex roots of (A.124) can cross the imaginary axis of the s plane only at the point $s = \pm j\omega = \pm j1$ no matter what the parameters α and β are.

The singular case of $J \equiv 0$, $\omega \neq 0$, has been discussed for a_k's as linear functions of α and β. The discussion, however, can be extended to the case when a_k's depend nonlinearily on the parameters α and β.

Derivation A.5. Sensitivity of Multiple Roots. In Section 1.7 it was pointed out that the root sensitivity of double-real roots cannot be obtained by treating double-real roots as real. The reason for this was that in an expression of real root sensitivity,

$$S^s_{j,r} = p_r \frac{\sum\limits_{k=0}^{n}(\partial a_k/\partial p_r)s_j^k}{\sum\limits_{k=0}^{n} kas_j^{k-1}}, \qquad (A.127)$$

the first derivative $F'(s)$ of the polynomial $F(s)$ appears in the denominator. Since for double-real roots both $F(s)$ and $F'(s)$ go to zero simultaneously, the value of $S^s_{j,r}$ is infinite [A.11].

The sensitivity of double-real roots should be found as the sensitivity of the pair of complex conjugate roots for which $\zeta_i = 1$ and $\omega_{ni} = -s_j$.

Then for small variations of parameters the double-real roots become two distinct real roots ($\zeta > 1$) or two complex conjugate roots ($\zeta < 1$), or they remain a double-real root ($\zeta = 1$). Therefore the sign of the sensitivity $S_{i,r}^{\zeta}$ indicates all three possible situations. It should be noted, however, that the sensitivity $S_{i,r}^{\zeta}$ for $\zeta = 1$ cannot be calculated from the sensitivities $S_{i,r}^{\sigma}$ and $S_{i,r}^{\omega}$ as shown in Section 1.7. This is true because the expressions for the latter sensitivities are obtained from equations

$$R \equiv \sum_{k=0}^{n} a_k X_k = 0 \qquad (A.90a)$$

$$I \equiv \sum_{k=0}^{n} a_k Y_k = 0, \qquad (A.90b)$$

in which (A.90b) for $\zeta = 1$ becomes identically zero. (For further explanation see equation A.69.)

The double-real root sensitivity should be calculated from equations

$$\sum_{k=0}^{n} a_k Y_{k-1}^{*} = 0 \qquad (A.72a)$$

$$\sum_{k=0}^{n} a_k Y_k^{*} = 0, \qquad (A.72b)$$

which, for double-real roots ($\zeta = 1$), represent the condition that $F(s)$ and $F'(s)$ simultaneously go to zero (see Derivation A.3).

By considering $a_k = a_k(p_r)$, $Y_k^{*} = Y_k(\omega_n, \zeta)$, and differentiating (A.72) with respect to p_r, we obtain

$$\frac{\partial \ln \omega_n}{\partial \ln p_r} \frac{\omega_n}{p_r} \sum_{k=0}^{n} a_k \frac{Y_{k-1}^{*}}{\partial \omega_n} + \frac{\partial \ln \zeta}{\partial \ln p_r} \frac{\zeta}{p_r} \sum_{k=0}^{n} a_k \frac{Y_{k-1}^{*}}{\partial \zeta} + \sum_{k=0}^{n} \frac{\partial a_k}{\partial p_r} Y_{k-1}^{*} = 0$$

$$\qquad (A.128)$$

$$\frac{\partial \ln \omega_n}{\partial \ln p_r} \frac{\omega_n}{p_r} \sum_{k=0}^{n} a_k \frac{\partial Y_k^{*}}{\partial \omega_n} + \frac{\partial \ln \zeta}{\partial \ln p_r} \frac{\zeta}{p_r} \sum_{k=0}^{n} a_k \frac{\partial Y_k^{*}}{\partial \zeta} + \sum_{k=0}^{n} \frac{\partial a_k}{\partial p_r} Y_k^{*} = 0,$$

where

$$\frac{\partial Y_k^{*}}{\partial \omega_n} = \frac{k-1}{\omega_n} Y_k^{*} \qquad (A.129)$$

and

$$(k-1)\frac{\partial Y_{k+1}^{*}}{\partial \zeta} + 2\zeta\omega_n k \frac{\partial Y_k^{*}}{\partial \zeta} + \omega_n^{2}(k+1)\frac{\partial Y_{k-1}^{*}}{\partial \zeta} = 0 \qquad (A.130)$$

with $\partial Y_0^{*}/\partial \zeta \equiv \partial Y_1^{*}/\partial \zeta \equiv 0$, $\partial Y_2^{*}/\partial \zeta = -2\omega_n$. Equation A.129 can be obtained from (A.71), whereas the relationship (A.130) can be proved by mathematical induction.

408   Basic Theorems and Derivations

Now, according to the definitions of sensitivities,

$$S_{i,r}^{\omega_n} \equiv \frac{\partial \ln \omega_n}{\partial \ln p_r} \tag{A.131}$$

$$S_{i,r}^{\zeta} \equiv \frac{\partial \ln \zeta}{\partial \ln p_r}, \tag{A.132}$$

(A.128) may be rewritten as

$$P_1 S_{i,r}^{\omega_n} + Q_1 S_{i,r}^{\zeta} + R_1 = 0$$
$$P_2 S_{i,r}^{\omega_n} + Q_2 S_{i,r}^{\zeta} + R_2 = 0, \tag{A.133}$$

in which

$$P_1 = \sum_{k=0}^{n} (k-2) a_k Y_{k-1}^*, \qquad P_2 = \sum_{k=0}^{n} (k-1) a_k Y_k^*$$

$$Q_1 = \zeta_i \sum_{k=0}^{n} \frac{\partial Y_{k-1}^*}{\partial \zeta}, \qquad Q_2 = \zeta_i \sum_{k=0}^{n} a_k \frac{\partial Y_k^*}{\partial \zeta} \tag{A.134}$$

$$R_1 = p_r^0 \sum_{k=0}^{n} \frac{\partial a_k}{\partial p_r^0} Y_{k-1}^*, \qquad R_2 = p_r^0 \sum_{k=0}^{n} \frac{\partial a_k}{\partial p_r^0} Y_k^*.$$

Equations A.134 are written for a complex pair of roots determined by $\omega_n = \omega_{ni}$, $\zeta = \zeta_i$, and $\partial a_k/\partial p_r^0 = (\partial a_k/\partial p_r)_{p_r=p_r^0}$.

Equations A.133 represent two equations in two unknowns, $S_{i,r}^{\omega_n}$ and $S_{i,r}^{\zeta}$, which can be solved to obtain

$$S_{i,r}^{\omega_n} = \frac{Q_1 R_2 - Q_2 R_1}{P_1 Q_2 - P_2 Q_1}$$
$$S_{i,r}^{\zeta} = \frac{P_2 R_1 - P_1 R_2}{P_1 Q_2 - P_2 Q_1}. \tag{A.135}$$

To illustrate the root sensitivity analysis and discuss the double-real root case, consider equation

$$s^3 + \alpha s^2 + \beta s + 1 = 0. \tag{A.136}$$

It has a double-real root $s_{1,2} = -2$ for $\alpha = p_1 = 4.25$, and $\beta = p_2 = 5$. To determine the sensitivities $S_{1,1}^{\omega_n}$ and $S_{1,1}^{\zeta}$, (A.134) and (A.135) are calculated for $\zeta_i = 1$, $\omega_{ni} = 2$, and $p_1 = 4.25$,

$$S_{1,1}^{\omega_n} = -0.163, \qquad S_{1,1}^{\zeta} = 1.21. \tag{A.137}$$

The positive sign of the sensitivity $S_{1,1}^{\zeta}$ indicates that for a small increase of the parameter $p_1 = \alpha$ from its nominal value $\alpha = 4.25$, the double-real roots $s_{1,2}$ become two distinct roots for which $\zeta > 1$.

Similarly, the sensitivities $S_{1,2}^{\omega_n}$ and $S_{1,2}^{\zeta}$ are found to be

$$S_{1,2}^{\omega_n} = 0.83, \qquad S_{1,2}^{\zeta} = -0.73, \qquad (A.138)$$

which indicates that for a small change in the parameter $\beta = p_2$ from its nominal value $\beta = 5$, while keeping $\alpha = 4.25$, the double-real root $s_{1,2} = -2$ becomes two complex conjugate roots. The sensitivity $S_{1,2}^{\zeta}$ is less than zero.

The above results can be generalized [A.12] if the algebraic equation $F(s) = 0$ is considered in the form

$$F(s, \mathbf{p}) = 0, \qquad (A.139)$$

where s_i, $(i = 1, 2, \ldots, n)$, are the roots (real or complex) and $\mathbf{p}$ is a vector with parameters p_r, $(r = 1, 2, \ldots, m)$. The nominal or unperturbed values of the parameters are denoted by p_r^0.

When s_i is a νth order root (that is, $(s - s_i)^\nu$ is a factor of F), the first $\nu - 1$ derivatives of F with respect to s_i are zero. In this case the total differential dF should be expanded to incorporate the first nonzero order term, so that

$$dF = \frac{1}{\nu!} \frac{\partial^\nu F}{\partial s_i^\nu} (ds_i)^\nu + \sum_{r=1}^{m} \frac{\partial F}{\partial p_r^0} dp_r. \qquad (A.140)$$

Equation A.140 can be solved for ds_i to give

$$ds_i = \left[-\frac{\nu!}{\frac{\partial^\nu F}{\partial s^\nu}(s)} \sum_{r=1}^{m} \frac{\partial F}{\partial p_r}(s) \, dp_r \right]_{\substack{s=s_i^0 \\ p=p_r}}^{1/\nu}. \qquad (A.141)$$

This relates differential changes in the characteristic root s_i due to differential changes in the parameters p_r. Then the sensitivity can be defined as

$$S_{i,r} \equiv -\nu! \left[\frac{\partial F/\partial p_r(s)}{\partial^\nu F/\partial s^\nu(s)} \right]_{\substack{s=s_i^0 \\ p=p_r}}, \qquad (A.142)$$

and the differential increment in the ith root s_i becomes

$$ds_i = \left(\sum_{r=1}^{m} S_{i,r} \, dp_r \right)^{1/\nu}. \qquad (A.143)$$

Note that sensitivity $S_{i,r}$ thus defined is a complex number if the root s_i is a complex root of equation $F(s) = 0$.

References

[A.1] W. S. Burnside and A. W. Panton, *The Theory of Equations*, 2d ed., Hodges, Figgis & Co., Dublin, Ireland, 1886.

[A.2] G. Chrystal, *Algebra*, Part I, 5th ed., Dover, New York, 1961.

[A.3] J. V. Uspensky, *Theory of Equations*, McGraw-Hill, New York, 1948.
[A.4] M. Marden, *The Geometry of the Zeros of a Polynomial in a Complex Variable*, American Mathematics Society, New York, 1949.
[A.5] E. Goursat, *A Course in Mathematical Analysis*, Vols. 1 and 2, Dover, New York, 1959.
[A.6] L. V. Ahlfors, *Complex Analysis*, McGraw-Hill, New York, 1953.
[A.7] Z. Nehari, *Conformal Mapping*, McGraw-Hill, New York, 1952.
[A.8] D. D. Šiljak, Analysis and Synthesis of Feedback Control Systems in the Parameter Plane, Pt. I—Linear Continuous Systems, *IEEE Trans.*, Pt. II (*Applications and Industry*), **83**, 449–458 (November 1964).
[A.9] Tables of Chebyshev Polynomials $S_n(x)$ and $C_n(x)$, National Bureau of Standards, Applied Mathematics Series No. 9, U.S. Government Printing Office, Washington, D.C., 1952.
[A.10] D. D. Šiljak, Generalization of the Parameter Plane Method, *IEEE Trans. Auto. Control*, **AC-11**, No. 1, 63–70 (January 1966).
[A.11] P. Kokotović and D. Šiljak, The Sensitivity Problem in Continuous and Sampled-Data Linear Control Systems by Generalized Mitrović's Method, *IEEE Trans.*, Pt. II (*Applications and Industry*), **83**, 324–328 (September 1964).
[A.12] R. L. Stapleford and D. T. McRuer, Sensitivity of Multiloop Flight Control System Roots to Open-loop Parameter Variations, *Proceedings of the 1966 JACC Conf.*, Seatle, Washington, 399–407 (August 1966).

APPENDIX B

Computer Applications

B.1 Introduction

Two problems are considered here: (1) plotting of the characteristic curves in the $\alpha\beta$ plane, and (2) the numerical solution of algebraic equations on both analog and digital computers. Both problems are of interest in system analysis and design. The plotting of the characteristic curves facilitates the analysis in the parameter plane, and the numerical solution of the algebraic equations enables a straightforward procedure for the inverse parameter mapping to obtain the root loci in the complex s plane. Thus, for design purposes both procedures can be programmed to obtain sufficient information about pole-zero configurations of relevant transfer functions as the system parameters are adjusted. This approach is utilized in the design of linear systems, which is considered in Chapter 2.

B.2 Plotting of the Characteristic Curves

To suggest a possible algorithm for plotting the characteristic curves by a digital computer, consider the case where the algebraic equation

$$F(s) \equiv \sum_{k=0}^{n} a_k s^k = 0, \tag{B.1}$$

has the coefficients a_k given as

$$a_k = b_k \alpha + c_k \beta + d_k. \tag{B.2}$$

By substituting

$$s = \sigma + j\omega \tag{B.3}$$

in (B.1), we obtain

$$\begin{aligned} B_1\alpha + C_1\beta + D_1 &= 0 \\ B_2\alpha + C_2\beta + D_2 &= 0, \end{aligned} \tag{B.4}$$

where

$$B_1 = \sum_{k=0}^{n} b_k X_k, \qquad B_2 = \sum_{k=0}^{n} b_k Y_k$$

$$C_1 = \sum_{k=0}^{n} c_k X_k, \qquad C_2 = \sum_{k=0}^{n} c_k Y_k \qquad (B.5)$$

$$D_1 = \sum_{k=0}^{n} d_k X_k, \qquad D_2 = \sum_{k=0}^{n} d_k Y_k.$$

In (B.5) the functions $X_k = X_k(\sigma, \omega)$, $Y_k = Y_k(\sigma, \omega)$ are given by the recurrence relations

$$\begin{aligned} X_{k+1} - 2\sigma X_k + (\sigma^2 + \omega^2) X_{k-1} &= 0 \\ Y_{k+1} - 2\sigma Y_k + (\sigma^2 + \omega^2) Y_{k-1} &= 0, \end{aligned} \qquad (B.6)$$

where $X_0 \equiv 1$, $X_1 \equiv \sigma$, $Y_0 \equiv 0$, $Y_1 \equiv \omega$.

Equations B.4 can be considered as two equations in two unknowns, α and β, which can be solved for α and β to get

$$\begin{aligned} \alpha &= \alpha(\sigma, \omega) \\ \beta &= \beta(\sigma, \omega), \end{aligned} \qquad (B.7)$$

provided that the Jacobian

$$J = \Delta = B_1 C_2 - B_2 C_1 \qquad (B.8)$$

is different from zero.

To obtain the values of α and β for given values of σ and ω and thus plot a desired characteristic curve, the following algorithm can be used.

1. Read the coefficients b_k, c_k, and d_k in the coefficients a_k.
2. Read in the required s plane contour and the number of points to be plotted.
3. Calculate the necessary X_k and Y_k functions for $k = 0, 1, \ldots, n$ for the first point of interest on the contour (that is, σ_0, ω_0) using (B.6).
4. Calculate the values of $B_1, C_1, D_1, B_2, C_2, D_2$ from (B.5). If $\omega_0 = 0$, go to step 8.
5. Solve the parameter plane equations B.4 for α and β and calculate the Jacobian using (B.8). If $J = 0$, go to step 7.[1]
6. Print $\sigma_0, \omega_0, \alpha, \beta, J$ and plot on the $\alpha\beta$ plane.
7. Calculate the next point on the s plane contour (that is, σ_1, ω_1) and set $\sigma_0 = \sigma_1$, $\omega_0 = \omega_1$ and repeat steps 3, 4, 5, 6, 7 for all the points of interest on the contour; thus the calculations are completed.

---

[1] There are singular cases when $J = 0$, but a section of the corresponding characteristic curve still exists (see Derivation A.4, Appendix A). A specific part of the computer program can be readily written to incorporate the singular cases.

8. For this step $\omega = 0$, and a range of values of α are specified and each α, β is calculated using (B.2); the $s = \sigma_0$ contour is plotted in the $\alpha\beta$ plane.

The algorithm can be readily extended to more complicated functions, $a_k(\alpha, \beta)$, discussed in Derivation A.2 of Appendix A.

B.3 Automatic Analog Solution of Algebraic Equations and the Plotting of Root Loci

Numerous methods for analog solution of algebraic equations and the plotting of root loci have been developed. Methods presented in the computer literature can be divided essentially into two categories: methods that employ special-purpose computers (cited in references B.1–5), and those that employ general-purpose computers (cited in references B.6–8). Application of the former techniques is significantly limited because electromechanical devices not usually available in the standard analog computing equipment are required. Therefore methods utilizing general-purpose computers have attracted greater attention in common engineering design. However, the disadvantage of such methods as developed so far is that they require considerable participation of the computer operator in the solution process; hence relatively complex schemes are employed in trial-and-error procedures.

In comparing references B.1–8, the computing technique described in reference B.9 represents an important advance, for it extends the method of steepest descent [B.10] to the analog solution of algebraic equations and the finding of root loci. It enables the closing of the feedback loop, and the computing system determines the root values automatically. But the disadvantage of the proposed computer setup is that special devices for harmonic synthesis cannot be avoided.

In this section a method is presented that applies the parameter plane method and the steepest descent approach [B.9–11] to the automatic solution of algebraic equations and the plotting of root loci. The method was first presented in reference B.12 and then extended to digital computer application in reference B.13.

The significance of the new procedure consists in its simplicity and its use of the general-purpose analog computing equipment, such as potentiometers, summers, and servomultipliers. Moreover, the corresponding computer circuit can be realized on extremely small-sized analog facilities. Since the solutions are obtained in the rectangular coordinates, no transformations of output data are necessary. Furthermore, all main computer circuits may be programmed on a removable patch-board for highest-degree equations expected in the computer application. Thus, the programming of lower-degree equations may be reduced to a minimum.

For the solution of higher-degree algebraic equations the use of a digital computer may be preferable [B.13]. This is discussed in Section B.4.

Steepest Descent Equations: For the application of the steepest descent approach [B.9–13], the problem of solving the algebraic equation

$$F(s) \equiv \sum_{k=0}^{n} a_k s^k = 0, \tag{B.1}$$

where a are real coefficients and $s = \sigma + j\omega$ is to be reformulated in terms of the corresponding set of differential equations

$$\frac{d\sigma}{dt} = -h \frac{\partial V}{\partial \sigma}$$
$$\frac{d\omega}{dt} = -h \frac{\partial V}{\partial \omega}. \tag{B.9}$$

In (B.9) the variables σ and ω are considered time-dependent variables, h is a positive constant, and $V(\sigma, \omega)$ is a function satisfying the following conditions:

1. V is nonnegative
2. The derivatives $\partial V/\partial \sigma$ and $\partial V/\partial \omega$ exist
3. The zeros of V are located at the roots of $F(s) = 0$
4. The zeros of V are the only minima of V
5. The derivative dV/dt is always negative.

With the application of steepest descent equations B.9, the search for the roots of (B.1) is reduced to the minimization of function V. The minima of V are the steady-state solutions of (B.9), and according to requirement (3), are located at the roots of (B.1). Therefore the computer, which is supposed to solve (B.9), will automatically seek out a particular root in dependence on the chosen initial conditions (σ_0, ω_0). The other stationary points are immaterial to the minimization process since they are the points of unstable equilibrium. As the transient solutions of (B.9) represent the steepest descent trajectories, the roots may be found at a rapid rate by this process.

When the coefficients a_k are given numerically, the polynomial $F = F(s)$ can be given as

$$F = R + jI, \tag{B.10}$$

where $R = R(\sigma, \omega)$, $I = I(\sigma, \omega)$. The selection of V as a function of R and I is not critical. In the following developments the function

$$V = |R| + |I| \tag{B.11}$$

will be used. It has been shown [B.9, 12, 14] that this form of V satisfies the requirements listed above and represents a suitable minimizing criterion with an appropriate selectivity.

After F is specified as a function of R and I, the problem consists in deriving the expressions of partial derivatives $\partial V/\partial \sigma$ and $\partial V/\partial \omega$ in a form suitable for the mechanization of the steepest descent equations B.9. Since, according to (B.10) and (B.11), V is a function of the variables R and I that, in turn, are functions of the independent variables σ and ω, the derivatives $\partial V/\partial \sigma$ and $\partial V/\partial \omega$ are given by

$$\frac{\partial V}{\partial \sigma} = \frac{\partial V}{\partial R}\frac{\partial R}{\partial \sigma} + \frac{\partial V}{\partial I}\frac{\partial I}{\partial \sigma}$$

$$\frac{\partial V}{\partial \omega} = \frac{\partial V}{\partial R}\frac{\partial R}{\partial \omega} + \frac{\partial V}{\partial I}\frac{\partial I}{\partial \omega}.$$
(B.12)

From (B.11), it follows that $\partial V/\partial R = \text{sign } R$, and $\partial V/\partial I = \text{sign } I$, including by convention that sign $0 \equiv 0$. Taking into account that the functions R and I are analytic, that is,

$$\frac{\partial R}{\partial \sigma} = \frac{\partial I}{\partial \omega}$$

$$\frac{\partial R}{\partial \omega} = -\frac{\partial I}{\partial \sigma},$$
(B.13)

equations B.12 can be reduced to

$$\frac{\partial V}{\partial \sigma} = \frac{\partial R}{\partial \sigma}\text{sign } R + \frac{\partial I}{\partial \sigma}\text{sign } I$$

$$\frac{\partial V}{\partial \omega} = \frac{\partial R}{\partial \sigma}\text{sign } I - \frac{\partial I}{\partial \sigma}\text{sign } R$$
(B.14)

with sign $0 \equiv 0$.

With relationship (B.14) the partial derivatives $\partial V/\partial \sigma$ and $\partial V/\partial \omega$ are expressed in terms of R, I, $\partial R/\partial \sigma$, and $\partial I/\partial \sigma$. Therefore the simplicity of the computer circuit, necessary for the instrumentation of (B.9), depends inherently upon the expressions for R, I, $\partial R/\partial \sigma$, and $\partial I/\partial \sigma$. Application of the parameter plane method enables these expressions to be obtained in a convenient form. While the following mathematical developments seem to be somewhat involved, it should be emphasized that the actual application of the derived results is particularly simple.

The real and imaginary parts of the polynomial $F(s)$ are

$$R \equiv \sum_{k=0}^{n} a_k X_k$$

$$I \equiv \sum_{k=0}^{n} a_k Y_k.$$
(B.15)

By using (1.49) of Chapter 1, we have

$$X_k = \sigma Y_k^* - (\sigma^2 + \omega^2) Y_{k-1}^*, \tag{B.16}$$

where

$$Y_{k+1}^* - 2\sigma Y_k^* + (\sigma^2 + \omega^2) Y_{k-1}^* = 0, \tag{B.17}$$

with $Y_0^* \equiv 0$, $Y_1^* \equiv 1$. From (1.47),

$$Y_k = \omega Y_k^*. \tag{B.18}$$

Applying (B.16), (B.17), and (B.18) to (B.15), we can readily derive

$$R = a_0 + \sigma \sum_{k=1}^{n} a_k Y_k^* - (\sigma^2 + \omega^2) \sum_{k=1}^{n} a_k Y_{k-1}^*$$

$$I = \omega \sum_{k=1}^{n} a_k Y_k^* \tag{B.19}$$

and

$$\frac{\partial R}{\partial \sigma} = \sum_{k=1}^{n} k a_k Y_k^* - \sigma \sum_{k=1}^{n} k a_k Y_{k-1}^*$$

$$\frac{\partial I}{\partial \sigma} = \omega \sum_{k=1}^{n} k a_k Y_{k-1}^*. \tag{B.20}$$

In deriving (B.20) the relationship

$$\frac{\partial^m Y_k}{\partial \sigma^m} = k(k-1) \cdots (k-m+1) Y_{k-m}^* \tag{B.21}$$

is used, which can be obtained directly from the expression $s^k = X_k + jY_k$ and $Y_k = \omega Y_k^*$.

Equations B.19 and B.20, through (B.14), enable the steepest descent equations B.9 to be easily solved on a general-purpose analog computer.

Computer Circuit and Operation: The analog computer, based on the procedure outlined in the foregoing section, is shown in Figure B.1. Three distinct parts of the computer circuit may be recognized: Y_k^* function generator, a_k coefficient potentiometers and summers, and the steepest descent circuit. Whereas in Figure B.1 the two former parts are given in block notation, the steepest descent circuit is represented in detail. For the sake of clarity, the circuit diagrams of Y^* function generator and a_k coefficient potentiometers and summers are shown separately in Figures B.2 and B.3, respectively.

The steepest descent circuit, as shown in Figure B.1, mechanizes (B.9) in regard to (B.14) by employing the operational amplifiers 1 through 6 and the relay amplifiers RA1 and RA2. The relays RL1 and RL2 are necessary

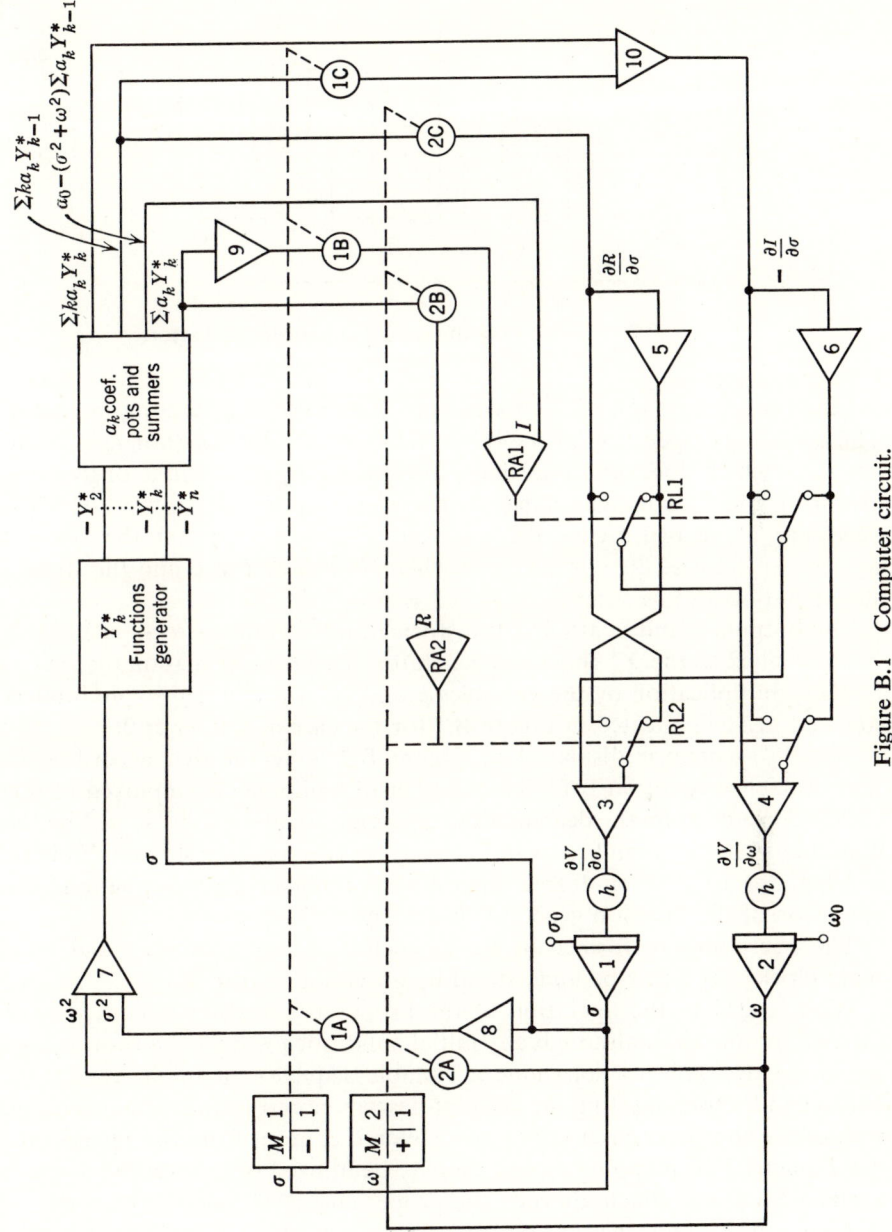

Figure B.1 Computer circuit.

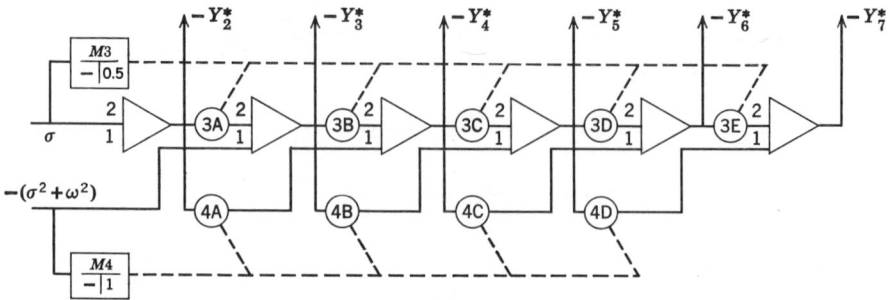

Figure B.2 Circuit diagram of Y_k^* functions generator.

for the implementation of the terms sign R and sign I appearing in (B.14). The oscillatory operation of the relays RL1 and RL2 when R or I, or both, are zero, which does not appreciably affect the outputs σ and ω of the integrators 1 and 2, is in accordance with the assumption sign $0 \equiv 0$, which is necessary in (B.14). Hence the transient solutions represent the steepest descent trajectories also in the cases where $R = 0$ or $I = 0$ and the steady-state solution will be reached as $R = I = 0$.

The outputs σ and ω are first transformed into σ and $-(\sigma^2 + \omega^2)$ before being applied to the Y_k^* functions generator. This transformation, as well as further multiplication by the variables σ and ω, is performed by multipliers $M1$ and $M2$ as indicated in Figure B.1 for the case $\sigma < 0$, $\omega > 0$.

The Y_k^* generator displayed in Figure B.2 is synthesized according to the recurrence equation B.17. The number of components employed in this circuit depends upon the degree of the algebraic equation to be solved by the computer. The circuit in Figure B.2, which uses two standard servomultipliers, corresponds to a seven-degree equation. For higher-degree equations the extension of the function generator is evident.

The operations performed by the a_k coefficient potentiometers and summers block may easily be understood by observing Figure B.3.

With regard to the operation of the proposed computer circuit, it is of interest to note that, although any initial conditions will yield a root, it may be found desirable to determine a suitable sequence of initial conditions beforehand. Disconnecting the steepest descent circuit enables the remaining part of the computer circuit to map any closed contour from the s plane into the F plane. This mapping procedure may be utilized to bracket the roots in s plane contours, which alleviate the proper choice of initial conditions.

Also worth noting here is the rate of descent along the gradient paths, which may vary, depending on the topology in the s plane. This rate, however, may easily be controlled manually, or even automatically, by adjusting the rate constant h in the steepest descent circuit shown in Figure B.1.

Automatic Analog Solution

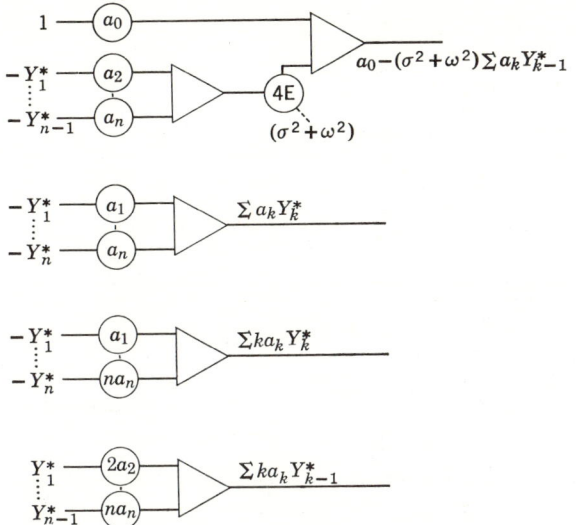

Figure B.3 Circuit diagram of a_k coefficient potentiometers and summers.

In order to illustrate the proposed method, the solution of the seven-degree algebraic equation

$$s^7 + 0.3s^6 + 1.2s^5 + 0.09s^4 + 0.52s^3 + 0.08s^2 + 0.16s + 0.16 = 0 \quad \text{(B.22)}$$

having the roots

$$s_1 = -0.5, \qquad s_{2,3} = 0.5 \pm j0.5$$
$$s_{4,5} = \pm j\sqrt{0.8}, \qquad s_{6,7} = -0.4 \pm j0.8$$

is demonstrated in Figure B.4, where only the upper half of the s plane is considered. The steepest descent trajectories, which begin at various initial conditions, converge to the roots s_1, s_3, s_5, and s_7. Figure B.4 shows the high accuracy of the roots obtained from the computer. In general the procedure enables more than 0.5% accuracy of the maximum root modulus to be achieved if the root values are read on the display unit of the digital voltmeter.

Note that the seven-degree algebraic equation was solved by using four servomultipliers. In order to arrive at the solution of twelve-degree equations, only two additional multipliers are required.

Root-Locus Plotting: In control engineering applications it is of interest to plot the root loci corresponding to an arbitrary parameter appearing in the coefficients of the characteristic equation, which is the algebraic equation with real coefficients. To illustrate the automatic plotting of such generalized root loci by the method outlined, the parameter p in the following equation

$$25s^4 + 35s^3 + 28.5s^2 + ps + 3.4 = 0 \quad \text{(B.23)}$$

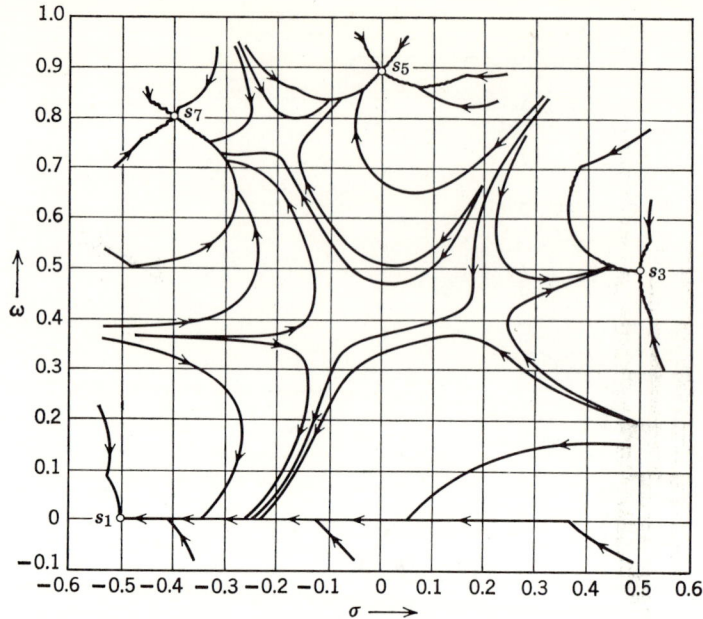

Figure B.4 Steepest descent trajectories and root locations.

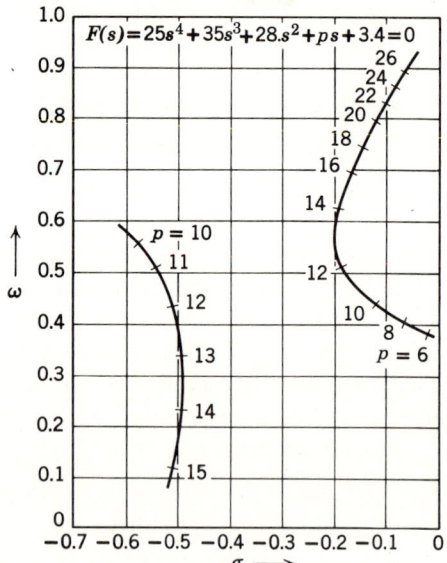

Figure B.5 Root locus plot.

is varied. By following the displacement of each root individually in accordance with the slow variations of the parameter p, the computer, which represents a closed-loop system, automatically plots the root loci, as can be seen in Figure B.5.

B.4 A Convergent Digital Computer Procedure for Solving Algebraic Equations

In many practical problems involving the solution of algebraic equations there is sufficient a priori knowledge about the roots to permit the application of the readily applied, rapidly converging synthetic division method based upon the relations outlined in Theorem A.2 of Appendix A. The limitation of the Newton-Raphson method, Lin's method, Bairstow's method, and other synthetic division methods is that convergence is heavily dependent on the initial conditions [B.15].

For problems in which the a priori information about the location of the roots is inadequate, the Lehmer-Schur method, Graeffe's root-squaring method [B.15], Bernoulli's method [B.15], or the methods of Lance [B.16], may be used and convergence is guaranteed. These methods are by no means simple and straightforward and are not rapidly convergent and thus, for efficiency, are usually used to calculate only the approximate root locations so that one of the more rapid synthetic division methods may then be applied. The need for a straightforward method that always converges and that has a rapid convergence on the region of the root is thus indicated.

In this section the steepest descent approach, outlined in the preceding analog procedure, is extended and adapted for the efficient digital computation of the zeros of analytic functions involving polynomials. The steepest descent approach is used again to minimize a nonnegative function, the minimal values of which are zero and correspond to the zeros of the analytic function under investigation. The magnitude of the increment in the direction of steepest descent for each iteration is calculated from equations similar to those used in the rapidly converging Newton-Bairstow method.

In particular, the rapid factorization of polynomials having either real or complex coefficients is achieved. By expressing polynomials in terms of X_k and Y_k functions, the methods are readily programmed on a digital computer.

Algorithm: For the numerical evaluation of the zeros of $F(s)$ the difference equations corresponding to the differential equations B.9 are considered:

$$\Delta\sigma = -h\frac{\partial V}{\partial \sigma} \quad \text{and} \quad \Delta\omega = -h\frac{\partial V}{\partial \omega}. \tag{B.24}$$

The direction of the steepest descent path, given by the ratio of $\Delta\sigma$ and $\Delta\omega$, is independent of h; that is,

$$\frac{\Delta\sigma}{\Delta\omega} = \frac{\partial V/\partial\sigma}{\partial V/\partial\omega}. \tag{B.25}$$

The magnitude of the increment in the direction given by $\Delta\sigma$ and $\Delta\omega$ is determined by h. The value of h should be chosen to minimize the function V along the direction of steepest descent. Consideration is now given to the calculation of a suitable h.

In the neighborhood of a zero of $F(s)$, the higher-order terms in a series expansion for R and I are negligible, and therefore the following approximations are valid:

$$-R \simeq \frac{\partial R}{\partial\sigma}\Delta\sigma + \frac{\partial R}{\partial\omega}\Delta\omega$$

$$-I \simeq \frac{\partial I}{\partial\sigma}\Delta\sigma + \frac{\partial I}{\partial\omega}\Delta\omega, \tag{B.26}$$

where $\Delta\sigma$ and $\Delta\omega$ are the distances in the σ and ω directions from a point in the neighborhood of the zero of $F(s)$ to the zero of $F(s)$. Using the Cauchy-Riemann equations and solving (B.26) for $\Delta\sigma$ and $\Delta\omega$ yields

$$\Delta\sigma \simeq \frac{-R(\partial R/\partial\sigma) - I(\partial I/\partial\sigma)}{(\partial R/\partial\sigma)^2 + (\partial I/\partial\sigma^2)}$$

$$\Delta\omega \simeq \frac{I(\partial R/\partial\sigma) - I(\partial R/\partial\sigma)}{(\partial R/\partial\sigma)^2 + (\partial I/\partial\sigma)^2}. \tag{B.27}$$

If the values of R, $\partial R/\partial\sigma$, $\partial I/\partial\sigma$ can be calculated, then successive application of equations B.27 in the region of a zero of $F(s)$ gives rapid convergence to the zero location for any required accuracy. It is to be noted that equations B.27 are similar to those used in the quadratically convergent Newton-Bairstow method [B.15]. The difference lies in the fact that equations B.27 are of the form of the steepest descent equations B.24, where

$$V = R^2 + I^2 \tag{B.28}$$

$$h = \frac{0.5}{(\partial R/\partial\sigma)^2 + (\partial I/\partial\sigma)^2} > 0. \tag{B.29}$$

This is evident since (B.27) may be derived from (B.24) and (B.28), and the function $R^2 + I^2$ satisfies the conditions 1–5 given in the preceding section.

(It is immediately apparent that the first three conditions, are satisfied; and by realizing that $R^2 + I^2$ is the square of the modulus of $F(s)$, the maximum modulus theorem may be applied to show that the fourth condition is satisfied.) It is concluded that the application of (B.24) and (B.28), or the equivalent equations B.27, indicates the direction of the steepest descent of V. (An exception to this exists where $\omega = 0$. For this case the direction indicated is along the ω axis.) For regions other than the neighborhoods of the zeros of $F(s)$, the application of (B.27) will give a poor estimate of the increment for which V is minimized in the direction of steepest descent of V. However, even if more than one calculation is necessary, an increment can be found in the direction of steepest descent for which the value of V is reduced, and so convergence may be guaranteed.

The following algorithm, which arises out of the above theoretical developments, is the basis for a convergent method for finding zeros of analytic functions.

1. Scale to have some roots within the unit circle.
2. Choose initial approximations other than on the real axis, that is, σ_0, ω_0.
3. Compute R, I, $\partial R/\partial \sigma$, $\partial I/\partial \sigma$, and V.
4. Compute the new approximations from (B.27): $\sigma_1 = \sigma_0 + \Delta\sigma$, $\omega_1 = \omega_0 + \Delta\omega$.
5. Take σ_1 as σ_0 and ω_1 as ω_0 and repeat steps 3, 4, 5, including the following step 6 until convergence occurs.
6. If the calculated value of V is greater than the value for the preceding iteration, reduce the increments $\Delta\sigma$ and $\Delta\omega$ used previously until the value of V is smaller than for the preceding iteration.

Once a zero of $F(s)$ is determined using synthetic division, a reduced polynomial is found and this may be used in the evaluation of the next zero. As in other procedures, a suitable first approximation may be read into the machine or calculated from either the first three or the last three coefficients of the polynomial; or, if frequency scaling is used, at least some of the zeros will be in the vicinity of the origin and a convenient first approximation may be $\sigma = 0$, $\omega = 1$.

The computer time for each iteration using Newton's method is about three-quarters of that for the above method. However, as mentioned previously, by using the methods discussed here, convergence is guaranteed and no a priori information about the root location is required. For the seventh-order equation of the preceding section, for example, all the roots were found in 17 iterations.

Transcendental equations involving polynomials can also be solved using the approach, and thus the method has a more general application.

References

[B.1] W. W. Soroka, *Analog Methods in Computation and Simulation*, McGraw-Hill, New York, 1954, pp. 127–158.

[B.2] N. N. Mikhailov, Electrical Devices for Solution of Algebraic Equations (in Russian), *Avtomatika i Telemekhanika*, **19**, No. 5, 477–490 (1958).

[B.3] M. L. Morgan and J. C. Looney, Design of the ESIAC Algebraic Computer, *Trans. Inst. of Radio Engrs.*, **EC-10**, 524–529 (September 1961).

[B.4] A. Lepschy, A Method for the Research of the Zeros of a Polynomial with an Analog Computer, *Alta Frequenza* (Milan, Italy), **30**, 216–218 (March 1961).

[B.5] F. E. Liethen, C. H. Houpis, and J. J. D'Azzo, An Automatic Root Locus Plotter Using an Analog Computer, *AIEE Trans.*, Pt. II (*Applications and Industry*), **79**, 1960, 523–527 (January 1961).

[B.6] G. Schernberg and J. F. Riordan, Analog Calculation of Polynomial and Trigonometric Expansions, *Mathematical Tables and Other Aids for Computers*, **7**, No. 41, 246–253 (January 1953).

[B.7] V. M. Eliasberg, Solution of Algebraic Equations by the Application of Analog Computers (in Russian), *Avtomatika and Telemekhanika*, **20**, No. 6, 756–761 (1959).

[B.8] A. A. Kosarev, A. V. Martinov, and L. I. Akunina, A Method for Computation of Complex Roots of Algebraic Equations by the Application of Analog Computers (in Russian), *Ibid.*, **23**, No. 2, 163 (1962).

[B.9] L. Levine and H. F. Meissinger, An Automatic Analog Computer Method for Solving Polynomials and Finding Root Loci, *National Convention Record, Inst. Radio Engrs.*, Pt. 4, 164–172 (March 1957).

[B.10] F. J. Murray, *The Theory of Mathematical Machines*, King's Crown Press, New York, 45–62 (1948).

[B.11] M. V. Rybashev, Analog Solution of Algebraic and Transcendental Equations by Gradient Method (in Russian), *Avtomatika and Telemekhanika*, **22**, No. 1, 77–88 (1961).

[B.12] P. Kokotović and D. D. Šiljak, Automatic Analog Solution of Algebraic Equations and Plotting of Root Loci by Generalized Mitrovic Method, *AIEE Trans.*, Pt. II (*Applications and Industry*), **83**, 324–328 (September 1964).

[B.13] J. Moore, A Convergent Algorithm for Solving Polynomial Equations, *J. Appl. Math.* **14**, No. 2, 182–185 (April 1967).

[B.14] J. A. Ward, The Down-Hill Method of Solving $F(z) = 0$, *J. Assoc. for Computing Machinery*, Baltimore, Maryland, **4**, 148–158 (1957).

[B.15] A. Ralston, *A First Course in Numerical Analysis*, McGraw-Hill, New York, 1965.

[B.16] G. N. Lance, Solution of Algebraic and Transcendental Equations on an Automatic Digital Computer, *J. Assoc. for Computing Machinery*, **6**, No. 1 97–101 (January 1959).

APPENDIX C
Algebraic Domain Versus Time Domain

C.1 Introduction. The Laplace Transform[1]

The ultimate goal of linear system analysis and design is the appropriate form of both the time- and the frequency-domain system responses. Thus the Laplace transform is used to interpret the design of linear systems as an adjustment of the pole-zero configuration of relevant transfer functions. The techniques are developed to adjust system parameters so that desired pole-zero locations are obtained. Then the correlation between the system response and pole-zero configurations is essential. Frequency-domain response can be simply obtained from the pole-zero locations, and the bandwidth curves can be plotted directly on the parameter plane, as shown in Section 2.3. The evaluation of the time response is discussed here.

Let the function $f(t)$ of the real argument t be given identically equal to zero for $t \leq 0$ and different from zero for all, or at least some, values of $t \geq 0$. The unilateral Laplace transform $F(s)$ of the function $f(t)$ is defined as

$$F(s) = \int_0^\infty f(t)e^{-st}\,dt, \qquad (C.1)$$

where $s = \sigma + j\omega$ is the complex variable. Under the conditions that the function $f(t)$ is piecewise-continuous on every finite interval of t, and that $|f(t)| \leq Me^{at}$ for some choice of the constants M and a, the integral in (A.15) converges for $s > a$. In addition, for the integral (C.1) to have a

---

[1] Here, only the definition of the Laplace transformation is given. For further properties of the transformation and its use in the linear system analysis, references C.1–3 are recommended.

simple interpretation, $f(t)$ must be single-valued almost everywhere,[2] in the range $t \geq 0$. In the linear system analysis treated here the time functions satisfy the above conditions in most common cases, and their corresponding Laplace transforms are readily obtained and given in tables [C.3].

The transformation (A.15) is called the *Laplace transformation* (abbreviated $\mathscr{L}$ transformation). The *inverse Laplace transformation* ($\mathscr{L}^{-1}$ transformation), which expresses the time function in terms of the associated Laplace transform $F(s)$, is given as

$$f(t) = \frac{1}{2\pi j} \int_{c-j\infty}^{c+j\infty} F(s) e^{ts}\, ds, \tag{C.2}$$

where c, the abscissa of absolute convergence, must be larger than the real parts of all singularities of $F(s)$.

In most common situations in linear system analysis, the open-loop transfer function $G(s)$ is a rational function of s, and the input function $r(t)$ is such that $R(s)$ is also a rational function of s. Then the system output

$$c(t) = \mathscr{L}^{-1}\left[\frac{G(s)}{1+G(s)} R(s)\right] \tag{C.3}$$

can be determined as the inverse Laplace transformation of a rational function $C(s)$

$$C(s) = \frac{Q(s)}{P(s)}, \tag{C.4}$$

where $Q(s)$ and $P(s)$ are polynomials in s. The degree n of the polynomial $P(s)$ is higher, or at least equal to the degree of the polynomial $Q(s)$.

Since the function $C(s)$ of (C.4) is a rational function with distinct poles (of arbitrary order), which are the roots of $P(s) = 0$, the inverse Laplace transform and the Cauchy integral formula [C.2] can be used to obtain the general solution $c(t)$ in the form

$$c(t) = \frac{1}{2\pi j} \int_{c-j\infty}^{c+j\infty} C(s) e^{st}\, ds = \sum_{k=1}^{p} \text{Res}\, [C(s)e^{st}]_{s=s_k}. \tag{C.5}$$

As known [C.2], the residue of a function $C(s)e^{st}$ at the pole s_k of an order m is

$$\text{Res}\, [C(s)e^{st}]_{s=s_k} = \frac{1}{(m-1)!} \lim_{s \to s_k} \frac{d^{m-1}[(s-s_k)^m C(s)e^{st}]}{ds^{m-1}}. \tag{C.6}$$

---

[2] "Almost everywhere" is a precise mathematical expression meaning "everywhere except for a set of points (here representing values of t) that can be covered by a set of line segments the sum of whose lengths is arbitrarily small." Such a set might contain infinitely many points. "Equals" almost everywhere implies in particular "equals at all points of continuity."

Introduction. The Laplace Transform

When the equation $P(s) = 0$ has simple roots ($m = 1$), the formula C.6 is valid, providing that $0! = 1$.

In case all roots of $P(s) = 0$ are simple and real the solution $c(t)$ is

$$c(t) = \sum_{k=1}^{n} \frac{Q(s_k)}{P'(s_k)} e^{s_k t}, \tag{C.7}$$

where $P'(s_k) = \left[\dfrac{dP(s)}{ds}\right]_{s=s_k}$.

In particular, when $P(s) = 0$ has one zero root—that is, $P(s) = sP_1(s)$—equation C.7 can be written

$$c(t) = \frac{Q(0)}{P_1(0)} + \sum_{k=1}^{n-1} \frac{Q(s_k)}{s_k P_1'(s_k)} e^{s_k t}. \tag{C.8}$$

If among the roots of $P(s) = 0$ there are two complex conjugate roots $s_{k,k+1} = \sigma_k \pm j\omega_k$, then

$$\frac{Q(s_k)}{P'(s)} = \rho_k e^{j\theta_k} \quad \text{and} \quad \frac{Q(s_{k+1})}{P'(s_{k+1})} = \rho_k e^{-j\theta_k}, \tag{C.9}$$

where ρ_k and θ_k are known real numbers. There the kth and $(k+1)$th terms in the summation of (C.7) can be joined together to obtain

$$\frac{Q(s_k)}{P'(s_k)} e^{s_k t} + \frac{Q(s_{k+1})}{P'(s_{k+1})} e^{s_{k+1} t} = \rho_k e^{j\theta_k} e^{(\sigma_k + j\omega_k)t} + \rho_k e^{-j\theta_k} e^{(\sigma_k - j\omega_k)t}$$
$$= 2\rho_k e^{\sigma_k t} \cos(\omega_k t + \theta_k). \tag{C.10}$$

Finally, the output $c(t)$ has the form

$$c(t) = \sum_{k=0}^{l} \frac{Q(s_k)}{P'(s_k)} e^{s_k t} + \sum_{k=0}^{n-l} 2\rho_k e^{\sigma_k t} \cos(\omega_k t + \theta_k), \tag{C.11}$$

where l denotes the number of real roots of $P(s) = 0$, and ρ_k and θ_k are given by (C.9).

In practical applications, the formula (C.11) is cumbersome to use. Thus the time responses are calculated for fundamental pole-zero configurations and given as charts.

One of the most fundamental pole-zero configurations is the one consisting of two conjugate complex poles. James, Nichols, and Phillips [C.4] presented a series of graphs showing the unit-step function response for such a system. Truxal [C.5] added a negative real zero to the conjugate complex pole pair and calculated some data for this configuration. Muligan [C.6] made an extensive study of the correlation between the pole-zero locations and the maxima and minima of the system transient response. His work has been further extended by Zemanian [C.7] to include a method of finding the value of time at which the response crosses the final value line. Elgerd and Stephens [C.8] extracted from the infinite number of pole-zero configurations the ones that have a practical significance. In that respect their work represents further extension of the concepts proposed in references C.4 and 5. When, after a careful sampling procedure, the fundamental cases have been selected, there remains the problem of bringing to a minimum the number of independent graphs needed to cover each case adequately. By means of these graphs, the designer is then able to correlate immediately by inspection the connection between pole-zero configuration and real-time response. The classification of the pole-zero locations is greatly simplified by a fundamental characteristic feature of linear systems: the *dominance* of certain poles and zeros. It is a well-known fact [C.5] that without losing much accuracy we might obtain the time response of a large class of systems by considering only the poles and zeros located closest to the origin. The introduction of real, large-magnitude poles of the system transfer function changes only slightly the form of the step input response. *The criterion commonly adopted in design is that any real poles that are to contribute negligibly to the step function response should be placed at least six times as far from the $j\omega$ axis as those poles governing the response. It is also largely irrelevant whether such poles far from the $j\omega$ axis are actually real or complex as long as the corresponding real parts of the poles are much larger than that of the significant poles.*

Other facts in favor of the Elgerd-Stephens approach are that in the system design it is often the intention to manipulate the system adjustable parameters in such a manner that the limited number of poles are located close to the $j\omega$ axis; these are the predominant poles and often called the control poles. All the other poles and the related zeros are forced as far to the left as possible. Additional poles and zeros close to the origin of the s plane are allowed if they form dipoles or if some special reason in the design requires their existence in the dominance region of the s plane. Thus the Elgerd-Stephens approach is an extremely practical accessory to the parameter plane design of linear systems. The system parameters are adjusted so that a convenient pole-zero configuration is obtained that guarantees the particular time response desired. In adjusting the system parameters and pole-zero locations, the graphs proposed by Elgerd and Stephens are used. Then, if the contribution of some additional poles and zeros are doubtful, the system time response may be computed by the inverse Laplace transform [C.5] or by computer simulation. If the obtained response is not satisfactory, the procedure can be repeated unless the desired time response characteristics are achieved. The graphs of the Elgerd-Stephens approach are presented in Section C.2.

Since the time-domain description of the linear systems in the *state space*

State Space Equations and Characteristic Equation   429

is suitable for various design techniques and methods [C.9–11], the relationship between the state space description and the algebraic domain is shown in Section C.3. A recursive system is given by which the coefficients of the characteristic polynomial are obtained from the state space equation describing the system. In Section C.3 a transition from the linear system description used in Section 2.2 of Chapter 2 and the state space description is also outlined.

C.2 Fundamental Pole-Zero Configurations and Time-Response Charts

The following pole-zero configurations are considered and the results are given in the graphs indicated:

Case A: One negative real pole—Figure C.1a.
Case B: Two negative real poles—Figure C.1b.
Case C: Two negative real poles and one negative real zero—Figure C.1c.
Case D: Two complex poles—Figure C.1d.
Case E: Two complex poles and one negative real zero—Figure C.2.
Case F: Two complex and one negative real pole—Figure C.3.
Case G: Two complex poles, one negative real pole, and one negative real zero—Figure C.4.
Case H: Two complex poles, one negative real pole, and two complex zeros—Figure C.5.
Case I: Two pairs of complex poles—Figure C.6.

All the given charts are based on a normalized time variable. As a consequence, they are applicable to any system independent of the speed of the response. The normalization procedure is based upon the well-known relationship: $\mathscr{L}[c(t/a)] = a\mathscr{L}[c(t)] = aC(s)$. In other words, if the pole-zero configuration in the s plane is shrunk uniformly in the ratio of a to 1, the corresponding time function is stretched in the ratio of 1 to a. The normalized time $t_n = t/a$ is defined in each case separately, as shown in the given diagrams. The application of the given charts is straightforward and needs no further explanation.

C.3 State Space Equations and Characteristic Equation

Before the characteristic equation is derived from the state space equations, it is of interest to consider the transition from the linear system description

$$\sum_{j=1}^{l} D_{ij}x_j = f_i, \quad (i = 1, 2, \ldots, l), \tag{C.12}$$

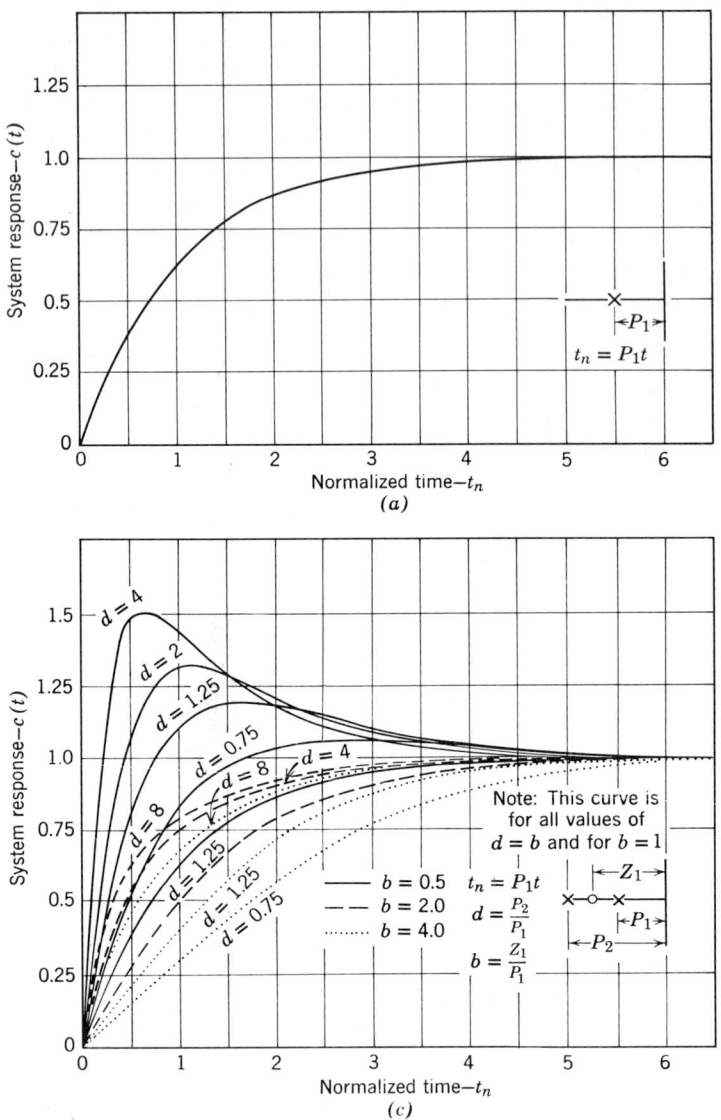

Figure C.1 Responses to a step input of systems: (a) with one negative real pole; (b) with two negative real poles; (c) with two poles and one zero, all real and negative; (d) with two complex poles.

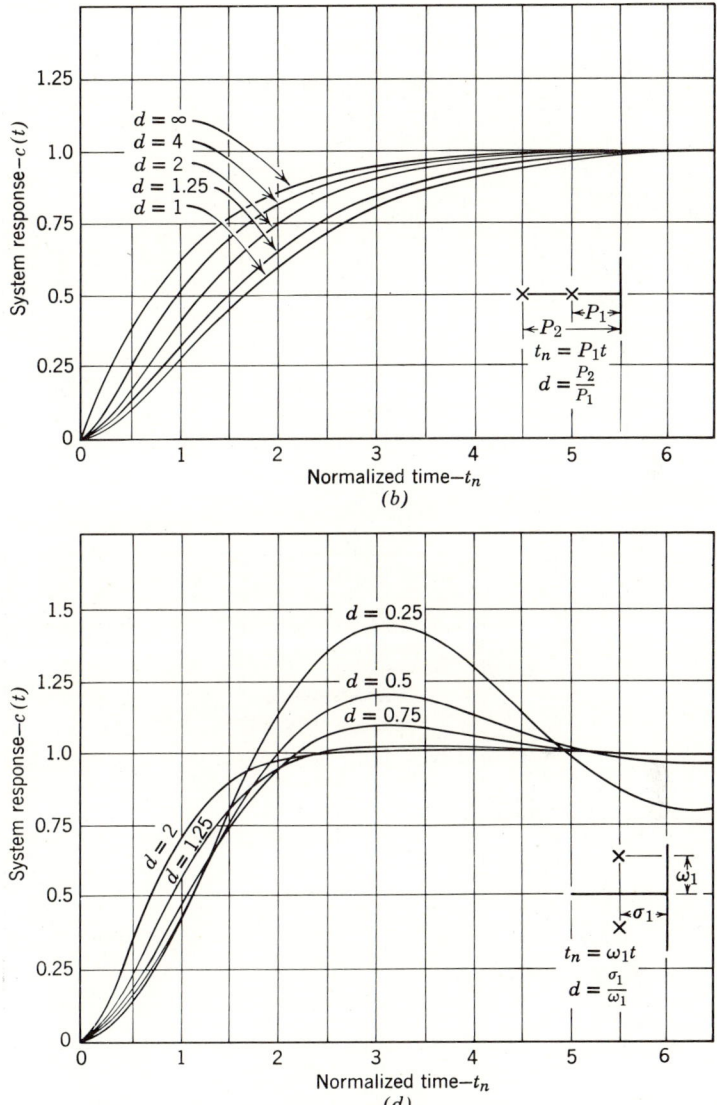

Figure C.1 (*continued*)

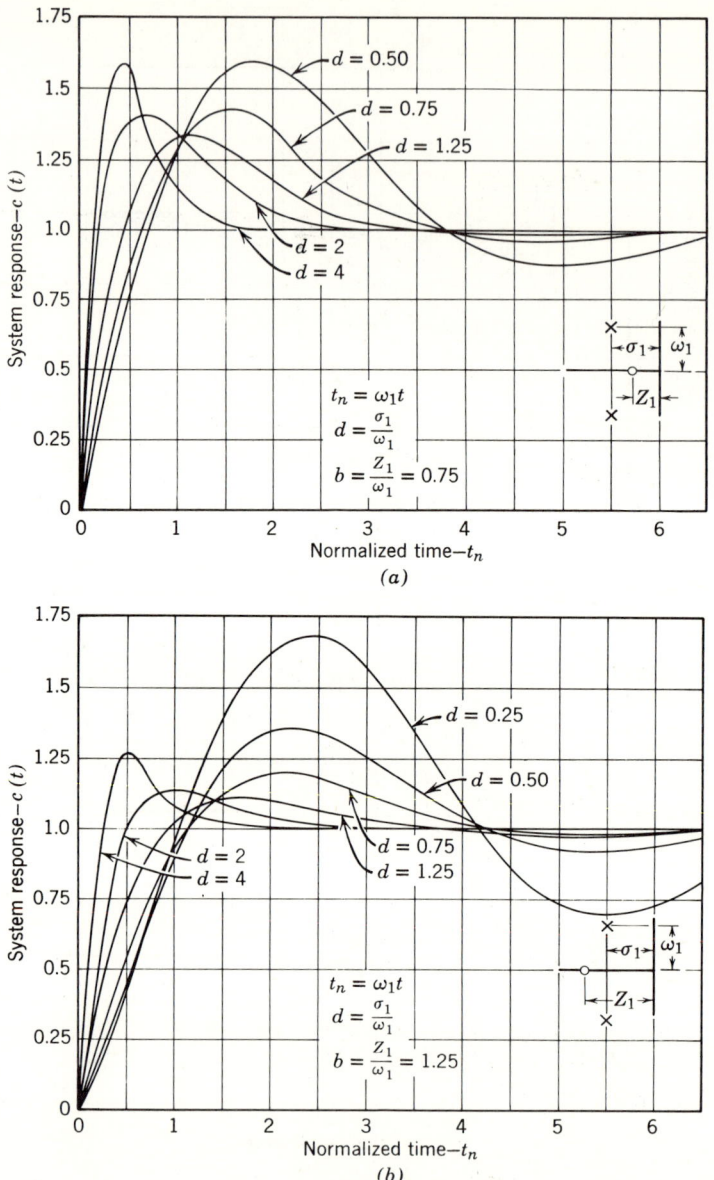

Figure C.2 Responses to step input for system with two complex poles and one negative real zero.

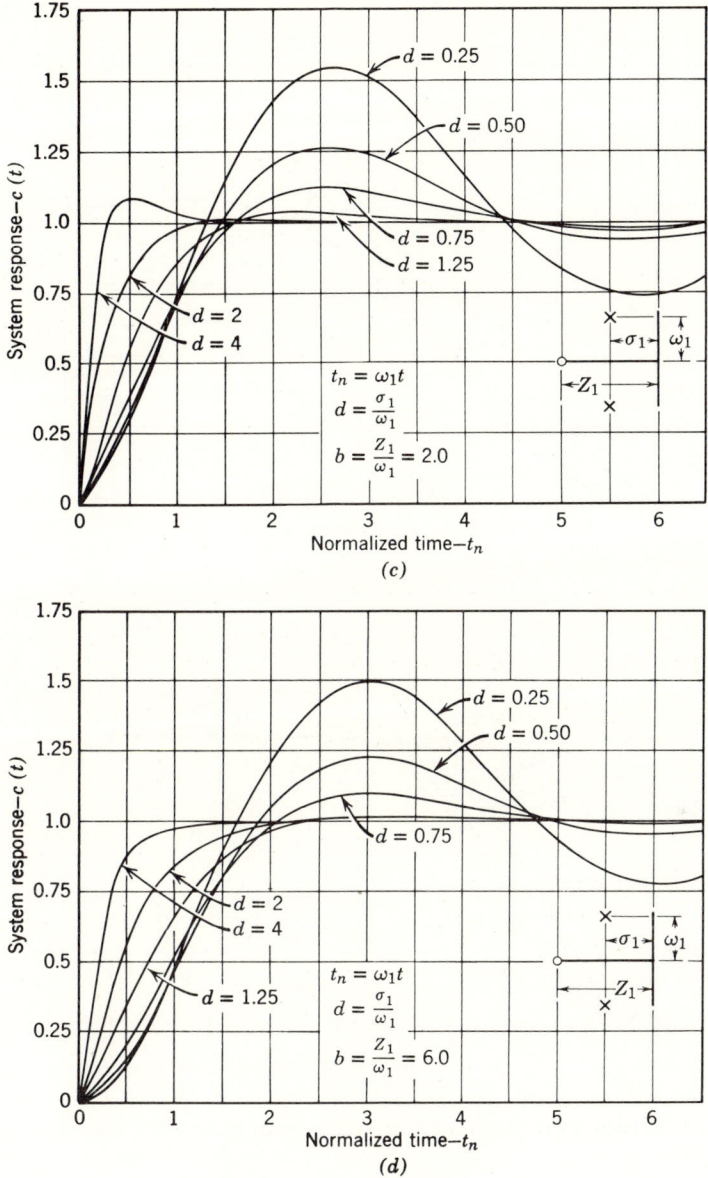

Figure C.2 (*continued*)

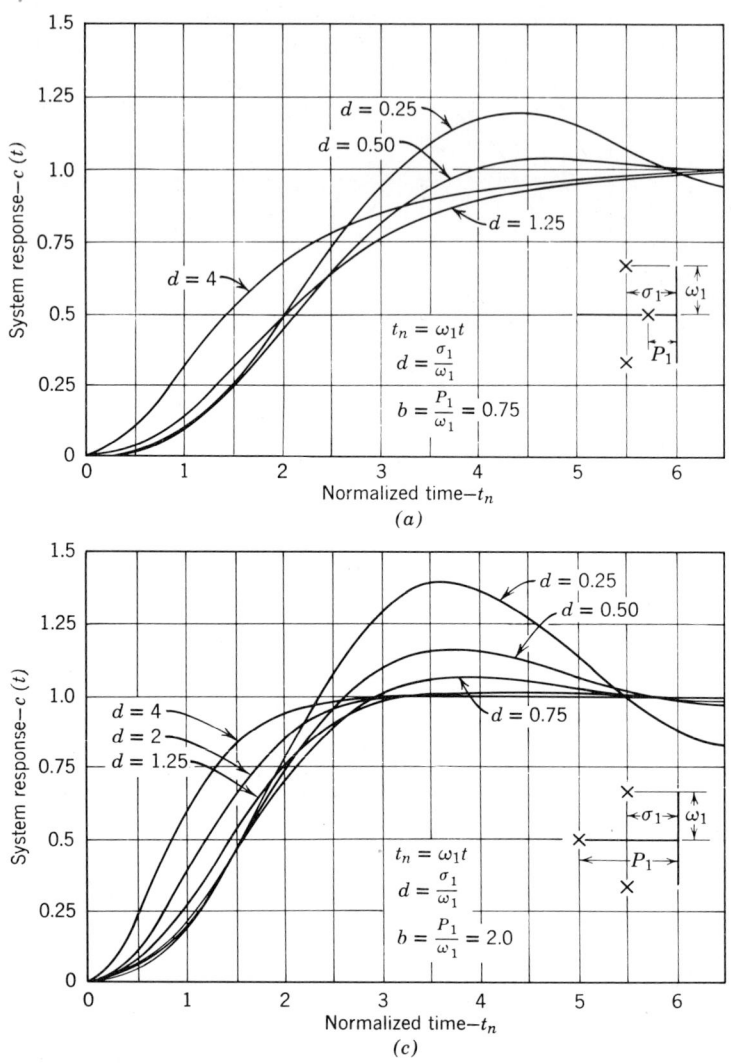

Figure C.3 Responses to step input for system with three poles, two complex and one real.

434

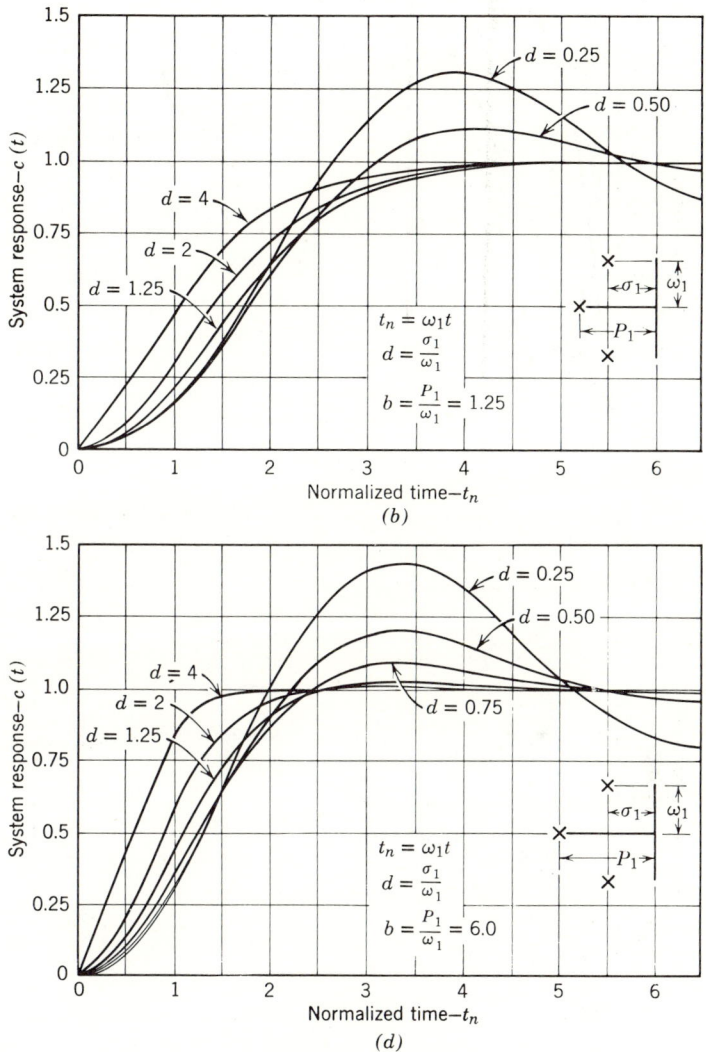

Figure C.3 (*continued*)

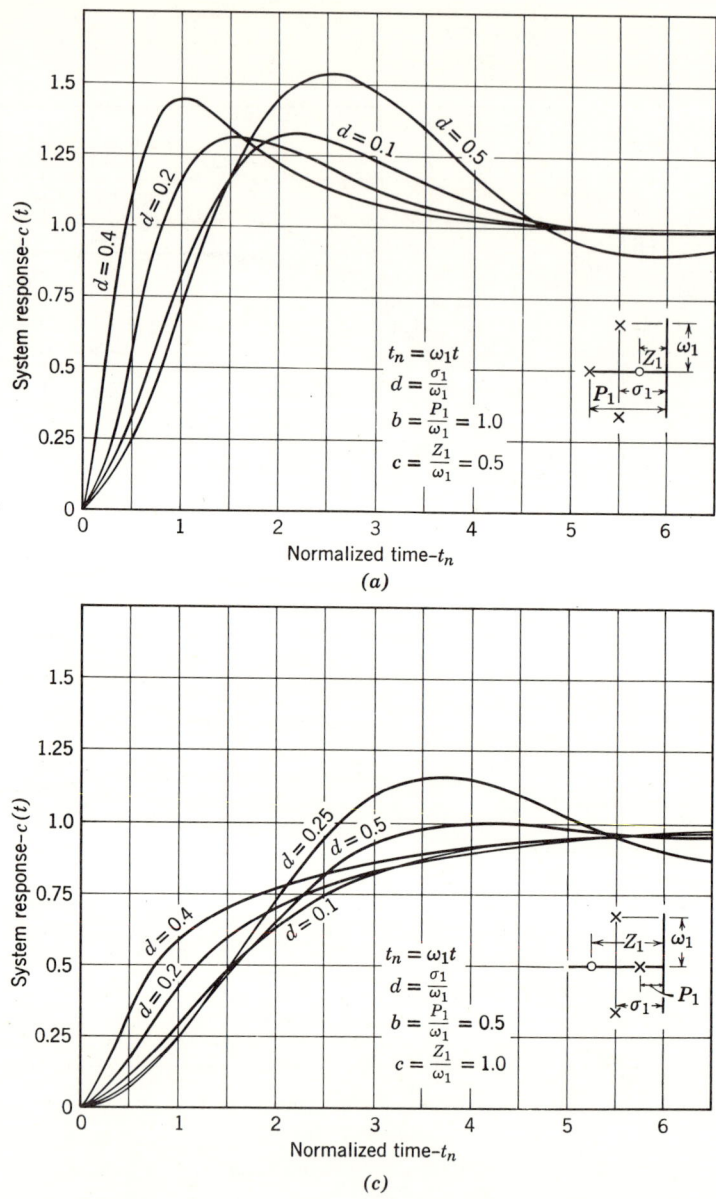

Figure C.4 Responses to step input for system with three poles, two complex and one real, plus one zero.

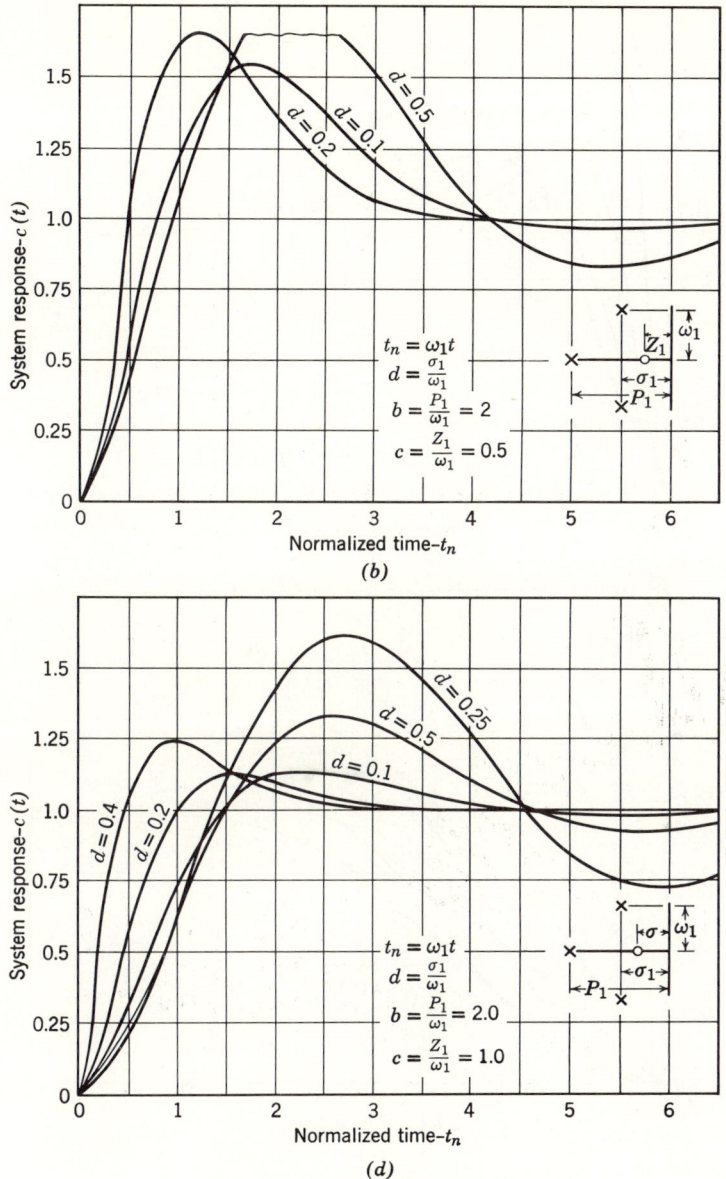

Figure C.4 (continued)

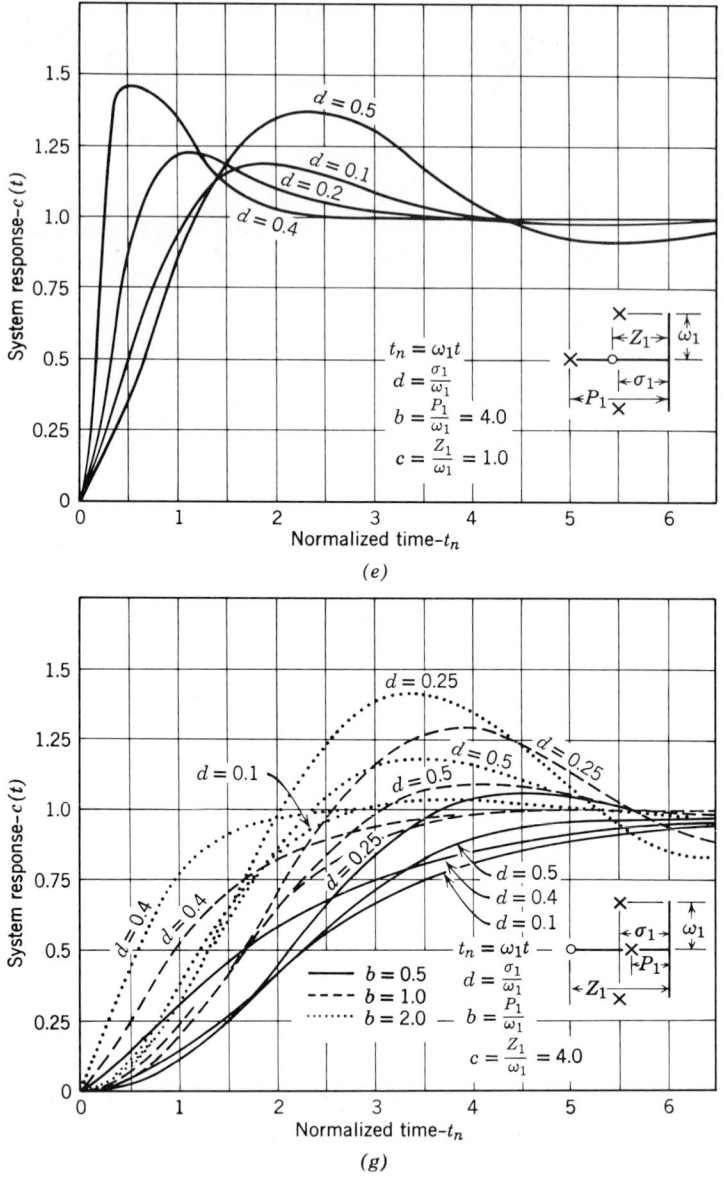

Figure C.4 (continued)

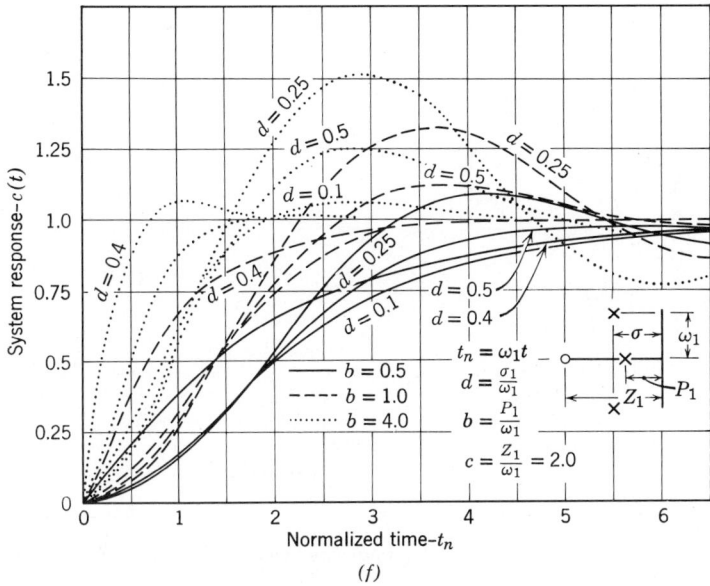

(f)

Figure C.4 (*continued*)

where

$$D_{ij} \equiv u_{ij}\frac{d^2}{dt^2} + v_{ij}\frac{d}{dt} + w_{ij} \qquad (C.13)$$

to the equivalent state space description

$$\dot{\mathbf{z}} = \mathbf{A}\mathbf{z} + \mathbf{B}\mathbf{u}, \qquad (C.14)$$

where $\mathbf{A}$ is an $n \times n$ matrix, $\mathbf{B}$ is an $n \times r$ matrix, $\mathbf{z}$ is the state n vector, and $\mathbf{u}$ is the input r vector.[3]

If the substitution

$$\dot{x}_j = y_j, \qquad (j = 1, 2, \ldots, l) \qquad (C.15)$$

is introduced in (C.12), we can rewrite (C.12) in a matrix form

$$\mathbf{U}\dot{\mathbf{y}} = \mathbf{f} - \mathbf{W}\mathbf{x} - \mathbf{V}\mathbf{y}, \qquad (C.16)$$

where $\mathbf{U} = (u_{ij})$, $\mathbf{V} = (v_{ij})$, $\mathbf{W} = (w_{ij})$ are all $l \times l$ matrices, and $\dot{\mathbf{y}} = \{\dot{y}_1, \dot{y}_2, \ldots, \dot{y}_l\}$, $\mathbf{y} = \{y_1, y_2, \ldots, y_l\}$, $\mathbf{x} = \{x_1, x_2, \ldots, x_l\}$, and $\mathbf{f} = \{f_1, f_2, \ldots, f_l\}$ are all l vectors.

---

[3] More general and rigorous treatment of the transition is given by E. Polak, An Algorithm for Reducing a Linear, Time-Invariant Differential System to State Form, *IEEE Trans. Auto. Control*, **AC–11**, No. 3, 577–580 (July 1966).

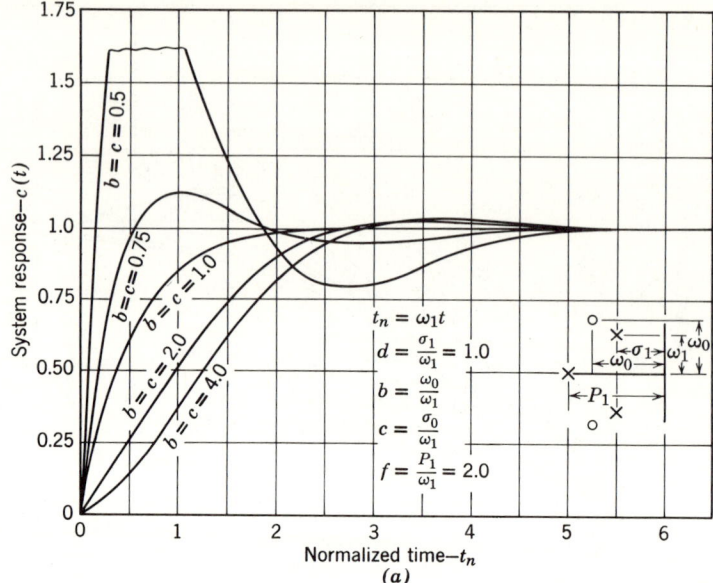

Figure C.5 Responses to step input for system with three poles, two complex and one real, plus two complex zeros.

By introducing the state vector $z = \{x, y\}$ and solving equation (C.16) for $\dot{y}$ (it is assumed that det $U \neq 0$), we can combine (C.15) and (C.16) to obtain the state space equation C.14 where

$$A = \begin{bmatrix} -W & -V \\ 0 & I \end{bmatrix}, \quad B = \begin{bmatrix} 0 \\ U^{-1} \end{bmatrix} \quad \text{(C.17)}$$

and $u = \{0, f\}$. In (C.17), 0 and I are the zero and identity $l \times l$ matrices, respectively. It can be observed that the number of scalar equations represented by the state space equation C.14 is $n = 2l$.

The relation between the state space equations and the characteristic polynomial is based upon a certain theorem, proved in reference C.12. It is also considered in reference C.9. The result of the theorem is a recursive system for calculating the coefficients of the characteristic polynomial from the state space matrix equation describing the system. The results of the theorem are outlined without the proof, which can be found in references C.9 and C.12.

Let us consider a linear system described by equation

$$\dot{z} = Az, \quad \text{(C.18)}$$

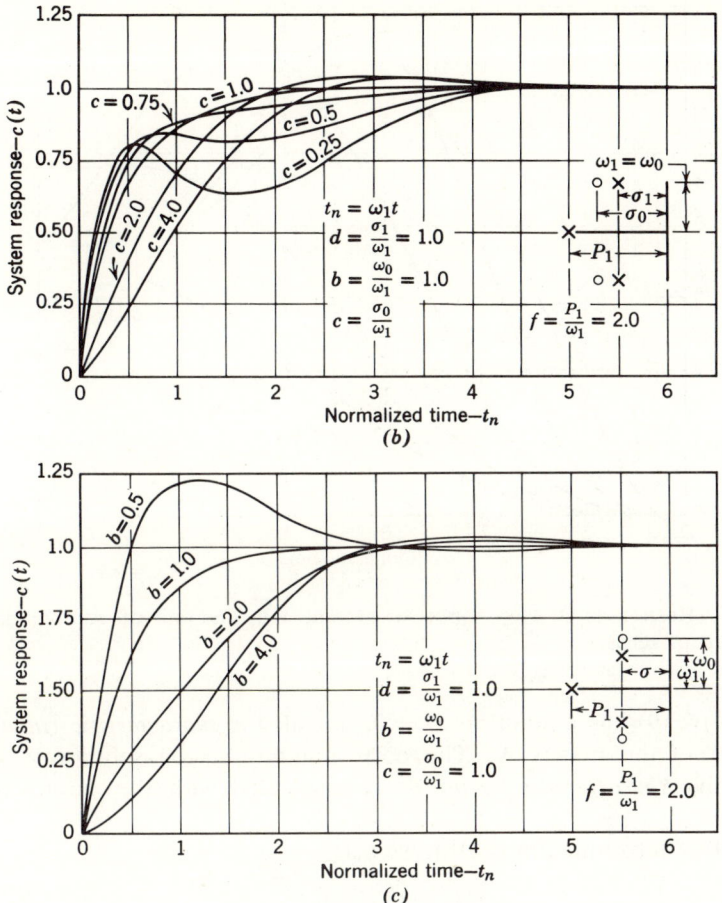

Figure C.5 (continued)

where $\dot{\mathbf{z}} = \{\dot{z}_1, \dot{z}_2, \ldots, \dot{z}_n\}$ and $\mathbf{z} = \{z_1, z_2, \ldots, z_n\}$ are column vectors and $A = (a_{ij})$ is a square $(n \times n)$ matrix. This equation can be viewed as a transformation of the vector $\mathbf{z}$ into the vector $\dot{\mathbf{z}}$. The question arises whether there exists a vector $\mathbf{z}$ so that the transformation $\mathbf{A}$ produces a vector $\dot{\mathbf{z}}$, which has the same direction in vector space as the vector $\mathbf{z}$—that is, does the equation

$$\dot{\mathbf{z}} = s\mathbf{z} \qquad (C.19)$$

exist where s is a scalar and represents a factor of proportionality? This is known as the *characteristic value problem*, and a value of s, for example, s_i,

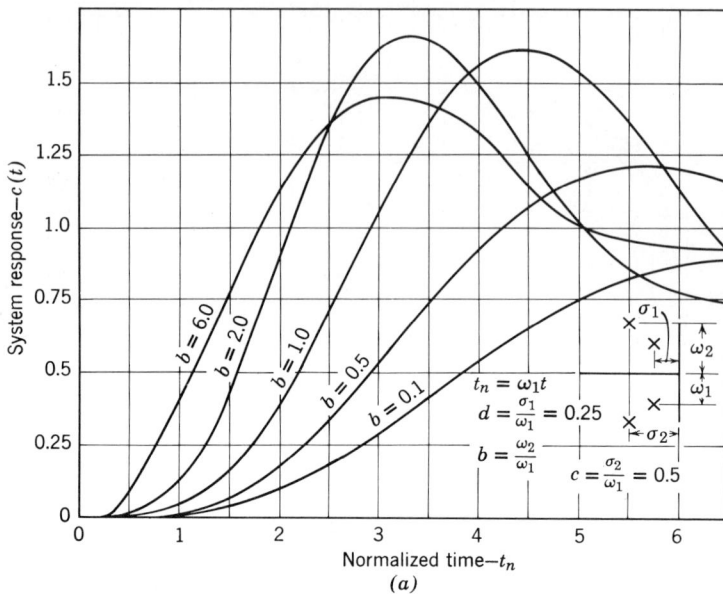

Figure C.6 Responses to step input for system with one pair of complex poles and one pair of zeros.

for which (C.19) has a solution $z_i \neq 0$, is called a *characteristic value (or eigenvalue)* of the matrix **A**. The corresponding vector solution $z_i \neq 0$ is called the *characteristic vector* of **A** associated with the characteristic value s_i.

If (C.19) is substituted in (C.18), we get

$$(s\mathbf{I} - \mathbf{A})\mathbf{z} = 0, \tag{C.20}$$

where **I** is the identity matrix. Equation C.20 has a nontrivial solution if and only if the determinant

$$\det(s\mathbf{I} - \mathbf{A}) = 0. \tag{C.21}$$

Equation C.21 is the *characteristic equation* of the linear system described by the state space equation, and has the common polynomial form

$$\sum_{k=0}^{n} a_k s^k = 0, \quad a_n \equiv 1. \tag{C.22}$$

It is of interest to give the expressions of the coefficients a_k of (C.22) in terms of the coefficients a_{ij} of the matrix $\mathbf{A} = (a_{ij})$ defined in (C.18).

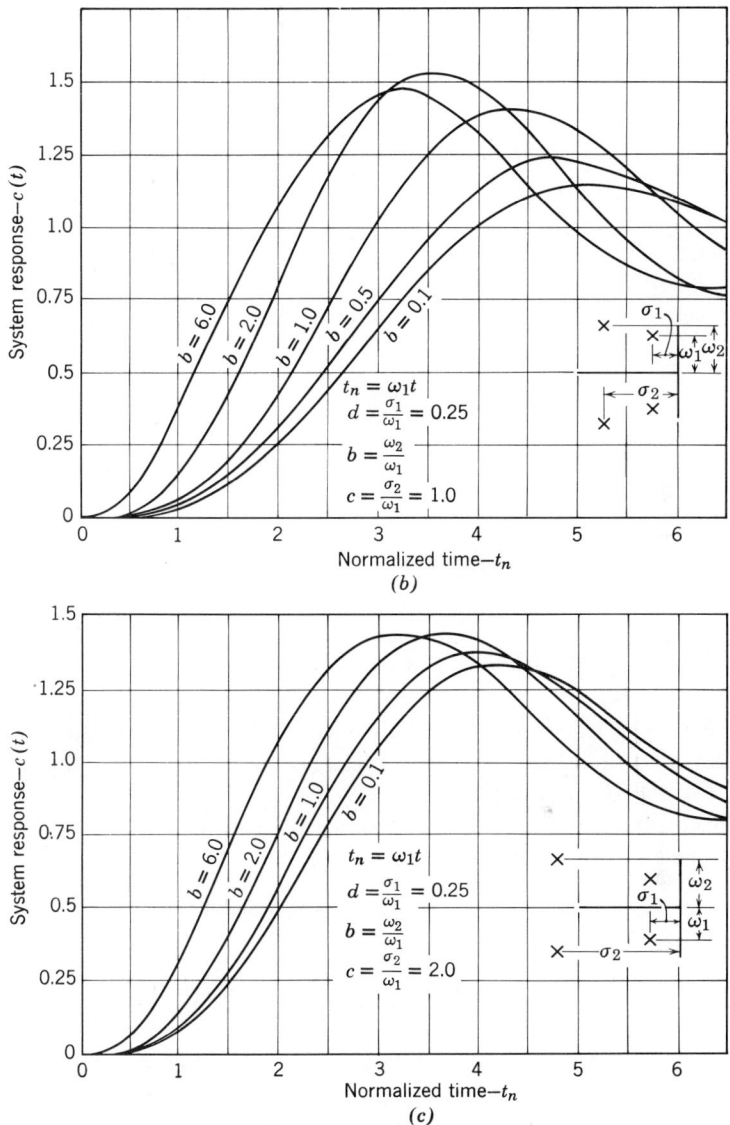

Figure C.6 (*continued*)

The result of the above-mentioned theorem is the recursive system

$$a_n \equiv 1$$
$$a_{n-1} = -T_1$$
$$a_{n-2} = -\tfrac{1}{2}(a_{n-1}T_1 + T_2)$$

$$\vdots$$

$$a_{n-i} = -\frac{1}{i}(a_{n-i+1}T_1 + a_{n-i+2}T_2 + \cdots + a_{n-1}T_{i-1} + T_i)$$

$$\vdots$$

$$a_1 = -\frac{1}{n-1}(a_2T_1 + a_3T_2 + \cdots + a_{n-1}T_{n-2} + T_{n-1})$$

$$a_0 = -\frac{1}{n}(a_1T_1 + a_2T_2 + \cdots + a_{n-1}T_{n-1} + T_n), \tag{C.23}$$

where T_i is the trace of $\mathbf{A}^i$ ($\mathbf{A}$ multiplied by itself i times). In practice, the application of the recursive system (C.23) is convenient if a computer is used in the calculation process.

It should be noted that the state space description of a linear system can be obtained directly from equation

$$D(s)x = D_i(s)f, \tag{C.24}$$

where

$$D(s) \equiv \sum_{k=0}^{n} a_k s^k, \qquad D_i(s) \equiv \sum_{k=0}^{m} b_k s^k \tag{C.25}$$

and $n \geq m$. Equations C.24 and 25 are derived in Section 2.2 as (2.12) and (2.13).

Since the forced function $f = f(t)$ is assumed as known function of time, the right-hand side of (C.24) can be considered as a known function of time $u = u(t)$, which is

$$u(t) = D_i f(t). \tag{C.26}$$

Thus (C.24) can be rewritten as

$$D(s)x = u. \tag{C.27}$$

State Space Equations and Characteristic Equation

In this case we may define state variables as follows. Let

$$x_1 = x$$

$$x_2 = \frac{dx}{dt} = \dot{x}_1$$

$$\vdots$$

$$x_n = \frac{d^{n-1}x}{dt^{n-1}} = \dot{x}_{n-1}. \tag{C.28}$$

Then nth-order linear differential equation C.27 can be rewritten as n first-order differential equations

$$\dot{\mathbf{x}} = \mathbf{Ax} + \mathbf{bu}, \tag{C.29}$$

where $\mathbf{A}$ is the $n \times m$ system matrix

$$\mathbf{A} = \begin{bmatrix} 0 & 1 & 0 & \cdots & 0 \\ 0 & 0 & 1 & \cdots & 0 \\ \cdot & \cdot & \cdot & & \cdot \\ \cdot & \cdot & \cdot & & \cdot \\ \cdot & \cdot & \cdot & & \cdot \\ 0 & 0 & \cdots & \cdots & 1 \\ -\dfrac{a_0}{a_n} & -\dfrac{a_1}{a_n} & \cdots & \cdots & -\dfrac{a_{n-1}}{a_n} \end{bmatrix} \tag{C.30}$$

and $\mathbf{x} = \{x_1, x_2, \ldots, x_n\}$, $\mathbf{u} = \{u_1, u_2, \ldots, u_n\}$, and $\mathbf{b} = \{0, 0, \ldots, 1/a_n\}$ are n vectors.

Before concluding this section, it is of interest to mention briefly the concept of *controllability and observability* introduced by Kalman [C.9–11].

Quite generally a linear invariant system can be described by state space equations

$$\begin{aligned} \dot{\mathbf{x}} &= \mathbf{Ax} + \mathbf{Bu} \\ \mathbf{y} &= \mathbf{Cx} + \mathbf{Du}, \end{aligned} \tag{C.31}$$

where $\mathbf{A}, \mathbf{B}, \mathbf{C}$, and $\mathbf{D}$ are, respectively, $n \times m, n \times r, p \times n, p \times r$ constant matrices. The n vector $\mathbf{x}$ is the state of the system, the r vector $\mathbf{u}$ is the input, and the p vector $\mathbf{y}$ is the output of the system. This is illustrated by the system block diagram of Figure C.7.

Intuitively speaking, the system described by (C.31) is *controllable* if, knowing the matrices $\mathbf{A}$ and $\mathbf{B}$ and the initial state $\mathbf{x}_0$ at $t = t_0$, we can construct an input $\mathbf{u}(t)$ that will bring the state of the system to zero state $\mathbf{0}$ at

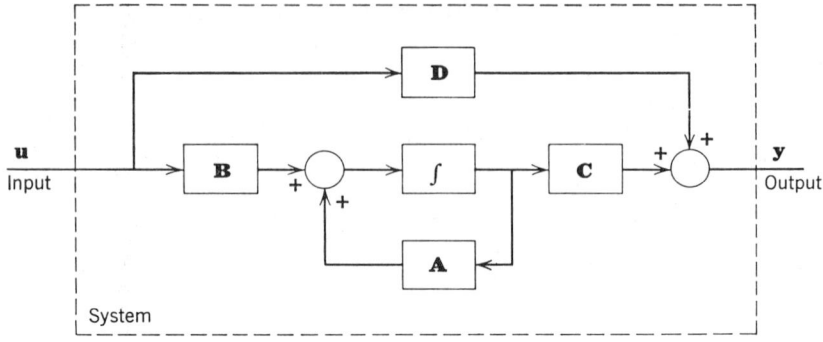

Figure C.7 System block diagram.

finite time. To express simply the conditions for controllability, equations C.31 are rewritten in the *canonical (or normal) form*.

Assume that the characteristic values of **A** are distinct. If the linear transformation $\mathbf{x} = \mathbf{Mz}$ is introduced, where **M** is the $n \times n$ *modal matrix* formed so that its columns are vectors $k_i \mathbf{x}_i$ (k_i are scalars, $\mathbf{x}_i$ are characteristic vectors, and the matrix **M** is assumed nonsingular so that its inverse $\mathbf{M}^{-1}$ exists), then (C.31) can be rewritten as

$$\dot{\mathbf{z}} = \boldsymbol{\alpha}\mathbf{z} + \boldsymbol{\beta}\mathbf{u}$$
$$\mathbf{y} = \boldsymbol{\gamma}\mathbf{z} + \boldsymbol{\delta}\mathbf{u}, \qquad (C.32)$$

where $\boldsymbol{\alpha} = \mathbf{M}^{-1}\mathbf{AM}$, $\boldsymbol{\beta} = \mathbf{M}^{-1}\mathbf{B}$, $\boldsymbol{\gamma} = \mathbf{CM}$, and $\boldsymbol{\delta} = \mathbf{D}$. Equations C.32 are known as the canonical (or normal) form for the state equations. In this form the differential equations in terms of the state variables $z_1, z_2, \ldots, z_n$ are uncoupled. That is, they are of the form $\dot{z}_i = \lambda_i z_i + f_i$ where f_i is the forcing function applied to the ith state variable z_i. This results from the fact that the matrix $\boldsymbol{\alpha}$ is a diagonal matrix composed of the characteristic values; that is, $\boldsymbol{\alpha} = \text{diag}(s_1, s_2, \ldots, s_n)$.

In the canonical form the conditions for controllability become quite clear. A particular state variable cannot be controlled in the above sense if the input is not coupled into this variable. Thus the condition for complete controllability is that the matrix $\boldsymbol{\beta}$ have not zero rows.

The idea of *observability* is closely related to that of controllability. A completely *observable* system has the property that all its dynamic modes of motion can be ascertained from measurements of the available outputs. That is, there are no zero columns of the matrix $\boldsymbol{\gamma}$ in (C.32).

In case **A** has repeated eigenvalues, we should consider the controllability in a more basic form, which is that there be an input $\mathbf{u}(t)$, $0 \leq t \leq T \leq \infty$, so that $\mathbf{x}(0) = \mathbf{x}_0$ can be forced to $\mathbf{x}(T) = \mathbf{0}$. This leads to the necessary and sufficient conditions for controllability that the vectors **B**, **AB**, . . . ,

$A^{n-1}B$ be linearly independent. Similarly, the basic requirement of observability is that for some $T > 0$ and all initial states x_0, knowledge of A, C, and $y(t)$, $0 \le t \le T$, is sufficient to determine the initial state x_0. This results in the necessary and sufficient conditions for observability that the vectors $C^*, A^*C^*, \ldots, A^{*(n-1)}C^*$, are linearly independent. (The asterisk indicates the conjugate transpose of the unstarred matrix.)

The concepts of controllability and observability are essential in the analysis of multivariable systems. They should be taken into account when a physical system is represented by the differential equations in either the operator form of (C.12) or the state space form (C.31), since further conclusions about the system behavior rely upon these concepts. A detailed discussion of the controllability and observability can be found in references C.9–11.

References

[C.1] G. Doetsch, *Handbuch der Laplace Transformation*, Vols. 1 to 3, Birkhauser, Basel, 1950.

[C.2] W. Kaplan, *Operational Methods for Linear Systems*, Addison-Wesley, Reading, Mass., 1962.

[C.3] P. A. McCollum and B. F. Brown, *Laplace Transform Tables and Theorems*, Holt, Rinehart, and Winston, New York, 1965.

[C.4] H. M. James, N. B. Nichols, and R. S. Phillips, *Theory of Servomechanisms* McGraw-Hill, New York, 1947.

[C.5] L. G. Truxal, *Control System Synthesis*, McGraw-Hill, New York, 1955.

[C.6] L. H. Muligan, Jr., The Effect of Pole and Zero Locations on the Transient Response of Linear Dynamic Systems, *IRE Proc.*, **37**, 516–529 (May 1949).

[C.7] A. H. Zemanian, Further Effects of the Pole and Zero Locations on the Step Response of Fixed Linear Systems, *AIEE Trans.*, **74**, Pt. II (*Applications and Industry*), 52–55 (March 1955).

[C.8] O. I. Elgerd and W. C. Stephens, Effect of Closed-loop Transfer Function Pole and Zero Locations on the Transient Response of Linear Control Systems, *AIEE Trans.*, **78**, Pt. II (*Applications and Industry*), 121–127 (May 1959).

[C.9] L. A. Zadeh and C. A. Desoer, *Linear System Theory: The State Space Approach*, McGraw-Hill, New York, 1963.

[C.10] P. M. DeRusso, R. J. Roy, and C. M. Close, *State Variables for Engineers*, Wiley, New York, 1965.

[C.11] S. C. Gupta, *Transform and State Variable Methods in Linear Systems*, Wiley, New York, 1966.

[C.12] V. N. Fadeeva, *Computational Methods of Linear Algebra*, Dover, New York, 1959.

APPENDIX D

Stability Criteria for Linear Systems

D.1 Introduction

A linear feedback system is stable if all the roots of the corresponding characteristic equation, which is an algebraic equation with real coefficients, have negative real parts. The necessary and sufficient conditions for indicating the number of roots of an algebraic equation, which have negative real parts, have been formulated by Hermité [D.1–3] in terms of the coefficients of the algebraic equation. These conditions were reformulated later by Hurwitz [D.3, 4] in a well-known and widely used form. Nyquist [D.5] and Mikhailov [D.6] proposed graphical solutions of the stability problem in the frequency domain. On the basis of the Vishnegradsky work [D.7] (see Section 1.4), Neimark [D.8–10] interpreted the absolute stability of linear systems in the parameter space (D-decomposition method).

The relative stability of a linear feedback system was considered after finding the conditions for the characteristic equation to have all roots with the absolute values of the real parts greater than a certain number specified in advance [D.11–14]. Such conditions were correlated with the study of the settling time of the system transient response. However, if other system characteristics are to be considered, such as speed of response, overshoot, and so on, it is necessary to investigate the relative damping coefficients of all roots of the characteristic equation, which is defined as a ratio between the real part of the root and its absolute value [D.14–18].

The extension of the Hurwitz, Nyquist, and Mikhailov stability criteria has been proposed [D.19] to investigate the relative stability of linear feedback systems in terms of the relative damping coefficients of the characteristic roots. The extension utilized the Chebyshev functions T_k and U_k. Later [D.20], by employing the functions X_k and Y_k, the analysis has been applied to the relative stability studies in which the absolute value of the real parts of characteristic roots is the measure of the degree of stability.

The D-decomposition method is briefly outlined and compared with the parameter plane concept as presented in Chapter 1. By using the concept of the parameter plane mapping, the D-decomposition method can be readily extended to the relative stability analysis.

D.2 Hurwitz Criterion

Consider the characteristic equation in the form

$$F(s) \equiv \sum_{k=0}^{n} a_k s^k = 0, \tag{D.1}$$

where the coefficients a_k are real and $a_n > 0$. If the complex variable s is substituted by a new variable w according to

$$s = w e^{j(\theta - \pi/2)}, \tag{D.2}$$

the half of the s plane that is to the left of the straight line in Figure D.1a is mapped into the left half of the w plane as shown in Figure D.1b.[1] If a root of (D.1) is located on the straight line l, it can be determined by

$$s = -\omega_n \zeta + j\omega_n \sqrt{1 - \zeta^2}, \tag{D.3}$$

where ω_n is an undamped natural frequency and $\zeta = -\cos \theta$ is a relative damping coefficient. After the application of the substitution (D.2), (D.1) becomes

$$F_1(w) \equiv \sum_{k=0}^{n} a_k e^{jk(\theta - \pi/2)} w^k = 0, \tag{D.4}$$

where the coefficients $a_k e^{jk(\theta - \pi/2)}$ are complex numbers. Then if a root of (D.1) is located on the straight line l of Figure D.1a, it would become a pure imaginary root of (D.1), and it would appear on the imaginary axis of the w plane as shown in Figure D.1b.

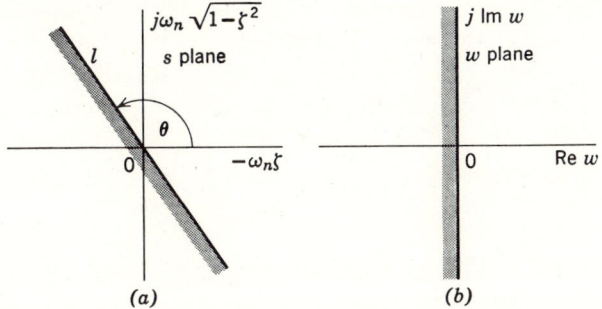

Figure D.1 Relative stability boundary in the s and w plane.

---

[1] This approach has been used in references D.21 and D.22 and repeated in reference D.23.

For the characteristic equation D.1 to have all roots with the relative damping coefficient greater than a prescribed value ζ, all the roots of (D.3) must have negative real parts and be located in the left half of the w plane. To check that this condition is being satisfied, apply the Hermité [D.1–3] results to (D.4). Thus consider a new equation that has the form

$$F_1(jw) \equiv \sum_{k=0}^{n} a_k e^{jk\theta} w^k = 0 \tag{D.5}$$

derived from (D.4). The coefficients $a_k e^{jk\theta}$ of (D.5) can be developed in the form

$$a_k e^{jk\theta} = b_k + jc_k, \tag{D.6}$$

where

$$b_k = a_k T_k(-\zeta) = (-1)^k a_k T_k(\zeta)$$
$$c_k = a_k \sqrt{1-\zeta^2}\, U_k(-\zeta) = (-1)^{k+1} a_k \sqrt{1-\zeta^2}\, U_k(\zeta). \tag{D.7}$$

Functions $T_k(\zeta)$ and $U_k(\zeta)$ are Chebyshev functions of the first and second kind, respectively. These functions are obtained according to the recurrence formulas

$$T_{k+1}(\zeta) - 2\zeta T_k(\zeta) + T_{k-1}(\zeta) = 0$$
$$U_{k+1}(\zeta) - 2\zeta U_k(\zeta) + U_{k-1}(\zeta) = 0, \tag{D.8}$$

with $T_0(\zeta) \equiv 1$, $T_1(\zeta) \equiv \zeta$, $U_0(\zeta) \equiv 0$, $U_1(\zeta) \equiv 1$. For pertinent values of ζ, the Chebyshev functions may be computed once and for all and put in tables (see Appendix A).

The number of roots of (D.4) that lie in the right half of the w plane may now be determined in terms of the coefficients b_k and c_k of (D.6) and (D.7). The number of sign changes in the sequence

$$1, \Delta_2, \Delta_4, \Delta_6, \ldots, \Delta_{2n} \tag{D.9}$$

is equal to the number of the roots of (D.4) that have positive real parts. The members of the train (D.9) are the even diagonal minors of the determinant

$$\Delta_{2n} = \begin{vmatrix} c_n & c_{n-1} & c_{n-2} & \cdots & c_1 & c_0 & 0 & \cdots & 0 & 0 \\ b_n & b_{n-1} & b_{n-2} & \cdots & b_1 & b_0 & 0 & \cdots & 0 & 0 \\ 0 & c_n & c_{n-1} & \cdots & c_2 & c_1 & c_0 & \cdots & 0 & 0 \\ 0 & b_n & b_{n-1} & \cdots & b_2 & b_1 & b_0 & \cdots & 0 & 0 \\ \cdot & \cdot & \cdot & & \cdot & \cdot & \cdot & & \cdot & \cdot \\ \cdot & \cdot & \cdot & & \cdot & \cdot & \cdot & & \cdot & \cdot \\ \cdot & \cdot & \cdot & & \cdot & \cdot & \cdot & & \cdot & \cdot \\ 0 & 0 & 0 & \cdots & c_n & c_{n-1} & c_{n-2} & \cdots & c_1 & c_0 \\ 0 & 0 & 0 & \cdots & b_n & b_{n-1} & b_{n-2} & \cdots & b_1 & b_0 \end{vmatrix}, \tag{D.10}$$

which is a determinant of order $2n (\Delta_{2n} \neq 0)$.

By expanding the above determinant in terms of the elements of the last column and bearing in mind that $T_0(\zeta) \equiv 1$ and $U_0(\zeta) \equiv 0$, it follows from (D.7) that

$$\Delta_{2n} = a_0 \Delta_{2n-1}. \tag{D.11}$$

Now the Hurwitz relative stability criterion is stated as follows: the necessary and sufficient conditions for all roots of equation $F(s) = 0$ to have relative damping coefficients greater than the prescribed value ζ are that $a_0, \Delta_2, \Delta_4, \ldots \Delta_{2n-2}, \Delta_{2n-1}$ all be positive numbers. In the special case where $\zeta = 0$, this criterion is reduced to the original Hurwitz criterion for absolute stability analysis [D.3, 4].

In the relative stability analysis, the following cases may occur:

1. All determinants in the sequence (D.9) and a_0 are positive numbers. The corresponding control system is relatively stable with respect to a specified damping coefficient, and all roots of (D.1) lie to the left of the straight line l of Figure D.1a.

2. Some of the determinants in sequence (D.9) have a negative sign. This means that several roots of (D.1) have damping coefficients smaller than the prescribed value ζ. The number of sign changes in sequence (D.9) indicates, then, the number of these roots that lie to the right of the straight line l.

3. All determinants in sequence (D.9) are positive numbers and $\Delta_{2n-1} = 0$. This signifies that (D.1) has roots that appear in complex conjugate pairs with relative damping coefficients equal to the specified ζ. Consequently these roots are located on the straight line l of Figure D.1a.

4. If $\Delta_{2n} \neq 0$ and some of the determinants in sequence (D.9) vanish, then, for each group of successive zeros,

$$\Delta_{2h} \neq 0, \Delta_{2h+2} = \cdots = \Delta_{2h+2p} = 0, \Delta_{2h+2p+2} \neq 0,$$

in the calculation of the sign changes in sequence (D.9), the following must be set [D.3]: sign $\Delta_{2h+2i} = (-1)^{i(i-1)/2}$ sign $\Delta_{2h}, (i = 1, 2, \ldots, p)$.

To illustrate the foregoing procedure, consider the characteristic equation of the form

$$F(s) \equiv s^3 + 3s^2 + 6s + 4 = 0 \tag{D.12}$$

and determine whether all roots of this equation have relative damping coefficients greater than $\tfrac{1}{2}$.

The necessary Chebyshev functions have the following values: $T_0(\tfrac{1}{2}) \equiv 1$, $T_1(\tfrac{1}{2}) \equiv \tfrac{1}{2}$, $T_2(\tfrac{1}{2}) = -\tfrac{1}{2}$, $T_3(\tfrac{1}{2}) = -1$, $U_0(\tfrac{1}{2}) \equiv 0$, $U_1(\tfrac{1}{2}) \equiv 1$, $U_2(\tfrac{1}{2}) \equiv 1$, $U_3(\tfrac{1}{2}) = 0$.

The determinant Δ_{2n} given in (D.10) and the corresponding diagonal minors are

$$\Delta_6 = \begin{vmatrix} 0 & -3\sqrt{\frac{3}{2}} & 6\sqrt{\frac{3}{2}} & 0 & 0 & 0 \\ 1 & -\frac{3}{2} & -3 & 4 & 0 & 0 \\ 0 & 0 & -3\sqrt{\frac{3}{2}} & 6\sqrt{\frac{3}{2}} & 0 & 0 \\ 0 & 1 & -\frac{3}{2} & -3 & 4 & 0 \\ 0 & 0 & 0 & -3\sqrt{\frac{3}{2}} & 6\sqrt{\frac{3}{2}} & 0 \\ 0 & 0 & 1 & -\frac{3}{2} & -3 & 4 \end{vmatrix} = 0$$

(D.13)

$$\Delta_4 = \begin{vmatrix} 0 & -3\sqrt{\frac{3}{2}} & 6\sqrt{\frac{3}{2}} & 0 \\ 1 & -\frac{3}{2} & -3 & 4 \\ 0 & 0 & -3\sqrt{\frac{3}{2}} & 6\sqrt{\frac{3}{2}} \\ 0 & 1 & -\frac{3}{2} & -3 \end{vmatrix} = \frac{81}{2} > 0$$

$$\Delta_2 = \begin{vmatrix} 0 & -3\sqrt{\frac{3}{2}} \\ 1 & -\frac{3}{2} \end{vmatrix} = 3\sqrt{\frac{3}{2}} > 0.$$

Consequently, these conditions of relative stability relate to the previously mentioned case 3, and one root of (D.12) is real whereas the other two form a complex conjugate pair with $\zeta = \frac{1}{2}$.

D.3 Nyquist Criterion

Consider a linear feedback system with the open-loop transfer function $KG(s)$ and the characteristic equation

$$1 + KG(s) = 0. \tag{D.14}$$

In relative stability analysis it is of interest to determine the conditions under which (D.14) has no roots inside the s plane contours shown in Figure D.2.

Consider the closed contour C of Figure D.2a, which consists of the constant σ, straight line AB and the infinite semicircle that encloses the part of the s plane to the right of the straight line AB. Next perform the mapping of the contour C into the $G(s)$ plane when C is traced out in a clockwise direction. Applying the Cauchy argument principle (Theorem A.6) to $G(s)$ plot, we may conclude that for (D.1) to have no roots inside the contour C of Figure D.2a, it is necessary and sufficient that, when ω takes on values along the σ constant line from $-\infty$ to $+\infty$, the plot of the function $G(s)$ enclose the $(-1/K, j0)$ point in the counterclockwise direction as many times as there are poles of the function $KG(s)$ in the specified contour.

Nyquist Criterion   453

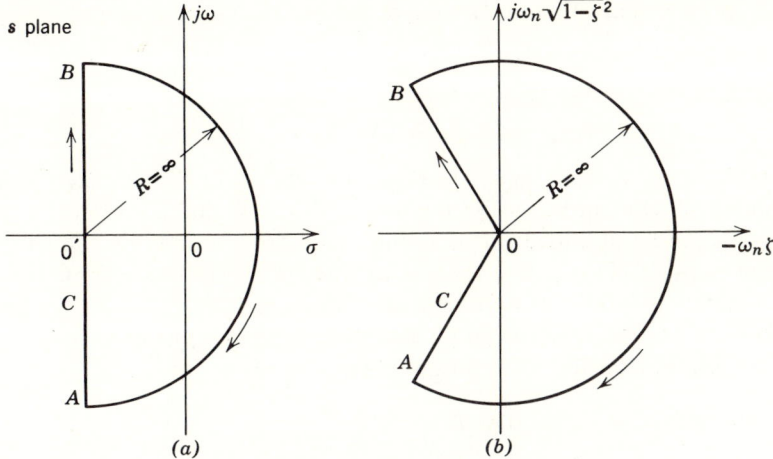

Figure D.2   The s plane contours.

On account of symmetry (see Theorem A.3), it is sufficient to map only the part $0'B$ of the contour C into the $G(s)$ plane, which corresponds to the variation of ω between 0 and $+\infty$, and then count the semiencirclements of the $G(s)$ plot about the $(-1/K, j0)$ point. It should also be noted that if the function $G(s)$ has singularities on the chosen σ constant line, they are handled in the usual fashion with the semicircles of infinitesimal radii that place the singularities outside the enclosed area.

In order to map the σ constant line from the s plane into the $G(s)$ plane, consider the open-loop transfer function in the form

$$KG(s) = K \frac{\sum_{k=0}^{m} c_k s^k}{\sum_{k=0}^{n} b_k s^k}, \qquad n \geq m \qquad (\text{D.15})$$

and substitute $s = \sigma + j\omega$ to obtain

$$KG(\omega, \zeta) = \frac{\sum_{k=0}^{m} c_k X_k + j \sum_{k=0}^{m} c_k Y_k}{\sum_{k=0}^{n} b_k X_k + j \sum_{k=0}^{n} b_k Y_k}, \qquad (\text{D.16})$$

where $X_k = X(\sigma, \omega)$ and $Y_k = Y(\sigma, \omega)$ are given by

$$\begin{aligned} X_k &= \sigma^k - \binom{k}{2}\sigma^{k-2}\omega^2 + \binom{k}{4}\sigma^{k-4}\omega^4 - \cdots \\ Y_k &= \binom{k}{1}\sigma^{k-1}\omega - \binom{k}{3}\sigma^{k-3}\omega^3 + \binom{k}{5}\sigma^{k-5}\omega^5 - \cdots \end{aligned} \qquad (\text{D.17})$$

454 Stability Criteria for Linear Systems

The polynomials X_k and Y_k thus defined in (D.17) can be obtained by the recurrence formulas

$$X_{k+1} - 2X_1 X_k + (X_1^2 + Y_1^2) X_{k-1} = 0$$
$$Y_{k+1} - 2X_1 Y_k + (X_1^2 + Y_1^2) Y_{k-1} = 0, \qquad \text{(D.18)}$$

where $X_0 \equiv 1$, $X_1 \equiv \sigma$, $Y_0 \equiv 0$, $Y_1 \equiv \omega$ as shown in Appendix A. Since the formulas (D.18) can be easily implemented in a digital computer, a general program can be obtained for mapping the σ straight line into the $G(s)$ plane on the basis of (D.16). For specified values of σ and ω the function $KG(s)$ is evaluated as a ratio of two complex numbers.

To illustrate the application of the generalized Nyquist criterion, consider the open-loop transfer function given as

$$KG(s) = K\,\frac{s+4}{s(s+10)(s^2+6s+10)} \qquad \text{(D.19)}$$

and determine the values of K for which all the roots of the corresponding characteristic equation D.14 have negative real parts $\sigma \leq -1$. By using (D.16), the curve $G(-1, \omega)$ is plotted in Figure D.3 for $-\infty < \omega < +\infty$. One of the poles ($s = 0$) of the transfer function $KG(s)$ in (D.19) is outside the contour C of Figure D.2a plotted for $\sigma = -1$. Therefore the segment $M_1 M_2$ of the real axis of the $G(s)$ plane corresponds to the condition $\sigma \leq -1$,

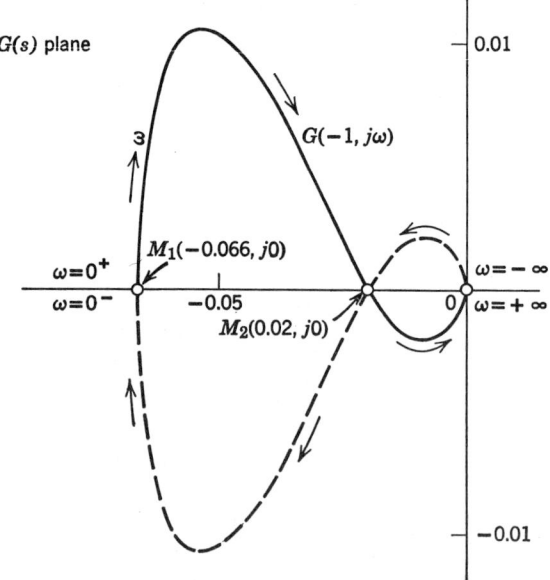

Figure D.3 Nyquist diagram.

for on that segment the point $(-1/K, j0)$ is enclosed once in the counter-clockwise direction by the $G(-1, \omega)$ curve. For all the characteristic roots to be with $\sigma \leq -1$, the gain constant K should be $15 \leq K \leq 50$, as calculated from the points M_1 and M_2 of Figure D.3.

In the case of the contour C in Figure D.2b, where the damping coefficients of the characteristic roots are of interest, (D.16) can again be used to plot the $G(\omega_n, \zeta)$ curve provided that the functions X_k and Y_k are computed from (D.18) by using the values $X_0 \equiv 1$, $X_1 \equiv -\omega_n \zeta$, $Y_0 \equiv 0$, $Y_1 \equiv \omega_n\sqrt{1-\zeta^2}$. The functions X_k and Y_k defined for the arguments ω_n and ζ are related to Chebyshev's functions

$$\begin{aligned} X_k &= (-1)^k \omega_n^k T_k(\zeta) \\ Y_k &= (-1)^{k+1} \omega_n^k \sqrt{1-\zeta^2}\, U_k(\zeta) \end{aligned} \quad (D.20)$$

as shown in Appendix A. Once the constant damping line $0B$ is mapped into the $G(s)$ plane as $G(\omega_n, \zeta)$ curve, the relative stability is interpreted by applying the generalized criterion to the obtained plot [D.19].

Note that the original Nyquist plot is obtained by mapping the imaginary axis of the s plane onto the $G(s)$ plane, which is related to the specific case $\sigma = \zeta = 0$.

D.4 Mikhailov Criterion

The Mikhailov stability criterion for linear systems represents a direct application of the Cauchy argument principle to the characteristic polynomial $F(s)$. This is discussed along with Theorem A.6 of Appendix A, by which the Mikhailov curve is plotted in Figure A.2.

D.5 D-Decomposition Method

The idea of interpreting the stability conditions for linear systems in the parameter plane as introduced by Vishnegradsky [D.7] has been generalized by Neimark [D.8–10] in his D-decomposition method for consideration of multiparameter problems in the parameter space. The D-decomposition method has been presented in English by Lanzkron and Higgins [D.24], and certain proofs and extension have been proposed in references D.25–28. The D-decomposition approach is briefly presented here and then compared with the parameter plane analysis proposed in Chapter 1.

Consider a real[2] polynomial

$$F(s) \equiv \sum_{k=0}^{n} a_k s^k, \quad (D.21)$$

---

[2] The real polynomials are considered here for simplicity. The complex polynomials can be treated in the same manner with minor formal modification as shown in reference [D.9].

456   Stability Criteria for Linear Systems

where the coefficients a_k are continuous functions of r system parameters p_i ($i = 1, 2, \ldots, l$), that is,

$$a_k = a_k(p_1, p_2, \ldots, p_l), \qquad (k = 0, 1, \ldots, n) \tag{D.22}$$

and $s = \sigma + j\omega$ is the complex variable. Now the l-dimensional vector space $\mathscr{P}$, of which the p_i are coordinates, can be decomposed into sets denoted by $D(m, n - m)$, which correspond to the polynomial (D.21) having m zeros with negative and $n - m$ zeros with positive real parts. Such a decomposition of the parameter space $\mathscr{P}$ into sets $D(m, n - m)$ is called the *D-decomposition*.

The boundaries of the sets $D(m, n - m)$ consist of those points for which the polynomial $F(s)$ has (finite) zeros with vanishing real parts. The boundaries consist of surfaces determined by

$$a_0 = 0 \tag{D.23}$$

$$a_n = 0 \tag{D.24}$$

and a surface determined by

$$\begin{aligned} R &\equiv \sum_{k=0}^{n} a_k X_k(0, \omega) = 0 \\ I &\equiv \sum_{k=0}^{n} a_k Y_k(0, \omega) = 0, \end{aligned} \tag{D.25}$$

where functions $X_k(\sigma, \omega)$ and $Y_k(\sigma, \omega)$ for $\sigma = 0$ are given by the recurrence formulas

$$\begin{aligned} X_{k+2} + \omega^2 X_k &= 0 \\ Y_{k+2} + \omega^2 Y_k &= 0, \end{aligned} \tag{D.26}$$

with $X_0 \equiv 1$, $X_1 \equiv 0$, $Y_0 \equiv 0$, $Y_1 \equiv \omega$.

The surface $a_0 = 0$ corresponds to a zero at the origin of the s plane, and the surface $a_n = 0$ corresponds to a zero at infinity of the s plane. The surface (D.25) corresponds to a pair of pure imaginary zeros of the polynomial $F(s)$.

Eliminating ω^2 from (D.25), we obtain

$$\Delta_{n-1} = \begin{vmatrix} a_{n-1} & a_n & 0 & 0 & \cdots & 0 \\ a_{n-3} & a_{n-2} & a_{n-1} & a_n & \cdots & 0 \\ \cdot & \cdot & \cdot & \cdot & & \cdot \\ \cdot & \cdot & \cdot & \cdot & & \cdot \\ \cdot & \cdot & \cdot & \cdot & & \cdot \\ 0 & 0 & a_0 & a_1 & \cdots & a_n \end{vmatrix} = 0, \tag{D.27}$$

where Δ_{n-1} is the next-to-the-last Hurwitz determinant Δ_n of $F(s)$. Thus the necessary and sufficient condition for the polynomial $F(s)$ to have a (finite)

zero with vanishing real part
$$a_0 a_n \Delta_{n-1} = 0. \tag{D.28}$$

The stability of a linear system with characteristic polynomial $F(s)$ is assured if a point in the space $\mathscr{P}$ is chosen from the set $D(n, 0)$.[3] Points on the boundary of the set $D(n, 0)$ apparently satisfy the condition (D.28).

As an example [D.29] consider a characteristic equation

$$s^5 + (\alpha + \beta + 10)s^4 + (10\alpha + \gamma + 200)s^3 + 200(\alpha + \beta + 10)s$$
$$+ 200(10\alpha + \gamma) + 10^5 = 0. \tag{D.29}$$

The Hurwitz conditions for $s = j\omega$ yield

$$(10\alpha + \gamma) > 0$$
$$(\alpha + \beta + 10) > 0 \tag{D.30}$$
$$(10\alpha + \gamma)(\alpha + \beta + 10)(200 - 10\alpha + \gamma) - 10^{-5} > 0.$$

The stability boundary corresponding to (D.30) is constructed in a three-dimensional parameter space in Figure D.4, where three points are indicated; M_1 and M_3 represent an unstable and a stable system, respectively, and M_2, which is on the boundary surface, represents a system on the verge of stability.

Although the D-decomposition is defined in a parameter space, it is convenient to apply only to two-parameter problems. The common conditions (D.25) for determining the boundary of $D(n, 0)$ provide two equations relating l parameters. Thus all parameters besides two should be given numerically.

There is a way, however, to improve this situation and extend the parameter mapping to multiparameter stability analysis. A technique is presented here for obtaining a convenient interpretation of the stability regions in the parameter space as it was proposed by George in reference D.30. The interpretation consists of imbedding a rectangular parallelepiped into a convex parameter domain of stability. The parallelepiped is assumed to have sides perpendicular to the coordinate axes and center at the known stable point $\bar{M}(\bar{p}_1, \bar{p}_2, \ldots, \bar{p}_l)$ located inside the domain $D(n, 0)$. Then the volume is maximized with respect to the constraints of the boundaries of $D(n, 0)$.

Let the coefficients a_k of the polynomial $F(s)$ in (D.21) be given as

$$a_k = p_1 b_k + p_2 c_k + d_k, \tag{D.31}$$

---

[3] Note that in general there are no necessary and/or sufficient conditions for the existence of the set $D(n, 0)$—it may be an empty set.

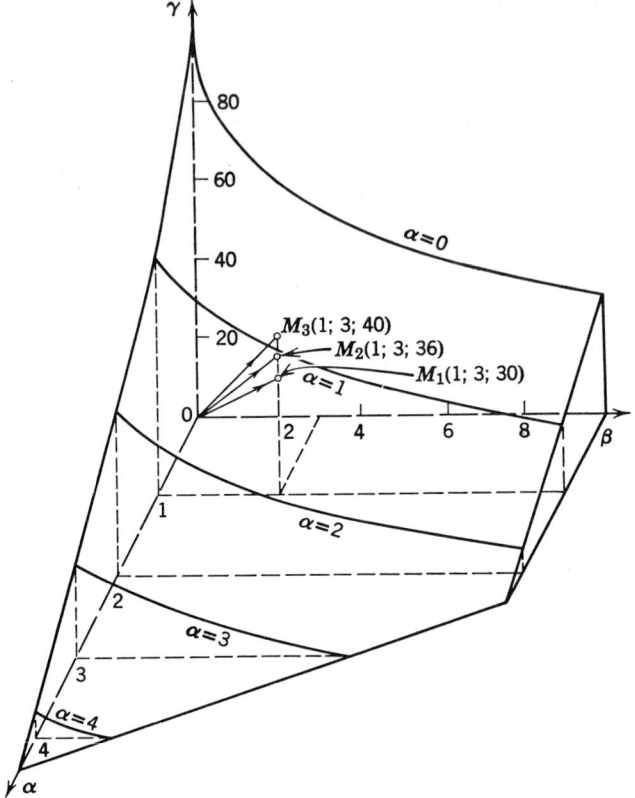

Figure D.4 A stability domain in a three-dimensional parameter space.

where
$$\begin{aligned} b_k &= b_k(p_3, p_4, \ldots, p_l) \\ c_k &= c_k(p_3, p_4, \ldots, p_l) \\ d_k &= d_k(p_3, p_4, \ldots, p_l). \end{aligned} \quad \text{(D.32)}$$

Then the *complex root boundaries of the domain* $D(n, 0)$ *in the parameter space* are determined by (D.25), which can be solved to yield

$$\begin{aligned} p_1 &= p_1(\omega, p_3, p_4, \ldots, p_l) \\ p_2 &= p_2(\omega, p_3, p_4, \ldots, p_l) \end{aligned} \quad \text{(D.33)}$$

provided that the corresponding Jacobian is different from zero (see Section 1.4).

The *real-root boundaries* of $D(n, 0)$ can be obtained directly from the characteristic equation D.21 for $s = 0$ and $s = \infty$, which yield $a_0 = 0$ and

$a_n = 0$. They can be represented by the equation

$$p_1 = p_1(p_2, p_3, \ldots, p_l). \tag{D.34}$$

Let the point $\bar{M}(\bar{p}_1, \bar{p}_2, \ldots, \bar{p}_l)$ be a stable point located inside $D(n, 0)$. Consider the volume v of the parallelepiped with sides

$$2(p_1 - \bar{p}_1), 2(p_2 - \bar{p}_2), \ldots, 2(p_l - \bar{p}_l)$$

that is given as

$$v = 2^l(p_1 - \bar{p}_1)(p_2 - \bar{p}_2) \cdots (p_l - \bar{p}_l). \tag{D.35}$$

Substituting (D.33) into (D.35) and extremizing, a necessary condition for $(\tilde{\omega}, \tilde{p}_3, \ldots, \tilde{p}_l)$ to occur at a maximum of v is that it be a solution of

$$\frac{\partial v}{\partial \omega} = 0, \quad \frac{\partial v}{\partial p_j} = 0, \quad (j = 3, 4, \ldots, l). \tag{D.36}$$

If (D.34) is substituted into (D.35) and extremized, a necessary condition[4] for $M^0(p_2^0, p_3^0, \ldots, p_l^0)$ to occur at a maximum of v is that it be a solution to

$$\frac{\partial v}{\partial p_j} = 0, \quad (j = 2, 3, \ldots, l). \tag{D.37}$$

Now let the parallelepiped P be

$$P = \{(p_1, p_2, \ldots, p_l) \,|\, |p_j - \bar{p}_j| \le \min\left[|\tilde{p}_j - \bar{p}_j|, |p_j^0 - \bar{p}_j|\right], \quad j = 1, 2, \ldots, l\}. \tag{D.38}$$

If each vertex point (corner) of P is contained in the stability domain containing $\bar{M}$, it follows that the region P is contained in $D(n, 0)$ of the polynomial $F(s)$ in (D.21).[5]

To illustrate the above procedure, consider the characteristic equation

$$F(s) \equiv s^3 + p_3 s^2 + p_2 s + p_1 = 0. \tag{D.39}$$

Let us choose the point $\bar{M}(0.5, 1, 1)$ inside the stable domain $D(3, 0)$.

---

[4] Standard sufficient conditions for these solutions to be maximal are given in footnote 58 in Chapter 8.

[5] If the domain $D(n, 0)$ is not convex, consider the euclidean norm $\delta = \|M - \bar{M}\| = \left[\sum_{j=1}^{l}(p_j - \bar{p}_j)^2\right]^{1/2}$. By substituting (D.33) or (D.34) into the norm, grad $\delta = 0$ can be solved for $M^*(p_1^*, p_2^*, \ldots, p_l^*)$. Let P^* be the rectangular parallelepiped with center at $\bar{M}$ and vertex at M^*. By now defining P to be a rectangular parallelepiped so that P is contained in every P^*, P yields a conservative estimate for the stability domain $D(n, 0)$. In (D.38), $\{p \,|\, P\}$ denotes a set of p's having a property P.

The complex root boundary of $D(3, 0)$ is given by equations

$$R \equiv p_1 - p_3\omega^2 = 0$$
$$I \equiv p_2\omega - \omega^3 = 0, \tag{D.40}$$

which can be solved to yield

$$p_1 = p_2 p_3. \tag{D.41}$$

Since (D.41) is the same type as (D.34), (D.37) becomes

$$2p_2 p_3 - p_2 - 0.5 = 0$$
$$2p_2 p_3 - p_3 - 0.5 = 0. \tag{D.42}$$

Solving (D.42) we obtain the point $\tilde{M}(0.655, 0.81, 0.81)$, which occurs at a maximum value of v.

The real root boundary is $s = 0$, which from (D.39) is given by

$$p_1 = 0. \tag{D.43}$$

Since (D.43) is also of the same type as (D.34), (D.37) yields the point $M^0(0, 1, 1)$ occurring at a maximum value of v.

From (D.38) a solution to the interpretation problem is given by P

$$|p_1 - 0.5| \leq 0.155$$
$$|p_2 - 1| \leq 0.19 \tag{D.44}$$
$$|p_3 - 1| \leq 0.19,$$

which is contained inside the stable domain $D(3, 0)$.

As seen above, the multiparameter stability analysis was formulated as an interpretation problem of the domains of stability that reduces to the solution of a system of algebraic equations. The simplicity of the interpretation is obtained at the expense of a considerable effort usually involved in solving the nonlinear algebraic equations.

Furthermore, it is of essential interest to note that the D-decomposition method is applicable only to stability problems and inherently belongs to frequency-domain methods of system analysis. Attempts to apply the method to the design of systems in terms of the transient response normally generate real difficulties.

An entirely different approach to system design based upon the parameter mapping is presented in Sections 2.4 and 2.5 of Chapter 2. It allows the multiparameter design with respect to both the frequency and the time response characteristics.

References

[D.1] C. Hermité, Le numbre des racines d'une équation algebraique comprises entre des limites données, *J. de Crelle*, Paris, **52** (1954).

[D.2] C. Hermité, Le numbre limite d'irrationalités auxquelle se réduisent les racines des équations a coefficients entiers complexes d'une degré et d'un discriminant données, *J. de Crelle*, Paris, **53** (1956).

[D.3] F. R. Gantmacher, *The Theory of Matrices*, Vol. 2, Chelsea, New York, 1960.

[D.4] A. Hurwitz, Über die Bedingungen, unter welchen eine Gleichung nur Wurzeln mit negativen reellen Teilen besitzt, *Mathematical Annual*, **46**, 273–284 (1895).

[D.5] H. Nyquist, Regeneration Theory, *Bell System Tech. J.*, **11**, 126–147 (1932).

[D.6] A. W. Mikhailov, Methods for Harmonic Analysis in the Automatic Control System Theory (in Russian), *Avtomatika i Telemekhanika*, No. 3 (1938).

[D.7] I. A. Vishnegradsky, Sur la théorie générale des régulateurs, *Compt. Rend.*, **83**, Paris, 318–321 (1876).

[D.8] Yu. I. Neimark, On the Problem of the Distribution of the Roots of Polynomials (in Russian), *Dokl. Akad. Nauk SSSR*, **58**, 357–360 (1947).

[D.9] Yu. I. Neimark, *Stability of Linearized Systems* (in Russian), LKVVIA, Leningrad, 1949.

[D.10] Yu. I. Neimark, D-decomposition of the Space of Quasipolynomials (On the Stability of Linearized Distributed Systems) (in Russian), *Appl. Math. Mech.*, **13**, No. 4, 349–380 (1949).

[D.11] Ya. Z. Tsypkin and P. W. Bromberg, Degree of Stability in Linear Systems (in Russian), *Contribution of NISO*, No. 9, Ed. VNTMAP (1946).

[D.12] A. A. Krasovskii, On the Degree of Stability of Linear Systems (in Russian), *Contributions of WWPA name of N. E. Zukovskii*, No. 281 (1948).

[D.13] N. G. Chetaev, On the Settling Time in Transients of Linear Systems (in Russian), *PMN*, **15**, No. 3 (1951).

[D.14] F. Strecker, *Praktische Stabilatatsprufung*, Springer, Berlin, 1950.

[D.15] A. Leonhard, Relative Damping as Criterion for Stability and as an Aid in Finding the Roots of a Hurwitz Polynomial, *Automatic and Manual Control*, Butterworths, London, 1952.

[D.16] W. Frey, A Generalization of the Nyquist and Leonhard Stability Criteria, *Brown Boveri Rev.*, No. 33 (1946).

[D.17] D. Mitrović, Graphical Analysis and Synthesis of Feedback Control Systems, Pt. I—Theory and Analysis, *AIEE Trans. Appl. Ind.*, **77**, 476–496 (January 1959).

[D.18] D. D. Šiljak, Analysis and Synthesis of Feedback Control Systems in the Parameter Plane, Pt. I—Linear Continuous Systems, *IEEE Trans. Appl. Ind.*, Pt. II, **83**, 449–458 (November 1964).

[D.19] M. R. Stojić and D. D. Šiljak, Generalization of Hurwitz, Nyquist, and Mikhailov Stability Criteria, *IEEE Trans. Auto. Control*, **AC-10**, No. 3, 250–255 (July 1965).

[D.20] D. D. Šiljak, A Note on the Generalized Nyquist Criterion, *IEEE Trans. Auto. Control*, **AC-11**, No. 2, 317 (April 1966).

[D.21] H. Bilharz, Bemerkung zu einem Satze von Hurwitz, *ZAMM*, **24**, 77–82 (1944).

[D.22] E. Frank, On the Zeros of Polynomials with Complex Coefficients, *Bull. Am. Math. Soc.*, **52**, 144–157 (1946).

[D.23] T. Takahashi, *Mathematics of Automatic Control*, Holt, Rinehart, and Winston, New York, 1966.

[D.24] R. W. Lanzkron and T. J. Higgins, D-decomposition Analysis of Automatic Control Systems, *IRE Trans. Auto. Control. Syst.*, **AC-4**, 150–171 (December 1959).

[D.25] T. Numakura and T. Miura, A New Stability Criterion of Linear Servomechanisms by a Graphical Method, *AIEE Trans. Appl. Ind.*, Pt. II, **76,** 40–48 (March 1957).

[D.26] M. V. Meerov, *Structural Synthesis of High-Accuracy Automatic Control Systems* (in Russian), State Press of Physics and Mathematical Literature, Moscow, 1959. (English translation by Pergamon, New York, 1965.)

[D.27] E. Polak, A Note on D-decomposition Theory, *IEEE Trans. Auto. Control*, **AC-9,** No. 1, 107–109 (January 1964).

[D.28] S. H. Lehnigk, *Stability Theorems for Linear Motion (With an Introduction to Liapunov's Direct Method)*, Prentice-Hall, Englewood Cliffs, New Jersey, 1966.

[D.29] K. W. Han and G. J. Thaler, Control System Analysis and Design Using a Parameter Space Method, *IEEE Trans. Auto. Control*, **AC-11,** No. 3, 560–563 (July 1966).

[D.30] J. H. George, On Parameter Stability Regions for Several Parameters Using Frequency Response, *IEEE Trans. Auto. Control*, **AC-12,** No. 2, 197–200 (April 1967).

APPENDIX E

Squared-Error Optimization with Stability Constraints

E.1 Introduction

In high-performance control systems the ultimate error is caused by both deterministic and stochastic input signals. The desired performance characteristics of linear control systems regarding deterministic inputs are expressed either by a prescribed degree of relative stability or, more precisely, in terms of system response specifications such as the overshoot, the settling time, the bandwidth, and so on. The behavior of control systems because of stochastic signals is designed by optimizing a performance index that describes certain average properties of the system [E.1–3].[1]

Numerous performance indexes have been proposed for optimization of control systems [E.3–7], and the proper index should be chosen in light of the nature of the control problem and specifications that have to be satisfied. However, although an appropriate performance index is selected, the optimization procedure may appear insufficient when applied to control system design. The performance indexes are average measures of the system behavior, and the system response as an exact function of time is unknown to the designer. The optimization procedures fail to reveal that the system is unstable or unrealizable, which may lead to serious limitations and errors in the control system design. Thus the optimization, if it is to be realistic and useful, must be compatible with the system stability and take into account the deterministic response specifications that are substantial characteristics of any high-quality control system [E.7].

---

[1] The error concept in feedback system optimization has been first proposed by I. Obradović [E.11] back in 1942, as pointed out by Schultz [E.5].

An illustration of the foregoing statement is the design of the aircraft autopilot in which a rapid elimination of effects of the atmospheric turbulence is required along with the stringent performance specification of the system response to deterministic input signals. Similar situations occur in radar tracking systems, in which the control system is to be designed by optimizing a statistical performance index over various situations that are weighted according to the associated probability of occurrence. However, such an optimization does not indicate the control system behavior in any particular situation and thus leaves the designer with an uncertainty about the stability and system response that are vital characteristics of radar tracking systems.

In order that the optimization procedure be generally applicable, various techniques have been proposed: weighting of the error signal with both deterministic and stochastic functions, model applications, standard forms of the system response, special penalty cases, constraints in the optimization in terms of additional performance index, and others. All these techniques heretofore described in the literature of the field provide an indirect measure of the actual system response. However, the quality of these measures depends on the particular control problem under investigation, and the designer has no possible way to check the quality during the optimization process. Thus the ambiguity about the system stability does not disappear.

In this appendix the optimization of linear control systems is performed with relative stability constraints in the parameter plane [E.7]. The design objective is to determine the free system parameters so that the mean-square error has its minimum under the constraints that all roots of the corresponding characteristic equation lie at certain locations. Therefore the optimization and stability considerations are performed simultaneously, enabling the design of both statistical performance specifications and the characteristics of the system response to deterministic input signals. The ability to incorporate the desired transient response specifications into the optimization process makes the proposed procedure a rather general method for the design of linear control systems.

The proposed graphical procedure is illustrated by the least-mean-square error optimization, although more general criteria can also be used. Linear continuous multiloop control systems with several adjustable parameters are designed by the optimization with relative stability constraints prescribed in advance. The approach can be extended to sampled-data systems and nonlinear systems, as well as to multivariable and adaptive control systems. The procedure is convenient for digital computer applications.

A limitation of the proposed method is that a form of the system structure and the compensation network must be chosen in advance. The design procedure is applied to control systems with inputs that are stationary, or may

E.2 Optimization

The optimization procedure with stability constraints will be best described by the following example.

Consider the control system shown in the upper-right corner of Figure E.1 with specifications

$$G_1(s) = \frac{K(s + \lambda\delta)}{s + \delta}$$

$$G_2(s) = \frac{1}{s(0.001s^2 + 0.025s + 0.25)}$$

$$\phi_S(s) = -\frac{\gamma_S}{\pi s^2} \quad \text{(E.1)}$$

$$\phi_N(s) = \gamma_N/\pi$$

$$u(t) = \text{unit step function,}$$

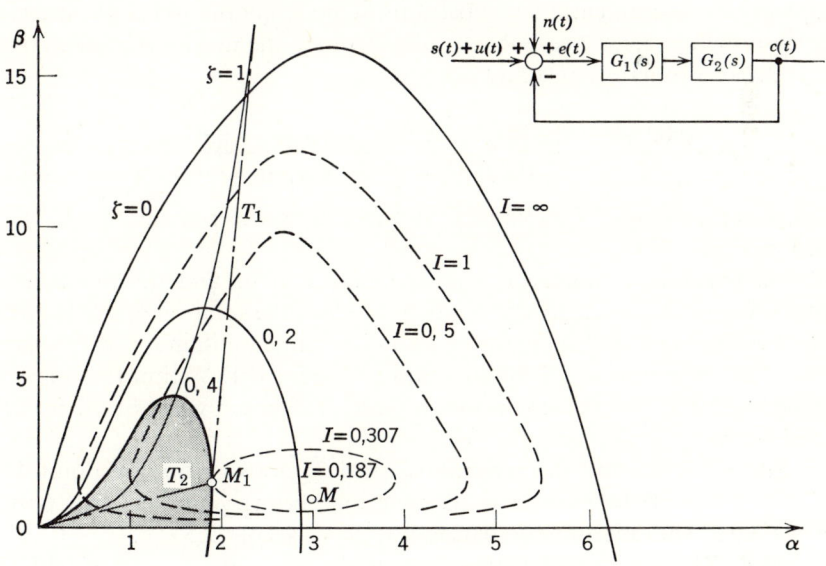

Figure E.1 Parameter plane diagram.

where $\phi_S(s)$ and $\phi_N(s)$ are power-density spectra of the stochastic input signal $s(t)$ and the noise $n(t)$, respectively.

It is required to determine the parameters K, λ, and δ of the integral compensator $G_1(s)$ so as to increase the system velocity constant and minimize the mean-squared error while maintaining the overshoot of the step function response below 30% of its steady-state value. At first, the noise component is not considered.

As known, the choice of the parameter $\delta(\ll 1)$ is not critical and the value 0.04 may be accepted. Then if the numerical value of γ_S is 2π, the power density of error $\phi_E(s)$, corresponding to the signal $s(t)$, is given as

$$\phi_E(s) = \frac{0.002s^3 + 0.05s^2 + 0.5s + 0.02}{0.001s^4 + 0.025s^3 + 0.25s^2 + (K + 0.01)s + 0.04K\lambda}. \quad \text{(E.2)}$$

The denominator of $\phi_E(s)$ is the characteristic polynomial of the system under investigation. By substituting $K + 0.01 = \alpha$ and $0.04K\lambda = \beta$ in (E.2), we may plot, in the usual fashion, the characteristic ζ curves as shown in Figure E.1. These curves determine in the $\alpha\beta$ plane the relative damping region that corresponds to the certain value of the relative damping coefficient ζ. Thus for $\zeta = 0.4$, the relative damping region is determined by the $\zeta = 0.4$ curve and is shown as shaded in Figure E.1. For the values of α and β lying in this region, all roots of the corresponding characteristic equation will have the relative damping coefficient greater than 0.4.

Applying the aforementioned substitution and using the well-known technique described in reference E.4, we may evaluate the mean-squared error I from (E.2) as a function of α and β

$$I(\alpha, \beta) = \frac{0.025\beta^2 - 0.375\alpha\beta - 1.55\beta + 10^{-4}\alpha - 0.65 \times 10^{-3}}{0.625\beta^2 + \alpha^2\beta - 6.25\alpha\beta}. \quad \text{(E.3)}$$

For different values of I, a family of the curves determined by (E.3) is plotted in Figure E.1.

Now in order to minimize the mean-squared error and simultaneously maintain the prescribed degree of relative stability, the solution of the control problem under investigation is found as a constrained minimum of I, which is located at the point M_1 of the diagram of Figure E.1. Within the relative damping region of $\zeta = 0.4$, the point M_1 has a minimum value of the mean-squared error.

The curve $\zeta = 0.4$ and the tangents T_1 and T_2 drawn from the point M_1 enable the roots of the characteristic equation corresponding to that point to be evaluated without any calculations: $s_{1,2} = -4.1 \pm j9.39$; $s_3 = -0.864$; $s_4 = -16.4$. The smaller real root s_3 and the zero of the corresponding closed-loop transfer function form a dipole whose effect may be neglected.

The other real root is large and its effect is also negligible. The step function response will, therefore, be governed only by the pair of complex roots $s_{1,2}$ whose value of the damping coefficient $\zeta = 0.4$ ensure that the overshoot is less than 30%.

From the coordinates of the point $M_1(\alpha = K + 0.01 = 1.89; \beta = 0.04K\lambda = 1.52)$, the values of the compensator parameters are $K = 1.88$ and $\lambda = 20.2$. The system velocity constant is 38 times greater than the velocity constant of the uncompensated system. The constrained minimum of the mean-squared error is $I = 0.307$. It is interesting to note from Figure E.1 that the absolute minimum $I = 0.187$, which is located at the point M, falls outside the relative damping region corresponding to $\zeta = 0.2$ and, therefore, results in a poorly damped system.

In a similar manner [E.7], the component of the mean-squared error corresponding to the noise can also be expressed as a function of α and β, which for $\gamma_N = 2\pi$ has the following form

$$J(\alpha, \beta) = \frac{\alpha\beta - 6.25\beta - 25\alpha^2}{0.625\beta + \alpha^2 - 6.25\alpha}. \tag{E.4}$$

Using the above equation, the same reasoning outlined above may be applied to the noise case. Furthermore, by the proposed techniques, it is also possible to introduce the quadratic constraint of the saturation signal into the optimization in the $\alpha\beta$ plane [E.4, 7].

The presented procedure can be readily extended to the multiple-input systems [E.9, 10].

References

[E.1] N. Wiener, *Extrapolation, Interpolation, and Smoothing of Stationary Time Series*, Wiley, New York, 1950.

[E.2] L. A. Zadeh and J. R. Ragazzini, An Extension of Wiener's Theory of Prediction, *J. Appl. Phys.*, **21**, No. 7, 645–655 (July 1950).

[E.3] J. H. Laning and R. H. Battin, *Random Processes in Automatic Control*, McGraw-Hill, New York, 1956.

[E.4] G. C. Newton, Jr., L. A. Gould, and J. F. Kaiser, Analytical Design of Linear Feedback Controls, Wiley, New York, 1957.

[E.5] W. C. Schultz and V. C. Rideout, Control System Performance Measures: Past, Present, and Future, *IRE Trans.*, **AC-6**, No. 1, 22–35 (February 1961).

[E.6] Z. V. Rekasius, A General Performance Index for Analytical Design of Control Systems, *IRE Trans.*, **AC-6**, No. 3, 217–222 (May 1961).

[E.7] D. D. Šiljak, Squared-Error Optimization with Stability Constraints, Ph.D. Thesis, University of Belgrade, Belgrade, Yugoslavia, March 1963.

[E.8] D. D. Šiljak, *Discussion of:* 'Dual-Input Systems with Saturation Constraint' by R. Gaylord, *Second IFAC Congress*, Basel, Switzerland, September 1963, Paper No. 411.

[E.9] C. W. Merriam, *Optimization Theory and the Design of Feedback Control Systems*, McGraw-Hill, New York, 1964.

[E.10] J. S. Tyler and F. B. Tuteur, The Use of a Quadratic Performance Index to Design Multivariable Control Systems, *IEEE Trans. Auto. Control*, **AC-11**, No. 1, 84–92 (January 1966).

[E.11] I. Obradović, The Deviation Area in Quick-Acting Regulation (in German), *Archiv Elektrotech.*, **36**, 382–390 (June 1942).

[E.12] W. J. Budurka, Sensitivity Constrained Optimal Control Synthesis, *IBM Res. Develop.*, **11**, No. 4, 427–435 (July 1967).

APPENDIX F

Harmonic Linearization of Typical Nonlinear Characteristics

F.1 Introduction

In this section harmonic linearization of common nonlinear characteristics $F(x, sx)$ is performed. The describing function $N = N_1 + jN_2$ is given in analytic form. For nonlinearities that most frequently occur in practical situations, the describing functions are given by diagrams.

The linearization is taken for both symmetrical and asymmetrical oscillations in the two separate sections. The process of evaluating the describing functions for specified nonlinear characteristics is described when the polynomial nonlinear characteristics are considered. The harmonic linearization of friction nonlinearities is performed and the advantages of the parameter plane analysis have been indicated. The nonlinearities with two input signals are also discussed as well as the systems with variable structures. A graphical method for calculating the describing function coefficients of certain classes of nonlinearities is also presented.

F.2 Piecewise Linear and Relay Characteristics. Symmetrical Oscillations

If the input $x = x(t)$ to the nonlinearity described by a nonlinear function $F(x, sx)$ is given as

$$x = A \sin \Omega t, \qquad (F.1)$$

then the function $F(x, sx) = F(A \sin \Omega t, A\Omega \cos \Omega t)$ can be linearized to obtain

$$F(x, sx) = N_1 x + \frac{N_2}{\Omega} sx. \qquad (F.2)$$

In (F.2), $N_1 = N_1(A, \Omega)$ and $N_2 = N_2(A, \Omega)$ are the coefficients of the describing function $N = N_1 + jN_2$, which are given by the integrals

$$N_1 = \frac{1}{\pi A} \int_0^{2\pi} F(A \sin \phi, A\Omega \cos \phi) \sin \phi \, d\phi$$

$$N_2 = \frac{1}{\pi A} \int_0^{2\pi} F(A \sin \phi, A\Omega \cos \phi) \cos \phi \, d\phi, \tag{F.3}$$

where $\phi = \Omega t$.

In the case when the nonlinearity is independent of the rate of the input signal—that is, the nonlinearity is characterized by a function $F(x)$—the describing function coefficients, $N_1 = N_1(A)$, $N_2 = N_2(A)$, are independent of frequency and are defined by

$$N_1 = \frac{1}{\pi A} \int_0^{2\pi} F(A \sin \phi) \sin \phi \, d\phi$$

$$N_2 = \frac{1}{\pi A} \int_0^{2\pi} F(A \sin \phi) \cos \phi \, d\phi. \tag{F.4}$$

When the nonlinearity is described by a single-valued nonlinear function, $F(x)$, then $N_2 \equiv 0$ and the harmonic linearization is $F(s) = N_1 x$, where N_1 is given in (F.4).

The harmonic linearization of nonlinearities commonly encountered in practical application, which has been performed in references F.1–3, is presented in Table F.1. First, a rather general characteristic representing the whole class of nonlinearities is considered and the coefficients N_1 and N_2 are evaluated. Then the other special cases are presented in two groups, namely, piecewise linear characteristics Nos. 2–11 and relay characteristics Nos. 12–18. The special cases can be obtained from the general characteristic No. 1 by properly manipulating the nonlinear parameters. For example, the characteristic No. 16 can be obtained from No. 1 by putting $E = D_1$, $s = T = R$, $a = b = c$, $k = \infty$, $k_1 = k_2 = 0$. Thus for other characteristics not included in Table F.1, but which belong to the same class described by the characteristic No. 1, the coefficients of the describing function can be obtained from the expressions N_1 and N_2 of the characteristic No. 1.

Besides the expressions for N_1 and N_2 in Table F.1, a restriction is given on the amplitude for the expressions to be valid. For example, in No. 4 the given expressions N_1 and N_2 are valid for $A \geq S$. If A is such that $D \leq A \leq S$, then the expression for N_1 and N_2 given in No. 3 should be used.

The process of calculating the coefficients of the describing function is outlined in the following sections.

Table F.1 Describing Function Coefficients for Symmetrical Oscillations

No.	Nonlinearity	Describing Function Coefficients
1	(see figure)	$N_1 = \dfrac{k}{\pi}\Bigg[-\arcsin\dfrac{D}{A} + \arcsin\dfrac{E}{A} - \arcsin\dfrac{R}{A} - \arcsin\dfrac{S}{A}$ $+ 2\arcsin\dfrac{T}{A} - \dfrac{D}{A}\left(1 - \dfrac{D^2}{A^2}\right)^{1/2} - \dfrac{E}{A}\left(1 - \dfrac{E^2}{A^2}\right)^{1/2} - \dfrac{R}{A}\left(1 - \dfrac{R^2}{A^2}\right)^{1/2}$ $+ \dfrac{S}{A}\left(1 - \dfrac{S^2}{A^2}\right)^{1/2} - 2\dfrac{T}{A}\left(1 - \dfrac{T^2}{A^2}\right)^{1/2} + 2\dfrac{R}{A}\bigg[-\left(1 - \dfrac{S^2}{A^2}\right)^{1/2} + 2\left(1 - \dfrac{T^2}{A^2}\right)^{1/2}\bigg]$ $+ 2\dfrac{D}{A}\left(1 - \dfrac{E^2}{A^2}\right)^{1/2} + \dfrac{k_1}{\pi}\bigg\{ -\arcsin\dfrac{E}{A} + \arcsin\dfrac{S}{A} + \dfrac{E}{A}\left(1 - \dfrac{E^2}{A^2}\right)^{1/2}$ $- \dfrac{S}{A}\left(1 - \dfrac{S^2}{A^2}\right)^{1/2} + 2\dfrac{bE - aS}{A(b-a)}\bigg[-\left(1 - \dfrac{E^2}{A^2}\right)^{1/2} + \left(1 - \dfrac{S^2}{A^2}\right)^{1/2}\bigg]\bigg\}$ $+ \dfrac{k_2}{\pi}\bigg[\pi - 2\arcsin\dfrac{T}{A} + 2\dfrac{T}{A}\left(1 - \dfrac{T^2}{A^2}\right)^{1/2}\bigg] - 4\dfrac{Tk_2 - c}{k_2 A}\left(1 - \dfrac{T^2}{A^2}\right)^{1/2},$ $N_2 = \dfrac{k}{\pi}\bigg[-\dfrac{D^2}{A^2} + \dfrac{E^2}{A^2} + \dfrac{R^2}{A^2} - \dfrac{S^2}{A^2} + 2\dfrac{R}{A}\left(\dfrac{R}{A} - \dfrac{S}{A}\right) + 2\dfrac{D}{A}\left(-\dfrac{D}{A} + \dfrac{E}{A}\right)\bigg]$ $+ \dfrac{k_1}{\pi}\bigg[-\dfrac{E^2}{A^2} + \dfrac{S^2}{A^2} + 2\dfrac{bE - aS}{A(b-a)}\left(-\dfrac{E}{A} + \dfrac{S}{A}\right)\bigg],\quad A \geq T$

Table F.1 (*continued*)

No.	Nonlinearity	Describing Function Coefficients
2		$N_1 = \dfrac{2k}{\pi}\left[\arcsin\dfrac{S}{A} + \dfrac{S}{A}\left(1 - \dfrac{S^2}{A^2}\right)^{1/2}\right],$ $N_2 \equiv 0, \qquad A \geq S$

Table F.1 (*continued*)

No.	Nonlinearity	Describing Function Coefficients
3	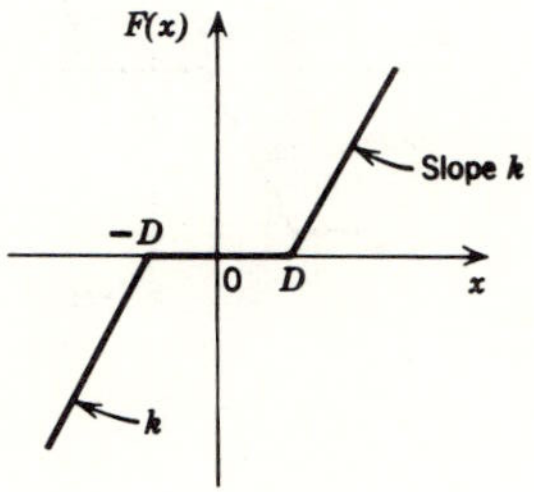 Dead-zone nonlinearity with slope k outside $[-D, D]$	$N_1 = k - \dfrac{2k}{\pi}\left[\arcsin\dfrac{D}{A} + \dfrac{D}{A}\left(1 - \dfrac{D^2}{A^2}\right)^{1/2}\right]$, $\quad A \geq D$ $N_2 \equiv 0$, Plot of N_1/k versus A/D 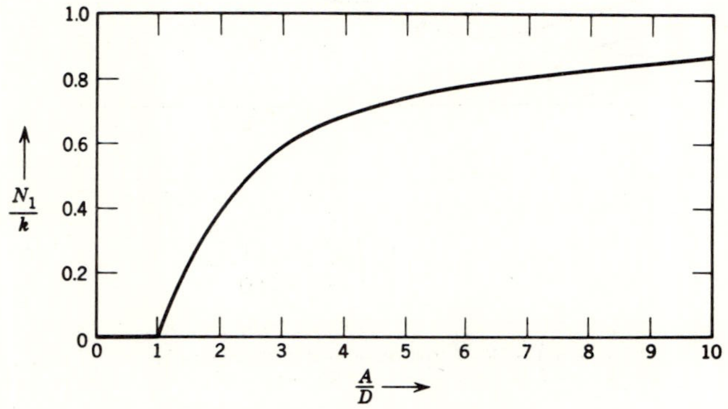

473

Table F.1 (*continued*)

No.	Nonlinearity	Describing Function Coefficients
4	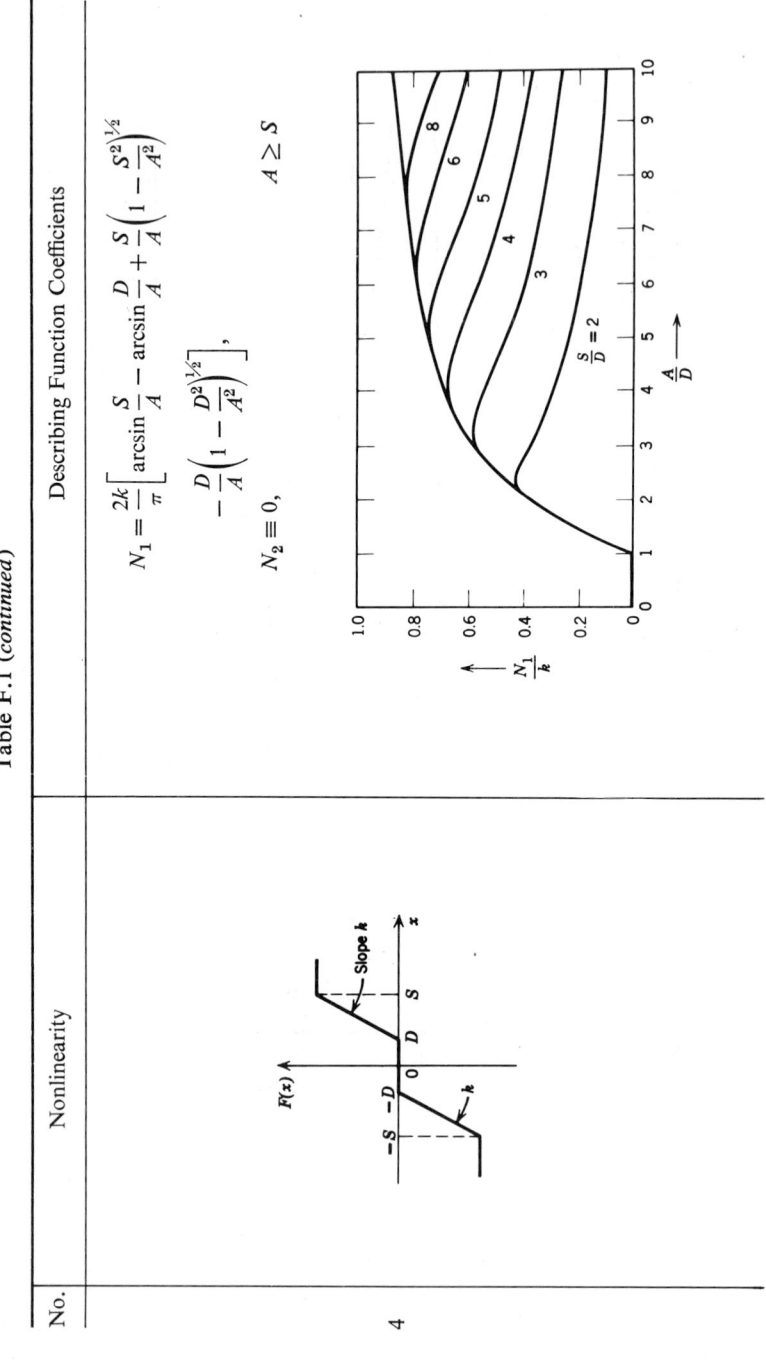	$N_1 = \dfrac{2k}{\pi}\left[\arcsin\dfrac{S}{A} - \arcsin\dfrac{D}{A} + \dfrac{S}{A}\left(1 - \dfrac{S^2}{A^2}\right)^{1/2} \right.$ $\left. - \dfrac{D}{A}\left(1 - \dfrac{D^2}{A^2}\right)^{1/2}\right],$ $N_2 \equiv 0, \qquad A \geq S$

474

Table F.1 (continued)

No.	Nonlinearity	Describing Function Coefficients
5	(piecewise linear characteristic $F(x)$ with slope k for $|x|\le S$ and slope k_1 for $|x|>S$)	$N_1 = k_1 - \dfrac{2}{\pi}(k_1 - k)\left[\arcsin\dfrac{S}{A} + \dfrac{S}{A}\left(1 - \dfrac{S^2}{A^2}\right)^{1/2}\right],$ $A \ge S$ $N_2 \equiv 0,$ (plot of N_1/k vs A/S for $\lambda = k_1/k = 2, 3, 4, 5, 6$)

Table F.1 (continued)

No.	Nonlinearity	Describing Function Coefficients
6	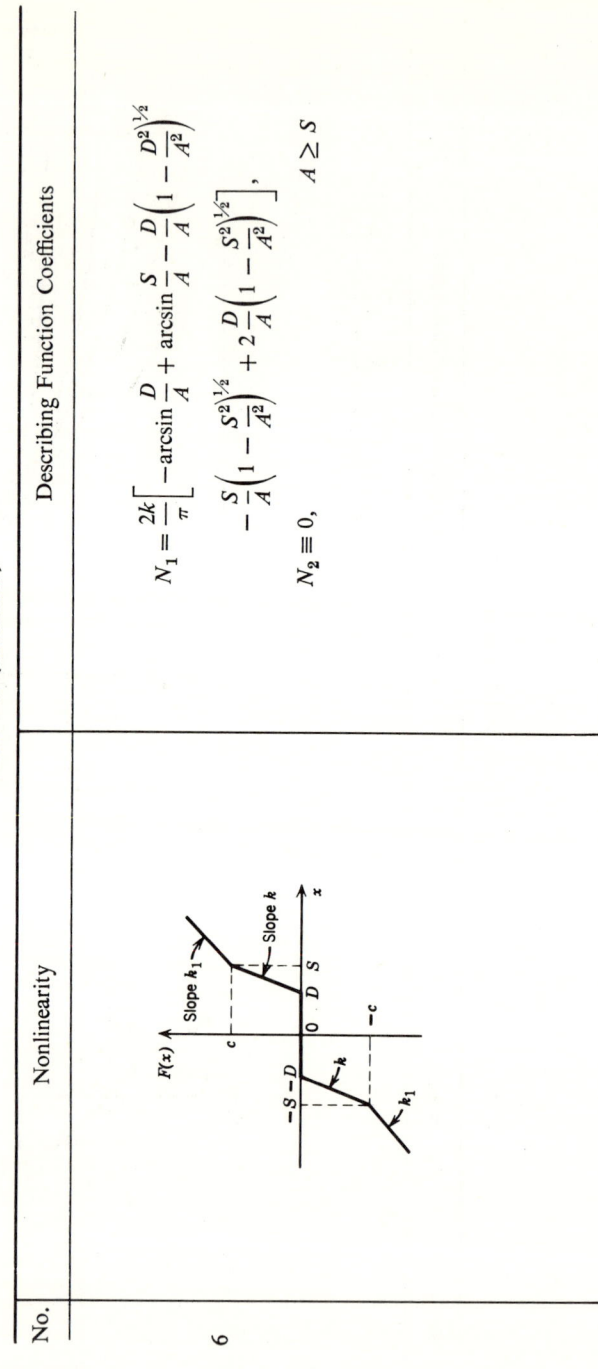	$N_1 = \dfrac{2k}{\pi}\left[-\arcsin\dfrac{D}{A} + \arcsin\dfrac{S}{A} - \dfrac{D}{A}\left(1 - \dfrac{D^2}{A^2}\right)^{1/2}\right.$ $\left. - \dfrac{S}{A}\left(1 - \dfrac{S^2}{A^2}\right)^{1/2} + 2\dfrac{D}{A}\left(1 - \dfrac{S^2}{A^2}\right)^{1/2}\right],$ $N_2 \equiv 0, \qquad A \geq S$

Table F.1 (*continued*)

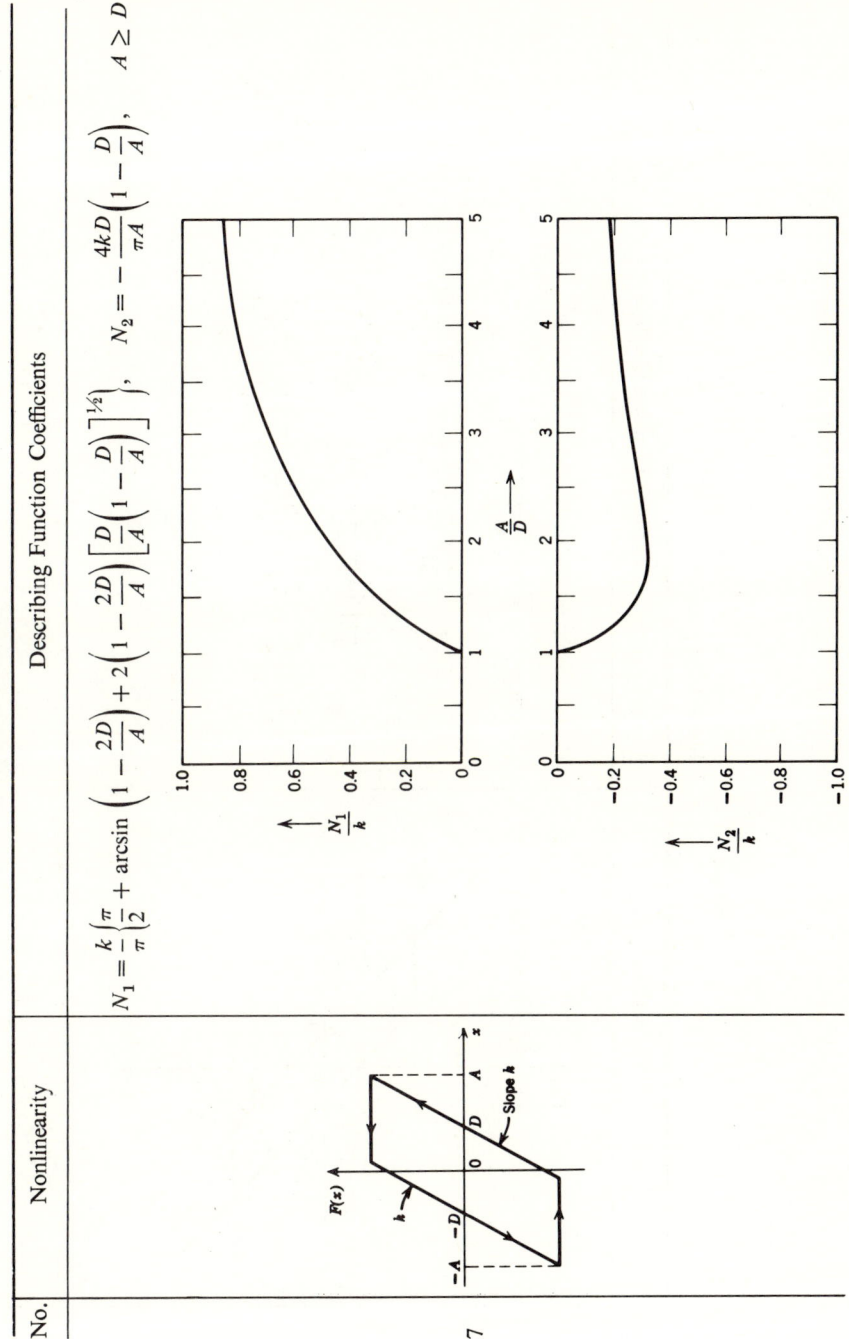

No.	Nonlinearity	Describing Function Coefficients
7		$N_1 = \dfrac{k}{\pi}\left\{\dfrac{\pi}{2} + \arcsin\left(1 - \dfrac{2D}{A}\right) + 2\left(1 - \dfrac{2D}{A}\right)\left[\dfrac{D}{A}\left(1 - \dfrac{D}{A}\right)\right]^{1/2}\right\}, \quad N_2 = -\dfrac{4kD}{\pi A}\left(1 - \dfrac{D}{A}\right), \quad A \geq D$

Table F.1 (*continued*)

No.	Nonlinearity	Describing Function Coefficients
8	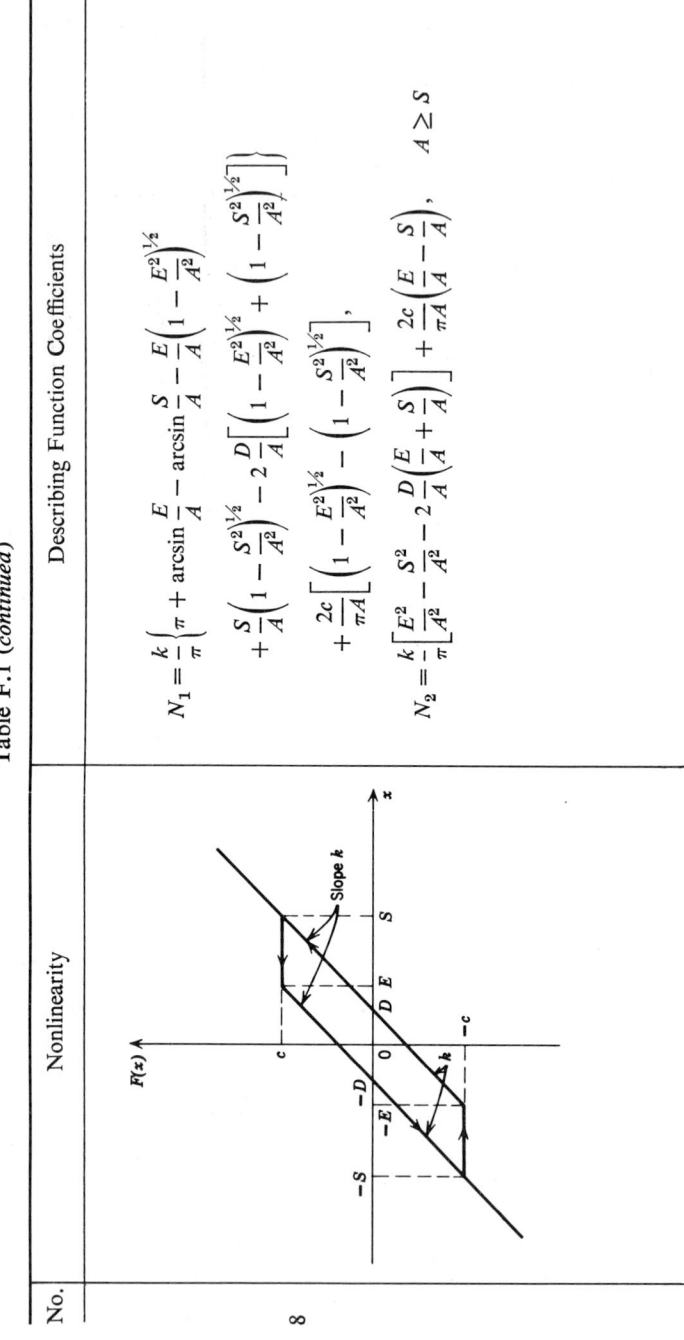	$N_1 = \dfrac{k}{\pi}\left\{\pi + \arcsin\dfrac{E}{A} - \arcsin\dfrac{S}{A} - \dfrac{E}{A}\left(1 - \dfrac{E^2}{A^2}\right)^{1/2}\right.$ $+ \dfrac{S}{A}\left(1 - \dfrac{S^2}{A^2}\right)^{1/2} - 2\dfrac{D}{A}\left[\left(1 - \dfrac{E^2}{A^2}\right)^{1/2} + \left(1 - \dfrac{S^2}{A^2}\right)^{1/2}\right]$ $\left. + \dfrac{2c}{\pi A}\left[\left(1 - \dfrac{E^2}{A^2}\right)^{1/2} - \left(1 - \dfrac{S^2}{A^2}\right)^{1/2}\right]\right\},$ $N_2 = \dfrac{k}{\pi}\left[\dfrac{E^2}{A^2} - \dfrac{S^2}{A^2} - 2\dfrac{D}{A}\left(\dfrac{E}{A} + \dfrac{S}{A}\right)\right] + \dfrac{2c}{\pi A}\left(\dfrac{E}{A} - \dfrac{S}{A}\right), \quad A \geq S$

Table F.1 (continued)

No.	Nonlinearity	Describing Function Coefficients
9	(hysteresis with deadzone and saturation, slope k, parameters c, D)	$N_1 = \dfrac{k}{\pi}\left\{\arcsin\dfrac{c+kD}{kA} + \arcsin\dfrac{c-kD}{kA}\right.$ $\left. + \dfrac{c+kD}{kA}\left[1 - \dfrac{(c+kD)^2}{k^2A^2}\right]^{1/2}\right.$ $\left. + \dfrac{c-kD}{kA}\left[1 - \dfrac{(c-kD)^2}{k^2A^2}\right]^{1/2}\right\},$ $N_2 = -\dfrac{4cD}{\pi A^2}, \qquad A \geq \dfrac{c+kD}{k}$
10	(hysteresis with deadzone, saturation, and offset R)	$N_1 = \dfrac{k}{\pi}\left\{\arcsin\dfrac{c+kD}{kA} + \arcsin\dfrac{c+kR}{kA} - \arcsin\dfrac{D}{A}\right.$ $\left. - \arcsin\dfrac{R}{A} + \dfrac{c+kD}{kA}\left[1 - \dfrac{(c+kD)^2}{k^2A^2}\right]^{1/2} - \dfrac{D}{A}\left(1 - \dfrac{D^2}{A^2}\right)^{1/2}\right.$ $\left. + \dfrac{c+kR}{kA}\left[1 - \dfrac{(c+kR)^2}{k^2A^2}\right]^{1/2}\right.$ $\left. - \dfrac{R}{A}\left(1 - \dfrac{R^2}{A^2}\right)^{1/2}\right\},$ $N_2 = -\dfrac{2c(R+D)}{\pi A^2}, \qquad A \geq \dfrac{c+kR}{k}$

Table F.1 (*continued*)

No.	Nonlinearity	Describing Function Coefficients
11	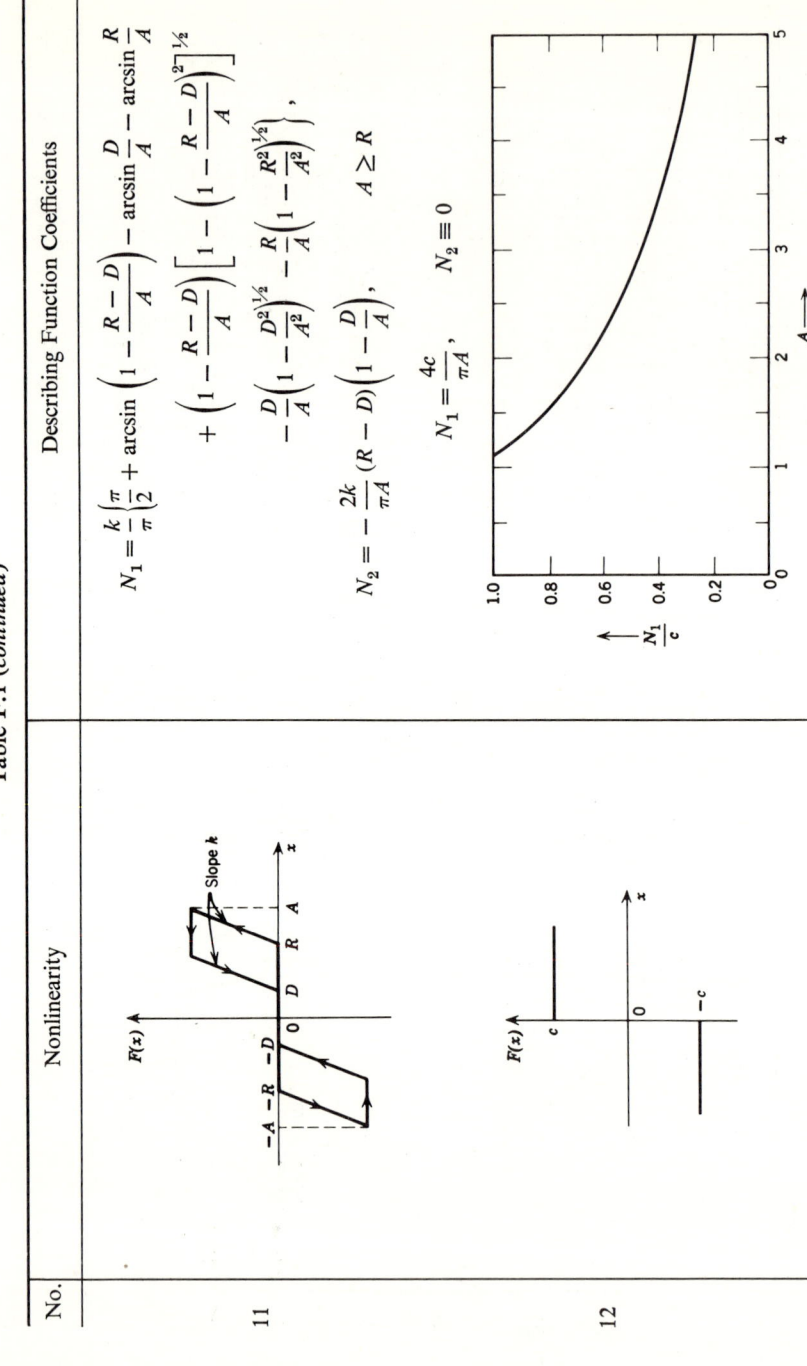	$N_1 = \dfrac{k}{\pi}\left\{\dfrac{\pi}{2} + \arcsin\left(1 - \dfrac{R-D}{A}\right) - \arcsin\dfrac{D}{A} - \arcsin\dfrac{R}{A}\right.$ $+ \left(1 - \dfrac{R-D}{A}\right)\left[1 - \left(1 - \dfrac{R-D}{A}\right)^2\right]^{1/2}$ $\left. - \dfrac{D}{A}\left(1 - \dfrac{D^2}{A^2}\right)^{1/2} - \dfrac{R}{A}\left(1 - \dfrac{R^2}{A^2}\right)^{1/2}\right\},$ $N_2 = -\dfrac{2k}{\pi A}(R-D)\left(1 - \dfrac{D}{A}\right),\qquad A \geq R$
12		$N_1 = \dfrac{4c}{\pi A},\qquad N_2 \equiv 0$

Table F.1 (*continued*)

No.	Nonlinearity	Describing Function Coefficients
13	(graph of $F(x)$: zero between $-D$ and D, then jumps to $-c$ for $x>D$ region shown as negative step, and to c for $x<-D$ region shown as positive step)	$N_1 = \dfrac{4c}{\pi A}\left(1 - \dfrac{D^2}{A^2}\right)^{1/2}, \quad A \geq D$ $N_2 \equiv 0,$ (plot of $\dfrac{D}{c}N_1$ versus $\dfrac{A}{D}$, peak near $\sqrt{2}$)

Table F.1 (*continued*)

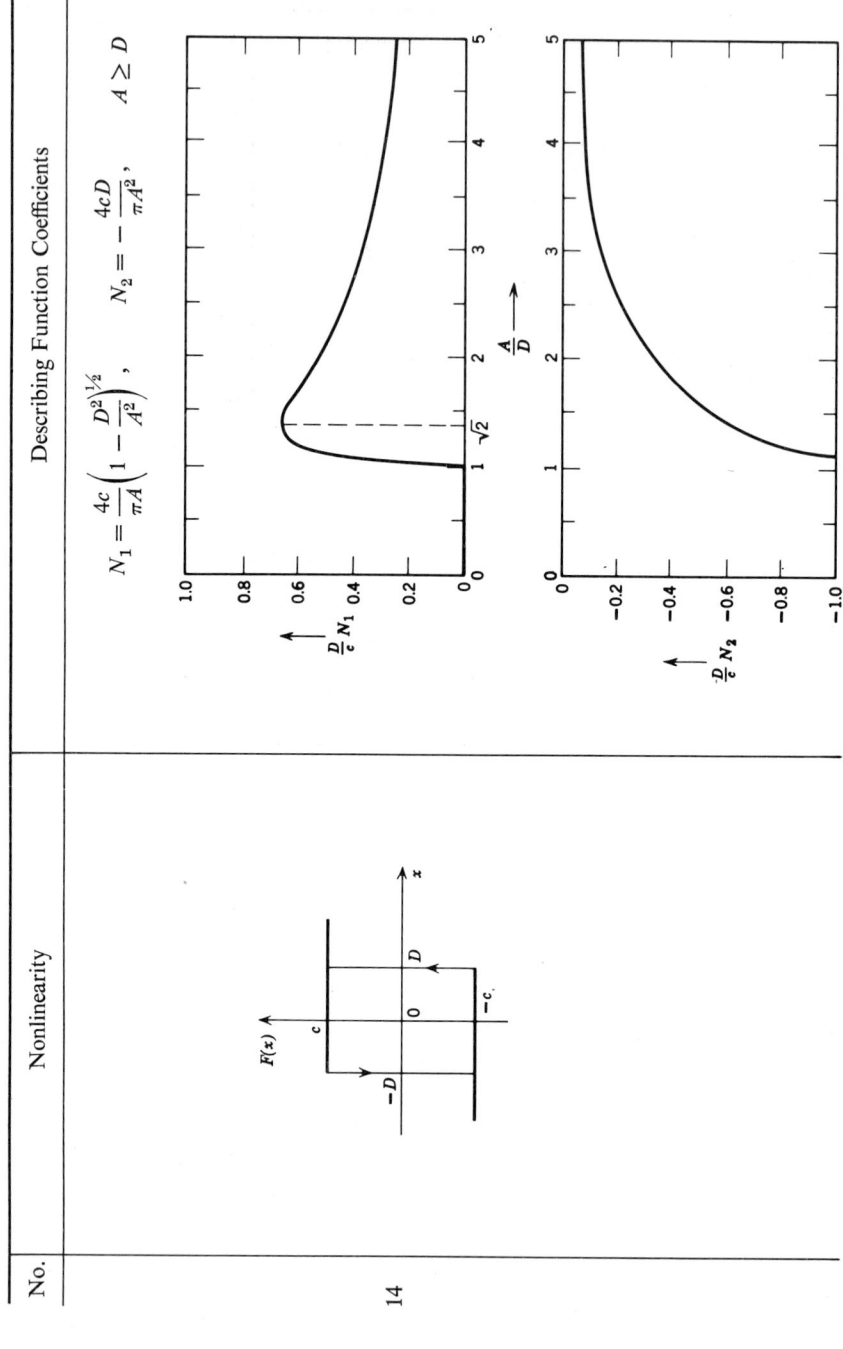

Table F.1 (*continued*)

No.	Nonlinearity	Describing Function Coefficients
15		$N_1 = \dfrac{2c}{\pi A}\left[\left(1-\dfrac{D^2}{A^2}\right)^{1/2}+\left(1-\dfrac{R^2}{A^2}\right)^{1/2}\right],$ $N_2 = -\dfrac{2c(R+D)}{\pi A^2},$ $A \geq D$
16		$N_1 = \dfrac{2c}{\pi A}\left[\left(1-\dfrac{D^2}{A^2}\right)^{1/2}+\left(1-\dfrac{R^2}{A^2}\right)^{1/2}\right],$ $N_2 = -\dfrac{2c(R-D)}{\pi A^2},$ $A \geq R$

Table F.1 (continued)

No.	Nonlinearity	Describing Function Coefficients
17	(hysteresis loop, width $2A$, height $2c$)	$N_1 \equiv 0,$ $N_2 = -\dfrac{4c}{\pi A}$
18	(staircase nonlinearity with levels $\pm c, \pm 2c, \pm 3c$ at breakpoints $\pm D_0, \pm D_1, \pm D_2$)	$N_1 = \dfrac{4c}{\pi A} \sum_{i=0}^{n} \sqrt{1 - \dfrac{D_i^2}{A^2}},$ $N_2 \equiv 0, \quad D_i \leq A \leq D_{i+1}$

F.3 Polynomial Characteristics

The polynomial type nonlinearities are referred to the functions

$$F(x) = kx^n, \quad n \text{ is an odd integer} \tag{F.5}$$

$$F(x) = kx^n \operatorname{sign} x, \quad n \text{ is an even integer,} \tag{F.6}$$

or combinations thereof [F.2, 3]. The polynomial characteristics are useful in approximating statical nonlinear characteristics obtained experimentally.

A general polynomial characteristic is shown in Figure F.1. The input sinusoidal signal and the corresponding output are also shown. Since the characteristic represents a single-valued function $F(x)$, the coefficient N_2 is identically zero, and the harmonic linearization is $F(x) = N_1 x$, where $N_1 = N_1(A)$ is given by

$$N_1 = \frac{1}{\pi A} \int_0^{2\pi} F(A \sin \phi) \sin \phi \, d\phi \tag{F.7}$$

according to (F.4) and $x = A \sin \phi$.

The integral (F.7) over $(0, 2\pi)$ may be replaced by four times the integral over $(0, \pi/2)$, this being the case for all single-valued symmetrical nonlinearities. This is illustrated by the output of the nonlinearity to be integrated, as shown in Figure F.1. For n odd, substitute $F(x) = kx^n$ into (F.7) to obtain

$$N_1 = \frac{4kA^{n-1}}{\pi} \int_0^{\pi/2} \sin^{n+1} \phi \, d\phi = \frac{2k}{\sqrt{\pi}} A^{n-1} \frac{\Gamma[(n+2)/2]}{\Gamma[(n+3)/2]}, \tag{F.8}$$

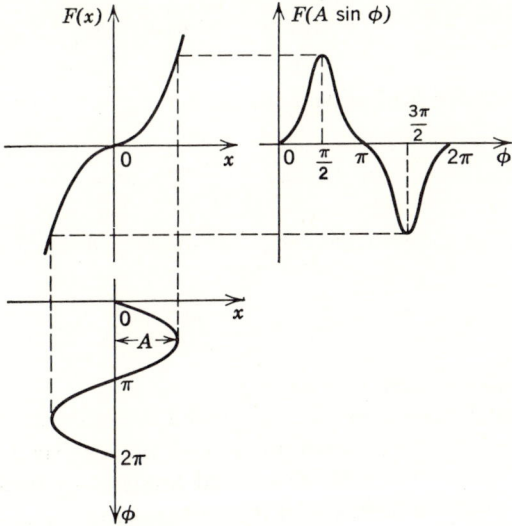

Figure F.1 Nonlinear polynomial characteristic.

486 Harmonic Linearization

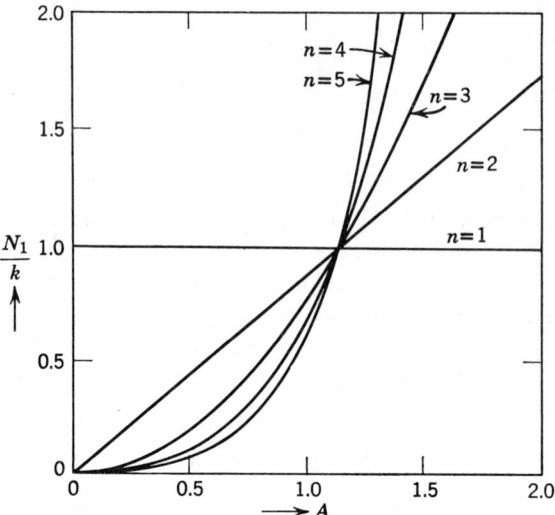

Figure F.2 Describing function for polynomial nonlinear characteristics.

where Γ is the gamma function. It is easy to show [F.3] that the same expression of N_1 in (F.8) is valid whether n is odd or even. For different values of n, the function $N_1(A)$ is given in the diagram of Figure F.2.

The obtained result can be directly extended to the case where

$$F(x) = \sum_{i=1}^{n} k_i x^i, \tag{F.9}$$

since, in general, for a nonlinear function $F_e(x, sx)$ given as

$$F_e(x, sx) = \sum_{i=1}^{n} F_i(x, sx), \tag{F.10}$$

the describing function coefficients, N_{1e} and N_{2e}, are calculated as

$$N_{1e} = \sum_{i=1}^{n} N_{1i}, \qquad N_{2e} = \sum_{i=1}^{n} N_{2i}, \tag{F.11}$$

N_{1i} and N_{2i} being related to $F_i(x, sx)$.

The polynomial nonlinearities are applied to obtain the describing function of symmetric multivalued polynomial nonlinearities such as hysteresis loops for magnetic materials [F.2]. A general analysis in these cases, however, becomes hopelessly unwieldy and it is better to obtain results for a specific case under investigation.

F.4 Static, Coulomb, and Viscous Friction

In this section the describing function of a nonlinear element with friction characteristics is given without derivations on the basis of the reference by Tou and Schultheiss [F.4] and its extension by Silberberg [F.5]. These friction characteristics are of particular interest here since the parameter plane analysis can be applied with obvious advantages.

Consider a rotating body with the moment of inertia J and the coefficient of viscous friction B, to which a torque τ_a is applied. If static, coulomb, and viscous frictions are present, then the nonlinear characteristic is shown in Figure F.3. Static friction is defined as the friction torque T_s necessary to initiate motion of the body from rest. Coulomb friction is defined as the friction torque T_c that opposes motion of the body and is constant, independently of speed. Viscous friction is defined as the friction torque that opposes motion of the body and is proportional to the speed. The torque, $T_f = F(\dot{\theta})$, as shown in Figure F.3, is therefore the total friction torque that is a function of the angular velocity $\dot{\theta}$ of the rotating body.

In the above mentioned references, the describing function $N' = N'(A, \Omega)$ is derived that relates the amplitude A_e of the effective torque

$$\tau_e = A_e \sin(\Omega t + \alpha)$$

and the amplitude A of the applied torque $\tau_a = A \sin \Omega t$ according to Figure F.4, in which τ_a, τ_e, $\dot{\Xi}$, Ξ represent the frequency-domain notations corresponding to τ_a, τ_e, $\dot{\theta}$, θ, respectively. The describing function N' is

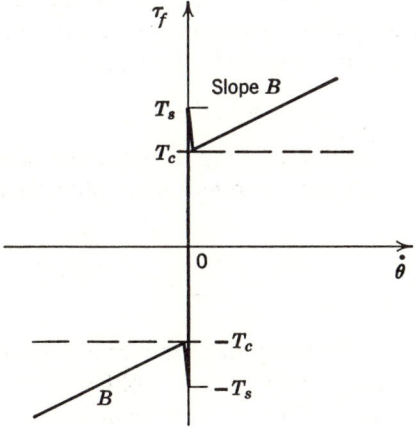

Figure F.3 Nonlinear characteristic of an element with static, coulomb, and viscous friction.

Harmonic Linearization

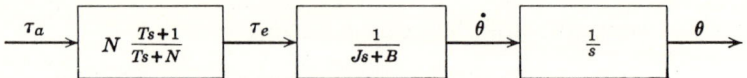

Figure F.4 Block diagram representing a rotating body with the moment of inertia J, viscous friction B, coulomb friction, and static friction.

given as

$$N'(A, \Omega) = N(A)\frac{Ts + 1}{Ts + N(A)}, \qquad (F.12)$$

where $N(A) = N_1(A) + jN_2(A)$ is presented in diagrams of Figure F.5.

As can be seen from Figure F.4, the equivalent transfer function $G(s, N)$ of the rotating element is

$$G(s, N) = \frac{N(Ts + 1)}{s(Js + B)(Ts + N)}. \qquad (F.13)$$

In applying conventional analysis techniques, such as the Nyquist diagram used in reference F.4, difficulties arise because the transfer function of the rotating element, which may be a part of a feedback system, is $G(s, N)$ rather than the simple product $NG(s)$. In the parameter plane analysis this entails no difficulty since $N = N_1 + jN_2$ can be considered as $N = \alpha + j\beta$, where $\alpha = N_1(A)$ and $\beta = N_2(A)$, given in the diagrams of Figure F.5. This case (F.13) is referred to as the *variable-pole describing function* and is treated in Section 4.8.

In Section 4.8 a feedback system is considered that exhibits the quadratic viscous friction usually present at high speed. The system has a block diagram (Figure F.6) that may represent the positional control system with a tachogenerator feedback. The components in the forward path of the inner control loop are the amplifier, generator, and servomotor. If the motor exhibits a quadratic friction, described by the function

$$F(\dot\theta) = k_1 \dot\theta^2 \, \text{sign} \, \dot\theta, \qquad (F.14)$$

where $\dot\theta$ represents the angular velocity of the motor shaft, then the differential equation representing the combination amplifier-generator-servomotor may be written as

$$(T_2 s + 1)\dot\theta = \frac{K_2}{T_1 s + 1} e - k_2 F(\dot\theta), \qquad (F.15)$$

in which

$e = e(t)$ is the error signal of the inner control loop;
K_2 is the gain constant of the amplifier;
T_1 is the time constant of the amplifier;
T_2 is the time constant of the servomotor;
k_2 is the proportionality factor.

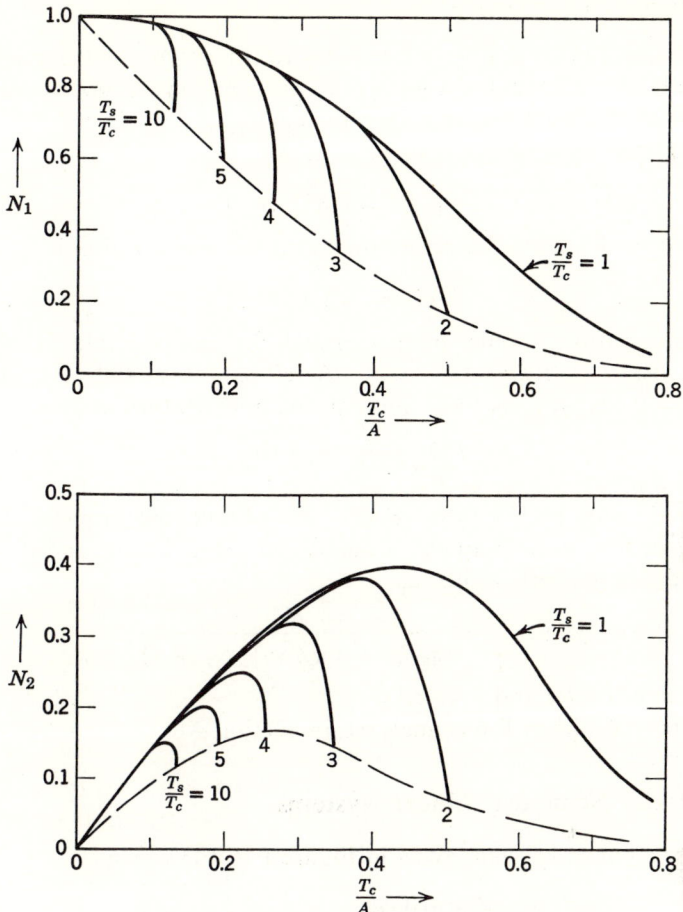

Figure F.5 Describing function diagrams.

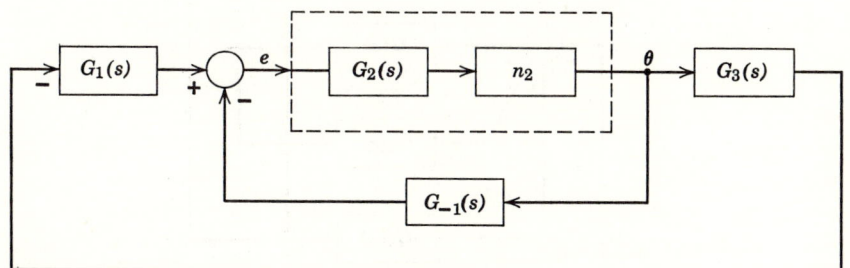

Figure F.6 Block diagram of a positional servo with quadratic friction in the forward path.

490   Harmonic Linearization

The function $F(\dot{\theta}) = k_1\dot{\theta}^2 \operatorname{sign} \dot{\theta}$ belongs to the polynomial characteristics of Section F.2 except that the argument of the function F is not the variable θ but its rate of change $\dot{\theta}$. Nevertheless, the corresponding describing function from (F.8) and $n = 2$, is given as

$$N_1'(A') = 0.85k_1 A', \tag{F.16}$$

for which $\dot{\theta} = A' \sin \Omega t$. Therefore the harmonic linearization is

$$F(\dot{\theta}) = N_1'(A')\dot{\theta}. \tag{F.17}$$

If the linearization should be performed for the variable θ, note that $\theta = A \sin \Omega t$, $\dot{\theta} = A\Omega \cos \Omega t$, and therefore $A' = A\Omega$. By denoting $N_1(A, \Omega) = 0.85k_1 A\Omega$, we have the required linearization as

$$F(\theta) = N_1(A, \Omega)s\theta, \tag{F.18}$$

where $s \equiv d/dt$.

By substituting (F.18) into (F.15), we obtain the transfer function $G_2(s, N_1)$ relating, in the frequency domain, the error signal e and the output shaft angle θ of the servomotor as

$$G_2(s, N_1) = \frac{K_2}{s(T_1 s + 1)(T_2 s + N_1 + 1)}, \tag{F.19}$$

where $N_1 = 0.85kA\Omega$ and $k = k_1 k_2$.

The system of Figure F.6 is analyzed in Section 4.8.

F.5 Variable-Structure Linear Systems

Consider a control system given in Figure F.7, for which

$$y = G_1(s)x, \qquad |x| < b \tag{F.20a}$$

$$y = G_2(s)x, \qquad |x| > b. \tag{F.20b}$$

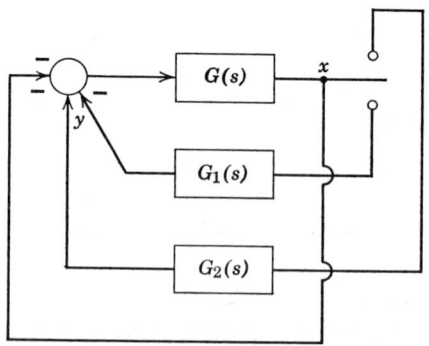

Figure F.7   Linear system with variable structure.

Assume that $x = A \sin \Omega t$. Then find the coefficients N_1 and N_2 relating x and y so that all the parts denoted by 1 in Figure F.8 are determined from (F.20a), and those with 2, from (F.20b). Thus according to reference F.2,

$$N_1 = \frac{1}{\pi A} \int_0^{2\pi} y(\phi) \sin \phi \, d\phi$$
$$N_2 = \frac{1}{\pi A} \int_0^{2\pi} y(\phi) \cos \phi \, d\phi, \tag{F.21}$$

or, because of symmetry in Figure F.8, we have

$$N_1 = \frac{2}{\pi A} \int_{-\phi_1}^{+\phi_1} y_1(\phi) \sin \phi \, d\phi + \frac{2}{\pi A} \int_{-\phi_1}^{+\phi_1} y_2(\phi) \sin \phi \, d\phi$$
$$N_2 = \frac{2}{\pi A} \int_{-\phi_1}^{+\phi_1} y_1(\phi) \cos \phi \, d\phi + \frac{2}{\pi A} \int_{-\phi_1}^{+\phi_1} y_2(\phi) \cos \phi \, d\phi, \tag{F.22}$$

where

$$\phi_1 = \arcsin \frac{b}{A} \tag{F.23}$$

and

$$y_1(\phi) = R_1(\Omega)A \sin \phi + I_1(\Omega)A \cos \phi + y_{1n}(\phi)$$
$$y_2(\phi) = R_2(\Omega)A \sin \phi + I_2(\Omega)A \cos \phi + y_{2n}(\phi), \tag{F.24}$$

and $\phi = \Omega t$. The transient components $y_{1n}(\phi)$ and $y_{2n}(\phi)$ in (F.24) exist if the transfer functions, $G_1(s)$ and $G_2(s)$, have time constants and the discontinuous jumps from parts 1 to 2 and vice versa are smoothed.

In specific cases, the calculations of the above integrals may be significantly facilitated. For example, if $G_1(s) = K_{-1}$ and $G_2(s) = K_{-2}$, then the harmonic linearization is reduced to the case of the nonlinearity given as No. 5 in Table F.1.

Similar reasoning to that outlined above can be applied to the system of Figure 3.10 shown in Section 3.2.

F.6 Systems with Two Nonlinearities

If, in a control system, there are two distinct nonlinearities

$$y_1 = F_1(x_1, sx_1), \quad y_2 = F_2(x_2, sx_2), \tag{F.25}$$

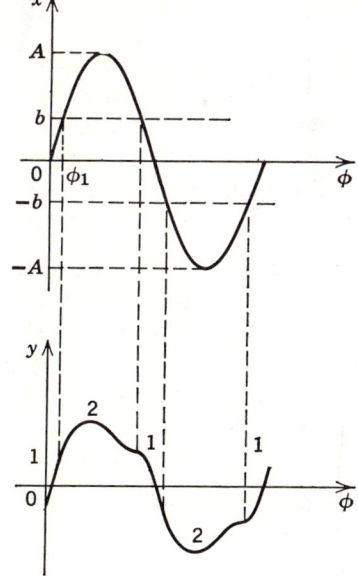

Figure F.8 Oscillations in variable-structure control system.

then, by assuming $x_1 = A_1 \sin \Omega t$ and $x_2 = A_2 \sin(\Omega t + \theta)$ and performing the harmonic linearization, we have

$$y_1 = N_{11}x_1 + \frac{N_{21}}{\Omega} sx_1, \qquad y_2 = N_{12}x_2 + \frac{N_{22}}{\Omega} sx_2, \qquad (F.26)$$

or, more often, they are

$$y_1 = N_{11}x_1, \qquad y_2 = N_{12}x_2. \qquad (F.27)$$

The coefficients N_{11}, N_{21}, N_{12}, and N_{22} are computed by using common formulas (F.3), except that for N_{11}, N_{21} and N_{12}, N_{22} there are $\phi = \Omega t$ and $\phi = \Omega t + \theta$, respectively. (The latter fact about the phase shift θ has no influence on the final results.)

In general, there are three cases to consider.

1. Two nonlinearities have common inputs, $x_1 = x_2 = x$, and $A_1 = A_2 = A$, $\theta \equiv 0$ (see Figure 3.7).
2. Two nonlinearities have inputs related by a linear differential equation

$$x_2 = G(s)x_1 \qquad (F.28)$$

(see, for example, Figure 3.8). Then, for $x_1 = A_1 \sin \Omega t$, $x_2 = A_2 \sin(\Omega t + \theta)$, we obtain

$$A_2 = |G_2(j\Omega)| A_1, \qquad \theta = \arg G(j\Omega). \qquad (F.29)$$

3. Two nonlinearities have their inputs related by a nonlinear differential equation. Such a case is illustrated by Figure 3.9. This case may be treated as similar to case 2 with (F.29) if it is possible to perform harmonic linearization of the nonlinearity n_2 relating the variables x_2 and y_2, and to include the frequency characteristic of that nonlinearity in the frequency characteristic $G_2(j\Omega)$. This concept has been utilized in reference F.2.

In case 2 when the generalized property for the transfer function $G_2(s)$ of Figure 3.9 is not fulfilled, the whole part, $y_2x_1y_1$, should be considered as one nonlinear element and we use (F.21) by substituting $y(\phi)$ with $y_1(\phi)$. The only variable that is considered as a harmonic oscillation is x_2. From $x_2 = A \sin \Omega t$, the variable y_2 can be determined either analytically or graphically from

$$y_2(\phi) = F_2(A \sin \phi, A\Omega \cos \phi). \qquad (F.30)$$

Then from the relation

$$x_1 = G_2(s)y_2, \qquad (F.31)$$

find exactly $x_1(\phi)$, $\phi = \Omega t$. The relationship F.31 is often easy to solve since, not being a sufficiently low-pass network, it is of a low order. Then, finally, the exact solution for $y_1(\phi)$ is

$$y_1(\phi) = F_1[x_1(\phi), sx_1(\phi)]. \tag{F.32}$$

The block diagram of Figure 3.9 is thereby reduced to that of Figure 3.6, for which

$$N_1 = \frac{1}{\pi A} \int_0^{2\pi} y_1(\phi) \sin \phi \, d\phi$$
$$N_2 = \frac{1}{\pi A} \int_0^{2\pi} y_1(\phi) \cos \phi \, d\phi, \tag{F.33}$$

instead of n and $G(s) = G_1(s)$.

F.7 Nonlinear Function $F(x_1, sx_1, x_2)$

In the case of a nonlinear function $F(x_1, sx_1, x_2)$, it is necessary to assume that $x_1 = A_1 \sin \Omega t$ and $x_2 = A_2 \sin (\Omega t + \theta)$. Then by using (F.29), reduce the function $F(x_1, sx_1, x_2)$ to $F_1(x_1, sx_1)$; that is,

$$x_2 = A_2 \sin \Omega t \cos \theta + A_2 \cos \Omega t \sin \theta$$
$$= \frac{A_2 \cos \theta}{A} x_1 + \frac{A_2 \sin \theta}{A\Omega} sx_1 \tag{F.34}$$

or, with (F.29), we have

$$x_2 = R_2(\Omega)x + \frac{I_2(\Omega)}{\Omega} sx_1, \tag{F.35}$$

where $R_2(\Omega)$ and $I_2(\Omega)$ are real and imaginary parts of the frequency characteristic $G_2(j\Omega)$ that relates x_2 and x_1. In (F.35), $R_2(\Omega)$ and $I_2(\Omega)$ represent constant coefficients dependent on the frequency Ω only. By substituting (F.35) in $F(x_1, sx_1, x_2)$ we obtain a function $F_1(x_1, sx_1)$ that can be linearized in the usual fashion (F.2).

To illustrate the procedure, consider a nonlinear element described by the function $F(x_1, x_2) = kx_2^2 \operatorname{sign} x_1$, where x_1 is the input variable and x_2 is the output variable of the element. Assume that $x_2 = A_2 \sin \Omega t$ and $x_1 = A_1 \sin (\Omega t - \theta)$. The relationship between the amplitudes A_1 and A_2 and the phase shift θ is determined by the linear part connecting the variables x_1 and x_2.

Under the above conditions, the nonlinear function $F(x_1, x_2)$ is a periodic function of the argument $\phi = \Omega t$ and can be rewritten as

$$F(x_1, x_2) = k(A_2 \sin \phi)^2 \operatorname{sign} [\sin (\phi - \theta)]. \tag{F.36}$$

The coefficients N_1 and N_2 of the describing function $N = N_1 + jN_2$ corresponding to $F(x_1, x_2)$ of (F.36) are functions of both the amplitude A_2 and the frequency Ω for different values of the phase shift θ. The coefficients

$N_1(A_2, \Omega)$ and $N_2(A_2, \Omega)$ are

$$N_1 = \frac{1}{\pi A_2} \int_0^{2\pi} F(x_1, x_2) \sin \phi \, d\phi$$

$$= \frac{2kA_2}{\pi} \int_0^{\pi} \sin^2 \phi \, \text{sign} \, [\sin(\phi - \theta)] \sin \phi \, d\phi$$

$$= \frac{2kA_2}{\pi} \left(-\int_0^{\theta} \sin^3 \phi \, d\phi + \int_{\theta}^{\pi} \sin^3 \phi \, d\phi \right)$$

$$= \frac{4kA_2}{\pi} (\cos \theta - \tfrac{1}{3} \cos^3 \theta), \qquad (F.37)$$

and, similarly, we have

$$N_2 = -\frac{4kA_2}{3\pi} \sin^3 \theta, \qquad (F.38)$$

in (F.37) and (F.38), $\theta = \theta(\Omega)$. Thus for $\theta = 0$ we obtain from (F.37) and (F.38)

$$N_1(A_2) = \frac{8kA_2}{3\pi} = 0.85 A_2, \qquad N_2(A_2) \equiv 0, \qquad (F.39)$$

which corresponds to the polynomial function $F(x_2) = kx_2^2 \, \text{sign} \, x_2$.

Similarly, when the nonlinear function $F(x_1, x_2) = kx_2^3 \, \text{sign} \, x_2 \, \text{sign} \, x_1$, it can be rewritten as

$$F(x_1, x_2) = k(A_2 \sin \phi)^3 \, \text{sign} \, (\sin \phi) \, \text{sign} \, [\sin(\phi - \theta)], \qquad (F.40)$$

for which $x_2 = A_2 \sin \phi$, $x_1 = A_1 \sin(\phi - \theta)$, and $\phi = \Omega t$. The corresponding coefficients, $N_1 = N_1(A_2, \Omega)$ and $N_2 = N_2(A_2, \Omega)$ of the describing function $N = N_1 + jN_2$, are derived as

$$N_1 = \frac{kA_2^2}{\pi} (\tfrac{3}{4}\pi - \tfrac{3}{2}\theta + \sin 2\theta - \tfrac{1}{8} \sin 4\theta)$$

$$N_2 = -\frac{kA_2^2}{\pi} \sin^4 \theta, \qquad (F.41)$$

where, again, $\theta = \theta(\Omega)$. The function $\theta(\Omega)$ is determined from the known linear part that separates the variables x_1 and x_2. In the special case $\theta = 0$, we have

$$N_1(A_2) = \frac{3kA_2^2}{4}, \qquad N_2(A_2) \equiv 0, \qquad (F.42)$$

which corresponds to the polynomial function $F(x_2) = kx_2^3$.

F.8 Approximate Graphical Procedure for Calculating the Describing Function

In case a nonlinear characteristic cannot be readily described analytically, reference F.6 presents a graphical method for calculating the describing function from the corresponding diagram of the characteristic that could be obtained experimentally. This method will be outlined in this section without proof, since the proof is presented in the reference. The method is applied to single-valued nonlinear functions. It may, however, be extended to multi-valued nonlinearities, as shown in reference F.6.

The exact formula for calculating the describing function $N_1(A)$, which is

$$N_1(A) = \frac{1}{\pi A} \int_0^{2\pi} F(A \sin \phi) \sin \phi \, d\phi, \tag{F.43}$$

can be approximately given as

$$N_1(A) \simeq \frac{2}{3A} \left[F(A) + F\left(\frac{A}{2}\right) \right]. \tag{F.44}$$

The formula (F.44) can be used to calculate $N_1(A)$ graphically from a given function $F(x)$.

In order to determine $N_1(A)$ by the formula (F.44), it is necessary to plot the right part of the function $F(x) = F(A)$ as shown in Figure F.9. By changing the scale on the x axis, the function $F(A/2)$ is added to the plot. Then, summing the ordinates of the two curves, the function $F(A) + F(A/2)$ is readily obtained. Two lines are drawn through the points $x = A_1$ and $x = -\frac{2}{3}$ parallel to the ordinate $F(x)$ axis. The corresponding value $N_1(A_1)$ is then determined as the part $\overline{cd}$ of the line $x = -\frac{2}{3}$, which is obtained by

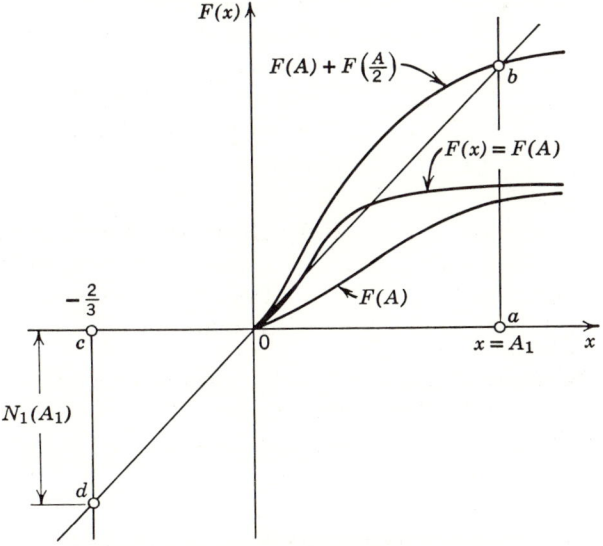

Figure F.9 Graphical evaluation of $N_1(A)$.

496 Harmonic Linearization

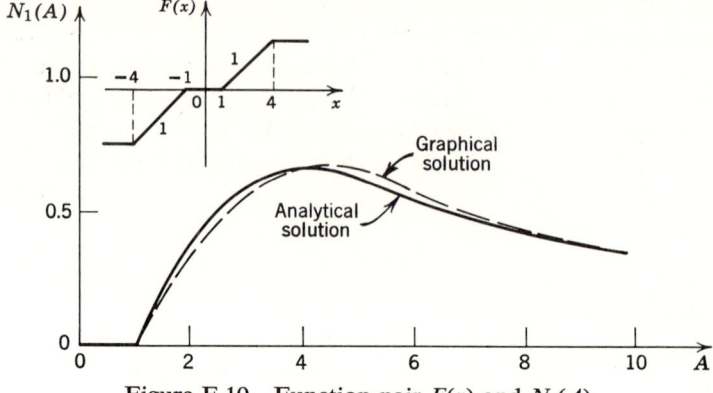

Figure F.10 Function pair $F(x)$ and $N_1(A)$.

drawing a straight line $x = A_1$ and function $F(A) + F(A/2)$, through the origin 0 up to the intersection d. From the similarity of the two triangles $0cd$ and $a0b$, we have

$$\overline{cd} = N_1(A_1) = \frac{2}{3} \frac{F(A_1) + F(A_1/2)}{A_1}.$$

(F.45)

Repeating the outlined process for several amplitudes, the curve $N_1(A)$ can be constructed.

If the above procedure is applied to the nonlinear characteristic of Figure F.10, the describing function $N_1(A)$ is obtained, as shown by the dotted line in Figure F.10. The analytic solution is plotted for comparison with respect to accuracy.

F.9 Harmonic Linearization of Nonlinear Characteristics in the Presence of Asymmetrical Oscillations

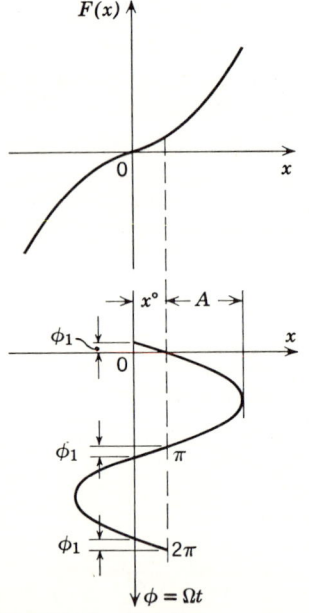

Figure F.11 Nonlinear characteristic $F(x) = kx^3$ and its input $x = x^0 + A \sin \phi$.

In the previous sections the harmonic linearization has been performed for symmetrical nonlinear characteristics when their input x is assumed to be a symmetrical oscillation, $x = A \sin \phi$, $\phi = \Omega t$. In this part of Appendix F the symmetrical and asymmetrical nonlinear characteristics $F = F(x, sx)$ are linearized when

Asymmetrical Oscillations

their input is
$$x = x^0 + x^*, \tag{F.46}$$
where $x^* = A \sin \phi$, $\phi = \Omega t$.

The nonlinear function F is then a periodic function of the argument ϕ with a constant term F^0, or
$$F = F^0 + \left(N_1 + \frac{N_2}{\Omega} s\right) x^*, \tag{F.47}$$
where
$$F^0 = F^0(A, \Omega, x^0), \quad N_1 = N_1(A, \Omega, x^0), \quad N_2 = N_2(A, \Omega, x^0), \tag{F.48}$$
which are determined by
$$F^0 = \frac{1}{2\pi} \int_0^{2\pi} F(x^0 + A \sin \phi, A\Omega \cos \phi)\, d\phi$$
$$N_1 = \frac{1}{\pi A} \int_0^{2\pi} F(x^0 + A \sin \phi, A\Omega \cos \phi) \sin \phi\, d\phi \tag{F.49}$$
$$N_2 = \frac{1}{\pi A} \int_0^{2\pi} F(x^0 + A \sin \phi, A\Omega \cos \phi) \cos \phi\, d\phi.$$

If the nonlinear function F does not depend on the rate of change of x—that is, $F = F(x)$—then
$$F^0 = F^0(A, x^0), \quad N_1 = N_1(A, x^0), \quad N_2 = N_2(A, x^0) \tag{F.50}$$
and
$$F^0 = \frac{1}{2\pi} \int_0^{2\pi} F(x^0 + A \sin \phi)\, d\phi$$
$$N_1 = \frac{1}{\pi A} \int_0^{2\pi} F(x^0 + A \sin \phi) \sin \phi\, d\phi \tag{F.51}$$
$$N_2 = \frac{1}{\pi A} \int_0^{2\pi} F(x^0 + A \sin \phi) \cos \phi\, d\phi.$$

To illustrate the calculations of the above integrals, consider the polynomial nonlinear characteristic $F(x) = kx^3$ as shown in Figure F.11. The input to the nonlinearity $x = x^0 + A \sin \Omega t$ is shown in Figure F.11. For $F^0 = F^0(A, x^0)$, we have from the first equation F.51
$$F^0 = \frac{k}{2\pi} \int_0^{2\pi} (x^0 + A \sin \phi)^3\, d\phi = k[(x^0)^3 + \tfrac{3}{2} x^0 A^2]. \tag{F.52}$$
For the coefficient $N_1 = N_1(A, x^0)$, we obtain from the second equation F.51
$$N_1 = \frac{k}{\pi A} \int_0^{2\pi} (x^0 + A \sin \phi)^3 \sin \phi\, d\phi = 3k\left[(x^0)^2 + \frac{A^2}{4}\right]. \tag{F.53}$$
Since the nonlinear function $F(x) = kx^3$ is a single-valued function of x, the coefficient $N_2 \equiv 0$.

The above coefficients, F^0, N_1, and N_2, for common nonlinearities are calculated and given in Table F.2.

Table F.2 Describing Function Coefficients for Asymmetrical Oscillations

No.	Nonlinearity	Describing Function Coefficients
1		$N_1 = \dfrac{4c}{\pi A}\left[1 - \dfrac{(x^0)^2}{A^2}\right]^{1/2},$ $N_2 \equiv 0,$ $F^0 = \dfrac{2c}{\pi}\arcsin\dfrac{x^0}{A},$ $A \geq x^0$
2		$N_1 = \dfrac{2c}{\pi A}\left\{\left[1 - \dfrac{(D+x^0)^2}{A^2}\right]^{1/2} + \left[1 - \dfrac{(D-x^0)^2}{A^2}\right]^{1/2}\right\},$ $N_2 \equiv 0,$ $F^0 = \dfrac{c}{\pi}\left(\arcsin\dfrac{D+x^0}{A} - \arcsin\dfrac{D-x^0}{A}\right),$ $A \geq D + |x^0|$

Table F.2 (*continued*)

No.	Nonlinearity	Describing Function Coefficients
3	(hysteresis relay with deadzone diagram)	$N_1 = \dfrac{2c}{\pi A}\left\{\left[1 - \dfrac{(D+x^0)^2}{A^2}\right]^{1/2} + \left[1 - \dfrac{(D-x^0)^2}{A^2}\right]^{1/2}\right\}$, $N_2 = -\dfrac{4cD}{\pi A^2}$, $F^0 = \dfrac{c}{\pi}\left(\arcsin\dfrac{D+x^0}{A} - \arcsin\dfrac{D-x^0}{A}\right)$, $A \geq D + |x^0|$
4	(hysteresis relay diagram)	$N_1 \equiv 0$, $N_2 = -\dfrac{4c}{\pi A}$, $F^0 \equiv 0$

Table F.2 (*continued*)

No.	Nonlinearity	Describing Function Coefficients
5	*F(x)* graph with parameters $-R, -D, 0, D, R$ on x-axis, levels $c, -c$, and x^0 offset	$N_1 = \dfrac{c}{\pi A}\left\{\left[1 - \dfrac{(R+x^0)^2}{A^2}\right]^{1/2} + \left[1 - \dfrac{(R-x^0)^2}{A^2}\right]^{1/2}\right.$ $\left. + \left[1 - \dfrac{(D-x^0)^2}{A^2}\right]^{1/2} + \left[1 - \dfrac{(D+x^0)^2}{A^2}\right]^{1/2}\right\},$ $N_2 = -\dfrac{2c}{\pi A}(R+D),$ $F^0 = \dfrac{c}{2\pi}\left(\arcsin\dfrac{R+x^0}{A} - \arcsin\dfrac{R-x^0}{A} - \arcsin\dfrac{D-x^0}{A}\right.$ $\left. + \arcsin\dfrac{D+x^0}{A}\right),$ $A \geq R + |x^0|$

Table F.2 (*continued*)

No.	Nonlinearity	Describing Function Coefficients
6	(graph: $F(x)$ vs x, deadband-saturation with slope k, thresholds $\pm S$, offset x^0)	$N_1 = \dfrac{k}{\pi}\left\{\arcsin\dfrac{S-x^0}{A} + \arcsin\dfrac{S+x^0}{A} + \dfrac{S-x^0}{A}\left[1 - \dfrac{(S-x^0)^2}{A^2}\right]^{1/2}\right.$ $\left. + \dfrac{S+x^0}{A}\left[1 - \dfrac{(S+x^0)^2}{A^2}\right]^{1/2}\right\},$ $N_2 \equiv 0,$ $F^0 = \dfrac{k}{\pi}\left[A\left\{\left[1 - \dfrac{(S+x^0)^2}{A^2}\right]^{1/2} - \left[1 - \dfrac{(S-x^0)^2}{A}\right]^{1/2}\right\}\right.$ $\left. + (S+x^0)\arcsin\dfrac{S+x^0}{A} - (S-x^0)\arcsin\dfrac{S-x^0}{A}\right],$ $A \geq S + |x^0|$

Table F.2 (*continued*)

No.	Nonlinearity	Describing Function Coefficients
7	(graph: $F(x)$ vs x, piecewise linear with deadband $\pm D$, slope k, offset x^0)	$N_1 = k \left\{ 1 - \dfrac{1}{\pi} \left(\arcsin \dfrac{D - x^0}{A} + \arcsin \dfrac{D + x^0}{A} \right) \right. $ $+ \dfrac{D - x^0}{A} \left[1 - \dfrac{(D - x^0)^2}{A^2} \right]^{1/2} + \dfrac{D + x^0}{A} \left[1 - \dfrac{(D + x^0)^2}{A^2} \right]^{1/2} \biggr\},$ $N_2 \equiv 0,$ $F^0 = \dfrac{kA}{\pi} \left\{ \left[1 - \dfrac{(D - x^0)^2}{A^2} \right]^{1/2} - \left[1 - \dfrac{(D + x^0)^2}{A^2} \right]^{1/2} \right.$ $+ kx^0 + \dfrac{k}{\pi} \left[D \left(\arcsin \dfrac{D - x^0}{A} - \arcsin \dfrac{D + x^0}{A} \right) \right.$ $\left. - x^0 \left(\arcsin \dfrac{D - x^0}{A} + \arcsin \dfrac{D + x^0}{A} \right) \right] \biggr\},$ $A \geq D + |x^0|$

Table F.2 (*continued*)

No.	Nonlinearity	Describing Function Coefficients
8	(plot of $F(x)$ with dead-zone and saturation: breakpoints at $\pm D$, $\pm S$, with small offset x^0)	$N_1 = \dfrac{k}{\pi} \Bigg\{ \arcsin\dfrac{S-x^0}{A} - \arcsin\dfrac{D-x^0}{A} + \arcsin\dfrac{S+x^0}{A} - \arcsin\dfrac{D+x^0}{A} + \dfrac{S-x^0}{A}\left[1 - \dfrac{(S-x^0)^2}{A^2}\right]^{1/2} - \dfrac{S-x^0}{A}\left[1 - \dfrac{(S-x^0)^2}{A^2}\right]^{1/2} + \dfrac{D+x^0}{A}\left[1 - \dfrac{(D+x^0)^2}{A^2}\right]^{1/2} - \dfrac{S+x^0}{A}\left[1 - \dfrac{(S+x^0)^2}{A^2}\right]^{1/2} + \dfrac{2kD}{\pi A}\left\{\left[1 - \dfrac{(S-x^0)^2}{A^2}\right]^{1/2} - \left[1 - \dfrac{(D-x^0)^2}{A^2}\right]^{1/2} + \left[1 - \dfrac{(D+x^0)^2}{A^2}\right]^{1/2} - \left[1 - \dfrac{(S+x^0)^2}{A^2}\right]^{1/2}\right\} + \dfrac{2k}{\pi A}(D-x^0)\left\{\left[1 - \dfrac{(D-x^0)^2}{A^2}\right]^{1/2} - \left[1 - \dfrac{(S-x^0)^2}{A^2}\right]^{1/2}\right\} + \dfrac{2k}{\pi A}(D+x^0)\left\{\left[1 - \dfrac{(D+x^0)^2}{A^2}\right]^{1/2} - \left[1 - \dfrac{(S+x^0)^2}{A^2}\right]^{1/2}\right\},$ $N_2 \equiv 0,$ $F^0 = \dfrac{kA}{\pi}\Bigg\{\left[1 - \dfrac{(D-x^0)^2}{A^2}\right]^{1/2} - \left[1 - \dfrac{(S-x^0)^2}{A^2}\right]^{1/2} - \left[1 - \dfrac{(D+x^0)^2}{A^2}\right]^{1/2} + \left[1 - \dfrac{(S+x^0)^2}{A^2}\right]^{1/2}\Bigg\} + \dfrac{k}{\pi}(D-x^0)\left(\arcsin\dfrac{D-x^0}{A} - \arcsin\dfrac{S-x^0}{A}\right) + \dfrac{k}{\pi}(D+x^0)\left(\arcsin\dfrac{D+x^0}{A} - \arcsin\dfrac{S+x^0}{A}\right),$ $A \geq S + |x^0|$

Table F.2 (*continued*)

No.	Nonlinearity	Describing Function Coefficients
9	(piecewise-linear characteristic with slopes k_1, k, k_1 and breakpoints at $\pm S$, with offset x^0)	$N_1 = k_1 - \dfrac{k_1 - k}{\pi A}\left\{(S - x^0)\left[1 - \dfrac{(S - x^0)^2}{A^2}\right]^{1/2}\right.$ $\left.+ (S + x^0)\left[1 - \dfrac{(S + x^0)^2}{A^2}\right]^{1/2}\right.$ $\left.+ A\left(\arcsin\dfrac{S - x^0}{A} + \arcsin\dfrac{S + x^0}{A}\right)\right\},$ $N_2 \equiv 0,$ $F^0 = k_1 x^0 + \dfrac{k_1 - k}{A}\left[A\left\{\left[1 - \dfrac{(S - x^0)^2}{A^2}\right]^{1/2} - \left[1 - \dfrac{(S + x^0)^2}{A^2}\right]^{1/2}\right\}\right.$ $\left.+ (S - x^0)\arcsin\dfrac{S - x^0}{A} + (S + x^0)\arcsin\dfrac{S + x^0}{A}\right],$ $A \geq S + |x^0|$

Table F.2 (*continued*)

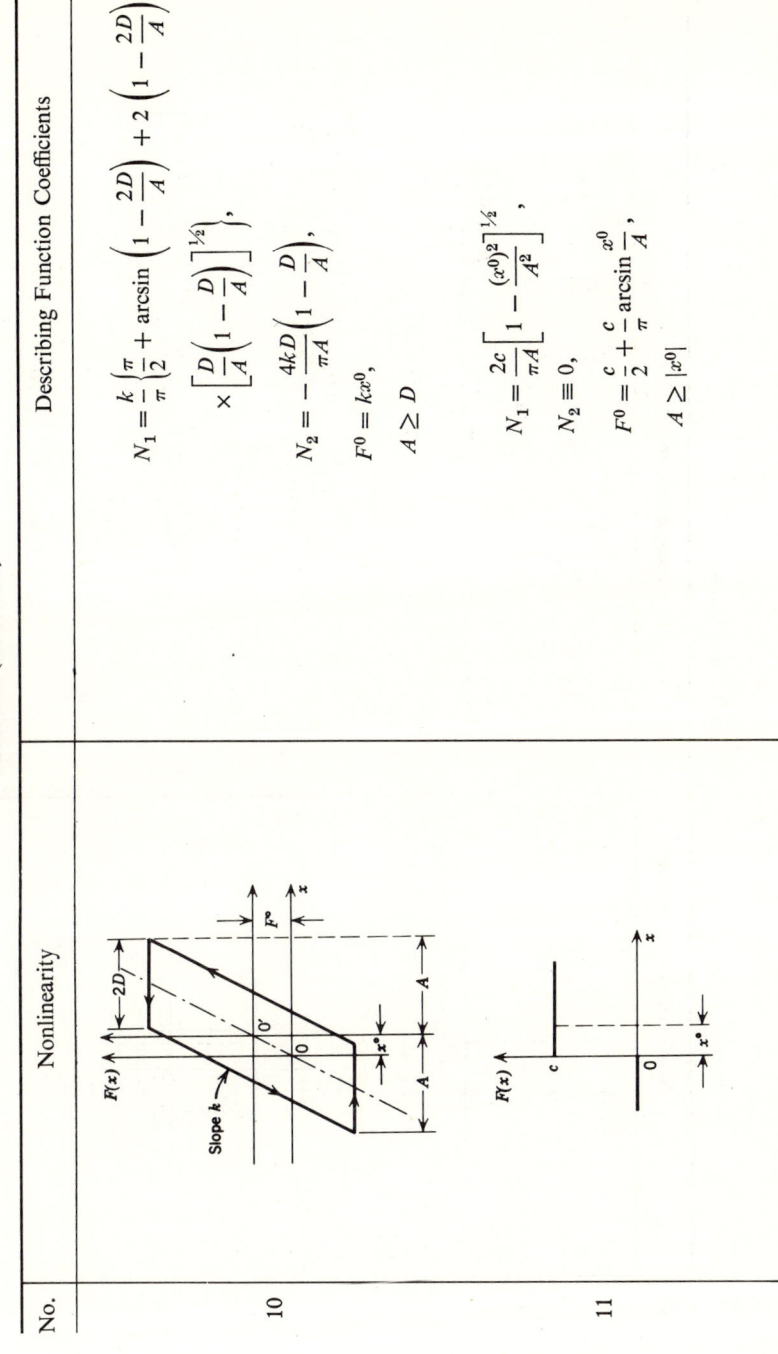

No.	Nonlinearity	Describing Function Coefficients
10		$N_1 = \dfrac{k}{\pi}\left\{\dfrac{\pi}{2} + \arcsin\left(1 - \dfrac{2D}{A}\right) + 2\left(1 - \dfrac{2D}{A}\right)\right.$ $\left.\times\left[\dfrac{D}{A}\left(1 - \dfrac{D}{A}\right)\right]^{1/2}\right\},$ $N_2 = -\dfrac{4kD}{\pi A}\left(1 - \dfrac{D}{A}\right),$ $F^0 = kx^0,$ $A \geq D$
11		$N_1 = \dfrac{2c}{\pi A}\left[1 - \dfrac{(x^0)^2}{A^2}\right]^{1/2},$ $N_2 \equiv 0,$ $F^0 = \dfrac{c}{2} + \dfrac{c}{\pi}\arcsin\dfrac{x^0}{A},$ $A \geq |x^0|$

Table F.2 (*continued*)

No.	Nonlinearity	Describing Function Coefficients
12		$N_1 = \dfrac{2c}{\pi A}\left[1 - \dfrac{(D-x^0)^2}{A^2}\right]^{1/2},$ $N_2 \equiv 0,$ $F^0 = \dfrac{c}{2} - \dfrac{c}{\pi}\arcsin\dfrac{D-x^0}{A},$ $A \geq |D-x^0|$
13		$N_1 = \dfrac{c}{\pi A}\left\{\left[1 - \dfrac{(D-x^0)^2}{A^2}\right]^{1/2} + \left[1 - \dfrac{(D+x^0)^2}{A^2}\right]^{1/2}\right\},$ $N_2 = -\dfrac{2cD}{\pi A^2},$ $F^0 = \dfrac{c}{2} + \dfrac{c}{2\pi}\left(\arcsin\dfrac{D+x^0}{A} - \arcsin\dfrac{D-x^0}{A}\right),$ $A \geq D + |x^0|$

Table F.2 (*continued*)

No.	Nonlinearity	Describing Function Coefficients
14		$N_1 = \dfrac{c}{\pi A}\left\{\left[1 - \dfrac{(R-x^0)^2}{A^2}\right]^{1/2} + \left[1 - \dfrac{(D-x^0)^2}{A^2}\right]^{1/2}\right\},$ $N_2 = -\dfrac{c}{\pi A^2}(R - D),$ $F^0 = \dfrac{c}{2} - \dfrac{c}{2\pi}\left(\arcsin\dfrac{R-x^0}{A} + \arcsin\dfrac{D-x^0}{A}\right),$ $A \geq |R - x^0|,\quad A \geq |x^0 - D|$
15		$N_1 = \dfrac{c}{\pi A}\left\{\left[1 - \dfrac{(R-x^0)^2}{A^2}\right]^{1/2} + \left[1 - \dfrac{(D+x^0)^2}{A^2}\right]^{1/2}\right\},$ $N_2 = -\dfrac{c}{\pi A^2}(R + D),$ $F^0 = \dfrac{c}{2} - \dfrac{c}{2\pi}\left(\arcsin\dfrac{R-x^0}{A} - \arcsin\dfrac{D+x^0}{A}\right),$ $A \geq |R - x^0|,\quad A \geq |x^0 + D|$

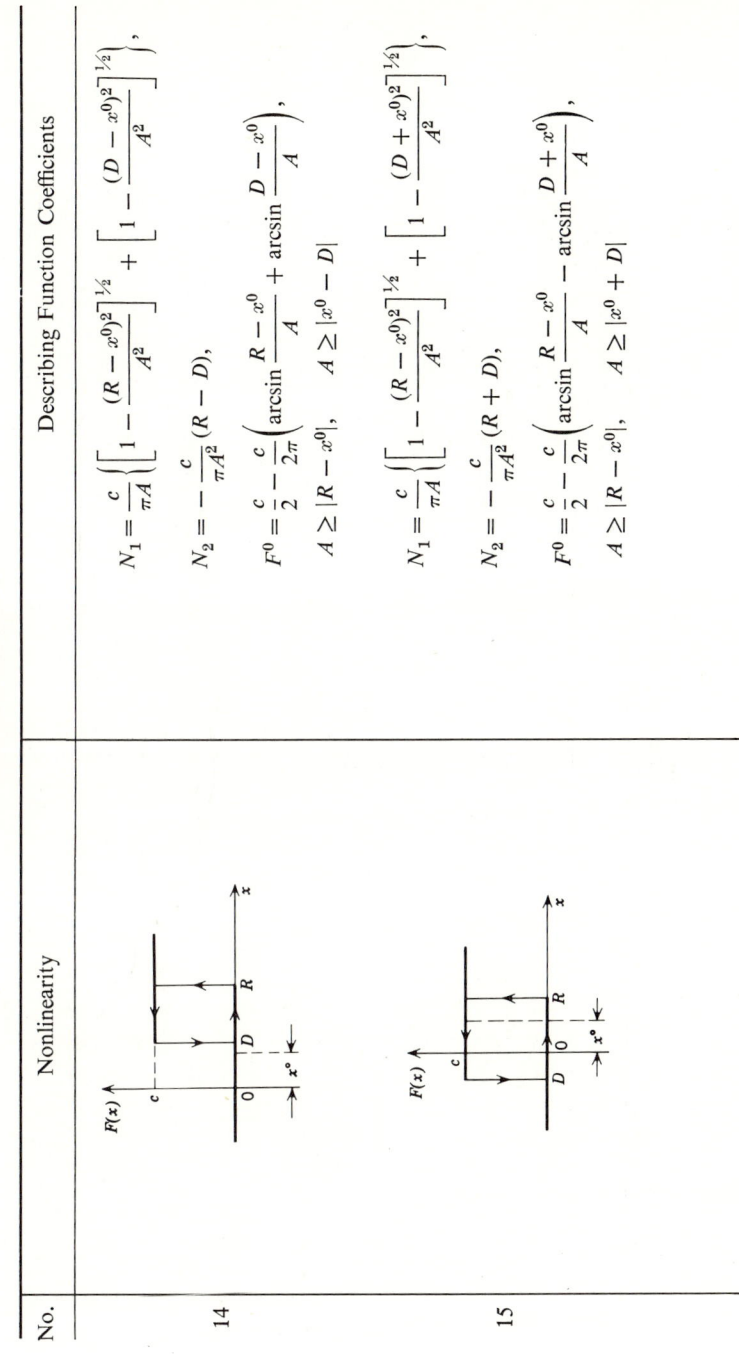

Table F.2 (*continued*)

No.	Nonlinearity	Describing Function Coefficients
16	$F(x)$ with Slope k_1 and Slope k_2, and x^0 marked	$N_1 = \dfrac{k_1 + k_2}{2} + \dfrac{k_1 - k_2}{\pi}\left\{\arcsin\dfrac{x^0}{A} + \dfrac{x^0}{A}\left[1 - \dfrac{(x^0)^2}{A^2}\right]^{1/2}\right\},$ $N_2 \equiv 0,$ $F^0 = \dfrac{k_1 + k_2}{2}x^0 + \dfrac{k_1 - k_2}{\pi}\left\{x^0 \arcsin\dfrac{x^0}{A} + A\left[1 - \dfrac{(x^0)^2}{A^2}\right]^{1/2}\right\}.$
17	$F(x)$ with Slope k, and x^0 marked	$N_1 = \dfrac{k}{2} + \dfrac{k}{\pi}\left\{\arcsin\dfrac{x^0}{A} + \dfrac{x^0}{A}\left[1 - \dfrac{(x^0)^2}{A^2}\right]^{1/2}\right\},$ $N_2 \equiv 0,$ $F^0 = \dfrac{k}{2}x^0 + \dfrac{k}{\pi}\left\{x^0 \arcsin\dfrac{x^0}{A} + A\left[1 - \dfrac{(x^0)^2}{A^2}\right]^{1/2}\right\}.$

Table F.2 (*continued*)

No.	Nonlinearity	Describing Function Coefficients
18	(graph: $F(x)$ vs x, with dead zone x^0 and slope k beyond)	$N_1 = \dfrac{k}{2} - \dfrac{k}{\pi}\left\{\arcsin\dfrac{x^0}{A} + \dfrac{x^0}{A}\left[1 - \dfrac{(x^0)^2}{A^2}\right]^{1/2}\right\},$ $N_2 \equiv 0,$ $F^0 = \dfrac{k}{2}x^0 - \dfrac{k}{\pi}\left\{x^0 \arcsin\dfrac{x^0}{A} + A\left[1 - \dfrac{(x^0)^2}{A^2}\right]^{1/2}\right\}.$

References

[F.1] R. Sridhar, A General Method for Deriving the Describing Functions for a Certain Class of Nonlinearities, *IRE Trans. Auto. Control*, **AC-5**, No. 2, 135–141 (June 1960).

[F.2] E. P. Popov and I. P. Palitov, Approximate Methods for Analysis of Nonlinear Automatic Systems (in Russian), State Press for Physics and Mathematical Literature, Moscow, 1960. (English translation: Foreign Technical Division, AFSC, Wright-Patterson AFB, Ohio, Report FTD-TT-62-910.)

[F.3] J. E. Gibson, *Nonlinear Automatic Control*, McGraw-Hill, New York, 1963.

[F.4] J. Tou and P. M. Schultheiss, Static and Sliding Friction in Feedback Systems, *J. Appl. Phys.*, **27**, No. 9, 1210–1217 (September 1953).

[F.5] M. Y. Silberberg, The Describing Function of an Element with Friction, *AIEE Trans.* Pt. II (*Applications and Industry*), **75**, 423–426 (January 1957).

[F.6] Y a. Z.Tsypkin, On the Relationship between the Equivalent Gain of Nonlinear Elements and their Characteristics (in Russian), *Avtomatika i Telemekhanika*, No. 4 (1956).

APPENDIX G

Accuracy Considerations in Describing Function Method

G.1 Introduction

The describing function, being an approximate method for solving nonlinear differential equations, can be justified by providing an estimate of the error caused by the approximations. Such an estimate was given in Section 3.4, where only a qualitative consideration of the error was outlined. Application of the describing function method, however, requires a more quantitative analysis of the accuracy to be expected.

Johnson [G.1] has presented an accuracy analysis based on the earlier studies of Bulgakov [G.2, 3]. The Johnson work is summarized and discussed by Truxal [G.4]. The somewhat extended analysis evaluates the second frequency correction and the fundamental component of the first amplitude correction. The results afford a method for evaluating the accuracy of the first approximation given by the describing function. On the basis of the accuracy evaluation, the first approximation can be improved. Similar results by a different approach have been obtained by Popov and Palitov [G.5]. In their study the higher harmonics of the periodic solution are calculated and, then, are used to correct the first approximation obtained in the first step of the analysis.

Essential progress in obtaining the indicated estimates was made by Rozenvasser [G.6, 7] based upon the work of Aizerman [G.8]. By reducing the differential equation to an integral one, Rozenvasser obtained more refined applicability conditions. In addition, an effective inequality was obtained for the error in the determination of the corresponding periodic solutions of piecewise linear systems.[1] Garber [G.9] extended the obtained

---

[1] The piecewise linear systems are systems with nonlinear characteristics made of straight line segments [G.8].

results to estimate the deviation of the periodic solutions from the harmonic oscillations in nonautonomous systems. Later Garber and Rozenvasser [G.10] reduced the estimation of the error in the describing function method to the solution of two basic problems: (1) the determination of a priori estimate of the higher harmonics of the periodic solutions, (2) the approximate determination of the amplitude and frequency of the periodic solutions from the solution of the first problem.

The first problem has been solved in [G.10] by some separate ideas suggested by Sagirow [G.11, 12]. In reference G.7 a solution of the second problem is obtained by treating an auxiliary variational problem and using reference G.14. Along these lines, Sandberg [G.13] derived an upper bound on the mean-square error incurred by applying the describing function technique. The expression for the error reflects the generalized filter property discussed in Section 3.4 and agrees with the results obtained by Rozenvasser and Garber.

The results of Rozenvasser and Garber are presented in this appendix.

G.2 Nonautonomous Systems

Consider the nonlinear differential equation

$$B(s)x + C(s)y = H(s)f, \qquad s \equiv \frac{d}{dt}, \tag{G.1}$$

where $B(s)$, $C(s)$, and $H(s)$ are polynomials in s and the degree of the polynomial $B(s)$ is higher than the degrees of the polynomials $C(s)$ and $H(s)$. Since the nonresonant case is considered, $B(jk\Omega) \neq 0$ for all integers k. The function $y = F(x)$, which represents the nonlinearity, is a symmetric, bounded, and piecewise continuous function of x.[2] The forcing function $f = f(t)$ is a periodic, piecewise continuous function with the frequency $\Omega = 2\pi/T$, so that

$$f(t) = f(t + T), \qquad f(t) = -f\left(t + \frac{T}{2}\right). \tag{G.2}$$

---

[2] Since the function $F(x)$ is a piecewise continuous function and it is necessary to perform the differentiation of $F(x)$ the necessary number of times, it is required to supplement (G.1) by *jump conditions* for the solution at the break points. The jump conditions are considered at the end of the appendix following reference G.8. This reference can be conveniently used as an introduction to the material presented in this appendix.

It should also be mentioned here that we can avoid the necessity of differentiating the nonsmooth function $F(x)$ by means of the substitution $H(s)f = -B(s)v$, $x + v = C(s)u$. According to (G.1), we obtain $B(s)u + F(x) = 0$ and the equation reduces to $B(s)u + F[C(s)u - v] = 0$.

Let $x = x(t)$ be a periodic solution of (G.1) that satisfies condition (G.2)

$$x(t) = x(t + T), \qquad x(t) = -x\left(t + \frac{T}{2}\right). \tag{G.3}$$

If the Fourier expansion of $y = y(t)$, which is

$$y(t) = \frac{2}{T} \sum_{k=-\infty}^{\infty} \int_0^{T/2} y(\tau) e^{(2k+1)j\Omega(t-\tau)} \, d\tau, \tag{G.4}$$

is substituted in (G.1), we obtain the solution $x(t)$ as the solution of the integral equation[3]

$$x(t) = \int_0^{T/2} \Phi(t - \tau) F[x(\tau)] \, d\tau + \psi(t), \tag{G.5}$$

where the kernel $\Phi(t - \tau)$ is given by its Fourier expansion

$$\Phi(t - \tau) = -\frac{2}{T} \sum_{k=-\infty}^{\infty} \frac{C[(2k+1)j\Omega]}{B[(2k+1)j\Omega]} e^{(2k+1)j\Omega(t-\tau)} \tag{G.6}$$

and the function $f(t)$ is represented by the Fourier expansion $\psi(t)$ given by

$$\psi(t) = \frac{2}{T} \sum_{k=-\infty}^{\infty} \int_0^{T/2} \frac{H[(2k+1)j\Omega]}{B[(2k+1)j\Omega]} f(t) e^{(2k+1)j\Omega(t-\tau)} \, d\tau. \tag{G.7}$$

The series (G.6) can be summed and expressed in the closed form as[4]

$$\Phi(t - \tau) = \begin{cases} -\sum_{i=1}^{n} \frac{C(s_i)}{B'(s_i)} \frac{\exp\,[s_i(t - \tau)]}{1 + \exp\,(\pi s_i/\Omega)}, & (0 < t - \tau < \pi/\Omega) \\ \sum_{i=1}^{n} \frac{C(s_i)}{B'(s_i)} \frac{\exp\,[s_i(t - \tau + \pi/\Omega)]}{1 + \exp\,(\pi s_i/\Omega)}, & (0 < \tau - t < \pi/\Omega) \\ \frac{1}{2} \sum_{i=1}^{n} \frac{C(s_i)}{B'(s_i)} \tanh\left(\frac{\pi s_i}{\Omega}\right), & (\tau - t = 0), \end{cases} \tag{G.8}$$

---

[3] Equation G.5 is of the Hammerstein type, since by the substitution $z = x - \psi$ we obtain

$$z(t) = \int_0^{T/2} \Phi(t - \tau) F[z(t) + \psi(t)] \, d\tau.$$

It should also be mentioned here that it is difficult to directly obtain the estimate of the error in the solution of (G.1) since the corresponding differential operator is unbounded. Therefore it is necessary to pass to the inverse operator—the integral operator.

[4] Let us consider the Fourier expansion of the function of the period T that coincides with $\Phi(t - \tau)$ for $0 < t - \tau < T/2$. Then for $\varphi_{2k+1} = j(2k+1)\Omega$,

$$-\frac{2}{T} \int_0^{T/2} \sum_{i=1}^{n} \frac{C(s_i)}{B'(s_i)} \frac{\exp\,(s_i\tau)}{1 + \exp\,(\pi s_i/\Omega)} \exp\,(-\varphi_{2k+1}\tau) \, d\tau = -\frac{2}{T} \sum_{i=1}^{n} \frac{C(s_i)}{B'(s_i)} \frac{1}{s_i - \varphi_{2k+1}}$$

$$= -\frac{2}{T} \frac{C(\varphi_{2k+1})}{B(\varphi_{2k+1})},$$

which proves (G.8).

provided that the zeros s_i ($i = 1, 2, \ldots, n$) of the polynomial $B(s)$ are assumed simple. The case where there are multiple zeros is considered in reference G.6.

The kernel $\Phi(t - \tau)$ represents the response of the linear part of the system to a periodic sequence of impulses. It can be calculated from the frequency characteristic $G(j\Omega) = C(j\Omega)/B(j\Omega)$ by summing first few terms in the summation (G.6).

In evaluating the periodic solution $x(t)$ of (G.1), the harmonic linearization method discards all the higher harmonics in (G.4). Then, the approximate solution is the corresponding fundamental component $x_1(t)$ of the Fourier series, which is determined as a solution of the approximate integral equation.

$$x_1(t) = \int_0^{T/2} \Phi_1(t - \tau) F[x_1(t)] \, d\tau + \psi_1(t), \tag{G.9}$$

where

$$\Phi_1(t - \tau) = \frac{\omega}{\pi} \left[\frac{C(j\Omega)}{B(j\Omega)} e^{j\Omega(t-\tau)} + \frac{C(-j\Omega)}{B(-j\Omega)} e^{-j\Omega(t-\tau)} \right] \tag{G.10}$$

is the fundamental component of the kernel $\Phi(t - \tau)$ and $\psi_1(t)$ is the first harmonic of the periodic function $\psi(t)$.

If the approximate solution is determined, the related accuracy can be simply evaluated from (G.5) and (G.9). The difference between the exact and approximate solution is transformed as

$$|x - x_1| \leq \left| \psi - \psi_1 + \int_0^{T/2} \Phi_h(t - \tau) F(x_1) \, d\tau \right.$$
$$\left. + \left| \int_0^{T/2} \Phi(t - \tau)[F(x) - F(x_1)] \, d\tau \right| \right| = |J_1| + |J_2|, \tag{G.11}$$

where $\Phi_h = \Phi - \Phi_1$ is the sum of the higher harmonics of the kernel Φ.

Since the approximate solution $x_1(t)$ is known, the integral J_1 can be evaluated or its upper bound can be determined. On the basis of certain assumptions concerning the characteristic $F(x)$, the integral J_2 can be majorized by a positive function ρ which depends only on max $|x - x_1|$. Thus, from (G.11), we have

$$|x - x_1| \leq |J_1| + \rho(\max |x - x_1|) \tag{G.12}$$

and the solution of this implicit inequality for $|x - x_1|$ yields the required quantitative estimate of the error made by the proposed approximation.

In the case where $F(x)$ satisfies the uniform Lipschitz condition

$$|F(\xi_1) - F(\xi_2)| \leq k |\xi_1 - \xi_2|, \quad (k = \text{const}), \tag{G.13}$$

we have $\rho(\eta) = kI\eta$, and

$$I = \int_0^{T/2} |\Phi(u)|\, du. \tag{G.14}$$

The accuracy estimate is of the form

$$|x - x_1| \leq \frac{\max |J_1|}{1 - kI}, \tag{G.15}$$

for which the applicability condition is

$$1 - kI > 0. \tag{G.16}$$

This inequality is satisfied when the linear part of the system has satisfactory filtering characteristics (see below).

Let us now discuss the quantities that enter in (G.15).

The function J_1, as can easily be shown, is the sum of the higher harmonics contained in the steady-state response of the system linear part to the periodic input signal $F(x_1) + [H(s)/C(s)]f$. It is not difficult to see that if the linear part of the system described by the transfer function $G(s) = C(s)/B(s)$ acts as a filter, I becomes small and so J_1 becomes smaller and the accuracy of the estimate (G.12) increases. It is also obvious that the applicability condition (G.16) is satisfied when the linear part confirms the generalized filter property. In a number of cases the integral J_1 can be found by direct integration of (G.1), using the known approximate solution $x_1(t)$. Sometimes it is easier to limit the value of max $|J_1|$ "from above," but then the accuracy estimates are weaker. The simplest way to estimate J_1 is to use the inequality

$$|J_1| \leq \max |\psi - \psi_1| + \max |F(x_1)| \left| \int_0^{T/2} \Phi_h(u)\, du \right|. \tag{G.17}$$

For piecewise differentiable nonlinear characteristics, the Lipschitz constant k is found as the maximum of the absolute value of the corresponding gain, that is, $k = \max |F'(x)|$. In practice this relationship can be used if max $|F'(x)|$ can be determined, which is the case in all piecewise linear continuous characteristics with finite number of segments.

The integral I in (G.14) can be calculated directly by using the expression for the kernel given in (G.8). Apparently, the quantity I is a function of the coefficients of polynomials $B(s)$ and $C(s)$, and the frequency Ω. When the transfer function $G(s)$ represents a low-pass filter, for sufficiently large values of Ω, the quantity $I(\Omega)$ satisfies the condition (G.16) which is necessary for the application of the accuracy estimate (G.15).

For most common types of transfer functions, the function $I(\Omega)$ can be tabulated [G.10].

Accuracy Considerations in Describing Function Method

For high-order systems, it is often simpler to estimate I by using the Bunyakovsky integral inequality[5]

$$I \leq \left[\frac{\pi}{\Omega} \int_0^{T/2} \Phi^2(u)\, du \right]^{1/2}. \tag{G.18}$$

Since the kernel is squared, the integral in (G.18) can be simply calculated by applying the Parceval formula

$$\int_0^{T/2} \Phi^2(u)\, du = \frac{\Omega}{\pi} \sum_{k=-\infty}^{+\infty} G[(2k+1)j\Omega] G[-(2k+1)j\Omega]$$

$$= \frac{2\Omega}{\pi} \sum_{k=0}^{\infty} |G[(2k+1)j\Omega]|^2, \tag{G.19}$$

which gives

$$I(\Omega) \leq \bar{I}(\Omega) = \sqrt{2 \sum_{k=0}^{\infty} |G[(2k+1)j\Omega]|^2}. \tag{G.20}$$

To calculate the summation within the brackets of (G.20), we start from (G.18) with

$$\bar{I}^2 = \int_0^{T/2} \Phi^2(u)\, du = \int_0^{T/2} \Phi^2(-u)\, du = \left[\int_0^{T/2} \Phi(t-u)\Phi(-u)\, du \right]_{t=0}. \tag{G.21}$$

In the second equation of (G.21), we used the periodicity of the kernel $\Phi(u)$. According to (G.21), the quantity $\bar{I}$ may be regarded as a consequence of the periodic impulse signal transformed by the transfer functions $G(-s)$ and $G(s)$ at $t = 0$. Therefore, in determining $\bar{I}$ we can apply the last expression of (G.8) where instead of $G(s)$ we use $G(s)G(-s)$. In doing so, we obtain

$$\frac{\Omega}{\pi} \sum_{k=-\infty}^{\infty} G[(2k+1)j\Omega] G[-(2k+1)j\Omega] = \sum_{i=1}^{n} \frac{C(s_i)C(-s_i)}{B'(s_i)B(-s_i)} \tanh\left(-\frac{\pi s_i}{2\Omega}\right). \tag{G.22}$$

Since to each zero s_i there correspond two equivalent terms in the summation of (G.22), we have $[B(s)B(-s)]'_{s=s_i} = B'(s_i)B(-s_i)$.

By applying (G.22), we obtain

$$\bar{I}(\Omega) = \left[\frac{\pi}{\Omega} \sum_{i=1}^{n} \frac{C(s_i)C(-s_i)}{B'(s_i)B(s_i)} \tanh\left(-\frac{\pi s_i}{2\Omega}\right) \right]^{1/2}. \tag{G.23}$$

---

[5] This inequality is sometimes called the Schwarz inequality [G.16] since it was discovered independently by Bunyakowsky in 1861 and Schwarz in 1885 [G.17].

Nonautonomous Systems

Suppose that the nonlinear characteristic $F(x)$ satisfies the inequality

$$k_1 \leq \frac{F(x)}{x} \leq k_2, \tag{G.24}$$

which is the case with most of the relay characteristics with a dead-zone. Then, using the transformation

$$F(x) - F(x_1) = \frac{F(x)}{x}(x - x_1) + \left[\frac{F(x)}{x} - \frac{F(x_1)}{x_1}\right]x_1 \tag{G.25}$$

in evaluation of $|J_2(t)|$, instead of (G.15), we get

$$|x - x_1| \leq \frac{\max |J_1| + (k_2 - k_1) \max \int_0^{T/2} |\Phi x_1|\, d\tau}{1 - k_2 I}. \tag{G.26}$$

The proposed accuracy estimates can be applied also to nonlinear characteristics $F(x, sx)$. If $F(x, sx)$ satisfies the uniform Lipschitz condition in both variables so that

$$|F(\xi_1, s\xi) - F(\xi_2, s\xi)| \leq k\, |\xi_1 - \xi_2|,$$
$$|F(\xi, s\xi_1) - F(\xi, s\xi_2)| < k_1\, |s\xi_1 - s\xi_2|, \tag{G.27}$$

where k and k_1 are constants, then we obtain

$$|x - x_1| \leq \max |J_1| + [k \max |x - x_1| + k_1 \max |sx - sx_1|]I,$$
$$|sx - sx_1| \leq \max |\dot{J}_1(t)| + [k \max |x - x_1| + k_1 \max |sx - sx_1|]I_1, \tag{G.28}$$

where

$$I_1 = \int_0^{T/2} |\Phi'(u)|\, du. \tag{G.29}$$

The estimate is

$$|x - x_1| \leq \frac{(1 - k_1 I_1) \max |J_1| - kI \max |\dot{J}_1|}{1 - kI - k_1 I_1}. \tag{G.30}$$

The applicability condition of the estimate (G.30) is

$$1 - kI - k_1 I_1 > 0, \tag{G.31}$$

which is again related to the filtering characteristics of the linear part of the system.

G.3 Autonomous Systems

The results obtained in the previous section cannot be extended to autonomous systems ($f \equiv 0$). It is not possible to satisfy applicability conditions (G.16) or (G.31) after a periodic solution is determined. In fact, from (G.5),

$$x(t) = \int_0^{T/2} \Phi(t - \tau) F[x(\tau)]\, d\tau = \int_0^{T/2} \Phi(t - \tau) \frac{F[x(\tau)]}{x(\tau)} x(\tau)\, d\tau. \quad \text{(G.32)}$$

If the nonlinear characteristic satisfies the Lipschitz condition (G.13), and specifically $\xi_2 = 0$, then

$$|F(\xi)| \leq k_0 |x|, \quad (k_0 = \text{const}). \quad \text{(G.33)}$$

From (G.32) and (G.33), we have

$$|x(t)| \leq k_0 \max |x| \int_0^{T/2} |\Phi(u)|\, du = k_0 I \max |x|. \quad \text{(G.34)}$$

If $k_0 I < 1$, then inequality (G.34) cannot be satisfied by nonzero periodic solutions.

To obtain the estimates for autonomous systems following reference G.10, we decompose the periodic solution $x(t)$ into the fundamental harmonic $x_1(t)$ and higher harmonics $x_h(t)$. Therefore, we have now two integral equations

$$x_1(t) = \int_0^{T/2} \Phi_1(t - \tau) F[x(\tau)]\, d\tau \quad \text{(G.35}a\text{)}$$

$$x_h(t) = \int_0^{T/2} \Phi_h(t - \tau) F[x(\tau)]\, d\tau, \quad \text{(G.35}b\text{)}$$

where $\Phi_1(u)$ is the first component of $\Phi(u)$ and $\Phi_h = \Phi - \Phi_1$.

Since the system is symmetric and autonomous,

$$x_1 = A \sin \Omega t = \frac{A}{2j}(e^{j\Omega t} - e^{-j\Omega t}). \quad \text{(G.36)}$$

By using (G.36) and (G.10) in (G.35a) and then equating the coefficients of $\exp(j\Omega t)$, we obtain

$$B(j\Omega) + C(j\Omega) N^*(A, x_h) = 0, \quad \text{(G.37)}$$

where

$$N^*(A, x_h) = \frac{2\Omega}{\pi A} \int_0^{T/2} F[A \sin \Omega t + x_h(t)](\sin \Omega t + j \cos \Omega t)\, dt. \quad \text{(G.38)}$$

When $x_h \equiv 0$, $N^*(A, 0) = N(A)$, equations (G.37) and (G.38) become the basic equations of the harmonic linearization (describing function) treated in Chapter 3. These basic equations yield approximate values A_0 and Ω_0

which are close to the exact values A and Ω if x_h is sufficiently small. Thus, the accuracy depends on the value of max $|x_h|$ which can be estimated from integral equation (G.35b).

Assuming that $F(x)$ satisfies (G.33), from (G.35b), we obtain

$$|x_h| \leq k_0(A + \max |x_h|) \int_0^{T/2} |\Phi_h(u)|\, du. \tag{G.39}$$

By denoting

$$I_h(\Omega) = \int_0^{T/2} |\Phi_h(u)|\, du, \tag{G.40}$$

and assuming that the condition

$$1 - k_0 I_h > 0 \tag{G.41}$$

is satisfied, we have

$$|x_h| \leq \frac{A k_0 I_h}{1 - k_0 I_h}. \tag{G.42}$$

Apparently, the accuracy of the obtained estimate increases for smaller values of I_h. The integral I_h is calculated in the same way as the integral I, and $I_h \ll I$ since the fundamental component is absent. By using the Bunyakovsky inequality we obtain

$$I(\Omega) \leq \left[2 \sum_{k=1}^{\infty} |G[(2k+1)j\Omega]|^2 \right]^{1/2}. \tag{G.43}$$

The quantity $I_h(\Omega)$ reflects the filtering properties of the system and decreases rapidly with increasing frequency.

If we know the value of max $|x_h|$ we can find estimates for the amplitude A and frequency Ω of the periodic solution $x(t)$. Let us rewrite equation G.37 as

$$\frac{1}{G(j\Omega)} + N(A, 0) = N(A, 0) - N(A, x_h), \tag{G.44}$$

where $G(j\Omega) = C(j\Omega)/B(j\Omega)$. We can now estimate the right-hand side of (G.44) to obtain

$$\delta N = N(A, x_h) - N(A, 0)$$

$$= \frac{2\Omega}{\pi A} \int_0^{T/2} [F(A \sin \Omega t + x_h) - F(A \sin \Omega t)] \sin \Omega t\, dt$$

$$+ j \frac{2\Omega}{\pi A} \int_0^{T/2} [F(A \sin \Omega t + x_h) - F(A \sin \Omega t)] \cos \Omega t\, dt. \tag{G.45}$$

The projection δN_φ of the vector δN on the ray with the polar angle φ is

$$\delta N_\varphi = \frac{2\Omega}{\pi A} \int_0^{T/2} [F(A \sin \Omega t + x_h) - F(A \sin \Omega t)] \cos (\Omega t - \varphi)\, dt. \quad \text{(G.46)}$$

By using the Lipschitz condition (G.13), we get

$$|\delta N_\varphi| \leq \frac{2\Omega}{\pi A} k_1 \max |x_h| \int_0^{T/2} |\cos(\Omega t - \varphi)|\, dt = \frac{4 k_1 \max |x_h|}{\pi A}. \quad \text{(G.47)}$$

Specifically, when $\varphi = 0$ and $\varphi = \pi/2$, we obtain

$$|\operatorname{Re} \delta N| = \left| \operatorname{Re} \left[\frac{1}{G(j\Omega)} + N(A, 0) \right] \right| \leq \rho(\Omega)$$

$$|\operatorname{Im} \delta N| = \left| \operatorname{Im} \left[\frac{1}{G(j\Omega)} + N(A, 0) \right] \right| \leq \rho(\Omega). \quad \text{(G.48)}$$

These inequalities can be solved for A and Ω to give estimates of the amplitude and frequency of the first harmonic. The solution can be obtained by a graphical procedure. After two loci $-1/G(j\Omega)$ and $N(A, 0)$ are plotted, the approximate values for the amplitude and frequency are found at their intersections. For every value of Ω along the locus $-1/G(j\Omega)$, a circle of the radius $\rho(\Omega)$ is drawn with the center on the curve. Tolerances for the amplitude and frequency of the corresponding periodic solution are determined by the extent of the plotted circle. By plotting several circles along the locus $-1/G(j\Omega)$, we can construct the envelopes of the circles as shown on Figure G.1. These envelopes determine the interval $[A_l, A_h]$ on the locus $N(A, 0)$ which in turn gives the accuracy estimate for the amplitude of the periodic solution. The estimate of the frequencies is determined by the interval $[\Omega_l, \Omega_h]$ where the limiting frequencies are at the centers of the circles tangent to the locus $N(A, 0)$.

After the estimates of the fundamental harmonic are obtained, we can return to (G.42) and determine the estimate $|x_h|$. Thus,

$$|x_h| \leq k_0 A_h \max_{\Omega_l \leq \Omega \leq \Omega_h} \frac{I_h}{1 - k_0 I_h}. \quad \text{(G.49)}$$

The obtained estimates are applicable only if the inequality (G.41) is satisfied, that is, when the filtering characteristics are sufficiently strong. These estimates can be improved, if further restrictions are placed on the nonlinear characteristic. A restriction $|F'(x)| \leq k$ where k is a constant, was used in references G.10 and G.11. Furthermore, nonautonomous systems can also be considered [G.7].

Autonomous Systems

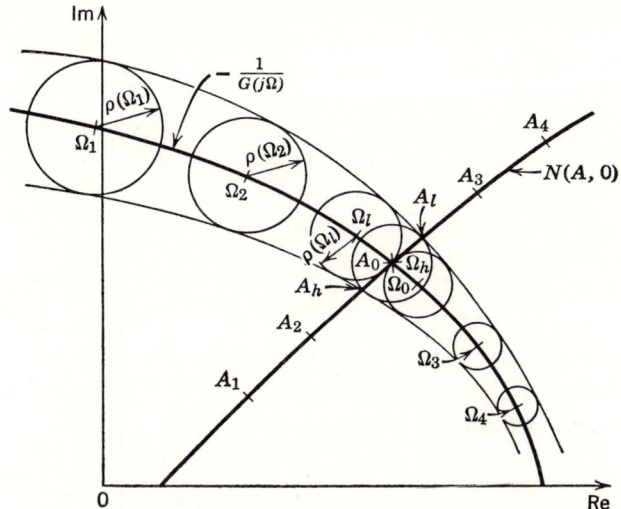

Figure G.1 Graphical interpretation of the inequalities.

To illustrate the practical application of the procedure, consider the nonlinear differential equation

$$(s^2 + 0.8s + 8)x - sF(x) = 0, \qquad (G.50)$$

where

$$F(x) = \begin{cases} x, & (|x| < 1) \\ \text{sign } x, & (|x| > 1). \end{cases} \qquad (G.51)$$

In this case $k_0 = 1$; the graph of $\rho(\Omega)$ is shown in Figure G.2. Condition (G.41) is satisfied for $\Omega > 1.15$. The loci of $-1/G(j\Omega)$ and $N_1(A, 0)$ are

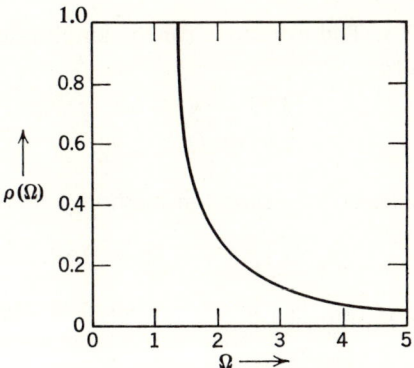

Figure G.2 Function $\rho(\Omega)$.

522 Accuracy Considerations in Describing Function Method

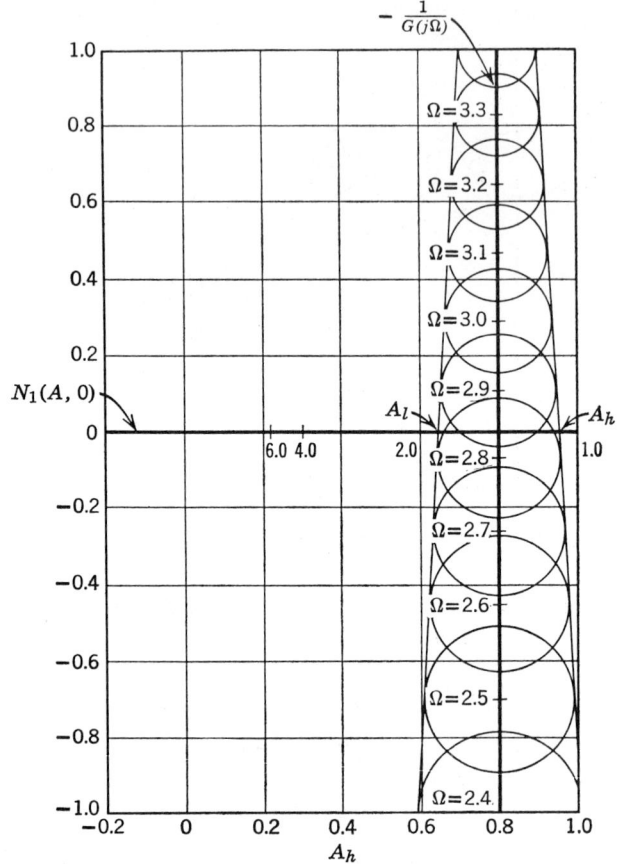

Figure G.3 Graphical solution of the estimates for A and Ω.

plotted in Figure G.3. The estimates for the amplitude of the first harmonic and the frequency are

$$1.23 \leq A \leq 1.85$$
$$2.75 \leq \Omega \leq 2.90. \tag{G.52}$$

The solution by the describing function method yields

$$A_0 = 1.55, \qquad \Omega_0 = 2.82. \tag{G.53}$$

From (G.49) we obtain $|x_h| \leq 0.29$. Therefore the maximum deviation of the variable x is

$$|x| \leq A_h + |x_h| = 2.14. \tag{G.54}$$

This example has been presented in references G.7 and G.10. In G.7 it has been solved by a variational approach that is an extension of the above results.

G.4   Jump Conditions

Consider (G.1) again

$$B(s)x + C(s)y = H(s)f, \qquad s \equiv \frac{d}{dt}, \qquad (G.1)$$

where $y = F(x)$ is a piecewise-linear, nonlinear characteristic. This means that the characteristic $F(x)$ is made up of straight line segments. Therefore it is of interest to consider the differentiation of the discontinuous functions such as relay characteristics, and clarify the term $C(s)y$. The consideration can then be directly extended to the term $H(s)f$ if $f = f(t)$ is a discontinuous function of time. The consideration follows reference G.8.

The symbol sy denotes the *ordinary derivative* of the function $y = F[x(t)]$ with respect to t, which exists everywhere apart from the points t_i, for which the function $y = y(t)$ has discontinuities; at the points for which $t = t_i$ the derivative does not exist and has a meaning only if we speak of the derivative from the right (for $t = t_i + \varepsilon$, $\varepsilon \to 0$, or simply $t = t_i + 0$) and of a derivative from the left (for $t = t_i - 0$).

At this point it is necessary to introduce the *Dirac delta functions*, or simple *delta functions*, which denote singular "functions" having the form of impulses. Strictly speaking, delta functions are not functions in the sense of classical mathematical analysis. They are extensively used in the theory of distributions, and a discussion of their properties can be found in reference G.15 and G.16. In this text a very limited and rather formalistic basic treatment of the delta function is given for the sake of clarity in presentation of jump conditions.

Intuitively, a delta function $\delta(t)$ is an idealization of a very narrow pulse, or impulse, occuring at a certain moment of time, for instance, say $t = 0$, with a finite total area that, for convenience, is normalized to 1. Thus

$$\delta(t) = \begin{cases} 0, & t \neq 0 \\ \infty, & t = 0 \end{cases} \qquad (G.55)$$

$$\int_{-\infty}^{+\infty} \delta(t)\, dt = 1.$$

Moreover, if $\delta(t)$ is a narrow pulse of area 1, then $g(t)\,\delta(t)$, where $g(t)$ is any "well-behaved" function in the neighborhood of $t = 0$, will be a narrow pulse of area $g(0)$. Therefore

$$\int_{-\infty}^{+\infty} \delta(t)g(t)\, dt = g(0). \tag{G.56}$$

This suggests that $\delta(t)$ be defined as a "limit" of an approximating sequence of functions $f(t, \lambda)$ so that, for any "well-behaved" function $g(t)$,

$$\lim_{\lambda \to 0} \int_{-\infty}^{+\infty} f(t, \lambda)g(t)\, dt = g(0). \tag{G.57}$$

There are many approximating sequences that can be used with (G.57) and they are discussed in reference G.15. One of the simplest is the sequence of pulses of width λ and height $1/\lambda$ centered on $t = 0$, in which case $f(t, \lambda)$ is

$$f(t, \lambda) = \frac{1}{\lambda}\left[u\left(t + \frac{\lambda}{2}\right) - u\left(t - \frac{\lambda}{2}\right)\right], \tag{G.58}$$

where $u(t)$ is the unit-step function. Now if $\delta(t) = \lim_{\lambda \to 0} f(t, \lambda)$, then formally

$$\begin{aligned}\delta(t) &= \lim_{\lambda \to 0} \frac{1}{\lambda}\left[u\left(t + \frac{\lambda}{2}\right) - u\left(t - \frac{\lambda}{2}\right)\right] \\ &= \frac{d}{dt} u(t).\end{aligned} \tag{G.59}$$

The derivative $du(t)/dt$ is a derivative in the *distribution sense* and is sometimes called the *generalized derivative*, which is denoted by $s*u(t)$. Then, more generally, $\delta^{(n)}(t)$, the delta function of order n, is the $(n+1)$th derivative of $u(t)$ in the distribution sense or, recursively, the first derivative of $\delta^{(n-1)}(t)$. Thus

$$\delta^{(n)}(t) = \frac{d^{n+1}}{dt^{n+1}} u(t) = \frac{d}{dt} \delta^{(n-1)}(t). \tag{G.60}$$

For the function $y(t)$ we have

$$s*y(t) = sy(t) + \sum_i \phi_{0i} \delta(t - t_i), \tag{G.61}$$

where the t_i are values of t at which $y(t)$ has discontinuities, and ϕ_{0i} is the magnitude of the discontinuity; that is,

$$\phi_{0i} = y(t_i + 0) - y(t_i - 0). \tag{G.62}$$

The summation in (G.61) is taken over all the discontinuities. Evidently, if the function $y(t)$ is continuous, the operations sy and $s*y$ are identical since all $\phi_{0i} \equiv 0$.

Jump Conditions

Applying repeatedly the operation s^*, we obtain a formula defining the generalized derivatives

$$s^*y(t) = sy(t) + \sum_i \phi_{0i} \delta(t - t_i)$$

$$(s^*)^2 y(t) = s^2 y(t) + \sum_i [\phi_{0i} \delta^{(1)}(t - t_i) + \phi_{1i} \delta(t - t_i)]$$

$$\vdots \qquad (G.63)$$

$$(s^*)^n y(t) = s^n y(t)$$
$$+ \sum_i [\phi_{0i} \delta^{(n-1)}(t - t_i) + \phi_{1i} \delta^{(n-2)}(t) + \cdots + \phi_{n-1,i} \delta(t - t_i)].$$

Here ϕ_{ji} ($j = 0, 1, \ldots, n$) are discontinuities of the functions $y(t), sy(t), \ldots, s^n y(t)$ at times $t = t_i$ when a discontinuity occurs for any one of these functions. The summation in formula (G.63) is taken over all the points t_i.

If we replace the operator s by s^* in (G.1), it becomes

$$B(s^*)x + C(s^*)y = H(s^*)f, \qquad (G.64)$$

where[6]

$$B(s^*) = \sum_{k=0}^{n} b_k s^{*k}, \qquad C(s^*) = \sum_{k=0}^{n} c_k s^{*k}, \qquad H(s^*) = \sum_{k=0}^{n} h_k s^{*k}. \qquad (G.65)$$

By substituting (G.63) into (G.64) and equating the coefficients by the equal order derivative of $\delta(t)$, we obtain

$$B(s)x + C(s)y = H(s)f \qquad (G.1)$$

with the supplementary conditions for each instant of time t_i

$$b_n \xi_0 + c_n \phi_0 = h_n \eta_0$$
$$b_n \xi_1 + b_{n-1} \xi_0 + c_n \phi_1 + c_{n-1} \phi_0 = h_n \eta_1 + h_{n-1} \eta_0$$
$$\vdots \qquad (G.66)$$
$$b_n \xi_{n-1} + \cdots + b_1 \xi_0 + c_n \phi_{n-1} + \cdots + c_1 \phi_0 = h_n \eta_{n-1} + \cdots + h_1 \eta_0,$$

which are called the *jump conditions*.

In (G.66), $\xi_0, \xi_1, \ldots, \xi_{n-1}$ and $\eta_0, \eta_1, \ldots, \eta_{n-1}$ are the discontinuities of the functions $x, sx, \ldots, s^{n-1}x$, and $f, sf, \ldots, s^{n-1}f$, respectively.

To determine the actual solution $x = x(t)$, either equation G.64 or equations G.1 supplemented with G.66 can be used since they are equivalent

---

[6] Normally the degrees of the polynomials $C(s^*)$ and $H(s^*)$ are less than the degree $B(s^*)$. However, this only means that $c_n, c_{n-1}, \ldots, h_n, h_{n-1}, \ldots$ are equal to zero.

in that respect. The solution procedure using the original equation, together with supplementary jump conditions, is called the *aprovision* and is described in reference G.8.

References

[G.1] E. C. Johnson, Sinusoidal Analysis of Feedback Control Systems Containing Nonlinear Elements, *AIEE Trans.*, Pt. II (*Applications and Industry*) **71**, 169–181 (1952).

[G.2] B. V. Bulgakov, Periodic Processes in Free Pseudo-Linear Oscillatory Systems, *J. Franklin Inst.*, **235**, 591–616 (June 1943).

[G.3] B. V. Bulgakov, On the Method of Van der Pol and Its Applications to Nonlinear Control Problems, *J. Franklin Inst.*, **241**, 31–54 (January 1946).

[G.4] J. G. Truxal, *Automatic Feedback Control System Synthesis*, McGraw-Hill, New York, 1955.

[G.5] E. P. Popov and I. P. Palitov, *Approximate Methods for Analysis of Nonlinear Automatic Systems* (in Russian), State Press for Physics and Mathematical Literature, Moscow, 1960. (English translation: Foreign Technical Division, AFSC, Wright-Patterson AFB, Ohio, Report FTD-TT-62-910.)

[G.6] E. N. Rozenvasser, The Accurate Determination of Periodic Regimes in Piece-Wise Linear Automatic Control Systems (in Russian), *Avtomatika i telemekhanika*, **21**, No. 9, 1279–1292 (September 1960).

[G.7] E. N. Rozenvasser, A Variational Approach to Estimation of the Harmonic Balance Method (in Russian), *Tekhn. Kibernetika*, No. 1, 111–123 (1964).

[G.8] M. A. Aizerman, Lectures on the Theory of Automatic Control (in Russian) FIZMATGIZ, Moscow, 1958. (English translation: Pergamon and Addison-Wesley, New York, 1963.)

[G.9] E. D. Garber, An Estimate of the Error of the Harmonic Balance Method (in Russian), *Avtomatika i Telemekhanika*, **24**, No. 4, 482–492 (April 1963).

[G.10] E. D. Garber and E. N. Rozenvasser, The Investigation of Periodic Regimes of Nonlinear Systems on the Basis of the Filter Hypothesis (in Russian), *Avtomatika i Telemekhanika*, **26**, No. 2, 277–287 (February 1965).

[G.11] P. Sagirow, On the Problem of Error Estimation in the Harmonic Balance (in German), *Z. Angew. Math. Mech.*, No. 10–11, 1960.

[G.12] P. Sagirow, On the Periodic Solutions of Nonlinear Differential Equations and on the Relation Between the Amplitude and Frequency (in German), *Z. Angew. Math. Mech.*, No. 3 (1961).

[G.13] I. W. Sandberg, On the Response of Nonlinear Control Systems to Periodic Input Signals, *Bell System Tech. J.*, **18**, No. 3, 911–927 (May 1964).

[G.14] L. S. Pontriyagin, V. P. Boltiansky, P. V. Gamkrelidze, and R. V. Mischenko, Mathematical Theory of Optimal Processes (in Russian), FIZMATGIZ, Moscow, 1961. (English translation: K. N. Trirogoff, ed. by L. W. Neustadt, Interscience, New York, 1962.)

[G.15] G. Doetsch, *Handbuch der Laplace Transformation*, Vols. 1–3, Birkhauser, Basel, Switzerland, 1950.

[G.16] L. A. Zadeh and C. A. Desoer, *Linear System Theory*, McGraw-Hill, New York, 1963.

[G.17] G. Sansone, *Orthogonal Functions*, Interscience, New York, 1959.

APPENDIX H
Proof of the Stability Theorems

H.1 Introduction

The purpose of this appendix is to bridge the Liapunov theory and Popov results by means of the Yakubovich-Kalman lemma. Therefore the Liapunov stability theorem is proved and then the proof of the lemma is given. On the basis of the Liapunov theorem and the Yakubovich-Kalman lemma, the Popov results can be derived. These results are given in terms of Yakubovich's stability theorems, the proof of which are given here in the section following the proof of the main lemma. This completes the material of Chapter 8 on stability analysis.

Since Yakubovich's stability theorems apply to systems with discontinuous characteristics, Section H.4 is devoted to this subject. After introducing the idea of the sliding motion ("chattering") by simple example, the definition of the solution of differential equations with discontinuous righthand sides is given in the sense of Filippov. The proofs of the stability theorems use this definition of the motion arising in discontinuous nonlinear systems.

In Section H.6 a brief survey of the "Lur'e problem" is outlined. Besides the history of the problem, it reviews the extensions of the results presented in Chapter 8 and gives short information about the related references.

A particular extension of the Lur'e approach to the estimation of finite stability domains proposed by Weissenberger and the author is treated separately in Section H.7.

H.2 Proof of Liapunov's Stability Theorem

Before we give a proof of the generalized version of the Liapunov stability theorem, Theorem 8.1, it is necessary to extend the notion of the sign-definite function to include the functions $V(\mathbf{x}, t)$, which are explicit functions of time.

Proof of the Stability Theorems

Functions $V(\mathbf{x}, t)$ are generally used to prove stability of nonautonomous systems [H.1]. In Chapter 8, however, the absolute stability of nonautonomous systems was analyzed on the basis of the simpler version $V(\mathbf{x})$ and it was not necessary to consider the functions $V(\mathbf{x}, t)$. The functions $V(\mathbf{x}, t)$, however, are needed later to prove Corollary C of total stability, which is related to parameter variations. The definition of positive definite function $V(\mathbf{x}, t)$ is the following.

Definition H.1. A function $V(\mathbf{x}, t)$ is positive definite if around the origin of the vector space $\mathscr{X}$ in some region $R = \{\|\mathbf{x}\| < h\}$ where $h > 0$ is a constant, the function $V(\mathbf{x}, t)$ satisfies the following conditions:
(a) $V(\mathbf{x}, t)$ has continuous first partial derivatives with respect to $\mathbf{x}$ and t;
(b) $V(\mathbf{0}, t) = 0$, for all t; and
(c) there exists a positive-definite scalar function $W(\mathbf{x})$ such that $V(\mathbf{x}, t) \geq W(\mathbf{x})$ for all t.[1]

In case of negative-definite function $V(\mathbf{x}, t)$, it is required that $-V(\mathbf{x}, t)$ be a positive-definite function. Hence a reformulation of the above definition is clear.

Now we are ready to outline the Liapunov stability theorem.[2] We shall consider uniform asymptotic stability in the large and define the class of Liapunov functions related to this case. By relaxing different conditions on the Liapunov function one obtains as consequences the weaker types of stability.

Theorem H.1. Liapunov Stability Theorem. Consider the free dynamic system $\dot{\mathbf{x}} = \mathbf{X}(\mathbf{x}, t)$ where $\mathbf{X}(\mathbf{0}, t) = \mathbf{0}$ for all t.

Suppose there exists a scalar function $V(\mathbf{x}, t)$ with continuous first partial derivatives with respect to $\mathbf{x}$ and t such that $V(\mathbf{0}, t) = 0$ and
(a) $V(\mathbf{x}, t)$ is a positive definite function, that is, there exists a continuous nondecreasing scalar function α such that $\alpha(0) = 0$ and, for all t and all $\mathbf{x} \neq \mathbf{0}$, $0 < \alpha(\|\mathbf{x}\|) \leq V(\mathbf{x}, t)$;
(b) There exists a continuous scalar function γ such that $\gamma(0) = 0$ and the derivative $\dot{V}(\mathbf{x}, t)$ of $V(\mathbf{x}, t)$ along the motion starting at t, $\mathbf{x}$ satisfies, for all t and $\mathbf{x} \neq \mathbf{0}$,[3]

$$\dot{V}(\mathbf{x}, t) \equiv \frac{dV(\mathbf{x}, t)}{dt} = \frac{\partial V}{\partial t} + \frac{\partial V}{\partial \mathbf{x}} \mathbf{X}(\mathbf{x}, t) \leq -\gamma(\|\mathbf{x}\|) < 0;$$

---

[1] A level surface $V(\mathbf{x}) = c$ is no longer cylindrical. However, for sufficiently small c it still separates the state space into two domains. For every c the cylindrical level surface $W(\mathbf{x}) = c$ contains in its interior the surface $V(\mathbf{x}) = h$. Because of this, for example, the function $V(\mathbf{x}) = \frac{1}{2}(x_1^2 + x_2^2)e^{-t}$ is not a positive-definite function.
[2] The outline of this theorem and the proof follows that of Kalman and Bertram [H.1].
[3] This is weaker than the usual requirement that $\dot{V}$ be negative definite, since we do not need the assumption that γ is nondecreasing.

(c) There exists a continuous, nondecreasing scalar function β such that $\beta(0) = 0$, and, for all t, $V(\mathbf{x}, t) \leq \beta(\|\mathbf{x}\|)$;[4] and

(d) $\alpha(\|\mathbf{x}\|) \to \infty$ with $\|\mathbf{x}\| \to \infty$.

Then the equilibrium state $\mathbf{x}_e = \mathbf{0}$ is uniformly asymptotically stable in the large; $V(\mathbf{x}, t)$ is called a Liapunov function.

The following corollary of the theorem is needed.

Corollary A. Weaker Stability Theorems. The following conditions are sufficient for the weaker types of stability:

(I) Uniform asymptotic stability: (a)–(c).
(II) Equiasymptotic stability in the large: (a)–(b), (d).
(III) Equiasymptotic stability: (a)–(b).
(IV) Uniform stability: (a), (c), and $\dot{V}(\mathbf{x}, t) \leq 0$ for all $\mathbf{x}, t$.
(V) Stability: (a), and $\dot{V}(\mathbf{x}, t) \leq 0$ for all $\mathbf{x}, t$.
(VI) No finite escape time: (a), (d), and $\dot{V}(\mathbf{x}, t) \leq cV(\mathbf{x}, t)$ for all $\mathbf{x}, t$; c being a positive constant.

The Liapunov stability theorem for autonomous systems, given as Theorem 8.1 in Section 8.5, is proved here as Corollary B.

Corollary B. Autonomous Systems. Given the autonomous dynamic system $\dot{\mathbf{x}} = \mathbf{X}(\mathbf{x})$ where $\mathbf{X}(\mathbf{0}) = \mathbf{0}$ for all t. In order for the equilibrium state $\mathbf{x}_e = \mathbf{0}$ of system to be equiasymptotically stable in the large it is sufficient that there exists a scalar function $V(\mathbf{x})$ with continuous first partial derivatives with respect to $\mathbf{x}$, such that $V(\mathbf{0}) = 0$ and:

(a_1) $V(\mathbf{x}) > 0$ for all $\mathbf{x} \neq \mathbf{0}$;
(b_1) $\dot{V}(\mathbf{x}) < 0$ for all $\mathbf{x} \neq \mathbf{0}$; and
(d_1) $V(\mathbf{x}) \to \infty$ with $\|\mathbf{x}\| \to \infty$.

Proof of the Liapunov Theorem. By using the condition (b) of the theorem, we see that V is decreasing along any motion $\mathbf{x}(t, \mathbf{x}_0, t_0)$,

$$V[\mathbf{x}(t, \mathbf{x}_0, t_0), t] - V(\mathbf{x}_0, t_0) = \int_{t_0}^{t} \dot{V}[\mathbf{x}(\tau, \mathbf{x}_0, t_0), \tau]\, d\tau < 0, \quad t \geq t_0. \quad \text{(H.1)}$$

(i) To show first uniform stability, we choose $\varepsilon > 0$, $\delta(\varepsilon) > 0$ so that $\beta(\delta) < \alpha(\varepsilon)$, which is always possible since β is continuous and $\beta(0) = 0$ as stated in (b). Now, by assumption, V does not increase in time and if $\|\mathbf{x}_0\| \leq \delta$, for all $t \geq t_0$ and arbitrary t_0, we have

$$\begin{aligned}\alpha(\varepsilon) > \beta(\delta) &\geq V(\mathbf{x}_0, t_0) \\ &\geq V[\mathbf{x}(t, \mathbf{x}_0, t_0), t] \\ &\leq \alpha[\|\mathbf{x}(t, \mathbf{x}_0, t_0)\|].\end{aligned} \quad \text{(H.2)}$$

---

[4] Condition (c) is sometimes restated as "V has an infinitely small upper bound."

Hence, (H.2) implies

$$\|\mathbf{x}(t, \mathbf{x}_0, t_0)\| < \varepsilon, \quad \text{for all } t \geq t_0, \tag{H.3}$$

for α is nondecreasing and positive as assumed in (a). This proves uniform stability.

(ii) Now let $\dot{V} < 0$ and show that $\|\mathbf{x}(t, \mathbf{x}_0, t_0)\| \to 0$ with $t \to \infty$ uniformly in t_0 and $\|\mathbf{x}_0\| < \nu$.

First, using (i) we conclude that for any constant c_1 and arbitrary t_0, ν can be found so that $\|\mathbf{x}\| \leq \nu$ implies $\|\mathbf{x}(t, \mathbf{x}_0, t_0)\| \leq c_1$ for $t \geq t_0$.

Let $0 < \rho \leq \|\mathbf{x}_0\|$ and $\theta = \theta(\rho)$ be chosen so that $\beta(\theta) < \alpha(\rho)$, and let $c_2(\rho, \nu)$ be the minimum of the continuous function $\gamma(\|\mathbf{x}\|)$ on the compact set $\theta(\rho) \leq \|\mathbf{x}\| \leq c_1(\nu)$. Then, from (b) and (H.1), we obtain

$$\begin{aligned} 0 < \alpha(\theta) &\leq V[\mathbf{x}(t, \mathbf{x}_0, t_0), t] \\ &\leq V(\mathbf{x}_0, t_0) - (t - t_0)c_2 \\ &\leq \beta(\nu) - (t - t_0)c_2. \end{aligned} \tag{H.4}$$

Let us define $T(\rho, \nu) = \beta(\nu)/c_2(\rho, \nu) > 0$. If the inequality $\|\mathbf{x}(t, \mathbf{x}_0, t_0)\| > \theta$ were true over the time interval $t_0 \leq t \leq t_0 + T$, then from (H.4) for $t = t_0 + T$ the relation $V[\mathbf{x}(t_0 + T, \mathbf{x}_0, t_0), t_0 + T] \leq \beta(\nu) - Tc_2 = 0$ which contradicts the assumption about positivity of V. Therefore, in the interval $[t_0, t_0 + T]$ there is a point t_1 for which $\|\mathbf{x}_1\| = \|\mathbf{x}(t_1, \mathbf{x}_0, t_0)\| = \nu$ and

$$\begin{aligned} \alpha[\|\mathbf{x}(t, \mathbf{x}_1, t_1)\|] &\leq V[\mathbf{x}(t, \mathbf{x}_1, t_1), t] \\ &\leq V(\mathbf{x}_1, t_1) \\ &\leq \beta(\theta) < \alpha(\rho), \end{aligned} \tag{H.5}$$

for all $t \geq t_1$. Then, $\|\mathbf{x}(t, \mathbf{x}_0, t_0)\| < \rho$ for all $t \geq t_0 + T(\rho, \nu) \geq t_1$, where T does not depend on t_0 and we have uniform asymptotic stability.

(iii) Finally, note that from (d) it follows that for any ν (however large) there is a constant $c_1(\nu)$ so that $\beta(\nu) < \alpha(c_1)$ and we have the uniform stability in the large. Furthermore, by choosing $N(\nu) = c_1(\nu)$ the uniform boundedness follows directly and we complete the proof.

Proof of Corollary A. (I) Parts (i) and (ii) of the proof.

(II) If we denote the maximum of $V(\mathbf{y}, t)$ for $\|\mathbf{y}\| < \|\mathbf{x}\|$ by $\beta_m(\|\mathbf{x}\|, t)$ and replace $\beta(\|\mathbf{x}\|)$ by β_m, the proof is the same save that δ and T depends on t_0.

(III) Same as (II) without part (iii) of the proof.

(IV) Part (i) of the proof.

(V) Same as (II) without parts (ii) and (iii) of the proof.

(VI) The additional requirement $\dot{V}(\mathbf{x}, t) \leq cV(\mathbf{x}, t)$ for all $\mathbf{x}, t$, implies $V[\mathbf{x}(t, \mathbf{x}_0, t_0)] \leq V(\mathbf{x}_0, t_0) \exp[c(t - t_0)]$. Then, from (a), $\alpha(\|\mathbf{x}(t, \mathbf{x}_0, t_0)\|) \leq V(\mathbf{x}_0, t_0) \exp[c(t - t_0)] < \infty$. From (d), $\|\mathbf{x}\| \to \infty$ implies $\alpha(\|\mathbf{x}\|) = \infty$, $t = \infty$. There is no finite escape time.

Proof of Corollary B. From (d), we conclude that for any number $a > 0$, there exists a number $b > 0$ so that $\|x\| > b$ implies $V(x) > a$. Define: $c(\|\mathbf{x}\|) = \min_{y} \{V(\mathbf{y}); \|\mathbf{y}\| = \|\mathbf{x}\|\}$, $\alpha(\|\mathbf{x}\|) = \min_{y} \{V(\mathbf{y}); \|\mathbf{x}\| \leq \|\mathbf{y}\| \leq b[c(\|\mathbf{x}\|)]\}$, $\beta(\|\mathbf{x}\|) = \max_{y} \{V(\mathbf{y}); \|\mathbf{y}\| \leq \|\mathbf{x}\|\}$, $\gamma(\|\mathbf{x}\|) = \min_{y} \{-\dot{V}(\mathbf{y}); \|\mathbf{y}\| = \|\mathbf{x}\|\}$. Since the defined functions α, β, and γ satisfy the assumptions of Theorem H.1, Corollary B is proved.

Let us now consider the case when there are perturbations in the system and it is described by the equation

$$\dot{\mathbf{x}} = \mathbf{X}(\mathbf{x}, t) + \mathbf{Y}(\mathbf{x}, t), \qquad (\text{H.6})$$

where we indeed have no information about $\mathbf{Y}$ except that it is "hopefully" not excessively large. Then the definition of the total stability is useful.

Let us assume that the right-hand side of (H.6) is such that the existence and uniqueness of the solutions $\mathbf{x}(t, \mathbf{x}_0, t_0)$, as well as their continuous dependence on the initial values is assured (see footnote 3 in Section 8.2). Furthermore, let $\mathbf{X}(\mathbf{0}, t) = \mathbf{0}$ for all $t \geq t_0$. Then we arrive at:

Definition H.2. Total Stability. The equilibrium $\mathbf{x} = \mathbf{0}$ of (H.6) is called totally stable if, for every $\varepsilon > 0$, two positive numbers $\delta_1(\varepsilon)$ and $\delta_2(\varepsilon)$ can be found such that for every solution $\mathbf{x}(t, \mathbf{x}_0, t_0)$ of (H.6) the inequality $\|\mathbf{x}(t, \mathbf{x}_0, t_0)\| < \varepsilon$, $(t > t_0)$ holds, whenever $\|\mathbf{x}_0\| < \delta_1$ and $\|\mathbf{Y}(\mathbf{x}, t)\| < \delta_2$ on the domain $D(\varepsilon, t_0)$: $\|\mathbf{x}\| < \varepsilon$, $t > t_0$.

To guarantee the total stability it is sufficient to prove that the unperturbed system $\dot{\mathbf{x}} = \mathbf{X}(\mathbf{x}, t)$ is uniformly asymptotically stable. The following corollary of the Liapunov stability theorem was proved by Malkin [H.73].

Corollary C. Malkin. If a positive-definite Liapunov function $V(\mathbf{x}, t)$ exists whose partial derivatives are bounded in a domain D and whose total derivative $\dot{V}_1$ for unperturbed system $\dot{\mathbf{x}} = \mathbf{X}(\mathbf{x}, t)$ is negative-definite, the equilibrium $\mathbf{x} = \mathbf{0}$ is totally stable.

According to the theorem there exist three real continuous nondecreasing scalar functions $\alpha(\|\mathbf{x}\|)$, $\beta(\|\mathbf{x}\|)$, and $\gamma(\|\mathbf{x}\|)$ defined on a closed interval $0 \leq \|\mathbf{x}\| \leq \varepsilon$ so that $\alpha(0)$, $\beta(0)$, $\gamma(0)$ are equal to zero and $\alpha \leq V \leq \beta$, $\dot{V}_1 \leq -\gamma$, where $\dot{V}_1$ corresponds to the unperturbed system ($\mathbf{Y} \equiv \mathbf{0}$). Suppose $\varepsilon < 0$ is given and let $0 < \xi < \alpha(\varepsilon)$. Then there is a number $\eta = \eta(\xi) < 0$ such that $V(\mathbf{x}^0, t) = \xi$ implies $\eta < \|\mathbf{x}^0\| < \varepsilon$ and, in addition, $\dot{V}_1(\mathbf{x}^0, t) < -\gamma(\eta)$. The function $\dot{V}_2$ of the perturbed system (H.6) differs from $\dot{V}_1$ by the term $\sum_{i=1}^{n} Y_i(\partial V/\partial x_i)$, which, because of the bounded partial derivatives of V, can be made arbitrarily small by appropriate choice of δ_2 in the above definition. Thus, for sufficiently small δ_2, we also have

$\dot V_2 < 0$ in the domain $\|\mathbf{x}\| < \delta_2$. Now, choose $\delta_1 = \delta_1(\varepsilon)$ so that $\delta_1 < \varepsilon$ and $V(t_0) = V(\mathbf{x}_0, t_0) < \xi$ when $\|\mathbf{x}_0\| < \delta_1$. Then $\|\mathbf{x}(t, \mathbf{x}_0, t_0)\| < \varepsilon$ for $t > t_0$. If this last inequality were not true for $t_1 > t_0$, we would have $V(t_1) > \xi$ since $\xi < \alpha(\varepsilon)$. However, $V(t_0) < \xi$, and because $\dot V_2 < 0$, V decreases monotonically this proves Malkin's result.

H.3 Yakubovich-Kalman Lemma

On the basis of the above theorem and the corollaries it may be concluded that an appropriate Liapunov function guarantees the stability of a given system. A major factor holding back the application of the theorem is the lack of methods for finding such a function in the general case. For a class of systems, however, that can be described by the nonlinear differential equations

$$\dot{\mathbf{x}} = \mathbf{P}\mathbf{x} + \mathbf{q}F(y), \qquad y = \mathbf{r}^T\mathbf{x}, \tag{H.7}$$

as discussed in Chapter 8, an appropriate type of Liapunov function $V = V(\mathbf{x}, y)$ proposed by Lur'e is available in the form

$$V = \mathbf{x}^T \mathbf{H} \mathbf{x} + q \int_0^y F(y)\,dy, \tag{H.8}$$

which is called "a quadratic form plus an integral involving the nonlinearity." Lur'e [H.2, 3] obtained a finite system of quadratic equations that may be written directly from the given differential equation, and showed that if this system of equations, which he called the *resolving equations*, has real roots, then this is a sufficient condition for the absolute stability of the system (H.7). This Lur'e approach has been extended to various cases of (H.7) by different authors, as discussed in the survey in Section H.6.

A significant change in the course of developing solution method for the "Lur'e problem" occurred when V. M. Popov [H.4–9] proposed a solution of the problem in the frequency domain. Popov's main results are given in Section 8.6, where it is pointed out that an essential role in concluding absolute stability is played by the inequality

$$\pi(\omega) \equiv \frac{1}{k} + \text{Re}\,(1 + jq\omega)G(j\omega) > 0, \qquad \text{for all real } \omega \geq 0, \tag{H.9}$$

where $G(s) = \mathbf{r}^T(\mathbf{P} - s\mathbf{I})^{-1}\mathbf{q}$, and $s = j\omega$. Inequality H.9 was derived without use of Liapunov functions. Popov also proved that inequality H.9 is satisfied if there exists a positive-definite Liapunov function ($V > 0$) of the form (H.8) whose derivative along the solutions of (H.7) is negative definite ($\dot V < 0$).

Several authors have proved partial converses to Popov's theorem on the existence of Liapunov functions. By strengthening a theorem of Yakubovich

[H.10] based on $\pi(\omega) > 0$, Kalman [H.11] succeeded in essentially completing Popov's theorem to a necessary and sufficient condition for the case $\pi(\omega) \geq 0$ by requiring that the system be completely controllable and observable. This last requirement of Kalman's can be avoided with some care, as shown by Meyer [H.12].

In the following discussion the proof of the Yakubovich lemma that establishes the converse to the Popov theorem or existence of Liapunov functions will be given for the principal case of (H.7) using the Lefschetz [H.13] version of the proof. That lemma is used in Chapter 8 to conclude that $\pi(\omega) > 0$ plus a very simple restriction provide necessary and sufficient conditions for V and $-\dot{V}$ both positive-definite and also absolute stability. The simplest particular case ($\pi \geq 0$, $\dot{V} \leq 0$) is solved by Kalman and can be found in reference H.11. Other particular cases can be treated as suggested in Chapter 8 or in references H.14, and 15.

Yakubovich-Kalman Lemma H.1. Let P be a real $n \times n$ matrix all of whose characteristic roots have negative real parts; let $\mathbf{Q}$ be a real symmetric positive definite $n \times n$ matrix ($\mathbf{Q} > 0$); let $\mathbf{q} \neq 0$ and $\mathbf{h}$ be two real n vectors; and let $\varepsilon > 0$, $\gamma \geq 0$ be two real scalars. Then a necessary and sufficient condition for existence of a solution as a real $n \times n$ matrix $\mathbf{H}$ (necessarily > 0) and a real n vector $\mathbf{g}$ of the system

$$\mathbf{HP} + \mathbf{P}^T\mathbf{H} = -\mathbf{g}\mathbf{g}^T - \varepsilon \mathbf{Q} \tag{H.10a}$$

$$\mathbf{Hq} + \mathbf{h} = -\sqrt{\gamma}\,\mathbf{g} \tag{H.10b}$$

is that ε be small enough and that the inequality

$$\gamma + 2\,\mathrm{Re}\,\mathbf{h}^T(\mathbf{P} - j\omega\mathbf{I})^{-1}\mathbf{q} > 0 \tag{H.11}$$

be satisfied for all real ω.

With $-\mathbf{G} = \mathbf{HP} + \mathbf{P}^T\mathbf{H}$ the condition (H.10a) can be readily identified as the quadratic inequality $\mathbf{G} - \mathbf{g}\mathbf{g}^T > 0$ used in the theorems of Chapter 8. In addition, note that (H.10a) and the Hurwitz property of $\mathbf{P}$ imply that $\mathbf{H}$ is symmetric, positive-definite matrix [H.13]. Later, inequality H.11 will be identified as Popov inequality H.9.

To prove the lemma, start by setting $\mathbf{n}(j\omega) = \mathbf{P}_{j\omega}^{-1}\mathbf{q}$, $\mathbf{P}_{j\omega}^{-1} = (\mathbf{P} - j\omega\mathbf{I})^{-1}$, where $\mathbf{n}(j\omega)$ is a complex vector function of ω. Using this notation, we may rewrite (H.11) in the form

$$\gamma + \mathbf{n}^*\mathbf{h} + \mathbf{h}^T\mathbf{n} > 0, \tag{H.12}$$

where $\mathbf{n}^*$ is the conjugate transpose of $\mathbf{n}$ (that is, for $\mathbf{n} = (n_{ik})$ we have $\mathbf{n}^* = (\bar{n}_{ki})^T$ as the conjugate transpose of $\mathbf{n}$).

The identity

$$\mathbf{HP}_{j\omega} + \mathbf{P}_{j\omega}^*\mathbf{H} = \mathbf{HP} + \mathbf{P}^T\mathbf{H} \tag{H.13}$$

is needed first. This can be derived by adding and subtracting $j\omega \mathbf{H}$ to $\mathbf{HP} + \mathbf{P}^T\mathbf{H}$, that is,

$$\mathbf{HP} + \mathbf{P}^T\mathbf{H} = \mathbf{HP} + \mathbf{P}^T\mathbf{H} + j\omega\mathbf{H} - j\omega\mathbf{H} = \mathbf{H}(\mathbf{P} - j\omega\mathbf{I}) + (\mathbf{P}^T + j\omega\mathbf{I})\mathbf{H}$$
$$= \mathbf{HP}_{j\omega} + \mathbf{P}^*_{j\omega}\mathbf{H}.$$

Premultiplying (H.13) by $\mathbf{q}^T\mathbf{P}^{*-1}_{j\omega}$ and postmultiplying by $\mathbf{P}_{j\omega}^{-1}\mathbf{q}$, then taking into account (H.10), we obtain the relationship

$$\mathbf{n}^*\mathbf{Hq} + \mathbf{q}^T\mathbf{Hn} = -(\mathbf{n}^*\mathbf{gg}^T\mathbf{n} + \varepsilon\mathbf{n}^*\mathbf{Qn}), \tag{H.14}$$

which will be used immediately.

Proof of Necessity. By replacing $\mathbf{Hq}$ from (H.10b) and using the identities $\mathbf{n}^*\mathbf{h} + \mathbf{h}^T\mathbf{n} = 2\,\mathrm{Re}\,\mathbf{h}^T\mathbf{n}$ and $\mathbf{n}^*\mathbf{g} + \mathbf{g}^T\mathbf{n} = 2\,\mathrm{Re}\,\mathbf{g}^T\mathbf{n}$, equation (H.14) can be rewritten as

$$2\,\mathrm{Re}\,\mathbf{h}^T\mathbf{n} = |\mathbf{g}^T\mathbf{n}|^2 - 2\sqrt{\gamma}\,\mathrm{Re}\,\mathbf{g}^T\mathbf{n} + \varepsilon\mathbf{n}^*\mathbf{Qn}, \tag{H.15}$$

where $|\mathbf{g}^T\mathbf{n}| = \mathbf{n}^*\mathbf{gg}^T\mathbf{n}$.

If we consider $\mathbf{Q}$ as a Hermitian matrix ($\mathbf{Q}^* = \mathbf{Q}$), then $\mathbf{Q}_1 = \mathbf{P}^{*-1}_{j\omega}\mathbf{Q}\mathbf{P}_{j\omega}^{-1}$ is also a Hermitian matrix derived from $\mathbf{Q}$ by the change of coordinates $\mathbf{x} \to \mathbf{P}_{j\omega}^{-1}\mathbf{x}$. Then $\mathbf{Q}_1 > 0$ as $\mathbf{Q} > 0$, and thus since $\mathbf{Q} \neq 0$,

$$\delta = \varepsilon\mathbf{q}^T\mathbf{Q}_1\mathbf{q} = \varepsilon\mathbf{n}^*\mathbf{Qn} > 0. \tag{H.16}$$

From (H.15) and (H.16) we have

$$2\,\mathrm{Re}\,\mathbf{h}^T\mathbf{n} = |\mathbf{g}^T\mathbf{n}|^2 - 2\sqrt{\gamma}\,\mathrm{Re}\,\mathbf{g}^T\mathbf{n} + \delta. \tag{H.17}$$

Adding γ to both sides of (H.17) gives

$$\gamma + 2\,\mathrm{Re}\,\mathbf{h}^T\mathbf{n} = |\mathbf{g}^T\mathbf{n} - \sqrt{\gamma}|^2 + \delta > 0, \tag{H.18}$$

which is (H.11). This proves the necessity proper.

Proof of Sufficiency. At the beginning it is necessary to establish a preliminary result. If $\mathbf{u}$ is a real constant vector so that $\mathrm{Re}\,\mathbf{u}^T\mathbf{n}(j\omega) = 0$ for any ω, then necessarily $\mathbf{u} = \mathbf{0}$.

Note that

$$G(s) = \mathbf{r}^T(\mathbf{P} - s\mathbf{I})^{-1}\mathbf{q} = \frac{C(s)}{B(s)} \tag{H.19}$$

is the transfer function of the linear part of the system. For $|s|$ large, the following expansion holds:

$$G_u(s) = \mathbf{u}^T(\mathbf{P} - s\mathbf{I})^{-1}\mathbf{q} = -\mathbf{u}^T\left(\frac{\mathbf{q}}{s} + \frac{\mathbf{Pq}}{s^2} + \frac{\mathbf{P}^2\mathbf{q}}{s^3} + \cdots\right), \tag{H.20}$$

from which it is easy to see that $G_u(s) \equiv 0$ implies $\mathbf{u} \equiv \mathbf{0}$ if and only if the vectors

$$\mathbf{q}, \mathbf{Pq}, \ldots, \mathbf{P}^{n-1}\mathbf{q} \qquad (\text{H}.21)$$

are linearily independent. Since the pair $(\mathbf{P}, \mathbf{q})$ is completely controllable, the linear independence in (H.21) is established.[5]

Moreover, it follows from equations (H.19) and (H.20) that the coefficients c_k of the polynomial $C(s)$ can be expressed as linear combination of the components u_i of the vector $\mathbf{u}$. Since $G_u(s) \equiv 0$, then $C(s) \equiv 0$ and $\mathbf{u} = \mathbf{0}$. These relationships are also reversible. Given a completely controllable system, it is possible to associate with any nondegenerate type (no common factors among the polynomials $B(s)$ and $C(s)$ in equation H.19) rational function

$$G_u(s) = \mathbf{u}^T(\mathbf{P} - s\mathbf{I})^{-1}\mathbf{q} = \frac{C(s)}{B(s)} \qquad (\text{H}.22)$$

an appropriate vector $\mathbf{u}$.

If it is known for a completely controllable system that Re $G_u(s) \equiv 0$, then $G_u(j\omega) + G_u(-j\omega) \equiv 0$, and the rational function (H.22) is odd. Since the polynomial $B(s)$ is Hurwitz, this takes place only if $C(s) \equiv 0$, which implies $\mathbf{u} = \mathbf{0}$.

We now proceed to prove the sufficiency proper. Since both $\mathbf{n}^*\mathbf{h} + \mathbf{h}^T\mathbf{u}$

---

[5] Popov [H.8] calls a system (H.7) "nondegenerate" if the corresponding transfer function $G(s)$ of (H.19) is such that the polynomials $C(s)$ and $B(s)$ have no common zeros.

If $\mathbf{P}$ and $\mathbf{q}$ are fixed and $\mathbf{r}$ is considered as a variable vector $\mathbf{u}$, then as $\mathbf{u}$ changes, so will the transfer function $G_u(s)$ relating the input $F(y)$ to the output $-y = -\mathbf{r}^T\mathbf{x}$ of the linear part of the system. The system is "completely controllable" if the identity $G_u(s) \equiv 0$ is possible only if $\mathbf{u} \equiv 0$.

A system (H.7) is "completely observable" if for fixed $\mathbf{P}$ and $\mathbf{r}$ with the substitution of variable vector $\mathbf{v}$ for fixed vector $\mathbf{q}$, the identity $G_\mathbf{v}(s) \equiv 0$ implies $\mathbf{v} \equiv 0$. Here the transfer function $G_\mathbf{v}(s)$ relates the input $F(y)$ with the output $-\mathbf{q}^T\mathbf{x}$. In analogy to the above expansion (H.20), we can show that the system is completely observable if and only if the vectors

$$\mathbf{r}, \mathbf{P}^T\mathbf{r}, \ldots, \mathbf{P}^{T(n-1)}\mathbf{r} \qquad (i)$$

are linearly independent.

Therefore [H.8], the system is nondegenerate if and only if both the vectors (H.21) and vectors (i) are linearly independent.

A simple reasoning can support the foregoing statement. If $C(s)$ and $B(s)$ of (H.19) are of degrees m and n ($m < n$) and they have common factors that can cancel, then the system becomes of order $n - k$ ($0 < k \leq m$), which is lower than n. Such a system is not completely controllable, as can be seen from the following. Suppose that the system is completely controllable. It means that there always exists an input that can force the system to move from the set of all initial states to the origin of the state space. This set has a dimension less than n. This contradicts the assumption that the system is of order n, which proves the above statement. A similar argument can be applied to prove the corresponding statement concerning the observability of the system.

Proof of the Stability Theorems

and $\mathbf{n}^*\mathbf{Q}\mathbf{n}$ are real rational functions of ω with numerators of degree $\leq n - 1$ and denominator of degree n, both go to zero when $\omega \to \infty$. The functions are also continuous for ω finite, and hence they have finite upper and lower bounds. Let μ be the upper bound of $\mathbf{n}^*\mathbf{Q}\mathbf{n}$ and ν the lower bound of $\mathbf{n}^*\mathbf{h} + \mathbf{h}^T\mathbf{n}$. From $\mathbf{n}^*\mathbf{Q}\mathbf{n} < 0$ for all finite ω, we have $\mu > 0$. Therefore

$$\gamma + \mathbf{n}^*\mathbf{h} + \mathbf{h}^T\mathbf{n} - \varepsilon \mathbf{n}^*\mathbf{Q}\mathbf{n} \geq \gamma + \nu - \varepsilon\mu. \tag{H.23}$$

From (H.11), $\gamma + \nu > 0$. By choosing $\varepsilon < \tfrac{1}{2}[(\gamma + \nu)/\mu]$, we get

$$\gamma + \mathbf{n}^*\mathbf{h} + \mathbf{h}^T\mathbf{n} - \varepsilon \mathbf{n}^*\mathbf{Q}\mathbf{n} > 0. \tag{H.24}$$

Let us now express the left-hand side of (H.24) as

$$\gamma + \mathbf{n}^*\mathbf{h} + \mathbf{h}^T\mathbf{n} - \varepsilon \mathbf{n}^*\mathbf{Q}\mathbf{n} = \frac{\xi(j\omega)}{B(j\omega)}$$

$$= \frac{\xi(j\omega)B(-j\omega)}{B(j\omega)B(-j\omega)}$$

$$= \frac{\eta(j\omega)}{B(j\omega)B(-j\omega)}, \tag{H.25}$$

where $\eta(s)$ is a polynomial with real coefficients, of degree $\leq 2n$, which has positive real values for any ω, that is, $\eta(j\omega) > 0$ for $-\infty < \omega < +\infty$. Then $\eta(s)$ is an even polynomial, that is, $\eta(-s) = \eta(s)$. Moreover, the polynomial $\eta(s)$ has pure imaginary roots $j\omega_0$ only, with even multiplicity k. To see that, let $\eta(s) = (s^2 + \omega_0^2)^k \theta(s)$, where $\theta(j\omega_0) \neq 0$. Then $\eta(j\omega) = (\omega_0^2 - \omega^2)^k \theta(j\omega)$, and the inequality $\eta(j\omega) > 0$ ($-\infty < \omega < +\infty$) follows only if the number k is even.

By using this fact and the fact that $\eta(j\omega) > 0$ for all real ω, we decompose the polynomial $\eta(s)$ into

$$\eta(s) = \phi(s)\phi(-s), \tag{H.26}$$

where $\phi(s)$ is a polynomial with real coefficients and degree $\leq n$ (for a more detailed discussion see reference H.8, Lemma 1).

The identity (H.25) then becomes

$$\gamma + \mathbf{n}^*\mathbf{h} + \mathbf{h}^T\mathbf{n} - \varepsilon \mathbf{n}^*\mathbf{Q}\mathbf{n} = \left|\frac{\phi(j\omega)}{B(j\omega)}\right|^2. \tag{H.27}$$

Since the left-hand side of this identity tends to γ as $\omega \to \infty$, we have

$$\frac{\phi(j\omega)}{B(j\omega)} = \sqrt{\gamma} + \frac{\phi_1(j\omega)}{B(j\omega)}, \tag{H.28}$$

where $\phi_1(j\omega)$ is a real polynomial of degree $\leq n - 1$. Since $\phi_1(j\omega)/B(j\omega)$ is a nondegenerate function, it can be uniquely expressed as

$$\frac{\phi_1(j\omega)}{B(j\omega)} = G_u(j\omega). \qquad (H.29)$$

Now, we can write

$$\gamma + \mathbf{n}^*\mathbf{h} + \mathbf{h}^T\mathbf{n} - \varepsilon\mathbf{n}^*\mathbf{Q}\mathbf{n} = \left(\frac{\phi_1(j\omega)}{B(j\omega)} + \sqrt{\gamma}\right)^*\left(\frac{\phi_1(-j\omega)}{B(-j\omega)} + \sqrt{\gamma}\right). \qquad (H.30)$$

If $g_1, g_2, \ldots, g_n$ are the real coefficients of $\phi_1(j\omega)$, choose the vector $\mathbf{g} = \{-g_1, -g_2, \ldots, -g_n\}$. Once $\mathbf{g}$ is found, the matrix $\mathbf{H}$ is obtained by (H.10a) of the lemma. Since $\mathbf{P}$ is a Hurwitz matrix, and $\mathbf{G} = \mathbf{gg}^T + \varepsilon\mathbf{Q}$ is positive-definite, then the matrix $\mathbf{H}$ is necessarily positive-definite ($\mathbf{H} > 0$).

The above choice of $\mathbf{g}$ may seem rather arbitrary. However, this $\mathbf{g}$ is now shown to also satisfy (H.10b) of the lemma. Referring to (H.29), it is seen that

$$\frac{\phi_1(j\omega)}{B(j\omega)} = -\mathbf{g}^T\mathbf{n}. \qquad (H.31)$$

Therefore, (H.30) and (H.14) yield for the chosen $\mathbf{g}$

$$\begin{aligned}\mathbf{n}^*\mathbf{h} + \mathbf{h}^T\mathbf{n} - \varepsilon\mathbf{n}^*\mathbf{Q}\mathbf{n} &= (\mathbf{n}^*\mathbf{g} - \sqrt{\gamma})(\mathbf{g}^T\mathbf{n} - \sqrt{\gamma}) - \gamma \\ &= \mathbf{n}^*\mathbf{gg}^T\mathbf{n} - \sqrt{\gamma}\,(\mathbf{n}^*\mathbf{g} + \mathbf{g}^T\mathbf{n}).\end{aligned} \qquad (H.32)$$

Thus, for any ω

$$\mathbf{n}^*(-\mathbf{Hq} - \mathbf{h} - \sqrt{\gamma}\,\mathbf{g}) + (-\mathbf{Hq} - \mathbf{h} - \sqrt{\gamma}\,\mathbf{g})^T\mathbf{n}$$
$$= 2\,\mathrm{Re}\,(-\mathbf{Hq} - \mathbf{h} - \sqrt{\gamma}\,\mathbf{g}) = 0. \qquad (H.33)$$

The vector in parentheses is real and it must vanish. Hence, (H.10b) is satisfied. In other words, a solution $(\mathbf{H}, \mathbf{g})$ has been constructed for equations (H.10) and the sufficiency of the inequality H.11 is thereby proved. This completes the proof of the Yakubovich-Kalman lemma.

It is important to note that, in case $\gamma = 0$, the inequality (H.11) should be supplemented with the requirement

$$\lim_{\omega \to \infty} \omega^2 \,\mathrm{Re}\, \mathbf{r}^T(\mathbf{P} - j\omega\mathbf{I})^{-1}\mathbf{q} > 0. \qquad (H.34)$$

Premultiplying identity H.13 by $[(\mathbf{P} - j\omega\mathbf{I})\mathbf{q}^{-1}]^T$, postmultiplying by $(\mathbf{P} - j\omega\mathbf{I})\mathbf{q}^{-1}$, and taking into account (H.10b) with $\gamma = 0$, we have $2\,\mathrm{Re}\,\mathbf{h}^T(\mathbf{P} - j\omega\mathbf{I})^{-1}\mathbf{q} \equiv [(\mathbf{P} - j\omega\mathbf{I})\mathbf{q}^{-1}]^T\mathbf{G}[(\mathbf{P} - j\omega\mathbf{I})\mathbf{q}^{-1}]$, where $-\mathbf{G} = \mathbf{HP} + \mathbf{P}^T\mathbf{H}$. Thus, $\mathbf{h}^T\mathbf{Pq} = \lim_{\omega \to \infty} \omega^2 \,\mathrm{Re}\,\mathbf{h}^T(\mathbf{P} - j\omega\mathbf{I})^{-1}\mathbf{q} = \frac{1}{2}\mathbf{q}^T\mathbf{Gq}$ and the necessity of the supplement (H.34) in the Yakubovich-Kalman lemma when $\gamma = 0$ is evident.

Proof of the Stability Theorems

It is now necessary to show that the inequality

$$\gamma + 2 \operatorname{Re} \mathbf{h}^T(\mathbf{P} - j\omega \mathbf{I})\mathbf{q} > 0 \qquad \text{(H.11)}$$

is actually the Popov inequality

$$\frac{1}{k} + \operatorname{Re}(1 + jq\omega)G(j\omega) > 0. \qquad \text{(H.9)}$$

Define

$$\mathbf{h} = \tfrac{1}{2}(q\mathbf{P}^T + \mathbf{I})\mathbf{r} \qquad \text{(H.35)}$$

as required by (8.93), and from the same equations (8.93) notice that

$$\gamma = -q\mathbf{r}^T\mathbf{q} + \frac{1}{k}. \qquad \text{(H.36)}$$

Substituting (H.35) and (H.36) in (H.11), we have

$$\frac{1}{k} - q\mathbf{r}^T\mathbf{q} + \operatorname{Re}[q\mathbf{r}^T\mathbf{P}(\mathbf{P} - j\omega\mathbf{I})^{-1}\mathbf{q} + \mathbf{R}^T(\mathbf{P} - j\omega\mathbf{I})^{-1}\mathbf{q}] > 0. \qquad \text{(H.37)}$$

Postmultiplying $\mathbf{P}$ by $(\mathbf{P} - j\omega\mathbf{I})^{-1}$ yields

$$\mathbf{P}(\mathbf{P} - j\omega\mathbf{I})^{-1} = \mathbf{I} + j\omega(\mathbf{P} - j\omega\mathbf{I})^{-1}. \qquad \text{(H.38)}$$

Thus from (H.37) we have

$$\frac{1}{k} + \operatorname{Re}[(1 + jq\omega)\mathbf{r}^T(\mathbf{P} - j\omega\mathbf{I})^{-1}\mathbf{q}] > 0. \qquad \text{(H.39)}$$

But $G(j\omega) = \mathbf{r}^T(\mathbf{P} - j\omega\mathbf{I})^{-1}\mathbf{q}$ and (H.39) becomes the Popov inequality H.9.

Since the Popov inequality, if satisfied, guarantees that V and $-\dot{V}$ are both positive definite, all that is now needed to conclude the absolute stability using the Liapunov stability theorem is the verification of condition (d) of Theorem 8.1, which is usually called the "Barbashin-Krassovsky complement" [H.16]. It is, therefore, left to show that $V \to \infty$ with $\|\mathbf{x}\| + |y| \to \infty$. Owing to $\mathbf{H} > 0$ and the property of the function $F(y)$ stated in (8.75) that the integral in (H.8) diverges when $y \to \pm\infty$, this complement is satisfied.

In the *simplest particular case* of (H.7), it was shown by Kalman [H.11] that the inequality sign ">" in (H.11) can be replaced by "$\geq$." This leads to $V > 0$ and $\dot{V} \leq 0$ and to assure absolute stability of system (H.11) for any $F(y)$ satisfying $0 < yF(y) < ky^2$, ($k \leq \infty$), Kalman showed that every solution of (H.7) is bounded and that the only invariant set of the Liapunov function V in $\dot{V} = 0$ is the origin of the state space—the unique equilibrium position of (H.7). Thus all solutions $\mathbf{x}(t) \to 0$ as $t \to \infty$, and system (H.7) is absolutely stable. It is, however, only fair to say that most of the difficulties in Kalman's extension of the Yakubovich lemma [H.10] were caused by weakening ">" to "$\geq$" ($\pi \geq 0$, $\dot{V} \leq 0$).

In the *general particular case* we rely on the results of Aizerman and Gantmacher [H.14], who extended the Popov criterion to that case, as shown in Section 8.7. These results were refined by Yakubovich [H.15] and the connection between the Liapunov function and the Popov criterion was established in the similar manner as for the principal case.

H.4 Discontinuous Nonlinearities

Discontinuous nonlinear characteristics give rise to new phenomena not observed in the systems with continuous characteristics. Of particular interest is the *sliding motion*, sometimes called the *chattering*. This motion gives rise to a more general definition of the solution $\mathbf{x}(t)$ of the equation

$$\dot{\mathbf{x}} = \mathbf{X}(\mathbf{x}, t), \tag{H.40}$$

referred to as *the solution in the Filippov sense* [H.17, 18]. This redefinition of the solution of (H.40) is necessary since it is employed in the absolute stability analysis of discontinuous systems considered in the following section of this appendix.

Before the solution is redefined, let us illustrate through some simple examples the sliding motion phenomena.

Consider the system equations:

$$\begin{aligned} \dot{x}_1 &= x_2 \\ \dot{x}_2 &= -x_1 - F(y), \qquad y = -x_1 - x_2, \end{aligned} \tag{H.41}$$

where

$$F(y) = \operatorname{sign} y \tag{H.42}$$

(see Figure 3.3a) and

$$\operatorname{sign} y = \begin{cases} +1, & \text{for } y > 0 \\ -1, & \text{for } y < 0 \end{cases} \tag{H.43}$$

sign 0 is undefined.

The motion is simply determined by integrating (H.41) with (H.42) and (H.43),

$$(x_1 + 1)^2 + x_2^2 = C^2, \qquad D^+\{y > 0\} \tag{H.44a}$$

$$(x_1 - 1)^2 + x_2^2 = C^2, \qquad D^-\{y < 0\}. \tag{H.44b}$$

In the phase $x_1 x_2$ plane, equations (H.44) represent the two family of circles crossed by the *switching line*

$$y = -x_1 - x_2 = 0 \tag{H.45}$$

on which the function $F(y)$ switches from $+1$ to -1 and vice versa. This line $y = 0$ divides the phase plane into domains D^+ and D^-, in which the system motion is well defined by (H.44), as shown in Figure H.1.

540 Proof of the Stability Theorems

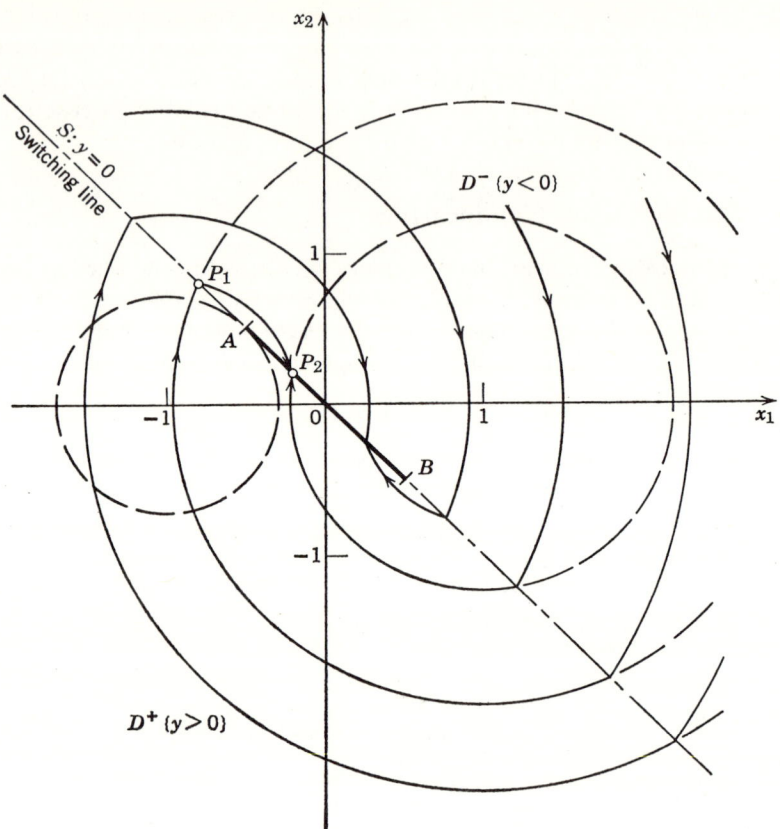

Figure H.1 Phase-plane portrait.

On the switching line S, however, there are two distinct situations illustrated by points P_1 and P_2, which need special attention. At the point P_1, the system switches from one type of motion D^+ to D^- and P_1 is called the *transition point*. At point P_2 such transition is impossible since the trajectories from both domains D^+ and D^- are directed toward the switching line. The point (x_1, x_2) cannot leave the line S. This will apparently happen whenever y and $\dot{y}$ have opposite signs on both sides of the switching line S. The point P_2 is called the *end point*.

In the specific system under consideration, the set of end points on the line S is determined by the condition

$$|x_1 - x_2| < 1 \tag{H.46}$$

and is represented by the segment AB in Figure H.1.

It is of interest to investigate the "after-end-point motion" once the point

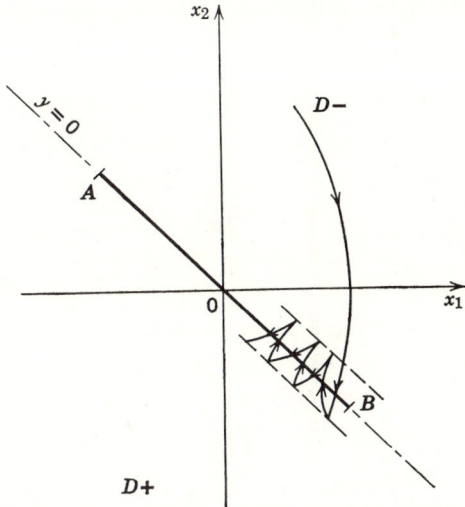

Figure H.2 Sliding motion—"chattering."

(x_1, x_2) falls on the line segment AB. It is impossible for the point to continue the motion along the line $y = 0$ since it is necessary that y change its sign by crossing the zero value. This is impossible and the remaining motion is undefined. The motion must continue since the only equilibrium point is at the origin of the $x_1 x_2$ plane. Indeed, the actual motion in the real system continues because of imperfections of the switching device. Namely, $F(y) = \text{sign } y$ will switch some time after $y = 0$, and the relay will start to "chatter" along the segment AB, as shown in Figure H.2. The motion has a relatively high frequency and tends, on the average, toward the origin. This motion is determined by the equation

$$y = -x_1 - x_2 = 0, \tag{H.47}$$

which, since $x_2 = \dot{x}_1$, has the solution

$$x_1 = Ce^{-t}, \tag{H.48}$$

where C is the initial value $x_{10} = x_1(0)$ on the segment AB.

A somewhat different situation arises when the nonlinear characteristic $F(y)$ is changed to

$$F(y) = \begin{cases} \text{sign } y, & \text{for } |y| > 1 \\ 0, & \text{for } |y| < 1 \end{cases} \tag{H.49}$$

and for $y = \pm 1$ it is not defined. The function $F(y)$ of (H.49) represents a relay with the dead zone.

Proof of the Stability Theorems

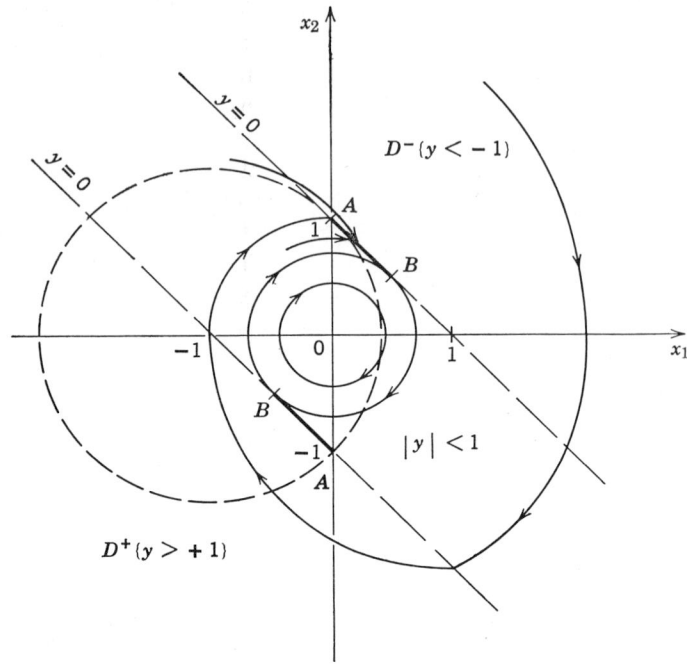

Figure H.3 Relay with the dead zone—phase portrait.

The phase portrait of the modified system is given in Figure H.3 according to

$$(x_1 + 1)^2 + x_2^2 = C^2, \quad y > 1 \quad \text{(H.50}a\text{)}$$

$$(x_1 - 1)^2 + x_2^2 = C^2, \quad y < 1 \quad \text{(H.50}b\text{)}$$

$$x_1^2 + x_2^2 = C^2, \quad y = 1. \quad \text{(H.50}c\text{)}$$

The chattering motion occurs on the segments AB. The chattering ends on a trajectory instead of at the singular point as previously.

Now, the idea of the solution of

$$\dot{\mathbf{x}} = \mathbf{X}(\mathbf{x}, t) \quad \text{(H.40)}$$

can be generalized to incorporate the sliding motion when the right-hand side $\mathbf{X}(\mathbf{x}, t)$ is a discontinuous function in $\mathbf{x}$.

Consider (H.40) in the domain D of the state space $\mathscr{X}$ consisting of all vectors $\mathbf{x}$, and let the domain D be divided by a continuous differentiable surface S into two domains D^+ and D^-. In the domains $D^+ + S$ and $D^- + S$ let there be given vector functions $\mathbf{X}^+$ and $\mathbf{X}^-$ that are continuous in t and continuously differentiable with respect to $\mathbf{x}$; for equation H.40, $\mathbf{X} = \mathbf{X}^+$ in

Discontinuous Nonlinearities 543

D^+ and $\mathbf{X} = \mathbf{X}^-$ in D^-. In this case the solution of (H.40) passing through domain D is continuous in a well-defined manner until it reaches the surface S. If the vector $\mathbf{X}$ in domain D^- is directed towards surface S, and in the domain D^+ from surface S, then the solution passes from D^- to D^+ through the transition point P on S (Figure H.4a).

In the case shown in Figure H.4b, we have the starting point P, which can be interpreted without any further difficulties.

If the vector $\mathbf{X}$ is directed toward S in both the domains D^+ and D^-, then we have the end point P on S (Figure H.4c). In this case we assume that for the real system the transition from $\mathbf{X}^+$ to $\mathbf{X}^-$ does not take place at the moment of reaching surface S but after some delay, owing to which the solution $\mathbf{x}(t)$ oscillates about the surface S as illustrated in Figure H.2. It can be shown

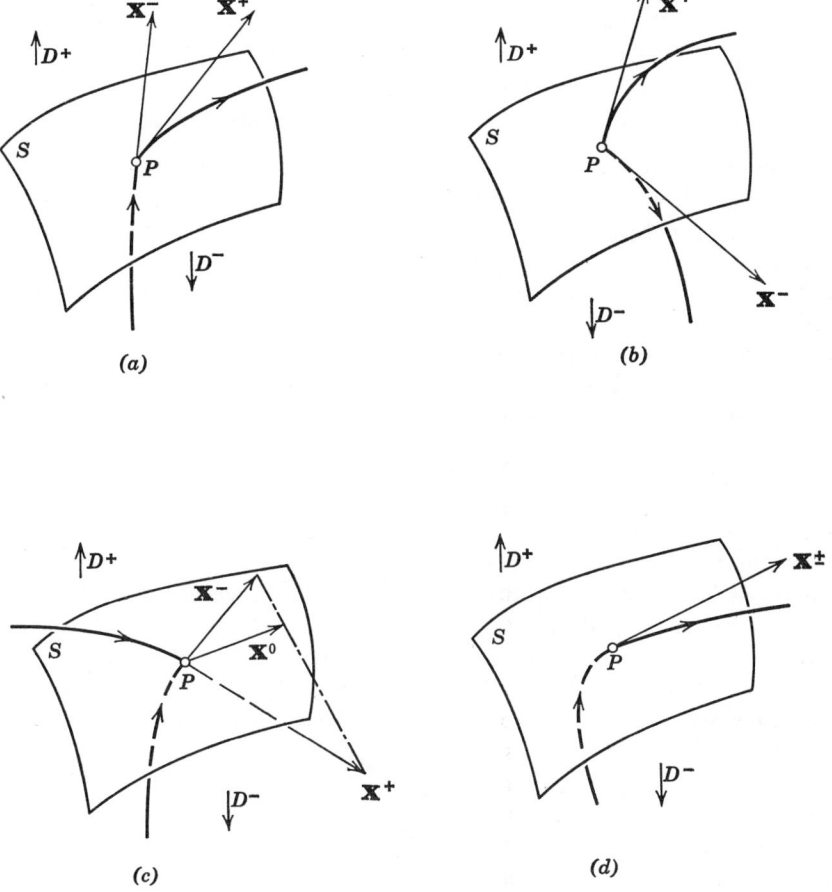

Figure H.4 Intersecting the switching surface.

[H.17, 18] that as the delay tends to zero, limit motion and velocity along surface S are

$$\dot{\mathbf{x}} = \mathbf{X}^0(\mathbf{x}, t) \equiv \alpha \mathbf{X}^+(\mathbf{x}, t) + (1 - \alpha)\mathbf{X}^-(\mathbf{x}, t), \qquad \text{(H.51)}$$

where $\mathbf{x} \in S$, and the number $\alpha (0 \leq \alpha \leq 1)$ so that the vector $\mathbf{X}^0$ is tangent to the surface S at point $\mathbf{x}$. The vector $\mathbf{X}^0$ in (H.51) can be constructed as follows. At the point $\mathbf{x} \in S$ we construct the vectors $\mathbf{X}^+$ and $\mathbf{X}^-$ and join their ends by a segment. The point of intersection with tangent plane to S at point $\mathbf{x}$ is the end point P of the required vector $\mathbf{X}^0$ (Figure H.4c). The function $\mathbf{x}(t)$ that satisfies (H.40) in D^+ and D^- and (H.51) on surface S is called *the solution of equation* $\dot{\mathbf{x}} = \mathbf{X}(\mathbf{x}, t)$ *in the Filippov sense*.

Before a precise definition of the solution is given, note that there is another possibility of sliding as shown in Figure H.4d. In this case the trajectory after *tangent point P* lies entirely on the switching surface S, that is, it coincides with the surface S. This case is particularly significant in optimal systems [H.19]. Other situations may be constructed, but they are essentially a variation of those presented in Figure H.4.

Now the precise definition of the solution in the Filippov sense is as follows.

Definition H.3. Solution in the Filippov Sense. Let the vector function $\mathbf{X}(\mathbf{x}, t)$, where $\mathbf{X}$ and $\mathbf{x}$ are n vectors, be measurable in domain Δ of space $\{\mathbf{x}, t\}$ and for any closed bounded domain[6] $\Delta \subset D$ let there exist a summable function $M(t)$ so that almost everywhere in Δ we have $|\mathbf{X}(\mathbf{x}, t)| \leq M(t)$. The absolutely continuous vector function $\mathbf{x}(t)$ is called the solution of equation $\dot{\mathbf{x}} = \mathbf{X}(\mathbf{x}, t)$ if for almost all t the vector $\dot{\mathbf{x}}$ belongs to the least convex closed set containing all the limiting values of vector $\mathbf{X}(\mathbf{x}^0, t)$, where $\mathbf{x}^0$ tends toward $\mathbf{x}$ in an arbitrary manner, and the values of function $\mathbf{X}(\mathbf{x}^0, t)$ on a set of zero-measure in the space $\mathscr{X}$ are ignored.

In particular, for the case shown in Figure H.4c, the values of function $\mathbf{X}(\mathbf{x}, t)$ on surface S must be ignored. Then the limiting values of vector $\mathbf{X}(\mathbf{x}^0, t)$ as $\mathbf{x}^0 \to \mathbf{x} \in S$ are simply the vectors $\mathbf{X}^+$ and $\mathbf{X}^-$, whereas the least convex closed set containing them is the segment joining their ends as given in Figure H.4c. Since the solution cannot leave S, the vector $\dot{\mathbf{x}}$ must be tangent to S. Hence its end lies at the point of intersection of the tangent to S and the above segment, that is, $\dot{\mathbf{x}} = \mathbf{X}^0$. The justification of such a definition of the solution is based essentially upon the introduction of a delay present in the real systems. The actual motion is then as shown in Figure H.2.

Similarly the vector $\dot{\mathbf{x}}$ can be defined in cases when there are several intersecting switching surfaces of discontinuity of the function $\mathbf{X}$, but then it is not always defined uniquely [H.18].

More about the differential equations with discontinuous right-hand sides can be found in references H.20–23.

---

[6] The symbol "$\subset$" means, in $\Delta \subset D$, that Δ is a subset of D.

H.5 Proofs of Yakubovich's Stability Theorems

In this section the absolute stability theorems 8.6–8.8 and H.1 are proved following Yakubovich [H.24]. Other absolute stability Theorems 8.2–8.6 in Chapter 8 are either direct consequence of these four theorems or are readily proved in Chapter 8 following the results of this section.

Before the theorems are proved, certain auxiliary propositions must be discussed.

Let us consider the system

$$\dot{\mathbf{x}} = \mathbf{X}(\mathbf{x}, t) \tag{H.40}$$

with function $\mathbf{X}(\mathbf{x}, t)$ bounded, when the arguments vary in a bounded region in which $\mathbf{X}(\mathbf{x}, t)$ is a measurable function of $\mathbf{x}$. The solution of the system (H.40) will be considered in the sense of Filippov, which allows $\mathbf{X}(\mathbf{x}, t)$ to be a discontinuous function. The following two lemmas due to Yakubovich [H.24] will incorporate the discontinuity property of $\mathbf{X}(\mathbf{x}, t)$ with respect to $\mathbf{x}$ and will serve as a basis for proving the stability theorems of Chapter 8. The two lemmas are based upon the work of Luzin [H.25], Demidovich [H.26, 27], and Yoschizawa [H.28].

Yakubovich Lemma H.2. Let us assume that there is a function $V(\mathbf{x})$ which satisfy the global Lipschitz condition in every bounded region of the state space $\mathscr{X}$, and:

(a) There is a constant ν and a continuous function $\gamma(\|\mathbf{x}\|) > 0$ given for $\|\mathbf{x}\| \geq \nu$ so that along any solution $\mathbf{x}(t, \mathbf{x}_0, t_0)$ of (H.40) we have almost everywhere the derivative $\dot{V}(\mathbf{x}) \leq -\gamma(\|\mathbf{x}\|)$ whenever $\|\mathbf{x}\| \geq \nu$. Let us also choose a positive constant β and define a bounded region

$$R = \{V(\mathbf{x}) \leq \beta\} \supset \{\|\mathbf{x}\| \leq \nu\}; \tag{H.52}$$

(b) $V(\mathbf{x}) \to \infty$ when $\|\mathbf{x}\| \to \infty$.

Then:

(1) R is an invariant region for the system (H.40), that is, $\mathbf{x}_0 \in R$ implies $\mathbf{x}(t, \mathbf{x}_0, t_0) \in R$ for all $t \geq t_0$; and

(2) Every solution $\mathbf{x}(t, \mathbf{x}_0, t_0)$ starting at any fixed $\mathbf{x}_0, t_0$ reaches the region R.

Since the invariant region R is bounded, the lemma provides the *ultimate boundedness* property of all the solutions of (H.40).

Proof of Lemma H.2. (1) Let $\mathbf{x}_0 \in R$ for arbitrary t_0, and $\mathbf{x}(t_1, \mathbf{x}_0, t_0) \notin R$ for some $t_1 > t_0$.[7] That is, $V[\mathbf{x}(t_1, \mathbf{x}_0, t_0)] = \beta_1 > \beta$. From (H.52) we have that $\|\mathbf{x}(t_1, \mathbf{x}_0, t_0)\| > \nu$. Let t_m be the largest value in the interval $[t_0, t_1]$ for which $\|\mathbf{x}(t_m, \mathbf{x}_0, t_0)\| = \nu$. If there is no such a value in $[t_0, t_1]$ we choose $t_m = t_0$. Now, from (H.52), we have $V(\mathbf{x}_0) = \beta_0 \leq \beta$. From (a), $\dot{V} \leq 0$

---

[7] The symbol "$\notin$" means "does not belong."

almost everywhere, and in the interval $[t_m, t_1]$ we have $\|\mathbf{x}\| > \nu$. Now, since $V(\mathbf{x})$ is continuous (it satisfies the global Lipschitz condition), we derive $\beta_1 = V[\mathbf{x}(t_1, \mathbf{x}_0, t_0)] \leq V(\mathbf{x}_0) = \beta_0$ which is a contradiction. Therefore, $\mathbf{x}(t_1, \mathbf{x}_0, t_0) \in \mathbf{R}$.

(2) Let us show that when $\mathbf{x}_0 \in \mathbf{R}$, which implies $\|\mathbf{x}_0\| < \nu$, then $\mathbf{x}(t_1, \mathbf{x}_0, t_0) \in \mathbf{R}$ for any $t_1 > t_0$. If it were not so, we would have $V(\mathbf{x}) > \beta$ and $\|\mathbf{x}\| > \nu$ for all t in the interval $[t_0, t_e]$ where the solution $\mathbf{x}(t, \mathbf{x}_0, t_0)$ exists. Note that $\dot{V} \leq 0$ almost everywhere on $[t_0, t_e]$, and $V(\mathbf{x}_0) > V[\mathbf{x}(t_e, \mathbf{x}_0, t_0)]$, that is, $\mathbf{x}(t, \mathbf{x}_0, t_0)$ is bounded on $[t_0, t_e]$ which implies $t_e = \infty$. But then, V decreases along any solution,

$$V[\mathbf{x}(t, \mathbf{x}_0, t_0)] = V(\mathbf{x}_0) + \int_{t_0}^{t} \dot{V}[\mathbf{x}(\tau, \mathbf{x}_0, t_0)]\, d\tau$$

$$\leq V(\mathbf{x}_0) - (t - t_0)c, \qquad (\text{H.53})$$

where $c > 0$ is the lower bound of the continuous function $\gamma(\|\mathbf{x}\|)$ on the bounded region $\{\|\mathbf{x}\| \geq \nu,\ V(\mathbf{x}) \leq V(\mathbf{x}_0)\}$, and $V(\mathbf{x}) \to \infty$ for $t \to \infty$, which is a contradiction since $V(\mathbf{x})$ is continuous. This proves the lemma.

Yakubovich Lemma H.3. If there is a bounded and closed set $\mathbf{R}$ which satisfies the conditions of Lemma H.2, then there exists a bounded solution $\mathbf{x}(t, \mathbf{x}_0, t_0) \in \mathbf{R}$ for t in the interval $(-\infty, +\infty)$.

Yakubovich [H.24] proves this lemma on the basis of some important results obtained in references [H.17, 29]. For a detailed proof, these references are needed. We will, however, only briefly discuss the idea of the proof.

Proof of Lemma H.3. Let $\mathbf{R} = \mathbf{R}_0$, $t_0 = 0$, and define $\mathbf{R}_1$ as a set

$$\mathbf{R}_1 = \{\mathbf{x}_0 \mid \mathbf{x}(-1, \mathbf{x}_0) \in \mathbf{R}\}. \qquad (\text{H.54})$$

That is, $\mathbf{R}_1$ is a set of all $\mathbf{x}$ with a property that for all solutions $\mathbf{x}(t, \mathbf{x}_0)$ of (H.40), $\mathbf{x} = \mathbf{x}_0$ implies $\mathbf{x}(-1, \mathbf{x}_0) \in \mathbf{R}_0$. Similarly, define

$$\mathbf{R}_j = \{\mathbf{x}_0 \mid \mathbf{x}(-j, \mathbf{x}_0) \in \mathbf{R}_0\} \qquad (j = 2, 3, \ldots).$$

From Lemma H.5, $\mathbf{R}_0$ is invariant and hence $\mathbf{R}_1 \subset \mathbf{R}_0$. Consequently,

$$\mathbf{R}_0 \supset \mathbf{R}_1 \supset \mathbf{R}_2 \supset \cdots, \qquad (\text{H.55})$$

as illustrated on Figure H.5. Since all $\mathbf{R}_j$ are closed, from (H.55) it follows that there exists a point $\mathbf{a}_0 \in \cap\, \mathbf{R}_j$.[8] Invariancy of $\mathbf{R}_0$ implies that any solution $\mathbf{x}(t, \mathbf{a}_0) \in \mathbf{R}_0$ for all $t \geq 0$.

Now, we should show that there is a solution $\mathbf{x}(t, \mathbf{a}_0)$ such that $\mathbf{x}(t, \mathbf{a}_0) \in \mathbf{R}_0$ for all $t \leq 0$. We note that since $\mathbf{a}_0 \in \mathbf{R}_j$, there exist solutions $\mathbf{x}(t, \mathbf{a}_0)$ such that $\mathbf{x}(-j, \mathbf{a}_0) \in \mathbf{R}_0$ for $t \geq -j$. By using the fact that the sets $\mathbf{R}_j$ are closed,

---

[8] The symbol "$\cap$" denotes the intersection of the sets $\mathbf{R}_j$.

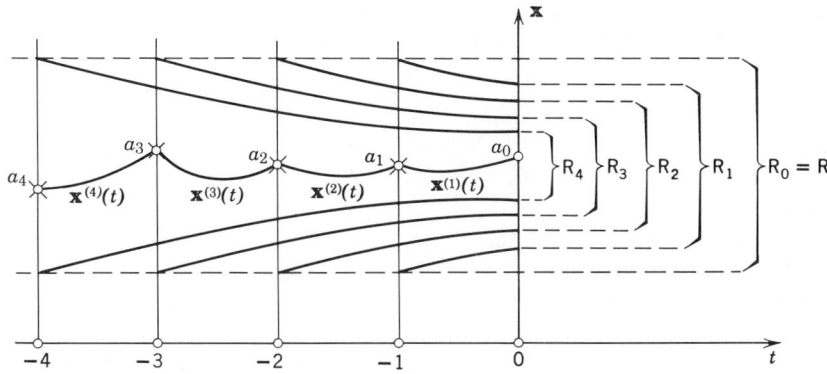

Figure H.5 State-time space trajectory.

we can show that there exist solutions $\mathbf{x}^{(m)}(t, \mathbf{a}_0)$ in each time interval $[-m, -m+1]$ such that $\mathbf{x}^{(m)}(-m, \mathbf{a}_m) = \mathbf{a}_m$ and $\mathbf{x}^{(m)}(-m+1, \mathbf{a}_m) = \mathbf{a}_{m-1}$ ($m = 1, 2, \ldots$) as illustrated in Figure H.5. Since $\mathbf{x}(-m, \mathbf{a}_m) \in R_0$, we have $\mathbf{x}(t, \mathbf{a}_m) \in R_0$ for all $t \geq -m$. Since m is an arbitrary integer, we actually have $\mathbf{x}(t, \mathbf{a}_0) \in R_0$ for all $t \leq 0$.

We proceed now to prove an auxiliary theorem of Yakubovich [H.24] which will serve as a basis for proving the stability theorems of Chapter 8.

Theorem H.2. Yakubovich. Given a dynamic system described by equations

$$\dot{\mathbf{x}} = \mathbf{P}\mathbf{x} + \mathbf{q}F(y) + \mathbf{f}(t), \qquad y = \mathbf{r}^T\mathbf{x}, \tag{H.56}$$

where $\mathbf{P}$ is a real $n \times n$ constant Hurwitz matrix; $\mathbf{q}$ and $\mathbf{r}$ are real constant n vectors; the nonlinear function $F(y)$ is generally discontinuous and satisfies the inequality

$$0 \leq \frac{F(y_1) - F(y_2)}{y_1 - y_2} \leq k, \tag{H.57}$$

where $0 < k \leq +\infty$, y_1 and y_2 are points of continuity of $F(y)$, and $-\infty < y_1 < y_2 < +\infty$; and $\mathbf{f}(t)$ is n vector function bounded in the entire interval $(-\infty, +\infty)$. The transfer function $G(s) = \mathbf{r}^T(\mathbf{P} - s\mathbf{I})^{-1}\mathbf{q}$, $s = \sigma + j\omega$, of the linear part of the system for $\sigma = 0$ satisfies

$$\pi(\omega) \equiv \frac{1}{k} + \text{Re}\,(1 + jq\omega)G(j\omega) > 0, \qquad \text{for all real } \omega \geq 0, \tag{H.58}$$

and

$$\lim_{\omega \to \infty} \pi(\omega)\omega^2 > 0, \qquad \text{if} \quad \lim_{\omega \to \infty} \pi(\omega) = 0. \tag{H.59}$$

Then the system is *dissipative*; this has three results: (1) in the state space $\mathscr{X}$ there is a bounded region R so that $\mathbf{x}_0 \in$ R implies $\mathbf{x}(t, \mathbf{x}_0, t_0) \in$ R for all $t \geq t_0$; (2) any solution $\mathbf{x}(t, \mathbf{x}_0, t_0)$ starting at any time t_0 outside the region R will enter the region at some time t_1 and stay there for all $t \geq t_1$; and there is at least one bounded solution $\mathbf{x}^0(t, \mathbf{x}_0, t_0) \in$ R in the interval $(-\infty, +\infty)$.

Before proving this theorem, a few remarks will be made to clarify some points of interest.

Since the nonlinear characteristic $F(y)$ is generally discontinuous, the solutions of (H.56) are well-defined everywhere but in the neighborhood of a discontinuity where the sliding motion can exist. To include the sliding in the solution of (H.56), it is necessary to consider the solutions in the Filippov sense: *A solution of* (H.56) *is any vector function* $\mathbf{x}(t)$ *such that, for any function* $\mathscr{F}(t) \equiv F[y(t)]$ *and all time t*,

$$\dot{\mathbf{x}} = \mathbf{P}\mathbf{x} + \mathbf{q}F(y) + \mathbf{f}(t) \qquad \text{(H.60a)}$$

$$F_-[y(t)] \leq \mathscr{F}(t) \leq F_+[y(t)], \qquad y = \mathbf{r}^T\mathbf{x}(t). \qquad \text{(H.60b)}$$

The function $\mathscr{F}(t)$ is uniquely determined by the solution $\mathbf{x}(t)$ and is called the *complementary function* of $F(y)$.

In the case of equations H.56, the sliding motion can be determined explicitely. Let us discuss the unforced case $\mathbf{f}(t) \equiv 0$ and without loss of generality let us assume that a discontinuity of $F(y)$ occurs at $y = 0$ where the discontinuity segment is determined by $F_+(0) = -F_-(0) = 1$.

The sliding motion can be readily studied in terms of y. First form the gradient $\nabla y = [\partial y/\partial x_1, \partial y/\partial x_2, \ldots, \partial y/\partial x_n]$ of the switching surface $y = 0$, and derive

$$\dot{y} = \nabla y \dot{\mathbf{x}} = \mathbf{r}^T\mathbf{P}\mathbf{x} + \mathbf{r}^T\mathbf{q}F(y). \qquad \text{(H.61)}$$

For values of $\mathbf{x}$ which satisfy the inequality

$$|\mathbf{r}^T\mathbf{P}\mathbf{x}| > |\mathbf{r}^T\mathbf{q}|, \qquad \text{(H.62)}$$

$\dot{y}$ has the same sign on both sides of the switching surface $y = 0$ in the state space $\mathscr{X}$, and the motion continues from region with $y > 0$ to region with $y < 0$ intersecting the switching surface $y = 0$. This is the *regular switching* which does not produce any new phenomena.

For those values of $\mathbf{x}$ satisfying the inequalities

$$|\mathbf{r}^T\mathbf{P}\mathbf{x}| < |\mathbf{r}^T\mathbf{q}|, \qquad \mathbf{r}^T\mathbf{q} < 0, \qquad \text{(H.63)}$$

from (H.61) we conclude that $\dot{y}$ is opposite in sign to y on both sides of the switching surface $y = 0$ on which $\dot{y} \doteq 0$. Thus, the motion is forced to stay on the switching surface as long as (H.63) holds.

To determine the sliding motion on the switching surface $y = 0$, note that for sliding the output $\mathscr{F}$ of the nonlinearity is evaluated from (H.61) as

$\mathscr{F} = -\mathbf{r}^T\mathbf{Px}/\mathbf{r}^T\mathbf{q}$. Now, the sliding motion is determined from (H.56) by a linear differential equation

$$\dot{\mathbf{x}} = \left[\mathbf{I} - \frac{\mathbf{q}\mathbf{r}^T}{\mathbf{r}^T\mathbf{q}}\right]\mathbf{Px}, \qquad \mathbf{r}^T\mathbf{x} = 0. \tag{H.64}$$

This discussion of the sliding motion can be readily extended to forced systems described by (H.56). For instance, when $\mathbf{f}(t) \equiv \mathbf{f}^0 = \text{const}$ and $\det \mathbf{P} \neq 0$, the stationary solutions $\mathbf{x} \equiv \mathbf{x}^0 = \text{const}$ are evaluated from (H.56) which yields $y^0 = \mathbf{r}^T\mathbf{x}^0$ and $y^0 + G(0)\mathscr{F}^0 + \mathbf{r}^T\mathbf{P}^{-1}\mathbf{f}^0 = 0$, where $\mathscr{F}^0 = F(y^0)$ is the complementary function. Then, if $G(0) \neq 0$, the necessary and sufficient condition for existence of a stationary solution $\mathbf{x}^0$ is that the function $F(y)$ complemented by the discontinuity segment, intersects the straight line $y + G(0)F + \mathbf{r}^T\mathbf{P}^{-1}\mathbf{f}^0 = 0$. The solution is $\mathbf{x}^0 = -\mathbf{P}^{-1}(\mathscr{F}^0\mathbf{q} + \mathbf{f}^0)$. Similarly, the periodic solutions of the sliding motion can be evaluated when $\mathbf{f}(t)$ is a periodic forcing function [H.24].

It is also necessary to note that the inequality (H.57) of the theorem requires that the function $F(y)$ is a non-decreasing function. This is stronger than necessary [H.24], but it is quite satisfactory for our purpose. Note that if $F(y)$ is a continuous function and $F'(y)$ exists, then inequality (H.57) means $0 \leq F'(y) \leq k$. For discontinuous $F(y)$ we have obviously $k = \infty$.

Proof of Theorem H.2. To prove the above theorem, let us use the function V in the form

$$V = \mathbf{x}^T\mathbf{H}\mathbf{x} + q\int_0^y F(y)\,dy, \tag{H.8}$$

where the matrix $\mathbf{H} > 0$ and q is the parameter which appears in (H.58). Note that the chosen V satisfies the global Lipschitz condition in every bounded region of the state space as required by Lemma H.2.

Differentiating, we obtain

$$-\dot{V} = [\mathbf{x}^T(\mathbf{G} - \mathbf{g}\mathbf{g}^T)\mathbf{x} + (\sqrt{\gamma}\mathscr{F} + \mathbf{g}^T\mathbf{x})^2] + \Omega_1 + \Omega_2, \tag{H.65}$$

where

$$-\mathbf{G} = \mathbf{HP} + \mathbf{P}^T\mathbf{H}, \qquad -\sqrt{\gamma}\,\mathbf{g} = \mathbf{Hq} + \tfrac{1}{2}(q\mathbf{P}^T + \mathbf{I})\mathbf{r}$$

$$-\gamma = q\mathbf{r}^T\mathbf{q} - \frac{1}{k}, \qquad -\Omega_1 = 2\mathbf{x}^T\mathbf{Hf} + q\mathscr{F}\mathbf{r}^T\mathbf{f} \tag{H.66}$$

$$\Omega_2 = \left(y - \frac{\mathscr{F}}{k}\right)\mathscr{F},$$

which is valid for both the regular and the sliding motion ($\dot{y} = 0$).

Now, we proceed to show that the function V of (H.8) is a Liapunov function as specified in Lemma H.2. From conditions (H.58), (H.59), and

Lemma H.1, we conclude that there is a matrix $\mathbf{H} > 0$ such that the bracketed expression in (H.65) is positive definite form in $\mathbf{x}$ and $\mathscr{F}$ (if $\gamma = \lim_{\omega \to \infty} \pi(\omega) > 0$), or only $\mathbf{x}$ (if $\gamma = 0$). In other words, for some constant $c > 0$, we have

$$-\dot{V} \geq c(\|\mathbf{x}\|^2 + \gamma \mathscr{F}^2) + \Omega_1 + \Omega_2. \tag{H.67}$$

From (H.66), and the restrictions (H.57), (H.60b) on the nonlinear characteristic, we can readily show that $\Omega_2 \geq 0$, and

$$\begin{aligned}\Omega_1 &\geq -c_1(\|\mathbf{x}\| + |\mathscr{F}|), \quad \text{for } \gamma \neq 0 \\ \Omega_1 &\geq -c_2\|\mathbf{x}\|, \quad \text{for } \gamma = 0,\end{aligned} \tag{H.68}$$

where c_1 and c_2 are positive constants. In deriving (H.68), we use the fact that $\mathbf{f}(t)$ is bounded and $|y| \leq \|\mathbf{r}\|\|\mathbf{x}\|$. Therefore, we conclude from above that for some $\nu > 0$ we have

$$-\dot{V} \geq c\|\mathbf{x}\|^2, \quad \text{for } \|\mathbf{x}\| \geq \nu, \tag{H.69}$$

and V satisfies condition (a) of Lemma H.2.

Let us now show that the condition (b) of Lemma H.2 ($V \to \infty$, when $\|\mathbf{x}\| \to \infty$) is satisfied. For $q \geq 0$ this is obvious. When $q < 0$ we first use the condition (H.57) and $F(y) = ky$ to show that the integral term of (H.8) is less or at most equal to $ky^2/2$. Then, we have

$$V \geq \mathbf{x}^T(\mathbf{H} + \tfrac{1}{2}q k \mathbf{r}\mathbf{r}^T)\mathbf{x}. \tag{H.70}$$

Let us now consider a linear system which we obtain from (H.56) for $\mathbf{f}(t) \equiv 0$ and $F(y) = Ky$, $0 \leq K \leq k$. For that system, $\Omega_1 \equiv 0$, $\Omega_2 \geq 0$ and, therefore, $\dot{V} < 0$ for $\mathbf{x} \neq \mathbf{0}$. Popov's conditions (H.58) and (H.59) are satisfied by the assumptions of the theorem and the linear system is stable, that is, the matrix $\mathbf{Q}(K) = \mathbf{P} + K\mathbf{q}\mathbf{r}^T$ is Hurwitz for all $K \in [0, k]$. This means that

$$\mathbf{H}_1 = \mathbf{H} + \tfrac{1}{2}qK\mathbf{r}\mathbf{r}^T > 0, \quad \text{for } 0 \leq K \leq k. \tag{H.71}$$

From (H.70) and (H.71) we conclude that $V(\mathbf{x}) \to \infty$ when $\|\mathbf{x}\| \to \infty$.

By Lemmas H.2 and H.3, the proof of Theorem H.2 is automatic.

It can be shown that if $\mathbf{f}(t) \equiv \mathbf{f}^0(t)$ is a T-periodic vector function and $0 \leq yF(y) \leq ky^2$, then Theorem H.2 implies that there is a T-periodic solution $\mathbf{x}^0(t)$. However, in general, there are solutions $\mathbf{x}(t)$ which do not tend to $\mathbf{x}^0(t)$ when $t \to \infty$.

Furthermore, Theorem H.2 applies to the system described by

$$\dot{\mathbf{x}} = \mathbf{P}\mathbf{x} + \mathbf{q}F(y) + \mathbf{f}(t, \mathbf{x}), \tag{H.72}$$

where $\mathbf{f}(t, \mathbf{x})$ is a bounded and continuous function in t and $\mathbf{x}$, and the following limiting process

$$\lim_{\|\mathbf{x}\|\to\infty} \frac{\|\mathbf{f}(t, \mathbf{x})\|}{\|\mathbf{x}\|} = 0 \tag{H.73}$$

is uniform in t. In this case, instead of (H.68) we should use the following estimates

$$\Omega_1 \geq -c_1(\|\mathbf{x}\|^2 + \mathscr{F}^2), \quad \text{for } \gamma \neq 0, \|\mathbf{x}\| \geq \nu, \tag{H.74}$$

or

$$\Omega_1 \geq -c_2\|\mathbf{x}\|^2, \quad \text{for } \gamma = 0, q = 0, \tag{H.75}$$

where $c_1 > 0$ and $c_2 > 0$ are sufficiently small and ν is sufficiently large [H.24].

Proof of Theorem 8.6. Let us now show that the Popov condition

$$\pi(\omega) \equiv \frac{1}{k} + \mathrm{Re}\,(1 + jq\omega)G(\sigma + j\omega) > 0, \quad \text{for all real } \omega \geq 0 \tag{H.76}$$

supplemented by (H.59) guarantees the exponential stability of the system

$$\dot{\mathbf{x}} = \mathbf{P}\mathbf{x} + \mathbf{q}F(y), \quad y = \mathbf{r}^T\mathbf{x} \tag{H.7}$$

for which $\mathbf{P}_\sigma = \mathbf{P} - \sigma\mathbf{I}$ is a Hurwitz matrix and nondecreasing $F'(y)$ satisfies

$$0 \leq F'(y) \leq k, \quad (0 \leq k \leq \infty). \tag{H.77}$$

We consider again the function V of (H.8) and derive

$$-\dot{V} + 2\sigma V = [\mathbf{x}^T\mathbf{G}_\sigma\mathbf{x} + 2\mathscr{F}\mathbf{x}^T\mathbf{g} + \gamma\mathscr{F}^2] + (y - \mathscr{F}k^{-1})\mathscr{F}$$

$$- 2\sigma q\left[\tfrac{1}{2}\mathscr{F}y - \int_0^y F(y)\,dy\right], \tag{H.78}$$

where

$$-\mathbf{G}_\sigma = \mathbf{H}\mathbf{P}_\sigma + \mathbf{P}_\sigma^T\mathbf{H}, \quad -\sqrt{\gamma}\,\mathbf{g}_\sigma = \mathbf{H}\mathbf{q} + \tfrac{1}{2}(q\mathbf{P}_\sigma^T + \mathbf{I})\mathbf{r},$$
$$-\gamma = q\mathbf{r}^T\mathbf{q} - 1/k. \tag{H.79}$$

By Yakubovich-Kalman lemma, we can conclude that there is a matrix $\mathbf{H} > 0$ such that the expression in the square brackets in (H.78) is positive-definite. So also is $(y - \mathscr{F}k^{-1})\mathscr{F}$. Now, let $q > 0$ (remember $\sigma < 0$) and we see that when $F'(y)$ is nondecreasing and $0 \leq F'(y) \leq k$, the last term in (H.78) is positive-definite. Thus, we can find a number $\mu > 0$ such that

$$-\dot{V} + 2\sigma V > 2\mu V. \tag{H.80}$$

By separating the variables and integrating with respect to time, we derive for $t \geq t_0$,

$$V[\mathbf{x}(t)] \leq V[\mathbf{x}(t_0)]\exp\,[2(\sigma - \mu)(t - t_0)]. \tag{H.81}$$

Since we interpreted V as a distance from the origin, (H.81) gives an estimate of how fast the states of the system approach the equilibrium. Actually, $[2(\mu - \sigma)]^{-1}$ can be regarded as the largest time constant everywhere in the state space $\mathscr{X}$ and may serve as a performance index of the system.

In fact, we conclude that

$$\mathbf{x}^T \mathbf{H} \mathbf{x} \leq V \leq \mathbf{x}^T \mathbf{H}_1 \mathbf{x} \tag{H.82}$$

where $\mathbf{H}_1 = \mathbf{H} + \tfrac{1}{2} q k \mathbf{r} \mathbf{r}^T$ as derived above. Since $\mathbf{H} > 0$, we can always find a constant $c > 0$ so that $\mathbf{x}^T \mathbf{H} \mathbf{x} \geq c \|\mathbf{x}\|^2$. Then, from (H.81), we conclude that there are positive constants M and μ such that for any fixed $\mathbf{x}_0$, t_0 and all $t \geq t_0$,

$$\|\mathbf{x}\| \leq M \|\mathbf{x}_0\| \exp\left[(\sigma - \mu)(t - t_0)\right]. \tag{H.83}$$

This proves exponential stability for $q > 0$.

Let us discuss now the case when $q < 0$. We use the transformation

$$F(y) = ky - F_{tr}(y), \tag{H.84}$$

$$\dot{\mathbf{x}} = \mathbf{P}_{tr} \mathbf{x} + \mathbf{q}_{tr} F_{tr}(y), \qquad y = \mathbf{r}^T \mathbf{x}, \tag{H.85}$$

where $\mathbf{P}_{tr} = \mathbf{P} + k\mathbf{q}\mathbf{r}^T$ and $\mathbf{q}_{tr} = -\mathbf{q}$. Then, $G_{tr}(s) = G(s)/[1 + kG(s)]$ and from (H.76) we obtain

$$\pi_1(\omega) \equiv \frac{1}{k} + \operatorname{Re}(1 - jq\omega)G_{tr}(\sigma + j\omega) > 0, \tag{H.86}$$

(see footnote 32 in Chapter 8). Now, we can repeat the above proof for the transformed system (H.85) using inequality (H.86) in which $q < 0$. We find that the assertion of Theorem 8.6 is true for the transformed system. Then, we note that the validity of the obtained results extends directly to the original system since it differs from the transformed system only in notation. This completes the proof of Theorem 8.6.

It is now important to note from (H.78) that when $\sigma = 0$, we can relax the condition on the nonlinear characteristic,

$$0 \leq yF(y) \leq ky^2, \qquad (0 \leq k \leq \infty), \tag{H.87}$$

and show readily that (H.76) with (H.59) guarantees absolute stability of the system (H.7). Thus, Popov's Theorem 8.2 is a direct consequence of Theorem 8.6.

Note also that we can use the weaker inequality (H.87) when $q = 0$. This is again based upon the expression in (H.78).

Proof of Theorems 8.7 and 8.8. To Theorem 8.7 we apply Theorem H.2 to conclude that the system

$$\dot{\mathbf{x}} = \mathbf{P}\mathbf{x} + \mathbf{q}F(y) + \mathbf{f}(t), \qquad y = \mathbf{r}^T \mathbf{x}, \tag{H.56}$$

has a bounded solution $\mathbf{x}^0(t)$ in the entire interval $(-\infty, +\infty)$.

Now, we take an arbitrary solution and define
$$z(t) = x(t) - x^0(t). \tag{H.88}$$
Let us choose the function
$$V(z) = z^T H z. \tag{H.89}$$
Differentiating with respect to time and using (H.56), we obtain
$$-\dot{V} + 2\sigma V = \left\{ z^T G_\sigma z + 2(\mathscr{F} - \mathscr{F}^0) + \frac{1}{k}(\mathscr{F} - \mathscr{F}^0)^2 \right\}$$
$$+ (\mathscr{F} - \mathscr{F}^0)^2 \left[\frac{y - y^0}{\mathscr{F} - \mathscr{F}^0} - \frac{1}{k} \right], \tag{H.90}$$
where
$$-G_\sigma = HP_\sigma + P_\sigma^T H, \qquad -g = Hq + \tfrac{1}{2}r, \tag{H.91}$$
$y^0 = r^T x^0(t)$, and $\mathscr{F}^0 = F[y^0(t)]$ is the complementary function for $F(y^0)$.

On the basis of the inequality (H.57) and condition (H.60b) we conclude that the expression in the square brackets of (H.90) is nonnegative. It remains to show that there is a matrix $H > 0$, so that
$$\begin{array}{ll} G - kgg^T > 0, & \text{for } k \neq \infty \\ G > 0, \ g = 0, & \text{for } k = \infty. \end{array} \tag{H.92}$$
That such a matrix exists, we conclude by using the Yakubovich-Kalman Lemma and the conditions of Theorem 8.7,
$$\pi(\omega) \equiv \frac{1}{k} + \text{Re } G(\sigma + j\omega) > 0, \qquad \text{for all real } \omega \geq 0, \tag{H.93}$$
and if $k = \infty$, then also
$$\lim_{\omega \to \infty} \omega^2 \text{ Re } G(\sigma + j\omega) > 0. \tag{H.94}$$
Since $G > 0$ and $P - \sigma I$ is a Hurwitz matrix, the expression in the curly brackets of (H.90) is a positive-definite form in z and $\mathscr{F}$ for $k \neq \infty$, or only z for $k = \infty$. Therefore, there is a number $\mu > 0$ such that we have $-\dot{V} + 2\sigma V > 2\mu V$ almost everywhere. That is,
$$V[z(t)] \leq V[z(t_0)] \exp\left[2(\sigma - \mu)(t - t_0)\right], \tag{H.95}$$
for all $t \geq t_0$. This leads to
$$\|x(t) - x^0(t)\| \leq M\|x(t_0) - x^0(t_0)\| \exp\left[(\sigma - \mu)(t - t_0)\right], \tag{H.96}$$
for a constant $M > 0$. Uniqueness of the solution $x^0(t)$ is now a direct consequence of (H.96). Actually, if we consider $x(t)$ as an arbitrary solution and let $t_0 \to -\infty$ for fixed t, we find that $x(t) \equiv x^0(t)$. Now, if $f(t)$ is a T-periodic vector function, that is, $f(t + T) \equiv f(t)$ where T is a constant, then

$\mathbf{x}^0 \equiv \mathbf{x}^0(t+T)$ is a solution of the system (H.56) bounded in $(-\infty, +\infty)$. When $\mathbf{f}(t)$ is an almost-periodic vector function, we define

$$\mathbf{z} = \mathbf{x}^0(t+T) - \mathbf{x}^0(t), \tag{H.97}$$

where T is an arbitrary δ-almost period of $\mathbf{f}(t)$. By differentiating V, we obtain

$$-\dot{V} + 2\sigma V = \mathbf{z}^T \mathbf{G}_\sigma \mathbf{z} + 2\mathscr{F}\mathbf{z}^T \mathbf{g} + \frac{1}{k}\mathscr{F}^2 + \mathscr{F}^2\left[\frac{y^0(t+T) - y^0(t)}{\mathscr{F}} - \frac{1}{k}\right]$$
$$+ 2\mathbf{z}^T \mathbf{H}[\mathbf{f}(t+T) - \mathbf{f}(t)], \tag{H.98}$$

where $\mathscr{F} = \mathscr{F}^0(t+T) - \mathscr{F}^0(t)$.

Since there is a number $c > 0$ so that

$$\mathbf{z}^T \mathbf{H}[\mathbf{f}(t+T) - \mathbf{f}(t)] \geq -c\|\mathbf{f}(t+T) - \mathbf{f}(t)\|V^{1/2} \geq -c\,\delta V^{1/2}, \tag{H.99}$$

by repeating the above reasoning, we again conclude that there exists a number $\mu > 0$ so that

$$-\dot{V} + 2(\sigma - \mu)V > 2\mu c V^{1/2} \tag{H.100}$$

almost everywhere. So, (H.100) can be rewritten to give

$$V[\mathbf{z}(t)]^{1/2} \leq V[\mathbf{z}(t_0)]^{1/2} e^{2(\sigma-\mu)(t-t_0)} + \frac{\delta}{\mu - \sigma}. \tag{H.101}$$

Now, letting $t_0 \to -\infty$, we have $V[\mathbf{z}(t)]^{1/2} \leq \delta/(\mu - \sigma)$, that is, T is a $\delta/(\mu - \sigma)$-almost period of the vector function $\mathbf{x}^0(t)$ with metric $\|\mathbf{x}\| = V(\mathbf{x})^{1/2}$ (remember $\sigma < 0$). Therefore $\mathbf{x}^0(t)$ is an almost-periodic vector function and the proof of Theorem 8.7 is completed.

The proof of Theorem 8.8 is now a direct consequence of the proof of Theorem 8.7. It should only be noted that the inequality

$$0 \leq \frac{F(y_1) - F(y_2)}{y_1 - y_2} \leq k \tag{H.57}$$

was needed to show that the expression in the square brackets of (H.90) is nonnegative, as well as, to conclude the existence of a solution $\mathbf{x}^0(t)$ in the interval $(-\infty, +\infty)$. If we know that such a solution exists, we can replace inequality (H.57) with

$$0 \leq \frac{F(y) - \mathscr{F}^0(t)}{y - y^0(t)} \leq k \tag{H.102}$$

and conclude that the mentioned expression in (H.90) remains nonnegative. Theorem 8.8 is proved.

H.6 Survey

In this section a brief historical review of the "Lur's problem" is given and some extensions of the material of Chapter 8 are outlined without proof. For more detailed survey, the references H.14, H.30, and H.31 are recommended.

The *absolute stability problem* was first formulated by Lur'e and Postnikov [H.2] for the system

$$\dot{\mathbf{x}} = \mathbf{P}\mathbf{x} + \mathbf{q}F(y), \qquad y = \mathbf{r}^T\mathbf{x}, \tag{H.7}$$

where the nonlinear characteristic is situated inside the first and third quadrants of the yF plane ($k = \infty$). Both the principal case ("direct control") and the simplest particular case ("indirect control") of the system in (H.7) were treated, and for the first time a Liapunov function of the type "a quadratic form plus an integral involving the nonlinearity" was proposed. Lur'e [H.3] extended the results and formulated a system of quadratic equations that may be obtained directly from the given differential equations. He proved that if this system of equations, which he called the *resolving equations*, has real roots, then the system (H.7) is absolutely stable.

The Lur'e approach was further extended by many authors. Significant results were obtained by Yakubovich [H.33], Malkin [H.32], and Letov [H.34], and the most general form of the approach was proposed by Rozenvasser [H.35]. It is also important to note that Letov [H.36] suggested an extension of the Lur'e problem for systems that contain several nonlinear functions. A complete treatment of the Lur'e method is given in reference H.14 by Aizerman and Gantmacher.

The problem of absolute stability of (H.7) in a finite sector (k_1, k_2) was formulated by Aizerman [H.37]. The principal task was to find the most widely separated numbers k_1 and k_2 for which absolute stability is preserved. This led to the conjecture known as *Aizerman's conjecture*. Namely, the sector (k_1, k_2) for obvious reasons must be contained in the "Hurwitz sector," that is, the largest sector inside which the linear system derived from (H.7) by substituting Ky for $F(y)$ is stable. The question was asked: Do these two sectors coincide? If characteristics that asymptotically approach the straight lines bounding the "Hurwitz sector" are excluded [H.38, 39], it was shown that the answer is affirmative for second-order systems and for some systems of higher order [H.40]. However, Pliss [H.41] constructed a third-order system for which the largest absolute stability sector (k_1, k_2) (this sector is later refered to as "Popov's sector"—see Section 8.15), is only a portion of the "Hurwitz sector." Another simple counterexample to Aizerman's conjecture of the third-order system has been produced by Dewey and Jury [H.42]. In Section 8.15 an example by Fitts [H.43] is given that represents a counterexample for both Aizerman's and Kalman's [H.44] conjectures, the latter being proposed for incremental rather than dc gains in the sector (k_1, k_2). Thus only the general problem of identifying the class of systems for which the answer is affirmative remained unsolved. Singling out such a class of systems would reduce, in their case, the problem of absolute stability to a linear problem.

An entirely different approach to the "Lur'e problem" of absolute stability was proposed by V. M. Popov [H.4-9]. He, as distinct from all the preceding authors, expressed his sufficient conditions not in terms of the Lur's resolving equations, but in terms of the *frequency response* of the linear part of the system. The graphical interpretation of the Popov criterion is expressed in terms of linear control theory, and is therefore convenient for practical control problems. Moreover, Popov proved that all the results obtained previously and related to the Liapunov function "a quadratic form plus an integral of the nonlinearity" are encompassed by his criterion.

Making use of his results on certain matrix inequalities, Yakubovich [H.10] established the converse proposition. He proved that if the Popov criterion together with some simple conditions is satisfied, then there exists a Liapunov function of the form

$$V = \mathbf{x}^T \mathbf{H} \mathbf{x} + q \int_0^y F(y)\, dy. \tag{H.8}$$

In a series of papers Yakubovich applied his results to various important problems of absolute stability. He examined in details the general particular case of system (H.7) in reference H.15 and solved [H.24] the forced case

$$\dot{\mathbf{x}} = \mathbf{P}\mathbf{x} + \mathbf{q}F(y) + \mathbf{f}(t), \qquad y = \mathbf{r}^T \mathbf{x} \tag{H.56}$$

as shown in Section H.5 by Theorem H.2. In reference H.46, Yakubovich extended the results of Popov [H.8] for systems with multivalued (hysteresis) nonlinearities. Yakubovich [H.47] also strengthened the Popov criterion by applying additional constraints to nonlinear characteristics. This last result was given in terms of the following theorem.

Theorem H.3. Monotonic Nonlinearity. Consider system (H.7) in the principal case where matrix $\mathbf{P}$ is Hurwitz and the differentiable nonlinear function $F(y)$ satisfies the condition

$$-k_1 \leq F'(y) \leq k_2, \tag{H.103}$$

where $k_1 \neq \infty$, $k_2 \neq \infty$ and, without loss of generality, $k_1 \geq 0$, $k_2 \geq k$.

If for some $\tau_1 > 0$, $\tau_2 \geq 0$ satisfying the condition

$$+\infty \geq \tau_2 \lim_{y \to +\infty} \frac{q}{y^2} \left[\int_0^y F(y)\, dy - \frac{yF(y)}{2} \right] \geq 0 \tag{H.104}$$

and all real $\omega \geq 0$, the condition

$$\pi(\omega) \equiv \tau_1[k^{-1} + \text{Re } G(j\omega)] + q \text{ Re}[j\omega G(j\omega)]$$
$$+ \tau_2\omega^2[1 + (k_2 - k_1)\text{Re } G(j\omega) - k_1k_2 |G(j\omega)|^2] > 0 \quad \text{(H.105)}$$

is also satisfied for some q, then the system (H.7) is absolutely stable.

This theorem is proved in reference H.47. Extensions are obtained by Dewey and Jury [H.48, 49], Brockett and Williams [H.45], and Zames [H.50].

The fundamental lemma of Yakubovich [H.10] was refined by Kalman [H.11] for the simplest particular case ("indirect control" in the Lur'e terminology) and $k = \infty$, by employing the controllability and observability theory. These two results constitute what is known as the Yakubovich-Kalman lemma. This lemma is used frequently in Chapter 8 and is proved in Section H.3. Further refinement of the lemma was performed by Meyer [H.12], who showed that the controllability and observability restrictions can be eliminated with some care. An extended form of the lemma for the principal and general particular case was proved recently by Yakubovich [H.51].

Absolute stability with prescribed damping coefficient was investigated by Naumov and Tsypkin [H.52] and by Naumov [H.53], and was precisely formulated by Yakubovich [H.24]. This is called exponential absolute stability and is outlined in Section 8.10. This represents a generalization of the Popov criterion for contours in the complex plane other than the imaginary axis. This is an important extension for practical application of absolute stability analysis. This applies to forced systems also [H.24].

It is of essential interest to note that the Popov criterion was extended to systems with a nonunique equilibrium state and discontinuous nonlinearities by Tsypkin [H.54] and Halanay [H.55] and in a refined form by Gelig [H.56]. The results obtained by Gelig are discussed briefly here.

Derivations of the original Popov criterion were essentially based on the continuity of the function $F(y)$. In the case of discontinuous nonlinearities, therefore, the criterion, cannot be used to guarantee absolute stability. In fact, when $y = 0$ is a point of discontinuity of the function $F(y)$, and $G(0) = 0$, the system (H.7) has a "rest segment"

$$\mathbf{x} = -\mathbf{P}^{-1}\mathbf{q}\mathscr{F}, \quad \mathscr{F} \in [F_-(0), F_+(0)], \quad \text{(H.106)}$$

where $\mathscr{F}(t) = F[y(t)]$ is the complementary function satisfying inequality H.60b. Therefore the absolute stability of the equilibrium state $\mathbf{x}_e \equiv \mathbf{0}$ is meaningless, even the Popov criterion is verified. It is therefore necessary to define absolute stability of the set of equilibrium states called the "stationary set."

There are several possible definitions of the stability of stationary sets: an arbitrary solution may approach any point of the stationary set, or may tend to the set, that is, asymptotically approach the set without approaching any point of the set. Therefore precise definitions are necessary [H.56].

Consider a system

$$\dot{\mathbf{x}} = \mathbf{X}(\mathbf{x}), \qquad (H.107)$$

where $\mathbf{x}$ and $\mathbf{X}(\mathbf{x})$ are real n vectors and $\mathbf{X}(\mathbf{x})$ is a measurable and bounded function in every portion of the state space $\mathscr{X}$. The solution $\mathbf{x}(t)$ is understood in the sense of Filippov, and the stationary set is assumed as bounded, connected, and closed.

Stationary set S is called stable in the large, if it is stable "in the small" in the sense of Liapunov and

$$\rho[\mathbf{x}(t), S] = \min_{\mathbf{z} \in S} \|\mathbf{x}(t) - \mathbf{z}\| \to 0, \quad \text{for } t \to \infty. \qquad (H.108)$$

If S is stable in the large and, in addition every solution $\mathbf{x}(t)$ for $t \to \infty$ approaches a certain equilibrium state $\mathbf{z} \in S$, then we say that the stationary set S is pointwise stable in the large.

Gelig [H.56] proved the following auxiliary proposition of Liapunov type.

Gelig's Lemma H.4. Suppose that for arbitrary $\mathbf{c} \in S$ and arbitrary vector $\mathbf{x}$ there exists a continuous function $V(\mathbf{x}, \mathbf{c})$ satisfying the following conditions.

(a) $V(\mathbf{x}, \mathbf{c}) > 0$ for $\mathbf{x} \notin S$, $\lim_{\mathbf{x} \to \mathbf{c}} V(\mathbf{x}, \mathbf{c}) = 0$,

and $\lim_{\|x\| \to \infty} V(\mathbf{x}, \mathbf{c}) = \infty$.

(b) There is an ε neighborhood of the set S in which for any $\mathbf{c}$ the function $V(\mathbf{x}, \mathbf{c})$ does not increase along any trajectory of the system (H.107).

(c) There is such $\mathbf{c}_0 \in S$, so that $V(\mathbf{x}, \mathbf{c}_0)$ does not increase along any trajectory, and on any surface $V(\mathbf{x}, \mathbf{c}_0) = \text{const}$ there are no positive half-trajectories of the system (H.107) besides the equilibrium state.

Then the stationary set S of the system (H.107) is stable in the large.

In addition to (a)–(c), let the function $V(\mathbf{x}, \mathbf{c})$ satisfy the following conditions.

(d) $V(\mathbf{x}, \mathbf{c}) = 0$ for any $\mathbf{c} \in S$ only if $\mathbf{x} = \mathbf{c}$.

Then the stationary set S is pointwise stable in the large.

The application of this lemma and the Yakubovich-Kalman Lemma gives the following result [H.56].

Theorem H.4. *Nonunique Equilibrium Position.* Consider the system (P.7) with a Hurwitz matrix **P** (the principal case).

(a) The function $F(y)$ has for $y = 0$ a discontinuity of the first kind, and for the points of continuity the inequality $F(y)y \geq 0$ is satisfied. In addition, there exists such $\varepsilon < 0$ that $F(y) \geq F_+(0)$ for $0 < y < \varepsilon$, and $F(y) \leq F_-(0)$ for $-\varepsilon < y < 0$;

(b) $G(0) = 0$, $G(j\omega) \neq 0$ for $\omega \neq 0$;

(c) The system is completely controllable and observable, that is, the rank of $[\mathbf{q}, \mathbf{Pq}, \ldots, \mathbf{P}^{n-1}\mathbf{q}]$ and $[\mathbf{r}, \mathbf{P}^T\mathbf{r}, \ldots, \mathbf{P}^{T n-1}\mathbf{r}]$ is equal to n; and

(d) There exists $q \geq 0$ so that for all real $\omega \geq 0$, $\pi(\omega) \equiv \text{Re}[(1 + jq\omega) \times G(j\omega)] > 0$ and, in case $q > 0$ the number q^{-1} is not an eigenvalue of matrix $\mathbf{P}$.

Then, the "rest segment" (H.105) is pointwise stable in the large.

In reference H.15 the general critical case was considered where $\mathbf{x} = 0$ is the unique stationary position.

The Popov criterion was also extended to *time-varying systems*, by Rozenvasser [H.57]. It can be shown [H.58] that the results [H.59–63] obtained by other approaches can be reduced to that of Rozenvasser. This is treated in Section 8.12, and the Yakubovich version of the problem is given in Section H.5.

Significant results were obtained in formulating the Popov conditions for nonlinear sampled-data systems. Fundamental results were obtained by Tsypkin [H.64] and they are now briefly reviewed.

A large class of nonlinear sampled-data systems can be described by difference equations of the form

$$\mathbf{x}_{t+1} = \mathbf{P}\mathbf{x}_t + \mathbf{q}F(y_t), \quad y_t = \mathbf{r}^T\mathbf{x}_t. \quad \text{(H.109)}$$

The transfer function of the linear part of the system from the input F to the output $(-y)$ is $G(e^s)$, where $G(s) = \mathbf{r}^T(\mathbf{P} - s\mathbf{I})^{-1}\mathbf{q}$. We say that the principal case is that in which all the roots of $\det(\mathbf{P} - e^s\mathbf{I}) = 0$ are inside the unit circle. Tsypkin obtained [H.64] the following result.

Theorem H.5. Sampled-Data Systems. Suppose that the system (H.109) represents a principal case and the nonlinear function $F(y)$ satisfies the condition $0 \leq yF(y) \leq ky^2$.

(a) If for all real ω, $0 \leq \omega \leq 2\pi$, the condition

$$\pi(\omega) \equiv \frac{1}{k} + \text{Re } G(e^{j\omega}) > 0 \quad \text{(H.110)}$$

is satisfied, then the system (H.109) is exponentially asymptotically stable: there are constants $M > 0$, $\mu > 0$, that depend only on $\mathbf{P}, \mathbf{q}, \mathbf{r}, k$, so that for every solution $\mathbf{x}_t(t)$ the condition

$$\|\mathbf{x}_t(t)\| < M\|\mathbf{x}_t(t_0)\|e^{-\mu(t-t_0)}, \quad t, t_0 = 0,1,2,\ldots, t \geq t_0, \quad \text{(H.111)}$$

is fulfilled.

(b) If the rank of matrices $[\mathbf{q}, \mathbf{Pq}, \ldots, \mathbf{P}^{n-1}\mathbf{q}]$ and $[\mathbf{r}, \mathbf{P}^T\mathbf{r}, \ldots, \mathbf{P}^{T n-1}\mathbf{r}]$ is equal to n, and the weakened condition (H.110) is satisfied, that is, ">"

is changed to "$\geq$", then the solution $\mathbf{x}_t(t) \equiv \mathbf{0}$ of the system (H.109) is absolutely stable.

(c) Statement (a) remains valid if in (H.109) $F(y_t)$ is an arbitrary function of $\mathbf{x}_t$, t, and perhaps some system parameters, which is related to y_t by the relationship $0 \leq yF(y) \leq ky^2$

This result was further developed by Szegö and Kalman [H.65, 66] and by Jury and Lee [H.67], who extended the theorem to multiple nonlinearities.

The time-delay system can also be investigated by a modified Popov criterion for absolute stability. It was shown by Popov and Halanay [H.7], and later by Desoer [H.68].

Multiple nonlinearities can also be considered in the framework of the Popov conditions. That was shown by Jury and Lee [H.67], Tokumaru and Saito [H.69], and Anderson [H.70]. Yakubovich's results in this direction are outlined in connection with Theorem H.2 and the vector function $\mathbf{f}(t, \mathbf{x})$ (see equation H.76). It should be noted, however, that in the multiple nonlinearity case the Popov criterion loses much of its simplicity.

The Popov condition can be reformulated in terms of the "circle criterion." Along these lines, the work of Sandberg, Zames, Brockett, Williams, and others is reviewed in reference H.30. It is only fair to say that the circle criterion is a direct reformulation of the Popov criterion, since some useful results have been obtained by this approach.

A significant reformulation of the Popov condition was given by Naumov and Tsypkin [H.52, 53] in terms of the Bode diagrams. This enables a design of absolutely stable system with desired degree of stability.

Another reformulation of the absolute stability criterion of Popov was given in the parameter space [H.71, 72]. This gave a possibility to study the influence of system parameters on the stability of nonlinear systems. This treatment is outlined in Section 8.14.

H.7 Finite Stability Domains

It is obvious that asymptotic stability in the large and, therefore, absolute stability are never possible in the real world, where all states of the system are bounded. Consequently, we are often interested in making sure that solutions starting inside a finite domain[9] of the state space $\mathscr{X}$ tend to the equilibrium position. The basic problem is: Given an asymptotically stable equilibrium point, find a set of all initial states in $\mathscr{X}$ that result in solutions going to the equilibrium point.

---

[9] If this domain is arbitrarily small, then the question can be resolved directly by linearization in all but critical cases. When the domain is finite, the problem of estimating the domain becomes difficult [H.73–76].

Suppose that the equilibrium point $\mathbf{x} = \mathbf{0}$ of the linear part of differential equation

$$\dot{\mathbf{x}} = \mathbf{P}\mathbf{x} + \mathbf{q}F(y), \qquad y = \mathbf{r}^T\mathbf{x} \tag{H.7}$$

is asymptotically stable. Let the Liapunov function $V(\mathbf{x})$ be constructed for the linear part, and its derivative $\dot{V} = W(\mathbf{x})$ be obtained for (H.7). If $W(\mathbf{x})$ is negative-definite throughout the state space $\mathscr{X}$, then from Theorem 8.1 it follows that the equilibrium point is asymptotically stable in the large. But if $W(\mathbf{x})$ vanishes for certain points $\mathbf{x} \neq \mathbf{0}$, we cannot conclude this. However, as long as one of the hypersurfaces $V(\mathbf{x}) = \text{const}$ is located entirely inside the domain specified by $W(\mathbf{x}) = 0$, then it definitely belongs to the domain of attraction of the point $\mathbf{x} = \mathbf{0}$. Thus it is necessary to find those surfaces $V = \text{const}$ that contact the surface $W = 0$ from the inside, and to single out from all these surfaces the one that corresponds to the lowest value V_0 of the constant. Then the selected domain $V(\mathbf{x}) < V_0$ is a subdomain of the *domain of attraction*. At a contact point $\mathbf{z}$ of those surfaces, grad V is parallel to grad W. Hence

$$W(\mathbf{z}) = 0, \quad \text{and} \quad \text{rank} \left.\frac{\partial(V, W)}{\partial(x_1, x_2, \ldots, x_n)}\right|_{\mathbf{x}=\mathbf{z}} = 1. \tag{H.112}$$

Certain freedom is involved in the construction of the Liapunov function. There is no general method, however, for systematically constructing Liapunov functions to improve the stability-domain estimates. The most general result is that of Zubov, elaborated in reference H.74. Numerical techniques to determine stability boundaries have been developed by Rodden [H.75] for a wide variety of nonlinear control systems. Efficient procedures for numerical Liapunov-function generation that are suitable for computer implementation and yield the largest stability domains were developed by Weissenberger [H.76, 77].

It is of interest here to show how the particular Liapunov function

$$V = \mathbf{x}^T \mathbf{H} \mathbf{x} + q \int_0^y F(y)\, dy \tag{H.8}$$

arising in the absolute stability can be used to establish domains of asymptotic stability. This was proposed by Weissenberger [H.77] as shown in the following.

Let us assume that a Liapunov function (H.8) has been determined so that the system (H.7) is absolutely stable in the sector $[0, k]$, that is, for all nonlinear characteristics $F(y)$ so that

$$0 \leq yF(y) \leq ky^2. \tag{H.113}$$

Suppose now that for a particular nonlinearity the inequality H.113 holds only for a limited range of $y(\mathbf{x})$, for instance,

$$|y(\mathbf{x})| < y_m \tag{H.114}$$

and, in fact, the domain of asymptotic stability in the state space is of finite extent. An estimate of this stability domain may be readily computed as follows: V will be a Liapunov function for (H.7) as long as inequality H.114 is satisfied. Hence the region

$$V(\mathbf{x}) < V_m, \qquad V_m = \min_{\mathbf{x} \in \{\mathbf{x} \mid |y(\mathbf{x})| = y_m\}} \{V\} \tag{H.115}$$

will be contained in the stability domain. Since a trajectory of (H.7) never leaves the domain $V(\mathbf{x}) \leq V_m$ while (H.114) is satisfied, and since (H.114) is always satisfied in this domain, it is a domain of asymptotic stability. This domain is determined by conditions

$$\operatorname{grad} V = \lambda \operatorname{grad} y, \qquad y = y_m \tag{H.116}$$

or

$$2\mathbf{H}\mathbf{x} + qF(y)\mathbf{r} = \lambda \mathbf{r}, \qquad \mathbf{r}^T \mathbf{x} = y_m. \tag{H.117}$$

Equation H.117 gives

$$\mathbf{A}\mathbf{z} = \mathbf{b}, \tag{H.118}$$

where

$$\mathbf{A} = \begin{bmatrix} 2\mathbf{H} & -\mathbf{r} \\ \mathbf{r}^T & 0 \end{bmatrix}, \qquad \mathbf{z} = \begin{bmatrix} \mathbf{x} \\ \lambda \end{bmatrix}, \qquad \mathbf{b} = \begin{bmatrix} -qF(y_m)\mathbf{r} \\ y_m \end{bmatrix}. \tag{H.119}$$

Equation H.117 and H.118 also give

$$V_m = \min_{c = y_m, -y_m} \left(q \int_0^c F(y)\, dy + \frac{2^{n-1} y_m^2 \det \mathbf{H}}{\det \mathbf{A}} \right). \tag{H.120}$$

Note that, in general, the nonlinear function $F(y)$ is not symmetrical and, consequently, the integral term in equation H.120 may have different values for $c = y_m$ and $c = -y_m$. A noteworthy special case occurs when $q = 0$. In this case, (H.120) becomes simply

$$V_m = \frac{2^{n-1} y_m^2 \det \mathbf{H}}{\det \mathbf{A}}. \tag{H.121}$$

This special result is useful in two ways: first, the domain depends explicitly on y_m; second, the domain is valid for all nonlinearities contained in the augmented sector shown in Figure H.6.

To illustrate the procedure, let us consider a system (H.7) with

$$G(s) = \frac{2s + 1}{s^2 + 2ps + p - 1}, \qquad F(y) = \operatorname{sgn} y - py. \tag{H.122}$$

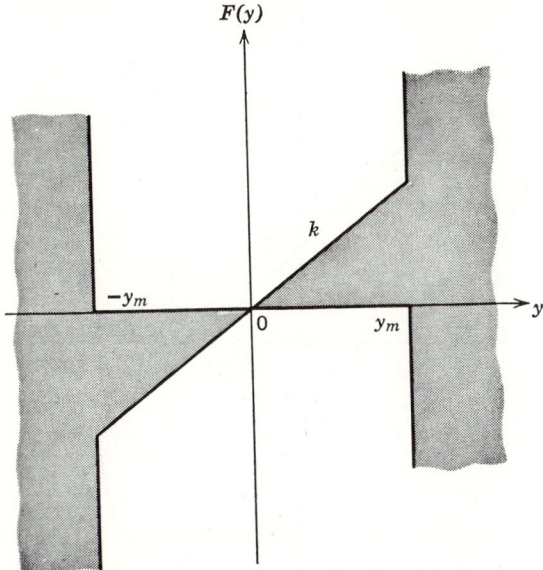

Figure H.6 Augmented sector.

For all $p > 0$, $G(s)$ satisfies the simplified Popov condition

$$\text{Re } G(j\omega) > 0, \quad \forall \omega \geq 0. \tag{H.123}$$

Consequently, $k = \infty$ and a quadratic Liapunov function

$$V = \mathbf{x}^T \mathbf{H} \mathbf{x} \tag{H.124}$$

exist and may be found by satisfying (H.95), that is,

$$-(\mathbf{HP} + \mathbf{P}^T \mathbf{H}) = \mathbf{G} > 0, \quad \mathbf{Hq} + \mathbf{r} = 0. \tag{H.125}$$

It follows from (H.125) that

$$\mathbf{H} = \begin{bmatrix} h_{11} & 1 \\ 1 & 2 \end{bmatrix} \tag{H.126}$$

subject to the condition

$$\mathbf{G} = \begin{bmatrix} 2\varepsilon & 2 + \varepsilon - h_{11} \\ 2 + \varepsilon - h_{11} & 6 + \varepsilon \end{bmatrix} > 0, \quad p = 1 + \varepsilon, \quad \varepsilon > 0. \tag{H.127}$$

The choice $h_{11} = 1.25$, $\varepsilon = 0.06$ gives, with $y_m = 1/p$, the stability domain estimate

$$1.25 x_1^2 + 2 x_1 x_2 + 2 x_2^2 < 0.44, \tag{H.128}$$

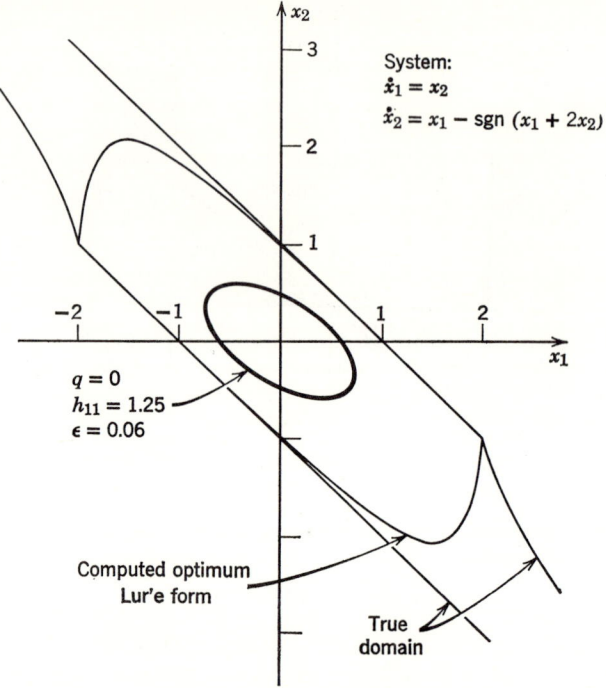

Figure H.7 Stability domains.

which is shown in Figure H.7, together with the true stability region and an estimate found by the numerical Liapunov-function generation procedure used by Weissenberger [H.76], which yields a Lur'e form ($q \neq 0$) that is optimum in the sense of producing the domain of greatest area. It is clear that the results for a particular nonlinearity are not likely to be as good as found by a systematic and effective procedure such as that of reference H.76, although we should expect distinctly improved results for $q \neq 0$. Furthermore, a strict comparison of the obtained results is unfair, since the quadratic domain is valid for any nonlinear characteristic $F_1(y)$ replacing the relay sgn y that satisfies the condition $yF_1(y) > y$, $|y| < 1$.

In spite of the relative conservativeness of the result, there are three unique advantages of this method: (1) the algebra is simple and allows systems of all orders to be treated with relative ease, a feature absent in other approaches, in which the algebraic difficulty is significant and increases with system order [H.75, 76]; (2) results may be obtained for a class of nonlinearities; and (3) the stability domain may be found as an explicit function of y_m.

These results can be immediately extended to estimate finite regions of exponential stability, thus providing additional information about how fast the solutions approach the equilibrium point. Furthermore, the property of exponential stability makes it possible to consider a class of nonautonomous differential equations and to determine a bound on the forcing function that guarantees that all the solutions remain bounded inside the computed region [H.79].

Let us consider (H.7) once again:

$$\dot{\mathbf{x}} = \mathbf{P}\mathbf{x} + \mathbf{q}\,F(y), \qquad y = \mathbf{r}^T\mathbf{x}, \tag{H.7}$$

for which the sector conditions (H.113) for continuous nonlinear characteristics $F(y)$ and Popov inequality

$$\pi(\omega) \equiv \frac{1}{k} + \mathrm{Re}\,G(\sigma + j\omega) > 0, \qquad \forall \omega \geq 0, \tag{H.96}$$

supplemented with condition (H.97) if $k = \infty$, are satisfied. Then, by Yakubovich's Theorem 8.6, there exist two positive constants M and μ such that for any solution $\mathbf{x}(t)$ of (H.7) and any $t \geq t_0$ we have

$$\|\mathbf{x}(t)\| \leq M\|\mathbf{x}(t_0)\| \exp\left[(\sigma - \mu)(t - t_0)\right]; \tag{H.129}$$

that is, the equilibrium point $\mathbf{x} = \mathbf{0}$ of (H.7) is globally exponentially stable with degree of stability σ for all continuous nonlinear characteristics $F(y)$ that satisfy sector condition (H.113).

According to the Yakubovich-Kalman lemma, the Popov condition (H.96) is necessary and sufficient for the existence of a Liapunov function $V(\mathbf{x}) = \mathbf{x}^T\mathbf{H}\mathbf{x}$ with the derivative along solutions of (H.7) as

$$-\dot{V} = \left[\mathbf{x}^T\mathbf{G}_\sigma\mathbf{x} + 2\mathbf{x}^T\mathbf{g}\mathscr{F} + \frac{1}{k}\mathscr{F}^2\right] + \left(y - \frac{1}{k}\mathscr{F}\right)\mathscr{F} - 2\sigma V, \tag{H.130}$$

where

$$-\mathbf{G}_\sigma = \mathbf{H}\mathbf{P}_\sigma + \mathbf{P}_\sigma^T\mathbf{H}, \qquad -\mathbf{g} = \mathbf{H}\mathbf{q} + \tfrac{1}{2}\mathbf{r}, \tag{H.91}$$

$\mathbf{P}_\sigma = \mathbf{P} - \sigma\mathbf{I}$, and $\mathscr{F} = F[y(t)]$. The matrix $\mathbf{H} > 0$ satisfies the following matrix inequalities:

$$\begin{aligned} \mathbf{G}_\sigma - k\mathbf{g}\mathbf{g}^T > 0, &\quad \text{for} \quad k \neq \infty \\ \mathbf{G}_\sigma > 0,\, \mathbf{g} = 0, &\quad \text{for} \quad k = \infty. \end{aligned} \tag{H.92}$$

If the nonlinearity leaves the sector for $|y| \geq y_m$, that is,

$$0 \leq y F(y) \leq ky^2, \qquad |y| < y_m, \qquad y \neq 0, \qquad F(0) = 0, \tag{H.131}$$

the property of exponential stability will be of finite extent in the state space $\mathscr{X}$ and the following definition is useful:

Definition H.4. Domain of Exponential Stability. A domain D_σ of exponential stability is defined as the set of all points $\mathbf{x}_0 = \mathbf{x}(t_0)$ in $\mathscr{X}$ for which solutions $\mathbf{x}(t)$ of (H.7) starting at $\mathbf{x}_0$ are exponentially stable with degree σ.

We assume that the Popov inequality (H.96) is satisfied for some σ and k and that $F(y)$ lies in the augmented sector of Fig. H.6; that is, $F(y)$ satisfies conditions (H.131). Then we are interested to find the largest domain $\bar{D}_\sigma$ ($\bar{D}_\sigma \subset D_\sigma$) defined by

$$\bar{D}_\sigma = \{\mathbf{x} \mid |y| < y_m; V(\mathbf{x}) < V_m\}, \tag{H.132}$$

where the constant V_m is determined as in (H.115); that is,

$$V_m = \min_{|y(\mathbf{x})|=y_m} V(\mathbf{x}). \tag{H.133}$$

Since $y = \mathbf{r}^T\mathbf{x}$ and $V(\mathbf{x}) = \mathbf{x}^T\mathbf{H}\mathbf{x}$, it can be shown from (H.133) that

$$V_m = y_m^2 (\mathbf{r}^T\mathbf{H}^{-1}\mathbf{r})^{-1}. \tag{H.134}$$

Therefore, with (H.134), (H.133) reduces to

$$\bar{D}_\sigma = \{\mathbf{x} \mid \mathbf{x}^T\mathbf{H}\mathbf{x} < V_m\}. \tag{H.135}$$

Now the problem is to find the matrix $\mathbf{H} > 0$ in (H.135)

A search for an appropriate matrix $\mathbf{H}$ can proceed in two different directions, based on the results obtained in [H.77] and [H.78]. A *specific* matrix $\mathbf{H}$ can be found [H.78] from

$$\mathbf{H}\mathbf{P}_\sigma + \mathbf{P}_\sigma^T\mathbf{H} = \mathbf{u}\mathbf{u}^T, \tag{H.136}$$

if $\pi(\omega)$, given in (H.96), is rewritten as

$$\pi(\omega) = \frac{\phi(j\omega)\phi(-j\omega)}{|\mathbf{P}_\sigma - j\omega\mathbf{I}| \, |\mathbf{P}_\sigma + j\omega\mathbf{I}|}, \tag{H.137}$$

as suggested in Section H.3, and the vector $\mathbf{u}$ is chosen from

$$\mathbf{u}^T(\mathbf{P}_\sigma - s\mathbf{I})^{-1}\mathbf{q} = \frac{\phi(s)}{|\mathbf{P}_\sigma - s\mathbf{I}|} - k^{-1/2}. \tag{H.138}$$

Another approach is to generate a *set* of appropriate $\mathbf{H}$ matrices directly from the matrix inequalities (H.95) as suggested in [H.79]. By applying the Sylvester inequalities we reduce (H.95) to a system of algebraic inequalities involving the elements of **H**. This approach allows a certain freedom in choosing the corresponding Liapunov function $V(\mathbf{x}) = \mathbf{x}^T \mathbf{H} \mathbf{x}$. In general, each Liapunov function produces a different domain $\bar{D}_\sigma$ with respect to extent and orientation, and it is desirable to select one that produces in some sense the best estimate of the domain $\bar{D}_\sigma$. A way to find a "best" estimate $\tilde{D}_\sigma$ is to maximize the volume of $\bar{D}_\sigma$ (a simple matter for quadratic domains) on the set of generated **H** matrices.

To illustrate the analysis of the second approach let us consider the equation

$$\begin{bmatrix} \dot{x}_1 \\ \dot{x}_2 \end{bmatrix} = \begin{bmatrix} 0 & 1 \\ -4 & -4 \end{bmatrix} \begin{bmatrix} x_1 \\ x_2 \end{bmatrix} - \begin{bmatrix} 0 \\ 1 \end{bmatrix} F(x_1 + x_2). \qquad (H.139)$$

The Popov conditions (H.96) and (H.97) are satisfied for $k = \infty$ and $-1 \leq \sigma \leq 0$. From (H.7), $\mathbf{g} = \mathbf{0}$,

$$\mathbf{H} = \begin{bmatrix} h_{11} & \frac{1}{2} \\ \frac{1}{2} & \frac{1}{2} \end{bmatrix} \qquad (H.140)$$

and the domain $\bar{D}$ is

$$2h_{11}x_1^2 + 2x_1 x_2 + x_2^2 < y_m^2 \qquad (H.141)$$

for all $F(y)$ inside the augmented sector (Fig. H.8). The condition $\mathbf{G}_\sigma > 0$ yields the Sylvester inequalities

$$h_{11} < \frac{-2}{\sigma},$$

$$|h_{11} - (2-\sigma)^2| < [(2-\sigma)^4 - \sigma^2 - 4\sigma + 4]^{\frac{1}{2}}. \qquad (H.142)$$

The area of $\bar{D}_\sigma$ is $\pi y_m^2 (2h_{11} - 1)^{-\frac{1}{2}}$ and the maximum-area domain is produced by the least h_{11} satisfying (H.142). Apparently, the extent of the domain $\tilde{D}_\sigma$ is a function of σ; for instance, the area of $\tilde{D}_0$ is more than three times the area of $\tilde{D}_{-1}$.

Now consider the nonautonomous equation

$$\dot{\mathbf{x}} = \mathbf{P}\mathbf{x} + \mathbf{q} F(y) + \mathbf{f}(t), \qquad y = \mathbf{r}^T \mathbf{x} \qquad (H.56)$$

with the forcing function $\mathbf{f}(t)$. It was shown before that if the Popov condition holds then for every $\mathbf{f}(t)$ bounded on $(-\infty, +\infty)$ and every continuous characteristic $F(y)$ that satisfies sector conditions (H.113) there exists a bounded domain D such that the solutions $\mathbf{x}(t)$ of (H.56), which start in D, remain there for all future time. In addition, there exists on $(-\infty, +\infty)$ a unique solution $\mathbf{x}^0(t) \in D$, which is exponentially stable in the domain D with the degree σ, that is, there exist two positive constants M and μ such that for any $t \geq t_0$ and any solution $\mathbf{x}(t) \in D$, we have

$$\|\mathbf{x}(t) - \mathbf{x}^0(t)\| \leq M \|\mathbf{x}(t_0) - \mathbf{x}^0(t_0)\| \exp\left[(\sigma - \mu)(t - t_0)\right]. \quad \text{(H.143)}$$

We now proceed to compute a particular bound on $\mathbf{f}(t)$ which will guarantee that $D \subset \bar{D}_\sigma$ for all $F(y)$ inside the augmented sector of Fig. H.6.

We seek a bound ξ on $\mathbf{f}(t)$:

$$|\mathbf{f}(t)| < \xi, \quad t \in (-\infty, +\infty), \quad \text{(H.144)}$$

which implies that no solution of (H.56) leaves $\bar{D}_\sigma$. This property of solutions is assured by requiring that the derivative of $V(\mathbf{x}) = \mathbf{x}^T \mathbf{H} \mathbf{x}$ along solutions of (H.56), which is denoted by $\dot{V}_{(H.56)}$, be negative on $V(\mathbf{x}) = V_m$. According to (H.56),

$$\dot{V}_{(H.56)} = \dot{V}_{(H.7)} + 2\mathbf{x}^T \mathbf{H} \mathbf{f}, \quad \text{(H.145)}$$

where $\dot{V}_{(H.7)}$ is given in (H.130). For $\mathbf{x} \in \bar{D}_\sigma$ and $F(y)$ inside the augmented sector

$$\dot{V}_{(H.7)} < 2\sigma V. \quad \text{(H.146)}$$

Also,

$$V = \mathbf{x}^T \mathbf{H} \mathbf{x} \leq \eta \|\mathbf{x}\|^2 \quad \text{(H.147)}$$

and

$$\mathbf{x}^T \mathbf{H} \mathbf{f} < \xi \eta \|\mathbf{x}\|, \quad \text{(H.148)}$$

where

$$\eta = \max_i \{s_i(\mathbf{H})\}; \quad \text{(H.149)}$$

$s_i (i = 1, 2, \ldots, n)$ are eigenvalues (characteristic roots) of the matrix $\mathbf{H}$. Combining (H.147) and (H.148), we conclude that $\dot{V}_{(H.56)} < 0$ for all $\|\mathbf{x}\| > -\xi/\sigma$ and $\mathbf{x} \in \bar{D}_\sigma$. Consequently, if the sphere $\|\mathbf{x}\| = -\xi/\sigma$ is contained inside $\bar{D}_\sigma$, then all solutions which start inside $\bar{D}_\sigma$ remain there for all future time. The largest value of ξ which ensures this property of $\bar{D}_\sigma$ is given by

$$\xi = -\sigma \left(\frac{V_m}{\eta}\right)^{1/2}. \quad \text{(H.150)}$$

It is of interest to note that the bound ξ on the forcing function $\mathbf{f}(t)$ in (H.150) depends directly on the degree ξ of exponential stability.

A significant extension of these results would be to consider Liapunov functions of the complete Lur'e type (H.8); that is, "a quadratic form plus an integral of the nonlinearity." The additional term in the Liapunov function will allow improvement of the estimates of stability domains. Future work should also be devoted to the computational aspect of the problem.

The presented results can also be extended to cases in which the differential equations have discontinuous nonlinear characteristics. Moreover, forcing terms that depend on both the states and time should be considered.

References

[H.1] R. E. Kalman and J. E. Bertram, Control System Design via the Second Method of Liapunov—Pt. I: Continuous-time Systems; Pt. II: Discrete Time Systems, *J. Basic Eng., Trans. ASME*, **82,** 371–400 (1960).

[H.2] A. I. Lur'e and V. N. Postnikov, On the theory of Stability of Control Systems (in Russian), *Prikl. Mat. i Mekhan.*, **8,** No. 3, 246–248 (1944).

[H.3] A. I. Lur'e, *Some Nonlinear Problems in the Theory of Automatic Control* (in Russian), GOSTEKHIZDAT, Moscow, 1951. (English translation: Her Majesty's Stationery Office, London, 1957.)

[H.4] V. M. Popov, Criterii de stabilitate pentra sistemele neliniare de neglare automata, basate pe utilizarea transformatei Laplace (in Rumanian), *Studii si Cercetari de Energetica, Acad., R.P.R.*, **9,** No. 1, 119–135 (1959).

[H.5] V. M. Popov, Criterii suficiente de stabilitate asimptotica in mare pentru sistemele automate neliniare cu mai multe organe de executie (in Rumanian), *Studii si Cercetari de Energetica, Acad. R.P.R.*, **9,** No. 4 (1959).

[H.6] V. M. Popov, On the Absolute Stability of Nonlinear Control Systems (in Russian), *Avtomatika i Telemekhanika*, **22,** No. 8, 961–979 (August 1961).

[H.7] V. M. Popov and A. Halanay, On the Stability of Nonlinear Control Systems with a Delayed Variable (in Russian), *Avtomatika i Telemakhanika*, **23,** No. 7, 783–786 (February 1962).

[H.8] V. M. Popov, Solution of a New Stability Problem for Control Systems (in Russian), *Avtomatika i Telemekhanika*, **24,** No. 1, 7–26 (January 1963).

[H.9] V. M. Popov, Criterii de stabilitate pentru sistemele automate continuind elemente neunivoce, *Probleme de Automatizare, Edituza Academici R.P.R.*, October 13, 1960.

[H.10] V. A. Yakubovich, Solution of Certain Matrix Inequalities Encountered in the Theory of Automatic Control (in Russian), *Dokl. Akad. Nauk SSSR*, **143,** No. 6, 1304–1307 (1962).

[H.11] R. E. Kalman, Lyapunov Functions for the Problem of Lur'e in Automatic Control, *Proc. Nat. Acad. Sci. US*, **49,** No. 2, 201–205 (February 1963).

[H.12] K. R. Meyer, Liapunov Functions for the Problem of Lur'e, *Proc. Nat. Acad. Sci. US*, **53,** 501–503 (1965).

[H.13] S. Lefschetz, *Stability of Nonlinear Control Systems*, Academic, New York, 1965.

[H.14] M. A. Aizerman and R. F. Gantmacher, *Absolute Stability of Regulator Systems*, Holden-Day, San Francisco, 1964. (Translation of the 1963 Russian edition by USSR Academy of Sciences.)

[H.15] V. A. Yakubovich, Absolute Stability of Nonlinear Control Systems in Critical Cases (in Russian), *Avtomatika i Telemekhanika*, Pt. I: **24,** No. 3, 293–303 (March 1963); Pt. II: **24,** No. 6, 717–731 (June 1963); Pt. III: **25,** No. 5, 601–612 (May 1964).

[H.16] E. A. Barbashin and N. N. Krassovsky, On the Existence of a Liapunov Function in the Case of Asymptotic Stability in the Large (in Russian), *Prikl. Mat. i Mekham.*, **18,** 345–350 (1954).

[H.17] A. G. Filippov, Differential Equations with Discontinuous Right-Hand Sides (in Russian), *Mat. Sb.*, **51(93),** No. 1 (1960).

[H.18] A. G. Filippov, Application of the Theory of Differential Equations with Discontinuous Right-Hand Sides to Nonlinear Problems in Automatic Control, *Proc. of the 1st IFAC Congress*, Moscow, **1,** 1098–1100 (1960).

[H.19] A. A. Fel'dbaum, *Introduction to the Theory of Optimal Automatic Systems* (in Russian) FIZMATGIZ, Moscow, 1963. (English translation: Academic, New York.)

[H.20] I. Flügge-Lotz, *Discontinuous Automatic Control*, Princeton University Press, Princeton, New Jersey, 1953.
[H.21] J. André and P. Seibert, After End-Point Motions of General Discontinuous Control Systems and Their Stability Properties, *Proc. of the 1st IFAC Congress*, Moscow, **1**, 919–922 (1960).
[H.22] M. A. Aizerman and F. R. Gantmacher, Some Aspects of the Theory of a Nonlinear Automatic Control System with Discontinuous Characteristics, *Proc. of the 1st IFAC Congress*, Moscow, **1**, 913–918 (1960).
[H.23] S. Lefschetz, *Stability of Nonlinear Control Systems*, Academic, New York, 1965.
[H.24] V. A. Yakubovich, The Method of Matrix Inequalities in the Stability Theory of Nonlinear Control Systems, Pt. I: The Absolute Stability of Forced Oscillations (in Russian), *Avtomatika i Telemekhanika*, **25**, No. 7, 1017–1029 (July 1964).
[H.25] N. N. Luzin, The Qualitative Investigation of Motion of a Train (in Russian), *Mat. Sb.*, **39**, No. 3 (1932).
[H.26] B. P. Demidovich, On the Dissipativity of a Nonlinear System of Differential Equations (in Russian), *Vestn. Mosk. Gos. Univ. Ser. Mat. i Mekhan.*, No. 6 (1961).
[H.27] B. P. Demidovich, On the Dissipativity of a Nonlinear System of Differential Equations (in Russian), *Vestn. Mosk. Gos. Univ. Ser. Mat. i Mekhan.*, No. 1 (1962).
[H.28] T. Yoschizawa, Liapunov's Functions and Boundedness of Solutions, *Funkcial. Ekvac.*, **2**, 95–142 (1959).
[H.29] B. M. Budak, The Idea of Motion in a Generalized Dynamic System, *Uch. Zap. Mosk. Gos. Univ., Mathematika*, No. 5, 155 (1952).
[H.30] R. W. Brockett, The Status of Stability Theory for Deterministic Systems, *IEEE Trans. Auto. Control*, **AC-11**, No. 3, 596–607 (July 1966). (A survey paper.)
[H.31] R. E. Gantmacher and V. A. Yakubovich, Absolute Stability of Nonlinear Regulator Systems (in Russian) *Proceedings of Trudi II Vsesoyuznogo Sezda po Teoreticheskoi i Prikladnoy Mekhanike*, Nauka, Moscow, January 29–February 5, 1965
[H.32] I. G. Malkin, On the Theory of Stability of Control Systems (in Russian), *Prikl. Mat. i Mekhan.*, **15**, No. 1 59–66 (1951).
[H.33] V. A. Yakubovich, On Nonlinear Differential Equations for Control Systems with a Single Regulator (in Russian), *Vestn. Leningr. Univ.*, **2**, No. 7 (1960).
[H.34] A. M. Letov, Stability of Nonlinear Regulating Systems (in Russian), GOSTEKHIZDAT, Moscow, 1955. (English translation: Princeton University Press, Princeton, New Jersey, 1961.)
[H.35] E. N. Rozenvasser, On the Construction of a Liapunov Function for a Class of Nonlinear Systems (in Russian), *Izv. Akad. Nauk SSSR, Otd. Tekh. Nauk.*, No. 2 (1960).
[H.36] A. M. Letov, Stability of Control Systems with Two Power Elements (in Russian), *Prikl. Mat. i Mehkan.*, **17**, No. 4 (1953).
[H.37] M. A. Aizerman, On the effect of Nonlinear Functions of Several Variables on the Stability of Automatic Control Systems (in Russian), *Avtomatika i Telemekhanika*, **8**, No. 1, 64–72 (January 1947).
[H.38] N. P. Yerugin, On Some Questions of Global Stability of Motion in the Qualitative Theory of Differential Equations (in Russian), *Prikl. Mat. i Mekhan.*, **14**, No. 5, 459–512 (May 1950).
[H.39] N. N. Krassovsky, Theorems on the Stability of a Motion Defined by a System of Two Equations (in Russian), *Prikl. Mat. i Mekhan.*, **16**, No. 5, 547–554 (May 1952).
[H.40] N. N. Krassovsky, On the Stability of the Solutions of a System of Two Differential Equations (in Russian), *Prikl. Mat. i Mekhan.*, **17**, No. 6, 651–672 (June 1953).

[H.41] V. A. Pliss, Certain Problems in the Theory of Stability in the Whole (in Russian), Leningrad University Press, Leningrad, 1958.
[H.42] A. G. Dewey and E. I. Jury, A Note on Aizerman's Conjecture, *IEEE Trans. Auto. Control*, **AC-10**, No. 4, 482–483 (October 1965).
[H.43] R. E. Fitts, Two Counterexamples to Aizerman's Conjecture, *IEEE Trans. Auto. Control*, **AC-11**, No. 3, 553–556 (July 1966).
[H.44] R. E. Kalman, Physical and Mathematical Mechanisms of Instability in Nonlinear Automatic Control Systems, *J. Basic Eng.*, *Trans. ASME*, **79**, 553–556 (April 1957).
[H.45] R. W. Brockett and J. L. Williams, Frequency Domain Stability Criteria—Pts. I and II, *IEEE Trans. Auto. Control*, **AC-10**, Nos. 3 and 4, 255–261, 407–413 (July October 1965).
[H.46] V. A. Yakubovich, Method of Matrix Inequalities in the Stability Theory of Nonlinear Control Systems; Pt. III—Absolute Stability of Systems with Hystheresis Nonlinearities (in Russian), *Avtomatika i Telemekhanika*, **26**, No. 5, 753–764 (May 1965).
[H.47] V. A. Yakubovich, Method of Matrix Inequalities in the Stability Theory of Nonlinear Control Systems; Pt. II—Absolute Stability in a Class of Nonlinearities with a Condition on the Derivative (in Russian), *Avtomatika i Telemekhanika*, **26**, No. 4, 577–590 (April 1965).
[H.48] A. G. Dewey and E. I. Jury, A Stability Inequality for a Class of Nonlinear Feedback Systems, *IEEE Trans. Auto. Control*, **AC-11**, No. 1, 54–62 (January 1966).
[H.49] A. G. Dewey, On the Stability of Feedback Systems with One Differentiable Nonlinear Element, *IEEE Trans. Auto. Control*, **AC-11**, No. 3, 485–491 (July 1966).
[H.50] G. Zames, On the Input-Output Stability of Time-Varying Nonlinear Feedback Systems—Part I: Conditions Derived Using Concepts of Loop Gain, Conicity, and Positivity, *IEEE Trans. Auto. Control*, **AC-11**, No. 2, 228–239 (April 1966).
[H.51] V. A. Yakubovich, Solution of Certain Special Matrix Inequalities Occurring in the Theory of Automatic Control (in Russian), *Dokl. Akad. Nauk SSSR*, **143**, No. 6, 1304–1307 (1962).
[H.52] B. N. Naumov and Ya. Z. Tsypkin, A Frequency Criterion for Absolute Process Stability in Nonlinear Automatic Control Systems (in Russian), *Avtomatika i Telemekhanika*, **25**, No. 6, 852–867 (June 1964).
[H.53] B. N. Naumov, An Investigation of Absolute Stability of the Equilibrium State in Nonlinear Automatic Control Systems by Means of Logarithmic Characteristics (in Russian), *Avtomatika i Telemekhanika*, **26**, No. 4, 591–600 (April 1965).
[H.54] Ya. Z. Tsypkin, On Stability "in the Large" of Relay Automatic Systems (in Russian), *Izv. Akad. Nauk SSSR, Otd. Tekh. Nauk, Mekhan. i Mashinostr.*, No. 3 (1963).
[H.55] A. Halanay, Theoria calitativa a ecatiilor differenntiale (in Rumanian), *Editura Acad. R.P.R.*, 1963.
[H.56] A. Kh. Gelig, Investigation of Stability of Nonlinear Discontinuous Automatic Control Systems with a Nonunique Equilibrium State (in Russian), *Avtomatika i Telemekhanika*, **25**, No. 2, 153–160 (February 1964).
[H.57] E. N. Rozenvasser, The Absolute Stability of Nonlinear Systems (in Russian), *Avtomatika i Telemekhanika*, **24**, No. 3, 304–313 (March 1963).
[H.58] W. T. Higgins, Jr. and D. G. Schultz, The Stability of Certain Nonlinear, Time-Varying Systems, Report No. CSL-66-2, Control System Laboratory, The University of Arizona, May 1966.
[H.59] J. J. Bongiorno, Jr., An Extension of the Nyquist-Barkhausen Stability Criterion to Linear Lumped-Parameter Systems with Time-Varying Elements, *IEEE Trans. Auto. Control*, **AC-8**, No. 2, 166–170 (April 1963).

[H.60] K. S. Navendra and R. M. Goldwyn, A Geometrical Criterion for the Stability of Certain Nonlinear Nonautonomous Systems, *IEEE Trans. Circuit Theory*, **CT-11**, No. 3, 406–408 (September 1964).

[H.61] I. W. Sandberg, A Frequency-Domain Condition for the Stability of Feedback Systems Containing a Single Time-Varying Nonlinear Element, *Bell System Tech. J.*, **43**, Pt. 2, 1601–1608 (July 1964).

[H.62] Z. V. Rekasius and J. R. Rowland, A Stability Criterion for Feedback Systems Containing a Single Time-Varying Nonlinear Element, *IEEE Trans. Auto. Control*, **AC-10**, No. 3, 352–354 (July 1965).

[H.63] G. Zames, On the Input-Output Stability of Time-Varying Nonlinear Feedback Systems—Part II: Conditions Involving Circles in the Frequency Plane and Sector Nonlinearities, *IEEE Trans. Auto. Control*, **AC-11**, No. 3, 465–476 (July 1966).

[H.64] Ya. Z. Tsypkin, Absolute Stability of Equilibrium Position and Transients in Nonlinear Sampled-Data Systems (in Russian), *Avtomatika i Telemekhanika*, **24**, No. 12, 1601–1614 (December 1963).

[H.65] G. Szegö and R. E. Kalman, Sur la stabilité absolue d'un système d'équations aux differences finies, *Compt. Rend. Acad. Sci.*, **257**, 338–390 (1963).

[H.66] G. Szegö, On the Absolute Stability of Sampled-Data Control Systems, *Proc. Nat. Acad. Sci., US.*, **49**, No. 9, 558–560 (1963).

[H.67] E. I. Jury and B. W. Lee, The Absolute Stability of Systems with Many Nonlinearities (in Russian), *Avtomatika i Telemekhanika*, **26**, No. 6, 945–965 (June 1965).

[H.68] C. A. Desoer, A Generalization of the Popov Criterion, *IEEE Trans. Auto. Control*, **AC-10**, No. 2, 182–185 (April 1965).

[H.69] H. Tokumaru and N. Saito, On the Stability of Automatic Control Systems with Many Nonlinear Characteristics, *Mem. Fac. Eng., Kyoto University*, **27**, Pt. 3, 347–379 (July 1965).

[H.70] B. D. O. Anderson, Stability of Control Systems with Multiple Nonlinearities, *J. Franklin Inst.*, **3**, No. 6, 535–540 (1966).

[H.71] D. D. Šilijak, Popov Inequality via Parameter Plane, *Proc. 1st Princeton Conf. Information Sciences and Systems*, Princeton University, Princeton, New Jersey, March 30–31, 1967, pp. 183–187.

[H.72] D. D. Šiljak, Absolute Stability in the Parameter Space, *Proc. 1st Asilomar Conf. Circuit and System Theory*, Asilomar, 627–632 (November 1–3, 1967).

[H.73] W. Hahn, *Theory and Application of Liapunov's Direct Method*, Prentice-Hall, Englewood Cliffs, N.J., 1967.

[H.74] S. G. Margolis and W. G. Vogt, Control Engineering Applications of V. I. Zubov's Construction Procedure for Liapunov Functions, *IRE Trans. Auto. Control*, **8**, No. 1, 104–113 (1963).

[H.75] J. J. Rodden, Numerical Applications of Liapunov Stability Theory, *Proc. 1964 Joint Auto. Control Conf.*, Stanford, June 1964, pp. 261–268.

[H.76] S. Weissenberger, Stability-Boundary Approximations for Relay-Control Systems Via a Steepest-Ascent Construction of Lyapunov Functions, *J. Basic Eng., Trans. ASME*, Series D, **88**, 419–428 (June 1966).

[H.77] S. Weissenberger, Application of Results from the Absolute Stability Problem to the Computation of Finite Stability Domains, to appear in IEEE Transactions on Automatic Control.

[H.78] J. A. Walker and N. H. McClamrock, Finite Regions of Attraction for the Problem of Lur'e, *Int. J. Control*, **6**, No. 4, 331–336 (1967).

[H.79] D. Šiljak and S. Weissenberger, Regions of Exponential Stability for the Problem of Lur'e, *Inter. Symp. Network Theory*, Belgrade, Sept. 4–7, 1968.

Problems

Chapter 1

Problem 1.1. Given the algebraic equation

$$s^2 + \alpha s + \beta = 0. \tag{P1.1}$$

(a) What should be the values of α and β for equation P1.1 to have the roots $s_{1,2}$ with $\zeta = 0.5$, $\omega_n = 100$? What are the corresponding values of σ and ω?
(b) What is the expression for the Jacobian $J = J(R, I/\alpha, \beta)$?
(c) Express the functions $\alpha = \alpha(\sigma, \omega)$ and $\beta = \beta(\sigma, \omega)$ in terms of X_k and Y_k functions. Then plot the $\Sigma = 0$ and $\zeta = 0.5$ curves. From the obtained plot, calculate by inspection the approximate root values for the point $M(72; 5000)$.
(d) Write a general computer program for calculating $X_k(\sigma, \omega)$ and $Y_k(\sigma, \omega)$ functions from their recurrence formulas.

Problem 1.2. The algebraic equation has the form

$$F(s) \equiv \sum_{k=2}^{n} a_k s^k + \alpha s + \beta = 0, \quad a_n \neq 0. \tag{P1.2}$$

(a) Find the values of σ and ω for which the Jacobian is zero.
(b) Express the functions $\alpha = \alpha(\sigma, \omega)$ and $\beta = \beta(\sigma, \omega)$ in terms of the X_k and Y_k functions.
(c) Derive the conditions for the extrema of the Σ curves with respect to α and express them in terms of X_k functions only.
(d) Derive the equations for the $\zeta = 1$ curve. In the case under investigation, it is known that the $\zeta = 1$ curve has cusps. Derive the equations for calculating the locations of the cups along the $\zeta = 1$ curve.
(e) What are the Σ curves and what is their slope in the $\alpha\beta$ plane? Apply the conclusions about the σ curves to calculate approximately the roots of equation

$$s^3 + s^2 + 0.25s + 0.017 = 0 \tag{P1.3}$$

from the parameter plane diagram of Figure P1.1. What, approximately, are the roots of the corresponding algebraic equation at the point M_2 of the diagram in Figure P1.1?
(f) Write a general computer program for plotting the parameter plane curves in case of (P1.2).

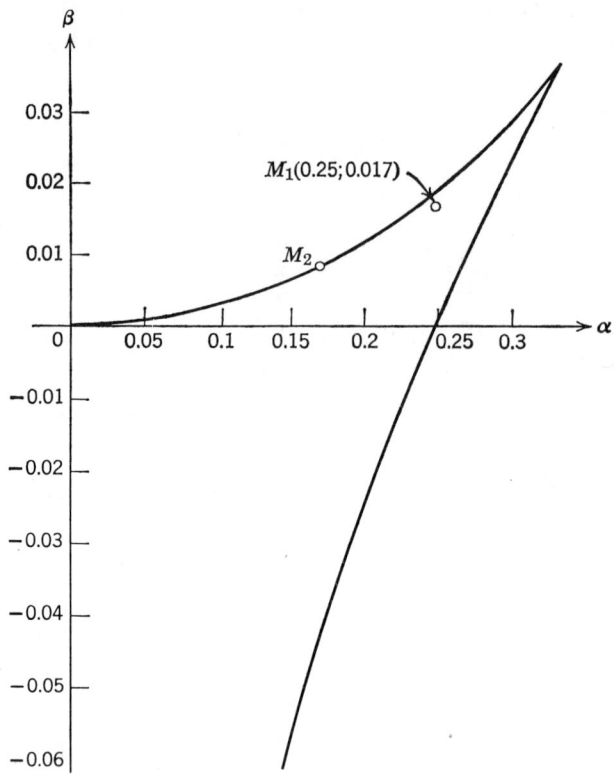

Figure P1.1 Parameter plane diagram.

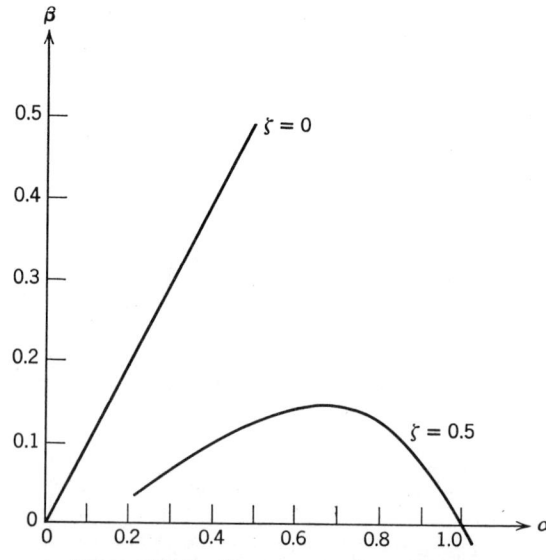

Figure P1.2 Parameter plane diagram.

Problem 1.3. Given the algebraic equation in the form

$$F(s) \equiv \sum_{\substack{k=0 \\ k \neq p,q}}^{n} a_k s^k + \alpha \beta s^p + \beta s^q = 0, \qquad a_n \neq 0. \tag{P1.4}$$

(a) Express the Jacobian $J = J(R, I/\alpha, \beta)$.
(b) What are the equations of the $\zeta = 1$ curve? Determine the slope of the tangents at the points on the $\zeta = 1$ curve. What are these tangents?
(c) Apply the expression of J obtained in (a) to equation

$$s^3 + s^2 + \alpha \beta s + \beta = 0 \tag{P1.5}$$

and shade the curves $\zeta = 0$ and $\zeta = 0.5$ plotted in Figure P1.2. Determine the number of roots at various regions of the plot with respect to both the regions $\zeta \geq 0$ and $\zeta \geq 0.5$. Before the $\zeta = 0.5$ curve is shaded, complete the curve approximately for the root distribution test without calculating additional points.
(d) Sketch the $\zeta = 0$ ($\omega > 0$) curve for equation

$$s^4 + \alpha s^3 + \beta s^2 + 5s + 2 = 0 \tag{P1.6}$$

by finding the behavior of the curve for $\omega \to 0$, $\omega \to \infty$ and the extrema with respect to α and β. Then shade the curve and determine the related root distribution evaluating all the roots at one of the extrema. (Hint: In evaluating all the roots at the extrema, two roots are known and use the synthetic division method of Theorem A.2 to evaluate the remaining two roots.)

Problem 1.4. For equation

$$s^4 + 7s^2 + 3qs^2 + \alpha s + \beta = 0, \tag{P1.7}$$

the parameter plane diagram of $\zeta = 0.5$ curve is shown in Figure P1.3.

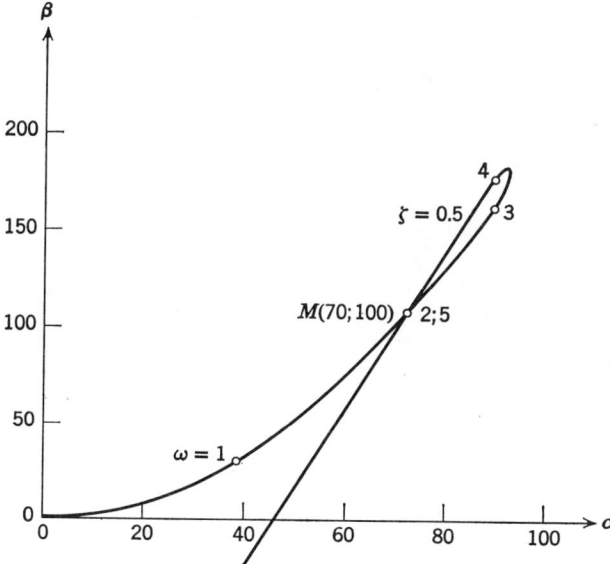

Figure P1.3 Parameter plane diagram.

Problems

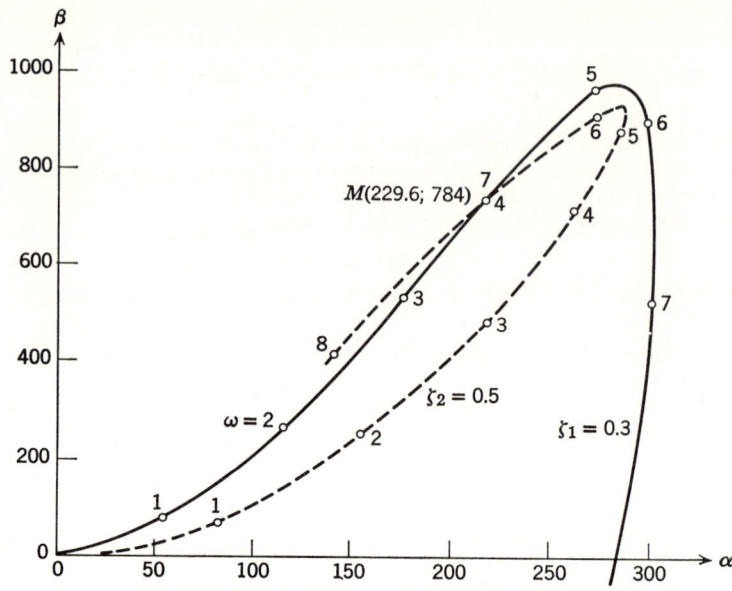

Figure P1.4 Parameter plane diagram.

(a) Derive the equations for the boundaries of the $\zeta \geq 0.5$ region R_4.
(b) Shade the boundaries and determine the root distribution in the $\alpha\beta$ plane by calculating all the roots at the point $M(70; 100)$.
(c) Repeat the same investigation on the plot of Figure P1.4, which corresponds to equation

$$s^4 + 9.4s^3 + 81.8s^2 + \alpha s + \beta = 0. \tag{P1.8}$$

(d) Plot the $F(s)$ curve for a chosen point at the region with maximum number of roots on Figure P1.3. Then prove that the region is $R_n = R_4$. Repeat the same investigation for the case of equation P1.8 and Figure P1.4. (For the procedure, refer to Theorem A.6, Appendix A, which states the Cauchy Argument Principle.)

Problem 1.5. Given the algebraic equation

$$s^5 + 7s^4 + 18s^3 + \alpha s^2 + \beta s + 6 = 0. \tag{P1.9}$$

(a) Determine the equations for $\zeta = 0$ curve and sketch the curve estimating the behavior of the curve for $\omega \to 0$, $\omega \to \infty$, and the extrema with respect to α and β.
(b) Determine the root distribution for the condition $\zeta \geq 0$.
(c) Use the plot of the $F(s)$ curve to determine the number of the roots relative to a chosen contour at a certain convenient point of the $\alpha\beta$ plane (Theorem A.6, Appendix A).

Problem 1.6. Given the algebraic equation

$$\alpha\beta s^2 + (\alpha\beta + 1 - \beta)s + 1 = 0. \tag{P1.10}$$

(a) Plot the root boundaries for the condition $\zeta \geq 0$ and determine the corresponding root distribution.

(b) Repeat the whole procedure when the condition is that all the roots have the real part less than -1.
(c) Use the $F(s)$ curve to determine the number of roots at certain points of the $\alpha\beta$ plane (Theorem A.6, Appendix A).
(d) Write a general computer program for plotting the parameter plane curves when $a_k = b_k\alpha + c_k\beta + d_k\alpha\beta + h_k$.

Problem 1.7. Given the algebraic equation

$$s^4 + (3 + \alpha)s^3 + (2 + 3\alpha)s^2 + (10 + 2\alpha)s + 10\beta = 0. \quad (P1.11)$$

(a) Write the computer program for plotting the $\zeta = 0$ curve (Appendix B) and plot the curve.
(b) Determine the corresponding root distribution. (Refer to the origin of the $\alpha\beta$ plane for the calculation of all the roots at one point in the $\alpha\beta$ plane.)
(c) Use the $F(s)$ curve to determine the number of the left-half plane roots at a certain point of the $\alpha\beta$ plane (Theorem A.6, Appendix A). Use the computer program to plot the $F(s)$ curve (Appendix B).
(d) Write a general computer program for plotting the curve $F(s)$ in the $F(s)$ plane. The value of σ is constant and ω varies from 0 to ∞.

Problem 1.8. Given the algebraic equation

$$s^3 + 3s^2 + (2 + \alpha)s + \beta = 0 \quad (P1.12)$$

and the corresponding parameter plane diagram in Figure P1.5.
(a) It is required to determine approximately all the roots at the points M_1 and M_2. Then check the results by comparing the coefficients of the equation obtained by multiplying the approximate root values with the coefficients of (P1.12).
(b) Perform the inverse parameter plane mapping along the $\zeta = 0.3$ curve as shown in Figure P1.5.

Problem 1.9.
(a) Prove the following statement: If the Jacobian $J = J(R, I/\alpha, \beta)$ defined in (1.6) is positive (negative), then facing the direction in which ω_n increases, the left (right) side of the ζ curve corresponds to a positive increment $\Delta\zeta$ and, therefore, should be shaded.

The proof should be based upon the direction of the vector

$$\mathbf{J}_1 = J_1\mathbf{k} \quad \text{which is defined as} \quad \mathbf{J}_1 = \Delta\boldsymbol{\omega}_n \times \Delta\boldsymbol{\zeta}, \quad (P1.13)$$

where

$$\Delta\boldsymbol{\omega} = \frac{\partial \omega_n}{\partial \alpha}\mathbf{i} + \frac{\partial \omega_n}{\partial \beta}\mathbf{j}$$

$$\Delta\boldsymbol{\zeta} = \frac{\partial \zeta}{\partial \alpha}\mathbf{i} + \frac{\partial \zeta}{\partial \beta}\mathbf{j}. \quad (P1.14)$$

(b) Then apply the above statement to the $\zeta = 0.3$ curve of the equation

$$0.0005s^4 + (0.005\alpha + 0.06)s^3 + (0.06\alpha + 1)s^2 + (\alpha + \beta)s + 0.8\alpha\beta = 0. \quad (P1.15)$$

Plot the $\zeta = 0.3$ curve using a computer.
(c) Finally, determine the corresponding root distribution.

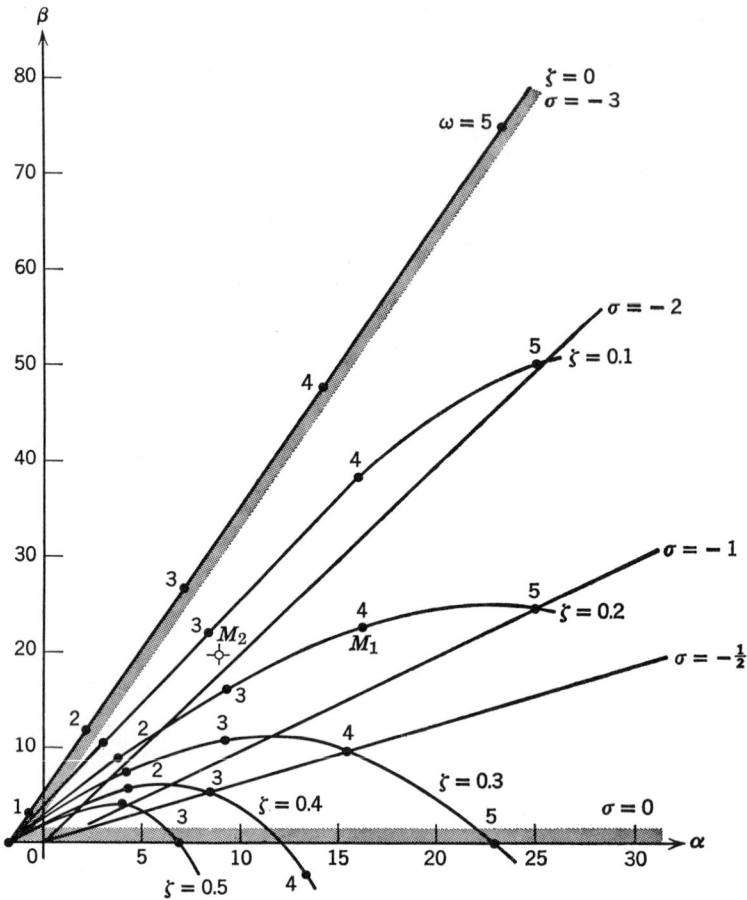

Figure P1.5 Parameter plane diagram.

Problem 1.10. A transcendental equation[1] is given as

$$F(s) \equiv P(s)e^{sT_t} + Q(s) = 0, \qquad (P1.16)$$

which can be rewritten as

$$F(s) \equiv \sum_{k=0}^{n} a_k(s)s^k = 0, \qquad (P1.17)$$

where

$$a_k(s) = (b_k + c_k e^{sT_t})\alpha + (d_k + e_k e^{sT_t})\beta + f_k + g_k e^{sT_t}. \qquad (P1.18)$$

(a) Derive the parameter plane equations $\alpha = \alpha(\sigma, \omega)$, $\beta = \beta(\sigma, \omega)$ for this case assuming $s = \sigma + j\omega$ and

$$e^{sT_t} = e^{\sigma T_t}(\cos \omega T_t + j \sin \omega T_t). \qquad (P1.19)$$

Repeat the procedure for $\alpha = \alpha(\omega_n, \zeta)$, $\beta = (\omega_n, \zeta)$.

---

[1] Transcendental equations arise in linear systems having a pure time delay sometimes called transport lag and denoted by T_t. See, for example, references 1.22, 50.

(b) Plot the $\Sigma = 0$ curve for the case

$$F(s) \equiv e^s T_t s^2 + Ks + KT = 0 \qquad (P1.20)$$

when $T_t = 1$ sec, and determine the region of stability in the plane of $\alpha(\alpha = K)$ and $\beta(\beta = KT)$.

(c) Plot the $\zeta = 0.2$ curve and show that there exists no values of α and β that permit all the roots of (P1.20) to have $\zeta \geq 0.2$.

Problem 1.11. Consider the equation

$$s^5 + (42 + p_1)s^4 + (588 + 42p_1)s^3 + (2744 + 588p_1 + p_2)s$$
$$+ (2744p_1 + 2p_3)s + p_4 = 0, \quad (P1.21)$$

in which the coefficients are expressed as linear functions of four system parameters, $p_1, p_2, p_3,$ and p_4.

(a) Find the values of the parameters so that the complex conjugate root pair is $s_{1,2} = -5 \pm j6$, and these roots become $s'_{1,} = -2 \pm j2$ when $p'_1 = p_1$, $p'_2 = 10p_2$, $p'_3 = 10p_3$, and $p'_4 = 10p_4$. (This problem is related to the "large root sensitivity" which is discussed in Chapter 2.)

(b) Write a general computer program for calculating $\partial X_k/\partial \omega$ and $\partial Y_k/\partial \omega$ from the X_k, Y_k recurrence relationship.

Chapter 2

Problem 2.1. Given a system with the block diagram of Figure P2.1. Determine the values of the parameters K_1, K_{-1}, and K_{-2} so that $K_v \geq 10$, $\zeta \simeq 0.5$, $\omega_n \simeq 15$. The remaining pair of characteristic roots should have the real part less than $-4\omega_n\zeta$.

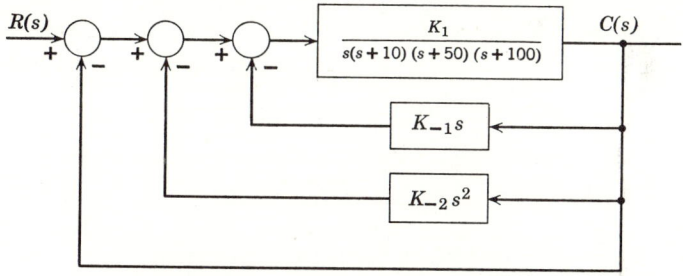

Figure P2.1   System block diagram.

Problem 2.2. Given a one-loop unity-feedback control system with an open-loop transfer function

$$G_2(s) = \frac{1}{s(s+4)}. \qquad (P2.1)$$

Introduce a lead cascade compensator with the transfer function

$$G_1(s) = K_1 \frac{s+Z}{s+P}. \qquad (P2.2)$$

Determine the parameters K_1, Z, and P so that the bandwidth is less than 80 rad/sec, velocity error constant is 250, the damping coefficient of the control poles is approximately $\zeta = 0.5$, and the overshoot of the step function response is less than 30%. (Use the third-degree equation charts of Section 2.5.)

Problem 2.3. Given a one-loop unity-feedback control system with an open-loop transfer function

$$G_2(s) = \frac{1}{s(s^2 + 2s + 1.2)}. \tag{P2.3}$$

Introduce an integral cascade compensator with the transfer function

$$G_1(s) = K_1 \frac{s + \lambda\delta}{s + \delta}, \tag{P2.4}$$

where $\lambda\delta \ll 1$ and $\lambda \simeq 10\delta$. Determine the compensator parameters K_1, λ, and δ so that the overshoot of the step function response is less than 30% and the velocity error constant $K_v \geq 5$. (To facilitate the design, the terms in the characteristic equation that contain only δ should be neglected so that only the two last coefficients of the equation have the adjustable compensator parameters. Plotting of the $\zeta = 1$ and $\zeta = 0.3$ curves is sufficient to meet the design specifications.)

Without neglecting the terms with δ, repeat the solution procedure without plotting any curves. The values $\zeta = 0.3$ and $\omega_n = 0.6$ of the dominant characteristic roots are given in advance.

Problem 2.4. Given the feedback control system in Figure P2.2, with transfer functions

$$G_1(s) = \frac{K_1}{s(s + 1)(s + 2)}, \quad G_{-1}(s) = 1 + K_{-1}s. \tag{P2.5}$$

Determine the parameters K_1 and K_{-1} so that the velocity error constant is $K_v = 0.5$ and the damping factor of the complex roots is $\zeta = 0.7$. (Use third-degree equation charts of Section 2.5.)

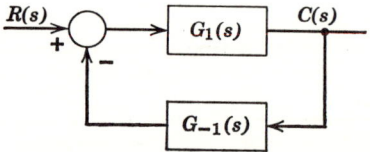

Figure P2.2 System block diagram.

If the transfer function $G_{-1}(s)$ is changed to

$$G_{-1}(s) = 1 + K_{-1}s + K_{-2}s^2 \tag{P2.6}$$

and $K_1 = 8$, calculate K_{-1} and K_{-2} so that $\zeta = 0.5$ and (a) $K_v = 0.5$, (b) $K_v = 1$. Perform the solution procedure in both the coefficient and the parameter plane.

Problem 2.5. Given the unsymmetrical parallel-T network of Figure P2.3. Study the pole-zero locations and the bandwidth frequency as functions of the parameters α and β.

Plot the constant Q loci in the $\alpha\beta$ plane. (Note that $Q = Q(\alpha, \beta)$ can be expressed as function of the adjustable parameters α and β.)

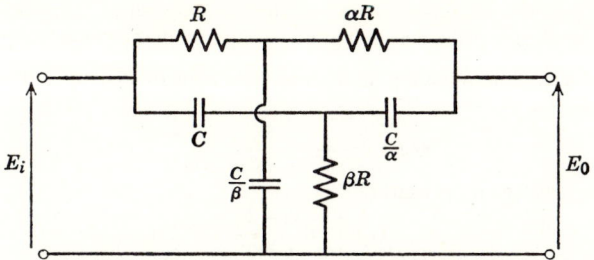

Figure P2.3 Unsymmetrical parallel-T network.

Problem 2.6. A bridged-T network is shown in Figure P2.4, which can be used to generate complex conjugate zeros in the left half of the s plane. Investigate the pole-zero configuration of the network transfer function as function of the parameters α and β.

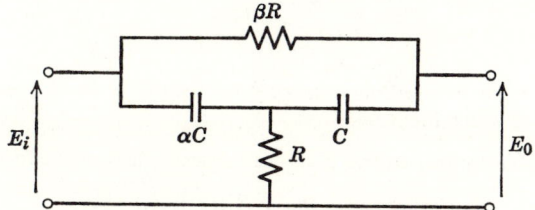

Figure P2.4 Bridged-T network.

Problem 2.7. Given the forward loop transfer function

$$G_2(s) = \frac{K_2}{s(s+0.5)(s+7)} \tag{P2.7}$$

of a one-loop unity-feedback control system. Introduce a cascade compensator with the transfer function

$$G_1(s) = \frac{s+Z}{s+P} \tag{P2.8}$$

so that the control characteristic roots are given by $\zeta = 0.5$ and $\omega_n = 6$. The minimum acceptable value of K_v is 4.15. Perform the solution process analytically.

If the above analytical calculations are performed, the obtained bandwidth frequency is $\omega_b \simeq 6.5$ rad/sec. Redesign the system so that the frequency ω_b is reduced to 4 rad/sec.

Problem 2.8. Given the system of Figure P2.5. Determine the parameters K_1, K_{-1}, P

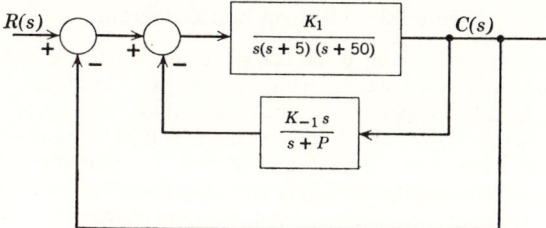

Figure P2.5 System block diagram.

582 Problems

so that $K_v \geq 5$ and the dominant characteristic roots are set at the locations for which $\zeta = 0.5$, $\omega_n = 10$. Perform the design without using any graphical technique.

Problem 2.9. Given a one-loop unity-feedback control system with the open-loop transfer function

$$G_2(s) = \frac{K_1}{s(s^2 + 2s + 1.2)}.$$ (P2.9)

By using the cascade lag compensator

$$G_2(s) = \frac{s + \lambda \delta}{s + \delta},$$ (P2.10)

satisfy the requirements $K_v \geq 5$, $\zeta = 0.3$, $\omega_b = 0.9$, where K_v is the velocity error constant, ζ is the damping factor of the dominant characteristic roots, and ω_b is the bandwidth specification.

Problem 2.10. Given a servosystem with forward transfer function

$$G(s) = \frac{1000}{s(s + 10)}.$$ (P2.11)

It is required that $K_v \geq 100$; settling time $T_s \leq 0.5$ sec ($T_s \simeq 4/\omega_n \zeta$); and almost critically damped response to a step input.

Hint: A reasonable compensator is a lag-lead cascade filter of the form

$$G_c(s) = \frac{K(s + Z_1)(s + Z_2)}{1000(s + P_1)(s + P_2)}$$ (P2.12)

with parameters K, Z_1, Z_2, P_2 to be determined (the pole of the lag section P_1 is small and it may be assumed as $P_1 \simeq 0$). As is common practice, let $Z_1 Z_2/P_1 P_2 = 1$.

Problem 2.11. A system is given in Figure P2.6 in which a combined cascade feedback compensation is used. Four variables, P, Z, K_1, and K_{-1}, should be determined in the design procedure. Select these parameters for a chosen set of performance characteristics.

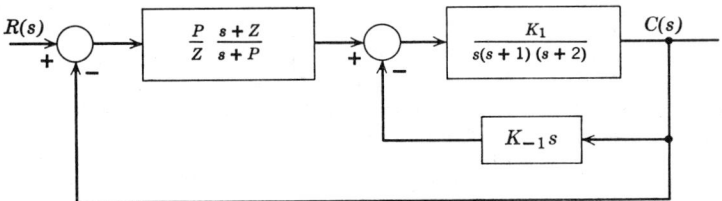

Figure P2.6   System block diagram.

It should be noted that parameter plane diagrams are plotted for only two variables, and to perform the design by a purely graphical technique, one should do the following.
  (a) Restrict the cascade compensator to be a lead filter, and define $\gamma = Z/P = 0.1$, where the numerical value is based on common engineering practice.
  (b) Consider K_1 as a "third parameter," and select a sequence of numerical values for K_1. A family of parameter plane curves can be prepared for each value of K_1 on a digital computer until a satisfactory solution is obtained. (Note that interpolation between K_1 values is not difficult.)

Problem 2.12. Given the system of Figure P2.7. Determine the parameters K, γ, P so that the following specifications are met: (a) $K_v \geq 28$, (b) step response overshoot less than 30%.

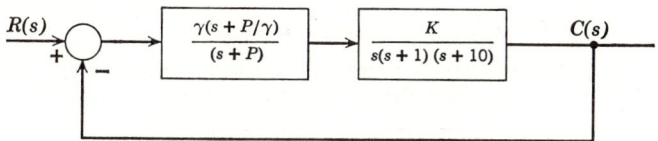

Figure P2.7 System block diagram.

Since such a design will reveal a value of $\gamma \simeq 25$, which is greater than ten, this indicates a multiple lead compensator with the transfer function

$$G_c(s) = \frac{\gamma_1 \gamma_2 (s + P_1/\gamma_1)(s + P_2/\gamma_2)}{(s + P_1)(s + P_2)} \qquad (P2.13)$$

or a tachometer feedback compensation. Try both possibilities to improve the previous design.

Problem 2.13. For a control system with the forward transfer function

$$G(s) = \frac{50}{s(s + 5)}, \qquad (P2.14)$$

it is possible to locate characteristic roots at $s_{1,2} = -5 \pm j5$ ($\zeta = 0.707$, $\omega_n = 5\sqrt{2}$) by using either tachometer feedback ($K_{-1} = 0.1$) or a cascade compensator ($Z = 5$, $P = 10$). In each case, $K_v = 5$. Design a system using both feedback and series compensator to satisfy the same specifications.

Reconsider the above problem in view of a specification requiring minimum sensitivity of the specified roots with respect to the forward gain K appearing in (P2.14) as $K = 50$. In other words, K is now considered as variable with its nominal value $K = 50$. The sensitivity index S to be minimized is

$$S = [(S_K^\zeta)^2 + (S_K^{\omega_n})^2]^{1/2}. \qquad (P2.15)$$

The parameter Z of the cascade compensator is $Z = 5$. How does the parameter P influence the design? What is the value of the parameter K_{-1} that satisfies the minimum of S?

Problem 2.14. A portion of a control system has the forward transfer function

$$G(s) = \frac{K}{s(s + 2)(s + 4)}. \qquad (P2.16)$$

Dominant closed-loop transfer function poles with $\zeta = 0.5$ and $\omega_n = 5$ are required for the desired response. Overshoot is of prime importance; therefore, ζ should be maintained within 1% accuracy for the anticipated gain variations of 5%. Velocity and acceleration signals are required elsewhere in the system and are therefore available for use as feedback signals.

Note that the problem as stated gives three specifications (pole pair and S_K^ζ) and three parameters. Additional specification concerning $S_K^{\omega_n}$ would normally require the insertion of an additional parameter, whereas a specification regarding the total pole sensitivity would not. Also note that the specifications permit ω_n to vary, if necessary, in order to maintain ζ approximately constant.

Problem 2.15. Consider the system of Figure P2.8 with transfer functions

$$G_1(s) = \frac{K_1(s+Z)}{s+0.2}, \qquad G_2(s) = \frac{K_2}{s(s^2+0.8s+1)} \qquad (P2.17)$$

$$G_{-1}(s) = K_{-1}s, \qquad G_{-2}(s) = K_{-2}s^2.$$

By using the truncation technique, determine the unspecified parameters of (P2.17) so that the characteristic roots are set so that the dominant complex roots $s_{1,2}$ are determined

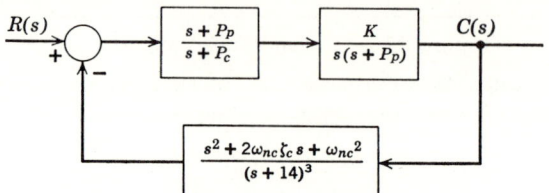

Figure P2.8 System block diagram.

by $\zeta = 0.5$, $\omega_n = 1$ and the double real root by $s_{3,4} = -10$. Then use the calculated parameters to calculate the exact root values (perform this calculation on a digital computer using the method given in Appendix B) and compare them with the approximate ones.

To further illustrate the accuracy of the approximation involved in the truncation, specify $K_1K_2 = 80.2$ and $Z = 1$ and plot the root locus of the inverse parameter mapping corresponding to $\zeta = 0.3$ and $\alpha = K_2K_{-2}$, $\beta = K_2K_{-1}$. The root locus should be plotted from the original characteristic equation and the truncated equations. Discuss the obtained root loci in view of the accuracy of the truncation.

Problem 2.16. Consider the control system in Figure P2.8. The feedback compensation poles are set at $s = -14$. The zeros of the feedback compensator (ζ_c, ω_{nc}), the pole of the series compensator P_c, and the system gain K are to be determined so that the dominant roots vary from point $A(-5 \pm j6)$ to point $B(-2 \pm j2)$ on the s plane as the gain decreases by a factor of 10.

Study the system by both the parameter plane mapping and the inverse parameter mapping for $\zeta = 0.6$.

Chapter 3

Problem 3.1. Choose transfer functions $G_1(s)$ and $G_2(s)$ in the system of Figure 3.7 and write the differential equation with respect to the variable x. The forcing functions f_1 and f_2 may be first considered as identically zero.

Then the transfer function $G_2(s)$ is given as

$$G_2(s) = \frac{0.5(s+1)}{0.2s+1} \qquad (P3.1)$$

and the nonlinearity n_2 is not present in the system—the signal x is applied directly to the block $G_2(s)$. If

$$(0.2s+1)c = 0.5(s+1)x = f_2, \qquad (P3.2)$$

where c is the output of the block $G_2(s)$ and $f_1(t) \equiv 0$, write the differential equation describing the system with respect to the signal x.

(Of course, assume that $F_1(x)$ and $F_2(x)$ are the nonlinear characteristics of the nonlinearities n_1 and n_2, respectively.)

Repeat the procedure for the system in Figure 3.8 with arbitrary choice of the transfer functions involved.

Problem 3.2. Derive the expressions for the describing function when the nonlinear characteristic is piecewise-linear with the dead-zone D and saturation S. Plot the corresponding diagrams N_1/k versus A/D when $S = 0$, and N_1/k versus A/S when $D = 0$. Then plot on the same diagrams the amplitude ratio of the first and third harmonic. In which case ($D = 0$ or $S = 0$) is the low-pass nature of the linear part of the system following the nonlinearity not critical when the amplitude of the input signal to the nonlinearity is large (small)?

Problem 3.3. Derive the expressions for the describing functions of nonlinearities No. 5, No. 13, No. 14, and No. 17 of Table F.1, Appendix F.

Write a computer program for calculating the describing function No. 14 and verify the corresponding diagram in Table F.1.

Problem 3.4. By using the approximate formula

$$N_1(A) \simeq \frac{2}{3A}\left[F(A) + F\left(\frac{A}{2}\right)\right] \tag{P3.3}$$

calculate $N_1(A)$ for the nonlinear characteristic No. 13 of Table F.1, in Appendix F, and compare the obtained results with the corresponding diagram given in Table F.1.

Repeat the procedure for a polynomial nonlinearity and use the diagram of Figure F.2 of Appendix F for comparison.

Problem 3.5. Differential equation describing a physical pendulum is

$$\ddot{x} + \dot{x} + \frac{g}{l}\sin x = 0. \tag{P3.4}$$

For small deviations $\sin x$ in (P3.4) may be substituted by the linear and cubic term of the corresponding Taylor series. Show that the solution in the first approximation of the Krylov-Bogoliubov method is

$$x = a_0 e^{-\delta t} \sin\left\{\omega_0\left[t + \frac{a^2}{32\delta}(e^{-2\delta t} - 1)\right] + \theta\right\}, \tag{P3.5}$$

where $\delta = \lambda/2$ and $\omega_0^2 = g/l$.

Use a computer to simulate (P3.4) and study the solution for different values of the parameters λ and ω_0.

If $\lambda = 0$ and only the linear term in the Taylor series is considered, the solution of (P3.4) becomes $x = A\sin(\Omega t + \theta)$. Express A and Ω as functions of the initial position $x(0) = x_0$ and the initial velocity $\dot{x}(0) = \dot{x}_0$, $t_0 = 0$.

Problem 3.6. Consider the linear differential equation

$$x + \lambda\dot{x} + \omega_0^2 x = 0, \tag{P3.6}$$

where the coefficient λ is small. Show that the second approximation of the solution in the Krylov-Bogoliubov method is given as

$$x = a\sin\phi \tag{P3.7}$$

$$\dot{a} = -\frac{\lambda}{2}$$

$$\phi = \omega_0\left[1 - \frac{1}{8}\left(\frac{\lambda}{\omega_0}\right)^2\right]. \tag{P3.8}$$

Then calculate the solution in (P3.7) using (P3.8) and compare it with the exact one obtained by direct integration of (P3.6).

Take $\lambda/\omega = \ln 2/\pi$ and calculate the error in frequency caused by using the expression $\omega = \phi$ in (P3.8) instead of the exact frequency obtained by the integration.

Problem 3.7. Derive the expressions for F^0, N_1, and N_2 for the asymmetric nonlinearities No. 1, No. 2, No. 11, No. 12, and No. 17 of Table F.2.

Use the computer to check the diagrams in Figures 5.7c, 5.11, and 5.14.

Problem 3.8. The dual-input describing function (DIDF) is defined for two sine waves at the input to the nonlinearity, with one wave at a multiple frequency of the first:

$$x = A, \quad \sin(\Omega t + \theta) + A_2 \sin n\Omega t. \tag{P3.9}$$

The magnitude of the DIDF is defined as

$$|\text{DIDF}| \triangleq \frac{\text{amplitude of desired frequency component in output}}{\text{amplitude of same frequency component in input}}. \tag{P3.10}$$

The phase shift of the DIDF is defined as the shift from input to output of the nonlinearity of the same frequency component. Since there are two components in the input (P3.9), a DIDF may be defined for each.

Derive the general expressions for DIDF assuming the function $F(x) = x^3$ for the first and third harmonic.

Problem 3.9. Bendixon's theorem states: If a half-trajectory[2] T remains in a closed bounded domain D without approaching a singular point (equilibrium position), then the trajectory T is either a closed trajectory or approaches a closed trajectory. This theorem gives sufficient conditions for existence of a closed trajectory-limit cycle. Its principal limitation is the difficulty of determining the domain D satisfying the requirements of the theorem. In the case of a ring-shaped domain D bounded by two closed curves C_1 and C_2 (see Figure 3.13) it is sufficient for the existence of at least one limit cycle that (a) trajectories enter (leave) D through every point of C_1 and C_2; (b) there are no singular points either in D or on C_1 and C_2. There is also a test developed by Bendixon, sometimes called the negative criterion of Bendixon, which gives sufficient conditions for the nonexistence of limit cycles. It says: if the expression $\partial X_1/\partial x_1 + \partial X_2/\partial x_2$ of equations[3]

$$\begin{aligned} \dot{x}_1 &= X_1(x_1, x_2) \\ \dot{x}_2 &= X_2(x_1, x_2) \end{aligned} \tag{P3.11}$$

does not change its sign within a simply connected domain[4] D, then no closed trajectory can exist in D.

The proof is immediate if the divergence theorem[5] of Gauss is used. Suppose that a

---

[2] By a half-trajectory we mean the portion of a trajectory associated with a solution $x(t)$ over the interval $t_0 \leq t < +\infty$ or $-\infty < t \leq t_0$ where t_0 is arbitrary but finite.

[3] This is the state variable form of differential equations that is discussed in Section 8.2.

[4] A simply connected domain has a property that any two points can be joined by a broken line lying wholly within the domain.

[5] See, for example, L. A. Pipes, *Applied Mathematics for Engineers and Physicists*, 2nd ed. McGraw-Hill, New York, 1958.

closed curve such as C exists in D. If D' is the domain bounded by C, then

$$\iint_{D'} \left(\frac{\partial X_1}{\partial x_1} + \frac{\partial X_2}{\partial x_2} \right) dx_1\, dx_2 = \oint_C (X_1\, dx_1 - X_2\, dx_2) \neq 0. \tag{P3.12}$$

However, along the path C equations 2.67 hold, and so $X_1\, dx_1 - X_2\, dx_2 = 0$. Hence the simple integral is zero. This contradiction proves the criterion.

As an application of Bendixon's negative criterion, consider the van der Pol equation

$$\begin{aligned} \dot{x} &= x_2 \\ \dot{x}_2 &= -x_1 + \varepsilon(1 - x_1^2)x_2 \end{aligned} \tag{P3.13}$$

and determine if it is possible to have a limit cycle within the strip $-1 < x_1 < 1$ when $\varepsilon \neq 0$.

Chapter 4

Problem 4.1. A dynamic system is described by the following differential equations:

$$\begin{aligned} b\dot{x} + c\dot{y} &= 0 \\ d\ddot{y} + e\dot{y} + f\dot{x} + gF(x) &= 0 \end{aligned} \tag{P4.1}$$

where $be \ll cf$, and $b < 0$. The nonlinear characteristic is $F(x) = k\,\text{sign}\,x$.

Study the possible periodic solutions: (a) directly solving the characteristic equation of the linearized system; (b) by using the parameter plane method; and (c) by utilizing the Nyquist diagram or root-locus technique.

Simulate equations P4.1 on the computer and compare the graphoanalytical results of the harmonic linearization (describing function) obtained in (a), (b), and (c) with the computer solutions.

Problem 4.2. Repeat the solution procedure of the previous problem, P4.1, when the equations P4.1 are modified as

$$\begin{aligned} a\ddot{x} + b\dot{x} + c\dot{y} &= 0 \\ d\ddot{y} + e\dot{y} + f\dot{x} + gF(x) &= 0, \end{aligned} \tag{P4.2}$$

where $be \ll cf$, and $a, b < 0$. The nonlinear characteristic $F(x)$ is

$$F(x) = \begin{cases} x - 1, & x > 1 \\ 0, & -1 \leq x \leq 1 \\ x + 1, & x < -1. \end{cases} \tag{P4.3}$$

Problem 4.3. Given a control system with a block diagram on Figure P4.1a, where

$$G(s) = \frac{K}{(T_1 s + 1)(T_2^2 s^2 + T_3 s + 1)} \tag{P4.4}$$

and $F(x) = 10\,\text{sign}\,x$, $(\beta = \infty)$, represents the nonlinearity.

(a) By applying the parameter plane method and the describing function technique, show that for $K = 50$, $T_1 = 0.02$ sec., $T_2^2 = 0.001$ sec^2, and $T_3 = 0.01$ sec the periodic solution $x = A \sin \Omega t$ has the values $A \simeq 370$ and $\Omega \simeq 39$ rad/sec.

(b) Determine the stability of the above solution using the sensitivity analysis.

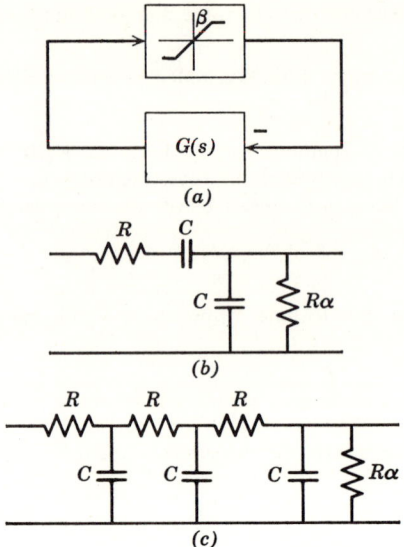

Figure P4.1 The oscillator.

(c) Find the sensitivity of the amplitude and frequency of the solution to small changes in the parameters K and T_3.

(d) Simulate the system and indicate the accuracy of the graphoanalytical solution obtained in (a).

Problem 4.4. To build an oscillator, one may use the system of Figure P4.1a with a network given in Figure P4.1b as the linear part of the system $G(s)$.

(a) By using the describing function and the parameter plane, determine the relationship between the parameters α and β (the slope of the linear part of the nonlinear characteristic) and the normalized frequency $\nu = RC\Omega$. Then derive the sensitivity S_α^ν of the normalized frequency ν with respect to the parameter α defined as

$$S_\alpha^\nu \triangleq \frac{\partial \ln \nu}{\partial \ln \alpha}. \tag{P4.5}$$

(b) Repeat the content in (a) but with the network shown in Figure P4.1c.

(c) Build such an oscillator on the analog computer and verify the results obtained in (a) and (b).

Problem 4.5. Given the system in Figure P4.2 with

$$G_1(s) = \frac{K_1}{s(s+1)}, \quad G_2(s) = \frac{1}{s+2}, \quad G_{-1}(s) = K_{-1}s, \tag{P4.6}$$

and

$$F(x) = \begin{cases} x, & |x| < 1 \\ \pm 1, & |x| \geq 1. \end{cases} \tag{P4.7}$$

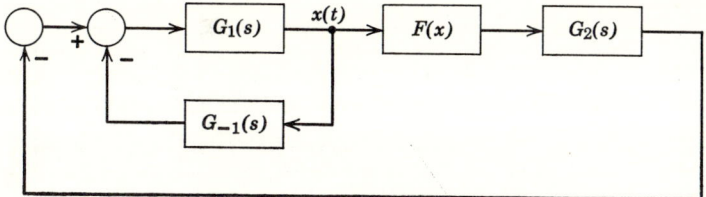

Figure P4.2 System block diagram.

By using the describing function technique, determine the values of the parameters K and K_{-1} for which the system is stable. Show that for $K = 10$, $K_{-1} = 0.02$ the system can exhibit a limit cycle with the amplitude $A \simeq 1.6$ and frequency $\Omega \simeq 1.54$ rad/sec.

Simulate the system on a computer and verify the analytic results.

Problem 4.6. Use the same system as in Problem 4.5 but with the nonlinearity $F(x)$ given in Figure 4.31a. Then choose the linear parameters K, K_{-1}, and the nonlinear parameters c, D, so that the system (a) has a periodic solution, and (b) is stable. Simulate the system on a computer and compare the results with those obtained analytically. Discuss case (b).

Problem 4.7. The van der Pol equation is

$$\ddot{x} - \varepsilon(1 - x^2)\dot{x} + x = 0. \qquad (P4.8)$$

Study the periodic solutions for different values of the parameter ε by using the harmonic linearization (describing function) and the parameter plane method.

Calculate the sensitivity of the frequency and amplitude of the oscillations with respect to ε.

Simulate the equation on the computer and compare the above analytical results with the computer solutions.

Problem 4.8. Consider the system of Figure P4.2 with

$$G_1(s) = \frac{1}{s+2}, \quad G_2(s) = \frac{K_2}{s(s+1)}, \quad G_{-1}(s) = \frac{K_{-1}}{s+0.1} \qquad (P4.9)$$

and the nonlinear characteristic $F(x)$ given in equation P4.7. Another nonlinear characteristic $F_{-1}(x)$ given as

$$F_{-1}(x) = \begin{cases} x - 1, & x > 1 \\ 0, & -1 \leq x \leq 1 \\ x + 1, & x < -1 \end{cases} \qquad (P4.10)$$

is placed inside the inner feedback loop associated with the block $G_{-1}(s)$.

By using the describing function technique and the parameter plane method, study the stability of the system as a function of system parameters K_2, K_{-1} when (a) the nonlinearity $F_{-1}(x)$ precedes $G_{-1}(s)$, and (b) the nonlinearity $F_{-1}(x)$ follows $G_{-1}(s)$.

Problem 4.9. To build an oscillator in which a simple control of the amplitude is possible without affecting the frequency of oscillation, the system in Figure P4.3 is proposed with

$$G(s) = \frac{K_1}{s(s+0.5)(s+1)(s+2)}. \qquad (P4.11)$$

Problems

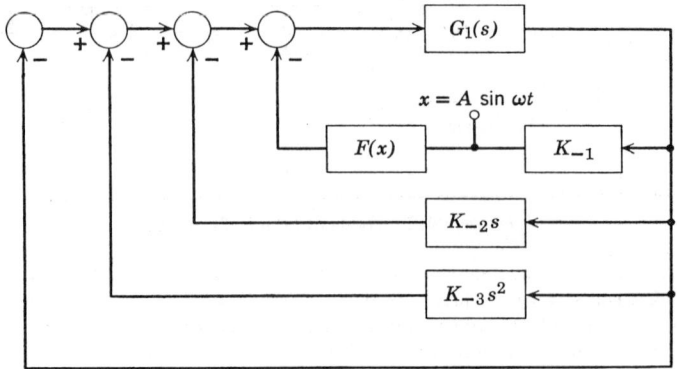

Figure P4.3 Oscillator block diagram.

By choosing
$$\alpha = KK_{-3}, \qquad \beta = K_1 K_{-1} N_1, \qquad \text{(P4.12)}$$
where the describing function $N_1 = N_1(A)$ corresponds to $F(x)$, show that the frequency
$$\omega_0^2 = \frac{1}{3.5}(1 + K_1 K_{-2}) \qquad \text{(P4.13)}$$
does not depend on β (this is the singular case treated in Derivation A.4, Appendix A, for which the curves $\zeta = 0$ and $\omega = \omega_0$ are identical).

Then simulate the system on the computer and show that for $K_1 = 5$, $K_{-1} = 0.4$, $K_{-2} = 0.5$, the frequency $\omega = 1$ will be constant, while the amplitude is varied by controlling the value of K_{-3}. (Choose the nonlinear element to perform the task. Use the parameter plane method to plot $\zeta = 0$, $\omega = 1$ curve.)

Chapter 5

Problem 5.1. Repeat the analysis of the example in Section 5.3 but with nonlinear characteristic No. 11 of Table F.2 in Appendix F. (Note: $m = 0$.)

Problem 5.2. Derive the expressions of N_1 and F^0 of the following cases of Table F.2 in Appendix F:
(a) No. 1
(b) No. 4
(c) No. 6
(d) No. 17

Problem 5.3. A system is described by the differential equations
$$Ts^2 y + 2sy - 400sx = F(x)$$
$$2s^2 x + sx + 400sy = f_0, \qquad \text{(P5.1)}$$
where $s \equiv d/dt$ and $0 \le f_0 \le 500$. The nonlinearity $F(x)$ is a pure relay with symmetrical characteristic as in the example in Section 5.3. Examine the asymmetrical oscillations for various values of f_0 and $T(T \simeq 2)$. Perform the analysis both analytically and in the parameter plane.

Simulate the system on the computer and compare the results with those obtained by the graphical analysis.

Problem 5.4. Given the system of Figure 5.1 with the following transfer functions:

$$G_1(s) = 1, \quad G_2(s) = \frac{K}{(s+1)(s+2)}, \quad G_3(s) = \frac{1}{s}, \quad G_{-1}(s) = 1. \quad \text{(P5.2)}$$

A constant signal $f_0(0 \le f_0 \le 100)$ is applied at the return point between the transfer functions $G_2(s)$ and $G_3(s)$. The nonlinearity is No. 13 of Table F.2 in Appendix F, with $c = 10$, $D = 5$.

Study the asymmetrical oscillations as f_0 is varied for several values of K and K_{-1}. Repeat the analysis when the nonlinearity $F(x)$ is backlash (characteristic No. 10 of Table F.2, Appendix F) with $D = 5$, $k = 1$.

Problem 5.5. Given the system in Figure 5.1 with transfer functions of (5.17). The nonlinearity n is given in Table F.2, No. 12, of Appendix F. Investigate the asymmetrical oscillations under various conditions on the nonlinearity and variable system parameters K_1, K_2, K_3, K_{-1}. Simulate the system on the computer and discuss the obtained graphical results.

Chapter 6

Problem 6.1. Given the differential equation

$$\ddot{x} - 2p\dot{x} + F(x) = 0, \quad \text{(P6.1)}$$

where

$$F(x) = \begin{cases} 1, & x > 0 \\ -1, & x < 0 \end{cases} \quad \text{(P6.2)}$$

and $p = \frac{1}{2}$.

It is required to calculate approximately the transient solution $x(t)$ for the initial condition $x(0) = 1$, using both the functions $\sigma(a)$, $\omega(a)$ and the functions $\sigma_1(a)$, $\omega_1(a)$. Compare the obtained results with the computer solution given in Figure P6.1.

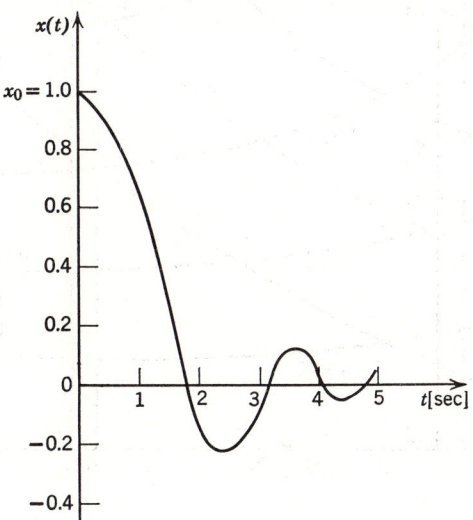

Figure P6.1 Computer solution.

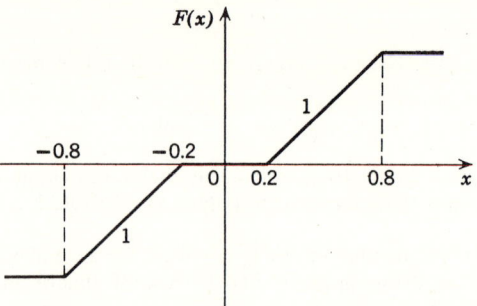

Figure P6.2 Nonlinear characteristic.

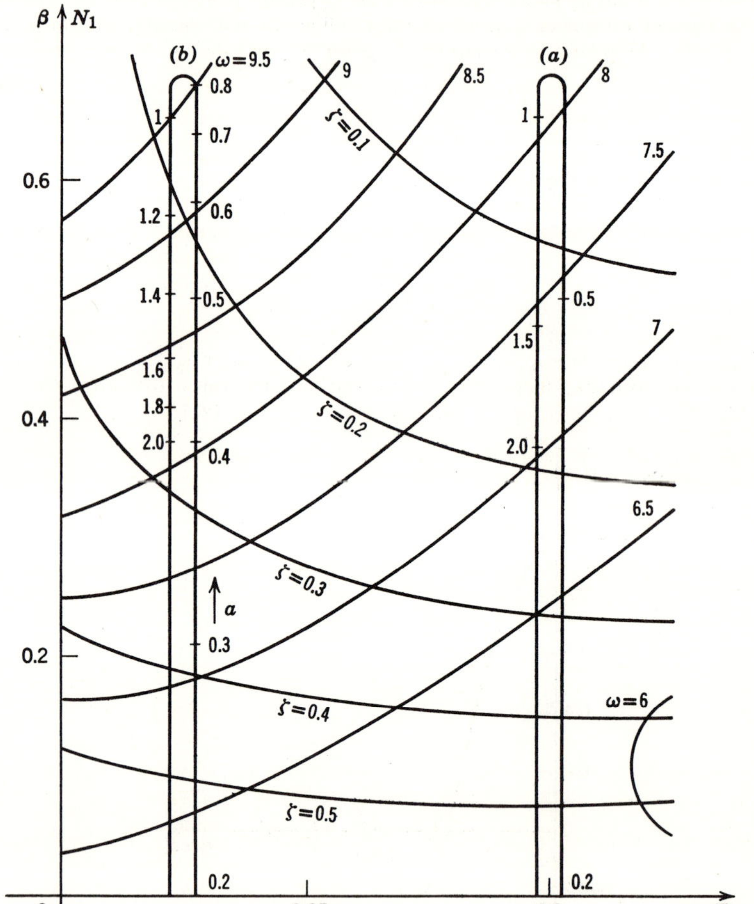

Figure P6.3 Parameter plane diagram.

Problem 6.2. Calculate the transient response of the van der Pol equation

$$\ddot{x} - p(1 - x^2)\dot{x} + x = 0 \tag{P6.3}$$

for $p = 1$ and an arbitrary chosen initial condition $x(0) = x_0$. Use both the functions $\sigma(a)$, $\omega(a)$ directly from (6.62) and the functions $\sigma_1(a)$, $\omega_1(a)$. Simulate the equation on the computer for comparison of the approximate solutions.

Problem 6.3. Given a control system with a block diagram shown in Figure 6.1 and the transfer functions given in (6.76). The nonlinearity n_{-1} has a characteristic shown in Figure P6.2. It is required to sketch the transient oscillations for two different values of the system parameter: (a) $T_{-1} = 0.1$ and (b) $T_{-1} = 0.025$. The initial condition is $x(0) = 2$ for both cases.

The parameter plane diagram in terms of the ζ and ω_n curves is shown in Figure P6.3. After the functions $\zeta(a)$ and $\omega_n(a)$ are calculated from the diagram, they should be used to construct the envelope curve $a(t)$ and the frequency curve $\omega(t)$. To simplify the analysis it should be assumed that $a_0 = x_0$ and $\phi_0 = \pi/2$. The computer solutions are given in Figures P6.4 and P6.5 for comparison.

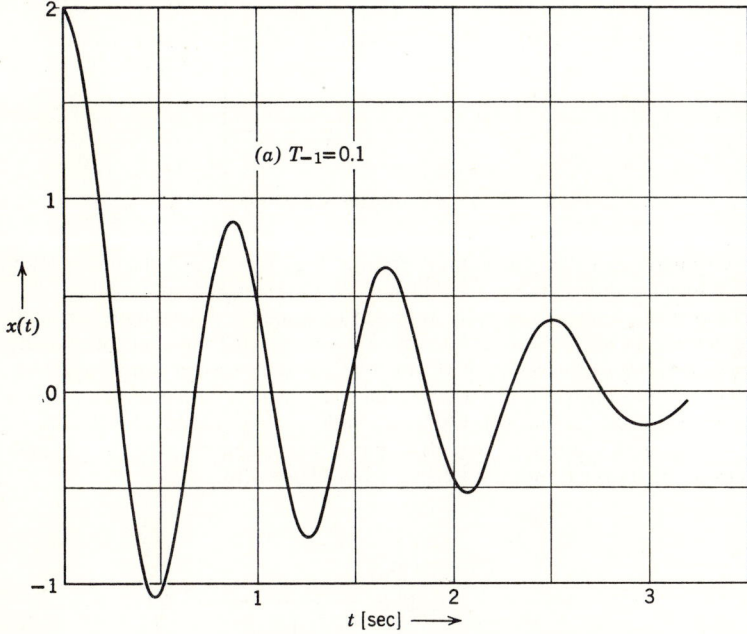

Figure P6.4 Computer solution.

Problem 6.4. Consider again the same system of Problem 6.3, but with the nonlinear characteristic shown in Figure P6.6. It is required to consider two parameter values: (a) $T_{-1} = 0.1$ and (b) $T_{-1} = 0.025$ with the initial condition $x_0 = 3$, and to determine (a) the settling time T_s for the amplitude a to become $a = a_s = 1.1$; (b) the overshoot x_m; (c) the number of oscillations k in the interval $(a_0, a_s) = (3, 1.1)$. In the analysis, consider three intervals ($n = 3$) and use equations 6.85, 6.88, and 6.91. To facilitate the calculation, assume that $a_0 = x_0 = 3$, $\phi_0 = \pi/2$.

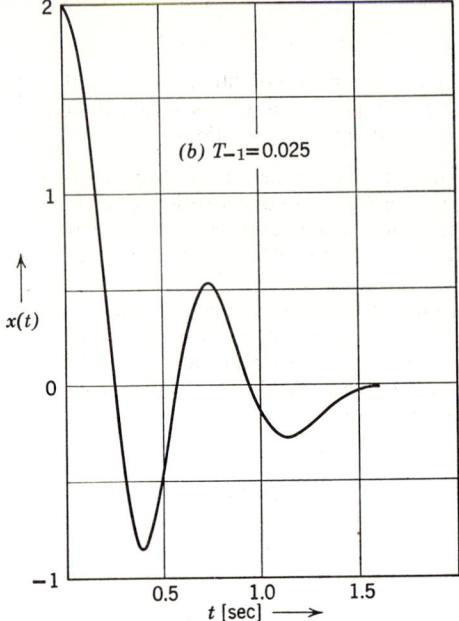

Figure P6.5 Computer solution.

The necessary parameter plane diagram is given in Figure P6.7, and the corresponding computer solutions are shown in Figures P6.8 and P6.9 for comparison.

The conclusions about the nature of the transient responses in this example is related to the stability considerations of Section 4.5, where it was pointed out that the stability is not a sufficient measure of the system performance. The same system is investigated here by the transient response analysis and it was possible to distinguish the nature of the transient process between the cases (a) and (b), which are both stable as indicated in Figure 4.17. The method used in this problem enables a quick but rough estimate of the salient characteristics of the transient responses displayed in Figure 4.18.

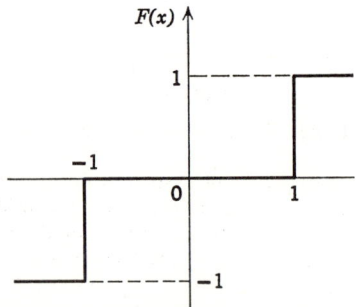

Figure P6.6 Nonlinear characteristic.

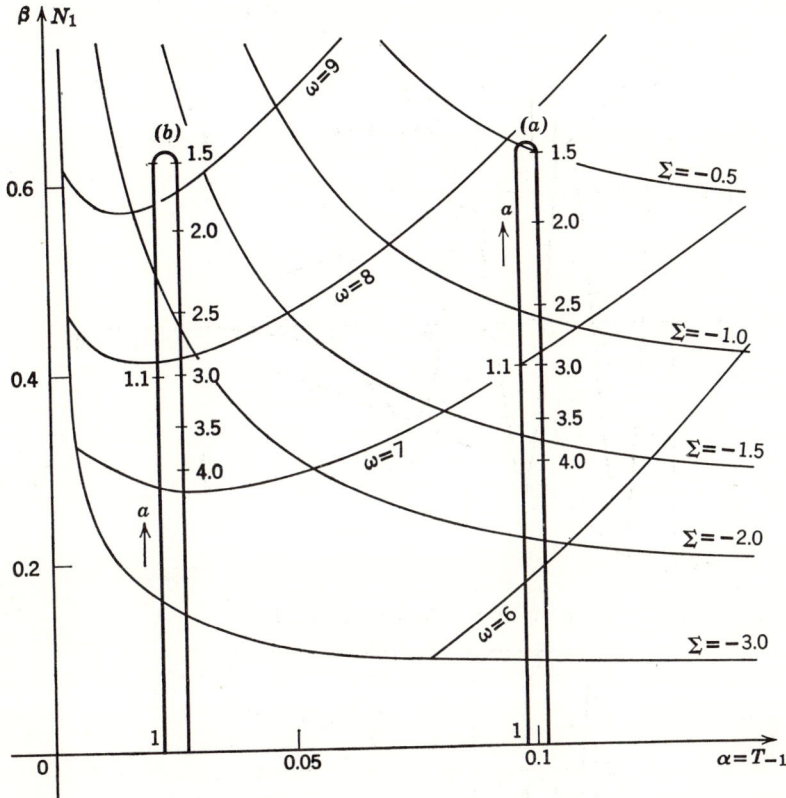

Figure P6.7 Parameter plane diagram.

Problem 6.5. Given the system in Figure P6.10. The characteristic equation of the linearized system is

$$s^3 + (3 + 60N_{12})s^2 + (2 + 60N_{11})s + 60 = 0, \qquad (P6.4)$$

where $N_{11} = N_{11}(a_1)$ and $N_{12} = N_{12}(a_2)$. It is assumed that

$$x = a \sin \omega t$$
$$x_1 = a_1 \sin(\omega t + \pi/2) \qquad (P6.5)$$
$$x_2 = a_2 \sin(\omega t + \pi),$$

where

$$a_1 = a\omega, \qquad a_2 = a\omega^2. \qquad (P6.6)$$

Thus $N_{11} = N_{11}(a, \omega)$, $N_{12} = N_{12}(a, \omega)$ are

$$N_{11} = \frac{2}{\pi}\left\{\arcsin\frac{3}{a\omega} + \frac{3}{a\omega}\left[1 - \left(\frac{3}{a\omega}\right)^2\right]^{1/2}\right\}.$$

$$N_{12} = \frac{2}{\pi}\left\{\arcsin\frac{3}{a\omega^2} + \frac{3}{a\omega^2}\left[1 - \left(\frac{3}{a\omega^2}\right)^2\right]^{1/2}\right\}. \qquad (P6.7)$$

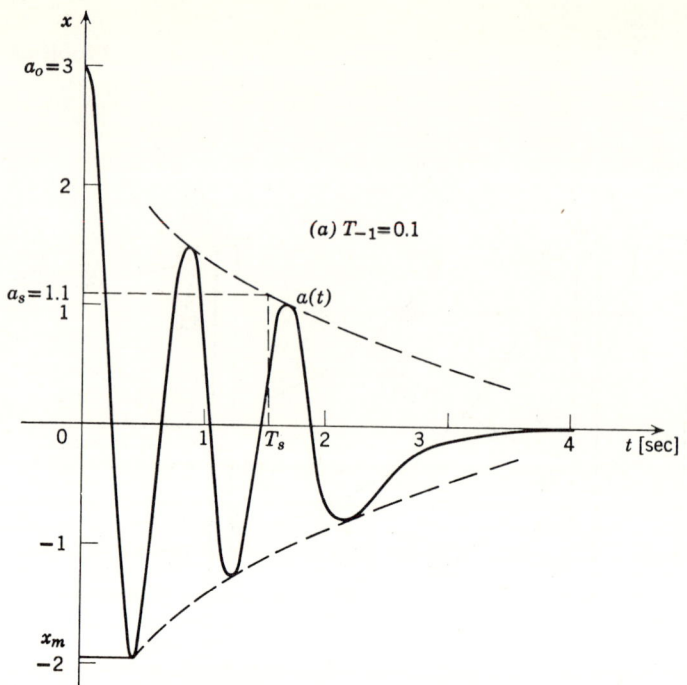

Figure P6.8 Computer solution.

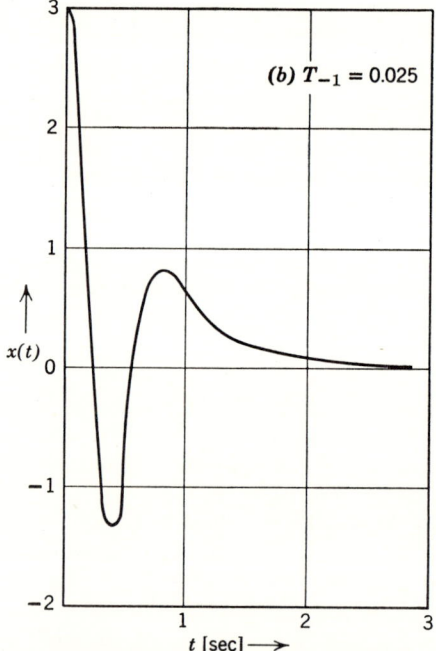

Figure P6.9 Computer solution.

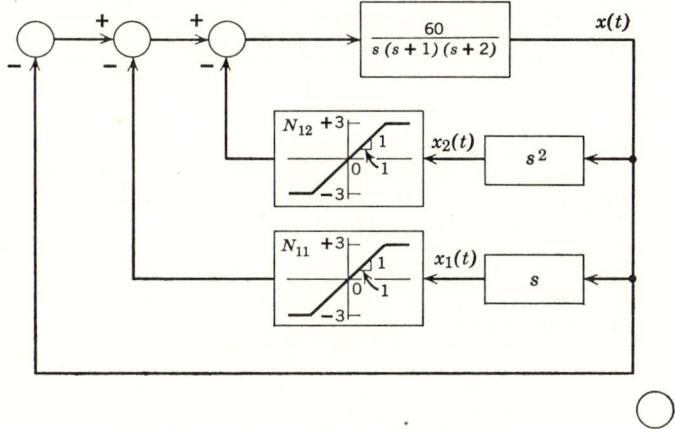

Figure P6.10 System block diagram.

If α and β denote N_{12} and N_{11}, respectively, a grid of $M(\alpha; \beta)$ loci can be plotted with a and ω as loci parameters (similar to the grid of Figure 4.29). Then this grid can be superimposed on the σ and ω curves in the parameter $\alpha\beta$ plane. The actual motion of the M point during the transient process will be along the intersections of the grid and ω curves where the values of ω coincides. Once the motion of the M point is drawn, the corresponding values of a and σ can be read from the plot. This yields $\sigma(a)$ and $\omega(a)$, which can be used to reconstruct the transient response $c(t)$. Perform this calculation for $x_0 = a_0 = 5$ and $x_0 = a_0 = 9$, and compare them with the computer simulation given in Figure P6.11a and b. Is there a stable or unstable limit cycle? What are the values of A and Ω?

Problem 6.6. The analysis of nonlinear transient oscillations is essentially based on the fact that under certain conditions the motion of a nonlinear system can be represented by a motion of an equivalent linear time varying system. This fact enables known linear methods to be applied quite effectively to nonlinear transients. That leads to a possibility of applying the proposed analysis to the analysis of transient oscillations in originally linear time varying systems [6.9, 10].

Now an originally time varying system can be analyzed by the same techniques by extending the Krylov-Bogoliubov method (the method is outlined in Section 3.5, Chapter 3).

The Krylov-Bogoliubov method applies to systems described by the second-order nonlinear differential equation

$$\ddot{x} + \varepsilon f(x, \dot{x}) + \omega_0^2 x = 0, \tag{P6.8}$$

where $\varepsilon > 0$ is the small parameter and ω_0 is a constant. The first approximation of the solution of (P6.8) is

$$x(t) = a(t) \sin [\omega_0 t + \phi(t)], \tag{P6.9}$$

where $a(t)$ and $\phi(t)$ are determined by the equations

$$\dot{a} = \frac{-\varepsilon}{2\pi\omega_0} \int_0^{2\pi} f(a \sin \psi, a\omega_0 \cos \psi) \cos \psi \, d\psi \tag{P6.10}$$

and

$$\dot{\phi} = \frac{\varepsilon}{2\pi a \omega_0} \int_0^{2\pi} f(a \sin \psi, a\omega_0 \cos \psi) \sin \psi \, d\psi, \tag{P6.11}$$

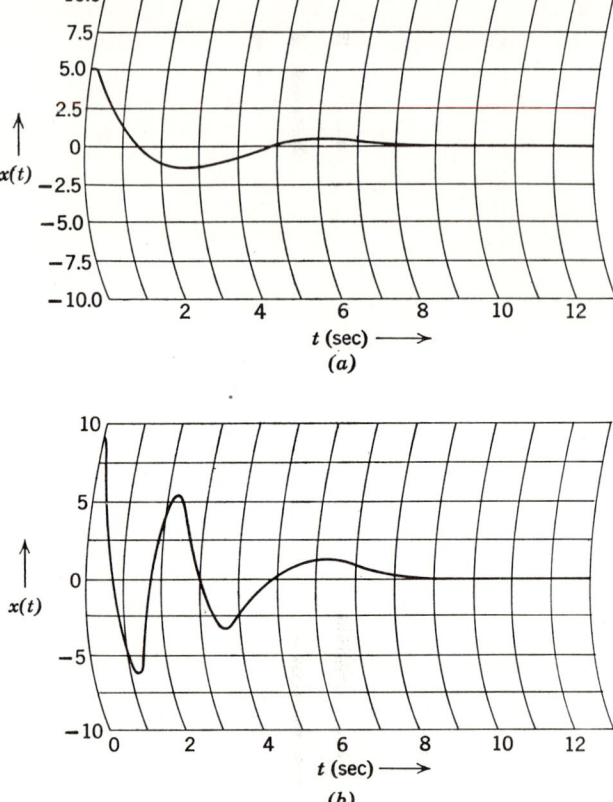

Figure P6.11 Computer solutions.

where $\psi = \omega_0 t + \phi$. The solution procedure is to solve (P6.10) for $a(t)$ and (P6.11) for $\phi(t)$. Then (P6.9) yields the approximate solution of (P6.8).

In the linear time varying case

$$\ddot{x} - 2\sigma\dot{x} + (\sigma^2 + \omega^2)x = 0, \qquad (\text{P6.12})$$

where $\sigma(t)$ and $\omega(t)$ are functions of time, the analogy with (P6.8) yields

$$\ddot{x} + [(\sigma^2 + \omega^2 - \omega_0^2)x - 2\sigma\dot{x}] + \omega_0^2 x = 0 \qquad (\text{P6.13})$$

and

$$\varepsilon f(x, \dot{x}) = (\sigma^2 + \omega^2 - \omega_0^2)x - 2\sigma\dot{x}. \qquad (\text{P6.14})$$

Evaluating the integrals (P6.10) and (P6.11), we obtain

$$\dot{a} = \sigma a$$

$$\dot{\phi} = \tfrac{1}{2}(\sigma^2 + \omega^2 - \omega_0^2). \qquad (\text{P6.15})$$

After integration of (P6.15), the approximate solution of (P6.9) becomes

$$x(t) = a_0 \exp\left(\int_0^t \sigma_1 \, dt\right) \sin\left(\int_0^t \omega_1 \, dt + \phi_0\right), \tag{P6.16}$$

where a_0, ϕ_0 are integration constants and σ_1, ω_1 stand for σ and $\tfrac{1}{2}(\sigma^2 + \omega^2 - \omega_0^2)$. Differentiating (P6.16) twice and substituting into (P6.12), we obtain two equations:

$$\dot{\sigma}_1 + \sigma_1^2 - 2\sigma_1\sigma - \omega_1^2 + \sigma^2 + \omega^2 = 0 \tag{P6.17}$$

$$2\sigma_1\omega_1 + \dot{\omega}_1 - 2\omega_1\sigma = 0.$$

If σ and ω are slowly varying functions of time, then $\dot{\sigma} \simeq \dot{\omega} \simeq 0$ and $\sigma_1 = \sigma$, $\omega_1 = \omega$. In other words, the approximate solution of (P6.12) is

$$x = a_0 \exp\left(\int_0^t \sigma \, dt\right) \sin\left(\int_0^t \omega \, dt + \phi_0\right), \tag{P6.18}$$

where $\sigma(t)$ and $\omega(t)$ are slowly varying functions of time.

The above procedure can now be extended to high-order linear differential equations of the form

$$\sum_{k=0}^{n} a_k s^k x = 0, \quad s \equiv \frac{d}{dt}, \tag{P6.19}$$

where

$$a_k = a_k(\alpha, \beta), \quad (k = 0, 1, \ldots, n) \tag{P6.20}$$

and $\alpha = \alpha(t)$, $\beta = \beta(t)$ are known functions of time.

The solution $x(t)$ of (P6.19) can be assumed in the form (P6.18) if the *dominant pair of roots* in the characteristic equation

$$F(s) \equiv \sum_{k=0}^{n} a_k s^k = 0 \tag{P6.21}$$

can be determined by

$$s^2 - 2\sigma s + (\sigma^2 + \omega^2) = 0, \tag{P6.22}$$

where $\sigma = \sigma(t)$, $\omega = \omega(t)$ are slowly varying in time. Note that the rate of change is not constrained in the variations of the parameters $\alpha(t)$ and $\beta(t)$, but rather in the changes of $\sigma(t)$ and $\omega(t)$. This allows a study of fast parameter variations, provided that the constraints on $\sigma(t)$ and $\omega(t)$ are not violated.

Once the functions $\alpha(t)$ and $\beta(t)$ are given, the parameter analysis yields the functions $\sigma(t)$ and $\omega(t)$. Then a tangential approximation (see Section 6.4) can be used to reconstruct the solution $x(t)$ determined by (P6.18). In majority of cases, however, the functions $\sigma(t)$ and $\omega(t)$ contain sufficient information about the solution $x(t)$ and it is not necessary to perform the construction of $x(t)$.

In a system design with computer simulation, the parameter variations may be chosen on the basis of the parameter plane analysis so that the functions $\sigma(t)$ and $\omega(t)$ result in a relatively fast response $x(t)$. This will be considered in the next section.

Now the above reasoning should be applied to the system of Figure P6.12 with

$$G_1(s) = \frac{K_1}{s(s+1)}, \quad G_2(s) = \frac{K_2}{s+2}, \quad G_{-1}(s) = K_{-1}s. \tag{P6.23}$$

600 Problems

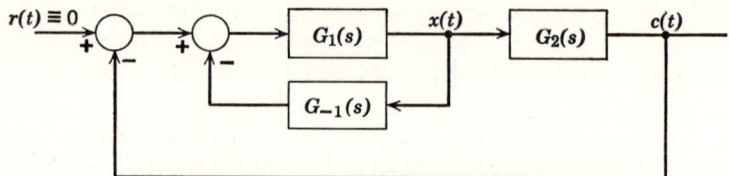

Figure P6.12 System block diagram.

The parameters $\alpha = K_1 K_{-1}, \beta = K_1 K_2$ are chosen so that $K_1 = 1$, $K_2 = 6$, and $K_{-1}(t) = at$, $a = \text{const}$.

(a) Plot the $\alpha\beta$ plane diagram ($0 \leq \alpha \leq 6$, $0 \leq \beta \leq 12$, $-1 \leq \sigma \leq 0$, $0.1 \leq \omega \leq 1.5$; choose $a = 1$, $x_0 = 10$; and compute the transient oscillation using the tangential approximation. Compare the obtained result with the computer solution in Figure P6.13.

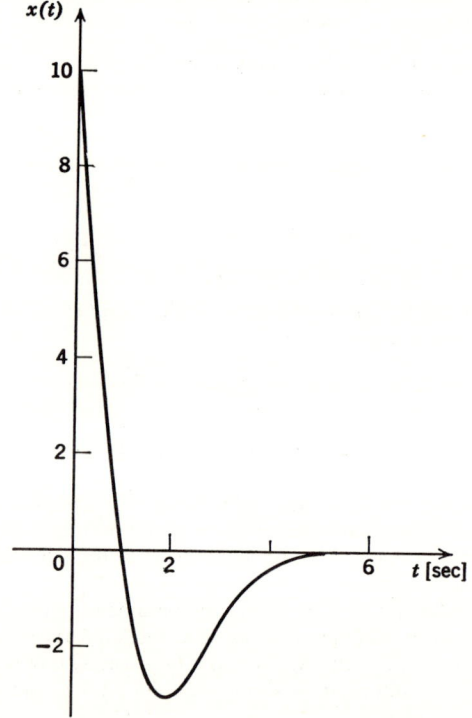

Figure P6.13 Computer solution.

(b) Simulate the system on the computer, change the constant a in $K_{-1}(t) = at$ from 1 to 5, and choose the value of a that yields the fastest response $x(t)$ for $x_0 = 10$. Explain the result by parameter plane analysis.

(c) Repeat the procedure of (b), but with $K_{-1}(t) = at^2$, $1 \leq a \leq 10$, and choose the fastest response.

Chapter 7

Problem 7.1. Given the nonlinear control system shown in Figure P7.1. Show that the amplitude A and the phase φ of the forced oscillations are $A = 21$, $\varphi = 35°$. Can the jump resonance take place?

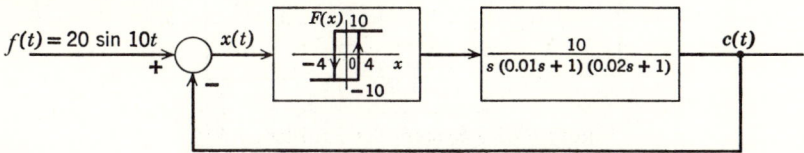

Figure P7.1 Forced nonlinear system.

Problem 7.2. Repeat the analysis of Problem 7.1 where the nonlinear characteristic is a pure relay—that is, $F(x) = \text{sign } x$—and determine the values of A and φ. Plot a relationship between A and A_f for fixed $\Omega_f = 10$.

Problem 7.3. If in the system of Figure P7.1 the nonlinear characteristic is linear with saturation, that is,

$$F(x) = \begin{cases} x, & |x| < 1 \\ \pm 1, & |x| \geq 1, \end{cases} \quad (\text{P7.1})$$

study the forced oscillations in the NA plane (see Figure 7.6) for different values of Ω_f. Is there a jump resonance phenomenon?

Problem 7.4. Given a system with the block diagram in Figure P7.1 but with

$$F(x) = \begin{cases} \dfrac{10}{0.4} x, & |x| < 0.4 \\ \pm 10, & |x| \geq 0.4 \end{cases} \quad (\text{P7.2})$$

and the transfer function of the linear part

$$G(s) = \frac{10}{s(s+1)}. \quad (\text{P7.3})$$

If $A_f = 5$ and Ω_f varies in the range $2 \leq \Omega_f \leq 6$, by plotting a diagram in the $\alpha\beta$ plane (see Figure 7.2) determine the periodic motions under forced conditions.

Indicate the jump resonance and plot the curves A versus A_f, and φ versus A_f for $0 \leq A_f \leq 10$.

Problem 7.5. Given the feedback system in Figure P7.2. Display the jump resonance phenomenon by plotting the parameter plane diagram for $A_f = 2, 20, 40$, and $1 < \Omega_f \leq 15$. Construct the functions $A(\Omega_f)$ and $\varphi(\Omega_f)$ for the same values of A_f.

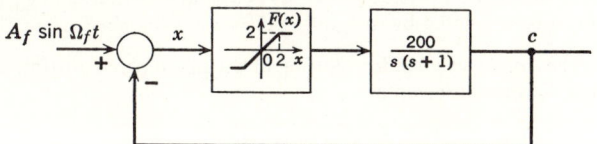

Figure P7.2 Feedback system for Problem 7.5.

Problem 7.6. Given the nonlinear system of Figure P7.3. Display the jump resonance for frequencies $2 \leq \Omega_f \leq 10$.

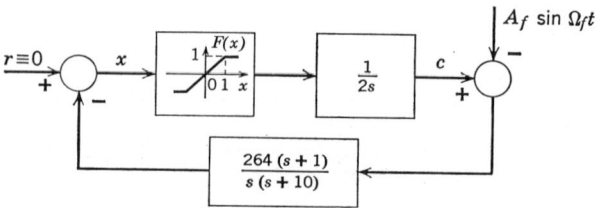

Figure P7.3 System for Problem 7.6.

Chapter 8

Problem 8.1. A nonlinear differential equation is

$$\dot{x}_1 = x_1^2. \tag{P8.1}$$

(a) Are the requirements of the existence and uniqueness theorem satisfied? (See the footnote following (8.6) in Section 8.2.)

(b) Examine (P8.1) with respect to finite escape time. Can equation P8.1 represent a physical system?

Problem 8.2. The motion of a free-falling body is described by

$$\ddot{x} = -g, \tag{P8.2}$$

where g is constant. By directly integrating equation P8.2, plot the trajectories in the $x_1 x_2$ plane (phase plane with $x_1 = x$, $x_2 = \dot{x}$) for different initial velocities x_{20} directed vertically upward, if the body starts from the surface of the earth ($x_{10} = 0$).

Problem 8.3. The types of equilibrium points for the second-order linear system

$$\ddot{x} + 2\zeta\omega_n \dot{x} + \omega_n^2 x = 0 \tag{P8.3}$$

are given in Table P8.1. For the case $\zeta = 0(\omega_n = \omega)$, denote $x_1 = x$, $x_2 = \dot{x}$ and integrate equation

$$\dot{x}_2 + \omega^2 x_1 = 0, \tag{P8.4}$$

to show that the equilibrium point is a center and the trajectories in the phase plane are a family of ellipses.

Calculate the time necessary for the point $P(x_1, x_2)$ to move along the trajectory from a point P_1 to a point P_2. Then verify the well-known fact that the period of oscillations in a *harmonic oscillator* (P8.4) does not depend on their amplitude.

Hint: Before integration, multiply equation P8.4 by x_2 and note that $x_2 = \dot{x}_1$. Time interval along the trajectories is calculated by integrating $dt = dx_1/\dot{x}_1 = dx_1/x_2$.

Problem 8.4. The *isocline method* for plotting of the phase plane portrait is general and applies to equations of motion given as

$$\begin{aligned}\dot{x}_1 &= X_1(x_1, x_2) \\ \dot{x}_2 &= X_2(x_1, x_2).\end{aligned} \tag{P8.5}$$

The slope of the phase plane trajectories at every point $P(x_1; x_2)$ of the phase plane is

$$\frac{dx_2}{dx_1} = \frac{X_2}{X_1}, \qquad (P8.6)$$

which is obtained by eliminating time from (P8.5). The locus of points in the $x_1 x_2$ plane that correspond to constant values of the slope $dx_2/dx_1 = c$ is termed an *isocline*. To plot the phase portrait, the isoclines are first constructed for different values of constant c. Once the isocline is traced, we draw along it small line segments with the slope c. After repeating the process for several isoclines covering a region of interest in the phase plane, the slope of the segments determine the tangents of the trajectories passing through the specific region. When the initial values $P_0(x_{01}; x_{02})$ are given, a continuous curve can be traced by always following the direction of line segments. The curve is clearly the trajectory starting at the initial point $P_0(x_{01}; x_{02})$.

(a) By using the isocline method, plot the phase plane portraits for four stable types of equilibrium points: center, focus, node, and saddle point (Table P8.1).

Show also that in the case of the node equilibrium point there are straight line trajectories $x_2 = s_1 x_1$, $x_2 = s_2 x_1$, where s_1 and s_2 are real roots of the *characteristic equation*

$$s^2 + 2\zeta\omega_n s + \omega_n^2 = 0 \qquad (P8.7)$$

corresponding to the state space (phase plane) equations

$$\dot{x}_1 = x_2$$
$$\dot{x}_2 = -\omega_n^2 x_1 - 2\zeta\omega_n x_2 \qquad (P8.8)$$

of system (P8.3). (Hint: Starting from $x_2 = kx_1$ and expressing $k = x_2/x_1 = \dot{x}_2/\dot{x}_1$, show that k is a root of the characteristic equation P8.7. Make use of equations P8.8.)

(b) By the isocline technique plot the phase-plane portrait for the van der Pol equations

$$\dot{x} = x_2$$
$$\dot{x}_2 = -x_1 - \varepsilon(1 - x_1^2)x_2 \qquad (P8.9)$$

when $\varepsilon = 1$ and the region of interest in the phase plane is determined by inequalities $-2.5 < |x_1| < 2.5$, $-3.5 < |x_2| < 3.5$. What is the type of the equilibrium point?

(c) Given the equation $\ddot{x} + \dot{x}x + x = 0$. What is the equilibrium point? Show the trajectories in the neighborhood of the point.

Problem 8.5. *Liénard Construction* of the phase plane trajectories applies to equations of the form

$$\ddot{x} + F(\dot{x}) + x = 0 \qquad (P8.10)$$

or, in the state-space notation,

$$\dot{x}_1 = x_2$$
$$\dot{x}_2 = -x_1 - F(x_2) \qquad (P8.11)$$

the procedure is the following. First plot on the phase-plane the curve $x_1 = -F(x_2)$ that is the zero slope isocline ($c = 0$, see Problem P8.4) and is usually referred to as the *Liénard characteristic*. Then, to determine the slope of a trajectory passing through any point $P(x_1; x_2)$, a straight line is drawn from P to the point R on the Liénard Characteristic

Table P8.1.[6]

Equilibrium point	Characteristic roots	Phase plane portrait
Center $\zeta = 0$, $\omega_n^2 > 0$	roots s_1, s_2 on imaginary axis	closed elliptical orbits around origin
Stable focus $0 < \zeta < 1$, $\omega_n^2 > 0$	complex roots s_1, s_2 in left half-plane	inward spiral to origin
Unstable focus $-1 < \zeta < 0$, $\omega_n^2 > 0$	complex roots s_1, s_2 in right half-plane	outward spiral from origin
Stable node $1 < \zeta$, $\omega_n^2 > 0$	real negative roots s_1, s_2	trajectories into origin along $x_2 = s_2 x_1$ and $x_2 = s_1 x_1$
Unstable node $\zeta < -1$, $\omega_n^2 > 0$	real positive roots s_1, s_2	trajectories out of origin along $x_2 = s_2 x_1$ and $x_2 = s_1 x_1$

---

[6] This table is adapted from reference 8.19 which can be used for a detailed study of the listed phase plane portraits.

Table P8.1. (*continued*)

Equilibrium point	Characteristic roots	Phase plane portrait
Saddle point $\omega_n^2 < 0$ (figure is for $\zeta > 0$)	s_1, s_2 real, opposite signs	$x_2 = s_2 x_1$; $x_2 = s_1 x_1$
$\zeta = 1$ stable	s_1, s_2 equal negative real	$x_2 = -\omega_n x_1$
$\zeta = -1$ unstable	s_1, s_2 equal positive real	$x_2 = x_1$
$\omega_n^2 = 0$ $\zeta \omega_n > 0$	$-2\zeta\omega_n$, 0	$x_2 = -2\zeta\omega_n x_1 + c$
$\omega_n^2 = 0$ $\zeta \omega_n < 0$	0, $2\zeta\omega_n$	$x_2 = 2\zeta\omega_n x_1 + c$

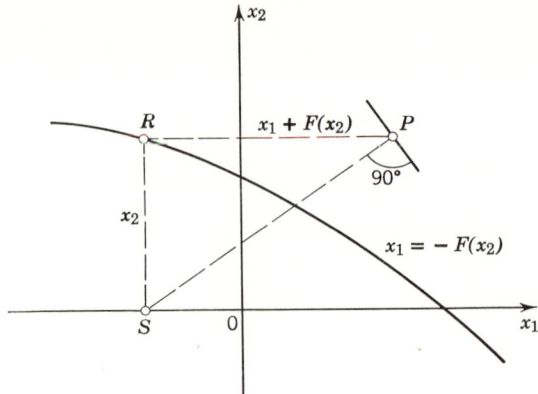

Figure P8.1   Liénard construction.

parallel to the x_1 axis, as shown in Figure P8.1. The point R is projected vertically to the x_1 axis to obtain point S. The slope of the trajectory at P is orthogonal to the straight line SP. Moreover, an arc through P with center at S represents a better approximation of the trajectory. A short segment of arc can be extended to a neighboring point P_1. Then a new center S_1 may be located and the process repeated. Of course, the shorter the segments of arc the more accurate the trajectory.

(a) Prove the Liénard construction by finding the isocline segment at P in Figure P8.1 from (P8.11), and comparing it with the geometry used in Figure P8.1 to obtain the slope of the trajectory at P.

(b) Consider a mass, spring, and damper system (Figure 8.4) described by

$$\dot{x}_1 = x_2$$
$$\dot{x}_2 = -Kx_1 - \frac{1}{M} F(x_2), \qquad (P8.12)$$

where M is the mass, K is the spring constant, and $F = F(x_2)$ is the friction characteristic given in Figure F.3 of Appendix F. Plot the phase plane portrait using the Liénard construction. What is the equilibrium position of the system?

Problem 8.6. Time scaling of the phase plane trajectories is necessary if we are interested in obtaining the states of the system as explicit functions of time. Of course, the scaling should be performed without the general solution of the differential equations and directly on the phase plane portrait.

Time appears implicitly in (P8.5), relating the state variables as $x_2 = dx_1/dt$. Thus for small increments Δt, we have $\tilde{x}_2 = \Delta x_1 / \Delta t$, where $\tilde{x}_2$ is the average value of x_2 in the time interval Δt. If Δt is sufficiently small, it can be approximated as $\tilde{x}_2 \simeq x_2(0) + \Delta x_2/2$, from which we obtain $2/\Delta t = [2x_2(0) + \Delta x_2]/\Delta x_1$. If the increments in time are equal along the trajectory, a construction of Figure P8.2 can be used to determine the corresponding points. Figure P8.2 is self-explanatory if it is noted that the angle $\theta = \tan^{-1}(2/\Delta t)$ is calculated from $2/\Delta t = [2x_2(0) + \Delta x_2]/\Delta x_1$ derived above.

(a) Draw a phase plane trajectory for the linear equation $\ddot{x} + \dot{x} + x = 0$ and find the time necessary for the point $P(x_1; x_2)$ to travel from the point $P_0(-1, 0)$ to the point where $x = x_1$ has a maximum.

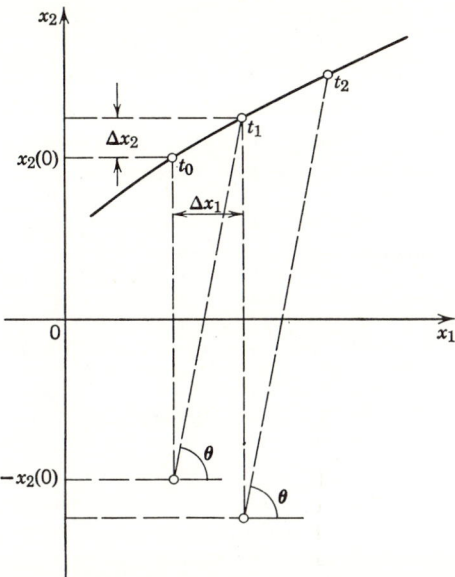

Figure P8.2 Time scaling.

(b) By using the above scaling method, calculate the time necessary for the point $P(x_1; x_2)$ to pass through the first quadrant of the phase plane if the system is the harmonic oscillator: $\ddot{x} + \omega x = 0$. Check the result by direct integration of $dt = dx_1/\dot{x}_1 = dx_1/x_2$.

Problem 8.7. Given the equation
$$\dot{x} = A(t)x, \tag{P8.13}$$
where $A(t)$ is a function given for $t \geq 0$ and piecewise-continuous in the interval $[0, \infty]$.

(a) Show that the boundedness of $\int_{t_0}^{t} A(\tau)\, d\tau$ implies stability in the sense of Liapunov.

(b) Show that the stability is not always sufficient for a satisfactory system behavior when
$$A(t) = \begin{cases} \ln 10, & 0 \leq t \leq 10 \\ 0, & t > 10. \end{cases} \tag{P8.14}$$

(Take, for example, the initial condition $x(0) = 10^{-5}$ and calculate x at $t = 10$.)

Problem 8.8. Given a system described by the equations
$$\dot{x}_1 = x_2 - cx_1(x_1^2 + x_2^2)$$
$$\dot{x}_2 = -x_1 - cx_2(x_1^2 + x_2^2), \tag{P8.15}$$
where c is a positive constant. Use the square of the Euclidean norm (equation 8.28) as a Liapunov function and determine the stability of the system.

Problem 8.9. Suppose that the trajectories of the second-order system have the form given in Figure P8.3. What can be said about the behavior of every solution for $t \to +\infty$? Is the origin an asymptotically stable solution of the system? If not, why?

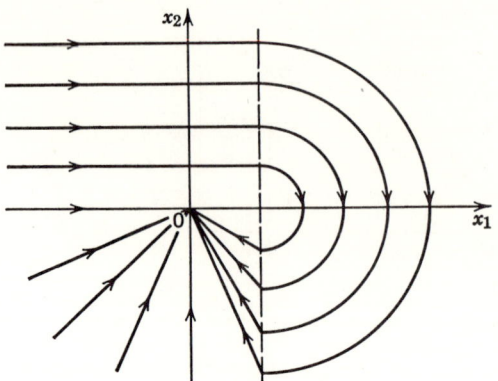

Figure P8.3 Phase plane portrait.

Problem 8.10. A system is described by the equations

$$x_1 = px_1^3 - x_2$$
$$x_2 = x_1 + px_2^3, \qquad \text{(P8.16)}$$

where p is a constant parameter. By using the Liapunov function $V = \tfrac{1}{2}(x_1^2 + x_2^2)$, show that a linearization of (P8.16) does not provide information about stability of the equilibrium $x_1 = x_2 = 0$. Discuss all three cases: $p < 0, p > 0, p = 0$.

Problem 8.11.
(a) Write a general computer program for plotting the envelope for the inequality

$$\pi(\omega) \equiv \frac{1}{k} + \operatorname{Re}(1 + jq\omega)G(\sigma + j\omega) > 0, \qquad \text{for all real } \omega \geq 0 \qquad \text{(P8.17)}$$

as it was proposed in Section 8.11, and check the curves on Figures 8.19 and 8.20.
(b) Choose a transfer function $G(s)$ and apply the program to calculate the regions of the absolute stability in the $\alpha\beta$ plane where $\alpha = q$ and $\beta = 1/k$.

Problem 8.12.
(a) Write a general program for plotting the envelope of the inequality

$$\pi(\omega) \equiv \frac{1}{k} + \operatorname{Re} G(\sigma + j\omega) > 0, \qquad \text{for all real } \omega \geq 0, \qquad \text{(P8.18)}$$

where

$$G(s) = \frac{C(s) + \beta D(s)}{B(s)}. \qquad \text{(P8.19)}$$

(b) Given the transfer function $G(s)$ of the system linear part as

$$G(s) = \frac{s^2 + \beta}{s^3 + 2s^2 + s + 1}. \qquad \text{(P8.20)}$$

If the inequality

$$\pi(\omega) \equiv \frac{1}{k} + \text{Re } G(\sigma + j\omega) > 0, \quad \text{for all real } \omega \geq 0 \tag{P8.21}$$

is considered, determine the value of β so that k is at maximum for absolute stability of the system with the exponent $\sigma = 0$. (Note that the envelope has three branches and three asymptotes.)

Problem 8.13. By using the results of the generalized Nyquist criterion (see Section D.2 of Appendix D) and the functions $X_k(\sigma, \omega)$, $Y_k(\sigma, \omega)$, make a general computer program to check the Popov inequality

$$\pi \equiv \frac{1}{k} + (1 + jq\omega)G(\sigma + j\omega) > 0, \quad \text{for all } \omega \geq 0. \tag{P8.22}$$

Then apply the computer program to determine the sector of the nonlinearity when the linear part is described by

$$G(s) = \frac{1}{(s+1)(s+2)(s+3)}. \tag{P8.23}$$

The system should be absolutely stable with $\sigma = 0$. In the case of (P8.23), is the Aizerman conjecture true?

Repeat the procedure for $\sigma = 0.5$ and find again the maximum value of k.

Problem 8.14. Given the system of equations

$$\begin{aligned}\dot{x}_1 &= -x_1 + x_2 - F(x_1) \\ \dot{x}_2 &= -x_1 + x_3 \\ \dot{x}_3 &= -x_1 + 2F(x_1)\end{aligned} \tag{P8.24}$$

in which $F(x) \in (k_1, k_2)$. Investigate the absolute stability in terms of the parameters k_1 and k_2.

Problem 8.15. A nonlinear, time varying system is given by

$$\begin{aligned}\dot{\mathbf{x}} &= \mathbf{P}\mathbf{x} - \mathbf{q}z \\ z &= F(y, t) \\ y &= \mathbf{r}^T\mathbf{x} - \rho z,\end{aligned} \tag{P8.25}$$

where $\mathbf{P}$ is Hurwitz $n \times n$ constant matrix; $\mathbf{q}$, $\mathbf{r}$ are constant n vectors; $F(y, t)$ is a continuous function that satisfies

$$F(0, t) = 0, \quad 0 < yF(y, t) < ky^2 \tag{P8.26}$$

and the scalar $\rho > 0$ (also $\rho \neq \mathbf{r}^T\mathbf{P}^{-1}\mathbf{q}$).

Show that the absolute stability of the system is guaranteed if the Popov inequality

$$\pi(\omega) \equiv \frac{1}{k} + \text{Re } G(j\omega) > 0 \tag{P8.27}$$

is satisfied for all real $\omega \geq 0$, where

$$G(j\omega) = \mathbf{r}^T(\mathbf{P} - j\omega\mathbf{I})\mathbf{q} + \frac{\rho}{j\omega}. \tag{P8.28}$$

In the proof use the function
$$V = \mathbf{x}^T \mathbf{P} \mathbf{x} + \alpha(y - \mathbf{r}^T\mathbf{x})^2 \tag{P8.29}$$
and $2\alpha\rho = 1$, $\alpha > 0$.

Problem 8.16. Prove that for the given nonlinearity the describing function variation is always less or at most equal to the total gain variation.

Problem 8.17. In Section 8.15, on Aizerman's conjecture, a feedback system with the linear part
$$G(s) = \frac{Ks(s+a)}{[(s+b)^2 + 0.9^2][(s+b)^2 + 1.1^2]} \tag{P8.30}$$
and the nonlinearity
$$F(y) = y^3 \tag{P8.31}$$
is shown to exhibit a limit cycle. Consider K and Ka as the parameters α and β, and apply the procedure in the parameter space (Section 8.14) to determine the stability of the system. Can the results of the stability analysis ensure the instability of the system concluded on the basis of experimental investigations?

Problem 8.18. Consider the linear time varying system
$$\dot{x}_1 + x_1/t = f_1(t) \tag{P8.32}$$
for values of $0 < t_0 \le t$.

(a) Show that for $t_0 < 0$, equation P8.32 has a finite escape time.
(b) By considering the impulse response ϕ_{11} (which in this case is a 1×1 matrix) of this system, show that the system is asymptotically stable, but not uniformly asymptotically stable.
(c) Now suppose that the initial state at some $t_0 > 0$ is zero. Let the input $f_1(t)$ be a unit-step function. By evaluating the unit-step response, show that the system is not BIBO stable.

Problem 8.19. Given a free ($\mathbf{f} = 0$) nonlinear system with the block diagram of Figure 8.13 and the transfer function
$$G(s) = \frac{1}{s(s^2 + \alpha s + \beta)}. \tag{P8.33}$$

Show that the Popov condition (8.112) verifies Aizerman's conjecture when $q = \alpha/\beta$, and $\sigma = 0$.
(Hint: First find the condition on K, α, β for the system to be stable as linear, $F(y) = K$; $s = j\omega$. Then show that the obtained condition represents the envelope of the absolute stability region, that is, the condition satisfies the corresponding equations $\pi(\omega) = 0$, $\pi'(\omega) = 0$ of the envelope; $\pi(\omega)$ derived from (8.112) and (P8.33) has no positive real zeros.)

Problem 8.20. Find a finite stability domain for the system represented by
$$\begin{aligned}\dot{x}_1 &= x_2 \\ \dot{x}_2 &= -7x_1 - 7x_2 - F(x_1 + x_2),\end{aligned} \tag{P8.34}$$
where $F \in [0, \infty]$, $(k = \infty, q = 0)$, and y_m is given.

Index

Aizerman, M. A., 292, 293, 368, 511, 539, 555, 556
  aprovision method of, 526
  conjecture of, 365, 555
Algebraic equation, 376
  root of, 376
Almost everywhere, 426
Amplitude, 122
  first correction of, 511
Anderson, B. D. O., 560
Antecedent, 17
Antosiewicz, H. A., 292
Approximate methods, 109
Aprovision, 526
Arc, 18
  continuous, 18
  differentiable, 18
  Jordan, 18
  simple, 18
Athans, M., 263
Atherton, D. P., 203
Auto-resonance, 133

Bairstow method, 421
Bandwidth, 8, 76
Barabashin-Krasovsky complement, 538
Barker, A. C., 78
Bendixon, I., 143, 586
  criterion of, 586
Bernoulli method, 421

Bertram, J. E., 292, 293, 308, 314, 528
Block diagrams, 68, 119
Bode, H. W., 61
  diagrams of, 560
Bogoliubov, N. N., 111, 121, 137, 138, 142, 145, 146, 150, 160, 162, 178, 254
  method of, 137
Bonenn, Z., 162
Bongiorno, J. J., Jr., 342
Booton, R. C., Jr., 277
Boundedness, 313
  equibounded, 313
  ultimate, 545
  uniform, 313
Boyer, R. C., 203
Brockett, R. W., 292, 557, 560
Brown, R. G., 2
Bulgakov, B. V., 109, 136, 146, 511
Bunyakowsky, V., 516
  inequality of, 516

Canonical form, 446
Cartesian product, 344
Cauchy, integral formula, 426
  Lipschitz condition, 294, 344, 351
  principle of argument, 392, 452
  Riemann conditions, 34, 53
Cesari, L., 292
Characteristic curves, 19, 411
Characteristic equation, 68, 154, 298, 442

611

Characteristic polynomial, 68
Characteristic roots, 68, 321
Characteristic value, 442
  problem of, 441
Characteristic vector, 442
Chattering, 539, 541
Chebyshev functions, 19, 385
Chetaev, N. G., 320
Circle criterion, 342, 560
Close, R., 292
Coddington, E. A., 344
Cofactor, 67, 117
Compensator, 61
  integral, 466, 580
  lead, 583
  multiple lead, 583
Complementary function, 323
Complex coefficients, 28, 449
Complex root boundary, 7
Computer, analog solution of roots, 413
  digital solution of roots, 421
  plotting of curves, 411
  simulation, 11, 62
Constraint, design with, 87
  equations of, 89
  inequality, 89
  stability, 84
Contour, 40, 453
Controllability, 322, 445, 535
Correctly set problem, 343
Cosgriff, R. L., 162
Cramer rule, 66
Curve, characteristic, 19
  constant bandwidth, 13, 76
  constant ω_n, 13, 23
  constant σ, 13, 26
  constant ζ, 6, 13, 23
  $\zeta = 1$, 401
  simple closed Jordan, 19

D'Alembert, J. le R., 376
D-decomposition, 2, 62, 456
Degree of freedom, 63
Delta function, 130, 298
Demidovich, B. P., 545
DeRusso, P., 292
Describing function, 122, 146, 470
  accuracy consideration of, 127, 511
  applicability conditions of, 127, 515
  approximate calculation of, 494
  for asymmetrical input, 205
  coefficients of, 470–485, 497–509
  dual-input, 277, 586
  frequency dependent, 198
  polynomial nonlinearities, 485, 497
  two nonlinearities, 492
  variable pole, 488
  variable structure, 491
Desoer, C. A., 292, 314, 560
Dewey, A. G., 109, 368, 555, 557
  counterexample of, 368
Differential equations, 65
  canonical form of, 446
  linear, 64, 116, 440
  nonlinear, 117, 293
  operator, 65, 68
  scalar, 293
  set of, 66
  vector, 293
Dirac delta function, 298, 523
Direct control, 321
Direct path, 69
Discontinuous nonlinearities, 113, 539
Dither effect, 202
Domain, 32
  of attraction, 561
  of exponential stability, 566
  in parameter space, 457
  simply connected, 586
  in state space, 564
Dominancy, 61, 248
  criterion of, 428

Eigenvalues, 321, 442
Elgerd, O, I., 428
Elliot, D. W., 3
End point, 540
Envelope, 357
  criterion, 354
  definition, 361
  equations of, 356
  of oscillations, 255
Equilibrium, 143, 177, 297
  nonunique, 558
  stability of, 308
Equivalent linearization, 142, 148
Error, acceleration constant, 83
  mean-squared, 466
  optimization, 465
  positional constant, 82
  signal, 69
  specification, 465

velocity constant, 13, 82
Euclidean distance, 303
Euclidean space, *see* Vector
Evans, W. R., 61

Falb, P. L., 263
Feedback path, 69
Filippov, A. G., 321
  solution of, 323, 539
Finite escape time, 302, 320, 529, 530
Fitts, R. E., 368, 369, 370, 555
  example of, 368
Forbidden zone, 327
Forcing function, 65, 204, 296
  almost periodic, 339
  constant, 211
  periodic, 278
  as perturbation, 345
  T-periodic, 339
Fourier series, 123
Frequency characteristic, 327
  modified, 328
Frequency response, 14
Friction, coulomb, 114, 487
  quadratic, 114, 487
  viscous, 114, 487
Function, of Chebyshev, *see* Chebyshev functions
  complementary, 323
  Dirac delta, 523
  entire, 376
  forcing, *see* Forcing function
  of Liapunov, 304, 317, 529
  of Lur'e, 532
  negative-definite, 317, 528
  negative-semidefinite, 317
  nondegenerate, 322, 535
  positive-definite, 317, 528
  positive-semidefinite, 317
  rational, 426
  transfer, 68, 322
  X_k and Y_k, 384
Fundamental implicit theorem, 381
Fundamental theorem of algebra, 376

Gain, incremental, 368
  total, 368
Gantmacher, F. R., 292, 293, 539, 555
Garber, E. D., 111, 511, 512
Gauss, K. F., 376
  divergence theorem of, 586

Gelb, A., 203
Gelig, A. Kh., 557, 558
Generalized coordinates, 63, 116
Generalized filter property, 133
George, J. H., 362, 457
Gibson, J. E., 136, 203, 277, 286, 292
Global behavior, 107, 301
Goldfarb, L. C., 121, 142
Goldwyn, R. M., 342
Goursat, E., 355
Gradient, 548
Graeffe method, 421
Grensted, P. E. W., 232, 233, 239, 240, 250
Gumowski, I., 345

Hahn, W., 292, 313
Halanay, A., 557, 560
Hale, J. K., 109, 136, 300, 351
Hammerstein equation, 513
Hancock, H., 363
Harmonic balance method, 111
Harmonic linearization, 110, 469, 518
Harmonic oscillator, 602
Harmonics, damped, 240, 247
  estimate of, 520
  first, 124
  higher, 132
Hermité, C., 448, 450
Heseltine, J. C. W., 3
Higgins, T. J., 2, 455
Homogeneity principle, 64
Hurwitz, A., 1, 60, 62, 74, 290, 443
  characteristic equation, 331
  criterion, 449
  determinant, 456
  matrix, 321
  polynomial, 335
  sector, 368, 555
Hysteresis, 113, 486

Image, 17
Impulse response, 298
Indirect control, 321
Initial conditions, 68, 254
Initial state, *see* State
Input, 63, 296, 426, 446
  deterministic, 463
  stochastic, 463
Instability, 320, 329, 370
Integral manifolds, 351
Integration constants, 71

Isocline, 603
 method of, 602

Jacobian, 16, 34, 382
James, H. M., 427
Johnson, E. C., 511
Jump conditions, 525
Jump resonance, 282
Jury, E. I., 109, 368, 555, 557, 560
 counterexample of, 368

Kalman, R. E., 291, 292, 293, 308, 314, 336, 368, 445, 528, 533, 538, 556, 557, 560
 lemma, see Yakubovich, V. A.
Kantorovich, L. V., 109, 123
Kochenburger, R. J., 121
Kokotović, P., 3, 343
Krylov-Bogoliubov method, 137
 Popov extension of, 597
 for time-varying equations, 597
Krylov, N. M., 111, 121, 137, 138, 142, 145, 146, 160, 162
 method of, see Krylov-Bogoliubov method
Krylov, V. I., 109, 123

Lagrange, J. L., 290
 stability of, 313
Lance method, 421
Lanzkron, R. W., 2, 455
Laplace transformation, 425
 inverse of, 426
LaSalle, J. P., 292, 300
Lee, B. W., 560
Lefschetz, S., 292, 293, 533
Lehmer-Schur method, 421
Letov, A. M., 292, 555
Levinson, N., 276, 344
Liapunov, M. A., 72, 108, 160, 162, 178, 290, 292, 301, 304
 direct method, 304
 first method, 290, 301
 function, 304, 307, 529
 instability theorem, 320
 second method, 290, 304
 stability concept, 302
 stability theorem, 318, 527
Liénard, A., characteristic of, 603
 construction of, 603
Limit cycle, 121, 144
 stability of, 158

Linearization, 299, 301, 365
 harmonic, see Harmonic linearization
 validity of, 300
Lin method, 421
Liouville, J., 377
Lipschitz condition, see Cauchy, Lipschitz condition
Liu, R., 314
Local behavior, 107, 301
Loeb, J. M., 203
Low-pass filter, 133, 515
Lozier, J. C., 202
Lur'e, A. I., 291, 532, 555
 function, see Function, of Lur'e
 problem, 291, 326, 532
 resolving equations, 532, 555
Luzin, N. N., 545

MacColl, L. A., 202, 203
Malkin, I. G., 136, 292, 293, 347, 531, 555
 corollary of, 531
Mapping, 17
 conformal, 15, 17
 continuous, 17
 continuously differentiable, 17
 differentiable, 17
 inverse, 17, 249
 inverse parameter, 17
 parameter, 15
Massera, J. L., 292
Matrix, 293
 characteristic roots of, 321
 eigenvalues of, 321
 Hurwitz, see Hurwitz, matrix
 inequality, 330
 Jacobian, 299
 modal, 446
 state transition, 298
 of system, 445
 unit (identity), 293
Maxwell, J. C., 1, 290
Meerov, M. V., 2
Meyer, K. R., 291, 533, 557
Mikhailov, A. W., 1, 60, 62, 74, 448
 criterion, 455
Minorsky, N., 136
Mitropolsky, Yu. A., 150
Mitrović, D., 2, 3
M locus, 155, 269
Modified frequency characteristic, see Frequency characteristic

Moivre formula, 377, 387
Motion, bounded, 313
  deviations of, 302
  equations of, 63, 293
  equibounded, 313
  fixed, 302
  perturbed, 302
  sliding (chattering), 323, 539, 548
  stability of, 303
  in the state space, 293
  uniformly bounded, 313
Muligan, L. H., Jr., 427

Narendra, K. S., 342
Naumov, B. N., 557, 560
Nehari, Z., 17
Neighborhood, 17
Neimark, Yu. I., 2, 448, 455
Nemytskii, V. V., 292
Newton-Bairstow method, 421
Nichols, H. M., 61, 427
Noise, 467
Nonlinear asymmetrical characteristic, 116
Nonlinear characteristic, 112
Nonlinear continuous characteristic, 112
Nonlinear discontinuous characteristic, 113
Nonlinear element, 111, 118
Nonlinearity, 118
  sector of, 323
Nonlinear modified characteristic, 206
Nonlinear multivalued characteristic, 113, 194
Nonlinear piecewise linear characteristic, 523
Nonlinear polynomial characteristic, 485
Nonlinear single-valued characteristic, 113
Norm, 303
Normalized time, 429
Nyquist, H., 1, 60, 62, 74, 162, 290, 448
  criterion, 327, 452
  diagram, 327

Obradovic, I., 463
Observability, 322, 445, 535
Ogata, K., 277
Oldenburger, R., 203
Operating point, 5, 18
Operator, 63, 68
Oppelt, W., 121
Oscillations, envelope of, *see* Envelope
  forced, 278, 337
  free, 121
  nonlinear damped, 240
  nonlinear periodic, 240
  number of, 259
  self-excited, 122
  stability of, 158
  subharmonic, 277
  sustained, 122, 159
  transient, 257
Output, 63, 69, 426, 446
Overshoot, 260, 466

Palitov, I. P., 111, 121, 129, 203, 277, 511
Parameter, 4, 65, 343
  bifurcation values of, 346
  controllable, 343
  critical values of, 346
  linear, 173
  mapping, *see* Mapping
  nonlinear, 173
  ordinary values of, 346
  perturbations, 346
  plane, 5
  small, 137, 245, 351
  space, 353, 456
  uncontrollable, 343
  variations, 52, 345
  vector, *see* Vector
Parceval formula, 516
Particular case, 321, 331
  general, 539
  simplest, 321, 336, 342, 533, 538
Peak amplitude ratio, 8, 13
Phase plane, 108, 539
  Bendixon criterion for, 586
  isocline method for, 602
  Liénard construction for, 603
  of linear second-order systems, 605
  time scaling of, 606
Phillips, R. S., 427
Pipes, L. A., 586
Pliss, V. A., 368, 555
Poincaré, H., 108, 136, 143, 160, 290
Polak, E., 439
Pole-zero configurations, 429
Polynomial, 375
  characteristic, 68, 375
  coefficients of, 375, 398
  complete, 375
  Hurwitz, *see* Hurwitz, A.
  identically vanishing, 375

incomplete, 375
Popov, *see* Popov, V. M.
terms of, 375
zero of, 376
Popov, E. P., 111, 121, 129, 137, 162, 203, 232, 233, 277, 511
  extension of, *see* Krylov-Bogoliubov method
Popov, V. M., 291, 326, 336, 342, 532, 535, 556, 560
  inequality, 326, 532
  line, 329
  polynomial, 365
  sector, 369
  solution of Lur'e problem, 532
  theorem, 328
Postnikov, V. N., 555
Prince, L. T., 276
Principal case, 321, 328, 533
Principle of argument, 392

Q factor, 80
Quadratic form, 293, 330
Quasioptimization, 263
Quasipolynomial, 73

Rayleigh, J. W. S., Lord, 122
Real root boundary, 7
Recurrence formulas, 19
Region, 32
  of absolute stability, 357
  closed, 32
  convex, 356
  invariant, 340
  stable, 72, 159, 182
  of ultimate boundedness, 545
Relative damping coefficient, 6
  variable, 262
Relay characteristics, 114
Remainder, 378, 379
  theorem, 378
Rest segment, 557
Riemann sphere, 41
Rise-time, 262
Rodden, J. J., 561
Root, 5
  area, 51
  boundary, 7
  complex, 7
  computer solution of, 413, 421
  distribution, 4

  dominant, 248
  double-real, 401
  locus, 47, 274, 419
  real, 5
  sensitivity, 4, 90
Rosenstein, A. B., 78
Rouché theorem, 396
Routh, E. J., 290
  array, 367
  test, 366
Roy, R., 292
Rozenvasser, E. N., 111, 341, 511, 512, 555, 559
Rutman, R. S., 343

Sagirow, P., 512
Saito, N., 560
Sandberg, I. W., 342, 512, 560
Sansone, G., 123
Schultheiss, P. M., 487
Schultz, W. C., 463
Schwarz, H. A., 516
  inequality of, 516
Sector, augmented, 563
  of describing function, 368
  Hurwitz, *see* Hurwitz, A.
  of nonlinearity, 324
  Popov, *see* Popov, V. M.
Sensitivity, 342
  of amplitude, 174
  analysis, 343
  coefficients, 345
  of frequency, 165, 174
  function, 345
  of multiple roots, 406
  of oscillations, 164
  of roots, 4, 90, 165
  specifications, 90
  survey of, 343
Settling time, 259
Shading rule, 34
Signal stabilization, 203
Silberberg, M. Y., 487
Singular cases, 27, 404
Singular loci, 360
Small parameter, *see* Parameter
Solution, 294
  almost periodic, 339
  bounded, 313
  constant, 297
  continuity of, 294

existence of, 294
explicit notation of, 295
in the Filippov sense, *see* Filippov, A. G.
general, 70
generating, 138
particular, 70
periodic, 122, 158, 204
perturbed, 50
properties of, 294
stability of, 158
steady-state, 73, 204
T-periodic, 339
transient, 71
ultimately bounded, 545
uniqueness of, 294
Sridhar, R., 203
Stability, 308, 529
  absolute, 72, 291, 326
  ASIL, 309
  asymptotic, 309
  BIBO, 313
  bounded input-bounded output, 313
  complete, 309
  conditions (linear), 71
  constraints of, 84
  degree of, 72, 336
  domain (parameter space), 457
  domain (state space), 560
  equiasymptotic, 310, 529
  equiasymptotic in the large, 310, 529
  exponential, 311, 337, 566
  global, 309
  Lagrange, *see* Lagrange, J. L.
  in the large, 309
  in the limit, 331
  of linear systems, 71, 448
  orbital, 159, 316
  of oscillations, *see* Oscillations
  practical, 312
  region (parameter space), 32, 353, 358
  region (state space), 339, 545
  relative, 72, 149, 179, 448
  test, 365
  total, 346, 531
  uniform, 308, 529
  uniform asymptotic, 310, 529
  uniform asymptotic in the large, 310, 529
State, 294
  equilibrium, 297
  initial, 294
  space, 294

space equations, 293, 429
transition matrix, *see* Matrix
variables, 294
vector, *see* Vector
zero-state response, 314
Stationary set, 557
  stability of, 558
Steady-state error constants, *see* Error
Steady-state response, 11
Steady-state solution, *see* Solution
Steepest descent, 414, 422
  circuit of, 418
  trajectories of, 420
Stein, W. A., 277
Stepanov, V. V., 292
Stephens, W. C., 428
Superposition principle, 64
Switching, regular, 548
Switching line, 539
Switching surface, 543, 548
Symmetry principle, 401
Synthetic division, 378, 379
System, adaptive, 203
  autonomous, 66, 297, 529
  continuous-time, 294
  controllable, 445, 535
  dissipative, 548
  forced, 296, 337
  free (unforced), 66, 296
  linear, 64, 299, 446
  nonautonomous, 66, 297
  nonlinear, 64, 117, 293
  nonlinear time varying, 341
  observable, 446, 535
  order of, 294
  perturbations of, 535
  response, 63, 70
  structure, 61, 118
  time invariant, 64, 298
  time varying, 64, 297, 341, 559
  variable structure, 490
Szegö, G., 560

Tangent point, 544
Thaler, G. J., 2, 3, 277
Theodorchik, K. F., 142
Tokumaru, H., 560
Tou, J., 487
Towill, D. R., 9
Trajectory, 296
  half-trajectory, 586

Transfer function, 68, 322
  closed-loop, 69
  nondegenerate, 322
  open-loop, 69, 119
  parallel-T network, 77
Transition point, 540
Transition vector function, *see* Vector
Transport lag, 113
Triangle inequality, 303
Truncation, 61, 94
Truxal, L. G., 427, 511
Tsien, H. S., 136
Tsypkin, Ya. Z., 162, 203, 292, 557, 559, 560
Tustin, A., 121

Undamped natural frequency, 6

van der Pol, B., 121, 137, 142, 143, 144, 160, 244
  equation of, 142, 244, 587
  method, 137
Vander Velde, W. E., 203
Variational equation, 299, 345
Vector, 293
  characteristic, 442
  Euclidean space of, 303
  linear independence of, 447, 535

  normed space of, 303
  norm of, 303
  parameter, 343
  space, 294
  state, 294, 445
Vishnegradsky, I. A., 2, 9, 25, 448, 455

Weierstrass, K., 376
Weissenberger, S., 527, 561, 564
West, U. C., 162, 203, 277
Williams, J. L., 557, 560

Yakubovich, V. A., 291, 292, 293, 334, 335, 336, 337, 339, 342, 532, 539, 545, 546, 547, 555, 556, 557
  lemma (Kalman), 291, 533
  Lemma H.2, 545
  Lemma H.3, 546
  Theorem H.2, 547
  Theorem H.6, 337, 551
  Theorem H.7, 337, 552
  Theorem H.8, 338, 552
Yoschizava, T., 313, 545

Zadeh, L., 292
Zames, G., 342, 557, 560
Zemenian, A. H., 427
Zero-state response, *see* State
Zubov, V. I., 561

COUNTY COLLEGE OF MORRIS
LEARNING RESOURCES CENTER

R00029 20549